Topics in Heterocyclic Chemistry

Volume 58

The series Topics in Heterocyclic Chemistry presents critical reviews on present and future trends in the research of heterocyclic compounds. Overall the scope is to cover topics dealing with all areas within heterocyclic chemistry, both experimental and theoretical, of interest to the general heterocyclic chemistry community. The series consists of topic related volumes edited by renowned editors with contributions of experts in the field. All chapters from Topics in Heterocyclic Chemistry are published OnlineFirst with an individual DOI. In references, Topics in Heterocyclic Chemistry is abbreviated as Top Heterocycl Chem and cited as a journal.

More information about this series at http://www.springer.com/series/7081

Hiromasa Kiyota
Editor

Marine Natural Products

With contributions by

R. A. Abdelhamid · N. E. Avalon · B. J. Baker ·
E. M. Carreira · S. Dietrick · D. J. Dixon ·
A. H. H. El-Desoky · M. Enomoto · J. R. Frost · H. Fuwa ·
K. Herrera · Y. Igarashi · J. Ishihara · H. Ishizuka ·
A. Iwasaki · P. Jakubec · C. Jiménez · H. Kiyota ·
S. Kokkaliari · H. Konno · M. Kuse · S. Kuwahara ·
S. V. Ley · K. Nagasawa · K. Okano · R. Sakai ·
P. Sondermann · K. Suenaga · T. Sugai · S. Tsukamoto
A. R. Uria · T. Wakimoto · J. Welsch · B. Yang
R. M. Young

Springer

Editor
Hiromasa Kiyota
Okayama University
Okayama, Japan

ISSN 1861-9282 ISSN 1861-9290 (electronic)
Topics in Heterocyclic Chemistry
ISBN 978-981-16-4636-2 ISBN 978-981-16-4637-9 (eBook)
https://doi.org/10.1007/978-981-16-4637-9

This Springer imprint is published by the registered company Springer Nature Singapore Pte Ltd.
The registered company address is: 152 Beach Road, #21-01/04 Gateway East, Singapore 189721, Singapore

Preface

Humans have been fascinated by the diverse biological activities of natural products, and this has led to the extraction and structural modification of natural medicinal compounds for the benefit of humankind. However, the mass supply of natural products with complex structures is a difficult challenge even with chemical synthesis, and compounds with simplified planar structures have become the mainstream types of small molecular pharmaceuticals. Accordingly, the structural expansion of the current two-dimensional drugs is limited, and now that new drug candidates have been depleted, natural products with three-dimensionally expanded structures are attracting attention again. Under these circumstances, 15 years after the first edition, it is a well-timed opportunity to publish the second edition of marine natural products in the heterocyclic chemistry series. This 18-chapter edition also covers the recent progress of natural product chemistry, i.e. isolation, structure determination, biosynthesis, total synthesis, analog synthesis, probe synthesis, and bioactivity research for various types of natural products originating from marine animals, plants, and microorganisms. I hope that this book will serve as a future guide for the next generation of natural product chemists.

I would like to express my sincere gratitude to all the contributors to this volume, editors of this series, and the staff of Springer, especially to Mr. Shinichi Koizumi. I am also thankful to Prof. Christopher J. Vavricka for his advice on English expression. I also thank my wife Mrs. Ikuko Kiyota for her multifaceted cooperation.

Okayama, Japan
April 2021

Hiromasa Kiyota

Contents

Part I
Isolation, Synthesis and Bioorganic Studies on Marine Alkaloids

Top Heterocycl Chem (2021) 58: 3–22
https://doi.org/10.1007/7081_2020_42

Published online: 14 November 2020

Manzamines: Marine Bioactive Heterocycles

Check for updates

Ahmed H. H. El-Desoky and Sachiko Tsukamoto

Contents

Abstract After the first discovery of manzamine A from a marine sponge by Sakai and colleagues in 1986, its derivatives have attracted much attention among chemists and biologists due to their structural complexity and diverse biological activities. So far, more than 100 manzamine-related alkaloids have been identified, most of which have a fused 6-, 6-, 5-, 8-, and 13-membered ring system connected to a β-carboline as shown in manzamine A. In addition, a variety of bioactivities including cytotoxic,

A. H. H. El-Desoky
Pharmaceutical Industries Research Division, Pharmacognosy Department, National Research Centre, Giza, Egypt

Graduate School of Pharmaceutical Sciences, Kumamoto University, Kumamoto, Japan
e-mail: eldesoky@kumamoto-u.ac.jp

S. Tsukamoto (✉)
Graduate School of Pharmaceutical Sciences, Kumamoto University, Kumamoto, Japan
e-mail: sachiko@kumamoto-u.ac.jp

antimicrobial, antimalarial, anti-HIV, and insecticidal are reported. This review covers the recent reports of novel structures and biological activities of the manzamines.

Keywords Anti-atherosclerosis · Anti-HIV · Antimalarial · Cytotoxicity · Manzamine · Marine sponge · Proteasome inhibitor

1 Introduction

The manzamine family is a class of alkaloids isolated from marine sponges. More than 100 derivatives with novel structures and significant biological activities have been reported to date. This chapter focuses on recent reports of the manzamines.

1.1 Discovery

Manzamine A (**1**) (Fig. 1) was first isolated by Sakai et al. in 1986 from a marine sponge *Haliclona* sp. collected in Okinawa, Japan, as a cytotoxic compound against P388 mouse leukemia cells [1]. It is composed of a *β*-carboline conjugated with an aliphatic moiety with a high degree of complexity. The structure was unprecedently novel, with complicated fused and bridged 6-, 6-, 5-, 8-, and 13-membered rings. The absolute configuration was determined by X-ray analysis of its hydrochloride salt. Later on, manzamine derivatives were frequently isolated from marine sponges belonging to the genera *Pellina* [2], *Xestospongia* [3, 4], *Haliclona* [1, 4, 5], *Pachypellina* [6], *Amphimedon* [7–9], *Petrosia* [10], *Cribochalina* [10], and *Acanthostrongylophora* [11–14].

Fig. 1 Structure of manzamine A (**1**)

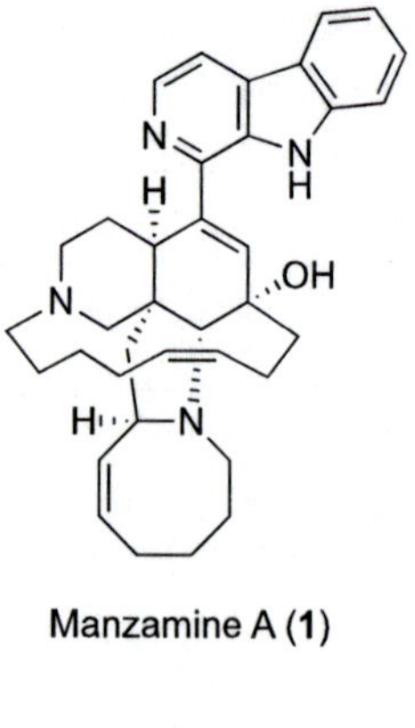

1.2 Biological Activities

A variety of biological activities of manzamine A (**1**) and its related alkaloids have been reported.

Cytotoxic activities were reported against mouse leukemia (P388) [1], human epidermoid carcinoma (KB) [6, 7], human colorectal adenocarcinoma (LoVo) [6], murine leukemia (L1210) [7], colorectal cancer (HCT116) [15], cervical cancer (HeLa, C33A, SiHa, and CaSki) [16], and pancreatic cancer (PANC-1, AsPC-1, MIA PaCa-2, and BxPC-3) [17] cells.

Manzamine alkaloids showed antibacterial activities against gram-positive and gram-negative bacteria, such as *Staphylococcus aureus* [2, 8, 18], *Sarcina lutea* [7, 8, 19], *Bacillus subtilis* [8, 18], *Corynebacterium xerosis* [19], *Escherichia coli* [18], and *Mycobacterium tuberculosis* [11].

Manzamine A (**1**) showed anti-protozoan activities against the chloroquine-sensitive (D6, Sierra Leone) and the chloroquine-resistant (W2, Indo-China) strains of *Plasmodium falciparum* with IC_{50} values of 4.5 and 6.0 ng/mL, respectively, in vitro [20]. Further, manzamine A (**1**) showed promising in vivo antimalarial activity against the rodent malaria parasite *P. berghei* in mice [21], which was indicated to be immune-mediated clearance [22].

Manzamine A (**1**) also exhibited in vitro and in vivo anti-toxoplasmosis activity against *Toxoplasma gondii* without significant toxicity to host cells [23] and anti-leishmanial activity against *Leishmania donovani*, the parasite causing visceral leishmaniasis [24]. Anti-HIV [20, 25] and anti-inflammatory [25] activities were also reported for manzamines.

In addition to the aforementioned reports, our group recently found that manzamines inhibited accumulation of cholesterol ester (CE) in macrophages and suppressed atherosclerosis in mice [12–14]. Further, our group found that manzamines showed inhibition of the proteasome [13, 14]. Although various activities have been reported for manzamines as promising drug leads, these activities have not yet been included. Details of these results are described in Sect. 3 of this chapter.

1.3 Biosynthesis

The biosynthesis of manzamine alkaloids has attracted much attention among a broad community of chemists. Kobayashi's group isolated ircinals A (**2**) and B (**3**) from the sponge *Ircinia* sp. [26] and suggested them to be plausible biogenetic precursors of manzamines A (**1**) and B (**4**) [5] (Scheme 1). Manzamine A (**1**) was chemically transformed from ircinal A (**2**) by a Pictet–Spengler reaction with tryptamine followed by oxidation with DDQ (2,3-dichloro-5,6-dicyano-*p*-benzoquinone) [26].

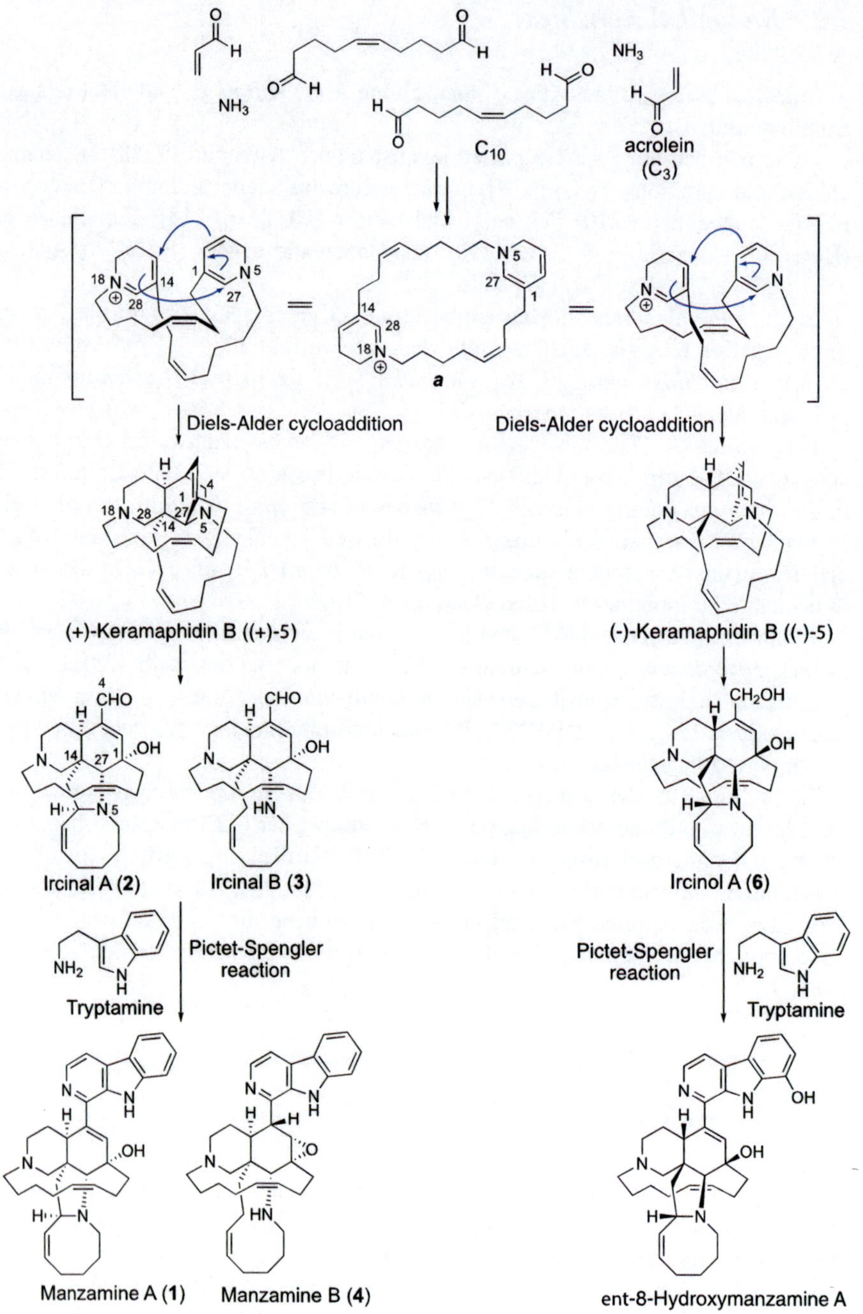

Scheme 1 A plausible biogenetic pathway of the manzamines

In 1992, Baldwin and Whitehead suggested an elegant biogenesis for the manzamines [27], which are derived from four units: ammonia, acrolein (C_3 unit), dialdehyde (C_{10} unit), and tryptophan (Scheme 1). A macrocyclic *bis*-dihydropyridine intermediate ***a*** may be converted to a pentacyclic intermediate such as keramaphidin B (**5**) through an intramolecular *endo*-Diels–Alder cycloaddition to afford manzamines A (**1**) and B (**4**) (Scheme 1).

Subsequently, in 1994, Kobayashi's group isolated ircinols A (**6**) and B, antipodes of the alcoholic forms of ircinals A (**2**) and B (**3**) and the first alkaloids in this class to have the opposite absolute configuration to that of the manzamines [28]. Furthermore, they isolated an enantiomeric mixture of keramaphidin B (**5**), and the (+)- and (−)-enantiomers were postulated to be the precursors of ircinal A (**2**) and ircinol A (**6**), respectively [8, 9]. As keramaphidin B (**5**) was revealed to be an enantiomeric mixture enriched with the (+)-isomer, the Diels–Alder cycloaddition would proceed under an incomplete stereoselective regulation.

Although the manzamines show unique chemical scaffolds and remarkable therapeutic potential, as described in Sect. 1.2, their use in drug development is hampered by the lack of a sustainable supply. In 2014, Hill and Hamann's group analyzed producers of the manzamines by MALDI-MS imaging and reported a sponge-associated bacterium *Micromonospora* sp. from the Indo-Pacific sponge *Acanthostrongylophora ingens* as a producer of manzamine A [29].

2 Recently Isolated Unique Manzamine Derivatives

To date, more than 100 manzamine congeners have been isolated, most of which share the common structural framework of manzamine A (**1**) or manzamine B (**4**) with some modifications. Reviews of the structures, biological activities, proposed biogenesis, and synthesis of manzamine alkaloids have been published already [30–33]. We describe below the unique structures and biological activities of the manzamines that have recently been reported.

2.1 Zamamidines A–D

Kobayashi and coworkers isolated two new manzamine-related alkaloids, zamamidines A (**7**) and B (**8**) [34], from the marine sponge *Amphimedon* sp. (SS-975) collected off Seragaki beach, Okinawa, Japan, and zamamidine C (**9**) from another specimen of the sponge (Fig. 2) [35]. Zamamidines A–C were the first examples of manzamines containing a second *β*-carboline moiety connected to N-2 of the typical manzamine framework through an ethylene unit. Their structures revealed the presence of the 6/6/11/13 ring system characteristic of manzamine B (**4**) for **7** and **8** and the 5/6/6/8/13 ring system of manzamine A (**1**) for **9**. Meanwhile, **7** and **9** showed positive Cotton effects [Δε +16.4 (221 nm) for **7** and Δε +16.9

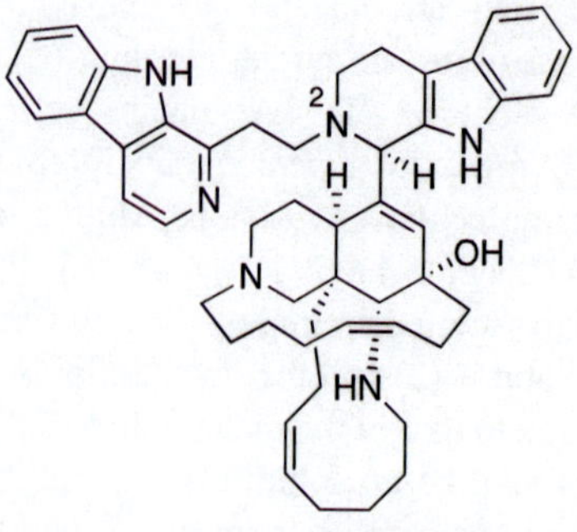

Zamamidine A (**7**)

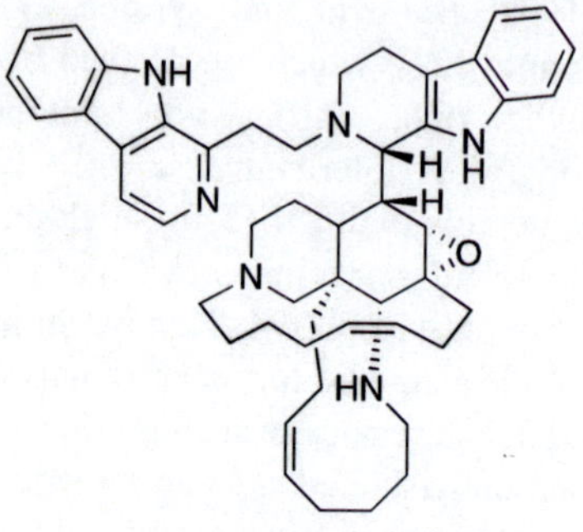

Zamamidine B (**8**)

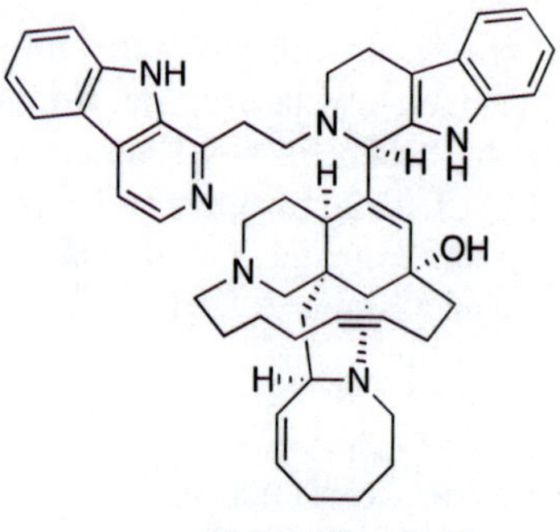

Zamamidine C (**9**)

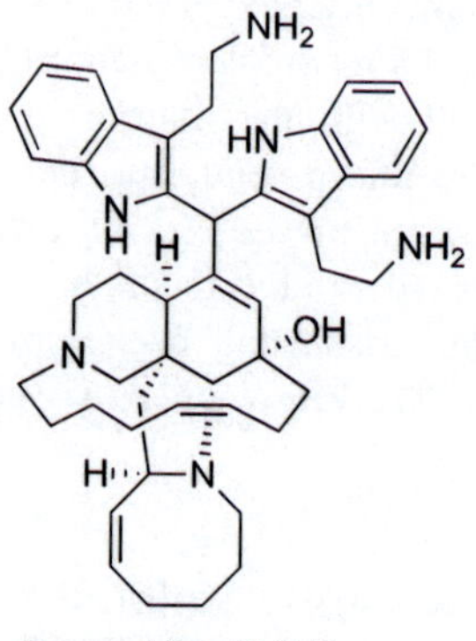

Zamamidine D (**10**)

Fig. 2 Structures of zamamidines A–D (**7**–**10**)

(223 nm) for **9**], which indicated a 1*R* configuration [36]. In contrast, **8** showed a negative Cotton effect [$\Delta\varepsilon$ −17.7 (223 nm)], indicating a 1*S* configuration. Zamamidines A–C showed cytotoxicity against P388 murine leukemia cells (IC_{50} 13.8, 14.8, and 14.1 μg/mL, respectively) and exhibited inhibitory activities against *Trypanosoma brucei brucei*, the parasite associated with sleeping sickness (IC_{50} 1.04, 1.05, and 0.27 μg/mL, respectively), and *Plasmodium falciparum* (IC_{50} 7.16, 12.20, and 0.58 μg/mL, respectively) in vitro.

Successively, they isolated zamamidine D (**10**) from the sponge *Amphimedon* sp. (SS-1231) collected at Zamami, Okinawa [18]. It was the first manzamine alkaloid possessing a 2,2′-methylenebistryptamine unit as the aromatic moiety instead of a *β*-carboline unit. Compound **10** may be derived from 3,4-dihydromanzamine A by oxidative cleavage of the C-1–N-2 bond, followed by attachment of tryptamine (Scheme 2). Compound **10** exhibited antimicrobial activity against gram-positive bacteria, *Staphylococcus aureus*, *Bacillus subtilis*, and *Micrococcus luteus* (MIC 8 μg/mL); a gram-negative bacterium, *Escherichia coli* (MIC 32 μg/mL); and fungi, *Aspergillus niger*, *Trichophyton mentagrophytes*, *Candida albicans*, and *Cryptococcus neoformans* (IC_{50} 16, 8, 16, and 2 μg/mL, respectively).

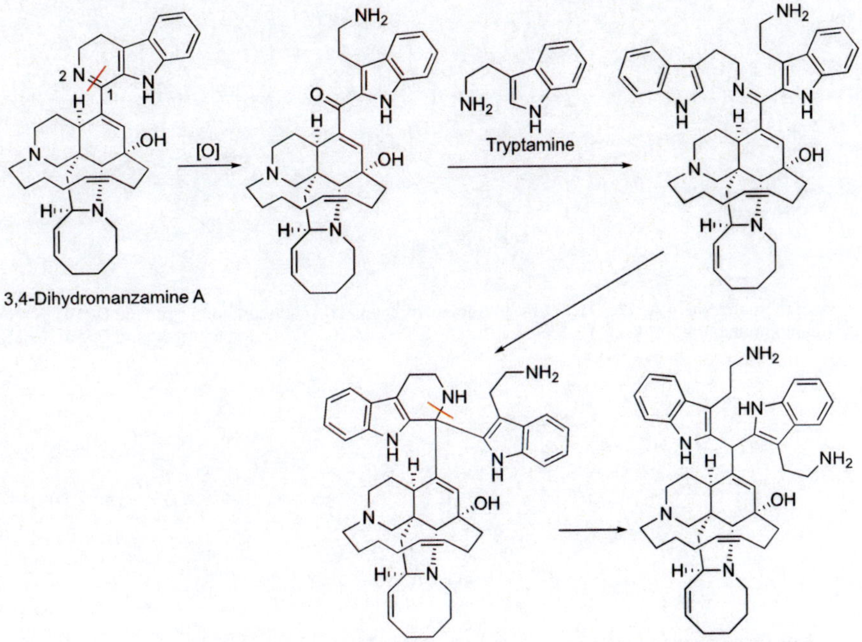

Scheme 2 A possible biogenetic pathway of zamamidine D (**10**) from 3,4-dihydromanzamine A

2.2 *Acanthomanzamines A–E*

In 2014, our group isolated acanthomanzamines A–E (**11–15**) from the marine sponge *Acanthostrongylophora ingens* (RMNH POR 6184) collected at Mantehage, North Sulawesi, Indonesia (Fig. 3a). Acanthomanzamines A (**11**) and B (**12**) were the first examples of manzamine alkaloids containing a tetrahydroisoquinoline moiety instead of a *β*-carboline ring [14]. An analysis of NMR spectra indicated that **11** and **12** possessed the same pentacyclic ring system as manzamine A (**1**), with a 1,2,3,4-tetrahydroisoquinole-6,7-diol ring attaching at C-10. The structure of **12** was indicated to be the same as **11**, except for the configuration of C-1. The ^{13}C chemical shifts of C-1 of **11** and **12** showed 6.5 ppm differences, suggesting that they are epimers at C-1. Although their C-1 configurations could not be determined by NOE experiments, the calculated ECD spectra indicated 1*S*- and 1*R*-configurations for **11** and **12**, respectively [14].

Acanthomanzamine C (**13**) possesses a unique 6/6/13/5/5/5 ring system, in which the classical eight-membered ring of manzamines is converted to bicyclic ring system composed of two five-membered rings fused at C-30–C-34. The configurations at C-30 and C-34 were analyzed by NOE correlations. Among the four possible configurations **13a–13d** (Fig. 3b), a NOE correlation between H-28 and H-33 excluded the possibility of the *trans*-fused rings, **13c** and **13d**, while NOE

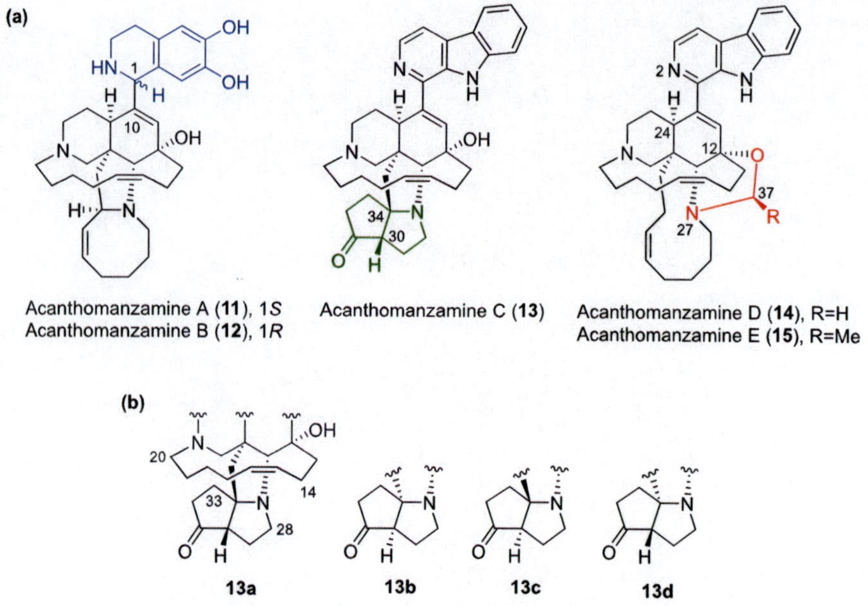

Fig. 3 (**a**) Structures of acanthomanzamines A–E (**11**–**15**). (**b**) Four possible structures of **13** (**13a**–**13d**) for study on the configuration of **13**

correlations from H-28 to H-14 and H-20 indicated that structure **13a** is more favorable than **13b**. In the proposed formation of **13**, the carbons at C-34 and C-31 of manzamine A (**1**) may be oxidized, followed by conversion to an iminium cation. Then, keto–enol tautomerization would generate a new link between C-30 and C-34 to afford **13** (Scheme 3) [14].

Acanthomanzamines D (**14**) and E (**15**) were the first examples of manzamines with oxazolidine and 2-methyloxazolidine rings, respectively, fused to manzamine skeleton. Ma'eganedin A, which was isolated from an *Amphimedon* sponge by Kobayashi's group, was the first example with a methylene group between N-2 and N-27 in manzamine B (**4**) [37]. In the cases of **14** and **15**, methylene and ethylidene groups, respectively, were inserted between N-27 and the oxygen at C-12. The relative configuration of the secondary methyl group at C-37 of **15** was determined to be β-oriented based on a H-24/H-37 NOE correlation.

The sponges containing manzamine alkaloids typically afforded high yields of manzamine A (**1**), but this sponge afforded **11** as a main metabolite (40 mg) along with **12**–**15** (2.0, 1.8, 2.9, and 3.7 mg, respectively) and **1** (3.5 mg). This is a rare case of **1** afforded as a minor compound.

Compounds **11** and **12** may be formed by a similar biosynthetic pathway to **1**, with dopamine instead of tryptamine at the step of a Pictet–Spengler reaction (Scheme 1).

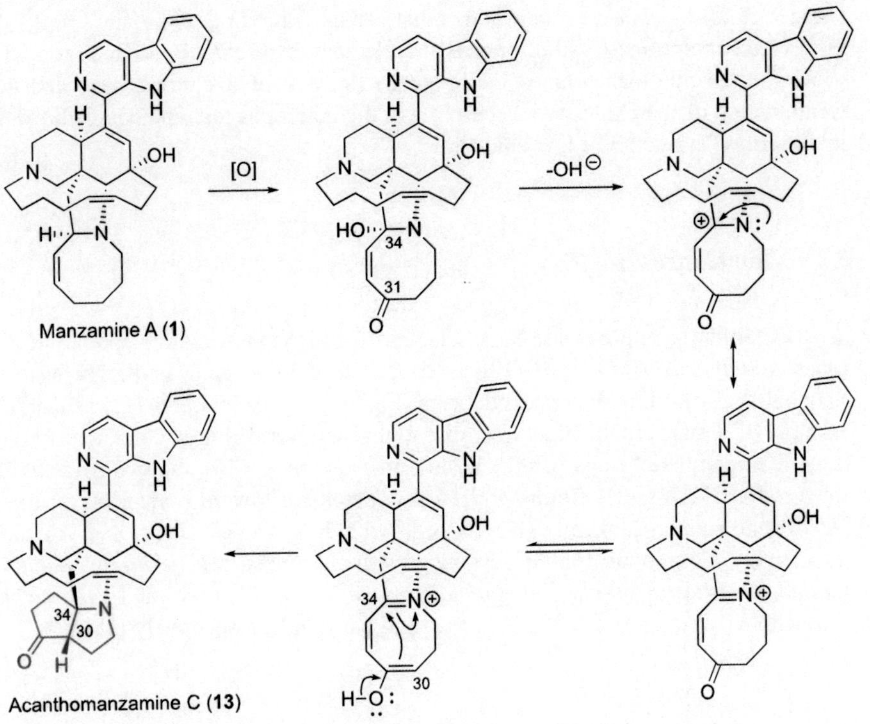

Scheme 3 A possible biogenetic pathway from manzamine A (**1**) to acanthomanzamine C (**13**)

Table 1 Inhibition of the proteasome and cytotoxicity[a]

Compounds	Inhibition of the proteasome[b] IC$_{50}$ (μM)	Cytotoxicity[c] IC$_{50}$ (μM)
Manzamine A (**1**)	2.0	13
Acanthomanzamine A (**11**)	4.1	4.2
Acanthomanzamine B (**12**)	7.8	5.7
Acanthomanzamine D (**14**)	0.63	15
Acanthomanzamine E (**15**)	1.5	>20
Acantholactam (**18**)	33	>50
Pre-*neo*-kauluamine (**19**)	0.34	16
neo-Kauluamine (**20**)	0.13	5.4

[a]Acanthomanzamine C (**13**) has not been subjected to biological tests due to the limited amount available
[b]Tested against the chymotrypsin-like activity of the proteasome
[c]Tested against HeLa cells

Compounds **11** and **12** showed cytotoxic activities against HeLa cells with IC$_{50}$ values of 4.2 and 5.7 μM, respectively, whereas **14** and **1** were less cytotoxic (IC$_{50}$ 15 and 13 μM, respectively), and the value of **15** was more than 20 μM (Table 1).

These results clearly indicated that the presence of the 1,2,3,4-tetrahydroisoquinoline-6,7-diol moiety accelerated cytotoxicity. In addition, we found that acanthomanzamines inhibited the activity of the proteasome and the accumulation of the cholesterol ester (CE) in the macrophages. The detailed results are described in Sect. 3 of this chapter.

2.3 *Kepulauamine A*

In 2017, Shin's group isolated kepulauamine A (**16**) from a marine sponge of the genus *Acanthostrongylophora* (12KL-15) collected in Kepulauan Seribu Marine National Park, north of Jakarta, Indonesia (Fig. 4) [38]. Compound **16** was another example of a manzamine alkaloid whose eight-membered ring is converted to two fused five-membered rings. Unlike acanthomanzamine C (**13**), the two rings of **16** are fused at N-27–C-31. Compound **16** was cytotoxic toward human lung cancer (A594) and human leukemia (K562) cells with IC_{50} values of 4.6 and 7.6 μM, respectively. Compound **16** inhibited the growth of *Staphylococcus aureus*, *Bacillus subtilis*, *Kocuria rhizophila*, *Salmonella enterica*, *Proteus hauseri*, and *Escherichia coli*, with MIC values of 8, 32, 16, 16, 8, and 64 ng/mL, respectively [38].

2.4 *Acantholactone and Acantholactam*

In 2012, Hamann's group reported the isolation of acantholactone (**17**), having unprecedented δ-lactone and ε-lactam rings, from a marine sponge of the genus *Acanthostrongylophora* collected in Indonesia (Fig. 5) [39]. Subsequently, in 2014, our group isolated acantholactam (**18**) from the sponge *Acanthostrongylophora ingens* (RMNH POR 8527) collected at Ti Toi, North Sulawesi, Indonesia (Fig. 5) [13]. Compound **18** was the first manzamine alkaloid containing a γ-lactam ring, whose nitrogen atom was substituted by a (Z)-2-hexenoic acid moiety.

Fig. 4 Structure of kepulauamine A (**16**)

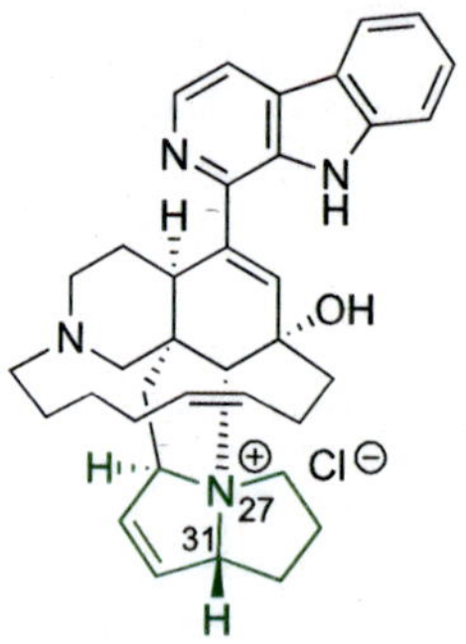

Kepulauamine A (**16**)

Fig. 5 Structures of acantholactone (**17**) and acantholactam (**18**)

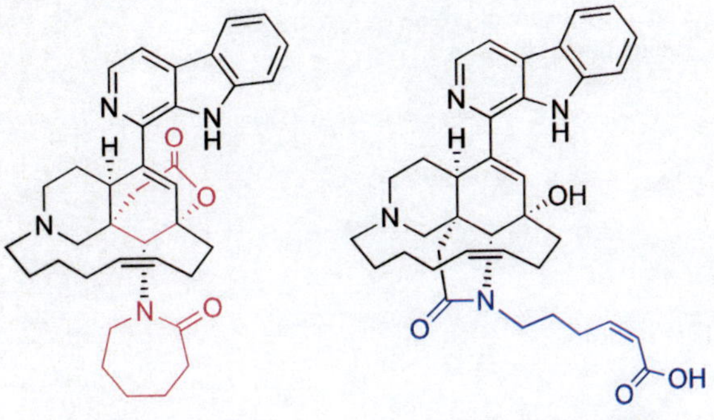

Acantholactone (**17**) Acantholactam (**18**)

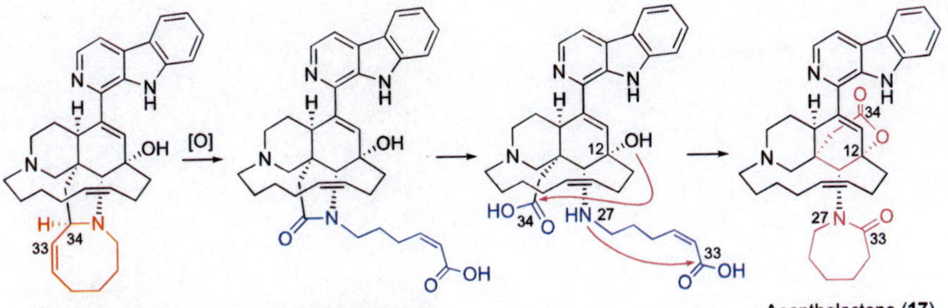

Manzamine A (**1**) Acantholactam (**18**) Acantholactone (**17**)

Scheme 4 A possible biogenetic pathway from manzamine A (**1**) to acantholactone (**17**) through acantholactam (**18**)

Biogenetically, acantholactam (**18**) would be derived from manzamine A (**1**) by oxidative cleavage of the C-33–C-34 bond of the eight-membered ring (Scheme 4). From the structural point of view, **17** could be biosynthesized from **18**, although these two alkaloids have not been isolated from the same sponge. Hydrolysis of the amide linkage of the γ-lactam ring in **18** may generate a carboxy group (C-34), which would form an ester linkage with 12-OH to yield a δ-lactone ring. Alternatively, amidation between N-27 and the C-33 carboxy group would afford an *N*-substituted ε-lactam ring to yield **17**.

2.5 *Pre-*neo*-kauluamine*

In 2014, our group isolated pre-*neo*-kauluamine (**19**) from the marine sponge *Acanthostrongylophora ingens* collected at Bajotalawaan (RMNH POR 3991), North Sulawesi, Indonesia (Fig. 6) [13]. Based on the ^{13}C chemical shifts, three carbons at C-30, C-31, and C-34 in the eight-membered ring were oxygenated, and

Fig. 6 Structure of pre-*neo*-kauluamine (**19**)

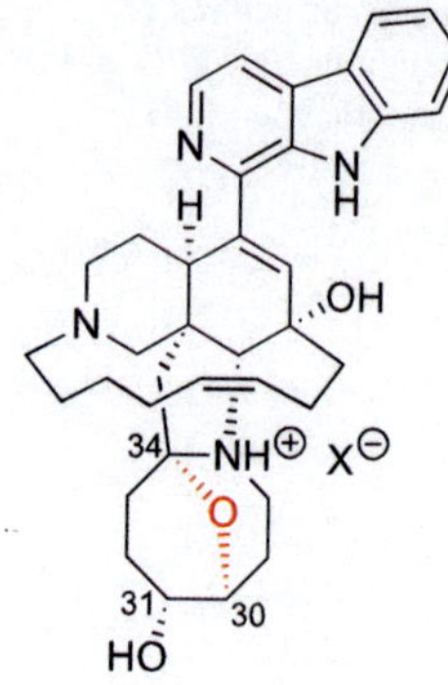

Pre-*neo*-kauluamine (**19**)

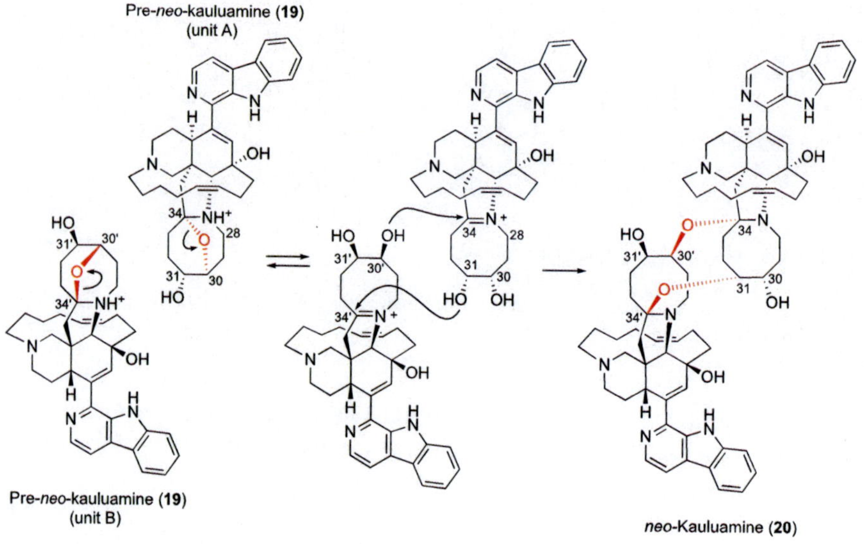

Scheme 5 A possible mechanism of the dimerization of pre-*neo*-kauluamine (**19**) to *neo*-kauluamine (**20**)

the HMBC correlation between H-30 and C-34 showed the presence of the ether linkage C-30–O–C-34.

Incidentally, after storage at −20°C for 2 months, pre-*neo*-kauluamine (**19**) was converted to the dimer *neo*-kauluamine (**20**) [23], which was isolated together with the monomer (**19**) from the same sponge. During storage, nucleophilic attacks occurred in two molecules of **19** that were equilibrated into a ring-opened diol form with a strained iminium carbon, to afford **20** (Scheme 5). Although the relative configurations of C-30′, C-31′, and C-34′ of **20** have yet to be determined by Hamann's group, our result that **19** dimerized to produce **20** clearly indicated the relative configuration of **20**. Moreover, the absolute configuration of **19** was indicated to be the same as that of manzamine A (**1**) on the basis of their ECD spectra, and therefore, the absolute configuration of **20** was also revealed to be the same.

3 Recently Reported Biological Activities

In previous reports, manzamine alkaloids have mainly been evaluated for biological activities related to cancer and infectious diseases including malaria, tuberculosis, and AIDS, with highly promising results [25, 30–33, 40]. During our search for pharmaceutical candidates among the marine invertebrates, we found several examples of proteasome inhibitory and anti-atherosclerotic activities. Bioassay-guided purification of sponges afforded new manzamine alkaloids, as described in Sects. 2.2, 2.4, and 2.5 of this chapter.

3.1 Inhibition of the Proteasome

The ubiquitin–proteasome system is an intracellular proteolytic mechanism responsible for selective protein degradation, which has essential roles in cell signaling, cell-cycle control, and transcription to development [41–45]. The proteasome is an intracellular high-molecular-weight protease subunit complex consisting of two subcomplexes, the 20S core particle (the 20S proteasome) and the 19S regulatory particle. Ubiquitin, composed of 76 amino acids, attaches to a target protein prior to degradation (Fig. 7). Ubiquitination requires the sequential actions of three enzymes: ubiquitin-activating enzyme (E1), ubiquitin-conjugating enzyme (E2), and ubiquitin-protein ligase (E3), and results in the formation of the polyubiquitin chain [42]. The formed polyubiquitin chain, attached to the target protein, is recognized by the proteasome, and the protein portion is degraded. Prior to the degradation, deubiquitinating enzymes (DUB) cleave the isopeptide bond at the C-terminus of ubiquitin and deconjugate the ubiquitin chain in target proteins [46–48]. Proteins enter the cavity of the barrel-shaped 20S proteasome to be degraded by

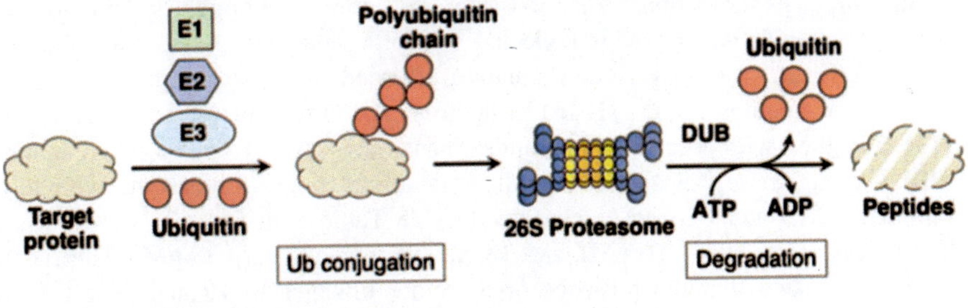

Fig. 7 The ubiquitin–proteasome system

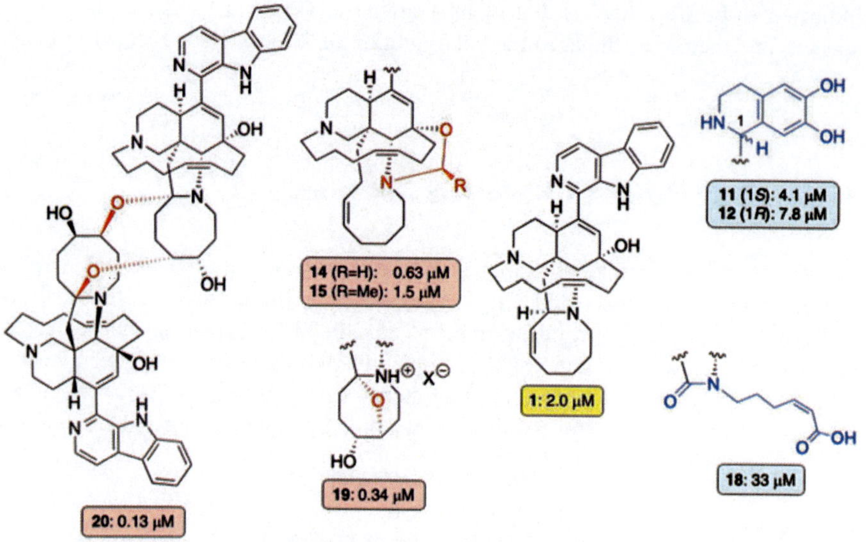

Fig. 8 Structures and proteasome inhibitory activities (IC_{50}) of manzamine derivatives. More potent and less potent compounds compared to **1** were shown in red and blue, respectively

the proteolytic active sites aligned in the cavity [49]. The 20S proteasome is classified as a threonine protease containing two pairs of three different sites that catalyze chymotrypsin-like (CT-L), trypsin-like (T-L), and caspase-like (C-L) activities. Since the level of the proteasome is increased in tumor cells, specific compounds inhibiting the proteasome-mediated proteolytic activity are suitable for use in cancer treatment.

In 2003, the proteasome inhibitor, Velcade® (bortezomib) [50], was approved as a powerful agent for the treatment of multiple myeloma by the FDA. Velcade shows antitumor activity against various tumor cells that are resistant to conventional chemotherapeutic agents [51]. After the approval of Velcade, Kyprolis® (carfilzomib) [52] and Ninlaro® (ixazomib) [53], which overcome the defects of Velcade, were approved in 2012 and 2015, respectively, and components of the ubiquitin–proteasome system also became attractive targets for cancer therapy [54–56].

In our experiments, the extracts of three sponge specimens of *Acanthostrongylophora ingens* collected in North Sulawesi, Indonesia, inhibited the proteasome. Bioassay-guided isolation afforded new manzamine derivatives, acanthomanzamines A–E (**11–15**), acantholactam (**18**), and pre-*neo*-kauluamine (**19**), together with known compounds, manzamine A (**1**) and *neo*-kauluamine (**20**). Among these, **19** and **20** significantly inhibited the proteasome with IC_{50} values of 0.13 and 0.34 μM, respectively (see Table 1 in Sect. 2.2) [13], while **1** was less active (IC_{50} 2.0 μM), and **18** was very weak (IC_{50} 33 μM). These data clearly indicated that the presence of the ether linkages in **19** and **20** intensely enhanced the activity (Fig. 8). Because **18**, the ring-opening metabolite, lacks proteasome inhibitory activity and cytotoxicity, the presence of the eight-membered ring in the manzamines is essential for the biological activities. Compounds **11**, **12**, **14**, and **15** inhibited the proteasome with IC_{50} values of 4.1, 7.8, 0.63, and 1.5 μM, respectively [14], indicating that the compounds containing *β*-carboline have more potent proteasome inhibitory activity than those containing 1,2,3,4-tetrahydroisoquinoline-6,7-diol moieties.

3.2 *Anti-atherosclerotic Activity*

The formation of foam cells in macrophages is an essential event in the progression of early phase atherosclerosis, and therefore its prevention is an effective intervention for atherosclerosis treatment. The accumulation of foam cells derived from macrophages in the subendothelial spaces is linked with early atherosclerotic lesions [57]. Foam cells produce various endogenous signaling molecules, such as cytokines, growth factors, and proteases, which play important roles in the progression of atherosclerotic lesions [57]. Chemically modified low-density lipoproteins (LDL), including oxidized LDL, acetylated LDL (Ac-LDL), and glycated LDL, are incorporated into macrophages through scavenger receptors (Fig. 9) [58]. In macrophages, the modified LDL releases cholesterol in a free form, and free cholesterol is toxic to cells. Cholesterol is then esterified to cholesterol esters (CE) by acyl-coenzyme A:cholesterol acyl-transferase (ACAT), located in the rough endoplasmic reticulum, and results in the formation of foam cells from macrophages [59].

In the screening, human monocyte-derived macrophages (HMDMs) were incubated with Ac-LDL in the presence of sample and BSA-conjugated [^{3}H]oleic acid at 37°C for 24 h (Fig. 10). Cellular lipids were extracted and the radioactivity of cholesteryl-[^{3}H]oleate was measured.

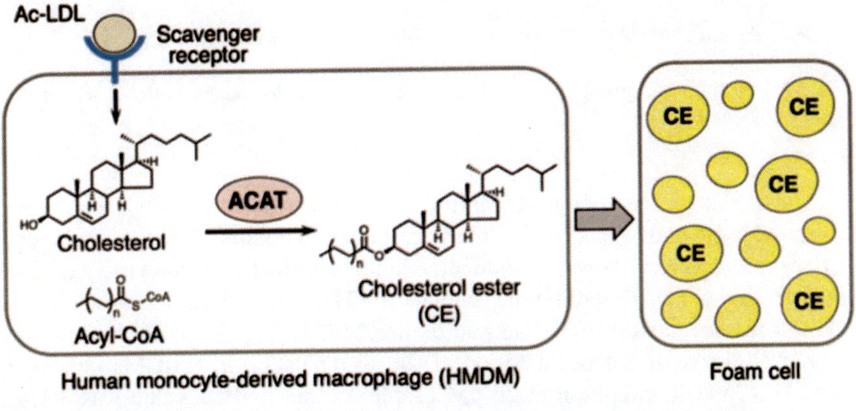

Fig. 9 Formation of foam cell from human monocyte-derived macrophage (HMDM)

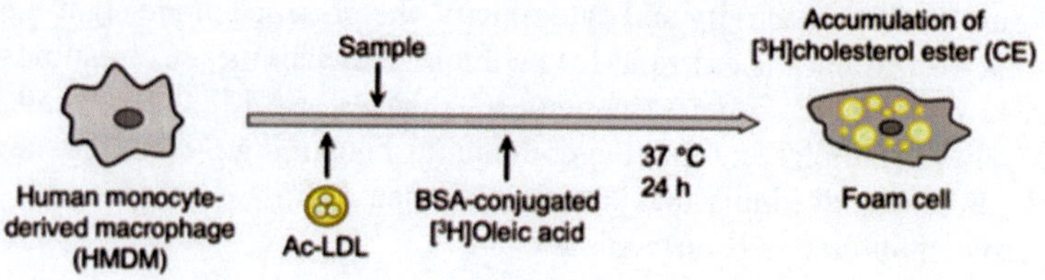

Fig. 10 Measurement of cholesterol ester (CE) formation

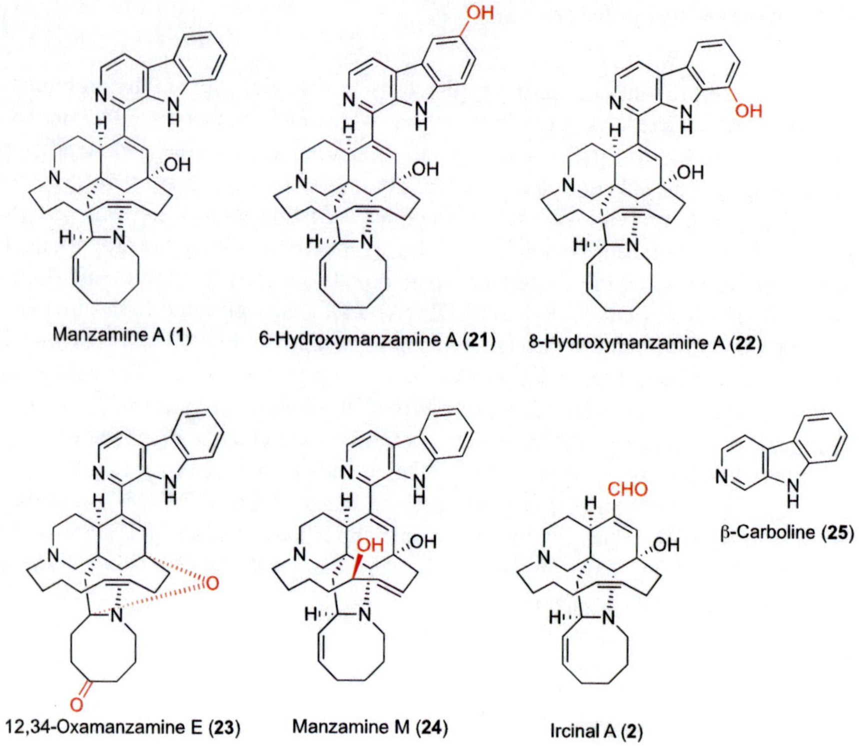

Fig. 11 Structures of manzamines (**1** and **21–24**) isolated from the sponge *A. ingens*, ircinal A (**2**), and β-carboline (**25**)

The extract of the sponge *Acanthostrongylophora ingens* (RMNH POR 3992) collected in Bajotalawaan, North Sulawesi, Indonesia, inhibited the accumulation of CE in macrophages. Bioassay-guided isolation afforded five known compounds: manzamine A (**1**), 6-hydroxymanzamine A (**21**), 8-hydroxymanzamine A (**22**), 12,34-oxamanzamine E (**23**), and manzamine M (**24**) (Fig. 11) [12]. Among these, **1** showed the most potent inhibition of CE accumulation in HMDMs induced by Ac-LDL by 80% inhibition at 20 μM (Table 2), and its IC_{50} value was 4.1 μM. Among the other manzamines, **21** (38%) and **22** (35%) were more effective than **23**

Table 2 Inhibition of accumulation of the cholesterol ester[a,b]

Compounds	Inhibition (%)	Compounds	Inhibition (%)
Manzamine A (**1**)	80	Acanthomanzamine A (**11**)	48
6-Hydroxymanzamine A (**21**)	38	Acanthomanzamine B (**12**)	73
8-Hydroxymanzamine A (**22**)	35	Acanthomanzamine D (**14**)	73
12,34-Oxamanzamine E (**23**)	4	Acanthomanzamine E (**15**)	61
Manzamine M (**24**)	12	Acantholactam (**18**)	0
Ircinal A (**2**)	16	Pre-*neo*-kauluamine (**19**)	91
β-Carboline (**25**)	24	*neo*-Kauluamine (**20**)	92

[a]Acanthomanzamine C (**13**) has not been subjected to biological testing due to the limited amount available
[b]Tested at a concentration of 20 μM

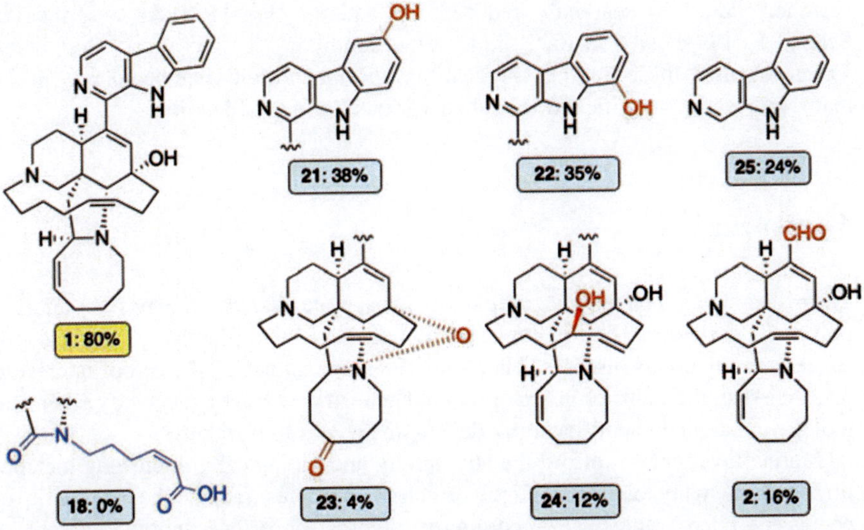

Fig. 12 Structures and inhibitory activities of the cholesterol ester accumulation of manzamine derivatives and **25** (20 μM). Much less potent compounds compared to **1** were shown

(4%) and **24** (12%). These data indicated that any structural modification of **1** would reduce the inhibitory activity (Fig. 12) and that modification at the pentacyclic portion would deteriorate the activity. Compound **1** is composed of β-carboline (**25**) and ircinal A (**2**); these compounds showed reduced inhibition, but addition of both compounds at a ratio of 1:1 did not recover the activity, clearly suggesting that the gross structure of **1** is necessary for the inhibitory effect on CE accumulation.

Manzamine A (**1**) showed the most significant inhibition of CE accumulation and was isolated in large amounts [323 mg from 300 g (wet weight) of sponge]. Detailed investigations were therefore carried out with **1**. The mechanism of action was found to be through the inhibition of ACAT activity, rather than any effect on the expression of the scavenger receptors or ACAT proteins [12]. Moreover, oral administration of **1** at a dose of 30 mg/kg/day in apoE-deficient mice, which are known to spontaneously undergo atherogenesis, significantly decreased total cholesterol, free cholesterol, LDL cholesterol, and triglycerides in plasma by approximately 40, 20, 40, and 50%, respectively. Meanwhile, **1** reduced the size of atherosclerotic lesions in apoE-knockout mice by 45%, whereas cross sections of the aortic sinus in the negative control mice exhibited a marked thickening of the intima filled with foam cells [12]. These results showed that **1** exhibited in vivo anti-atherosclerotic activity.

The extracts of other sponge specimens of *A. ingens* (RMNH POR 3991 and RMNH POR 8527) also inhibited accumulation of the cholesterol ester (CE) in macrophages. Pre-*neo*-kauluamine (**19**), *neo*-kauluamine (**20**), and acanthomanzamines A (**11**), B (**12**), D (**14**), and E (**15**), isolated from these sponges, exhibited similar inhibition levels to manzamine A (**1**) (Table 2). In contrast, acantholactam (**18**) markedly reduced the activity (Fig. 12). As described in Sect. 3.1, **18** showed neither proteasome inhibitory activity nor cytotoxicity. Taken together, these findings suggest that the intact eight-membered ring in the manzamines plays a critical role in their various biological activities.

4 Conclusions

The manzamines are a unique group of marine sponge alkaloids that possess unusual polycyclic skeletons. Their biological activities make them promising drug leads for cancer and infectious diseases. Therefore, the manzamines have been of interest to researchers in the fields of natural product chemistry, synthetic chemistry, chemical biology, medicinal chemistry, drug development, and biosynthesis.

Marine invertebrates afford the structurally and biologically interesting metabolites, but in many cases their real producers are presumed to be symbiotic or coexistent microorganisms. Accordingly, there is a serious supply problem of metabolites for the drug development. The recent discovery [29] of a sponge-associated bacterium that produces manzamine A (**1**) should lead to a sustainable supply and a new era in manzamine alkaloid research and exploitation.

References

1. Sakai R, Higa T, Jefford CW, Bernardinelli G (1986) J Am Chem Soc 108:6404
2. Nakamura H, Deng S, Kobayashi J, Ohizumi Y, Tomotake Y, Matsuzaki T, Hirata Y (1987) Tetrahedron Lett 28:621
3. Ichiba T, Sakai R, Kohmoto S, Saucy G, Higa T (1988) Tetrahedron Lett 29:3083
4. Kobayashi M, Chen Y-J, Aoki S, In Y, Ishida T, Kitagawa I (1995) Tetrahedron 51:3727
5. Sakai R, Kohmoto S, Higa T, Jefford CW, Bernardinelli G (1987) Tetrahedron Lett 28:5493
6. Ichiba T, Corgiat JM, Scheuer PJ, Kelly-Borges M (1994) J Nat Prod 57:168
7. Kobayashi J, Tsuda M, Kawasaki N, Sasaki T, Mikami Y (1994) J Nat Prod 57:1737

8. Kobayashi J, Tsuda M, Kawasaki N, Matsumoto K, Adachi T (1994) Tetrahedron Lett 35:4383
9. Tsuda M, Inaba K, Kawasaki N, Honma K, Kobayashi J (1996) Tetrahedron 52:2319
10. Crews P, Cheng X-C, Adamczeski M, Rodríguez J, Jaspars M, Schmitz FJ, Traeger SC, Pordesimo EO (1994) Tetrahedron 50:13567
11. Rao KV, Donia MS, Peng J, Garcia-Palomero E, Alonso D, Martinez A, Medina M, Franzblau SG, Tekwani BL, Khan SI, Wahyuono S, Willett KL, Hamann MT (2006) J Nat Prod 69:1034
12. Eguchi K, Fujiwara Y, Hayashida A, Horlad H, Kato H, Rotinsulu H, Losung F, Mangindaan RE, de Voogd NJ, Takeya M, Tsukamoto S (2013) Bioorg Med Chem 21:3831
13. El-Desoky AH, Kato H, Eguchi K, Kawabata T, Fujiwara Y, Losung F, Mangindaan REP, de Voogd NJ, Takeya M, Yokosawa H, Tsukamoto S (2014) J Nat Prod 77:1536
14. Furusato A, Kato H, Nehira T, Eguchi K, Kawabata T, Fujiwara Y, Losung F, Mangindaan REP, Voogd NJD, Takeya M, Yokosawa H, Tsukamoto S (2014) Org Lett 16:3888
15. Lin L-C, Kuo T-T, Chang H-Y, Liu W-S, Hsia S-M, Huang T-C (2018) Mar Drugs 16:252
16. Karan D, Dubey S, Pirisi L, Nagel A, Pina I, Choo Y-M, Hamann MT (2020) J Nat Prod 83:286
17. Kallifatidis G, Hoepfner D, Jaeg T, Guzmán EA, Wright AE (2013) Mar Drugs 11:3500
18. Kubota T, Nakamura K, Kurimoto S, Sakai K, Fromont J, Gonoi T, Kobayashi J (2017) J Nat Prod 80:1196
19. Watanabe D, Tsuda M, Kobayashi J (1998) J Nat Prod 61:689
20. Rao KV, Kasanah N, Wahyuono S, Tekwani BL, Schinazi RF, Hamann MT (2004) J Nat Prod 67:1314
21. Ang KKH, Holmes MJ, Higa T, Hamann MT, Kara UAK (2000) Antimicrob Agents Chemother 44:1645
22. Ang KKH, Holmes MJ, Kara UAK (2001) Parasitol Res 87:715
23. El Sayed KA, Kelly M, Kara UAK, Ang KKH, Katsuyama I, Dunbar DC, Khan AA, Hamann MT (2001) J Am Chem Soc 123:1804
24. Rao KV, Santarsiero BD, Mesecar AD, Schinazi RF, Tekwani BL, Hamann MT (2003) J Nat Prod 66:823
25. Yousaf M, Hammond NL, Peng J, Wahyuono S, McIntosh KA, Charman WN, Mayer AMS, Hamann MT (2004) J Med Chem 47:3512
26. Kondo K, Shigemori H, Kikuchi Y, Ishibashi M, Sasaki T, Kobayashi J (1992) J Org Chem 57:2480
27. Baldwin J, Whitehead R (1992) Tetrahedron Lett 33:2059
28. Tsuda M, Kawasaki N, Kobayashi J (1994) Tetrahedron 50:7957
29. Waters AL, Peraud O, Kasanah N, Sims JW, Kothalawala N, Anderson MA, Abbas SH, Rao KV, Jupally VR, Kelly M, Dass A, Hill RT, Hamann MT (2014) Front Mar Sci 1:54
30. Ashok P, Ganguly S, Murugesan S (2014) Drug Discov Today 19:1781
31. Ashok P, Lathiya H, Murugesan S (2015) Eur J Med Chem 97:928
32. Hu J-F, Hamann MT, Hill R, Kelly M (2003) Cordell GA (ed) The alkaloids: chemistry and biology, vol 60. Elsevier, San Diego, p 207
33. Hamann MT (2007) Curr Pharm Des 13:653
34. Takahashi Y, Kubota T, Fromont J, Kobayashi J (2009) Org Lett 11:21
35. Yamada M, Takahashi Y, Kubota T, Fromont J, Ishiyama A, Otoguro K, Yamada H, Ōmura S, Kobayashi J (2009) Tetrahedron 65:2313
36. Stöckigt J, Zenk MH (1977) J Chem Soc Chem Commun:646
37. Tsuda M, Watanabe D, Kobayashi J (1998) Tetrahedron Lett 39:1207
38. Kim C-K, Riswanto R, Won TH, Kim H, Elya B, Sim CJ, Oh D-C, Oh K-B, Shin J (2017) J Nat Prod 80:1575
39. Wahba AE, Fromentin Y, Zou Y, Hamann MT (2012) Tetrahedron Lett 53:6329
40. Radwan M, Hanora A, Khalifa S, Abou-El-Ela SH (2012) Cell Cycle 11:1765
41. Hershko A, Ciechanover A (1998) Annu Rev Biochem 67:425
42. Voges D, Zwickl P, Baumeister W (1999) Annu Rev Biochem 68:1015
43. Glickman MH, Ciechanover A (2002) Physiol Rev 82:373
44. Pickart CM (2001) Annu Rev Biochem 70:503

45. Finley D (2009) Annu Rev Biochem 78:477
46. Reyes-Turcu FE, Ventii KH, Wilkinson KD (2009) Annu Rev Biochem 78:363
47. Komander D, Clague MJ, Urbe S (2009) Nat Rev Mol Cell Biol 10:550
48. Fraile JM, Quesada V, Rodríguez D, Freije JM, López-Otín C (2012) Oncogene 31:2373
49. Demartino GN, Gillette TG (2007) Cell 129:659
50. Adams J (2003) Drug Discov Today 8:307
51. Sánchez-Serrano I (2006) Nat Rev Drug Discov 5:107
52. Thompson JL (2013) Ann Pharmacother 47:56
53. Muz B, Ghazaria RN, Ou M, Luderer MJ, Kusdono HD, Azab AK (2016) Drug Des Devel Ther 10:217
54. Bedford L, Lowe J, Dick LR, Mayer RJ, Brownell JE (2011) Nat Rev Drug Discov 10:29
55. Shen M, Schmitt S, Buac D, Dou QP (2013) Expert Opin Ther Targets 17:1091
56. Ciechanover A (2013) Bioorg Med Chem 21:3400
57. Ross R (1999) New Engl J Med 340:115
58. Steinberg D, Parthasarathy S, Carew TE, Khoo JC, Witztum JL (1989) New Engl J Med 320:915
59. Chang TY, Chang CC, Lin S, Yu C, Li BL, Miyazaki A (2001) Curr Opin Lipidol 12:289

Top Heterocycl Chem (2021) 58: 23–60
https://doi.org/10.1007/7081_2021_51

Published online: 19 February 2021

Total Synthesis of Manzamine Alkaloids

Darren J. Dixon and Pavol Jakubec

Contents

Abstract Manzamine alkaloids are a diverse and highly complex family of marine natural products. The unique macrocyclic structural features of these alkaloids inspired the development of many fascinating synthetic routes. However, the existing strategies typically target only individual alkaloids. Herein we present a detailed insight into a unified approach for the total synthesis of four complex manzamine alkaloids – namely nakadomarin A, manzamine A, ircinal A, and ircinol A – and towards keramaphidin B. The successful synthetic campaign over the past decade is based on stereoselective Michael nitro olefin addition, nitro-Mannich/ lactamization cascade, iminium ion/cyclization, stereoselective ring-closing

D. J. Dixon (✉)
Chemistry Research Laboratory, The Department of Chemistry, University of Oxford, Oxford, UK
e-mail: darren.dixon@chem.ox.ac.uk

P. Jakubec (✉)
Faculty of Chemical and Food Technology, Slovak University of Technology, Bratislava, Slovakia
e-mail: Pavol.jakubec@stuba.sk

metathesis, and cross-coupling reactions. The development of new catalytic systems for the Michael addition and ring-closing metathesis and unique, powerful route-shortening methodologies has been fundamental for the efficiency of the completed total syntheses.

Graphical Abstract

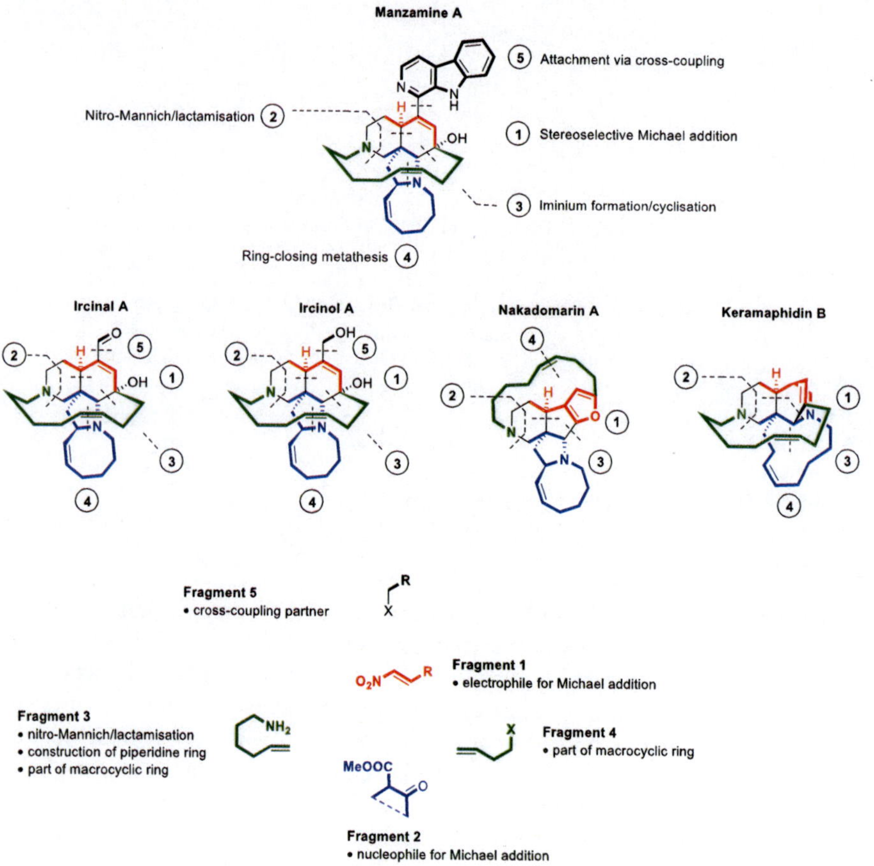

Keywords BIMP · Cascade · Iminium ion · Manzamine alkaloid · Metathesis · Michael addition · Nakadomarin A · Nitro-Mannich reaction · Organocatalysis · Stereoselective synthesis · Total synthesis

Abbreviations

AIBN	Azobisisobutyronitrile
BIMP	Bifunctional iminophosphorane
CSA	Camphorsulfonic acid
DABCO	1,4-Diazabicyclo[2.2.2]octane

DIBAL	Diisobutylaluminum hydride
DMAP	Dimethylaminopyridine
DMF	Dimethylformamide
DMSO	Dimethyl sulfoxide
Dr	Diastereomeric ratio
Er	Enantiomeric ratio
HWE	Horner-Wadsworth-Emmons
IBX	2-Iodoxybenzoic acid
KHMDS	Potassium bis(trimethylsilyl)amide
LHMDS	Lithium bis(trimethylsilyl)amide
MS	Mass spectrometry
pTSA	p-Toluenesulfonic acid
RCM	Ring-closing metathesis
THF	Tetrahydrofuran
TLC	Thin-layer chromatography

1 Introduction

Manzamine alkaloids are a class of marine alkaloids characterized by the presence of a macrocyclic ring, one or two nitrogen atoms, and for some members, by the presence of β-carboline moiety [1]. Manzamines may biogenically derive from ammonia, a C-10 unit (a symmetrical dialdehyde), tryptophan, and a C-3 unit (an acrolein equivalent), as it was proposed by Baldwin and Whitehead [2]. The novel family of marine alkaloids has been rapidly growing, already in 2008, Haman reported the existence of more than 80 manzamine alkaloids isolated from 16 different marine sponge species with genus *Amphimedon*, *Acanthostrongylophora*, *Haliclona*, *Xestospongia*, and *Ircinia* being the major source [3]. The family is characterized by a wide structural diversity, ranging from simple members such as keramaphidin C (**9**) [4] and manzamine C (**8**) [5, 6] to extremely complex alkaloids represented by dimeric neokauluamine (**6**) [7]. From a structural point of view, a typical manzamine alkaloid contains a complex array of rings of varying sizes, multiple stereogenic centers, and unusual structural motifs. Such characteristic features can be seen, for example, in madangamine D (**7**) [8], manzamine A (**1**) [9], nakadomarin A (**2**) [10], ircinal A (**3**) [11], ircinol A (**4**) [12], and keramaphidin B (**5**) [13]. Alongside their unique structural variety and complexity, manzamines have been fascinating the scientific community with their biological properties. The bioactivities include cytotoxicity, antimicrobial, pesticidal, and anti-inflammatory properties, and, perhaps most importantly, anti-malarial activities.

This review relays our journey on the development of a unified synthetic route to five members of the family – manzamine A (**1**), nakadomarin A (**2**), ircinal A (**3**), ircinol A (**4**), and keramaphidin B (**5**). Arguably, these molecules may be called – due to their breathtaking structural beauty, important biological properties, and the intense interest of the synthetic community – the most prominent manzamine alkaloids (Fig. 1).

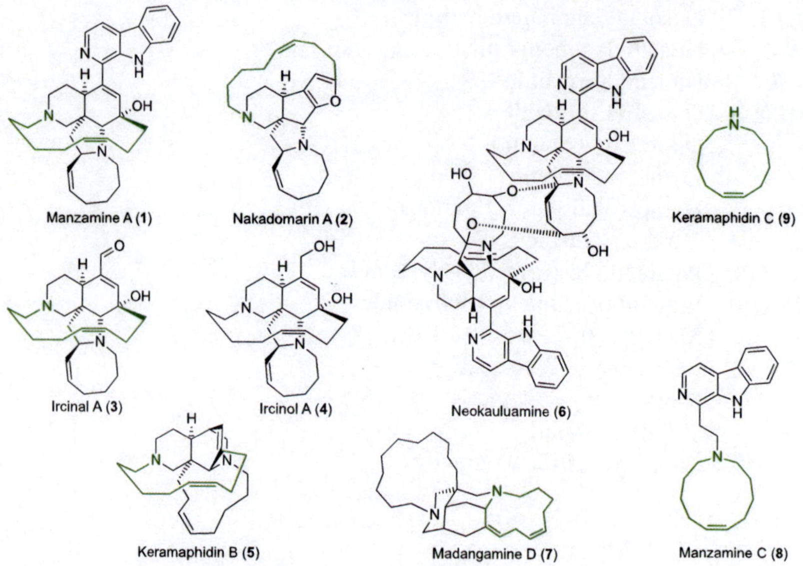

Fig. 1 Selected manzamine alkaloids

1.1 Isolation and Biological Properties of the Most Prominent Manzamine Alkaloids

Manzamine A (**1**), the prototype of the manzamine family, was first isolated by Higa in 1986 from a sponge in the Okinawa Sea [9]. Dissimilar to many other related alkaloids, a significant amount of manzamine A (**1**) was isolated from the natural source. From 735 g of a wet sponge, Higa separated 100 mg of the natural product as its hydrochloride salt. Moreover, in 2008 Hamann reported the development of "a method for kilogram-scale preparation of manzamines" yielding manzamine A (**1**) and other three members of the family [3]. The purification methods involved a large-scale extraction, chromatography, and crystallization. Manzamine A (**1**) has a pentacyclic (ABCDE) core comprising 6-, 6-, 5-, 13-, and 8-membered rings, two Z-configured double bonds, two tertiary amines, and five stereogenic centers. It displays a variety of important biological properties, including insecticidal, anti-bacterial, anti-inflammatory, and anti-cancer activities [9]. But perhaps the most noteworthy are its anti-malarial activities [9]. Manzamine A (**1**) remains a very promising lead for the treatment of malaria as it shows an improved activity over the clinically used drugs chloroquine and artemisinin both in vitro and in vivo [3, 14]. Its high importance as a potential lead for the development of novel drugs remains very clear due to regular discoveries of new promising biological properties [15–17].

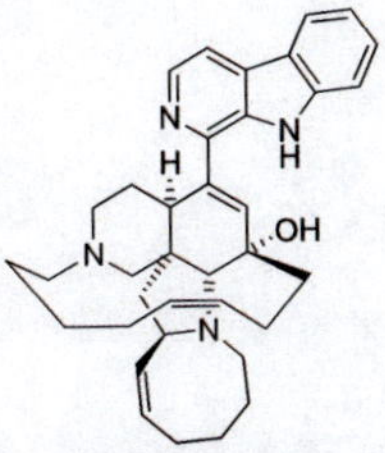

Manzamine A (**1**)

Nakadomarin A (**2**), unlike manzamine A (**1**), was isolated from a natural source only in a tiny amount. From 1 kg of a wet sponge, only 6 mg of the natural product was isolated by a combination of extraction and chromatographical methods [10]. Even the minute isolated quantity allowed the determination of its beautiful polycyclic structure comprising of an 8/5/5/5/15/6 ring system and four stereogenic centers, including one quaternary. Nakadomarin A (**2**) shows cytotoxicity against murine lymphoma L1210 cells, inhibitory activity against cyclin-dependent kinase 4, antimicrobial activity against a fungus *Trichophyton mentagrophytes,* and a Gram-positive bacterium *Corynebacterium xerosis* [10].

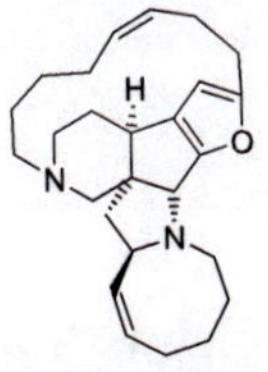

Nakadomarin A (**2**)

Both ircinal A (**3**) and ircinol A (**4**) were isolated for the first time by Kobayashi from marine sponges. Ircinal A (**3**) was isolated in 1992 from Okinawan marine sponge *lrcinia* sp. [11] followed by the isolation of ircinol A (**4**) in 1994 from the Okinawan marine sponge *Amphimedon* sp. [12] Interestingly, Kobayashi originally assigned the absolute configuration of ircinol A (**4**) to be opposite to ircinal A (**3**) and manzamine A (**1**) [12]. Only a personal communication with Winkler together with Winkler's independent total synthesis allowed determination of the absolute stereochemical configuration [18]. Ircinal A (**3**) is cytotoxic against L1210 murine leukemia cells and KB human epidermoid carcinoma cells in vitro [11]. Similar bioactivities were identified for ircinol A (**4**), it is cytotoxic against L1210 cells and KB cells and shows inhibitory activity against endothelin converting enzyme [12].

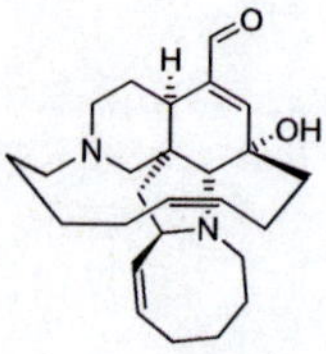
Ircinal A (**3**)

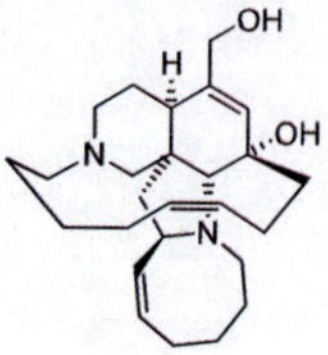
Ircinol A (**4**)

Kobayashi's fruitful isolation of novel natural products from marine sponges reached another landmark in 1994 when he isolated a complex, new alkaloid from the sponge *Amphimedon sp.* collected off the Kerama Islands, Okinawa. According to the origin of the alkaloid, it was named keramaphidin B (**5**) [13] and was found to show cytotoxicity against P388 murine leukemia and KB human epidermoid carcinoma cells [13].

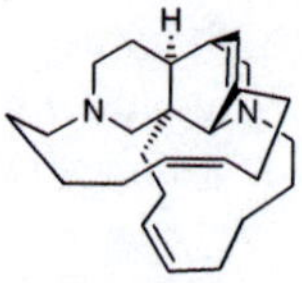
Keramaphidin B (**5**)

2 Synthetic Approaches to Manzamine Alkaloids

Although manzamine alkaloids are a relatively young family of natural products, they are already well known for their intriguing biological properties, wide molecular diversity, and a unique structural complexity containing unusual motifs. These key factors significantly contributed to the enormous interest of the scientific community, particularly of organic chemists and biochemists. As a result, since the discovery of manzamine A (**1**) in 1986, a surge of inventions resulted in many core syntheses. Multiple reports were dedicated to various approaches to manzamine A (**1**), nakadomarin A (**2**) ircinal A (**3**), ircinal A (**4**), and keramaphidin B (**5**) (Fig. 2) [110, 111, 113].

Eventually, all the most prominent manzamine alkaloids succumbed to total synthesis by various research groups. Accompanied by several groundbreaking discoveries in organic synthesis, manzamine A (**1**) was prepared synthetically for the first time by Winkler in 1998 [18]. Possessing the same pentacyclic core as manzamine A, (**1**) ircinal A (**3**) and ircinol A (**4**) were completed by Winkler at the same time [18]. In the following years, Martin accomplished a total synthesis of manzamine A (**1**), ircinal A (**3**), and ircinol A (**4**) [70, 71] and Fukuyama finished the preparation of unnatural enantiomer of manzamine A (**1**) [72]. In 2016 Nishida synthesized both ircinal A (**3**) and ircinol (**4**) and formally prepared manzamine A

Alkaloid	Isolation and properties[a]	Synthesis[b]
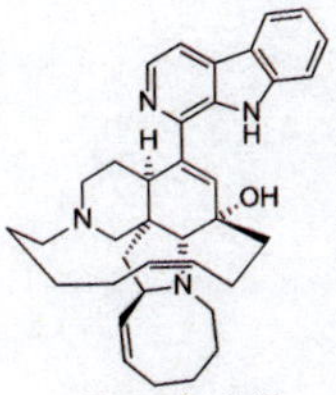 Manzamine A (**1**)	• Higa 1986 *Haliclona sp.* • Insecticidal • Anti-bacterial • Anti-inflammatory • Anti-cancer activities[9]	*Core:* • Pandit[19–24] • Hart[25–28] • Nakagawa, Hino 1990[29–31] • Nakagawa[32, 33] • Simpkins[34] • Martin[35, 36] • Winkler[37–38] • Winkler, Houk[39] • Winkler, Hamann[40] • Overman[41] • Leonard[42] • Yamamura[43-47] • Langlois[48–50] • Brands[51] • Coldham[52–55] • Clark[56, 57] • McIntosh[58] • Marazano[59, 60] • Magnus[61] • Markó[62] • Nishida[63–65] • Fukuyama[66] • Chau, Liu[67] • Dixon[68] *Formal:* • Nishida[69] *Total:* • Winkler[18] • Martin[70] • Martin[71] • Fukuyama[72] • Dixon[73]
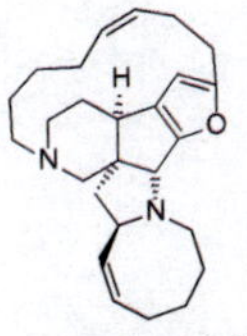 Nakadomarin A (**2**)	• Kobayashi 1997 *Amphimedon sp* • Cytotoxicity • Inhibitory activity against cyclin-dependent kinase 4 • Antimicrobial[10]	*Core:* • Fürstner[74, 75] • Nishida, Nakagawa[76, 77] • Magnus[61] • Tius[78] • Nishida[79] • Williams[80] • Kerr[81] • Zhai[82, 83] • Winkler[84] • Porter[85] • Takemoto[86] *Formal:* • Mukai[87] Stockman, Magnus[88] *Total:* • Nishida[89–91] • Kerr[92] • Dixon[93] • Funk[94] • Zhai[95] • Dixon[96, 97] • Dixon, Schrock, Hoveyda[98, 99] • Evans[100] • Boeckman[101] • Clark[102]
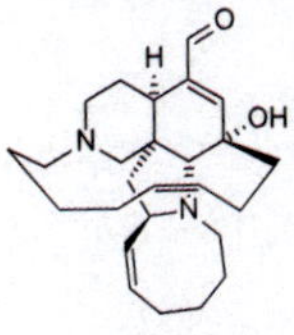 Ircinal A (**3**)	• Kobayashi 1992 *Ircinia sp* • Cytotoxic[11]	*Total:* • Winkler[18] • Martin[70] • Dixon[73] • Nishida[69]
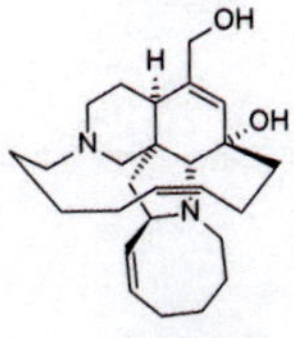 Ircinol A (**4**)	• Kobayashi 1994 *Amphimedon sp* • Cytotoxic • Inhibitory activity, endothelin converting enzyme[12]	*Total:* • Winkler[18] • Martin[70] • Dixon[73] • Nishida[69]
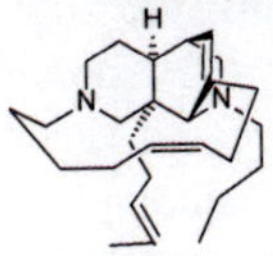 Keramaphidin B (**5**)	• Kobayashi 1994 *Amphimedon sp* • Cytotoxic[13]	*Core:* • Baldwin[103, 104] • Marazano, Das 1995[105] • Dixon[106] • Nakada[107] *Total:* • Baldwin[108, 109]

[a] Only the first reported isolation and selected properties referenced, synthesis of analogues excluded. For more details, see reviews by Tsukamoto, Hamann and Kubota[3, 110-112]

[b] Only the most relevant syntheses describing cores of a certain complexity are referenced

Fig. 2 Isolation, biological properties, and synthesis of the most prominent manzamine alkaloids

(**1**) [69]. Synthetic and theoretical studies on keramaphidin B (**5**) ultimately culminated in its biomimetic total synthesis achieved by Baldwin in 1998 [108, 109]. Although nakadomarin A (**2**) is one of the youngest family members, it became one of the most popular synthetic targets. To date it has already been successfully synthesized eight times, namely by Nishida in 2003 (unnatural enantiomer) [89] and 2004 (natural enantiomer) [90], Kerr in 2007 [92], Funk in 2010

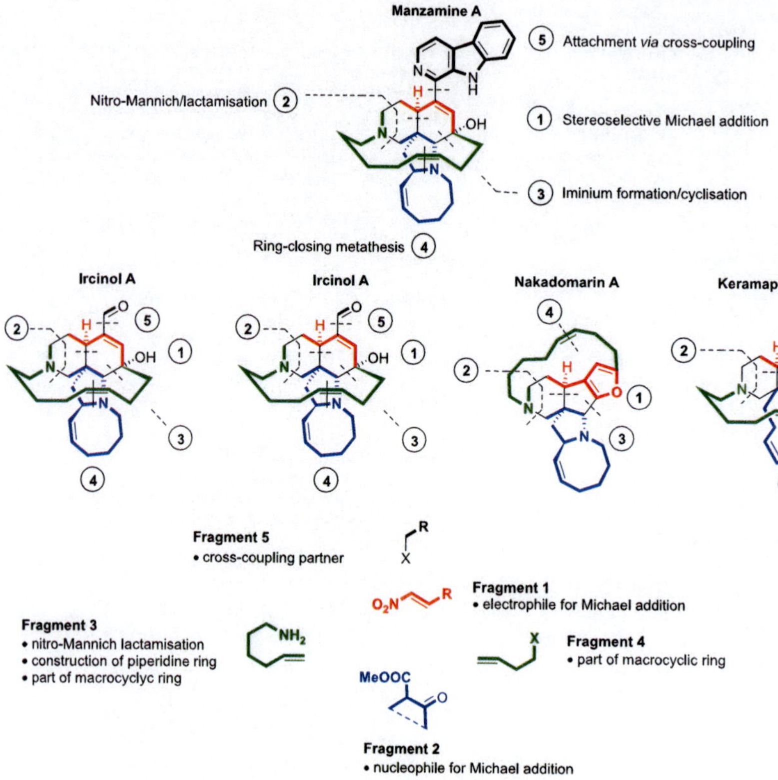

Fig. 3 Key structural units of the most prominent manzamine alkaloids and their corresponding building blocks

[94], Zhai in 2011 [95], Evans in 2013 [100], and most recently in 2016 by Boeckman [101] and Clark [102]. In 2010 two formal syntheses were reported by Mukai [87] and Stockman and Magnus [88]. Despite their elegance and very high efficiency all reported syntheses were designed to target individual alkaloids. Only Winkler's and Martin's approaches allowed the synthesis of manzamine A (**1**), ircinal A (**3**), and ircinol A (**4**) from the same intermediate. Interestingly, despite the structural similarities and the same biological origin, no general synthetic strategy for accessing all the most prominent manzamine alkaloids, e.g., manzamine A (**1**), nakadomarin A (**2**), ircinal A (**3**), ircinal A (**4**) and keramaphidin B (**5**) has been designed so far. This fact prompted us to analyze the structures further and identify key similarities from a synthetic point of view and possibly retrosynthetic disconnections (Fig. 3). The analysis revealed that three common key elements are present in all alkaloids – *Z*-configured double bond containing 10-carbon chain attached to a nitrogen atom (green fragment), mono- or bicyclic macrocyclic ring containing *Z*-configured double bond and a nitrogen atom (blue fragment), and a short carbon chain attached to a heteroatom (oxygen or nitrogen, red fragment).

Additionally, manzamine A (**1**), ircinal A (**3**), and ircinol A (**4**) contain an appendix of variable complexity attached to the pentacyclic core (black fragment). The structural resemblance between the alkaloids inspired a daring idea to construct all five complex alkaloids by the same synthetic strategy. Based on this idea and with suitable retrosynthetic analysis, we identified five essential chemical transformations that would assist with the stereoselective assembly of the alkaloids from 5 key fragments (Fig. 3). Thus, the general strategy applicable to the synthesis of the most prominent manzamine alkaloids would include stereoselective Michael addition, nitro-Mannich/lactamization cascade, iminium formation/cyclization, ring-closing metathesis, and a cross-coupling. Each of the essential transformations was selected to fulfill a very specific task. During the stereoselective Michael addition between a suitable 1,3-dicarbonyl nucleophile (fragment 2) and electrophilic nitro olefin (fragment 1) two consecutive stereogenic centers would be formed in a stereoselective manner. A nitro-Mannich/lactamization cascade was designed to form the piperidine ring present in all alkaloids. An iminium ion intermediate, formed from a corresponding amide, would allow a formation of the polycyclic structures via an individually designed cyclization. Ring-closing metathesis was selected to form the macrocyclic ring containing the *Z*-configured double bond. Importantly, the fruitful research on the field of metathesis offers a flexibility to construct the macrocyclic ring via either alkene or alkyne metathesis. Finally, a suitable cross-coupling reaction employing fragment 5 should enable a late-stage attachment of the "northern" appendix.

3 Nakadomarin A

3.1 Total Synthesis of Nakadomarin A: First-Generation, Alkyne Metathesis [96]

With the ambitious unified approach to manzamine alkaloids in mind, we began testing the hypothesis on nakadomarin A (**2**). In line with the general idea, we identified three key building blocks to assemble the polycyclic natural product – 5,5-bicycle **12** as a pro-nucleophile for the Michael addition, nitro olefin bearing tethered alkyne **11** as an electrophile, and amine **10** as the key component of the nitro-Mannich/lactamization (Fig. 4). Both the nitro olefin and the amine contained tethered alkynes, which would later form the macrocyclic ring via an alkyne ring-closing metathesis. The overall efficiency heavily relied on easy access to all building blocks via the short reaction sequences described in Schemes 1 and 2.

Amine **10** bearing the tethered internal alkyne was prepared from commercial chloride **13** in just four synthetic operations (Scheme 1). Firstly, the terminal alkyne was deprotonated with butyllithium and then methylated with methyl iodide yielding chloride **14**. Subsequently, the chloride underwent nucleophilic substitution with sodium azide in dimethyl sulfoxide, and the intermediate azide **15** was reduced to

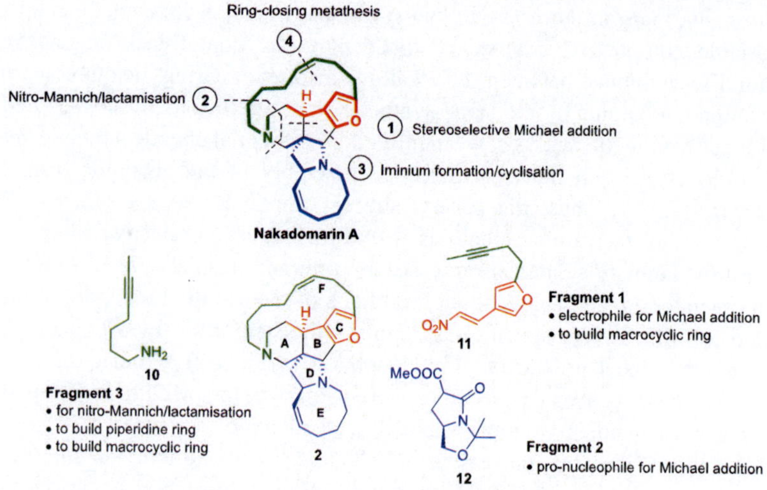

Fig. 4 Identification of the key structural units for the synthesis of nakadomarin A (**2**)

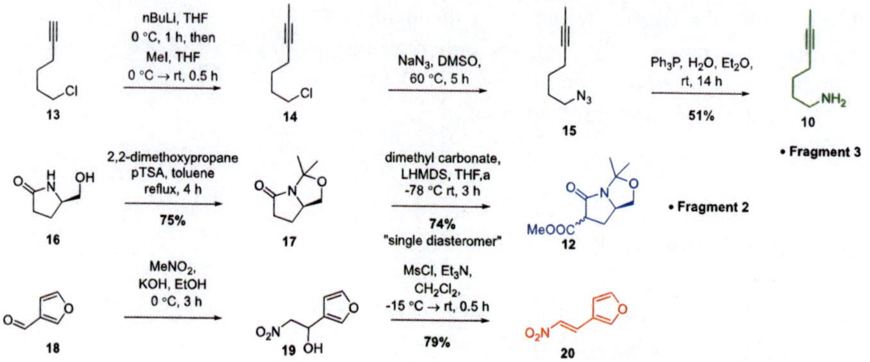

Scheme 1 Rapid synthesis of fragments **2** and **3** and model nitro olefin **20**

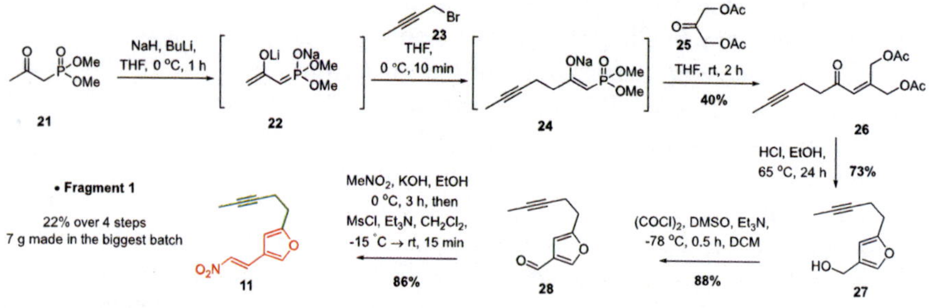

Scheme 2 Synthesis of fragment 1 – electrophilic nitroolefin **11** bearing tethered alkyne moiety

amine **10** using the mild Staudinger protocol. Even shorter was the synthesis of the bicyclic pro-nucleophile **12** (fragment 2), which consists of the protection of enantiomerically pure pyroglutamol (**16**) with 2,2-dimethoxypropane and the concomitant alkoxycarbonylation of lactam **17** with dimethyl carbonate (Scheme 1).

The remaining fragment 1, furanyl nitro olefin **11**, was constructed in four steps from commercially available ketophosphonate **21** (Scheme 2). A one-pot protocol initiated by double deprotonation of **21** followed by propargylation with bromide **23** and HWE reaction using diacetyl dihydroxyacetone (**25**) afforded enone **26**. Acid hydrolysis at elevated temperature then afforded alcohol **27**, which was subjected to Swern oxidation and 2-step Henry condensation to give the desired nitro olefin **11**. The whole sequence was performed multiple times on a multigram scale, and the biggest batch yielded almost 7 g of the valuable intermediate **11**.

Crucial to the success of our synthesis was the stereoselective Michael addition of chiral bicycle **12** and easily accessible nitro olefin **11**. Initially, we tested various conditions for the stereoselective construction of the C–C bond linking the quaternary stereocenter to the tertiary stereocenter on a model substrate **20** (Fig. 5). Preliminary results on model Michael acceptor **20**, prepared from furan carbaldehyde **18**, showed good levels of diastereocontrol when using bicyclic pro-nucleophile **12**.

The exploitation of achiral base DABCO (**30**) in toluene led to the formation of two of the possible four diastereoisomers **29** in 4.3:1 dr. Both diastereomers were separated and analyzed by single-crystal X-ray analysis. The analysis confirmed that the diastereomers were epimeric at the newly formed tertiary stereocenter and allowed us to draw important conclusions about the achieved stereocontrol. The stereoselectivity was controlled by the exclusive addition to the convex face of the enolized bicyclic pro-nucleophile. However, the facial selectivity of the addition to the nitro olefin was poor. To improve the poor diastereoselection, a range of chiral enantiomerically pure organocatalysts was tested. Pleasingly using a catalytic amount of cinchonine-derived urea **31a**, the reaction diastereoselectivity was significantly improved to 15:1:0:0 dr. The purity of nitro ester **29a** was further enhanced after column chromatography, and after crystallization **29a** was isolated in 80% yield as a single diastereomer. A reasonable reaction rate was achieved – the complete conversion was observed by TLC analysis already after 48 h. As expected, the use of the pseudoenantiomeric catalyst **32** derived from cinchonidine resulted in a poor and reversed diastereoselectivity (**29a**: **29b**: **29c**: **29d** = 4:12:1:1 dr) as well as reduced reaction rate (5 days). The model substrate study might have been a very short detour from the total synthesis but undoubtedly worthy of the efforts. The study provided essential information about the stability of both substrates, their inherent reactivity, catalytic activities of various catalytic systems, and shone light on the substrate and catalyst stereocontrol. We applied all the findings in the succeeding reaction of 5,5-bicyclic pro-nucleophile **12** with nitro olefin bearing the tethered alkyne **11** (Scheme 3).

Pleasingly, at slightly elevated temperature, the reaction was completed in 52 h producing nitro ester **33** in 81% yield as a single diastereomer (Scheme 3). In terms of our unified strategy, the importance of the successful, high yielding, and

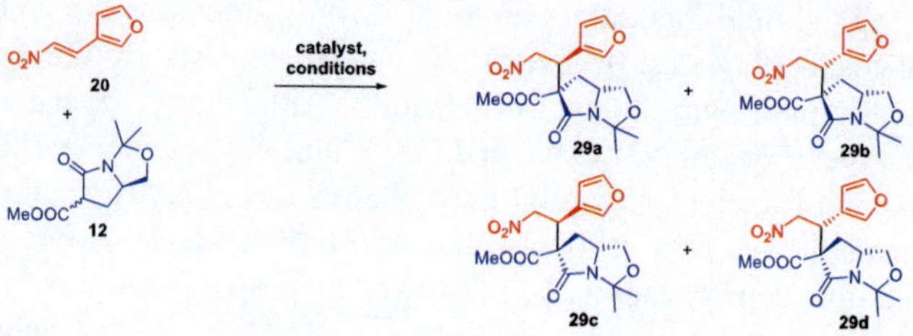

Catalyst (10 mol%)	Reaction conditions	Reaction time	Yield	Dr 29a:29b:29c:29d
 DABCO (30)	Toluene, 30 °C	7 days	38%	4.3:1:0:0
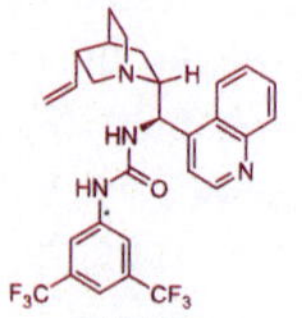 Catalyst 31a	Toluene, 30 °C	2 days	80% 1:0:0:0 After purification (single diastereomer)	Crude 15:1:0:0
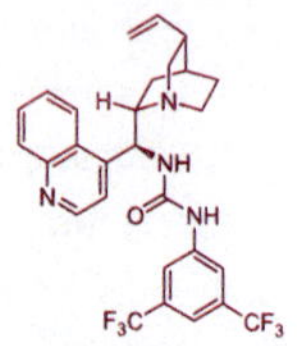 Catalyst 32	Toluene, 30 °C	5 days	61% Dr 6:19:1:1 After purification	Crude 4:12:1:1

Fig. 5 Michael addition of pro-nucleophile **12** and nitro olefin **20** – stereoselectivity model study

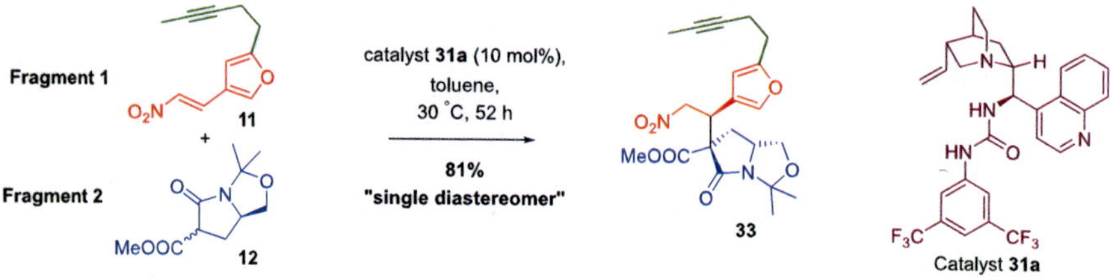

Scheme 3 Stereoselective Michael addition of fragment 1 and 2

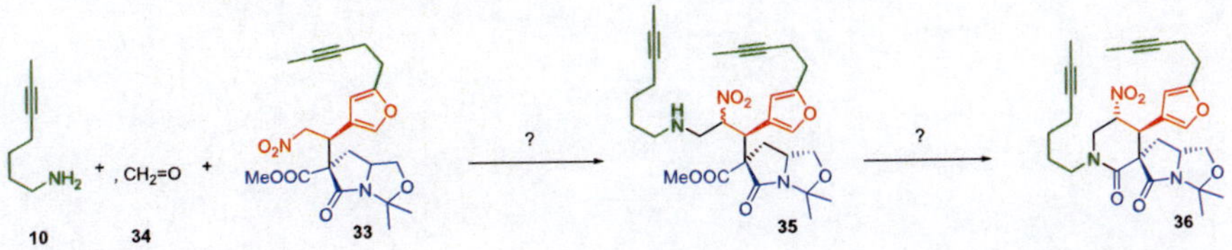

Scheme 4 Proposed multicomponent reaction for the formation of ring A of nakadomarin A (2)

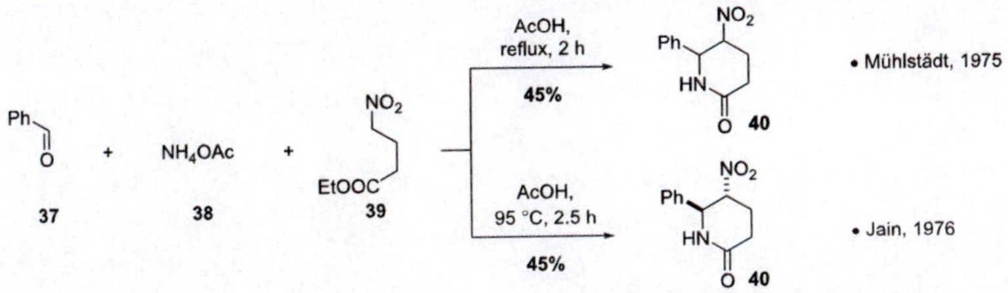

Scheme 5 Preliminary studies on the nitro-Mannich/lactamization cascade by Mühlstädt and Jain

stereoselective Michael reaction was immense. It allowed not only the immediate successful continuation of the nakadomarin A (**2**) first-generation synthesis but also became the cornerstone for the stereoselective assembly of all manzamine alkaloids. All our following total syntheses benefited, especially from the high stereoselectivity achieved by the combination of the catalyst and substrate control. With enantiomerically pure nitro ester **33** in hand, the piperidine containing ring A was the next challenge in our synthetic endeavors. We proposed to form the A ring in a three-component nitro-Mannich reaction between nitro ester **33**, amine **10** (fragment 3), and formaldehyde (**34**) followed by an intramolecular lactamization (Scheme 4).

In 1975 and 1976, Mühlstädt [114] and Jain [115], respectively, and independently reported a multicomponent synthesis of racemic nitro piperidinones **40** from ammonium acetate (**38**), simple nitro esters such as ethyl 4-nitrobutanoate (**39**) and carbonyl compounds (Scheme 5). Their seminal work, without a doubt, represented a breakthrough in the multicomponent construction of racemic nitro piperidinones and provided an attractive inspiration for our total synthesis. Surprisingly, despite its obvious huge synthetic potential, after the discovery, the method was rarely used and only then for specific syntheses [116].

Due to the higher complexity and high enantiomeric purity of nitro ester **33**, we decided to verify the seminal work and investigate the scope. To our delight, the Mühlstädt-Jain reaction proved to be far more general as it was demonstrated in a string of reports by us and others (Scheme 6, selected representative examples) [116–119]. The methodology applies to a broad range of heavily functionalized substrates **41** forming products **42** with different carbonyl moieties at C-3 and aromatic and heteroaromatic substituents at C-4 (**42a-c**). A variety of carbonyl components, including formaldehyde, glyoxylic acid, acetaldehyde, and anisaldehyde, successfully underwent the transformation yielding nitro piperidinone **42d-h** with the corresponding substituents at C-6. Further variability at position

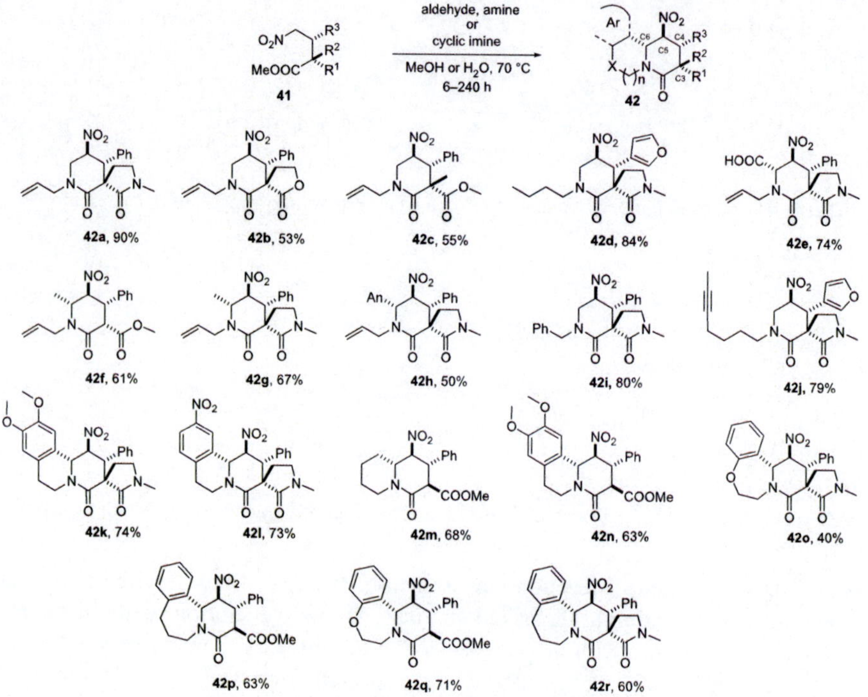

Scheme 6 The expanded scope of nitro-Mannich/lactamization cascade

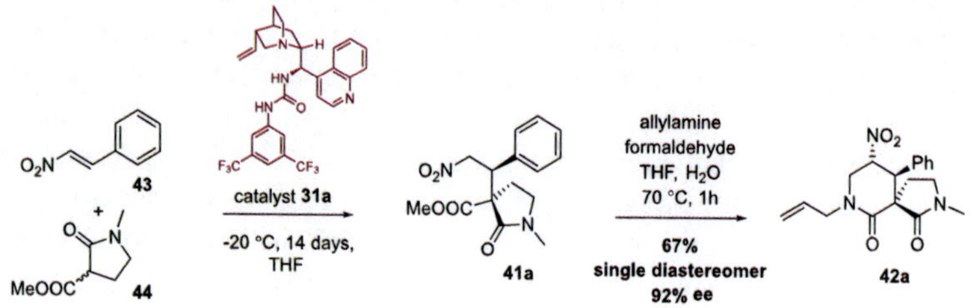

Scheme 7 One-pot Michael addition-nitro-Mannich/lactamization sequence

1 was secured by employing different amines such as allylamine, butylamine, benzylamine, or even moderately functionalized amines providing easy access to *N*-substituted lactams. Moreover, pre-formed cyclic imines were tolerated in the cascade reaction opening rapid access to diastereomerically pure polycyclic scaffolds such as **42 k-r**. Even stereoselective synthesis of substituted pyrrolidin-2-ones was achieved with functionalized 3-nitropropanoates as the starting materials (not shown) [119]. Overall, the original Mühlstädt-Jain reaction was significantly expanded in terms of its scope and transformed to a highly reliable tool for the formation of heavily substituted nitro piperidinones. Importantly, the high enantiomeric purity of nitro ester **41a** was not compromised during the reaction with

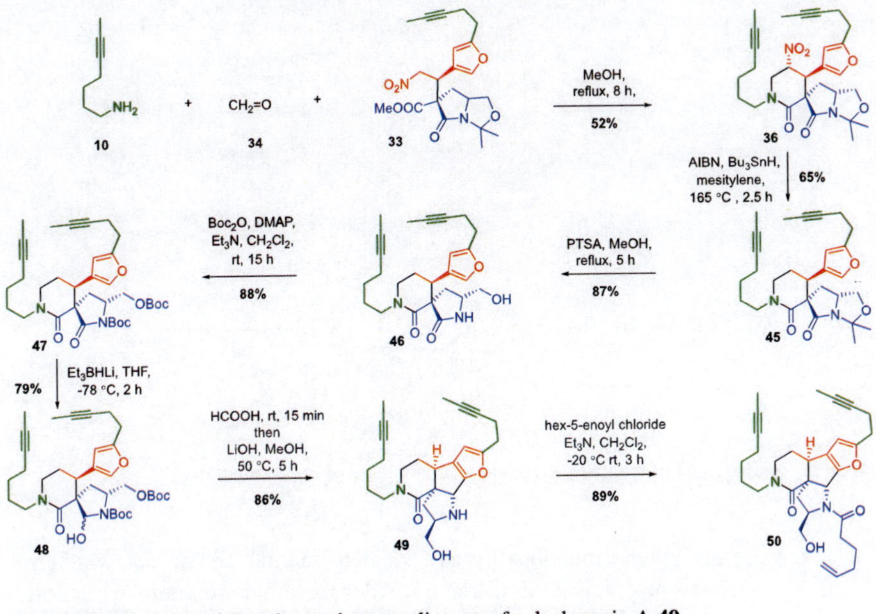

Scheme 8 Synthesis of the advanced tetracyclic core of nakadomarin A **49**

formaldehyde and allylamine (Scheme 7). The conservation of the stereochemical integrity (92% ee) during the one-pot procedure confirmed the suitability of the method for further elaboration of nakadomarin A (**2**).

Having optimized the nitro-Mannich/lactamization cascade, we confidently returned to the synthesis of ring A of nakadomarin A (**2**). Enantiomerically and diastereomerically pure nitro ester **33** was mixed with amine **10** and aqueous formaldehyde (**34**) and boiled in methanol for 8 h (Scheme 8). As a result of the cascade taking place, nitro piperidinone **36** was formed and isolated in acceptable yield as a single diastereomer. With the nitro group fulfilling its purpose in both the Michael addition and the nitro-Mannich reaction, it was removed with tributyltin hydride under standard radical conditions described by Ono [120]. To close ring B and construct the tetracyclic core **49** via an *N*-acyliminium ion cyclization, synthesis of suitable precursor **47** was required (Scheme 8).

This was achieved via a protecting group manipulation in **45** and its transformation to bis-Boc protected derivative **47**, which was perfectly poised for a selective hydride delivery achieved with Superhydride at low temperature. Next, intermediate **48** was subjected to neat formic acid for 15 h, followed by an alkaline hydrolytic workup affording the tetracyclic core of nakadomarin A **49**. The 11-step route to tetracyclic core **49** offered numerous possible endgame routes to complete the total synthesis. In fact our successful strategy involved a sequential building of macrocyclic rings E and F via alkene and alkyne metathesis, respectively. Suitable precursor for the metathesis endgame **52** was prepared from the tetracyclic core **49** in just three simple steps – hexenoylation, oxidation, and the Wittig reaction (Scheme 9). Alkyne RCM of **52** catalyzed by Schrock's catalyst was compatible

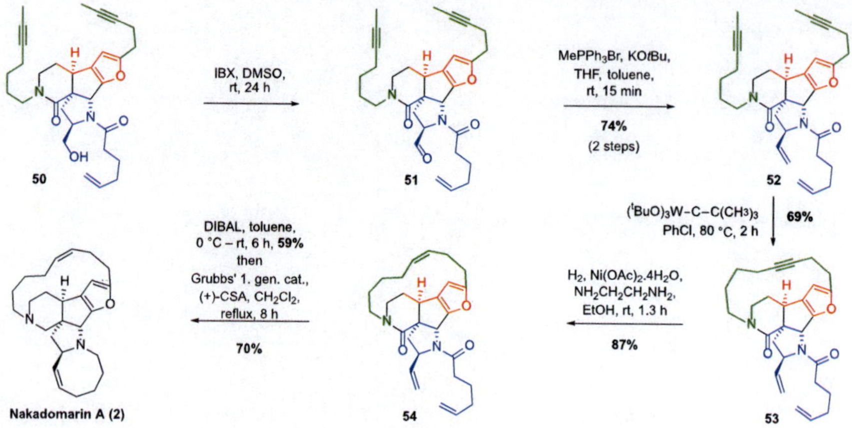

Scheme 9 Endgame of the first-generation total synthesis of nakadomarin A (**2**)

with the present alkene functionality and yielded internal alkyne **53**, which was readily reduced to macrocyclic *Z*-alkene **54**. After the amide to amine reduction in **54**, the remaining 8-membered ring D was closed by an alkene RCM enabled by Grubbs 1st generation catalyst. The structure of nakadomarin A (**2**) was confirmed by comparison of the collected analytical data with those of the literature. In summary, a 19-step route (longest linear sequence) to nakadomarin A (**2**) pivoting on the stereoselective construction of two consecutive stereocenters in a nitro olefin Michael addition has been developed. Several remarkable transformations were employed, including a highly diastereoselective furan/*N*-acyliminium cyclization and sequential alkyne and alkene ring-closing metathesis. Moreover, the proposed unified approach to manzamine alkaloids was successfully demonstrated for the first time. However, we believed the length of the first-generation synthesis of nakadomarin A (**2**) could be significantly shortened by employing a more convergent approach. Therefore, a second-generation route was sought.

3.2 Total Synthesis of Nakadomarin A: Second Generation, Alkyne Metathesis [97]

We scrutinized the first-generation route and identified the late-stage assembly of the 8-membered ring as significantly contributing to the high step count of the synthesis. Consequently, the 5,5 bicyclic pro-nucleophile **12** was replaced by 8,5 bicyclic pro-nucleophile **55** to slash the step count. While using the same key transformations, the utilization of the 8,5 bicyclic nucleophile instead of the 5,5 bicycle offered a much more convergent approach over the first-generation route (Fig. 6).

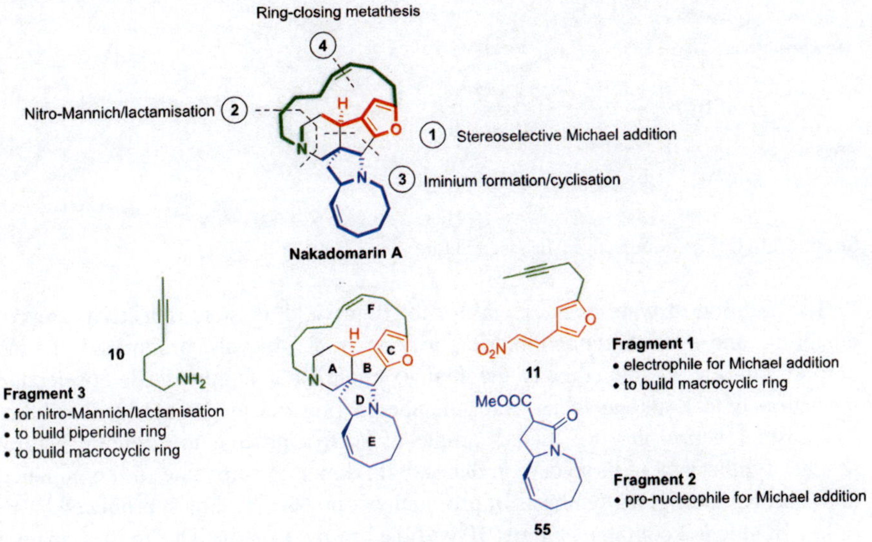

Fig. 6 Key structural units for the second-generation synthesis of nakadomarin A (**2**)

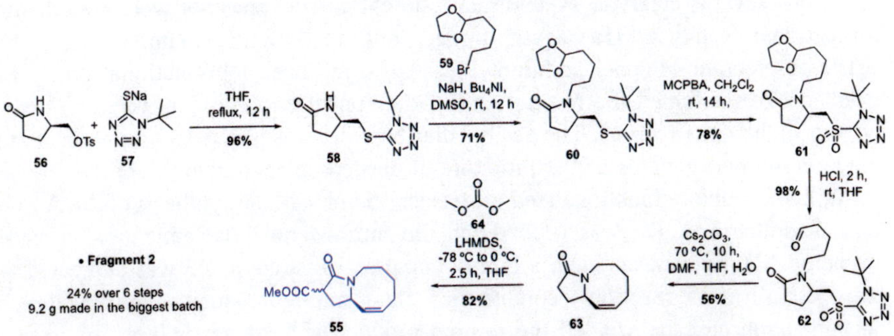

Scheme 10 Stereoselective synthesis of fragment 2, 8,5 bicyclic pro-nucleophile **55**

Additionally, the use of 8,5 pro-nucleophile **55** would remove several protective group manipulation steps and improve the overall atom economy of the process. Like the first-generation route, the synthesis again relied upon a rapid convergent assembly of 3 building blocks – the same nitro olefin **11** and amine **10** and modified fragment 2, 8,5-bicyclic pro-nucleophile **55** (Fig. 6). The most complex fragment 2 was prepared in six linear steps from the commercially available tosylate ester of pyroglutamol (**56**) (Scheme 10). In situ generated sodium thiolate **57** reacted with tosylate **56** and afforded sulfide **58**. *N*-Alkylation with bromide **59** followed by sulfide-sulfone oxidation and acetal deprotection generated aldehyde **62**, the precursor to the intramolecular Julia-Kocienski olefination. At moderate dilution on a multigram scale, this was efficiently carried out without racemization using cesium carbonate as the base in a THF-DMF-water mixture.

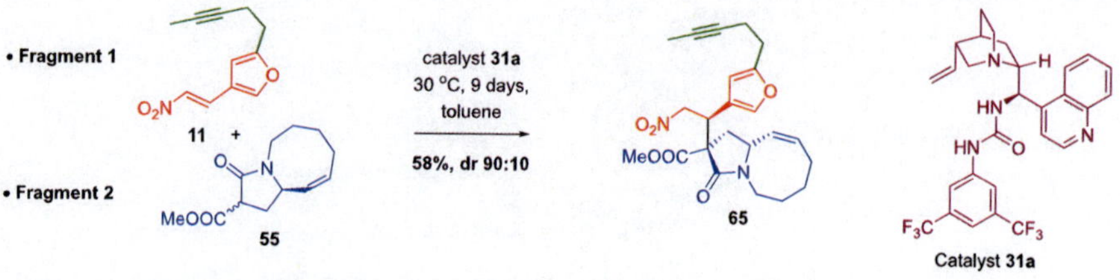

Scheme 11 Key stereoselective Michael addition

The addition of water was crucial for the high yield, diastereoselectivity, reproducibility, and even the enantiomeric purity of the 8,5-bicyclic product **63**. To the best of our knowledge, this is the first example of a highly diastereoselective formation of a *Z*-alkene in an eight-membered ring via an intramolecular Julia-Kocienski reaction and the first example of such a process in complex natural product synthesis. *C*-Alkoxycarbonylation with dimethyl carbonate (**64**) completed this practical multigram synthesis of pro-nucleophile **55**. The largest produced batch of key fragment 2 contained almost 10 g of the pro-nucleophile. Due to the apparent shortness and robustness of the building blocks synthesis, we rapidly proceeded to their stereoselective union via a conjugate addition and nitro-Mannich lactamization. Although several catalytic systems and stoichiometric reagents were tested, the bifunctional catalyst **31a** again turned out to be the champion for the diastereoselective Michael addition. Pleasingly, the use of bifunctional cinchona catalyst **31a** facilitated the reaction and formation of two consecutive stereogenic centers in the nitro ester **65**. The desired diastereomer was isolated after a 9-day long reaction in good yield as a 90:10 mixture of diastereomers (Scheme 11).

With the suitably functionalized nitro ester **65** in hand, the synthesis of the A ring of nakadomarin A (**2**) was probed via the nitro-Mannich/lactamization cascade (Scheme 12). Even when such a highly complex substrate as **65** was subjected to the established reaction conditions, the multicomponent nitro-Mannich lactamization cascade yielded the desired product **66** proving its high robustness and reliability. Spirocyclic nitro piperidinone **66** was formed after nitro ester **65** was briefly refluxed with formaldehyde (**34**) and amine **10** in methanol. Ono's reductive removal of the nitro group on **66** occurred smoothly using tributyltin hydride and AIBN as a radical initiator. To our delight, the denitrated piperidinone **67** was isolated as a single diastereomer. Further elaboration of **67** into the pentacyclic core of nakadomarin A **71** required a successful complete reduction of the δ-lactam to piperidine **68** and a partial reduction of the γ-lactam to hemiaminal **69**. Anticipating a similar low reactivity of both lactams and low stability of hemiaminal **69**, differentiation between both lactams represented one of the major synthetic challenges of the total synthesis. Fortunately, a significant reactivity difference between the lactams was observed. Remarkably, δ-lactam in **67** was fully reduced when exposed to lithium aluminum hydride followed by a formic acid treatment. The next treatment of the intact γ-lactam in **68** with DIBAL in toluene at −20°C afforded metallated hemiaminal **69**, which upon treatment with dilute aqueous

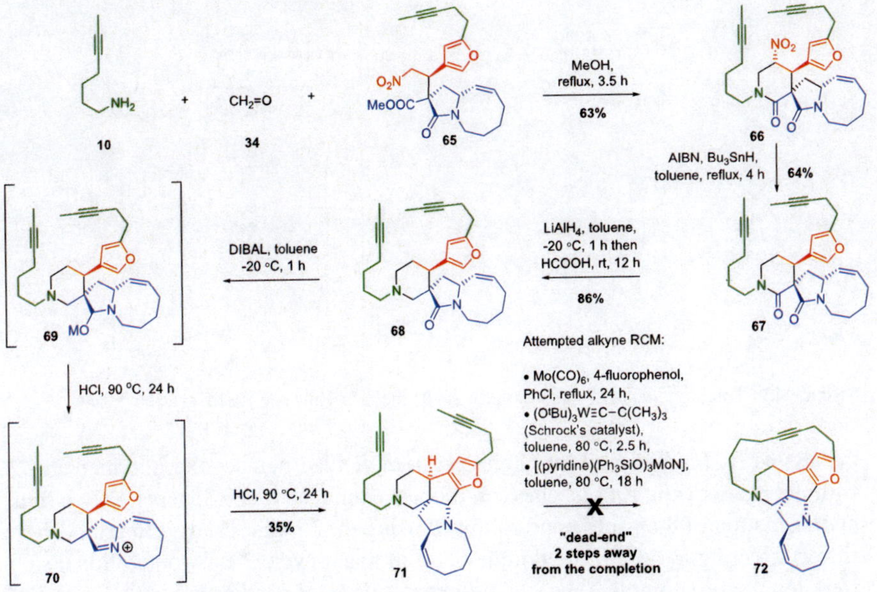

Scheme 12 Failed total synthesis of nakadomarin A including alkyne RCM – "dead-end" scenario

hydrochloric acid provided MS-detectable iminium ion **70** – the perfect substrate for an intramolecular furan-iminium cyclization. As a result of the anticipated, but unprecedented 5-*exo*-trig furan-iminium cyclization, the key pentacycle **71** was formed in a moderate yield. With the total synthesis almost completed, several commercially available alkyne ring-closing metathesis catalysts were screened on the pentacyclic compound with possessing alkynes. In particular, Mortreux's bench stable catalytic system [121] improved by Grela [122], Furstner's [123], and Schrock's catalyst [124] was amongst the compounds we tested. Disappointingly, none of the used catalysts yielded desired hexacyclic target product **72**, and only unreacted starting material **71** was recovered in most cases.

In-situ protection of both basic tertiary amines with camphor sulfonic acid (CSA) failed to provide the crucial reactivity of the catalysts, and no reaction was observed again. We speculated that the presence of the tertiary amines and the anticipated strain originating from the internal tripe bond in **72** significantly contributed to the lack of reactivity in the alkyne ring-closing metathesis. After reaching the synthetic "dead-end," we decided to redesign the original second-generation route and reshuffle the key steps. Such modification of the order of the steps would allow performing the alkyne metathesis on a different substrate possessing lactams rather than amines. Masking the basic amines seemed like a rather simple solution and a quick way out from the "dead-end" scenario of our second-generation synthesis of nakadomarin A (**2**). Subsequently, lactam **67** was selected as the most suitable substrate to perform the alkyne ring-closing metathesis. Indeed, the new reaction sequence turned out to be very fruitful (Scheme 13). Performed at high dilution (0.26 mM concentration),

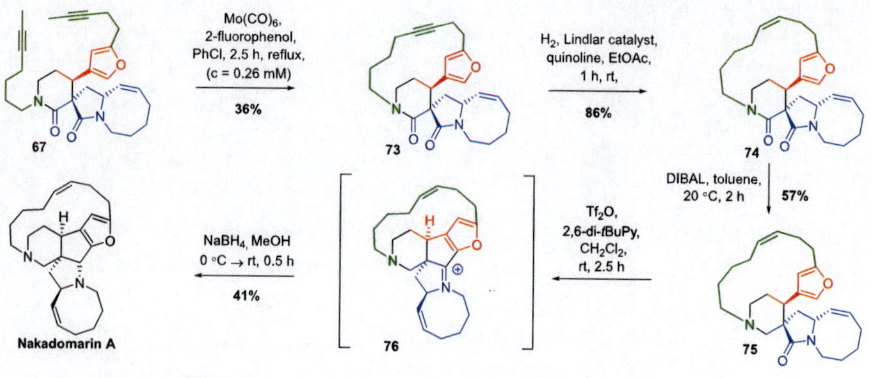

Scheme 13 Total synthesis of nakadomarin A (**2**) including alkyne RCM

the alkyne RCM catalyzed by Grela's system yielded pentacyclic intermediate **73**, with the macrocyclic ring in place, in modest chemical yield (Scheme 13). A semi-hydrogenation of the triple bond followed using the Lindlar catalyst to establish the correct *Z*-configuration of the double bond in macrocycle **74**. Two additional steps were required to complete the second-generation total synthesis. Firstly, even with the macrocyclic ring in place influencing the reactivity, the remarkable chemoselectivity enabled preferential reduction of the δ-lactam over the γ-lactam yielding spirocycle **75**. Activation of the remaining γ-lactam with triflic anhydride and 2,6-di-*tert*-butylpyridine triggered a cascade of reactions producing MS-detectable iminium species **76**, which upon stereoselective reduction with $NaBH_4$ provided nakadomarin A (**2**).

Despite the moderate chemical yield, the one-pot route-shortening transformation of lactam **75** to nakadomarin A (**2**) became one of the critical steps and highlights of the second-generation synthesis. Overall, the synthesis was completed in just 13 linear steps and significantly reduced the number of steps of the first-generation synthesis.

3.3 *Total Synthesis of Nakadomarin A, Third Generation, Alkene Ring-Closing Metathesis [93]*

Although both alkyne metathesis routes enabled an efficient synthesis of nakadomarin A (**2**), the necessary stereoselective reduction of the triple bond inherently prolonged the sequence. Therefore, as a part of our attempt to further improve the synthesis, a third-generation route employing an alkene metathesis was pursued [93]. The third-generation route to nakadomarin A (**2**) was built on the success of our previous first- and second-generation routes but addressed the major limitation of the second-generation route – 2-step construction of the macrocyclic ring. Straightforward stereoselective alkene RCM was proposed to solve this

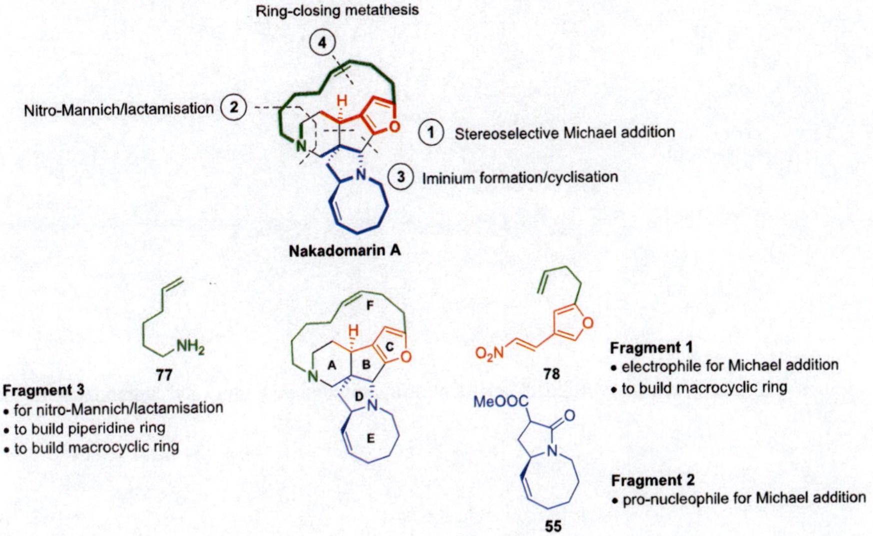

Fig. 7 Key fragments for the synthesis of nakadomarin A (**2**) employing stereoselective alkene RCM

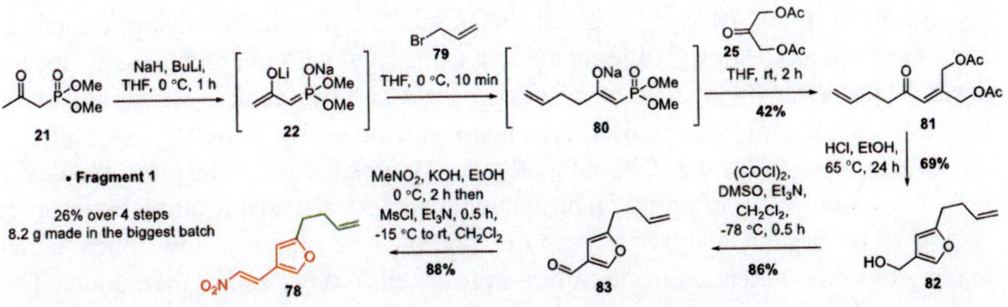

Scheme 14 Short preparation of fragment 1 for the third-generation synthesis of nakadomarin A (**2**)

significant limitation. Compared to the second-generation route, the overall synthetic strategy, concept, and key building blocks remained much unchanged except the side-chain terminal functional groups in fragments 1 and 3 (Fig. 7). Both fragments 1 and 3 contained tethered terminal alkenes instead of the internal alkyne. Their preparation required only a simple modification of starting materials because the reaction conditions were developed in our previous synthesis.

Nitro olefin **78** was constructed again in four steps from the same commercially available keto phosphonate **21** (Scheme 14). A one-pot, sequential allylation followed by HWE reaction using dihydroxyacetone diacetate (**25**) produced enone **81** (Scheme 14).

Acid hydrolysis at elevated temperature then afforded alcohol **82**, which was subjected to Swern oxidation and 2-step Henry condensation to give the desired nitro

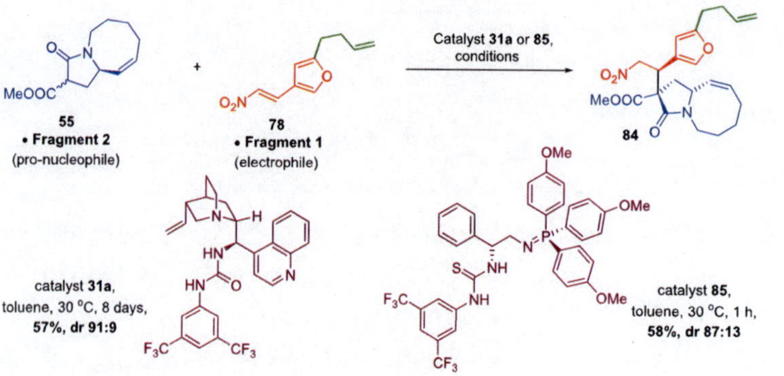

Scheme 15 Stereoselective Michael addition using bifunctional urea and iminophosphorane catalysts

olefin **78**. The whole sequence was performed on a multigram scale, and the biggest batch yielded more than 8 g of the stable intermediate suitable for the following diastereoselective Michael addition. In anticipation of similar results to those obtained in the second-generation route, the Michael addition of pro-nucleophile **55** and electrophile **78** was executed under identical conditions using a catalytic amount of organocatalyst **31a** (Scheme 15). Consistent with the previous results, the desired nitro ester **84** was isolated in good yield and respectable stereoselectivity. However, the prolonged reaction time again somehow diminished the overall efficiency of the key reaction. Consequently, the sluggish reaction inspired the development of a new class of powerful bifunctional organocatalysts. A problem observed during some organocatalyzed conjugate addition reactions is a low reaction rate leading to rather extended reaction times measured in days, rather than hours. This major problem is often attributed to the relatively low Brønsted basicity of common organocatalysts. This issue was addressed in a new class of potent bifunctional iminophosphorane BIMP organocatalysts containing a strongly basic moiety [125, 126]. The Michael addition of **55** and **78** was selected as a suitable showcase to demonstrate the power of BIMPs. A remarkable increase in the reaction rate was observed when BIMP organocatalyst **85** was employed, complete conversion and comparable diastereoselectivity were achieved already within 1 h (Scheme 15) [127]. Relying on the previously developed steps the following nitro-Mannich/lactamization (transformation **84** → **86**), protodenitration (**86** → **87**), chemoselective δ-lactam reduction (**87** → **88**), and furan-iminium cyclisation (**88** → **89**) occurred without incident giving rapid access to the tetracyclic core **89** (Scheme 16).

Our third-generation synthesis of nakadomarin A (**2**) was completed by a stereoselective alkene ring-closing metathesis. Initially, the macrocyclization promoted by Grubbs first-generation catalyst (**90**) in the absence of any additives showed a bias towards the *E*-isomer (*Z*:*E* 40:60). This poor and undesired stereoselectivity prompted us to investigate the effect of additives and different

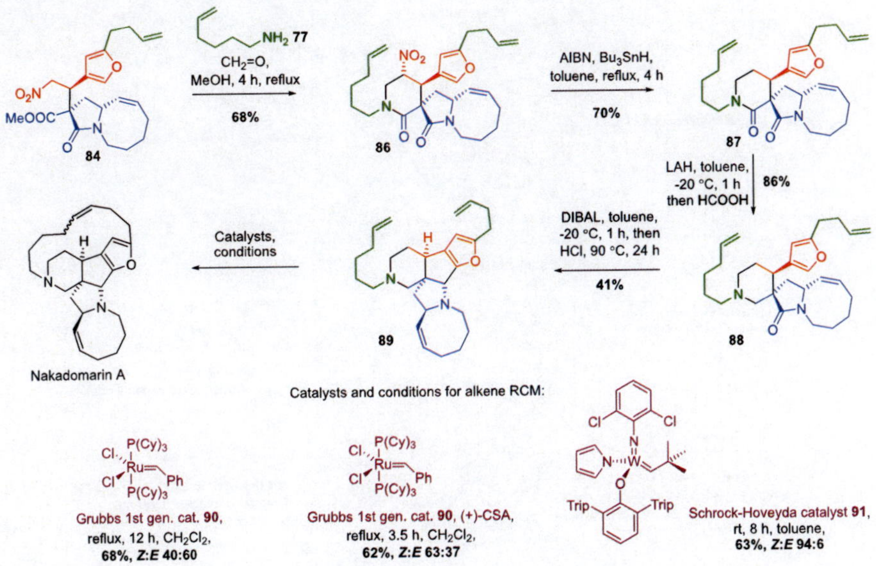

Scheme 16 Third-generation total synthesis of nakadomarin A (**2**) – endgame

catalytic systems. We found out that protonation of the tertiary amines with Brønsted acid ((+)-CSA) reversed the diastereoselectivity resulting in the formation of 63:37 *Z*:*E* mixture of isomers. The observed reversal of diastereoselectivity in alkene RCM caused by the amine protonation, allowed the separation of nakadomarin A (**2**) by chromatography and completion of the total synthesis. Nevertheless, the formation of a significant amount of *E*-isomer in the last step reduced the efficiency of the synthesis. The low stereoselectivity of the process was further studied and solved following the development of a series of new catalysts for *Z*-stereoselective RCM, by Schrock and Hoveyda [98, 99]. The same alkene ring-closing metathesis catalyzed by tungsten complex **91** provided remarkable results, nakadomarin A (**2**) was isolated in 63% yield and 94:6 *Z*:*E* ratio. In summary, the third-generation total synthesis of nakadomarin A (**2**) became the shortest and the most efficient amongst the developed routes. It features the facile stereoselective Michael addition catalyzed by the BIMP organocatalyst and the highly Z-stereoselective catalyst-controlled late-stage alkene RCM.

4 Manzamine A [73]

Manzamine A (**1**), arguably the most important member of the manzamine alkaloids, provides almost endless inspiration for drug development, the discovery of new methodologies, and advances in novel synthetic tactics.

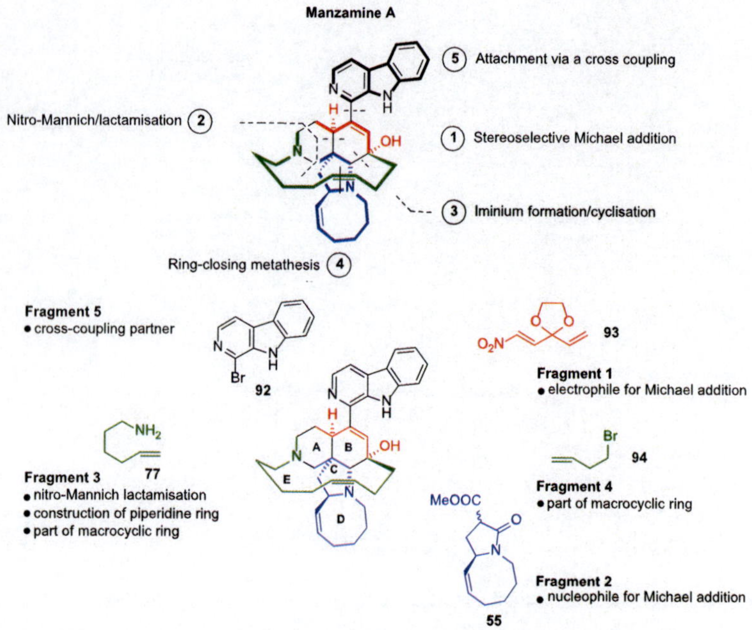

Fig. 8 Key fragments for the stereoselective synthesis of manzamine A (**1**)

Its fascinating polycyclic core comprises 6-, 6-, 5-, 13-, and 8-membered rings, two *Z*-olefins, two tertiary amines, and five stereocenters, including four contiguous and two quaternary centers. Due to its breathtaking molecular beauty and the unique biological profile, manzamine A (**1**) became an irresistible target for our unified synthetic strategy. All key steps of our general strategy towards manzamine alkaloids were thoroughly tested in the nakadomarin A (**2**) syntheses. The successful completion of all of them boosted our confidence in the general strategy and fueled its further application in the total synthesis of manzamine A (**1**). However, due to the overwhelming complexity of manzamine A (**1**) and the presence of unusual synthetic motifs, we expected that the general strategy would require a fine-tuning and still be put to the test. Especially challenging appeared to be the connection of the pentacyclic core with the β-carboline moiety. After considering multiple synthetic strategies and various disconnections, we proposed five key building blocks to take part in the stereoselective synthesis of manzamine A (**1**) (Fig. 8). Four of these building blocks, namely 8,5 bicyclic pro-nucleophile **55**, nitro olefin **93**, amine **77**, and bromide **94** containing tethered alkenes, would serve for the construction of the polycyclic core of manzamine A (**1**) and fit well into the proposed modular strategy. The fifth fragment, β-carboline derivative **92**, would serve for a late-stage attachment of the carboline moiety. The proposed synthesis was again reliant on a high convergency and easy access to the building blocks. Most of the necessary fragments were already available. 8,5 bicycle **55** was prepared as in the nakadomarin A (**2**) synthesis, and fragments 3, 4, and 5 were identified as commercially available. The

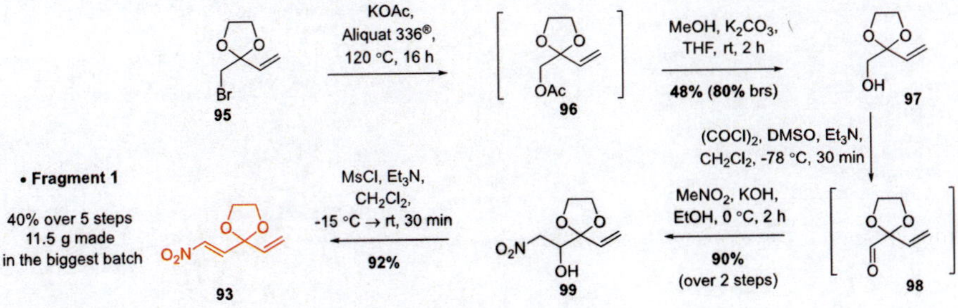

Scheme 17 Preparation of fragment 1 for the stereoselective synthesis of manzamine A (**1**)

actual preparation of building blocks was narrowed down to only nitro olefin **93**. The presence of the cyclic acetal in nitro olefin **93** predestined alcohol **97** as the key intermediate for its straightforward synthesis (Scheme 17). Despite the apparent challenges associated with the neopentylic structure, we believed that the alcohol could be accessed from readily available bromide **95** [128]. After significant optimization, Aliquat 336 was identified as a suitable reaction medium for the challenging nucleophilic substitution performed under very harsh reaction conditions. Potassium acetate was found as the most suitable nucleophile for the bromide replacement. Apart from the desired acetate **96**, unreacted bromide **95** was recovered from the crude reaction mixture as well. The bromide for acetate substitution was followed by a mild acetate deprotection mediated by potassium carbonate in methanol to produce neopentylic alcohol **97**. Alcohol **97** was oxidized to aldehyde **98** using the standard and reliable Swern protocol.

The aldehyde, which was used without any purification, was converted to the desired nitro olefin via a two-step Henry condensation. Excellent chemical yield for the last step and the robustness of the invented sequence allowed the preparation of fragment 1 on a multigram scale. The amount of material isolated from the biggest batch exceeded 10 g. After the rapid assembly of all essential building blocks, the Michael addition of pro-nucleophile **55** to nitro olefin **93** was tested under various conditions. Although the organocatalyzed approach had limited success, the KHMDS-promoted reaction proceeded rapidly with acceptable diastereoselectivity even at a very low temperature (Scheme 18). Pro-nucleophile **55** was deprotonated with KHMDS at −78°C, and the naked nucleophile **100** formed in the presence of 18-crown-6 reacted with electrophilic nitro olefin at −94°C to furnish nitro ester **101** in 86% yield as a 3:1 mixture of epimers at the tertiary stereogenic center. After the separation of these diastereomers by column chromatography, the major isomer was analyzed by X-ray diffraction. In an excellent agreement with the previous studies, the crucial analysis confirmed the desired configuration of both tertiary and quaternary stereogenic centers.

Aiming for the synthesis of ring A of manzamine A (**1**), the nitro ester **101** was further reacted with formaldehyde and amine **77** in refluxing methanol under previously established conditions (Scheme 19). The highly reliable nitro-Mannich/lactamization cascade proceeded without incident and afforded nitro piperidinone

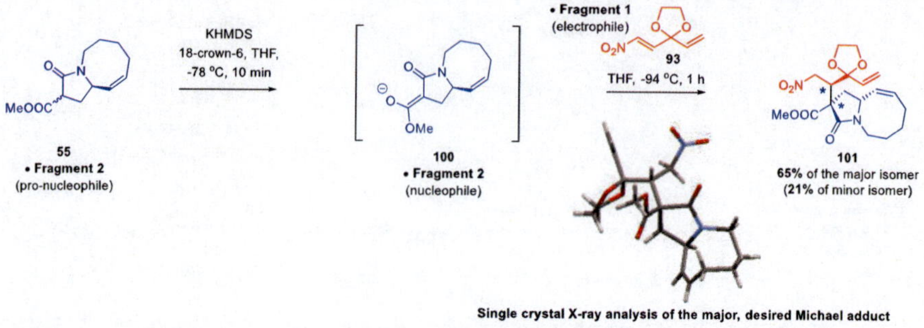

Scheme 18 Stereoselective Michael addition of pro-nucleophile **55** and nitro olefin **93**

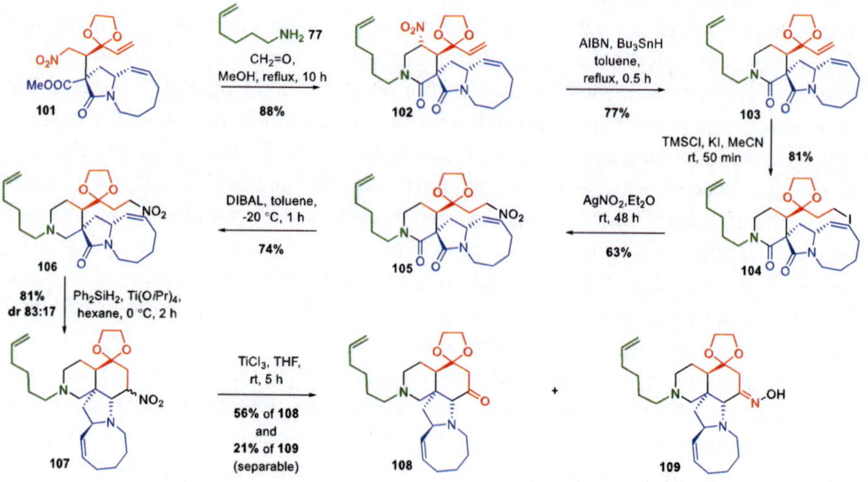

Scheme 19 Elaboration of the tetracyclic core of manzamine A **108**

102 in an excellent 88% yield as a single diastereomer. With suitably functionalized spirocycle **102** in hand synthesis of ring B via an intramolecular cyclization to the γ-lactam was considered. Such an approach required denitration of the nitro piperidinone **102**, followed by chemoselective functionalization of the double bond attached to the cyclic acetal. Ono's radical protodenitration of **102** proved to be highly reliable once again, yielding piperidinone **103**. Next, the chemo- and regioselective addition of hydrogen iodide was achieved applying the conditions developed by Feringa [129]. Noteworthy are not just the remarkable selectivities of the addition but also the high chemical yield for such an unusual transformation. The reactive iodide **104** then underwent nucleophilic substitution with silver nitrite performed with the exclusion of light. The resulting nitro lactam **105** possessed all key structural features for the intended cyclization – tethered nitro alkane as a pro-nucleophile and γ-lactam as a pro-electrophilic site of the molecule. Before any attempt to test the actual cyclization piperidinone **105** was reduced to piperidine **106** with DIBAL in toluene at −20°C. Substantial effort was then invested into the investigation of the partial reduction of lactam **106** with the concomitant intramolecular nitro-Mannich reaction. As a result, access to **107** proved to be straightforward using a mixture of diphenylsilane and titanium isopropoxide in hexane, introduced by Buchwald for reduction of amides to aldehydes [130]. Suitable functionalized tetracyclic core of manzamine A **107** was isolated in respectable 81% yield as a mixture of epimers in ratio 83:17. A broader synthetic potential of this single iminium/intramolecular nitro-Mannich annulation was recognized and further harvested in the development of a novel methodology [131]. Being epimeric at the stereogenic center bearing the nitro group, the intermediate **107** was perfectly poised for the next two transformations – the Nef reaction followed by the introduction of the last non-aromatic carbon chain. Although the Nef reaction is one of the best-developed transformations of the nitro group [132], the presence of a variety of sensitive functional groups significantly restricted the arsenal of available reagents.

A plethora of reagents is available for oxidative variants of the Nef reaction and modifications performed in highly acidic or basic media. These options were quickly ruled out from our screening due to the presence of tertiary amines, acid-labile acetal, and the epimerization prone product containing an α-amino ketone. On the other hand, McMurry's reductive variant of the Nef reaction [133] tolerated all functional groups present in the molecule and allowed the smooth transformation of the nitro alkane into a carbonyl compound. A mixture was ketone **108**, and oxime **109** was produced when nitro compound **107** was treated with titanium trichloride in tetrahydrofuran. Both products were separated by column chromatography, and oxime **109** was further analyzed by single-crystal X-ray analysis. The favorable outcome of the X-ray analysis (not shown) proved the desired configuration at all stereogenic centers of the tetracyclic core of manzamine A. Having executed the "umpolung" transformation of the nitroalkane to the ketone the stage was set for the introduction of the last non-aromatic carbon chain. The natural electrophilic character of the ketone **108** and its steric predispositions determined the choice of reagents for the envisioned stereoselective reaction. Initially, Grignard and organolithium organometallics were tested with very limited success. A thorough analysis of the crude reaction mixtures indicated base-induced elimination processes were occurring. Such an outcome leads to the moderation of the strongly basic nature of the nucleophiles achieved by transmetalation with anhydrous cerium chloride. The addition of organocerium derivative to ketone **108** proceeded smoothly with excellent stereoselectivity (Scheme 20). Moreover, the following acid-catalyzed acetal hydrolysis of the intermediate **110** performed in the same pot allowed the preparation of keto alcohol **111** in an overall excellent (92%) chemical yield.

The subsequent TMS-protection of the tertiary alcohol **111** and enol triflate formation proceeded according to plan and yielded diene **113**, which underwent an alkene RCM. Despite the presence of various functional groups, the macrocyclization catalyzed by first-generation Grubbs catalyst proceeded well and produced a mixture of the *Z*- and *E*-isomers, which were readily separated by

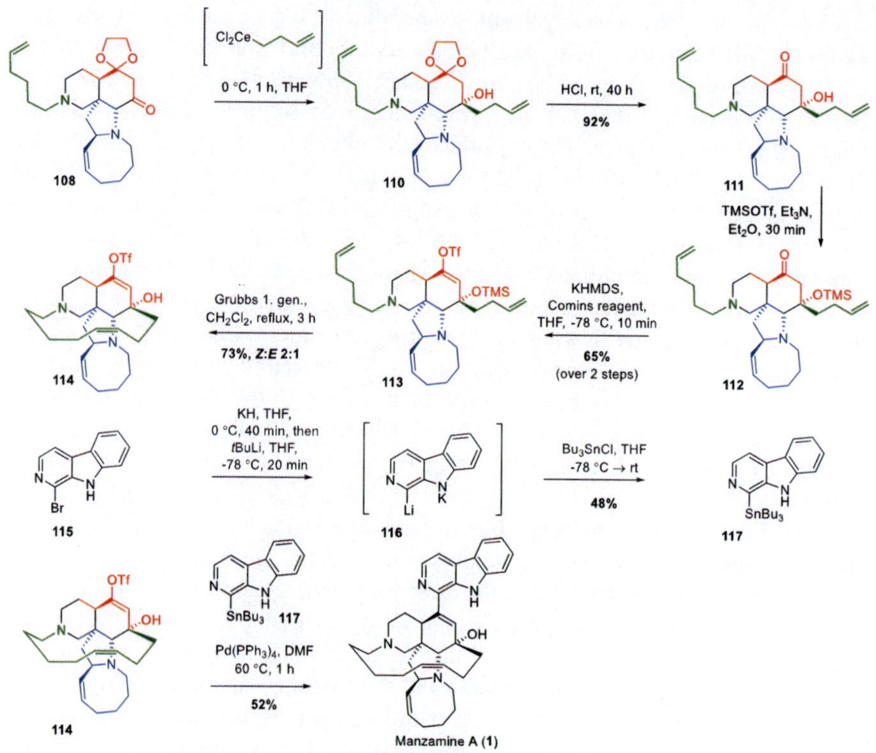

Scheme 20 Completion of the stereoselective total synthesis of manzamine A (**1**)

standard column chromatography. The desired *Z*-isomer **114** was isolated from the mixture already without the TMS protecting group. Therefore, it represented a highly valuable, direct precursor to manzamine A (**1**) and related natural products. Aligned with our strategy, the last step of the total synthesis required a cross-coupling of triflate **114** with a suitably functionalized β-carboline derivative. We considered organotin derivatives as suitable cross-coupling partners due to their higher stability towards the protodemetalation compared to their organoboron counterparts. Pleasingly, the synthesis of tributylstannylated β-carboline **117** was possible from commercially available bromide **115** via a one-pot protocol. Firstly, dianion **116** was formed by deprotonation of carboline **115** with potassium hydride followed by a lithium-halogen exchange mediated by *tert*-butyllithium. The highly reactive intermediate **116** was treated with tributyltin chloride and the reaction mixture subsequently filtered through a pad of basic alumina, which yielded the desired β-carboline **117**. Both rapid access to **117** and its excellent stability during storage at lower temperature enabled us to proceed to the endgame of manzamine A (**1**). The total synthesis of manzamine A (**1**) was completed by a Stille cross-coupling between triflate **114** and carboline derivative **117**. Considering the complexity of both coupling partners, the palladium-mediated reaction yielded the natural product in a respectable chemical yield. Pleasingly, all analytical data, including specific rotation, were in excellent agreement with reported data. Overall, we developed an 18-step long, stereoselective route to complex marine natural product manzamine A (**1**) featuring the stereoselective Michael addition and the two route-shortening nitro-Mannich reactions – the nitro-Mannich/lactamization cascade and the iminium/nitro-Mannich cyclization.

5 Ircinal A and Ircinol A

The development of the late-stage cross-coupling between triflate **114** and β-carboline derivative **117** allowed the formation of the last C–C bond in the manzamine A (**1**) total synthesis. Simultaneously, the Stille cross-coupling also inspired analogous syntheses of two other members of manzamine alkaloid family – ircinal A (**3**) and ircinol A (**4**).

In the last decades, various modifications of direct palladium-catalyzed carbonylation have been developed, allowing the straightforward transformation of an enol triflate to a conjugated aldehyde, ester, or amide. Employing Stille's original protocol [134] for reductive carbonylation, triflate **114** was directly turned into ircinal A (**3**) in a one-pot procedure (Scheme 21). The unreacted triflate **114** accompanying ircinal A (**3**) was recovered as well. Inspired by a protocol developed by Ortart [135], triflate **114** was exposed to carbon monoxide (**115**) and methanol under palladium catalysis. To our delight, a smooth alkoxycarbonylation took place, resulting in the formation of methyl ircinate **116**. The conjugate ester was readily reduced with DIBAL in toluene at low temperature, providing ircinol A (**4**), the fourth prepared member of the manzamine family accessed by our route [73]. The late-stage divergency enabled not only the synthesis of three natural products but also holds promise for future analogue production.

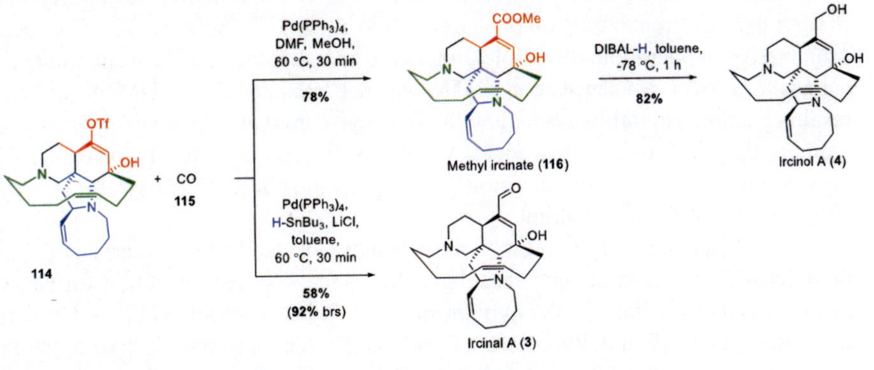

Scheme 21 Endgame of total synthesis of ircinal A (**3**) and ircinol A (**4**)

6 Keramaphidin B [106]

Keramaphidin B (**5**) belongs to a group of relatively younger and synthetically less explored members of the manzamine family. However, its discovery, synthesis, and biological origin remain one of the fascinating stories seen amongst the family members. At the time when manzamine A (**1**) was already well-known for its interesting biological properties and became a popular synthetic target, keramaphidin B (**5**) remained an unknown entity. At the same time, in the absence of any acceptable theory of the biological origin of manzamine alkaloids, intense theoretical studies were ongoing. Eventually, in 1992 Baldwin proposed that the manzamine family could derive from four building blocks: ammonia, a C-10 unit (a symmetrical dialdehyde), tryptophan, and a C-3 unit (an acrolein equivalent) [2]. Remarkably, the biogenetic route proposed by Baldwin included a structure closely related to keramaphidin B (**5**), which was discovered only 2 years later in 1994. Subsequently, in a series of papers, Baldwin experimentally validated his biomimetic synthesis of keramaphidin B (**5**) [108, 109].

Surprisingly, apart from the bio-inspired total synthesis by Baldwin, no other synthesis exists to date. We viewed this apparent lack of synthetic activities on the complex alkaloid as another opportunity to test our unified strategy. The close structural resemblance of keramaphidin B (**5**) with the other most prominent manzamine alkaloids and the completion of the previous syntheses amplified our interest. Accordingly, we further analyzed the structure of keramaphidin B (**5**) and identified four key building blocks for its potential synthesis (Fig. 9). Three fragments, namely, nitro olefin **11** and amines **10** and **77**, were already available. Dissimilar to our shortest syntheses of nakadomarin A (**2**) and manzamine A (**1**), we proposed that a nucleophile for the Michael addition would not contain the whole of the macrocyclic azocene ring (ring C, blue fragment). Instead, our proposal employed ester **117** as the pro-nucleophile and relied on a late-stage construction of the C-ring using amine **77**. Such strategic decision limited certain risks associated with the key stereoselective Michael addition as β-keto esters are excellent substrates in the stereoselective Michael addition with nitro styrenes. Initially, we engaged in a straightforward preparation of pro-nucleophile ester **117** (Scheme 22). This was achieved by an operationally simple protocol from valerolactone (**118**) and dimethyl carbonate. Firstly, δ-valerolactone (**118**) was deprotonated with LHMDS, and the resulting anion smoothly reacted with dimethyl carbonate yielding, upon acidic workup, ester **117** in excellent yield. With all key fragments readily available, we rapidly approached the first decision point of keramaphidin B (**5**) synthesis – the stereoselective Michael addition.

In conjunction with the nitro-Mannich/lactamization cascade, the diastereoselective nitro olefin Michael grew into the most significant transformation in our synthetic efforts. We hypothesized that the Michael addition between pro-nucleophile **117** and nitro olefin **11** could again secure stereoselective construction of the consecutive tertiary and quaternary stereocenters. In our previous syntheses of nakadomarin A (**2**), the diastereoselectivity of the reaction was controlled

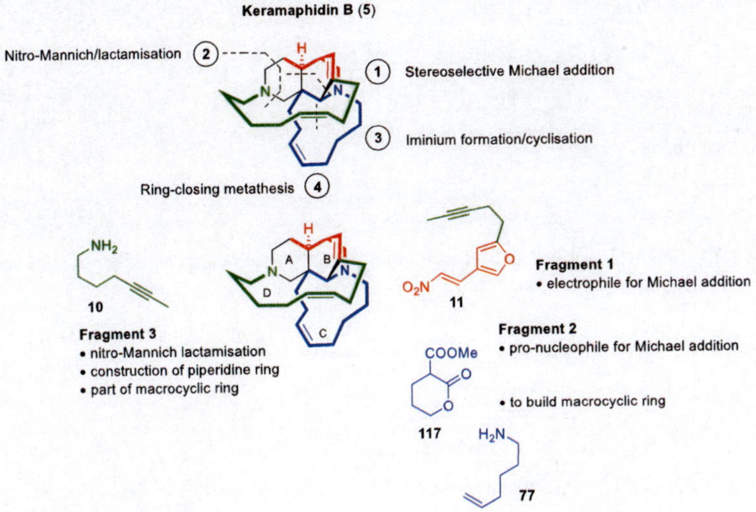

Fig. 9 Key building blocks for the proposed synthesis of keramaphidin B (**5**)

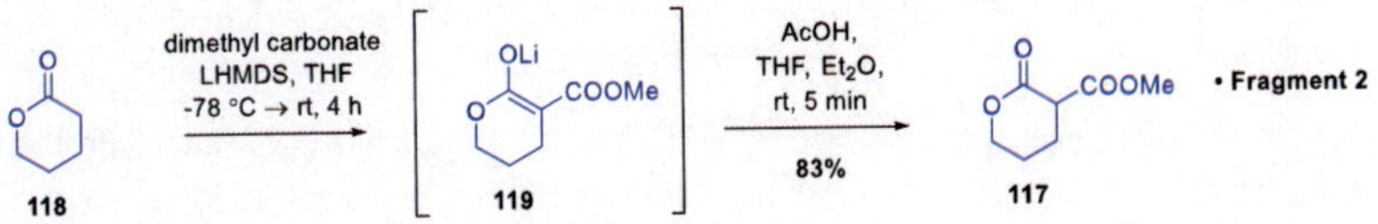

Scheme 22 Preparation of pro-nucleophile **117**, fragment 2

by a chiral, single enantiomer, cyclic β-amido ester **55**, and the bifunctional organocatalyst **31** (see Scheme 11). On the other hand, the envisaged synthesis of keramaphidin B (**5**) would use prochiral pro-nucleophile **117** and prochiral electrophile **11**. Both the tertiary and the quaternary stereocenters would be installed through an enantio- and diastereoselective organocatalyst-controlled Michael addition. Therefore, considering the different nature of the pro-nucleophiles, we decided to investigate the Michael addition on a model nitro olefin first. Simple nitro olefin **20** devoid of the tethered alkyne and organocatalyst **31b** was selected for these preliminary studies. Apart from the success of a cinchonine-derived catalyst **31a** in the nakadomarin A (**2**) syntheses, the selection of organocatalyst **31b** was heavily influenced by our preliminary studies on the enantioselective Michael addition to nitro olefins [136, 137].

In this study, a family of readily available organocatalysts possessing a range of hydrogen bond donor groups was evaluated for asymmetric organocatalytic activity in the dimethyl malonate Michael addition to β-nitrostyrene. Organocatalyst **31b** was identified to be the superior bifunctional catalyst to induce the excellent enantioselectivity in the Michael addition of dimethyl malonate (**121**) to nitro olefin **120** (Scheme 23). To our delight, utilization of the carefully selected cinchonine-derived bifunctional organocatalyst **31b** facilitated the Michael addition of ester **117**

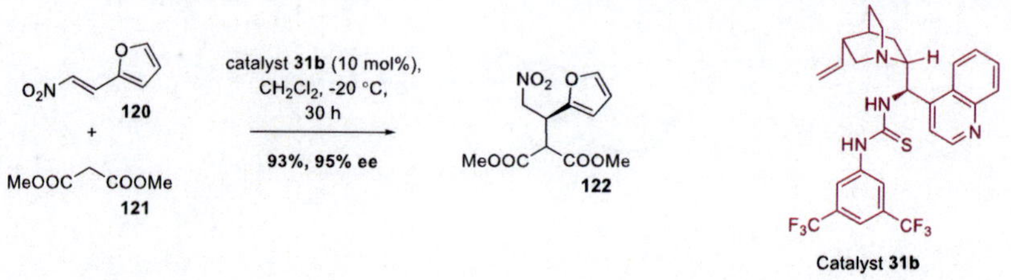

Scheme 23 Enantioselective Michael addition of dimethyl malonate controlled by organocatalyst **31b**

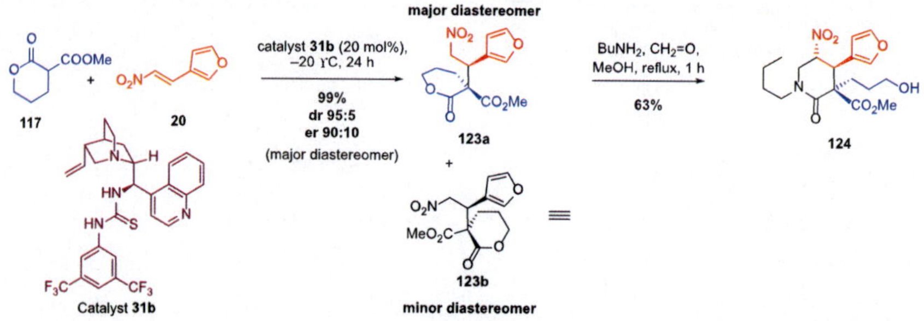

Scheme 24 Proof of principle studies on the Michael addition and nitro-Mannich/lactamization cascade

to nitro olefin **20**. It led to the isolation of nitro ester **123** in an excellent diastereoselectivity 95:5 (Scheme 24). Determination of the enantiomeric excess of the major diastereomer **123a** also revealed a similar level of the enantiocontrol (er 90:10). The observed outcome was in a good agreement with the previous total syntheses of nakadomarin A (**2**) and with the levels of enantiocontrol observed in the addition of dimethyl malonate to nitro olefin. Although the minor diastereomer **123b** formed during the addition possessing an undesired stereochemical configuration remained a side product, it played a crucial role in the progress of our synthesis efforts towards keramaphidin B (**5**). From its single-crystal X-ray analysis, we deduced the favorable configuration of the major diastereomer. Our deduction assumed a higher reactivity of the δ-lactone carbonyl in the presence of the methyl ester in the following lactamization. Indeed, the expected chemoselectivity was observed when nitro ester **123a** was treated with formaldehyde and butylamine under standard nitro-Mannich/lactamization conditions (MeOH, reflux, Scheme 24). The one-pot protocol yielded nitro piperidinone **124** possessing 3-hydroxypropyl chain and methoxycarbonyl functionality attached to the quaternary stereocenter – such an outcome results from an intramolecular attack of the amine to the lactone rather than the ester.

The model Michael addition and nitro-Mannich lactamization studies provided a highly valuable insight into both key reactions of the planned keramaphidin B (**5**) total synthesis. With the acquired knowledge about the stereochemical control and

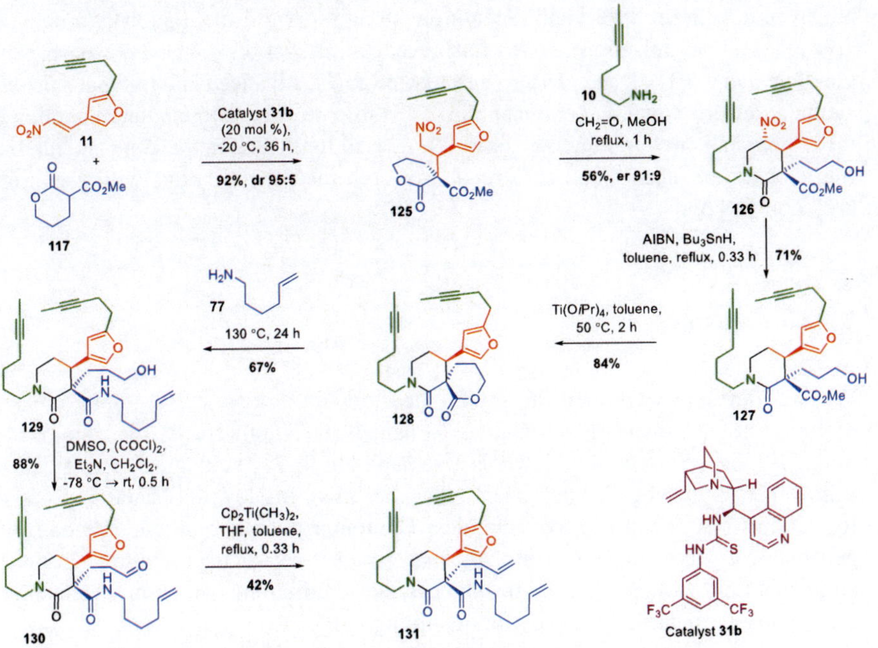

Scheme 25 Elaboration of the spirocyclic core of keramaphidin B **131**.

the chemoselectivity of the aminolysis, we confidently moved on and replaced the model nitro olefin for its functionalized analogue **11**. As expected, the Michael addition of **117** and nitroolefin **11** under the control of bifunctional organocatalyst **31b** proceeded smoothly. Compared to the model studies, a slightly extended reaction time (36 h) was necessary to achieve complete conversion. Nevertheless, the diastereo- and enantiocontrol and chemical yield remained satisfactory, nitro ester **125** was isolated in 92% yield and 95:5 dr. The exposure of nitro ester **125** to formaldehyde and amine **10** in boiling methanol for 1 h enabled the formation of nitro piperidinone **126** (Scheme 25). Our studies on model substrate **124** and the pioneering work from 2008 (Scheme 7) indicated excellent conservation of the enantiomeric purity during the nitro-Mannich lactamization cascade. We were delighted when such an anticipated outcome was confirmed again after 91:9 enantiomeric ratio had been determined in nitro piperidinone **126**. With the key highly enantiomerically enriched intermediate **126** in hand, our further activities were devoted to the elaboration of rings A and C of keramaphidin B (**5**). Firstly, the nitro group was removed after it had fulfilled its purpose in the Michael addition and the nitro-Mannich reaction yielding piperidinone **127**. After several unsuccessful direct attempts to transform ester **127** to amide **129**, a 2-step sequence was pursued. Titanium tetraisopropoxide-promoted lactonization of hydroxy ester followed by neat aminolysis of lactone **128** quickly reestablished the successful course of the synthesis. The Swern oxidation of alcohol **129** proceeded smoothly and provided aldehyde **130** in an 88% yield. Seemingly straightforward olefination of aldehyde **130** required optimization and was achieved by utilization of the Petasis reagent yielding diene **131**. Diene **131** represents the most advanced intermediate in our synthetic efforts towards keramaphidin B (**5**) to date. The intermediate containing the tethered alkenes and internal alkynes is poised for orthogonal alkene and alkyne ring-closing metatheses and construction of both macrocyclic rings with *Z*-configured double bonds.

7 Conclusion

We have developed a unified strategy for the short and stereoselective total synthesis of four complex manzamine alkaloids – namely nakadomarin A, manzamine A, ircinal A, and ircinol A – and towards keramaphidin B. The strategy is distinguished from other approaches by its general character as it employs the same key transformations for targeting all five alkaloids. The routes pivoted on the highly convergent union of five fragments through a stereoselective Michael addition, a three-component nitro-Mannich lactamization cascade, an iminium formation/cyclization, a ring-closing metathesis, and a cross-coupling.

References

1. Magnier E, Langlois Y (1998) Tetrahedron 54:6201
2. Baldwin JE, Whitehead RC (1992) Tetrahedron Lett 33:2059
3. Peng J, Rao KV, Choo Y-M, Hamann MT (2008) Manzamine alkaloids. In: Fattorusso E, Taglialatela-Scafati O (eds) Modern alkaloids. WILEY-VCH Verlag GmbH & Co. KGaA, Weinheim, p 189
4. Tsuda M, Kawasaki N, Kobayashi J (1994) Tetrahedron Lett 35:4387
5. Sakai R, Kohmoto S, Higa T, Jefford CW, Bernardinelli G (1987) Tetrahedron Lett 28:5493
6. MaGee DI, Beck EJ (2000) Can J Chem 78:1060
7. El Sayed KA, Kelly M, Kara UAK, Ang KKH, Katsuyama I, Dunbar DC, Khan AA, Hamann MT (2001) J Am Chem Soc 123:1804
8. Kong F, Graziani EI, Andersen RJ (1998) J Nat Prod 61:267
9. Sakai R, Higa T, Jefford CW, Bernardinelli G (1986) J Am Chem Soc 108:6404
10. Kobayashi J-I, Watanabe D, Kawasaki N, Tsuda M (1997) J Org Chem 62:9236
11. Kondo K, Shigemori H, Kikuchi Y, Ishibashi M, Sasaki T, Kobayashi J (1992) J Org Chem 57:2480
12. Tsuda M, Kawasaki N, Kobayashi J-I (1994) Tetrahedron 50:7957
13. Kobayashi J, Tsuda M, Kawasaki N, Matsumoto K, Adachi T (1994) Tetrahedron Lett 35:4383
14. Ang K, Holmes M, Kara U (2001) Parasitol Res 87:715
15. Ashok P, Lathiya H, Murugesan S (2015) Eur J Med Chem 97:928
16. Lin L-C, Kuo T-T, Chang H-Y, Liu W-S, Hsia S-M, Huang T-C (2018) Mar Drugs 16:252
17. Karan D, Dubey S, Pirisi L, Nagel A, Pina I, Choo Y-M, Hamann MT (2020) J Nat Prod 83:286
18. Winkler JD, Axten JM (1998) J Am Chem Soc 120:6425

19. Brands KMJ, Pandit UK (1989) Tetrahedron Lett 30:1423
20. Brands KMJ, Meekel AAP, Pandit UK (1991) Tetrahedron 47:2005
21. Pandit UK (1994) J Heterocyclic Chem 31:615
22. Pandit UK, Borer BC, Bieräugel H, Deerenberg S (1994) Pure Appl Chem 66:2131
23. Bieräugel H, Borer BC, Pandit UK (1995) Heteroc Commun 1:115
24. Pandit UK, Borer BC, Bieräugel H (1996) Pure Appl Chem 68:659
25. Hart DJ, McKinney JA (1989) Tetrahedron Lett 30:2611
26. Campbell JA, Hart DJ (1992) Tetrahedron Lett 33:6247
27. Bland D, Hart DJ, Lacoutiere S (1997) Tetrahedron 53:8871
28. Bland D, Chambournier G, Dragan V, Hart DJ (1999) Tetrahedron 55:8953
29. Nakagawa M, Lai Z, Torisawa Y, Hino T (1990) Heterocycles 31:999
30. Torisawa Y, Nakagawa M, Arai H, Lai Z, Hino T, Nakata T, Oishi T (1990) Tetrahedron Lett 31:3195
31. Torisawa Y, Nakagawa M, Hosaka T, Tanabe K, Lai Z, Ogata K, Nakata T, Oishi T, Hino T (1992) J Org Chem 57:5741
32. Nakagawa M, Torisawa Y, Uchida H, Nishida A (1999) J Synth Org Chem Jpn 57:1004
33. Uchida H, Nishida A, Nakagawa M (1999) Tetrahedron Lett 40:113
34. de Oliveira Imbroisi D, Simpkins NS (1991) J Chem Soc Perkin Trans 1:1815
35. Martin SF, Rein T, Liao Y (1991) Tetrahedron Lett 32:6481
36. Martin SF, Liao Y, Wong Y, Rein T (1994) Tetrahedron Lett 35:691
37. Winkler JD, Siegel MG, Stelmach JE (1993) Tetrahedron Lett 34:6509
38. Winkler JD, Stelmach JE, Siegel MG, Haddad N, Axten J, Dailey III WP (1997) Israel J Chem 37:47
39. Winkler JD, Axten J, Hammach AH, Kwak Y-S, Lengweiler U, Lucero MJ, Houk KN (1998) Tetrahedron 54:7045
40. Winkler JD, Londregan AT, Ragains JR, Hamann MT (2006) Org Lett 8:3407
41. Kamenecka TM, Overman LE (1994) Tetrahedron Lett 35:4279
42. Leonard J, Fearnley SP, Finlay MR, Knight JA, Wong G (1994) J Chem Soc Perkin Trans 1:2359
43. Li S, Kosemura S, Yamamura S (1994) Tetrahedron Lett 35:8217
44. Li S, Ohba S, Kosemura S, Yamamura S (1996) Tetrahedron Lett 37:7365
45. Li S, Kosemura S, Yamamura S (1998) Tetrahedron 54:6661
46. Li S, Yamamura S (1998) Tetrahedron 54:8691
47. Li S, Yamamura S, Hosomi H, Ohba S (1998) Tetrahedron Lett 39:2601
48. Magnier E, Langlois Y, Mérienne C (1995) Tetrahedron Lett 36:9475
49. Magnier E, Langlois Y (1998) Tetrahedron Lett 39:837
50. Urban D, Duval E, Langlois Y (2000) Tetrahedron Lett 41:9251
51. Brands KMJ, DiMichele LM (1998) Tetrahedron Lett 39:1677
52. Coldham I, Crapnell KM, Fernàndez J-C, Haxell TFN, Treacy AB, Coles SJ, Hursthouse MB, Moseley JD (1999) Chem Commun:1757
53. Coldham I, Crapnell KM, Fernàndez J-C, Moseley JD, Rabot R (2002) J Org Chem 67:6181
54. Coldham I, Pih SM, Rabot R (2005) Synlett 2005:1743
55. Pathak RB, Dobson BC, Ghosh N, Ageel KA, Alshawish MR, Saruengkhanphasit R, Coldham I (2015) Org Biomol Chem 13:3331
56. Clark JS, Hodgson PB, Goldsmith MD, Blake AJ, Cooke PA, Street LJ (2001) J Chem Soc Perkin Trans 1:3325
57. Clark JS, Townsend RJ, Blake AJ, Teat SJ, Johns A (2001) Tetrahedron Lett 42:3235
58. Lindsay HA, Salisbury CL, Cordes W, McIntosh MC (2001) Org Lett 3:4007
59. Herdemann M, Al-Mourabit A, Martin M-T, Marazano C (2002) J Org Chem 67:1890
60. Wypych J-C, Nguyen TM, Nuhant P, Bénéchie M, Marazano C (2008) Angew Chem Int Ed 47:5418
61. Magnus P, Fielding MR, Wells C, Lynch V (2002) Tetrahedron Lett 43:947
62. Solberghe GF, Markó IE (2002) Tetrahedron Lett 43:5061

63. Tokumaru K, Arai S, Nishida A (2006) Org Lett 8:27
64. Matsumura T, Akiba M, Arai S, Nakagawa M, Nishida A (2007) Tetrahedron Lett 48:1265
65. Mihara Y, Matsumura T, Terauchi Y, Akiba M, Arai S, Nishida A (2009) Bull Chem Soc Jpn 82:1520
66. Kita Y, Toma T, Kan T, Fukuyama T (2008) Org Lett 10:3251
67. Chau C-M, Liu K-M, Sha C-K (2011) Tetrahedron Lett 52:5068
68. Hawkins A, Jakubec P, Ironmonger A, Dixon DJ (2013) Tetrahedron Lett 54:365
69. Tokumaru K, Ohfusa T, Arai S, Nishida A (2016) J Antibiot 69:340
70. Martin SF, Humphrey JM, Ali A, Hillier MC (1999) J Am Chem Soc 121:866
71. Humphrey JM, Liao Y, Ali A, Rein T, Wong Y-L, Chen H-J, Courtney AK, Martin SF (2002) J Am Chem Soc 124:8584
72. Toma T, Kita Y, Fukuyama T (2010) J Am Chem Soc 132:10233
73. Jakubec P, Hawkins A, Felzmann W, Dixon DJ (2012) J Am Chem Soc 134:17482
74. Fürstner A, Guth O, Rumbo A, Seidel G (1999) J Am Chem Soc 121:11108
75. Fürstner A, Guth O, Düffels A, Seidel G, Liebl M, Gabor B, Mynott R (2001) Chem A Eur J 7:4811
76. Nagata T, Nishida A, Nakagawa M (2001) Tetrahedron Lett 42:8345
77. Nakagawa M, Uchida H, Ono K, Kimura Y, Yamabe M, Watanabe T, Tsuji R, Akiba M, Terada Y, Nagaki D, Ban S, Miyashita N, Kano T, Theeraladanon C, Hatakeyama K, Arisawa M, Nishida A (2003) Heterocycles 59:721
78. Leclerc E, Tius MA (2003) Org Lett 5:1171
79. Ono K, Nagata T, Nishida A (2003) Synlett:1207
80. Ahrendt KA, Williams RM (2004) Org Lett 6:4539
81. Young IS, Williams JL, Kerr MA (2005) Org Lett 7:953
82. Deng H, Yang X, Tong Z, Li Z, Zhai H (2008) Org Lett 10:1791
83. Wu F, Chen Y, Cheng B, Li Y, Zhai H (2013) Asian J Org Chem 2:561
84. Haimowitz T, Fitzgerald ME, Winkler JD (2011) Tetrahedron Lett 52:2162
85. Chavda JK, Procopiou PA, Horton PN, Coles SJ, Porter MJ (2014) Eur J Org Chem:129
86. Tsuji N, Stadler M, Kazumi N, Inokuma T, Kobayashi Y, Takemoto Y (2014) Org Biomol Chem 12:7919
87. Inagaki F, Kinebuchi M, Miyakoshi N, Mukai C (2010) Org Lett 12:1800
88. Stockman RA, McDermott PJ, Newton AF, Magnus P (2010) Synlett:559
89. Nagata T, Nakagawa M, Nishida A (2003) J Am Chem Soc 125:7484
90. Ono K, Nakagawa M, Nishida A (2004) Angew Chem Int Ed 43:2020
91. Nakagawa M, Nagata T, Ono K, Nishida A (2005) J Synt Org Chem Jpn 63:200
92. Young IS, Kerr MA (2007) J Am Chem Soc 129:1465
93. Jakubec P, Cockfield DM, Dixon DJ (2009) J Am Chem Soc 131:16632
94. Nilson MG, Funk RL (2010) Org Lett 12:4912
95. Cheng B, Wu F, Yang X, Zhou Y, Wan X, Zhai H (2011) Chem A Eur J 17:12569
96. Kyle AF, Jakubec P, Cockfield DM, Cleator E, Skidmore J, Dixon DJ (2011) Chem Commun 47:10037
97. Jakubec P, Kyle AF, Calleja J, Dixon DJ (2011) Tetrahedron Lett 52:6094
98. Yu M, Wang C, Kyle AF, Jakubec P, Dixon DJ, Schrock RR, Hoveyda AH (2011) Nature 479:88
99. Wang C, Yu M, Kyle AF, Jakubec P, Dixon DJ, Schrock RR, Hoveyda AH (2013) Chem A Eur J 19:2726
100. Bonazzi S, Cheng B, Wzorek JS, Evans DA (2013) J Am Chem Soc 135:9338
101. Boeckman RK, Wang H, Rugg KW, Genung NE, Chen K, Ryder TR (2016) Org Lett 18:6136
102. Clark JS, Xu C (2016) Angew Chem Int Ed 55:4332
103. Baldwin JE, Claridge TDW, Heupel FA, Whitehead RC (1994) Tetrahedron Lett 35:7829
104. Baldwin JE, Bischoff L, Claridge TDW, Heupel FA, Spring DR, Whitehead RC (1997) Tetrahedron 53:2271
105. Gil L, Bauchere X, Martin M-T, Marazano C, Das BC (1995) Tetrahedron Lett 36:6231

106. Jakubec P, Farley AJM, Dixon DJ (2016) Beilstein J Org Chem 12:1096
107. Shimoda H, Shibata T, Sekine D, Nakada M (2020) Heterocycles 100:3
108. Baldwin JE, Claridge TDW, Culshaw AJ, Heupel FA, Lee V, Spring DR, Whitehead RC, Boughtflower RJ, Mutton IM, Upton RJ (1998) Angew Chem Int Ed 37:2661
109. Baldwin JE, Claridge TDW, Culshaw AJ, Heupel FA, Lee V, Spring DR, Whitehead RC (1999) Chem A Eur J 5:3154
110. Hu J-F, Hamann MT, Hill R, Kelly M (2003) The manzamine alkaloids. In: The alkaloids: chemistry and biology, vol 60. Academic Press, New York, p 207
111. Kubota T, Kurimoto S, Kobayashi J (2020) The manzamine alkaloids. In: Knölker H-J (ed) The alkaloids: chemistry and biology, vol 84. Academic Press, New York, p 1
112. El-Desoky AHH, Tsukamoto S (2020) Manzamines: marine bioactive heterocycles. In: Topics in heterocyclic chemistry. Springer, Berlin, p 1
113. Jana S, Mekonnen HG (2018) ChemistrySelect 3:5198
114. Mühlstädt M, Schulze B (1975) J Prakt Chem 317:919
115. Bhagwatheeswaran H, Gaur SP, Jain PC (1976) Synthesis:615
116. Hynes PS, Stupple PA, Dixon DJ (2008) Org Lett 10:1389
117. Jakubec P, Helliwell M, Dixon DJ (2008) Org Lett 10:4267
118. Jakubec P, Cockfield DM, Helliwell M, Raftery J, Dixon DJ (2012) Beilstein J Org Chem 8:567
119. Pelletier SMC, Ray PC, Dixon DJ (2009) Org Lett 11:4512
120. Ono N, Miyake H, Tamura R, Kaji A (1981) Tetrahedron Lett 22:1705
121. Mortreux A, Blanchard M (1974) J Chem Soc Chem Commun:786
122. Grela K, Ignatowska J (2002) Org Lett 4:3747
123. Bindl M, Stade R, Heilmann EK, Picot A, Goddard R, Fürstner A (2009) J Am Chem Soc 131:9468
124. Schrock RR, Clark DN, Sancho J, Wengrovius JH, Rocklage SM, Pedersen SF (1982) Organometallics 1:1645
125. Núñez MG, Farley AJM, Dixon DJ (2013) J Am Chem Soc 135:16348
126. Formica M, Rozsar D, Su G, Farley AJM, Dixon DJ (2020) Acc Chem Res 10:2235
127. Farley AJM, Jakubec P, Goldys AM, Dixon DJ (2018) Tetrahedron 74:5206
128. Valentin M-L, Bolte J (1995) Bull Soc Chim Fr 11:1167
129. Feringa BL (1985) Synth Commun 15:87
130. Bower S, Kreutzer KA, Buchwald SL (1996) Angew Chem Int Ed 35:1515
131. Gregory AW, Chambers A, Hawkins A, Jakubec P, Dixon DJ (2015) Chem A Eur J 21:111
132. Ballini R, Petrini M (2015) Adv Synth Cat 357:2371
133. McMurry JE, Melton J (1973) J Org Chem 38:4367
134. Baillargeon VP, Stille JK (1986) J Am Chem Soc 108:452
135. Cacchi S, Morera E, Ortar G (1985) Tetrahedron Lett 26:1109
136. Ye J, Dixon DJ, Hynes PS (2005) Chem Commun:4481
137. Jakubec P, Cockfield DM, Hynes PS, Cleator E, Dixon DJ (2011) Tetrahedron Asymmetry 22:1147

Top Heterocycl Chem (2021) 58: 61–84
https://doi.org/10.1007/7081_2021_54

Published online: 10 March 2021

Total Syntheses of the Dictyodendrins

Kentaro Okano

Contents

Abstract The dictyodendrins, which were isolated from the Japanese marine sponge *Dictyodendrilla verongiformis* and the southern Australian marine sponge *Ianthella*, possess the unique 3,6-dihydropyrrolo[2,3-*c*]carbazole structure. They exhibit several biological activities such as telomerase inhibitory activity and β-secretase inhibitory activity. This review focuses on the methods to construct the characteristic pyrrolocarbazole skeleton of the dictyodendrins.

Keywords Cytotoxicity · Dictyodendrin · Marine sponge · Pyrrolocarbazole · Telomerase inhibitory activity · β-secretase inhibitory activity

K. Okano (✉)
Department of Chemical Science and Engineering, Kobe University, Kobe, Japan
e-mail: okano@harbor.kobe-u.ac.jp

Abbreviations

Ac	Acetyl
Ad	Adamantyl
BACE	β-site amyloid precursor protein cleaving enzyme (β-secretase)
Bn	Benzyl
Boc	*tert*-butoxycarbonyl
BrettPhos	2-(Dicyclohexylphosphino)-3,6-dimethoxy-2′,4′,6′-triisopropyl-1,1′-biphenyl
Bs	Benzenesulfonyl
Bu	Butyl
COD	1,5-Cyclooctadiene
CSA	Camphorsulfonic acid
DABCO	1,4-diazabicyclo[2.2.2]octane
DCB	Dichlorobenzene
DDQ	2,3-Dichloro-5,6-dicyano-1,4-benzoquinone
DMAP	4-(Dimethylamino)pyridine
DME	1,2-Dimethoxyethane
DMF	*N*,*N*-dimethylformamide
DMP	Dess–Martin periodinane
DMSO	Dimethyl sulfoxide
DMTMM	4-(4,6-Dimethoxy-1,3,5-triazin-2-yl)-4-methylmorpholinium chloride
DNA	Deoxyribonucleic acid
DPPF	1,1′-Ferrocenediyl-bis(diphenylphosphine)
Et	Ethyl
HPLC	High performance liquid chromatography
IC_{50}	Half maximal inhibitory concentration
LDA	Lithium diisopropylamide
LiHMDS	Lithium hexamethyldisilazide
Me	Methyl
MS	Molecular sieves
NBS	*N*-bromosuccinimide
NMO	*N*-methylmorpholine *N*-oxide
Ph	Phenyl
PIFA	Phenyliodine (III) bis(trifluoroacetate)
pin	Pinacol
Pr	Propyl
SEM	2-(Trimethylsilyl)ethoxymethyl
SPhos	2-(Dicyclohexylphosphino)-2′,6′-dimethoxybiphenyl
S-TCPTAD	*N*-tetrachlorophthaloyl-(*S*)-*tert*-leucinato
TBAF	Tetrabutylammonium fluoride
TBAI	Tetrabutylammonium iodide
TBS	*tert*-butyldimethylsilyl

TES	Triethylsilyl
Tf	Trifluoromethanesulfonyl
TFA	Trifluoroacetic acid
THF	Tetrahydrofuran
TMEDA	*N,N,N′,N′*-tetramethylethylenediamine
TMP	2,2,6,6-Tetramethylpiperidide
TMS	Trimethylsilyl
TosMIC	*p*-toluenesulfonylmethyl isocyanide
TPAP	Tetrapropylammonium perruthenate

1 Introduction

A number of marine natural products have been isolated and used for pharmaceutical purposes. Among them, several natural products consist of characteristic molecular structures, which have driven the synthetic chemists into a race for total synthesis. The dictyodendrins involve a 3,6-dihydropyrrolo[2,3-*c*]carbazole skeleton (Fig. 1). In 1993 Sato and co-workers reported isolation and structural elucidation of the related compounds **1a**, **1b**, and **2a** from the Japanese marine sponge *Dictyodendrilla*

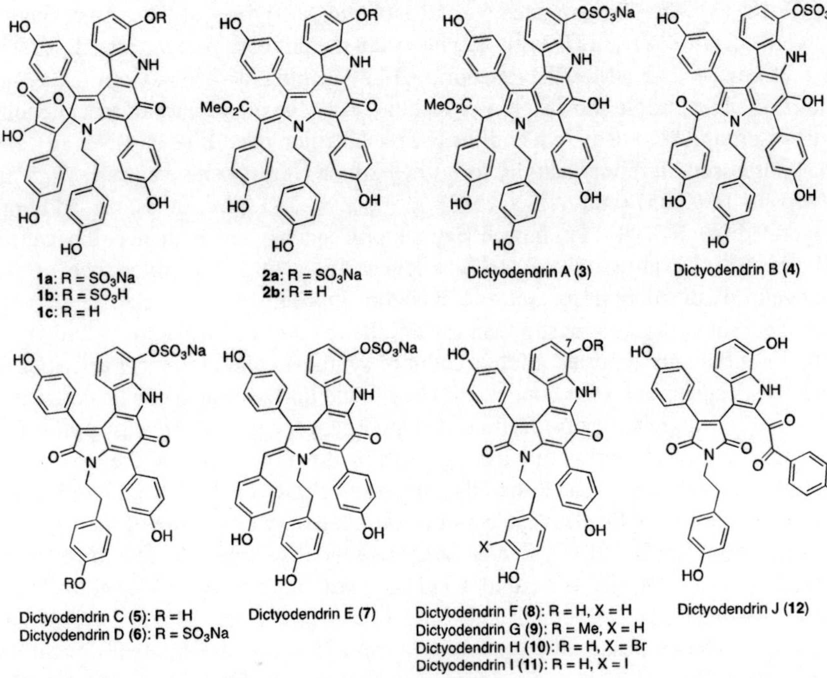

Fig. 1 Structures of the dictyodendrins and their congeners

verongiformis [1]. Ten years later, Fusetani and Matsunaga isolated five new compounds named as dictyodendrins A–E (**3–7**) and revealed that dictyodendrin A (**3**) exists as a mixture of racemates by separating each racemate using chiral HPLC [2]. Capon and co-workers isolated dictyodendrins F–J (**8–12**) from the southern Australian marine sponge *Ianthella* [3]. Syntheses of the dictyodendrins have been reported by Fürstner [4–6], Ishibashi [7], Tokuyama [8, 9], Jia [10, 11], Yamaguchi/Itami/Davies [12], Gaunt [13], Ready [14], Ohno [15, 16], and Guo/He [17]. Three reviews have been published on the total syntheses of the dictyodendrins [18–20].

2 Biological Activity

In 1993 Sato and co-workers reported the inhibitory activity of compounds **1a**, **1b**, and **2a** against bovine lens aldose reductase at IC_{50} values of 49 nM, 125 nM, and 112 nM, respectively [1]. They described that the sulfate group in compounds **1a** and **1b** was not necessary for the activity (**1c**: IC_{50} = 102 nM) and that the activity decreased (**2b**: IC_{50} = 567 nM) in the absence of the sulfate group. In 2003 Fusetani and Matsunaga reported **1a**, **2a**, and dictyodendrins A–E (**3–7**) showed 100% inhibition of telomerase activity at a concentration of 50 μg/L [2]. Based on the results of dictyodendrin F (**8**) showing no inhibition, the sulfate group was found to be necessary for the activity. In 2012, Capon and co-workers isolated dictyodendrins F–J (**8–12**) [3]. They investigated several biological activities of these compounds. Dictyodendrins F (**8**) and H–J (**10–12**) inhibited the protease β-secretase (BACE) at IC_{50} values of 1–2 μM. Dictyodendrin G (**9**) bearing a methoxy group at the 7 position of the indole was inactive, which indicates the importance of this phenolic hydroxy group. Among dictyodendrins F–J (**8–12**), dictyodendrins F (**8**), H (**10**), and I (**11**) possessed inhibitory activity for growth of the Gram positive bacteria *Bacillus subtilis* (ATCC6051 and ATCC6633) at IC_{50} = 1–3 μM levels. In addition, dictyodendrins F–I (**8–11**) showed cytotoxicity against the human colon cancer cell line SW620 at IC_{50} = 2–16 μM levels, indicating the importance of the tetracyclic oxidized pyrrolocarbazole skeleton. In contrast to the BACE activity, dictyodendrin G (**9**) was most potent toward the cancer cell line (IC_{50} = 2 μM).

In 2009 Fürstner reported a full account of syntheses and biological activities of the dictyodendrins and their congeners. They found that several synthetic congeners possessed low toxicity against HeLa and HepG2 cancer cells, which was appropriate for monitoring of inhibition of cancer growth by downregulation of the telomerase activity [6]. They investigated the DNA cleavage ability of the related compounds and found that the DNA cleavage could be prohibited by alkylation of the phenolic hydroxy groups. In 2016 Ready and co-workers reported cytotoxicity of dictyodendrins F (**8**), H (**10**), and I (**11**) against human colon cancer cell line (HCT116), and three non-small cell lung cancer cell lines (H157, H2073, and H2122) [14]. They observed that dictyodendrin I (**11**) was most potent toward the four cancer cell lines. In the case of human colon cancer cell line (HCT116), the IC_{50} values were 27 μM (dictyodendrin F), 13 μM (dictyodendrin H), and 8.8 μM (dictyodendrin I). In 2020 Ohno and co-workers reported cytotoxicity of dictyodendrin F (**8**) and its analogs against HCT116 using the colorimetric MTS assay and found that a pyrrolo[3,2-*c*]carbazole was more potent than pyrrolo[2,3-*c*] carbazole involved in the natural dictyodendrins [16]. In addition, the pyrrolo[3,2-*c*] carbazole also shows higher inhibitory activity against CDK2/CycA2 and GSK3β at $IC_{50} = 1–3\ \mu M$ levels, which indicates the potential of these pyrrole-fused carbazoles in the area of medicinal chemistry.

3 Syntheses of the Dictyodendrins

Synthetic approaches toward the tetracyclic pyrrolo[2,3-*c*]carbazole are categorized in Fig. 2. The structural formulae are the substrates and the reagents for constructing the pyrrolocarbazole. This review focuses on the detail of the syntheses as well as how this characteristic skeleton was formed.

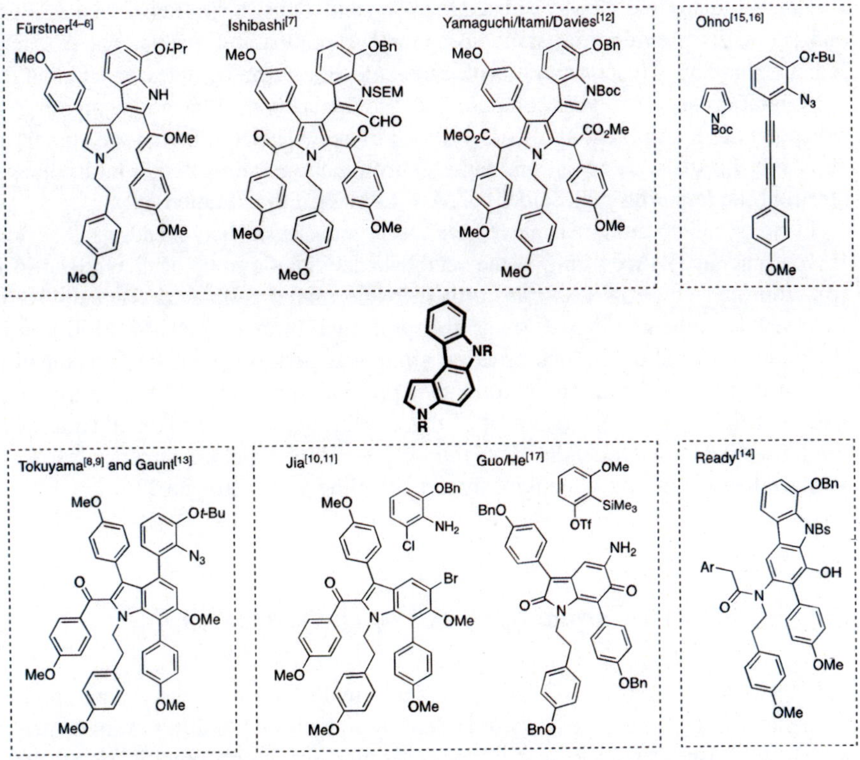

Fig. 2 Synthetic approaches toward construction of pyrrolo[2,3-*c*]carbazole

3.1 Fürstner's Total Syntheses of Dictyodendrins B, C, and E

The synthesis of dictyodendrin B (**4**) commenced with the preparation of the substituted pyrrole **13** through the reaction of TosMIC and chalcone **14**, which was synthesized from phenol **15** in two steps (Scheme 1). Reduction of the nitro group followed by acylation of the resultant amino group provided amide **16**. The ketoamide was then subjected to the titanium–graphite to provide the corresponding indole **17** in high yield. Subsequent 6π-electrocyclization led to the formation of the tetracyclic pyrrolo[2,3-*c*]carbazole **18**. This unique skeleton was found to undergo a Lewis acid-promoted skeletal rearrangement in Scheme 2, which forced to perform the stepwise introduction of the 4-methoxybenzoyl group. The isopropyl group was removed by BCl_3, and the resulting phenol **19** was converted to the mixed sulfate. Finally, the five methyl ethers were cleaved by BCl_3 with the aid of tetrabutylammonium iodide (TBAI) [21]. Removal of the trichloroethyl group led to the first total synthesis of dictyodendrin B (**4**).

The authors reported that the attempted Friedel–Crafts acylation of pyrrolocarbazole **20** resulted in migration of the aryl group (Scheme 2). The starting pyrrolocarbazole **20** was consumed by a combination of the benzoyl chloride and $SnCl_4$; however, none of the desired acylated compound **21** was observed. Instead, the undesired product **22** was isolated in 64% yield. Other Lewis acids such as $TiCl_4$ and $BF_3 \cdot OEt_2$ provided the same compound in comparable yields. These results indicate that $SnCl_4$ reacts with pyrrolocarbazole **20** to form complex **23** and that the 1,2-migration occurs to give compound **22** via the bridging Wheland complex. The exceptionally high reactivity of the pyrrolocarbazole skeleton forced the authors to carry out the stepwise acylation, namely, bromination, halogen–lithium exchange, electrophilic trap with *p*-anisaldehyde, and Ley–Griffith oxidation [22].

Fürstner and co-authors achieved the total syntheses of dictyodendrins C (**5**) and E (**7**) (Scheme 3). According to the established synthetic route of dictyodendrin B (**4**), isopropyl ether **18** was transformed to the mixed sulfate **24**. Demethylation followed by oxidation gave **25** as the protected form of dictyodendrin C (**5**). Reductive removal of the trichloroethyl group was performed with the combination of activated zinc and ammonium formate to provide a mixture of dictyodendrin C (**5**) and its reduced form **26**. Treatment of this mixture with oxygen in methanol led to the formation of dictyodendrin C (**5**) in 76% combined yield. Dictyodendrin E (**7**) was synthesized through Suzuki–Miyaura coupling with boronate **27**.

3.2 Ishibashi's Formal Synthesis of Dictyodendrin B

In 2009 Ishibashi and co-workers reported a formal synthesis of dictyodendrin B (**4**) (Scheme 4). They employed a symmetrical pyrrole diester **28** as a starting material, which was also used in their syntheses of other pyrrole alkaloids [23]. Stepwise introduction of the two different aryl groups proceeded to provide the fully

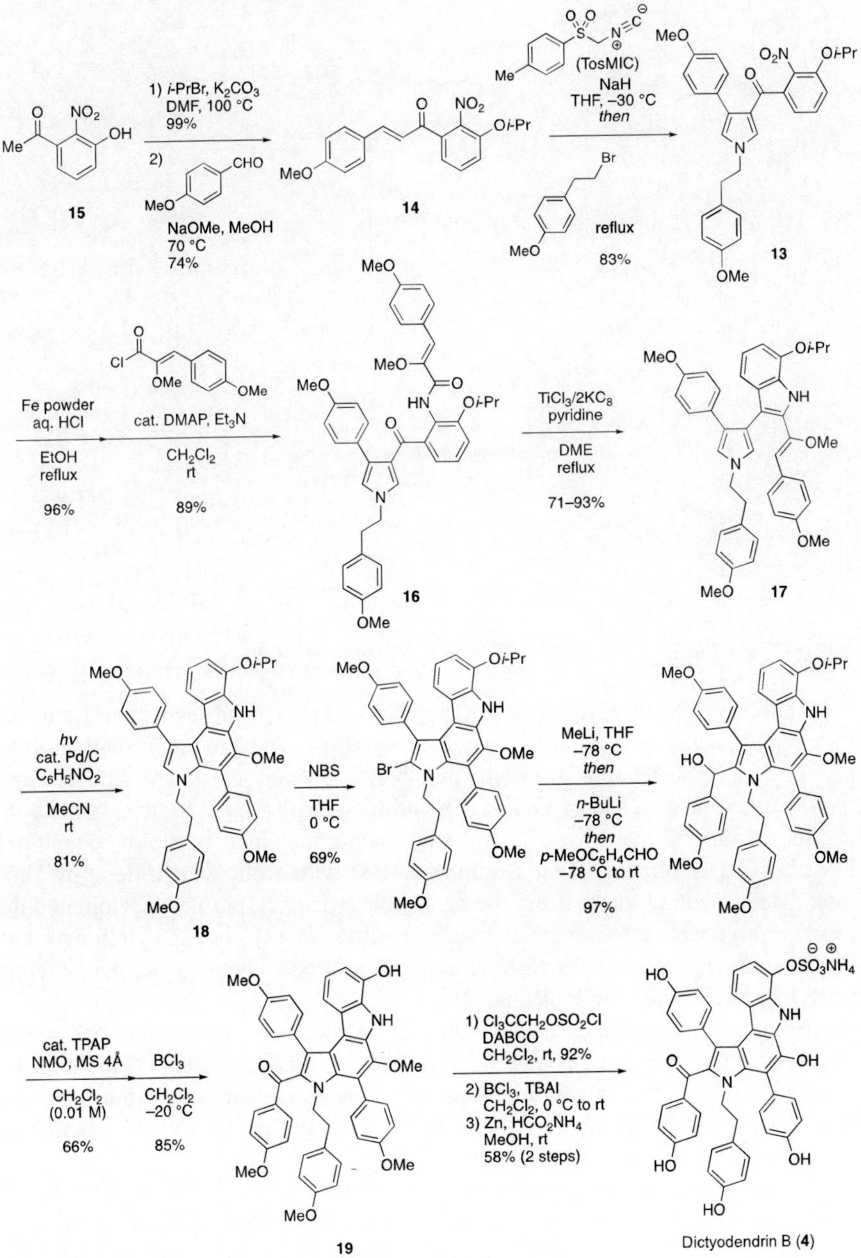

Scheme 1 Fürstner's total synthesis of dictyodendrin B

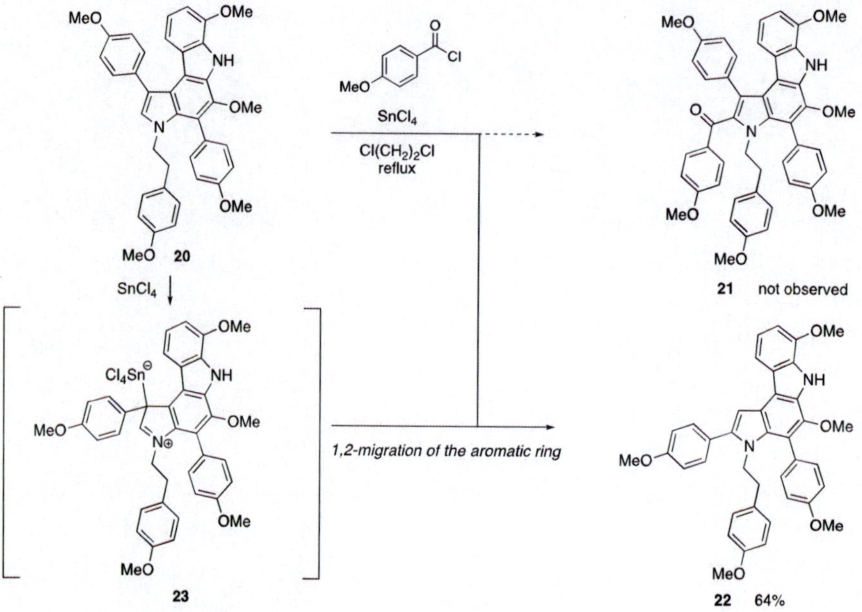

Scheme 2 Attempted Friedel–Crafts acylation of a pyrrolocarbazole

substituted pyrrole **29**. Hydrolysis of the methyl esters and subsequent activation of the resulting carboxylic acid moiety with DMTMM [24] led to the formation of the activated ester **30** bearing the seven-membered lactone. Treatment with the aryl Grignard reagent gave compound **31**. The primary alcohol was oxidized with DMP, and the resulting ketoaldehyde **32** was subjected to the pinacol coupling conditions using SmI_2. Dehydration of the resulting diol **33** with acetic anhydride in pyridine took place smoothly to construct the tetracyclic pyrrolocarbazole skeleton, and the resulting aryl acetate **34** was converted to methyl ether **35** in 91% yield over two steps. Acidic removal of the SEM group and hydrogenolysis of the benzyl ether provided the Fürstner's intermediate **19**.

The authors attempted dehydration of diol **33** under the acidic conditions, which led to migration of the aromatic group to provide **36** via intermediate **37** (Scheme 5). They gave a comment that this migration was promoted by the steric repulsion of the aryl group and the phenylethyl substituent. This undesired aryl migration was suppressed by employing the basic conditions.

3.3 *Tokuyama's Total Syntheses of Dictyodendrins A–E*

Tokuyama and co-workers achieved the total synthesis of dictyodendrin B (**4**) by developing a benzyne-mediated one-pot indoline formation/cross coupling (Scheme 6). The synthesis started with an iodine-selective halogen–lithium exchange of

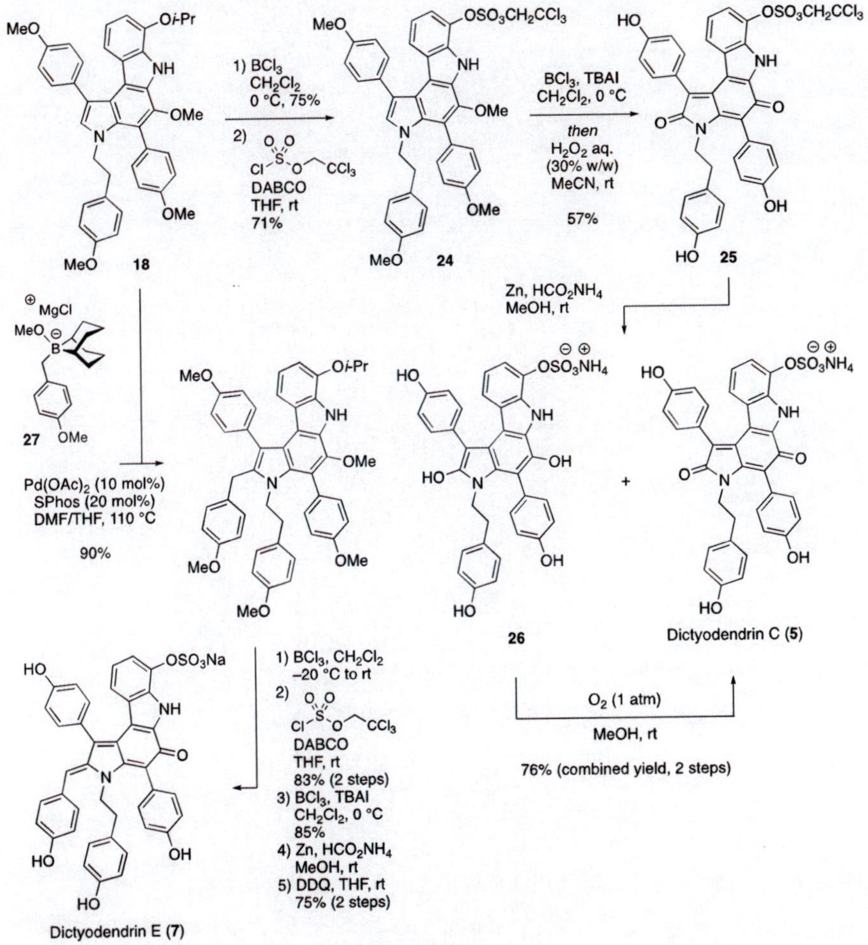

Scheme 3 Fürstner's total syntheses of dictyodendrins C and E

trihalobenzene **38** [25] followed by 1,4-addition to nitrostyrene **39** to provide nitroalkane **40** in 90% yield. The chemoselective reduction of the nitro group was performed using Fe/$FeCl_2$ in refluxing EtOH, and the resulting primary amine was protected as its Boc carbamate **41**. The compound was subjected to deprotonation conditions using $Mg(TMP)_2 \cdot 2LiBr$ [26] to generate benzyne **42** with the tethered nucleophilic nitrogen, which cyclized to construct the indoline skeleton. Subsequent transmetalation to copper followed by the palladium-catalyzed cross coupling provided multiply substituted indoline **43** in excellent yield, which was transformed to indole **44** by removal of the Boc group and DDQ oxidation. The phenylethyl moiety was introduced to the indole nitrogen to provide compound **45**, which underwent Friedel–Crafts acylation exclusively at the indole 2 position. The azidophenyl group was incorporated for the construction of the tetracyclic pyrrolocarbazole skeleton by

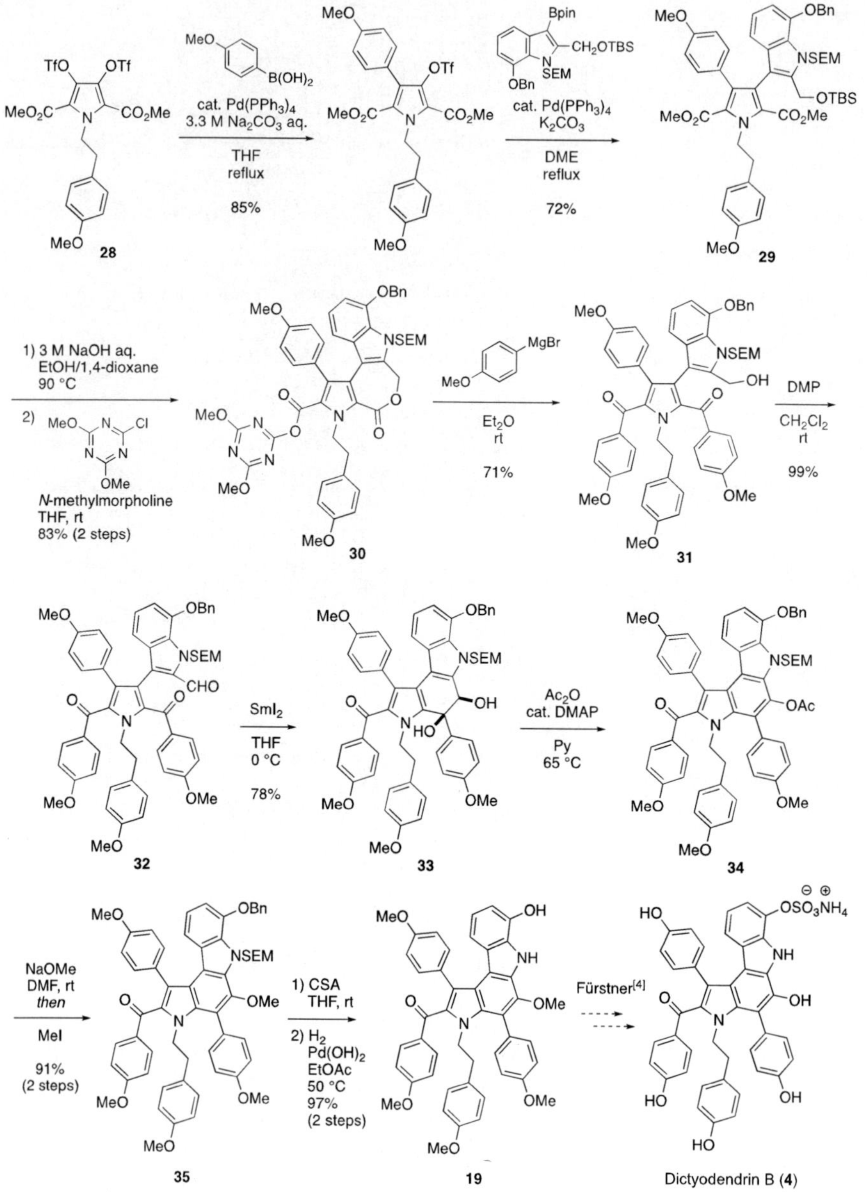

Scheme 4 Ishibashi's formal synthesis of dictyodendrin B

the palladium-catalyzed borylation and Suzuki–Miyaura coupling with azidophenyl iodide **46**. The tetracyclic framework was constructed by thermolysis of azide **47** by heating the reaction mixture at 180°C and subsequent insertion of the resulting nitrene into the proximal C–H bond. The total synthesis of dictyodendrin B (**4**)

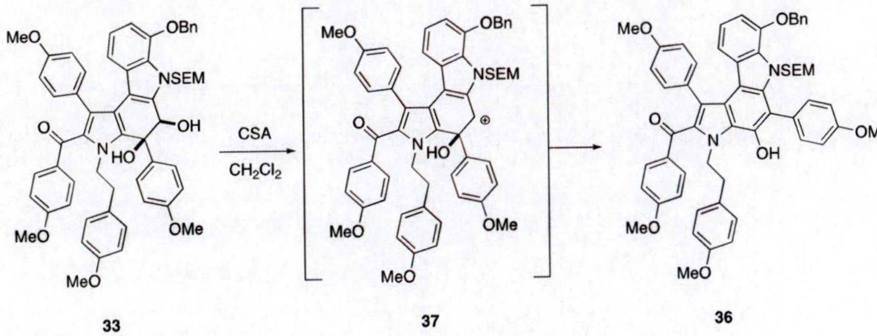

Scheme 5 Undesired acid-promoted migration of the aryl group

was achieved by modifying the method by Fürstner [4, 5]. Deprotection of the *tert*-butyl ether **48** was carried out by BCl_3 in the presence of pentamethylbenzene as a non-Lewis basic cation scavenger to prevent the undesired Friedel–Crafts alkylation [27]. Formation of the mixed sulfate and removal of the protecting groups led to the total synthesis of dictyodendrin B (**4**).

Dictyodendrins A (**3**) and C–E (**5–7**) were synthesized from indole **44** as the key synthetic intermediate (Scheme 7). The thermolytic construction of pyrrolocarbazole was also effective for the 2-unsubstituted indoles, and dictyodendrins C (**5**) and D (**6**) were successfully synthetized from the corresponding indoles **45** and **49**, respectively. Friedel–Crafts alkylation of indole **45** required further optimization of Lewis acids for the syntheses of dictyodendrins A (**3**) and E (**7**). Finally, the presented reaction conditions allowed to functionalize the 2-position of the indole to give the corresponding indoles **50** and **51**, which were converted into dictyodendrins A (**3**) and E (**7**).

3.4 *Jia's Total Syntheses of Dictyodendrins B and E and Formal Synthesis of Dictyodendrin C*

In 2013 Jia and co-workers reported the total syntheses of dictyodendrins B (**4**) and E (**7**) utilizing one-pot Buchwald–Hartwig amination/intramolecular C–H arylation (Scheme 8). The synthesis commenced with iodination of aminobiphenyl **52**, and the resulting 2-iodoaniline **53** was subjected to Larock indole synthesis [28] to give multiply functionalized indole **54** in 95% yield. N-Alkylation and regioselective bromination provided hexa-substituted indole **55**, which was transformed to pyrrolocarbazole **56** in 71% yield using stoichiometric amount of $Pd(OAc)_2$. Hydrogenolysis of the benzyl ether afforded the Fürstner's intermediate **19**. Reduction of the carbonyl group by a combination of $AlCl_3/LiAlH_4$ (1:2) [29] provided **57** for the synthesis of dictyodendrin E (**7**). The established reaction is a powerful

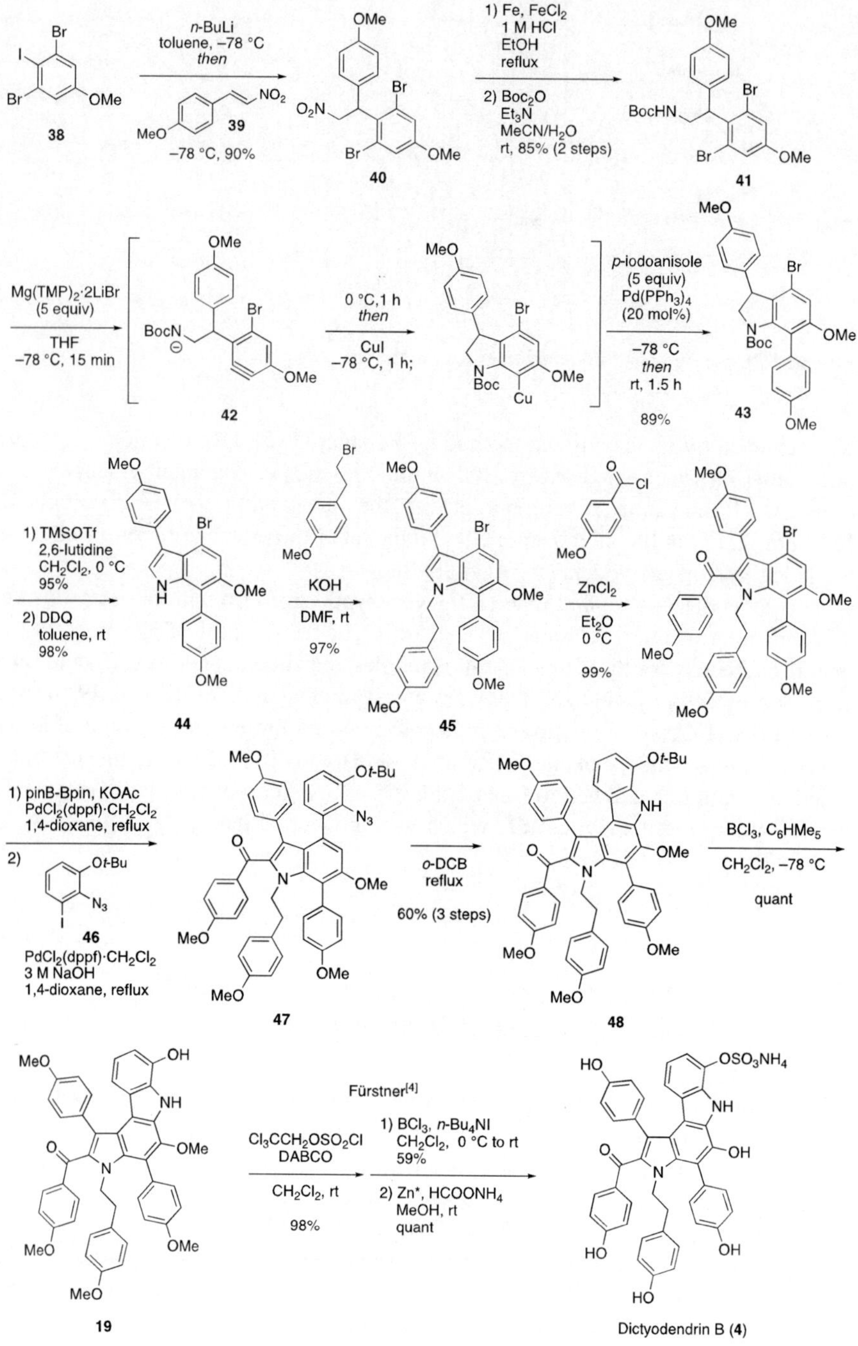

Scheme 6 Tokuyama's total synthesis of dictyodendrin B

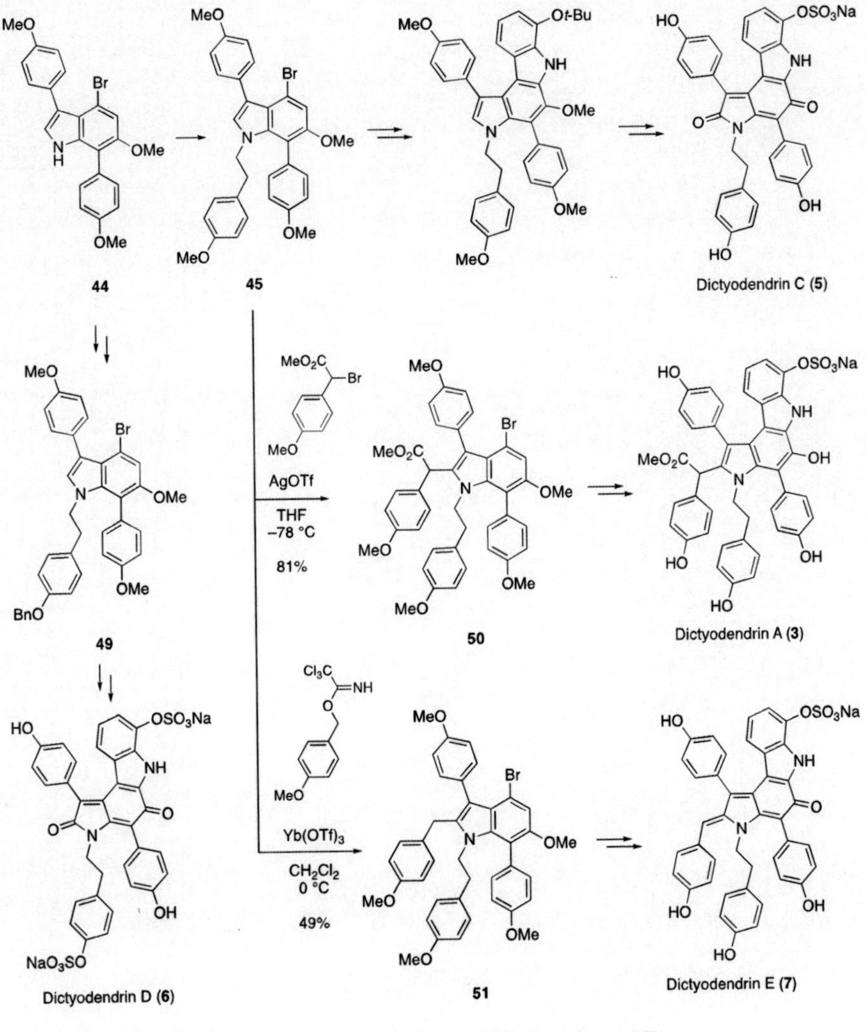

Scheme 7 Tokuyama's total syntheses of dictyodendrins A, C, D, and E

method for the construction of pyrrolocarbazole; however, the conditions were not applied to the synthesis of dictyodendrin A (**3**) bearing the ester moiety.

In 2014 Jia and co-workers achieved the formal synthesis of dictyodendrin C (**5**) (Scheme 9). Apart from their syntheses of dictyodendrins B and E, introduction of the bromo group at the later stage was difficult because more reactive indole 2 position was not substituted. Thus, they introduced the requisite bromo group onto the 2-iodoaniline **53**; however, subsequent Larock indole synthesis provided the desired indole **58** in low yield. Removal of the TES group and N-alkylation provided 5-bromoindole **59**, which smoothly underwent the key pyrrolocarbazole formation to give compound **60** for the synthesis of dictyodendrin C (**5**).

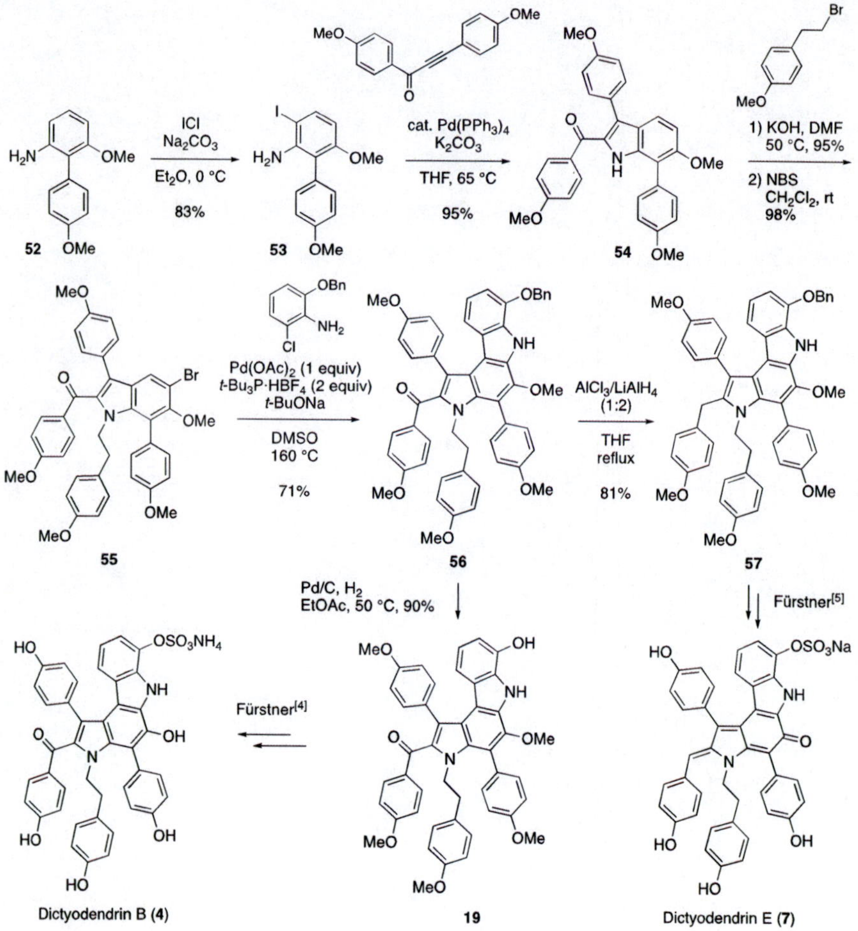

Scheme 8 Jia's total syntheses of dictyodendrins B and E

3.5 Yamaguchi/Itami/Davies' Total Syntheses of Dictyodendrins A and F

In 2015 Yamaguchi/Itami/Davies reported the total syntheses of dictyodendrins A and F using C–H functionalization (Scheme 10). The starting N-alkylated pyrrole **61** was subjected to the rhodium-catalyzed β-selective arylation [30]. The resulting arylated pyrrole **62** was converted to the fully substituted bromopyrrole **63** through the rhodium-catalyzed dialkylation with the diazoester. The bromo group was then transformed to the indole moiety, and the product was treated with LDA to provide pyrrolocarbazole **64** in 47% yield, through the 6π-electrocyclic reaction at the indole moiety. Further transformation led to the total synthesis of dictyodendrin A (**3**). The pyrrolocarbazole was oxidized to compound **65**, which was treated with 4 M HCl to

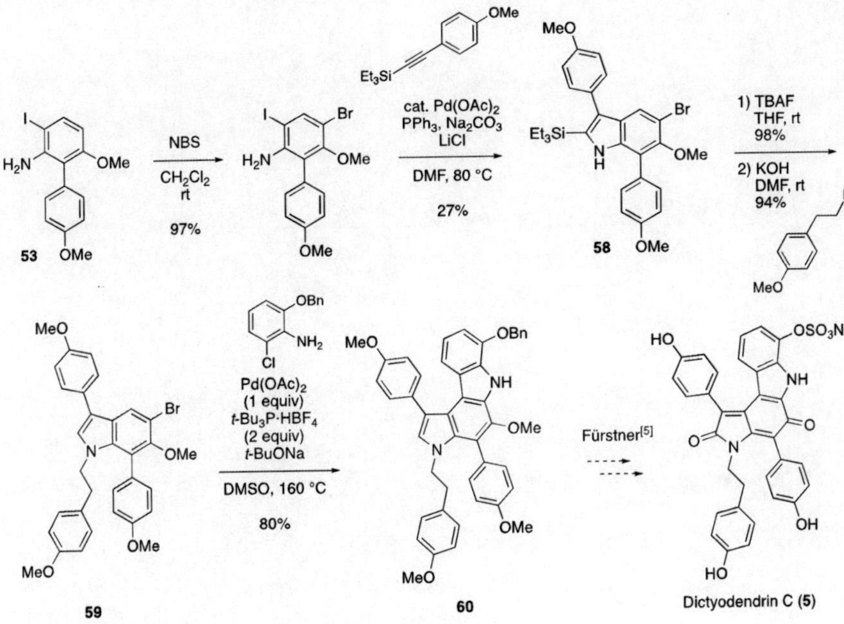

Scheme 9 Jia's formal synthesis of dictyodendrin C

furnish compound **66**. Global deprotection of **66** provided dictyodendrin F (**8**). This synthetic route was corresponding to the results, which were reported by Fusetani's isolation paper [2].

3.6 Gaunt's Total Synthesis of Dictyodendrin B

In 2015 Gaunt and co-workers reported total synthesis of dictyodendrin B (**4**) by utilizing C–H functionalization of 4-bromoindole (**67**) (Scheme 11). Arylation at position 3 and subsequent benzoylation at position 2 proceeded to provide 2,3-disubstituted indole **68**. Iridium-catalyzed C–H borylation at position 7 followed by Suzuki–Miyaura coupling with *p*-iodoanisole provided indole **69** with the bromo group at position 4 intact. N-alkylation and subsequent Suzuki–Miyaura coupling with arylboronic acid pinacol ester **70** provided compound **71**. They installed the requisite methoxy group in a two-step sequence, namely, regioselective monobromination and copper-catalyzed etheration to provide **72**. The regioselectivity was controlled by choosing the nitro group to deactivate the aromatic ring, which required to convert the nitro group to the azide **47** for constructing the pyrrolocarbazole. Further transformation led to the synthesis of dictyodendrin B (**4**). This synthetic route describes the synthetic potential of the C–H

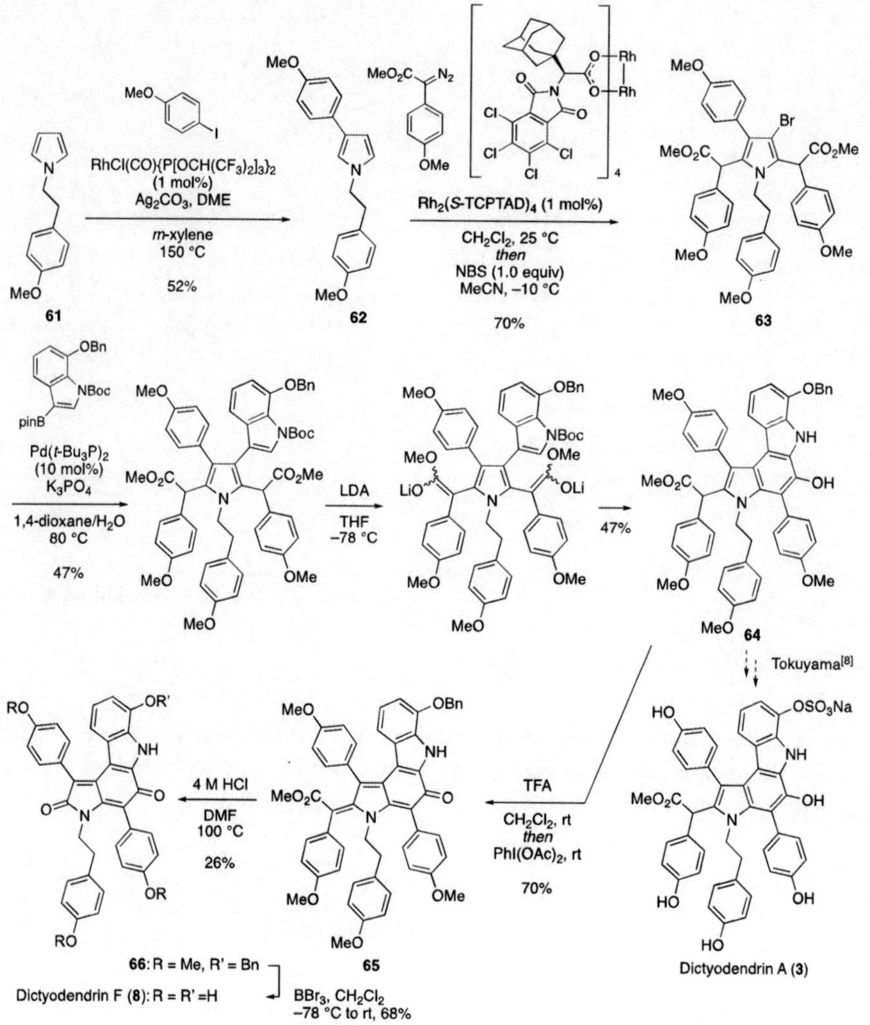

Scheme 10 Yamaguchi/Itami/Davies' total syntheses of dictyodendrins A and F

functionalization strategy by installing the bromo group to control the regioselectivity and utilize the bromo group itself for further transformation.

3.7 *Ready's Total Syntheses of Dictyodendrins F, H, and I*

In 2016 Ready and co-workers achieved concise total syntheses of dictyodendrins F (**8**), H (**10**), and I (**11**) by a unique approach (Scheme 12). The synthesis started with

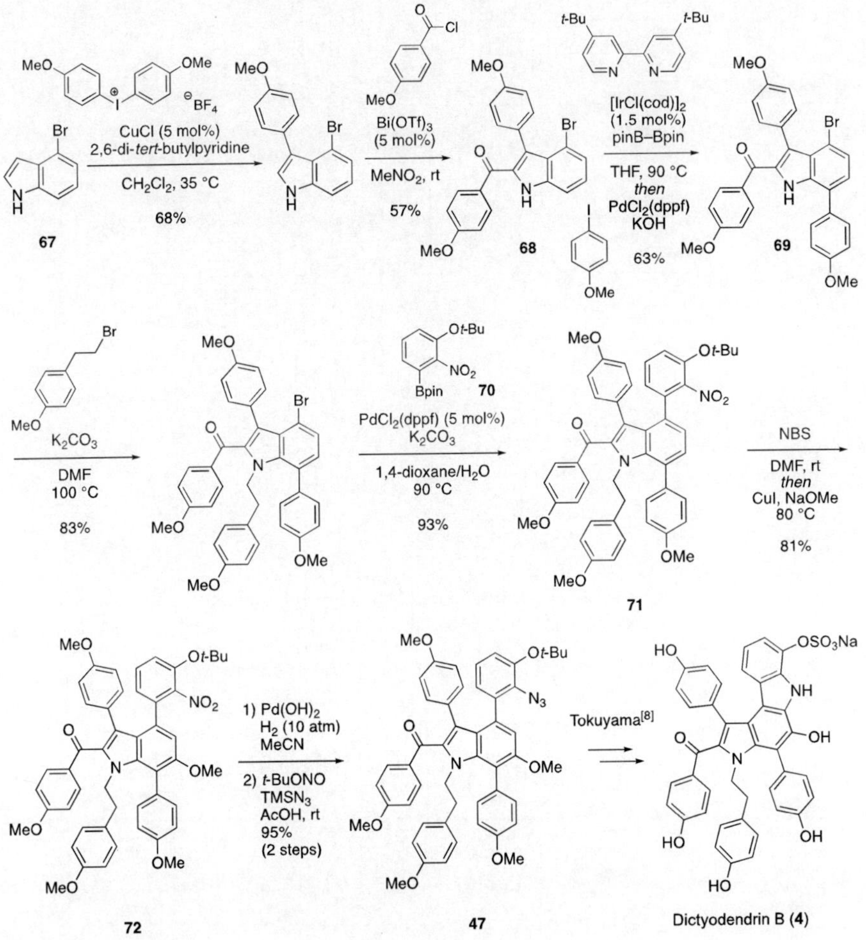

Scheme 11 Gaunt's total synthesis of dictyodendrin B

the [2 + 2] cycloaddition of adamantyl ynol ether **73** and ketene **74**, which was generated by heating *tert*-butyl ynol ether **75** to provide the four-membered vinylogous ester **76** in 56% yield. The use of adamantyl ynol ether **73** was crucial for the selective [2 + 2] cycloaddition. After introducing the amino functionality, compound **77** was heated at 80°C to promote retro [2 + 2] cycloaddition, which led to the formation of ketene **78**. Intramolecular cyclization provided aminocarbazole **79**. This aniline moiety was further acylated by ketene **80** derived from **81** whose rate of generation was controlled by using 3-pentyl substituent. The oxidized pyrrolocarbazole was constructed with LiHMDS followed by $PhI(OAc)_2$. Cleavage of benzyl ether **66** was carried out with BBr_3 to furnish dictyodendrin F (**8**). This strategy could be applied to the syntheses of dictyodendrins H (**10**) and I (**11**).

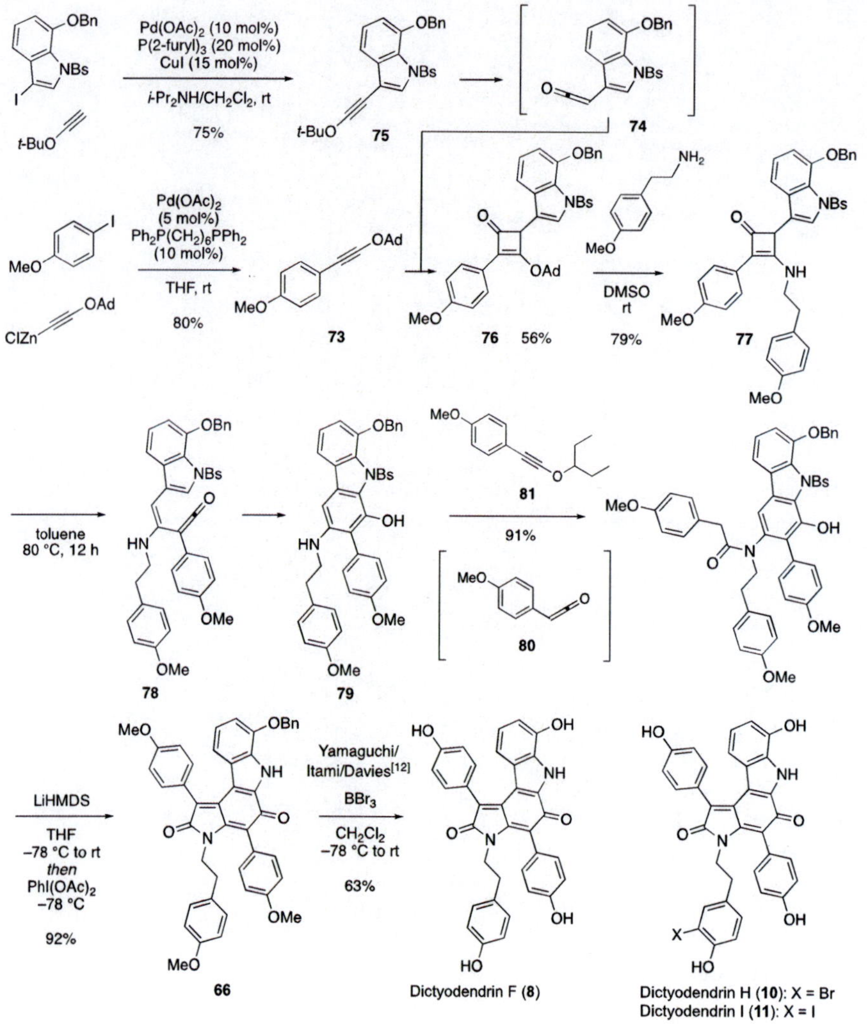

Scheme 12 Ready's total syntheses of dictyodendrins F, H, and I

3.8 *Ohno's Total Syntheses of Dictyodendrins B, C, and F and Formal Syntheses of Dictyodendrins A and D*

Ohno and co-workers reported total syntheses of dictyodendrins B (**4**), C (**5**), and F (**8**) by developing the gold-catalyzed formation of pyrrolocarbazole from a diyne bearing the different aryl groups at the terminal positions (Scheme 13). Previously they achieved a variety of gold-catalyzed annulation reactions to form the π-extended fused cyclic systems [31]. Apart from other total syntheses of the dictyodendrins, this method allowed one-pot construction of the pyrrolocarbazole

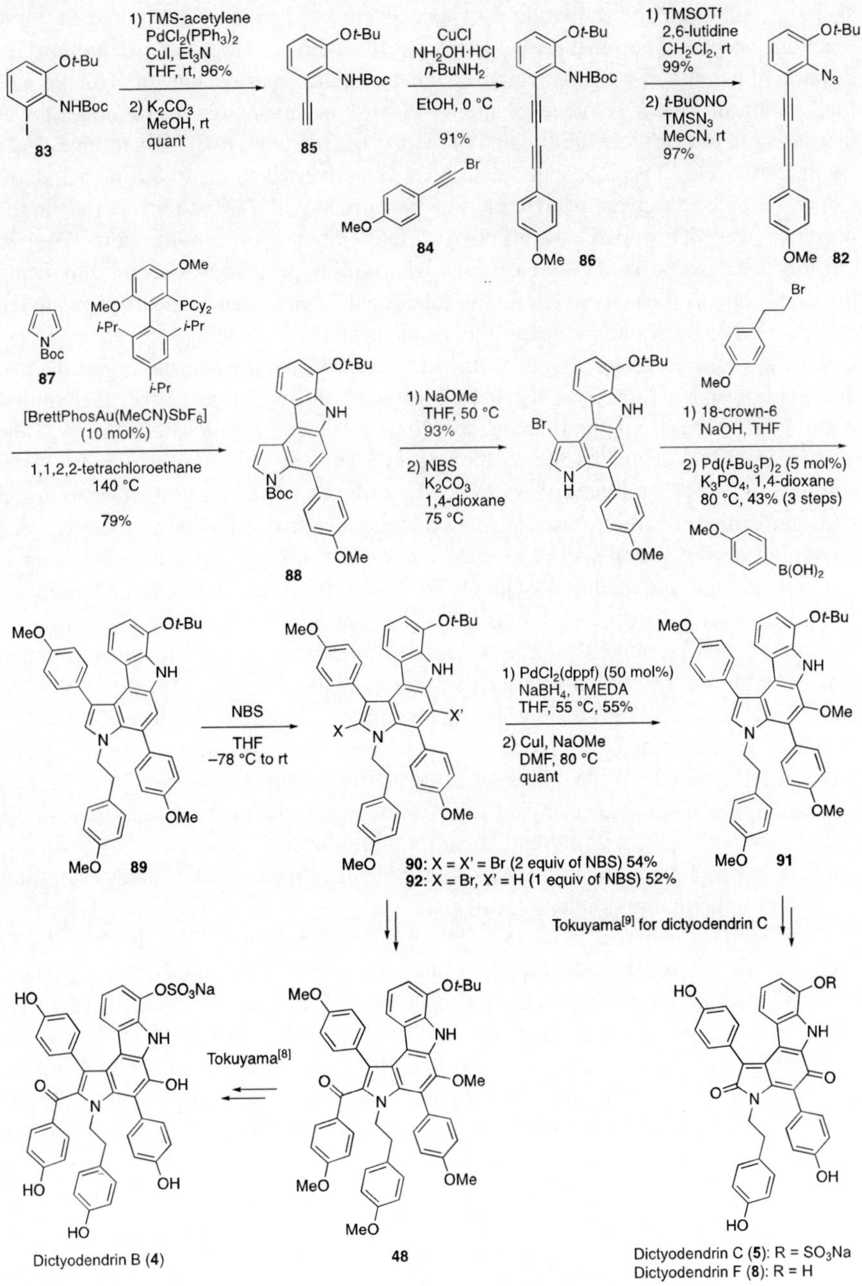

Scheme 13 Ohno's total syntheses of dictyodendrins B, C, and F

skeleton, although the preparation of azidephenylated diyne **82** required multiple reaction steps. The synthesis commenced with Sonogashira coupling of 2-iodoaniline derivative **83**, which is the synthetic intermediate of Tokuyama's total synthesis. After removal of the TMS group, anisylethynyl bromide **84** was incorporated to acetylene **85** by a combination of CuCl and *n*-$BuNH_2$ to give diyne **86** in 91% yield. The Boc carbamate **86** was converted to azide **82** in two steps. Construction of the pyrrolocarbazole was performed with *N*-Boc-pyrrole (**87**) in the presence of 10 mol% BrettPhosAu(MeCN)SbF_6 to provide the desired pyrrolocarbazole **88** in 79% yield. The high yielding process was secured by using the Boc group as the substituent. Other functional groups such as the tosyl group led to reduction of the yield and selectivity of the product. The Boc group was removed under the basic conditions, and treatment with NBS resulted in the regioselective bromination at position 3 of the indole skeleton. Introduction of the phenylethyl group and Suzuki–Miyaura cross coupling provided pyrrolocarbazole **89**. The remaining task is to install the methoxy group and the carbon functionality. Two equivalents of NBS provided dibromide **90**. Reductive removal of the bromo group was performed in the presence of substoichiometric palladium catalyst. The remaining bromo group was converted into the methoxy group using copper catalyst, according to the method by Gaunt. The resulting compound **91** is the Fürstner's synthetic intermediate of dictyodendrin C (**5**) and F (**8**). The use of equimolar amount of NBS resulted in the exclusive formation of the mono-brominated pyrrolocarbazole **92**, which was then subjected to the Fürstner's synthetic route to incorporate the benzoyl moiety for the synthesis of dictyodendrin B (**4**). The introduction of the requisite methoxy group proceeded smoothly in this case. The established synthetic route allowed construction of the unnatural pyrrolo[3,2-*c*]carbazole, which showed stronger biological activities than pyrrolo[2,3-*c*]carbazole that is involved in the natural dictyodendrins. This selectivity was controlled by the gold catalyst and the protecting group of the pyrrole nitrogen, and further exploration of such promising compounds is anticipated.

Ohno and co-workers achieved the formal synthesis of dictyodendrin A (**3**) (Scheme 14). They introduced the ester side chain in a multi-step sequence. Pyrrolocarbazole **89** reacted with oxalyl chloride followed by methanol provided ketoester **93** in 87% yield. Selective addition of Grignard reagent to the ketone carbonyl group was performed after hydrolysis of the methyl ester. Thus, ketoacid **94** was treated with the Grignard reagent led to the formation of compound **95**, which was subjected to the methyl ester formation followed by reduction of the alcohol to provide compound **96**.

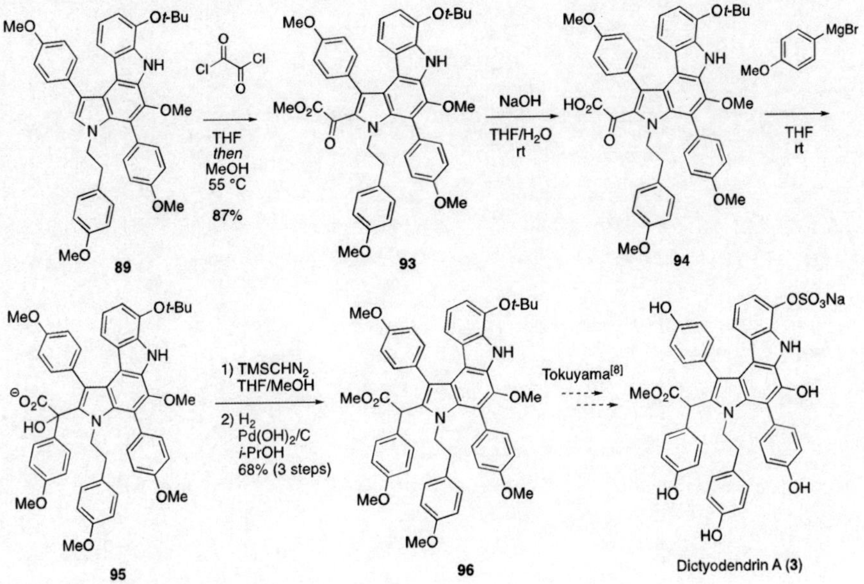

Scheme 14 Ohno's formal synthesis of dictyodendrin A

3.9 *Guo/He's Total Syntheses of Dictyodendrins F, G, H, and I*

Guo/He and co-workers reported total syntheses of dictyodendrins F (**8**), G (**9**), H (**10**), and I (**11**) from aminoquinone **97** (Scheme 15). The synthesis commenced with Suzuki–Miyaura cross coupling to provide arylated quinone **98** in 78% yield. Subsequent introduction of the tyramine moiety gave diaminoquinone **99** in 94% yield. Reduction/oxidation steps were necessary for introducing the acyl moiety; the amino functionality of **99** was not reactive toward acyl chloride **100** due to its electron deficiency. The resultant amino group attached to the dihydroquinone **101** underwent the smooth acylation, and PIFA oxidation [32] led to the formation of quinone **102**. Aldol condensation and removal of the Boc group provided compound **103**, which was converted to the oxidized pyrrolocarbazole **104** through the reaction with the benzyne derivative. Palladium-catalyzed hydrogenolysis of the benzyl ethers led to the first total synthesis of dictyodendrin G (**9**). This synthetic route was effective for the congeners, dictyodendrins F (**8**), H (**10**), and I (**11**).

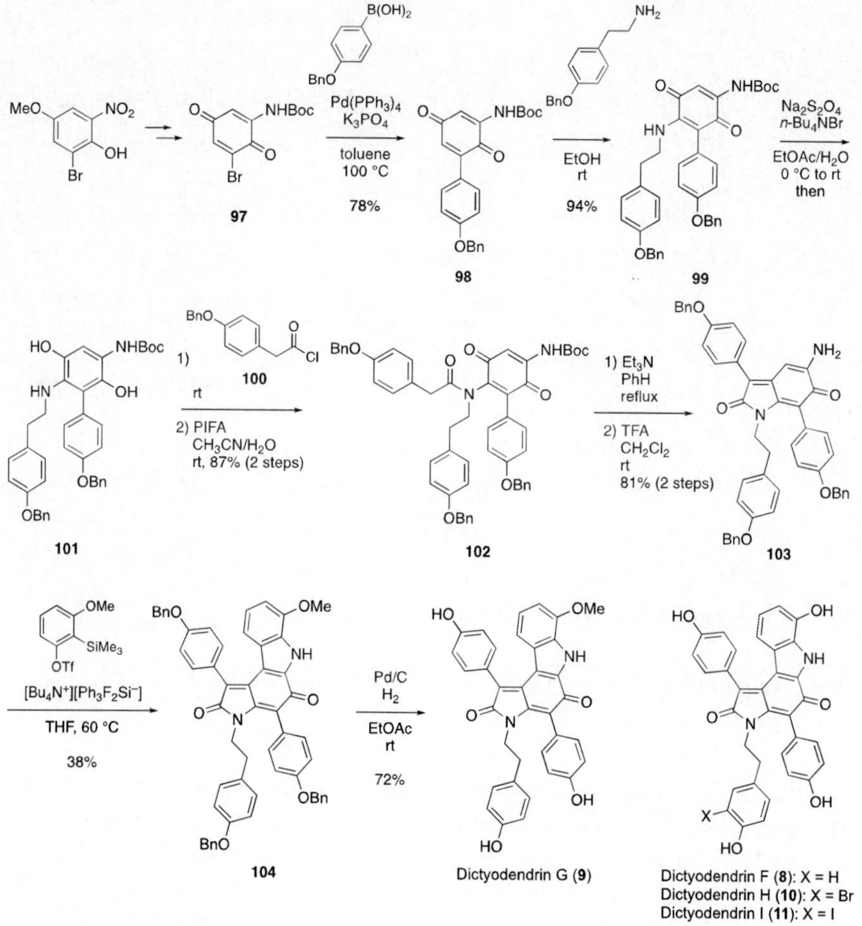

Scheme 15 Guo and He's total syntheses of dictyodendrins F, G, H, and I

4 Conclusion

This review has outlined an overview of isolation, biological activities, and chemical syntheses of the dictyodendrins. In addition to the wide range of the biological activities, the characteristic pyrrolocarbazole skeleton has attracted the synthetic chemists, which led to the discovery of the new synthetic methods. The development will become effective synthetic methods not only for the dictyodendrins and their congeners, but also for the structurally related pyrrole-containing natural products.

References

1. Sato A, Morishita T, Shiraki T, Yoshioka S, Horikoshi H, Kuwano H, Hanzawa H, Hata T (1993) Aldose reductase inhibitors from a marine sponge, *Dictyodendrilla* sp. J Org Chem 58:7632–7634
2. Warabi K, Matsunaga S, van Soest RWM, Fusetani N (2003) Dictyodendrins A–E, the first telomerase-inhibitory marine natural products from the sponge *Dictyodendrilla verongiformis*. J Org Chem 68:2765–2770
3. Zhang H, Conte MM, Khalil Z, Huang X-C, Capon RJ (2012) New dictyodendrins as BACE inhibitors from a Southern Australian marine sponge, *Ianthella* sp. RSC Adv 2:4209–4214
4. Fürstner A, Domostoj MM, Scheiper B (2005) Total synthesis of dictyodendrin B. J Am Chem Soc 127:11620–11621
5. Fürstner A, Domostoj MM, Scheiper B (2006) Total syntheses of the telomerase inhibitors dictyodendrin B, C, and E. J Am Chem Soc 128:8087–8094
6. Buchgraber P, Domostoj MM, Scheiper B, Wirtz C, Mynott R, Rust J, Fürstner A (2009) Synthesis-driven mapping of the dictyodendrin alkaloids. Tetrahedron 65:6519–6534
7. Hirao S, Yoshinaga Y, Iwao M, Ishibashi F (2010) A formal total synthesis of the telomerase inhibitor dictyodendrin B. Tetrahedron Lett 51:533–536
8. Okano K, Fujiwara H, Noji T, Fukuyama T, Tokuyama H (2010) Total synthesis of dictyodendrin A and B. Angew Chem Int Ed 49:5925–5929
9. Tokuyama H, Okano K, Fujiwara H, Noji T, Fukuyama T (2011) Total synthesis of dictyodendrins A–E. Chem Asian J 6:560–572
10. Liang JJ, Hu WM, Tao PY, Jia YX (2013) Total synthesis of dictyodendrins B and E. J Org Chem 78:5810–5815
11. Tao PY, Liang JJ, Jia YX (2014) Total synthesis of dictyodendrins B and E, and formal synthesis of dictyodendrin C. Eur J Org Chem:5735–5748
12. Yamaguchi AD, Chepiga KM, Yamaguchi J, Itami K, Davies HML (2015) Concise syntheses of dictyodendrins A and F by a sequential C–H functionalization strategy. J Am Chem Soc 137:644–647
13. Pitts AK, O'Hara F, Snell RH, Gaunt MJ (2015) A concise and scalable strategy for the total synthesis of dictyodendrin B based on sequential C–H functionalization. Angew Chem Int Ed 54:5451–5455
14. Zhang WH, Ready JM (2016) A concise total synthesis of dictyodendrins F, H, and I using aryl ynol ethers as key building blocks. J Am Chem Soc 138:10684–10692
15. Matsuoka J, Matsuda Y, Kawada Y, Oishi S, Ohno H (2017) Total synthesis of dictyodendrins by the gold-catalyzed cascade cyclization of conjugated diynes with pyrroles. Angew Chem Int Ed 56:7444–7448
16. Matsuoka J, Inuki S, Matsuda Y, Miyamoto Y, Otani M, Oka M, Oishi S, Ohno H (2020) Total synthesis of dictyodendrins A–F by the gold-catalyzed cascade cyclization of conjugated diyne with Pyrrole. Chem Eur J 26:11150–11157
17. Banne S, Reddy DP, Li WX, Wang CH, Guo J, He Y (2017) A unified modular synthetic strategy for dictyodendrins F, H, I, and G. Org Lett 19:4996–4999
18. Zhang WH, Ready JM (2017) Total synthesis of the dictyodendrins as an arena to highlight emerging synthetic technologies. Nat Prod Rep 34:1010–1034
19. Domingo V, Quilez del Moral JF, Barrero AF (2016) Chapter 1 – recent accomplishments in the total synthesis of natural products through C–H functionalization strategies. In: Atta-ur-Rahman (ed) Studies in natural products chemistry, vol 48. Elsevier, Amsterdam, pp 1–28
20. Matsuoka J (2020) Total synthesis of dictyodendrins. Yakugaku Zasshi 140:1313–1322
21. Brooks PR, Wirtz MC, Vetelino MG, Rescek DM, Woodworth GF, Morgan BP, Coe JW (1999) Boron trichloride/tetra-*n*-butylammonium iodide: a mild, selective combination reagent for the cleavage of primary alkyl aryl ethers. J Org Chem 64:9719–9721

22. Griffith WP, Ley SV, Whitcombe GP, White AD (1987) Preparation and use of tetra-*n*-butylammonium per-ruthenate (TBAP reagent) and tetra-*n*-propylammonium per-ruthenate (TPAP reagent) as new catalytic oxidants for alcohols. J Chem Soc Chem Commun:1625–1627
23. Iwao M, Takeuchi T, Fujikawa N, Fukuda T, Ishibashi F (2003) Short and flexible route to 3,4-diarylpyrrole marine alkaloids: syntheses of permethyl storniamide A, ningalin B, and lamellarin G trimethyl ether. Tetrahedron Lett 44:4443–4446
24. Kunishima M, Kawachi C, Morita J, Terao K, Iwasaki F (1999) Tani, S., 4-(4,6-Dimethoxy-1,3,5-triazin-2-yl)-4-methyl-morpholinium chloride: an efficient condensing agent leading to the formation of amides and esters. Tetrahedron 55:13159–13170
25. Yamada K, Kurokawa T, Tokuyama H, Fukuyama T (2003) Total synthesis of the duocarmycins. J Am Chem Soc 125:6630–6631
26. Eaton PE, Lee C-H, Xiong Y (1989) Magnesium amide bases and amido-Grignards. 1, ortho magnesiation. J Am Chem Soc 111:8016–8018
27. Okano K, Okuyama K, Fukuyama T, Tokuyama H (2008) Mild debenzylation of aryl benzyl ether with BCl_3 in the presence of pentamethylbenzene as a non-Lewis-basic cation scavenger. Synlett:1977–1980
28. Larock RC, Yum EK (1991) Synthesis of indoles via palladium-catalyzed heteroannulation of internal alkynes. J Am Chem Soc 113:6689–6690
29. MacDowell DWH, Springsteen AW (1976) A thiophene analog of 7,12-dihydropleiadene. J Org Chem 41:3046–3048
30. Yanagisawa S, Sudo T, Noyori R, Itami K (2006) Direct C−H arylation of (hetero)arenes with aryl iodides via rhodium catalysis. J Am Chem Soc 128:11748–11749
31. Matsuda Y, Naoe S, Oishi S, Fujii N, Ohno H (2015) Formal 4+2 reaction between 1,3-diynes and pyrroles: gold(I)-catalyzed indole synthesis by double hydroarylation. Chem Eur J 21:1463–1467
32. Tamura Y, Yakura T, Tohma H, Kikuchi K, Kita Y (1989) Hypervalent iodine oxidation of *p*-alkoxy- and related phenols: a facile and efficient synthesis of *p*-quinones. Synthesis:126–127

Top Heterocycl Chem (2021) 58: 85–104
https://doi.org/10.1007/7081_2020_41

Published online: 27 September 2020

Marine Bioluminescence with Dehydrocoelenterazine, an Imidazopyrazinone Compound

Masaki Kuse

Contents

Abstract Heterocycles play important roles in bioluminescence. Among them, the imidazopyrazinone compounds are utilized for light emission in some marine bioluminescent systems. Coelenterazine (CL) is the organic substance of aequorin, a photoprotein obtained from jellyfish, that possesses green fluorescent proteins (GFP). The dehydrogenative oxidation of CL affords dehydrocoelenterazine (DCL), which is also an organic substance for marine bioluminescence. DCL cannot react directly with oxygen. Therefore, it is converted to the chromophore, which is required for light emission in a photoprotein. The chromophore directly reacts with oxygen in the presence of reactive oxygen species (ROS) to emit visible light. Symplectin, a photoprotein derived from squids, and Pholasin, a photoprotein of mollusks, utilize the DCL-derived chromophore. Pholasin, a registered trademark of Knight Scientific Ltd., is widely utilized as a highly sensitive indicator of ROS, i.e., the detection of ROS that are produced from activated neutrophils by Pholasin is the biomarker for oxidative stresses. Since there are many excellent reviews regarding CL-dependent bioluminescence, this review focuses on the bioluminescent systems, which employ DCL as an organic substance, by describing the history of the research

M. Kuse (✉)
Graduate School of Agricultural Science, Kobe University, Kobe, Japan
e-mail: kuse@eagle.kobe-u.ac.jp

on these systems and the bioluminescent mechanism that is related to the DCL-derived chromophores.

Keywords Bioluminescence · Chromophore · Dehydrocoelenterazine · Luciferase · Luciferin · Pholasin · Photoprotein · Symplectin

Abbreviations

CL	Coelenterazine
DCL	Dehydrocoelenterazine
DTT	Dithiothreitol
GFP	Green fluorescent protein
GSH	Glutathione
GST	Glutathione *S*-transferase
HRP	Horseradish peroxidase
LC–MS	Liquid chromatography–mass spectrometry
MPO	Myeloperoxidase
NMR	Nuclear magnetic resonance
PBS	Phosphate-buffered saline
ROS	Reactive oxygen species
Tris	Tris(hydroxymethyl)aminomethane

1 Introduction

Beautiful bioluminescence, i.e., light emission from living organisms, has attracted attention since ancient times. Hence, many scientists have attempted to demystify bioluminescent systems to understand how luminous organisms emitted light. Many bioluminescent organisms are found in the ocean, and 90% of deep-sea organisms are suspected to be luminous [1, 2]. Light emission is essentially facilitated by the enzymatic/nonenzymatic oxidation of organic substances. The "burning" of organic molecules with oxygen to produce light emission in luminous organisms could be figuratively likened to the burning of a candle with fire. Although this metaphor is not scientifically correct, it would afford us an easy insight into what bioluminescence is. Heterocyclic compounds are utilized as organic substances in bioluminescence (Fig. 1) [3]. "Luciferin" means the substance of luciferase (a bioluminescent enzyme).

A glowing jellyfish, *Aequorea aequorea*, is one of the famous bioluminescent organisms in the ocean especially because it produces a green fluorescent protein (GFP) [4, 5]. GFP is widely utilized as an imaging material in biological components, in vivo and in vitro [6–8]. The energy required for GFP to emit light is supplied by aequorin, a photoprotein, which emits blue light [9]. After both proteins were discovered by Shimomura and Johnson in 1962 [10, 11], it was further reported

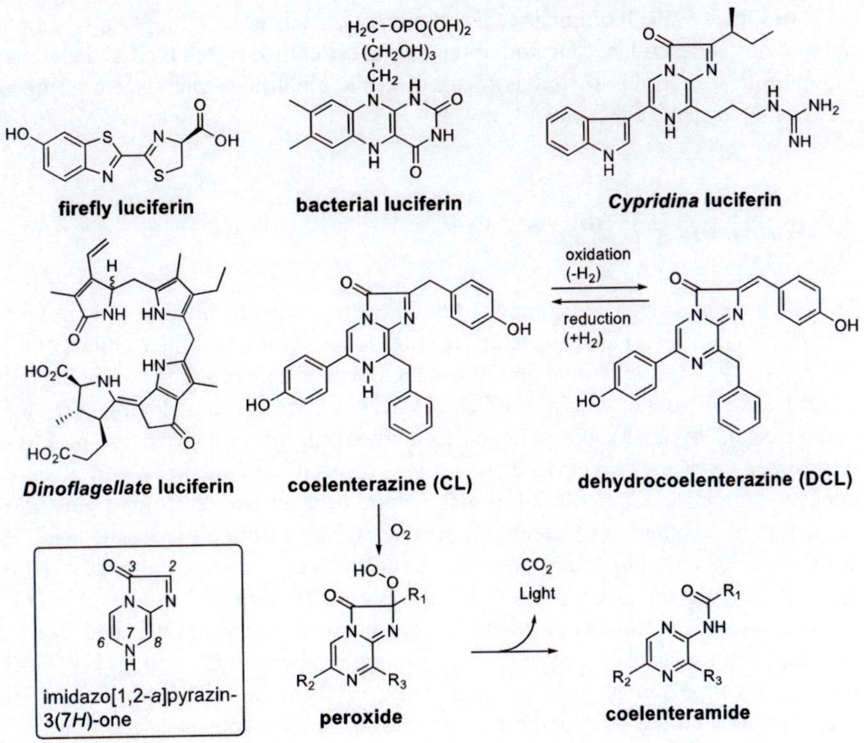

Fig. 1 Structures of heterocyclic organic substances utilized in bioluminescence

in 1975 that coelenterazine (CL) was the organic substance of aequorin [12]. CL was confirmed to exist as peroxide in the active site that is stabilized by hydrogen, which is bonded to Tyr184 [13]. Luminescence was initiated by binding calcium ions to aequorin to induce its conformational changes and cause the decomposition of the unstable peroxide. This decomposition affords an excited state of the oxidized compound (coelenteramide) of CL with the simultaneous production of CO_2. Light emission is obtained from the relaxation of the excited state to a ground state (Fig. 1).

Imidazopyrazinone is the core structure of CL, and its formal nomenclature is imidazo[1,2-*a*]pyrazin-3(7*H*)-one. CLs are found in many luminescent/nonluminescent marine organisms [14–17]; thus, they are expected to accumulate in the organisms because of the food chain processes [18]. Though the origin of biosynthesized CL is still unknown, Oba et al. demonstrated the biosynthesis of CL from a deep-sea copepod, *Metridia pacifica*, after feeding it with tyrosine and phenylalanine [19]. The amino acids were condensed to obtain CL. Dehydrocoelenterazine (DCL) is an oxidized compound of CL, which is also found in luminous organisms. CL and DCL are red–ox related; thus, the method of utilizing both compounds is different in the bioluminescent mechanism (Fig. 1).

In this review, the bioluminescence of marine organisms with imidazopyrazinone compounds is described. Since other reviews excellently covered the CL-dependent luciferase [20–22], this review is focused on the bioluminescent system, utilizing DCL as the organic substance [23].

2 *Watasenia scintillans*

DCL is the dehydrogenated compound of CL, which was identified from *Watasenia scintillans* (Japanese name is Hotaru-ika) by Inoue, Goto, and their colleagues in 1977 [24]. The synthetic route for CL and DCL was established as shown in Scheme 1 [25, 26]. 4-Methoxyacetophenone (**1**) was converted to aldoxime (**2**), and phenylacetaldehyde (**3**) was changed to aminonitrile (**4**). Condensation of **2** and **4** afforded N-oxide (**5**), which was then reduced and demethylated to give coelenteramine (**6**). Compound **6** is also natural product isolated from bioluminescent system. Another component (**11**) was synthesized from phenylacetic acid (**7**), which was converted to diazoketone (**8**). After **8** was converted to bromoketone (**9**), nitrone (**10**) was obtained from **9**. Hydrolysis of **10** yielded ketoaldehyde (**11**). Condensation of **6** and **11** afforded CL, which was oxidized to DCL with MnO_2. A literature summarizes the recent progress of the synthesis of CL analogs [27]. DCL itself is not involved in the bioluminescence of *Watasenia*. The disulfate of CL (Fig. 2) was reported as a luciferin (organic substrate for enzymatic reaction), and luciferase (an enzyme that catalyzes the oxidation of luciferin) requires ATP and magnesium ion as cofactors [28, 29]. Although a plausible mechanism for the

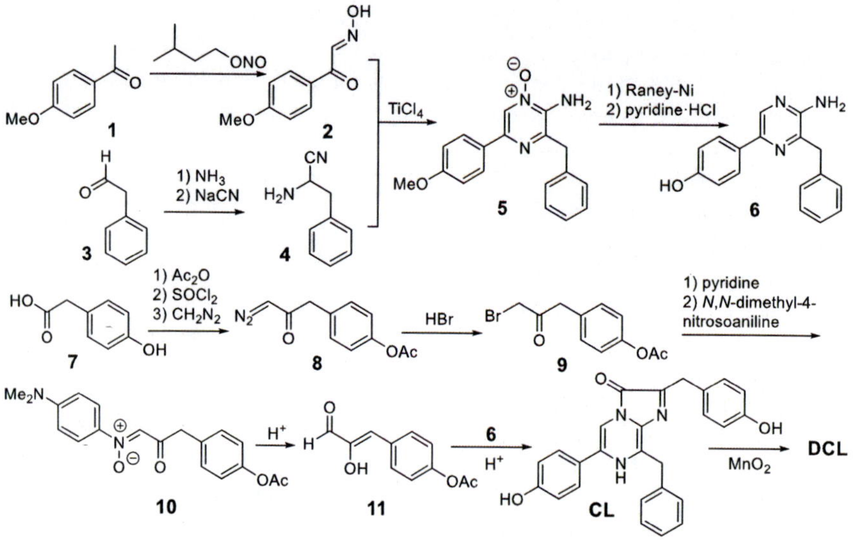

Scheme 1 Synthetic route for CL and DCL

Fig. 2 Structure of CL disulfate

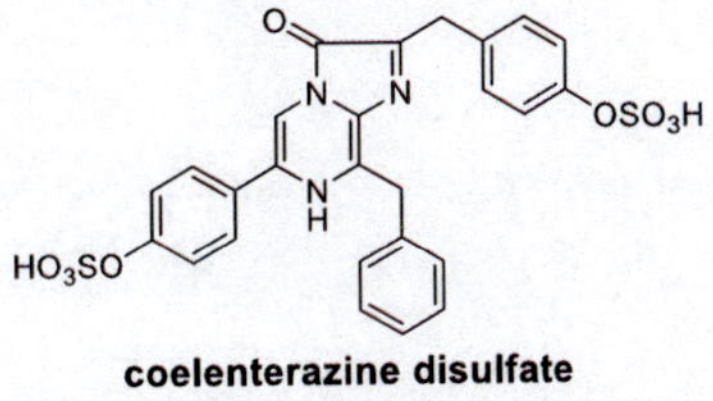

luciferin–luciferase (L–L) reaction was proposed to describe the involvement of adenylated CL disulfate [30], supportive data were not available to prove the involvement of adenylated CL disulfate [31], i.e., the detailed bioluminescent mechanism was unclear. In 2016, Sharpe's group reported the mass spectrometry (MS) analysis and transcriptome sequencing of a luminous organ [32]. They revealed that glowing squid crystal proteins are in the same superfamily as firefly luciferase, and *Watasenia* luciferase exhibited 19–20% sequence identity with firefly luciferase. The activity of the luciferase was not reported, hence the challenge in understanding the isolation of pure luciferase from *Watasenia* [33].

3 *Symplectoteuthis (Sthenoteuthis) oualaniensis*

The purple-back flying squid (*Symplectoteuthis/Sthenoteuthis oualaniensis*), Tobi-ika in Japanese, is distributed in the Pacific and Indian oceans [34]. There is an oval light organ on the surface of its mantle that emits blue light. The ink of the squid is utilized in luxury foods in Japan. Research on its bioluminescence was first reported by Tsuji and Leisman in 1981 [35]. A membrane-bound protein was found to be the origin of its light emission when it was stimulated by Na^+/K^+, at an optimum pH of 7.8. The organic substance for the bioluminescence was unknown until 1993 when Isobe's group discovered that DCL was the organic substance responsible for the bioluminescence [36]. An acetone adduct of DCL was isolated from the acetone powder of the light organs. The adduct was proven to be an artifact during the preparation of acetone powder from the light organs. DCL, existing in the light organ, was converted to the adduct by the conjugate addition of an enamine, derived from an amino group of protein and acetone, while the acetone powder was prepared (Fig. 3). After extensive efforts had been invested by the same group, the sulfhydryl group was confirmed to be the most active nucleophile toward DCL. Dithiothreitol (DTT) and glutathione (GSH) interacted with DCL to produce DCL–DTT and DCL–GSH, respectively (Fig. 4) [37].

Simultaneously, a photoprotein, later called "symplectin," was extracted in a 1.0 M KCl buffer. Both DCL–DTT and DCL–GSH, along with DCL, were also active substances in the luminescence of symplectin. There was equilibrium between DCL and these thiol adducts; thus, the liberated DCL was expected to be involved in the active site of symplectin to form a chromophore that was responsible for the light emission. From these results, a luminescent mechanism was proposed for

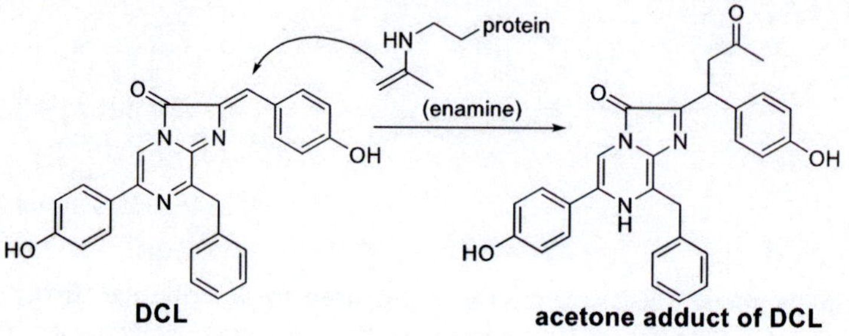

Fig. 3 Schematic representation of the production of the acetone adduct from DCL

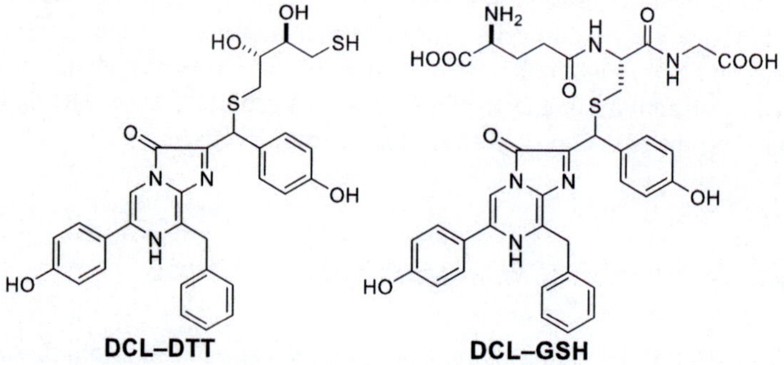

Fig. 4 Structures of the thiol adducts of DCL

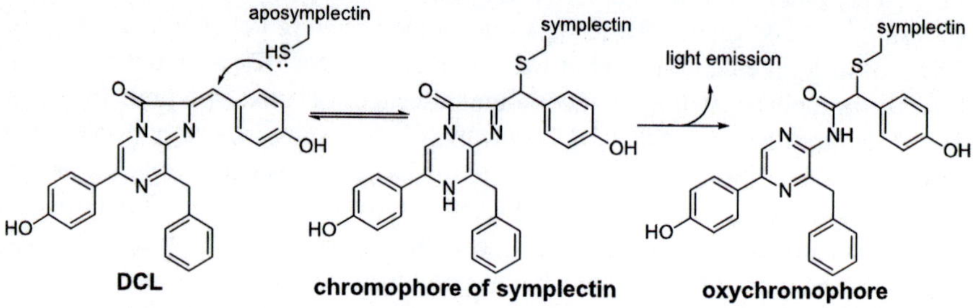

Fig. 5 Schematic representation of a plausible luminescent mechanism of symplectin

symplectin. A sulfhydryl group of the active site, cysteine, was connected to DCL by conjugate addition to form the chromophore of symplectin. The oxidation of the chromophore would result in the emission of light to afford an oxychromophore (Fig. 5).

The luminescent mechanism was demonstrated utilizing DTT (as an apo-symplectin model) and ^{13}C-labeled DCL. The structures of the model chromophore and oxychromophore were confirmed by nuclear magnetic resonance (NMR) [38, 39]. In 2002, an amino acid sequence of symplectin was reported. Symplectin is

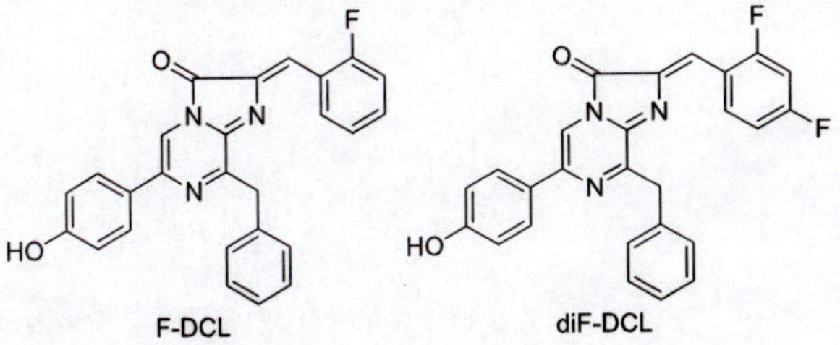

Fig. 6 Chemical structures of the fluorinated DCL analogs

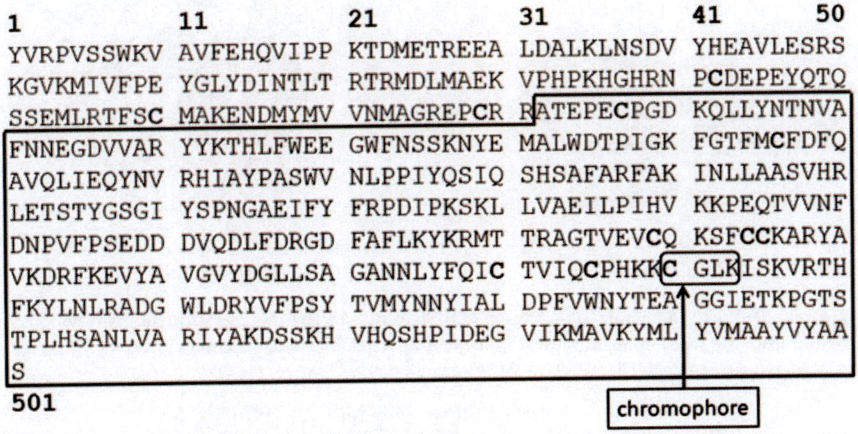

Fig. 7 Amino acid sequence of symplectin. The 40 kDa domain is highlighted with a box. The chromophore was expected to form at Cys390

a 60 kDa protein, which exhibits a fluorescent character due to the chromophore [40]. The partial digestion of symplectin with trypsin afforded a fluorescent 40 kDa fragment that also exhibited luminescent activity with DCL. The remaining *N*-terminal (20 kDa) of symplectin demonstrated some similarities with the mammalian carbon–nitrogen hydrolase domain (biotinidase/pantetheinase).

In 2008, the active site of symplectin was deduced to be Cys390, employing fluorinated DCL (F-DCL) as an organic substance [41, 42]. F-DCL afforded a stable chromophore probably because of the inductive effect of the fluorine atom (Fig. 6). Further, the inductive effect halted the equilibrium between the adduct and DCL. The reconstituted symplectin with F-DCL was digested with trypsin, after which the resulting peptides were analyzed by liquid chromatography–MS (LC–MS). A chromophore-containing peptide (Cys390–Lys393) plus F-DCL was observed, before and after the luminescence. From these results, Cys390 was expected to be the active site that formed the chromophore of symplectin (Fig. 7). Difluorinated DCL was later confirmed to be the most active non-natural substance of symplectin (Fig. 6) [43]. However, the detailed luminescent mechanism of symplectin is still unknown [44]. In 2017, Haddock's group reported their research on the occurrence and evolution of symplectin [45].

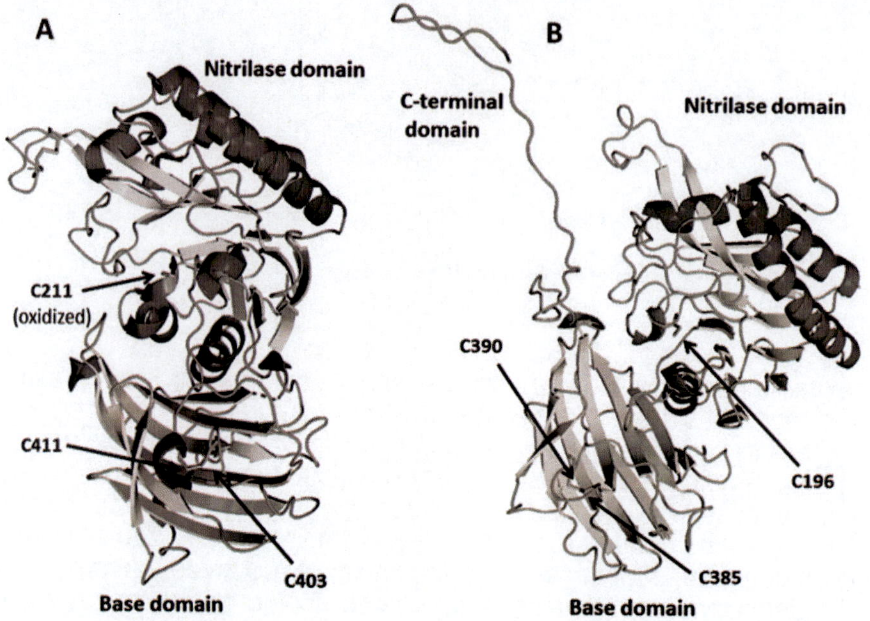

Fig. 8 (**a**) Crystal structure of vanin-1 (PDB:4CYF) and (**b**) modeled structure of symplectin, rebuilt from the reported data (https://bitbucket.org/wrf/squid-transcriptomes) [45]

They performed the transcriptome analysis to obtain five homologs of symplectin. Since the crystal structure of vanin-1 (pantetheinase) was reported in 2014 [46] and its sequence exhibited similarities with symplectin (30% identity), a modeling study was conducted to investigate the active site (cysteine) residue (Fig. 8). Vanin-1 consists of nitrilase and base domains, and Cys211 is located in the active site that is responsible for the hydrolysis of the carbon–nitrogen bond. Cys411 formed a disulfide bond with Cys403. In the modeled symplectin, Cys196 existed in the active site, which was like Cys211 (vanin-1), and Cys390 formed a disulfide bond with Cys385. From this modeling, a possible alternative in which Cys196 was the active site for bioluminescence was proposed.

4 *Eucleoteuthis luminosa*

A brief study on the bioluminescence of *Eucleoteuthis luminosa* (Suji-ika in Japanese) was reported by Shimomura [47]. The glowing squid belongs to the same genus as *Sthenoteuthis oualaniensis*. The photoprotein was partially extracted from the light organ, following the same method for symplectin. The light emission of the photoprotein (50 kDa) was increased by the pretreatment of DCL, and the photoprotein could be like or the same with symplectin. A large amount of DCL (600 μg) was also found in the liver (10 g). The most important finding was that the

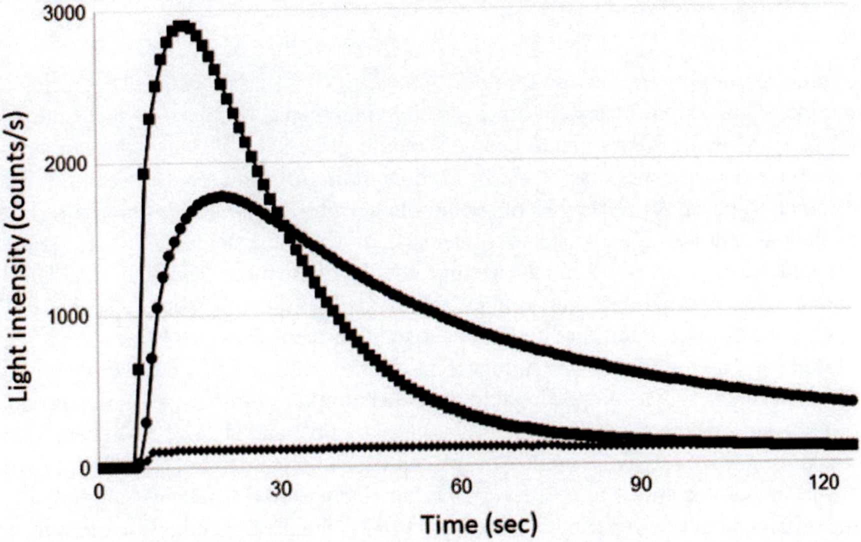

Fig. 9 Time course of luminescence of symplectin stimulated with a mixture of catalase–H_2O_2. At a time (5 s), symplectin (20 μL in 0.6 M KCl–PBS) was mixed with (■) catalase (20 μL, 1 mg/mL) and 3% H_2O_2 (200 μL) in 50 mM Tris with pH 8.3, (•) catalase (20 μL, 1 mg/mL) and 1% H_2O_2 (200 μL) in 50 mM Tris (pH 8.3), (◆) catalase (20 μL, 1 mg/mL) in 50 mM Tris, pH 8.3 (200 μL)

mixture of catalase and hydrogen peroxide (H_2O_2) strongly induced the light emission of the photoprotein. This phenomenon was also observed in the light emission of symplectin (Fig. 9). Some intermediates/compounds, derived from catalase–H_2O_2, could cause the induction of intense light emission, although the detailed mechanism is still unknown.

5 *Dosidicus gigas* (The Largest Luminescent Animal in the World)

In 2019, Oliveira's group reported the characterization of the bioluminescence of a Humboldt squid (*Dosidicus gigas*), which is one of the largest luminescent animals in the world [48]. They identified DCL in the luminous organ and succeeded in the purification and isolation of the photoprotein. The photoprotein was 60 kDa, as in symplectin. The isolated protein was analyzed by LC–MS and the transcriptome analyses to afford two gene products, which displayed high coverage (>80%) with symplectin and vanin-2. The photoprotein emitted blue light (470 nm) in the presence of DCL.

6 *Pholas dactylus*

The common glowing piddock, *Pholas dactylus* [49, 50], is a clam-like species of marine mollusks. From ancient times, its bioluminescence had been noticed at night. Further, chewing *Pholas* made people's mouth glow [51]. In 1887, the luciferin–luciferase reaction was discovered by Dubois in *Pholas* bioluminescence [52]. The research restarted in 1970 by Henry's and Michelson's group. *Pholas* luciferin (later named as Pholasin, a registered trademark of Knight Scientific Ltd. [53]) and luciferase were extracted from the acetone powder of the luminous organs of *Pholas*, and it was reported that the luminescence was significantly stimulated by Fe^{2+} [54]. Pholasin was confirmed to be a heat-stable protein (Mw. 46 kDa) after it was heated for 3 min at 60 °C. The luminescent reaction required molecular oxygen, and the light intensity was proportional to the concentration of Pholasin. Light emission of Pholasin was thereafter found to be activated by Fe^{2+} and H_2O_2 [55]. In the report, it was found that luciferase could be replaced with a mixture of Fe^{2+}–H_2O_2. In 1973, Pholasin was confirmed to be the photoprotein that was like aequorin. Subsequently, the involvement of a superoxide anion radical ($O_2^{\bullet-}$) in the light emission of Pholasin was demonstrated [56]. The xanthine–xanthine oxidase system could also induce the luminescence of Pholasin in the presence of oxygen. The sensitivity of Pholasin toward $O_2^{\bullet-}$ was superior to that of luminol. Thus, the report implied the important utility of Pholasin, as a sensor for reactive oxygen species (ROS). It was also reported that Pholasin did not recover after the luminescence, implying that an irreversible reaction occurred during the luminescence. They also tested the electrochemical oxidation of Pholasin and observed that oxidation only occurred at the anode, not at the cathode. The obtained experimental voltage for the oxidation was 23 kcal/mol, which was evidently much lesser than the corresponding voltage for light emission, at 495 nm (λ_{max} of the emission spectrum of Pholasin), which required >58 kcal/mol of energy. It further implied that there was a withdrawal of three electrons, the rate-limiting step, and that the following irreversible second step followed the luminescent reaction of Pholasin, i.e., it involved the interaction of molecular oxygen with a radical of luciferin. This reaction could be represented by Eq. (1):

$$\text{pholasin} + O_2^{\bullet-} \rightarrow \text{pholasin} - \text{OOH} \rightarrow \text{oxypholasin} + \text{Light} \qquad (1)$$

The luminescence of Pholasin was also induced by the addition of horseradish peroxidase (HRP) [57]. In the system employing HRP, H_2O_2 was not essential for the luminescence. They noticed the presence of a prosthetic group (chromophore) of Pholasin that was responsible for light emission. It was proposed that the oxidation of the chromophore with $O_2^{\bullet-}$ afforded H_2O_2, as an intermediate, after which the decomposition of the intermediate afforded the light emission. However, they could not isolate the chromophore from Pholasin.

Further, the same group also studied *Pholas* luciferase extensively [58, 59]. The luciferase, composed of two molecules of a subunit (Mw. 150 kDa) and possessing one copper atom per molecule, formed a 1:1 complex with one Pholasin molecule. The composition of the amino acids of the luciferase confirmed that 23% of it were aspartic and glutamic acids. These amino acids rendered the luciferase acidic (PI < 3.5). The luciferase was a glycoprotein that was composed of glucosamine, fucose, mannose, and galactose residues. The luciferase was a peroxidase, which catalyzes the oxidation of ascorbic acid in the presence of H_2O_2, and exhibited similar behavior as HRP. Subsequently, Henry et al. improved the method of isolating Pholasin and luciferase [60]. Pholasin was confirmed to be a glycoprotein, which possesses similar sugar chains as luciferase, and its revised Mw. was 34 kDa. Like luciferase, Pholasin was also an acidic protein, and the deduced quantum yield of Pholasin was 0.09 (based on the calibration with luminol [61]). They analyzed the protein–protein interactions between the luciferase and Pholasin to obtain the following parameters: the dissociation equilibrium constant (K_ds) for Pholasin–luciferase and oxypholasin–luciferase was 1.2×10^{-11} and 1.7×10^{-8}, respectively. Unfortunately, these important and extensive research in France halted abruptly in 1978 because of the extinction of *Pholas*. As mentioned by Michelson, this extinction might be caused by gastronomic delicacy; their utilization, as luminescent baits, by fishermen; and, perhaps, pollution by oil tankers. Michelson made a collective research review on *Pholas* bioluminescence, conducted by their research group [62]. The report revealed the importance of the surrounding environmental organisms, which was of interest to natural products chemists.

In 1987, Knight, Campbell, and Roberts continued the research in England. They reported the utilization of Pholasin in the detection of the activation of single neutrophils [53]. After they confirmed that the luminescence of Pholasin was induced by HRP and a neutrophil peroxidase (myeloperoxidase (MPO)) [63], Pholasin was confirmed to be a highly sensitive indicator of reactive oxygen metabolites, produced by the neutrophil. The sensitivity of Pholasin was 50- to 100-fold greater than that of luminol. From these results, it could be seen that Pholasin was successfully employed as the indicator of ROS [64, 65]. At that time, the exact chemical structure of Pholasin was unknown. Furthermore, Müller and Campbell investigated the chromophore of Pholasin [66]. CL and *Cypridina* (*Vargula*) luciferin [67, 68] did not activate the luminescence of Pholasin. From the fluorescent spectra, they proposed that a flavin compound could have been involved in the luminescence. In 2000, Campbell's group successfully cloned and expressed apopholasin (an apoprotein of Pholasin) [69]. Apopholasin contained 225 amino acids with a signal peptide of 20 amino acids, 3 predicted *N*-linked glycosylation sites, 1 *O*-linked site, 0 histidine, and 7 cysteines (Fig. 10). The recombinant apopholasin was expressed in insect cells to afford the fully processed glycosylated apopholasin (34 kDa) and the unprocessed apopholasin (26 kDa). After the apopholasin was mixed with acid methanol extracts of *Pholas*, the resulting apopholasin exhibited luminescence that was stimulated by sodium hypochlorite. The kinetics of the reactivated apopholasin was very different from that of the native Pholasin, thus the conclusion that native Pholasin could not be obtained from the recombinant apopholasin, employing the acid methanol extract, since the substance that constituted the chromophore of Pholasin was still unknown.

```
1          11         21         31         41
MACIVFVALV ALCLMQPGSG EEVQCAMNWT QANEYVFNVD WMTIFIYDYG
      Signal peptide   Matured pholasin
AQEQLYEDRA LGLCRIERAG PGTTKAVWIN WSNDTQSCVT RKTIFFEVGG

EIARLVDYRP QEDGTEKTFT RKFSSKMPGT YMLMDVCATR DADDKCIEGT

IVVTVRVSLY DEDNNGVMDE GKVIPSETIE DDIKDCGLLD QDVELDYTWT

QNECDLPDTV DEAEDTPSET GEFFW
201
```

Fig. 10 Amino acid sequence of Pholasin with a signal peptide (Met1–Gly20)

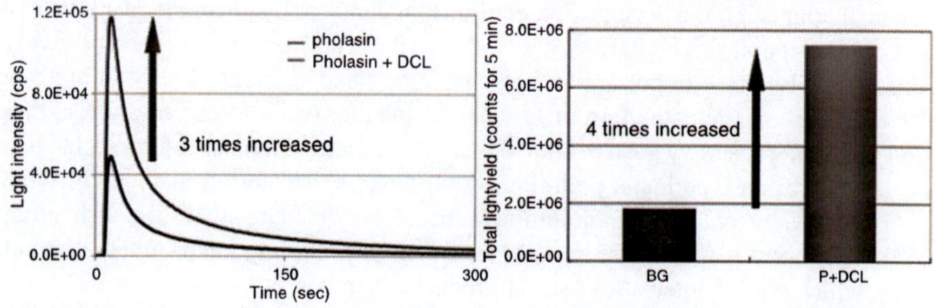

Fig. 11 Time course of luminescence and total light yield of Pholasin with/without DCL

The luminescent character of Pholasin triggered by ROS resulted in the establishment of Knight Scientific Limited by Jan Knight and Robert Knight, in 1990 [70]. They succeeded in cultivating *Pholas dactylus* through their 11-year research efforts, and the company started supplying the extracted and purified Pholasin from the cultivated *Pholas* as Pholasin and ABEL (Analysis By Emitted Light, a registered trademark of Knight Scientific Ltd.) antioxidant test kits for the detection of free radicals and oxidants with high sensitivity [71]. Pholasin is also widely utilized for the detection of free radicals and oxidants, such as ROS, chloramines, and bromamines [72–78].

In 2008, Kuse et al. observed that DCL could increase the light emission of Pholasin [79]. Pretreated Pholasin with DCL afforded threefold light intensity and fourfold total light yield for 5 min, and DCL was confirmed as an organic substance of Pholasin (Fig. 11). Although there was no proof that DCL was the true substance of the native Pholasin, this finding proposed that Pholasin afforded a similar chromophore as did symplectin.

Later, the same group reported the extraction and identification of the DCL–DTT adduct from Pholasin by a DTT treatment [80]. Since DCL was expected to form a stable chromophore with Pholasin, there could be an equilibrium between DCL and the chromophore (Fig. 12). Thus, an excess amount of DTT would convert chromophore into DCL–DTT. This hypothesis actually worked.

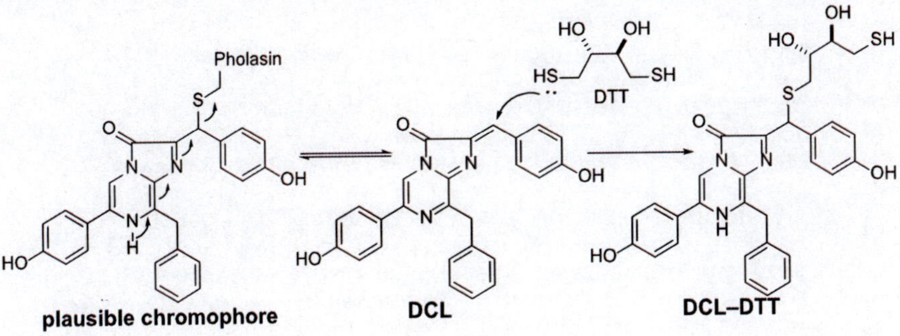

Fig. 12 Schematic representation of the conversion of the chromophore of Pholasin into a DCL–DTT adduct

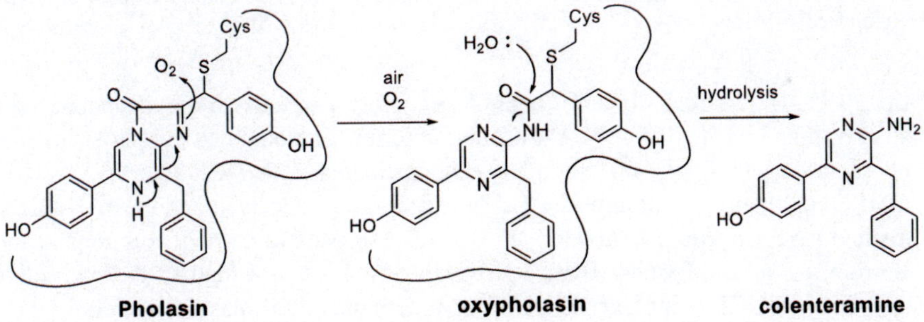

Fig. 13 Schematic representation of a plausible mechanism of the air oxidation of Pholasin to afford coelenteramine. During the storage of Pholasin, the prosthetic group of Pholasin could be oxidized to produce oxypholasin. Thereafter, the oxypholasin would be hydrolyzed to afford coelenteramine

The DCL–DTT adduct was prepared by treating Pholasin with DTT in methanol. The resulting solution was thereafter centrifugated to afford a fluorescent solution. The extract was analyzed by LC–MS by comparing it to the synthetic DCL–DTT. DCL–DTT was observed in the extract, and its MS spectra were identical to those of the authentic sample. As a supportive data, coelenteramine was also detected in the methanolic extract of Pholasin. Coelenteramine was expected to be produced by air oxidation during the storage of Pholasin (Fig. 13). Thus, DCL was confirmed to exist in Pholasin.

In 2020, Moriguchi et al. succeeded in cloning and expressing recombinant apopholasin (an apoprotein of Pholasin), utilizing a baculovirus–silkworm multigene expression system and DCL for the activation [81]. The recombinant apopholasin possessed C-terminal tags (histidine and FLAG® (DYKDDDDK) tags) with a cleavage site via the HRV 3C protease (Fig. 14).

Two recombinant apopholasins were expressed in the baculovirus–silkworm multigene expression system: apopholasin$_F$ contained a signal peptide at the N-terminal, and apopholasin$_M$ corresponded to the mature Pholasin beginning

```
1          11          21          31          41
MACIVFVALV ALCLMQPGSG EEVQCAMNWT QANEYVFNVD WMTIFIYDYG
     Signal peptide   Matured pholasin
AQEQLYEDRA LGLCRIERAG PGTTKAVWIN WSNDTQSCVT RKTIFFEVGG
EIARLVDYRP QEDGTEKTFT RKFSSKMPGT YMLMDVCATR DADDKCIEGT
IVVTVRVSLY DEDNNGVMDE GKVIPSETIE DDIKDCGLLD QDVELDYTWT
QNECDLPDTV DEAEDTPSET GEFFWGLEVL FQGPASHHHH HHGTDYKDDD
                                  HRV 3C cleavage  His Tag   FLAG® Tag
DK  (252 aa)          Matured pholasin  site
```

Fig. 14 Amino acid sequence of apopholasin with a cleavage site at HRV 3C protease, a histidine tag, and a Flag® tag at the C-terminal. Apopholasin possesses a signal peptide (Met1–Gly20), and mature apopholasin is called Glu21–Trp205. Seven cysteines in mature apopholasin were highlighted as bold

from Glu21. The recombinant apopholasins were extracted from the pupae of a silkworm and purified in a FLAG-affinity column. The obtained apopholasin$_F$ and apopholasin$_M$ (100 μL, 0.1 mg/mL) were incubated with DCL (2 μL, 2 mM in DMSO) to obtain apopholasin–DCLs (hereafter referred to as apopholasin–DCLs to distinguish them from *Pholas*-derived Pholasin). First, a mixture of solutions A and B from a Pholasin-based ABEL antioxidant test kit was added to the respective apopholasin–DCL complexes since the luminescence of Pholasin is generally initiated by the addition of this A and B mixtures from the ABEL kit. However, light emission was not observed at all. In attempting to obtain active apopholasin–DCLs, it was discovered that the addition of an HRP–H_2O_2 mixture would result in the activation of the apopholasin–DCL complexes (Fig. 15).

Both the DCL-incubated apopholasin$_F$ and apopholasin$_M$ emitted intense light when the peroxidase–H_2O_2 mixture was added, and the resulting luminescence was observed for 10 min before it gradually decreased. From these results, it was revealed that the luminescence depended on the presence of the apopholasins because DCL alone could not afford any light emission when the peroxidase–H_2O_2 mixture was added. This trend was also observed during the activation of Pholasin. From these results, it has been revealed that luminescence in Pholasin would be produced if similar activators were utilized, i.e., incubation with DCL and activation via an enzyme–peroxide mixture. Moreover, the catalase–H_2O_2 mixture is known to cause luminescence in symplectin [47]. However, this mixture failed to trigger luminescence strongly in the apopholasin–DCL complex. Another peroxide, *tert*-butyl hydroperoxide (TBHP), was also tested, but its light emission was poor, even in the presence of peroxidase or catalase activators (Fig. 16). From these results, it was clear that a peroxidase–hydrogen peroxide mixture specifically initiated the luminescent character of apopholasin–DCLs.

The luciferin–luciferase reaction in *Pholas dactylus* is known to be responsible for the light emission characteristics of mollusks since Pholasin is a type of luciferin and luciferase is a copper-containing peroxidase [62]. Reichl et al. reported that the

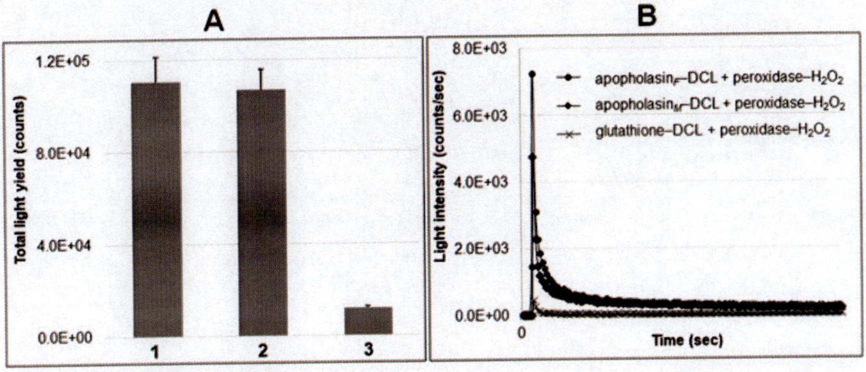

Fig. 15 (**a**) Total light yields and (**b**) time course of luminescence of apopholasin–DCLs for 10 min. (**a**): (1) apopholasin$_F$–DCL + peroxidase and H_2O_2. (2) apopholasin$_M$–DCL + peroxidase and H_2O_2. (3) GSH–DCL + peroxidase and H_2O_2. (**b**): time course of the light emission of apopholasin$_F$–DCL (–•–), apopholasin$_M$–DCL (–◊–), and GSH–DCL (–x–). Each sample was placed in a luminometer to quantify the amount of light emitted from the sample. After 5 s, a peroxidase–H_2O_2 mixture was quickly injected, and the resulting light emission was monitored for 10 min. Reprinted from Moriguchi et al. (2020) Bioorg Med Chem Lett 30:127177, with permission from Elsevier

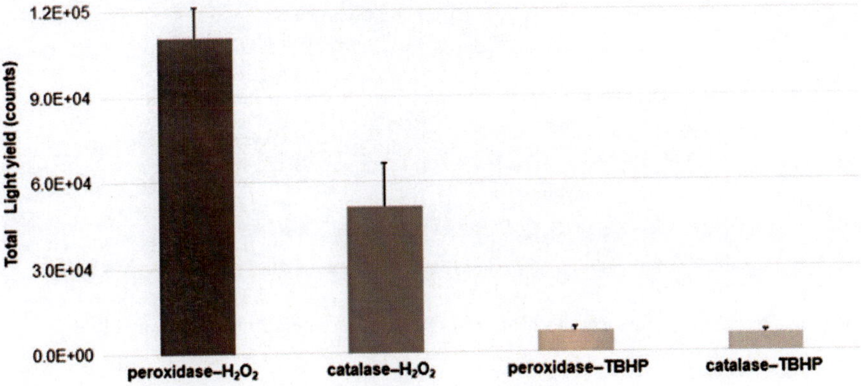

Fig. 16 Total light yields of apopholasin$_M$–DCL for 5 min: (1) apopholasin$_M$–DCL + peroxidase and H_2O_2, (2) apopholasin$_M$–DCL + catalase (5 mg/mL in PBS buffer, pH 8.5) and H_2O_2, (3) apopholasin$_M$–DCL + peroxidase and TBHP (5%), and (4) apopholasin$_M$–DCL + catalase (5 mg/mL in PBS buffer, pH 8.5) and TBHP (5%). Reprinted from Moriguchi et al. (2020) Bioorg Med Chem Lett 30:127177, with permission from Elsevier

oxidation of Pholasin was caused by Compound I or II, which was obtained from MPO and HRP in the presence of H_2O_2 to afford an intense light emission [73]. Compounds I and II are ferryl (iron-containing heme) intermediates, which were derived from the activation of oxygen in heme enzymes (a copper-containing metal center also afforded similar intermediates as *Pholas* luciferase) [82, 83]. Native ferric HRP is converted into Compound I when it is activated with H_2O_2. It is a green

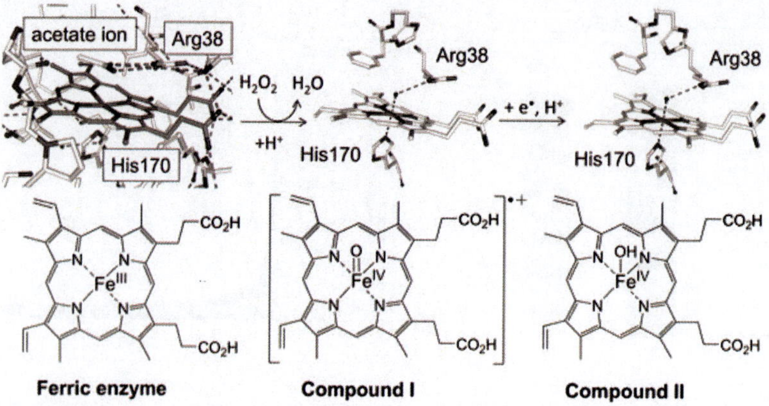

Fig. 17 Production of Compounds I and II from ferric HRP. The active sites of each structure are shown with a heme moiety. Ferric enzyme (PDB 1H5A), Compound I (PDB 1HCH), and Compound II (PDB 1H55) are shown from the X-ray data [82]. The radical cation is located in the heme moiety of Compound I

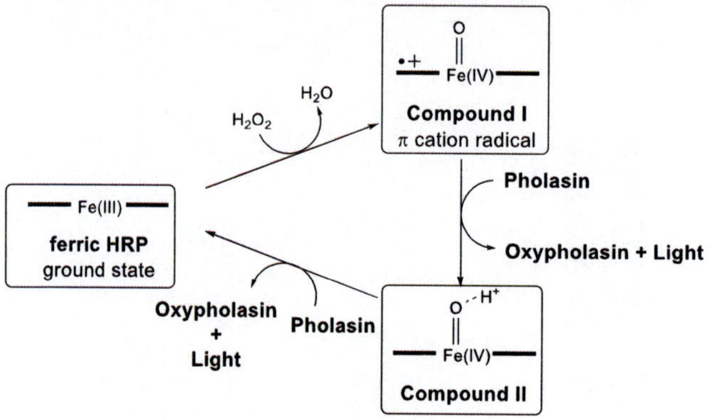

Fig. 18 Proposed luminescent mechanism of Pholasin, based on Compounds I and II

two-equivalent oxidized intermediate. Compound II is formed by a one-electron reduction of Compound I. It is a brown intermediate (Fig. 17). Both compounds could withdraw one proton and electron, and Reichl et al. proposed a luminescent mechanism (Fig. 18), resultantly. Compound I oxidizes Pholasin into oxypholasin in the presence of oxygen to afford Compound II. The obtained Compound II also oxidizes Pholasin into oxypholasin while emitting intense light to afford the native HRP. This phenomenon is consistent with the mechanism proposed by Michelson [62]. Thus, the utilization of recombinant apopholasin and DCL was simply a reproduction of this biological system that initiated light emission via the HRP–H_2O_2 mixture. However, a complete luminescent profile for the apopholasin–DCL complexes is yet to be determined.

Inoue et al. reported that a fusion protein consisting of glutathione *S*-transferase (GST) and apopholasin (GST–apopholasin) demonstrated luciferase-like activity for the oxidation of CL into coelenteramine while emitting blue light [84]. In their system, DCL did not afford any light emission. Afterward, they reported GST–apopholasin with DCL could afford light via the addition of catalase–H_2O_2 [85].

7 Conclusion

From the discovery of the luciferin–luciferase reaction in *Pholas dactylus* in 1887, extensive research on bioluminescence have been conducted ever since (Table 1). In this review, the detailed research results regarding the bioluminescence of *Pholas dactylus* and *Symplectoteuthis/Sthenoteuthis oualaniensis* are summarized. Two photoproteins, Pholasin and symplectin, were DCL-related bioluminescent, i.e., their chromophores were composed of DCL for light emission. Catalase–H_2O_2 and peroxidase–H_2O_2 induced the respective luminescence of symplectin and Pholasin. The exact and detailed chemical structures of the chromophores of Pholasin and symplectin are yet unknown. Although the precise luminescence mechanism of these DCL-dependent photoproteins is unknown, Pholasin is widely employed as detector of ROS derived from the cells. The detection has been applied to extracellular systems, although recent progress would apply these photoproteins as intracellular detectors of ROS.

Table 1 Important progresses related to DCL-dependent bioluminescence

Year	Description
1887	Discovery of luciferin–luciferase reaction in *Pholas dactylus*
1970	Isolation of *Pholas* luciferin and *Pholas* luciferase
1973	Fe^{2+}–H_2O_2 induced light emission of *Pholas* luciferin
1977	Discovery of DCL from *Watasenia scintillans*
1985	*Symplectoteuthis* photoprotein is stimulated by Na^+/K^+ ions
1987	*Pholas* luciferin was renamed as Pholasin
1990	Knight Scientific Ltd. was established and released Pholasin and ABEL test kits
1993	DCL is a substance of *Symplectoteuthis* photoprotein (symplectin)
1996	Research on the chromophore of Pholasin
2000	Cloning and expression of apopholasin
2000	Compounds I and II induced light emission in Pholasin
2008	Active site analysis of symplectin
2008	DCL worked as a substance of Pholasin
2009	DCL–DTT was isolated from Pholasin
2019	*Dosidicus* photoprotein emitted light in the presence of DCL
2020	Activation of the expressed recombinant apopholasin with DCL

References

1. Haddock SH, Moline MA, Case JF (2010) Annu Rev Mar Sci 2:443
2. Widder EA (2010) Science 328:704
3. Kaskova ZM, Tsarkova AS, Yampolsky IV IV (2016) Chem Soc Rev 45:6048
4. Shimomura O (2009) Angew Chem Int Ed 48:5590
5. Chalfie M (2009) Angew Chem Int Ed 48:5603
6. Pakhomov AA, Martynov VI (2008) Chem Biol 15:755
7. Zimmer M (2009) Chem Soc Rev 38:2823
8. Ozbakir HF, Anderson NT, Fan KC, Mukherjee A (2020) Bioconjug Chem 21:293
9. Webb SE, Karplus E, Miller AL (2015) Mol Reprod Dev 82:563
10. Shimomura O, Johnson FH, Saiga Y (1962) J Cell Comp Physiol 59:223
11. Johnson FH, Shimomura O, Saiga Y, Gershman LC, Reynolds GT, Waters JR (1962) J Cell Comp Physiol 60:85
12. Shimomura O, Johnson FH (1975) Proc Natl Acad Sci U S A 72:1546
13. Head JF, Inoue S, Teranishi K, Shimomura O (2000) Nature 405:372
14. Shimomura O, Inoue S, Johnson FH, Haneda Y (1980) Comp Biochem Physiol 65B:435
15. Shimomura O (1987) Comp Biochem Physiol 86B:361
16. Campbell AK, Herring PJ (1990) Mar Biol 104:219
17. Thomson CM, Herring PJ, Campbell AK (1997) J Biolumin Chemilumin 12:87
18. Haddock SH, Rivers TJ, Robison BH (2001) Proc Natl Acad Sci U S A 98:11148
19. Oba Y, Kato S, Ojika M, Inoue S (2009) Biochem Biophys Res Commun 390:684
20. Markova SV, Vysotski ES (2015) Biochemistry 80:714
21. Lee J (2016) Photochem Photobiol 93:389
22. Fleiss A, Sarkisyan KS (2019) Curr Genet 65:877
23. Kuse M (2014) Biosci Biotechnol Biochem 78:731
24. Inoue S, Taguchi H, Murata M, Kakoi H, Goto T (1977) Chem Lett 6:259–262
25. Kishi Y, Tanino T, Goto T (1972) Tetrahedron Lett 13:2747
26. Inoue S, Sugiura S, Kakoi H, Hashizume K, Goto T (1975) Chem Lett 4:141
27. Coutant EP, Janin YL (2015) Chem Eur J 21:17158
28. Inoue S, Kakoi H, Goto T (1976) Tetrahedron Lett 17:2971
29. Tsuji FI (1985) Proc Natl Acad Sci U S A 82:4629
30. Tsuji FI (2002) Biochim Biophys Acta 1564:189
31. Teranishi K, Shimomura O (2008) Biochim Biophys Acta 178:784
32. Gimenez G, Metcalf P, Paterson NG, Sharpe ML (2016) Sci Rep 6:27638
33. Shimomura O (2012) Bioluminescence, chemical principal and methods, Revised edn. World Scientific, Singapore, p 214
34. *Sthenoteuthis oualaniensis* is recent accepted nomenclature instead using *Symplectoteuthis*. http://www.marinespecies.org/
35. Tsuji FI, Leisman GB (1981) Proc Natl Acad Sci U S A 78:6791
36. Takahashi H, Isobe M (1993) Bioorg Med Chem Lett 3:2647
37. Takahashi H, Isobe M (1994) Chem Lett 23:848
38. Isobe M, Kuse M, Yasuda Y, Takahashi H (1998) Bioorg Med Chem Lett 8:2919
39. Kuse M, Isobe M (2000) Tetrahedron 56:2629
40. Fujii T, Ahn JY, Kuse M, Mori H, Matsuda T, Isobe M (2002) Biochem Biophys Res Commun 293:874
41. Isobe M, Fujii T, Kuse M, Miyamoto K, Koga K (2002) Tetrahedron 58:2117
42. Isobe M, Kuse M, Tani N, Fujii T, Matsuda T (2008) Proc Jpn Acad Ser B 84:386
43. Kongjinda V, Nakashima Y, Tani N, Kuse M, Nishikawa T, Yu CH, Harada N, Isobe M (2011) Chem Asian J 6:2080
44. Chou CM, Tung YW, Isobe M (2014) Bioorg Med Chem 22:4177
45. Francis WR, Christianson LM, Haddock SHD (2017) PeerJ 5:e3633

46. Boersma YL, Newman J, Adams TE, Cowieson N, Krippner G, Bozaoglu K, Peat TS (2014) Acta Cryst D70:3320
47. Shimomura O (2012) Bioluminescence, chemical principal and methods, Revised edn. World Scientific, Singapore, p 219
48. Galeazzo GA, Mirza JD, Dorr FA, Pinto E, Stevani CV, Lohrmann KB, Oliveira AG (2019) Photochem Photobiol 95:1179
49. Knight R, Thorne J (1982) Protistologica 18:53
50. Knight J, Knight R (1986) Phil Trans R Soc Lond B 313:509
51. Roda A (2011) Chemiluminescence and bioluminescence: past, present and future. The Royal Society of Chemistry, Cambridge, p 3
52. Dubois R (1887) Compt Rend Soc Biol 39:564
53. Roberts PA, Knight J, Campbell AK (1987) An Biol 160:139
54. Henry JP, Isambert MF, Michelson AM (1970) Biochim Biophys Acta 205:437
55. Henry JP, Michelson AM (1970) Biochim Biophys Acta 205:451
56. Henry JP, Michelson AM (1973) Biochimie 55:75
57. Henry JP, Michelson AM (1973) Biochimie 55:83
58. Michelson AM, Isambert MF (1973) Biochimie 55:619
59. Henry JP, Monny C, Michelson AM (1975) Biochemistry 14:3458
60. Henry JP, Monny C (1977) Biochemistry 16:2517
61. Lee J, Wesley AS, Ferguson JF, Seliger HH (1996) Bioluminescence in progress. Princeton University Press, Princeton, p 35
62. Michelson AM (1978) Methods Enzymol 57:385
63. Roberts PA, Knight J, Campbell AK (1985) Biochem Soc Trans 13:1139
64. Cotton B, Allshire A, Cobbold PH, Müller T, Campbell AK (1989) Biochem Soc Trans 17:705
65. Müller T, Davies EV, Campbell AK (1989) J Biolum Chemilum 3:105
66. Müller T, Campbell AK (1990) J Biolum Chemilum 5:25
67. Shimomura O, Goto T, Hirata Y (1957) Bull Chem Soc Jpn 30:929
68. Kishi Y, Goto T, Hirata Y, Shimomura O, Johnson F (1966) Tetrahedron Lett 7:3427
69. Dunstan SL, Sala-Newby GB, Fajardo AB, Taylor KM, Campbell AK (2009) J Biol Chem 275:9403
70. Knight Scientific Limited: http://www.knightscientific.com/
71. Knight J (1997) Immunol News 4:26
72. Witko-Sarsat V, Nguyen AT, Knight J, Descamps-Latscha B (1992) Free Radic Biol Med 13:83
73. Reichl S, Arnhold J, Knight J, Schiller J, Arnold K (2000) Free Radic Biol Med 28:1555
74. Reichl S, Vocks A, Petkovic M, Schiller J, Arnhold J (2001) Free Radic Res 35:723
75. Swindle EJ, Hunt JA, Coleman JW (2002) J Immunol 169:5866
76. Glebska J, Koppenol WH (2005) Free Radic Biol Med 38:1014
77. Jaffar NZ, Dan Z, Christopher S, John BD, Knight J, John H (2006) Clin Biochem 39:55
78. Sild E, Hõrak P (2010) Behav Ecol Sociobiol 64:2065
79. Kuse M, Tanaka E, Nishikawa T (2008) Bioorg Med Chem Lett 18:5657
80. Tanaka E, Kuse M, Nishikawa T (2009) ChemBioChem 10:2725
81. Moriguchi M, Takahashi R, Bubwoong K, Kuse M (2020) Bioorg Med Chem Lett 30:127177
82. Berglund GI, Carlsson GH, Smith AT, Szöke H, Henriksen A, Hajdu J (2002) Nature 417:463
83. Moody PCE, Raven E (2018) Acc Chem Res 51:427
84. Inoue S, Sahara-Miura Y, Nakamura M, Hosoya T (2020) Protein Expr Purif 171:105615
85. Inoue S, Sahara-Miura Y, Iimori R, Sakata Y, Hazama Y, Yoshida S, Nakamura M, Hosoya T (2020) Biochem Biophys Res Commun 526:404

Part II
Isolation and Synthesis of Water-Soluble Low Weight Molecular Compounds

Top Heterocycl Chem (2021) 58: 107–130
https://doi.org/10.1007/7081_2020_46

Published online: 22 December 2020

Chemical and Biological Aspects of Water-Soluble Heterocyclic Marine Natural Products

Ryuichi Sakai

Contents

Abstract Water-soluble marine natural products are interesting as they generally do not penetrate into the cells but exhibit biological activities through cell surface receptors, ion channels, and glycans. Some molecules, however, interact with cell membrane and disrupt it. In this review, discovery, structures, and biological activities of water-soluble marine natural heterocyclic molecules are summarized with special emphasis on their biological activity and functions.

Keywords Excitatory amino acid · Giant clam · Glutamate receptor · Marine natural products · Membrane · Mycosporine · Neuronal receptor · Peptide · Polyamine · Polycation · Sponge · Tunicate · Zoanthid

R. Sakai (✉)
Faculty of Fisheries Sciences, Hokkaido University, Hakodate, Japan
e-mail: ryu.sakai@fish.hokudai.ac.jp

1 Introduction

According to marine natural products (MNP) database MarineLit, more than 35,000 compounds are known from the sea to date. Most reported marine-derived compounds, especially bioactive compounds are extracted by organic solvents. The reason for this is that organic soluble compounds often exert potent biological activity including cytotoxicity and antimicrobial activity by interacting with intracellular targets upon crossing lipid bilayer membrane. Contrary, hydrophilic or ionic compounds such as sugars and amino acids cannot cross the cell membrane thus require specific transporters to enter the cell. Thus, natural products with such physico-chemical properties likely fail to show "bioactivity" in cell-based screening assays such as cytotoxicity and antimicrobial assays. Moreover, most of the conventional separation techniques such as silica gel chromatography or solvent partition used to separate organic molecule may not be applicable for highly polar or ionic molecules [1]. For these reasons, water-soluble compounds are left relatively untapped. However, they are potentially interesting source for new biologically active molecules, because they can interact with cellular targets on the cell surface, including ion channels, receptors, glycans, and membrane itself. Many interesting examples such as tetrodotoxin/saxitoxins that target voltage gated sodium channel; kainic acid/domoic acid that target ion channel glutamate receptors (iGluR); the *Conus* peptide toxins such as ω-conotoxin MVIII that target various neuronal ion channels and receptors including voltage gated calcium channel; lectins that target sugar chains, and polyalkylpyridiniums/saponins that target lipid bilayer [2]. The existence of these classical examples suggested that water-soluble marine metabolites are interesting source to explore molecules that potentially target cell surface machineries that often govern intercellular communications and tissue integration in multi-cellular animals and plants. Moreover, biosynthesis and origin of water-soluble molecules are often quite interesting but elusive. Because most biosynthetically well-known class of molecules such as polyketides, lipids, steroids, terpenes, non-ribosomal peptides are organic soluble in general but those of some water-soluble small molecules such as tetrodotoxin are puzzling. This suggests to us that water-soluble molecules are intriguing source to explore newer chemical space and biological activities in the science of marine natural products. In this review, I summarize works regarding water-soluble MNPs mainly conducted in my laboratory with special emphasis on discovery, structure, and biological functions of heterocyclic molecules.

2 Discovery of Water-Soluble Marine Heterocycles

To screen bioactive water-soluble MNP, we first employed in vivo assay in mice. Because central nervous system is a showcase of receptors and ion channels expressed in cell surface. An intracerebroventricular (i.c.v.) injection of water-soluble compounds in mice can deliver them directly to synaptic machineries. In the early stage in conotoxin research, mouse behavioral assay was used to discover potentially interesting molecule and it turn out to discover many important classes of *Conus* peptides [3]. Inspired by this finding, the in vivo-based behavioral assay was extended to aqueous extracts of various marine organisms. We screened more than 1,000 marine aqueous extracts to date and found various small to middle secondary metabolites as well as bioactive proteins.

3 Excitatory Amino Acids

3.1 Kainic Acid

Excitatory amino acid (EAA) is an acidic amino acid that triggers excitatory neuronal transmission by binding with ionotropic glutamate receptors (iGluRs). iGluRs are categorized structurally and functionally into NMDA (*N*-methyl-D-aspartate) and non-NMDA types, and later are further divided into kainite and AMPA (α-amino-3-hydroxy-5-methyl-4-isoxazole-propionate)-type iGluRs. As the names suggest, the classification was based on the specific ligands that activate each subtype. Kainic acid (kainite), a classical marine heterocyclic compound, has played a key role in earlier studies of iGluR research and became a standard tool to investigate kainite-type iGluRs. Kainic acid (**1**) is a product of red algae *Digenea simplex* that occurs in temperate to subtropical shallow water of Indo-Pacific and Atlantic seas. The alga has been traditionally used as anthelmintic medicine in human, and kainic acid was isolated as an active principal. Later, domoic acid (**2**), a higher congener of kainic acid was found from another red algae *Chondria armata* which has also been used as anthelmintic medicine in Japan. Domoic acid is produced not only by the red algae, but also by diatoms [4]. Thereby bloom of domoic acid-producing diatoms, a genus of *Pseudo-nitzschia* species, causes damage in health of marine animals as shellfish or fishes may accumulate domoic acid through food web [4]. Ingestions of contaminated foods by marine animals as well as human result in abnormal behaviors and death in the worst cases. Both **1** and **2**, and in addition mushroom toxin acromelic acids share the substructure of kainic acid and thus they are collectively called kainoids. Kainoids generally possess agonist activity for iGluR selective for KA/AMPA types. Because **1** is indispensable tool in neurophysiological researches, and has attractive structural feature, more than 40 stereoselective syntheses of this molecule have been reported [5, 6].

We are interested in this molecule as to biological aspect in the producing algae *D. simplex* because no direct relevance between the known physiological action of kainite and algal physiology seemingly exist. However, the fact that some glutamate related signaling are identified in plants may suggest some importance of kainic acid in the producing algae [7]. Towards further understanding of this aspect we choose to determine cellular localization of kainic acid using immunohistochemical (IHC) techniques [8]. A polyclonal antibody that specifically recognizes kainic acid was

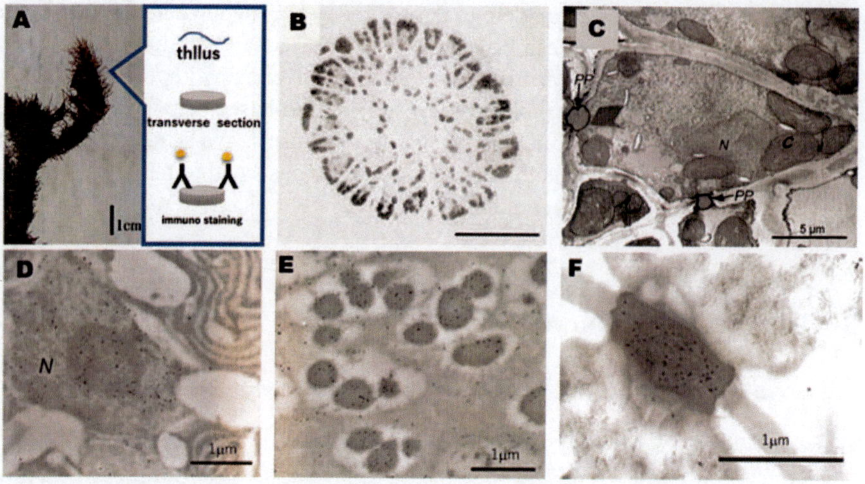

Fig. 1 Cellular and subcellular localization of **1** revealed by IHC experiments. (**a**) *Digenea simplex*. A hair-like thallus was cut and subjected to immuno-staining by anti-kainate antisera. (**b**) Light micrograph of kainic acid immunoreactivity in a transverse section of thallus labeled by gold nano particle. Dark parts are kainic acid immunoreactivity. (**c**) TEM of non-labeled *D. simplex* cell showing nucleus, N, and pit plugs, pp. TEM of (**c**) chloroplast, (**d**) nucleus, (**e**) granule body, (**f**) pit plug. Kainic acid immunoreactivity was labeled as black dots

successfully raised in rabbit. The immuno-staining of the section of algae clearly showed that kainic acid is localized most densely in the outer epithelial layer of fine cylindrical thallus. The cortical cells, but not the inner layers of the main axis, and cells of the rhizoid were also stained with this antibody. These observations suggested some defensive roles of kainic acid in the algae. Interestingly subcellular localization revealed by transparent electron microscopy (TEM) suggested that kainic acid is localized in nucleus, granule body (unannotated electron-dense compartment), and pit plug (red algae specific cell connection), however, no immunoreactivity was observed in the chloroplasts (Fig. 1).

Recently, biosynthetic pathway of kainic acid [9] was uncovered along with that of domoic acid [10] by genomic approaches. As proposed earlier, condensation of geranyl diphosphate and L-glutamic acid is a key step in domoic acid biosynthesis in *Pseudo-nitzschia*, followed by oxidation to form carboxylic acid and then oxidative cyclization to construct the pyrrolidine ring. In *C. armata*, isolation of putative biosynthetic intermediates allowed to propose DA biosynthesis in the red alga where *N*-geranylated *β*-hydroxy glutamate was found as key intermediate for pyrrolidine ring formation [11]. This general enzymatic conversion was conserved both in the diatom and in *D. simplex* [9] (Scheme 1). Because plant terpene synthesis generally takes place in the chloroplast through MEP pathway [12], our finding that chloroplast in *D. simplex* was devoid of kainic acid immunoreactivity suggests that biosynthesis of kainite takes place in cytosol rather than chloroplast using chloroplast-derived precursor.

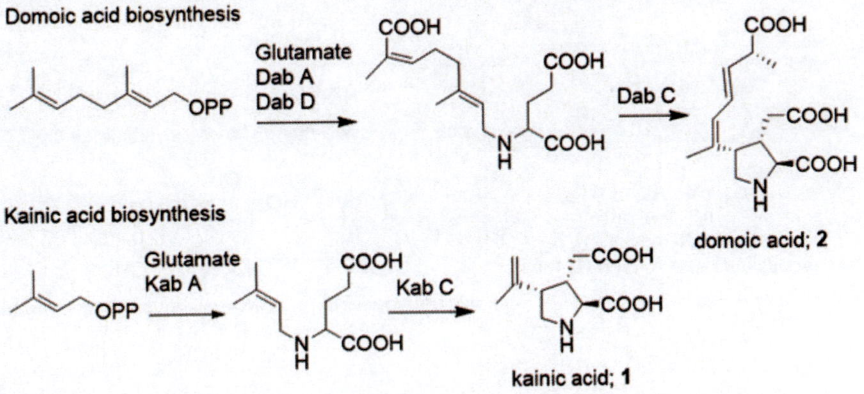

Scheme 1 Biosynthetic pathway of domoic acid and kainic acid

3.2 *Dysiherbaines and Related Compounds*

We found a Micronesian sponge initially identified as *Dysidea herbacea* [13] but revised later as *Lendenfeldia chondrodes*, exhibited potent pro-convulsant in mice in our in vivo screening. We isolated dysiherbaine (**3**) and neodysiherbaine-A (**4**) as the responsible compounds for excitatory activity in the extract. We also found several other compounds without neuronal activity in the same sponge. These compounds include dysibetaine (**5**), dysibetaine CPa (**6**) and CPb (**7**), PP (**8**) [14], and deoxynojirimycin-6-phosphate (**9a**) and its *N*-methyl derivative (**9b**) [15] (Scheme 2). Absolute stereochemistry for **6** and **8** were later confirmed by total syntheses [16, 17], and that for **7** was supported by synthesis of desmethyl analogue [18]. The structural diversity of small heterocyclic compounds found in the sponge demonstrated its rich biosynthetic capability. Interestingly, sponges collected only in Yap state Micronesia contained dysiherbaines, but not in other areas in central Pacific including Guam and Palau, thereby production of dysiherbaines not by sponge itself, but associated microorganisms are suspected. It has long been known that certain species of Pacific Dictyoceratid sponges, including *D. herbacea* (*Lammelodysidea herbacea*) and *L. chondrodes* heavily harbor symbiotic cyanobacteria *Oscillatoria spongeliae* up to 30–50% of sponge volume. Interestingly, however, *L. chondrodes* not only harbors *O. spongeliae* but also contains another cyanobacterial symbiont *Synechocystis* sp. [19].

We observed that both of these symbiotic cyanobacteria in mesohyl of the sponge. An immunohistochemical study combined with microscopic observations using dysiherbaine-specific antibody raised in rabbit revealed that immunoreactivity of dysiherbaine was localized specifically in the cells of *Synechocystis* sp., but not in *O. spongeliae*, eubacteria or sponge cells (Fig. 2). This result suggested that *Synechocystis* sp. plays important roles in production or storage of dysiherbaine [20]. *Synechocystis* sp. identified in the sponge is closely related to obligate symbionts *Prochloron didemni* and *S. trididemni* of tunicates. *Prochloron* is known to

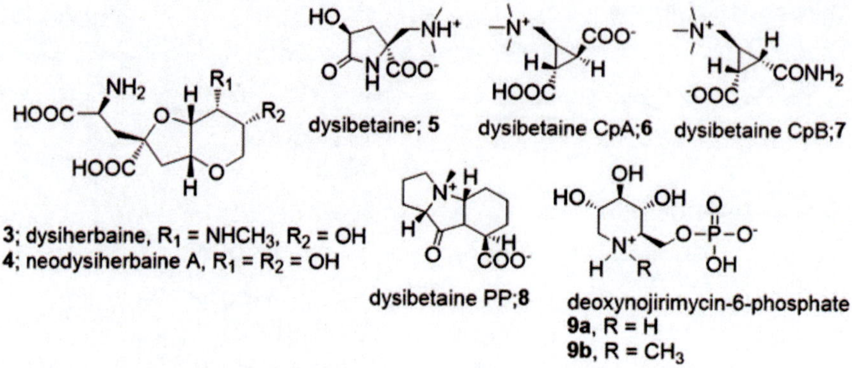

Scheme 2 Dysiherbaines and other water-soluble metabolites form *Lendenfeldia chondrodes*

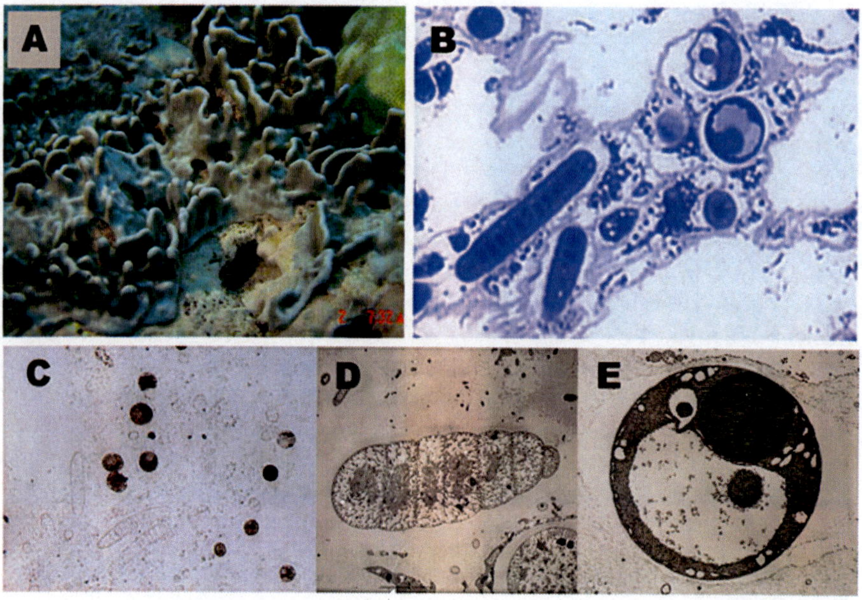

Fig. 2 Cellular and subcellular localization of dysiherbaine (**3**). (**a**) Underwater photography of *L. chondrodes*. (**b**) A Light micrograph of toluidine blue-stained section. (**c**) A section stained by anti-dysiherbaine antisera. TEM for (**d**) *Oscillatoria spongeliae* and (**e**) *Synechocystis* sp.

produce secondary metabolites represented by patellamides in the tunicate *Lissoclinum patella* [21], however, metabolic potential of symbiotic *Synechocystis* is left to be investigated.

Dysiherbaine (**3**) exhibited very potent excitotoxicity in mice, inducing long lasting convulsion with IC_{50} value of 6 pM/mouse, six times as potent as that of domoic acid, thus regarded as the most potent naturally occurring excitatory amino acid known to date [22]. Dysiherbaine (**3**) bind to GluK1 and GluK2 subtypes of kainite receptors very potently at *K*i values of 0.74 and 1.2 nM, respectively. Because of this potency, **3** selectively activates GluK1 site in the hetero tetrameric complex of GluK1/K4 receptor, showing new insight into behavior of heteromeric GluRs [23]. To gain structural insight into the potency and subtype selectivity, we performed structure-activity relationship study of dysiherbaines by synthesizing 11 analogues. We found that, besides importance of glutamate substructure of **3**, substituent of the perhydrofuro-pyrane ring plays significant roles in the activity. Further, crystal structures of ligand binding domain (LBD) of GulRs complexed with **3** indicated that binding of the ligand lead to conformational change in the receptor stabilizing "closed" conformation, typical of those observed in glutamate and other agonists. Interestingly, a truncated analogue, MSVIII-19 (**10**), synthesized by Sasaki's group that exhibited in vivo a transient coma-like sleep also bound to the LBD in the same fashion as **3**, however, it stabilized desensitized state of the receptor and inactivate it by functioning as antagonist [24]. This is an unexpected outcome because in AMPA receptors, closure of the LBD lobe has been thought to correlate to its channel opening [25, 26]. Thus the structure and functional observations with dysiherbaine and congeners posed complexity in GluR biophysics [27]. Because **10** was shown to be a functional antagonist for GluK1 and this receptor has been thought to be involved in pain transmission, it antinociceptive activities were tested in rat pain models revealed that it selectively reduced pain-related behaviors in neuropathic pain model while no effect was seen in the peripheral model [28]. This result showed potential of selective GluR1 antagonists as analgesic drugs (Scheme 3).

Inspired by the structure and activity of naturally occurring EAAs including kainic acid and dysiherbaine, Ikoma and Oikawa designed and synthesized a new molecule IKM-159 as racemic mixture (rac-**11**) using three component Ugi reaction coupled with olefin metathesis to construct tri hetero cyclic amino acids [29]. These compounds especially rac-**11** induced behavioral changes in mice in that the mice became totally flaccid after intracerebroventricular injection [30, 31]. Electrophysiological characterization indicated that these compounds attenuate excitatory neurotransmission through AMPA-type glutamate receptors [32]. Crystal structure of LBD of AMPA receptor complexed with IKM-159 indicated that the 2*R* congener from rac-**11** binds to the ligand binding cavity and change the conformation to "open state" which is typical of the conformation bound to GluR antagonist. These observations suggested that (2*R*)-IKM159 functions as antagonist for GluRA2

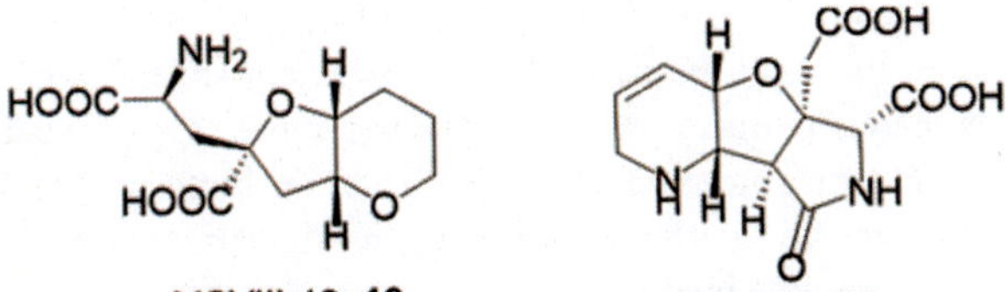

MSVIII-19, **10**

IKM-159, **11**

Scheme 3 Synthetic antagonists

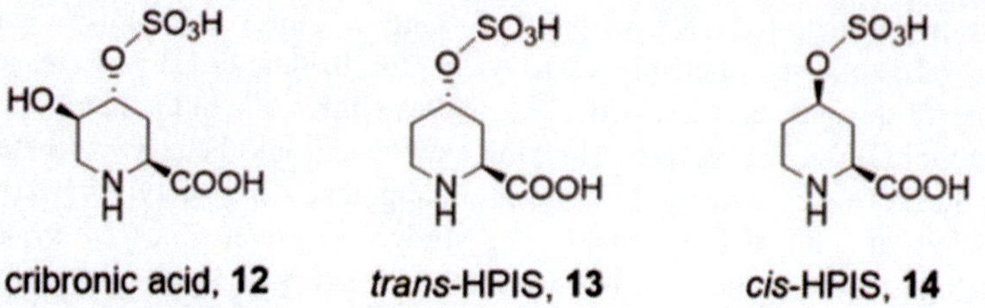

Scheme 4 NMDA agonists from sponges

receptor. Stereospecific synthesis of both (2*R*)- and (2*S*)-IKM159 followed by in vivo bioassay in mice showing the activity in only 2*R* isomer **11**, confirmed the above observation [33].

3.3 Sponge-Derived 4-Sulfooxypiperidine-2-Carboxylic Acids

During further investigation for neuroactive compounds we found that the aqueous sponge extracts induced convulsant behaviors characterized by violent but transient running and jumping behaviors. These behaviors are also observed after injection of NMDA agonists in our previous study. From two distinct sponges, *Stylotella aurantium* and *Axinella carteri* collected in Yap State, Micronesia, a known (2*S*,4*S*)-4-sulfooxypiperidine-2-carboxylic acid (*trans*-HPIS, **13**) was isolated as an active principal (Scheme 4). On the other hand, separation of a Palauan sponge *Cribrochalina olemda* afforded an isolation of cribronic acid [34], (2*S*,4*R*,5*R*)-5-hydroxy-4-sulfooxypiperidine-2-carboxylic acid (**12**) (Scheme 4). They both bind to NMDA site in the rat synaptic preparation at IC_{50} values of 214 and 83 nM, but not to AMPA or KA sites. *trans*-HPIS (**13**) was previously reported from the seed of the legume *Peltophorum africanum* and identified as an NMDA agonist [35, 36]. Of note, synthetic *cis*-HPIS (**14**) [36] did not show any activity in mice. Several other NMDA ligands are found from marine organisms and plants indicated that they are potential source of novel NMDA receptor-targeting drugs [2].

4 Neuroactive Purine Derivatives from Marine Sponges

Purines are one of the most frequently encountered classes of molecules during the separation of aqueous extracts. Plant methyl xanthines, such as caffeine (**15**) theophylline (**16**), and theobromine (**17**) are the most important class of bioactive purines. Needless to say, caffeine is widely used in recreational and medicinal purposes, and theophylline is clinically used as antiasthma drug. Marine organisms are rich source of bioactive purines exhibiting cytotoxicity, antimicrobial activity, enzyme inhibition, antiangiogenic activity, and neuronal activity [37] (Scheme 5).

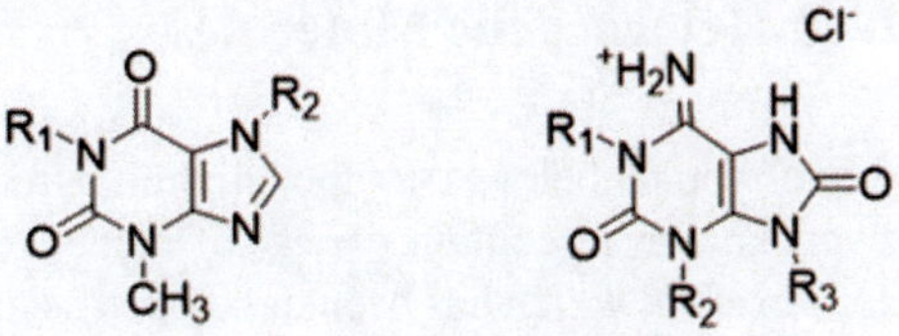

15, $R_1 = CH_3$, $R_2 = CH_3$
16, $R_1 = CH_3$, $R_2 = H$
17, Rr = H, $R_2 = CH_3$

18, $R_1 = CH_3$, $R_2 = H$, $R_3 = CH_3$
19, $R_1 = CH_3$, $R_2 = CH_3$, $R_3 = H$
20, $R_1 = H$, $R_2 = CH_3$, $R_3 = CH_3$
21, $R_1 = R_2 = H$, $R_3 = CH_3$

Scheme 5 Neuroactive purines

We found that aqueous extracts of Palauan Haplosclerida sponges, *Cribrochalina olemda*, *Haliclona* sp., and *Amphimedon* sp. induce transient running-jumping convulsion in mice via i.c.v. route. Bioactivity-guided separation of *C. olemda* resulted in isolation of 1,9-dimethyl-8-oxoisoguanine (**18**) as active principal [38]. Compound **18** and related purines **19–21** were also found in *Haliclona* sp. and *Amphimedon* sp. collected in the same region. All these purines induced convulsion in mice, and the behavioral phenotypes including transient running-jumping convulsion resembled to that of NMDA agonists, however, **18** did not interact directly with GluRs in the radioligand binding assay. Note that caffeine did not induce such convulsant activity in the mouse assay. It was shown, however, that **18** clearly showed hyperexcitability in cultured rat hippocampal neuron in an electrophysiological assay. Further physiological experiments indicated that **18** modify both inhibitory and excitatory neurotransmissions with higher impact on the inhibitory one. In a target screening panel, provided by Psychoactive Drug Screening Program (PDSP) showed moderate affinity for α-adrenoceptors (*K*i values of 885, 758, and 497 nM for each α1A, α1B, and α1D, respectively) and weak affinity for the serotonin receptor 5-HT_{1E} (2.9 μM), nicotinic receptor α3β2 (4.2 μM), and κ- and μ-opioid receptors (1.8 μM and 8.9 μM, respectively). The action of **18**, however, was not affected by inhibitors of α-adrenoceptors or $GABA_B$ receptors which regulate inhibitory neurotransmission [38]. These results together indicated that **18** modulates inhibitory neurotransmission to turn neuron into excitatory phase, however, actual molecular targets of the purine are remained to be identified. This study illustrated interesting and complex nature of neuronal modification by purines. As represented by caffeine, the multiple actions of the drug may be ascribed to its wide target selectivity [39]. Further investigation on this class of neuroactive molecules would result in next generation of purine-based drug discovery [37].

5 Tunicate-Derived Heterocyclic Molecules

Tunicates are marine invertebrate belongs to the phylum Urochordata, close to vertebrates. It is, however, known to contain variety of bioactive secondary molecules including ecteinascidins [40] which has been developed as first marine-derived anticancer drug. Because of this, tunicates have received intense attentions from marine natural product community and a number of biologically interesting compound are reported. Most of which, however, are lipophilic metabolites and biosynthesized likely by symbiotic microorganisms [41]. Aqueous extracts of tunicate are studied mainly in a context of physiological and ecological interests but are relatively untapped [42]. Here, we show our recent work looking into aqueous extract of tunicates using our assay systems.

5.1 *Mellpaladines*

A crude aqueous extract of Palauan Didemnidae tunicate suppressed motor activity of mice. This activity suggested a presence of compounds that interact with neuronal receptors. Separation of the extract guided by the mouse assay resulted in isolation of mellpaladines A-C (**22–24**), dopargimine (**25**), as active principals, along with known compounds guanidinobutyric acid (**26**), and dimeric polysulfur dopamines lissoclibadins 11 (**28**) and 12 (**29**) [43]. Dopargimine and mellpaladine A-C induced behavioral changes in mice. They induced tremor or convulsion at high dose (<50 nmol/mouse), but at lower dose (14 nmol/mouse) the motor activity of mice was largely suppressed by administration of these compounds. Compound **26** induced moderate behavioral change, but **28** and **29** did not show noticeable behavioral changes at 50 μg/mouse. Note that **26** is known to be a physiological substance in the brain and intracisternal injection of **26** in rabbit was convulsive at 5 mg/kg [44]. The molecular targets of the active compounds were screened in PDSP using radio ligand binding assay showing that **25** displace radioligands for δ-opioid receptor (DOR $Ki = 485$ nM) and α2B adrenergic receptor (5.9 μM). Mellpaladines A and B were also shown to bind to wide range of synaptic receptors including serotonin, α-adrenergic, dopamine, histamine, opioid, and Sigma receptors, and transporters such as dopamine and noradrenaline transporters. The structure of **25** is closely related to plant guanidine alkaloid dopargine (**27**) which is known as a key ingredient of black cohosh, a herbal medicine used to treat menopausal symptoms [45].

Other related guanidine alkaloids trypargine derivatives were found for various organisms including tunicates [46], a sponge [47], a flog [48], and a spider [49]. All these alkaloids are likely formed by Pictet–Spengler reaction between monoamines and an aldehyde derived from arginine. Isolation of putative precursor, **26** from this tunicate supported the above hypothesis. Mellpaladines are the first natural products that possess 1,2-dithoketal structure, although examples of dithioacetal compounds

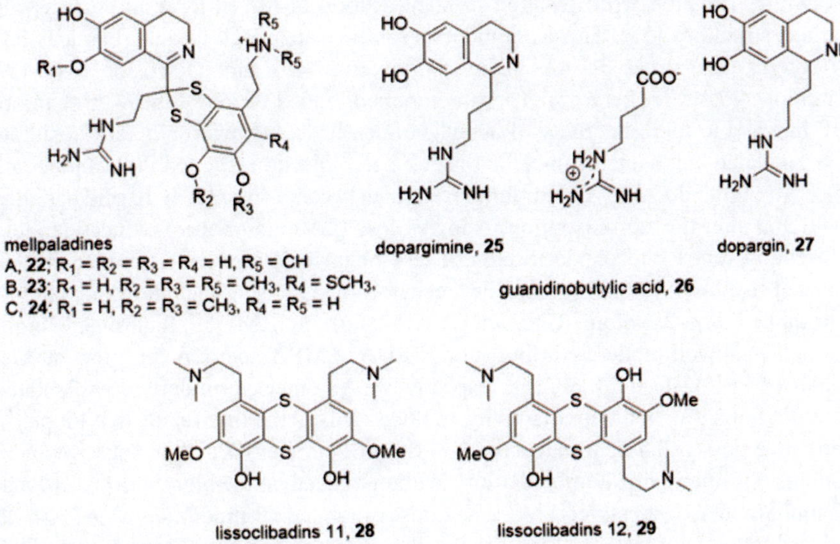

Scheme 6 Mellpaladines and related compounds

were reported in tunicates. Isolation of polysulfur dopamines **28** and **29** suggested that the tunicate is in a chemotype that produce polysulfur dopamines [50], thereby mellpaladines are speculated to be biosynthesized by condensation of dithiol or relevant derivatives of dopamine and dopargimine derivative. Biosynthesis especially origin of sulfur atoms in mellpaladines and polysulfur dopamines are of particular interest in further investigations [51] (Scheme 6).

5.2 *Aromatics from* Cnemidocarpa irene

Aqueous extract of the solitary tunicate *Cnemidocarpa irene* collected in Kojima inlet, off-Hakodate, Hokkaido was shown to be rich in small water-soluble aromatics. Because the extract inhibited acetylcholinesterase activity in vitro, further separation was carried out.

As a result, new β-carboline derivatives irenecarboline A (**30**) and B (**31**) along with *N*-methyl-β-carbolinium chloride (**32**) and 1,3,9-trimethyl-8-oxoisoguanine (**33**) were isolated as active compounds [52]. *N*-Methyl-β-carbolinium-3-carboxylate (**34**) was also isolated but was devoid of the activity. Irenecarboline B (**31**) showed the most potent inhibition with IC_{50} value of 0.47 μM which is comparable to galantamine a clinically used acetylcholinesterase inhibitor for treatment of Alzheimer's disease. Compound **33** is the first example of purines that exhibited this activity. Of note, **33** is closely related to a series of neuroactive purines isolated from sponge (See Sect. 4).

Further isolation work resulted in identification of fifteen more aromatics with various structures [53]. These compounds can be categorized into tauroamides **35–37**, pterin derivatives **38–42**, guanine derivative (**43**) and nucleoside (**44**), and tyramines **45–48** (Scheme 7). These compounds failed to inhibit growth of tumor cell line and to inhibit activity of acetylcholinesterase, however, biopterin disulfate **39** modulated behaviors of mice, in that, an i.c.v. injection resulted in flaccidness in mice. At higher doses (5–25 nmol/mouse), mice became completely flaccid and died 6–40 min after the administration. A lower dose (2.5 nmol/mouse) induced the loss of lighting reflex and ataxia motion. These behavioral profiles resembled to that induced by AMPA-type glutamate receptor antagonists, including 6-cyano-7-nitroquinoxaline-2,3-dione (CNQX), GYKI52466, and IKM-159. Radioligand binding assay showed moderate affinity for NMDA, AMPA, and KA receptors at IC_{50} values of 1.9, 0.91, and 0.68 μM, respectively. All other pterin derivatives isolated here did not show behavioral activity in mice. This is the first report that biopterin derivative possessing neuronal activity. Tyramine sulfates **47** and **48** caused violent convulsions upon i.c.v. administration, leading to death at the highest doses (17 and 92 nmol/mouse, respectively). At lower doses (3.4 and 23 nmol/mouse), **47** and **48** induced motor suppression. Target molecules of these compounds are not identified. Halogenated tyramines **45** and **46** induced minor behavioral changes, including loss of balance and frequent grooming, suggesting they modulate synaptic transmission. Notably, **45** is known as a ligand for the dopamine D2 receptor [54]. Structural and biological diversity of these aromatic metabolites suggested some relevance to the ecology and physiology of the tunicate. Interestingly it was found that the blood of the tunicate was fluorescent [52]. HPLC analysis showed that the *β*-carboliniums **30** and **32** were present in the blood serum while pterin **42** in the blood cells. LC-MS analysis for the juvenile tunicates found near the adult animal, and larvae found in aquarium-grown tunicate lacked these aromatic metabolites. These observations suggested that the above aromatic secondary metabolites were accumulated with growth after the metamorphosis of this animal.

In conclusion, water-soluble metabolites from tunicates provide us with insight into very unusual metabolism of tunicates. The isolates we show here may largely be of metabolites of tunicate rather than symbiotic bacterium, although origin of polysulfur compounds is debatable. The water-soluble metabolites are structurally close to primary metabolites, and some of them should even be categorized as primary metabolites. For instance, biopterin disulfate **39** isolated as neuroactive molecule here could exist in brain given that biopterin is known as a key cofactor in monoamine synthesis [55]. Presence of many neuroactive molecules suggest some physiological function of these compounds in tunicates.

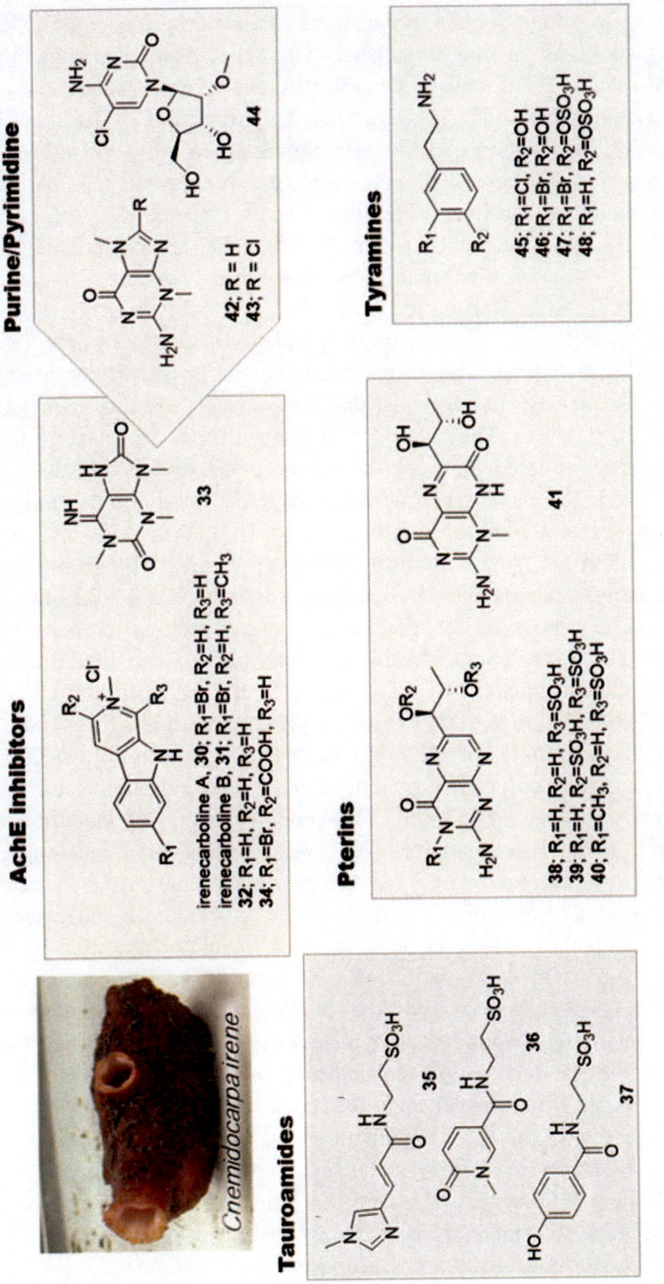

Scheme 7 Aromatic heterocycles from *Cnemidocarpa irene*

6 Marine Polycations

Marine sponges often contain polycationic molecules, presumably for defense against predators or fouling organisms. The most prominent example is alkyl pyridinium salts (APS) found in several different taxon of sponges on mainly in the order Haplosclerida [56–58]. Common structural unit of APS is *N*-alkylated pyridinium but the *N*-alkyl chain is interconnected to the 3-position of the other pyridinium unit to form cyclic dimers, oligomers to polymers [57]. An example of monomeric molecule and liner APS were recently reported [57, 59]. Length of the alkyl chain varies and double bonds or methyl branches are located in the alkyl chain in some APS [57]. Generally small APS are soluble in organic solvents, but polymers are water soluble. The molecular size of poly-APS from *Reniera sarai* was determined to be 5.5 and 19 kDa (corresponds to 26 and 100-mers, respectively) based on MALDI-TOFMS data, however, behaviors in gel filtration and dynamic light scattering experiment suggested that the molecule exist as aggregate with the size of 3×10^6 Da [60]. The poly-APS displays discrete biological activities from small APS as represented by potent hemolytic and acetylcholinesterase inhibition activity [61–63]. The membrane interaction of PAP resulted in the unique property to translocate certain molecules into the cells [63]. Poly-APS thus translocated GFP-plasmid and the gene was successfully expressed [63]. We often observe in the mouse assay, aqueous extract containing polymeric Poly-APS induces convulsion thus can be a nuisance constituent when one is seeking for novel neuroactive compounds. However, unique membrane activity, enzyme inhibition and other ecologically-related functions of poly-APS, as well as its distribution in the sponge and biosynthetic relation to other marine polycations are interesting, and it is worthy endeavor to explore more chemical and biological aspects in poly-APS [58].

We found another polycationic middle-sized molecule aculeines from the marine sponge *Axinyssa aculeata* [64–66]. The aqueous extract of this sponge induced convulsion in mice. Three aculeines (Acu) A-C were isolated, but further structure determination was carried out only for Acu A and B because Acu C was obtained as complex mixture of homologues [64]. Aculeins A (**49**) and B (**50**) were characterized as a peptide with 44-amino acid residues (AcuPep) modified by long chain polyamine (LCPA) in combination of spectrometric, spectroscopic, and *c*-DNA cloning [64]. The AcuPep is a ribosomal peptide, and the gene encodes the peptide originally containing a codon for tryptophan at the *N*-terminal of the mature peptide, however, in Edman degradation the first amino acid was undetected instead, the sequencing started from aspartic acid, the second amino acid residue in the *c*-DNA [64]. The structure of the *N*-terminal amino acid has long been elusive as no direct spectral interpretation on NMR was realistic due to broad spectral data. Isolation of free protoaculein B (pAcu-B, **51**) from the same extract, however, overcame this problem [65]. The spectral analysis of free pAcu-B leads to a proposal of *N*-terminal residue of Acu-B. The structure of **51** proposed here is unique as LCPA was attached to the unusual tricyclic tryptophan derivative. Detailed mass spectral analysis for both **49** and **50** indicated that both of those molecules contain AcuPep as a common

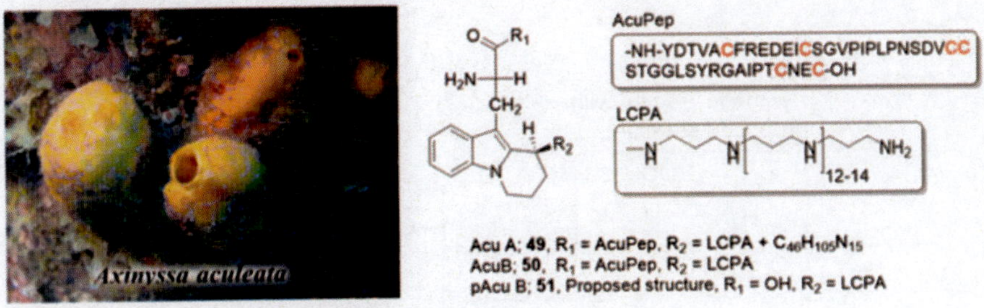

Scheme 8 Aromatic heterocycles From *Axinissa aculeata*

substructure but two LCPAs are attached to the *N*-terminal tryptophan-derived amino acid in the case of **49** instead of one in **50**, although detailed structure in the LCPA moiety of **49** is still unknown [65].

This structural characteristic, LCPA-modified polypeptide, is unique as no other examples besides diatom-derived silica mineralization protein silaffin [67], has been found in nature. Moreover, six cysteine residues in AcuPep were shown to form three disulfide bonds [65]. Although the pattern of disulfide bond formation is not determined, the interval of the cystatin residues is in a typical pattern found in cysteine knot peptide toxins including conotoxins [64]. Together the gross structures of **49** and **50** were proposed as novel LCPA-modified peptide (Scheme 8). Recently Oikawa's group attempted to synthesize **51**, they, however, encountered difficulty due to poor solubility of nitrobenzenesulfonyl (Ns)-protected LCPA [68]. This problem was overcome by using both *t*-butoxycarbonyl (Boc) and Ns as protecting groups for secondary amine and photoremovable 1-(2-nitrophenyl)ethoxycarbonyl (NPEC) group at the terminal primary amine. The fully protected pAcu-B was thus synthesized [68]. The stepwise deprotection of the fully protected pAcu-B was further elaborated and finally **51** with the proposed structure was synthesized [69]. The NMR data for the synthetic and natural **51**, however, were found to differ significantly [69]. Therefore the chemical reactivity of them was compared. Natural **51** and a synthetic model compound was reported to undergo dehydrogenation/ aromatization upon acid treatment in 6 M aqHCl. This hallmark reactivity in the natural product was not observed for the synthetic compound: namely, the same treatment resulted in de-amination first and then aromatization. These observation lead to the conclusion that the structure proposed for **51** was incorrect. The structural revision of **51** in now in progress (Scheme 9).

Aculeines are discovered as "neuroactive" compound because it induced convulsion in mice after i.c.v. administration, but they also exhibited cytotoxicity and hemolytic activity [64]. No antibacterial activity, however, was found. It was proposed that all the above bioactivities of aculeines is due to their action to disrupt cell membrane [64]. The deferential hemolytic activity of **49** was observed in the erythrocyte of several different animal species [64]. This observation suggested that the activity of **49** depends on phospholipid in the cell surface. Interestingly,

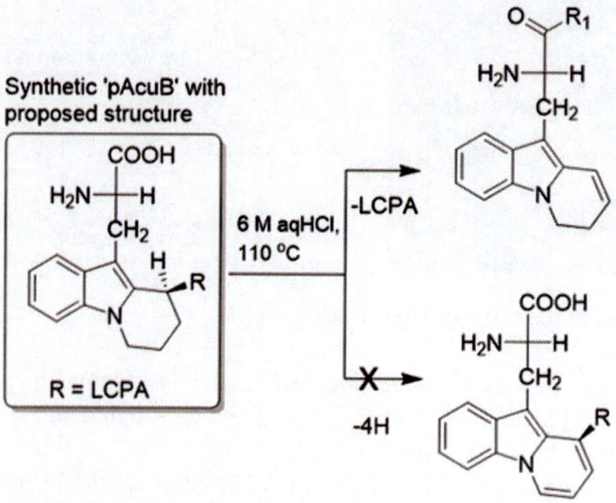

Scheme 9 Reactivity of proposed pAcu B

however, **49** failed to disrupt artificial membrane in POPC liposome, suggesting that **49** targets other phospholipids or some other cell surface molecules [64].

In conclusion, middle-sized water-soluble marine polycations such as poly-APS and aculeines both display interesting membrane activities. This action leads to wide variety of biological activities that have been reported especially for poly-APS. Membrane targeting is a common and widely used chemical defense strategy in nature as represented by antimicrobial peptides (AMP) that present in wide variety of organisms including mammals. Since chemical and biological research for AMPs have led to development of unique cell penetrating peptides that can translocate membrane-impermeable molecules into the cells. Thus, deeper insight into marine water-soluble polycations may lead to development of some useful drug delivery systems.

7 Polyguanidine Alkaloids from Zoantharians

Zoantharians, often referred to as zoanthids, belongs to hexacorallia, have been recognized as rich sources of bioactive molecules [70]. The most prominent example, palytoxins are complex polyols with potent toxicity in animals. In our screening study, we found the aqueous extract of *Epizoanthus illoricatus* collected in Palau induced convulsion in mice. Separation of the extract afforded a fraction that induce the convulsant activity, and a fraction that shows cytotoxicity. Further separation of the cytotoxic fraction resulted in isolation of a KB343 (**52**) [71]. The structure of **52**, determined by extensive analysis of NMR and mass spectral data to be a novel trisguanidine alkaloid where three guanidines are condensed in the southern hemisphere of the molecule. Although several polyguanidine alkaloids, such as Palau'amine [72]

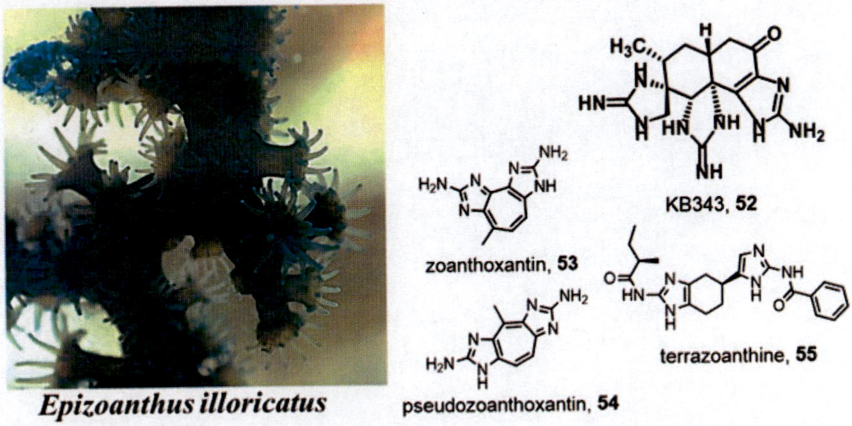

Scheme 10 Guanidine alkaloids from zoanthalian

are known in sponges, the compound that has three guanidino groups in one carbon skeleton is not precedented (Scheme 10).

KB 343 was toxic to both mice and to cultured cells. An i.c.v. injection of **52** in mice resulted in death without any noticeable neuronal symptoms with slow onset (1 to few days). Cytotoxicities of **52** against murine leukemia L1210, human tumor Hela and model neuronal cell SH-SY5Y were moderate with IC_{50} value of 1.96, 4.93, and 3.90 μM, respectively. Compound **52** also suppressed growth of baker's yeast. These observations collectively suggested that **52** permeate the cell membrane slowly and interact to putative cellular target to evoke biological activity. Given that the high-water solubility and *C*LogP value (−4.75), it is reasonable to assume that the molecule cannot permeate cell membrane by simple diffusion. Interestingly, however, guanidino group is known to play a key role in polar molecules that can penetrate cell membrane [73]. For example, cell penetrating peptides (CPP) generally contain several arginine residues and the guanidino group interact negatively charges species in the cell surface resulting in internalization through endocytosis or other undefined mechanisms [73]. Thus, a unique structural feature, three guanidines arranged in the bottom of the molecule, may play a key role in cell penetration of this molecule.

Several structurally related guanidino derivatives are reported in *Epizoanthus*, *Parazoanthus*, *Terrazoanthus*, and *Palythoa* [70]. These compounds commonly contain 2-amino imidazole groups as found in KB343. Zoanthoxantin (**53**) and pseudozoanthoxantin (**54**) form 14π aromatic system with liner and angular arrangement of two aminoimidazole via 7-membred ring, respectively. Terrazoanthine (**55**) also has two aminoimidazole units without fused aromatic system [74]. These molecules are hypothesized to be composed of arginine derived C_5N_3 building blocks. Cyclic addition of the building block lead to the formation of the above compounds. KB343 (**52**) is also envisioned as the biosynthetic product of the C_5N_3 building block, as tandem Diels-Alder type cycloadditions between three building blocks can formulate the skeletal structure of **52**. Thus, zoanthalians have

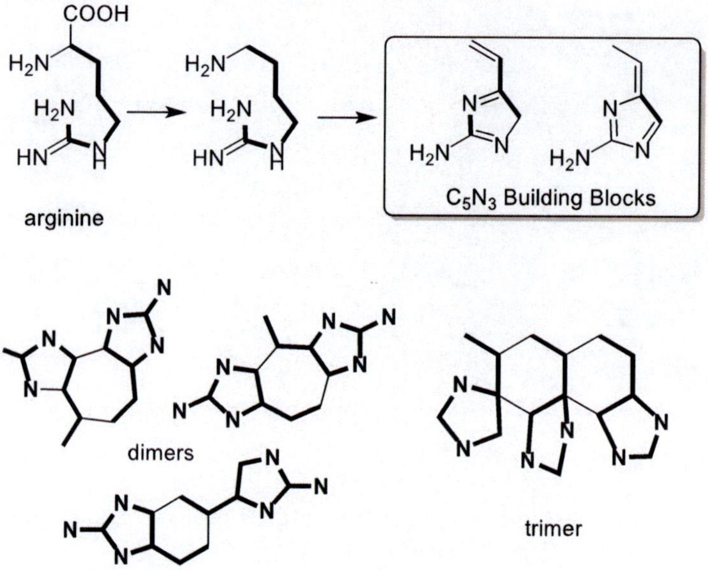

Scheme 11 The C_5N_3 building blocks

biosynthetic machinery to convert arginine to the C_5N_3 compounds and assemble them into structurally diverse polyguanidine alkaloids (Scheme 11).

8 Mycosporines

Mycosporines (MA) are water-soluble heterocyclic amino acid derivatives found widely in aquatic organisms [75–77]. They commonly derived from 4-deoxygadusol where one or two amino acid sidechains are incorporated by forming aminocyclohexenone or aminocycloheximine ring, respectively. MAs absorb harmful UV-A light very efficiently thus thought to play roles as natural sunscreen. Other additional functions beneficial to aquatic organisms including regulation of osmotic stress, heat or desiccating stress, antioxidation have also been found [77].

MAs are distributed widely in marine organisms, ranging from bacteria, microalgae, dinoflagellates, to red algae. They are also found both in invertebrates and vertebrates. This ubiquitous presence of MAs in aquatic organisms implicates its importance in marine ecosystem. Because MAs are found often in invertebrate species that associate symbiotic microorganisms, MA profiling, i.e. determining the presence of MA may help us to gain insight into the symbiotic relationship between the host animals and the symbionts.

This idea in mind, we isolated MAs in the aqueous extract of *L. chondrodes*, the sponge that produces dysiherbaine [75]. The major MA in the extract absorbed λmax

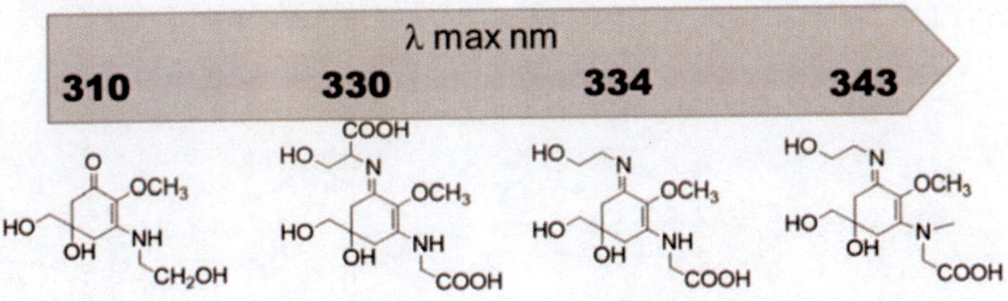

Scheme 12 Mycosporines from *L. chondrodes*

of 343 nm was named LC343 (**59**). We also isolated asterina 330 (**58**) and shinorine (**57**) as known MA and mycosporine ethanolamine **56** as a new MA. The physical properties of LC 343 were particularly interesting; first, it gave only broad NMR signals in water at room temperature and were tend to be broadened even at higher temperature while closely related **58** gave sharp NMR signal. Because the methyl signal was particularly broadened suggesting that the methyl substituent at the imine nitrogen drastically affected the dynamics of the molecule. Furthermore, introduction of the methyl group markedly shifted absorption maxima by 14 nm. Although the mechanism of the bathochromic shift is yet to be investigated, it is interesting that a small change in the sidechain structure of MAs can fine-tune the photochemical property of the molecule. Because *L. chondrodes* inhabits over shallow tropical coral reefs in Micronesia, it is constitutively exposed to strong sunlight. Therefore, the presence of MAs is the key to their survival (Scheme 12).

Because in the depth where the sponge thrives (5–10 m), UV-A (315–400 nm) light can penetrate sea water to reach to the benthic organisms, covering wavelength with various MA is reasonable strategy to maximize the protection efficacy.

Tropical marine invertebrate largely relies their survival on symbiosis. The most well-known example is symbiosis between coral and symbiotic symbiodiniacean cells (dinoflagellate family Symbiodiniaceae). It has been generally accepted that symbiodiniacean cells have ability to biosynthesize MA, and coral utilize these metabolites for protection. Recently, however, gene set for MA biosynthesis present in coral genome has suggested that coral itself can be a producer of MA [78]. These recent observations illustrated complex nature of MA dynamics in coral reef symbiotic ecosystem. Besides corals, giant clams are representative benthic invertebrates that harbor symbiodiniacean cells. We recently studied MA profiles in a species of giant clam, *Tridacna crocea* (*T. crocea*), commonly referred to as "boring clam." The clam, inhabiting in shallow coral reef, harbors symbiodiniacean cells in its mantel. The clam shell is open during the day and symbiodiniacean cells in the outer mantel were exposed to strong sunlight (Fig. 3). This life style maximize production of photosynthetic substance by symbiodiniacean cells while risk of UV exposure needs to be managed. This problem is thought to be compromised by MAs, however, exact mechanisms of protection were not known. We used metabolomics approach

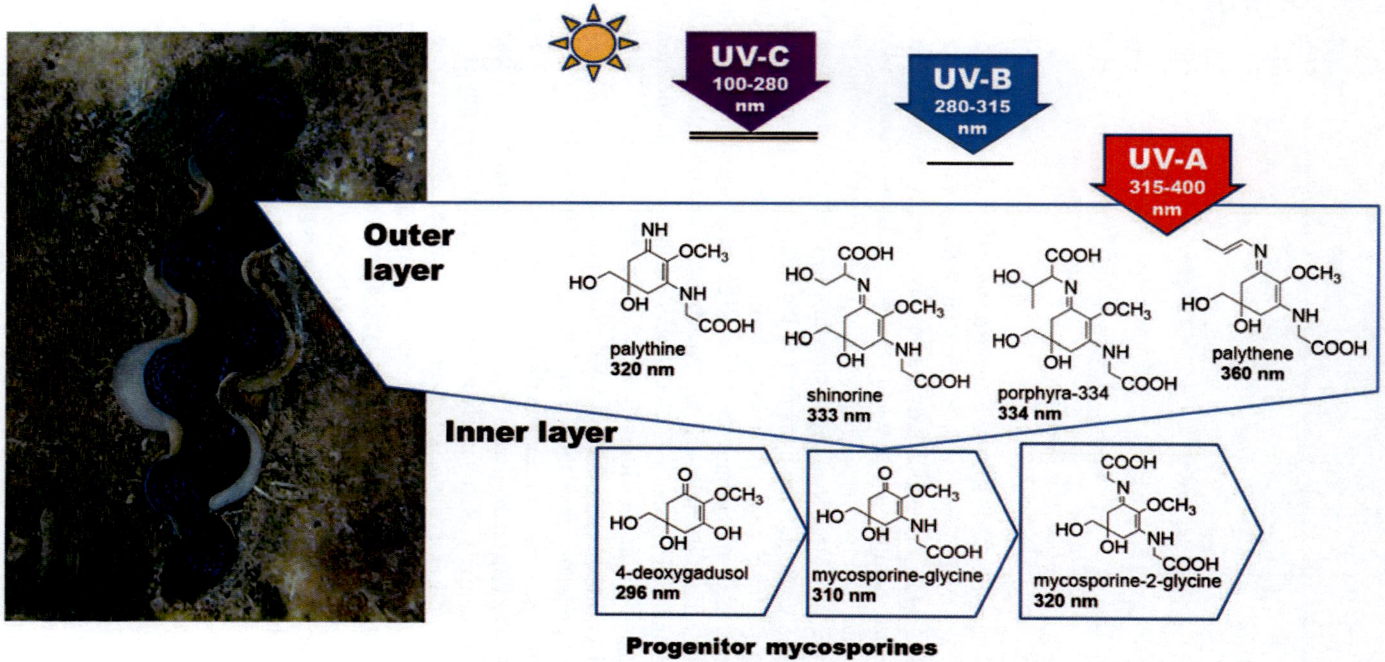

Fig. 3 "Smart use" of mycosporines in the giant clam. Mass-imaging experiments showed uneven distribution of MAs in the mantel tissue of *T. crocea*, where MA and putative precursors with UV-B absorbing capability were localized all over the mantel. Biosynthetically "processed" derivatives localized only in the outer layer of the mantel where is most severely suffered by UV-A penetration

as well as mass and UV imaging analysis to visualize localization of MAs in the mantel of the clam and identified eleven MAs with varying absorption properties [77]. The imaging experiment indicated that the localization of MA was "selective" as the outmost layer of the mantel had high density of processed MAs which absorbs light with 330–360 nm, while the inner layer was rich in precursor MAs that can be transformed to more complex derivatives. These finding suggested the "smart use" of precious sun screen in the clam where costly but most effective derivatives are transformed and accumulated at the site where the frontline defence from the UV light required the most.

9 Conclusion

We have demonstrated that aqueous extracts of marine organisms are attractive sources of biologically active and functional molecules with chemical space distinguishable from the lipophilic compounds. Most of the above compound possess new-class chemical structure and biosynthesis. As represented by dysiherbaine, molecule with unique structure lead to discovery of yet undescribed physiological functions of living organisms and that may lead to drug discovery. Marine ecosystem differs largely from that of terrestrial as it is in the water milieu. Water-soluble molecule like mycosporines can be useful markers to truck complex food web and symbiotic relationship of marine animals. Moreover, large production, either by chemical or biological means, mycosporines are of great commercial value in the area of skin care/cosmetic industries.

Acknowledgement Researches conducted by RS are supported by JSPS grants 22380114, 15H04546, 18H02270, 19H03040, and 20H04758.

References

1. Shimizu Y (1985) J Nat Prod 48:223–235
2. Sakai R, Swanson GT (2014) Nat Prod Rep 31:273–309
3. Olivera BM, Cruz LJ, Yoshikami D (1999) Curr Opin Neurobiol 9:772–777
4. Fritz L, Quilliam MA, Wright JL, Beale AM, Work TM (1992) J Phycol 28:439–442
5. Stathakis CI, Yioti EG, Gallos JK (2012) Eur J Org Chem 2012:4661–4673
6. Stonik VA, Stonik IV (2020) Molecules 25:3049
7. Toyota M, Spencer D, Sawai-Toyota S, Jiaqi W, Zhang T, Koo AJ, Howe GA, Gilroy S (2018) Science 361:1112–1115
8. Sakai R, Minato S, Koike K, Koike K, Jimbo M, Kamiya H (2005) Cell Tissue Res 322:491–502
9. Chekan JR, McKinnie SM, Moore ML, Poplawski SG, Michael TP, Moore BS (2019) Angew Chem 131:8542–8545
10. Brunson JK, McKinnie SM, Chekan JR, McCrow JP, Miles ZD, Bertrand EM, Bielinski VA, Luhavaya H, Oborník M, Smith GJ (2018) Science 361:1356–1358

11. Maeno Y, Kotaki Y, Terada R, Cho Y, Konoki K, Yotsu-Yamashita M (2018) Sci Rep 8:1–10
12. Lichtenthaler HK (2010) The non-mevalonate doxp/mep (deoxyxylulose 5-phosphate/methylerythritol 4-phosphate) pathway of chloroplast isoprenoid and pigment biosynthesis. In the chloroplast. Springer, Berlin, pp 95–118
13. Sakai R, Kamiya H, Murata M, Shimamoto K (1997) J Am Chem Soc 119:4112–4116
14. Sakai R, Suzuki K, Shimamoto K, Kamiya H (2004) J Org Chem 69:1180–1185
15. Sakai R, Kamiya H (2006) J Antibiot 59:507–511
16. IJzendoorn DR, Botman PN, Blaauw RH (2006) Org Lett 8:239–242
17. Sakai M, Ishikawa Y, Takamizawa S, Oikawa M (2013) Tetrahedron Lett 54:5911–5912
18. Tanaka K, Sakai M, Takamizawa S, Oikawa M (2015) Chem Lett 44:253–255
19. Ridley CP, Faulkner DJ, Haygood MG (2005) Appl Environ Microbiol 71:7366–7375
20. Sakai R, Yoshida K, Kimura A, Koike K, Jimbo M, Koike K, Kobiyama A, Kamiya H (2008) Chembiochem 9:543–551
21. Donia MS, Fricke WF, Partensky F, Cox J, Elshahawi SI, White JR, Phillippy AM, Schatz MC, Piel J, Haygood MG (2011) Proc Nat Acad Sci 108:E1423–E1432
22. Sakai R, Swanson GT, Shimamoto K, Green T, Contractor A, Ghetti A, Tamura-Horikawa Y, Oiwa C, Kamiya H (2001) J Pharmacol Exp Ther 296:650–658
23. Swanson GT, Green T, Sakai R, Contractor A, Che W, Kamiya H, Heinemann SF (2002) Neuron 34:589–598
24. Frydenvang K, Lash LL, Naur P, Postila PA, Pickering DS, Smith CM, Gajhede M, Sasaki M, Sakai R, Pentikaïnen OT (2009) J Biol Chem 284:14219–14229
25. Møllerud S, Frydenvang K, Pickering DS, Kastrup JS (2017) Neuropharmacology 112:16–28
26. Armstrong N, Gouaux E (2000) Neuron 28:165–181
27. Birdsey-Benson A, Gill A, Henderson LP, Madden DR (2010) J Neurosci 30:1463–1470
28. Qiu C-S, Lash-Van Wyhe L, Sasaki M, Sakai R, Swanson GT, Gereau IV RW (2011) Pain 152:1052–1060
29. Oikawa M, Ikoma M, Sasaki M, Gill MB, Swanson GT, Shimamoto K, Sakai R (2010) Bioorg Med Chem 18:3795–3804
30. Oikawa M, Kasori Y, Katayama L, Murakami E, Oikawa Y, Ishikawa Y (2013) Synthesis 45:3106–3117
31. Chiba M, Fujimoto C, Sakai R, Oikawa M (2015) Bioorg Med Chem Lett 25:1869–1871
32. Gill M, Frausto S, Ikoma M, Sasaki M, Oikawa M, Sakai R, Swanson GT (2010) Br J Pharmacol 160:1417–1429
33. Juknaite L, Sugamata Y, Tokiwa K, Ishikawa Y, Takamizawa S, Eng A, Sakai R, Pickering DS, Frydenvang K, Swanson GT (2013) J Med Chem 56:2283–2293
34. Sakai R, Matsubara H, Shimamoto K, Jimbo M, Kamiya H, Namikoshi M (2003) J Nat Prod 66:784–787
35. Evans SV, Shing TK, Aplin RT, Fellows LE, Fleet GW (1985) Phytochemistry 24:2593–2596
36. Mannaioni G, Alesiani M, Carlà V, Natalini B, Marinozzi M, Pellicciari R, Moroni F (1994) Eur J Pharmacol 251:201–207
37. Daly J (2007) Cell Mol Life Sci 64:2153–2169
38. Sakurada T, Gill MB, Frausto S, Copits B, Noguchi K, Shimamoto K, Swanson GT, Sakai R (2010) J Med Chem 53:6089–6099
39. Yu L, Coelho JE, Zhang X, Fu Y, Tillman A, Karaoz U, Fredholm BB, Weng Z, Chen J-F (2009) Physiol Genomics 37:199–210
40. Sakai R, Jares-Erijman EA, Manzanares I, Silva Elipe MV, Rinehart KL (1996) J Am Chem Soc 118:9017–9023
41. Schmidt EW, Donia MS (2010) Curr Opin Biotechnol 21:827–833
42. Taylor SW, Kammerer B, Bayer E (1997) Chem Rev 97:333–346
43. Uchimasu H, Matsumura K, Tsuda M, Kumagai K, Akakabe M, Fujita MJ, Sakai R (2016) Tetrahedron 72:7185–7193
44. Jinnai D, Sawai A, Mori A (1966) Nature 212:617–617

45. Godecke T, Lankin DC, Nikolic D, Chen S-N, van Breemen RB, Farnsworth NR, Pauli GF (2009) J Nat Prod 72:433–437
46. Van Wagoner RM, Jompa J, Tahir A, Ireland CM (1999) J Nat Prod 62:794–797
47. Davis RA, Duffy S, Avery VM, Camp D, Hooper JN, Quinn RJ (2010) Tetrahedron Lett 51:583–585
48. Akizawa T, Yamazaki K, Yasuhara T, Nakajima T, Roseghini M, Erspamer GF, Erspamer V (1982) Biomed Res 3:232–234
49. Cesar LM, Mendes MA, Tormena CF, Marques MR, De Souza BM, Saidemberg DM, Bittencourt JC, Palma MS (2005) Toxicon 46:786–796
50. Wang W, Takahashi O, Oda T, Nakazawa T, Ukai K, Mangindaan RE, Rotinsulu H, Wewengkang DS, Kobayashi H, Tsukamoto S (2009) Tetrahedron 65:9598–9603
51. Jiang C-S, Müller WE, Schröder HC, Guo Y-W (2012) Chem Rev 112:2179–2207
52. Tadokoro Y, Nishikawa T, Ichimori T, Matsunaga S, Fujita MJ, Sakai R (2017) ACS Omega 2:1074–1080
53. Miyako K, Yasuda Y, Shinada T, Fujita M, Sakai R (2020) J Nat Prod 83:3156–3165
54. Angelina E, Andujar S, Moreno L, Garibotto F, Párraga J, Peruchena N, Cabedo N, Villecco M, Cortes D, Enriz RD (2015) Mol Inform 34:28–43
55. Nagatsu T, Matsuura S, Sugimoto T (1989) Med Res Rev 9:25–44
56. Turk T, Frangež R, Sepčić K (2007) Mar Drugs 5:157–167
57. Laville R, Thomas OP, Berrue F, Reyes F, Amade P (2008) Eur J Org Chem 2008:121–125
58. Tribalat M-A, Marra MV, McCormack GP, Thomas OP (2016) Planta Med 82:843–856
59. Timm C, Volk C, Sasse F, Köck M (2008) Org Biomol Chem 6:4036–4040
60. Sepčić K, Guella G, Mancini I, Pietra F, Serra MD, Menestrina G, Tubbs K, Macek P, Turk T (1997) J Nat Prod 60:991–996
61. McClelland D, Evans R, Abidin I, Sharma S, Choudhry F, Jaspars M, Sepčić K, Scott RH (2003) Br J Pharmacol 139:1399–1408
62. Malovrh P, Sepčić K, Turk T, Maček P (1999) Comp Biochem Phys Part C Pharmacol Toxicol Endocrinol 124:221–226
63. McLaggan D, Adjimatera N, Sepčić K, Jaspars M, MacEwan DJ, Blagbrough IS, Scott RH (2006) BMC Biotechnol 6:6
64. Matsunaga S, Jimbo M, Gill MB, Wyhe LL, Murata M, Nonomura K, Swanson GT, Sakai R (2011) Chembiochem 12:2191–2200
65. Matsunaga S, Kishi R, Otsuka K, Fujita MJ, Oikawa M, Sakai R (2014) Org Lett 16:3090–3093
66. Matsunaga S, Sakai R, Jimbo M, Kamiya H (2007) Chembiochem 8:1729–1735
67. Kröger N, Deutzmann R, Sumper M (1999) Science 286:1129–1132
68. Shiozaki H, Miyahara M, Otsuka K, Miyako K, Honda A, Takasaki Y, Takamizawa S, Tukada H, Ishikawa Y, Sakai R (2018) Org Lett 20:3403–3407
69. Irie R, Miyahara M, Nakamura S, Honda A, Sakai R, Oikawa M (2020) J Nat Prod
70. Guillen PO, Jaramillo KB, Genta-Jouve G, Thomas OP (2020) Nat Prod Rep 37:515–540
71. Matsumura K, Taniguchi T, Reimer JD, Noguchi S, Fujita MJ, Sakai R (2018) Org Lett 20:3039–3043
72. Kinnel RB, Gehrken HP, Scheuer PJ (1993) J Am Chem Soc 115:3376–3377
73. Futaki S, Nakase I (2017) Acc Chem Res 50:2449–2456
74. Guillen PO, Jaramillo KB, Genta-Jouve G, Sinniger F, Rodriguez J, Thomas OP (2017) Org Lett 19:1558–1561
75. Oda Y, Zhang Q, Matsunaga S, Fujita MJ, Sakai R (2017) Chem Lett 46:1272–1274
76. Goto-Inoue N, Sato T, Morisasa M, Yamashita H, Maruyama T, Ikeda H, Sakai R (2020) Sci Rep 10:1–10
77. Oren A, Gunde-Cimerman N (2007) FEMS Microbiol Lett 269:1–10
78. Shinzato C, Shoguchi E, Kawashima T, Hamada M, Hisata K, Tanaka M, Fujie M, Fujiwara M, Koyanagi R, Ikuta T (2011) Nature 476:320–323

Top Heterocycl Chem (2021) 58: 131–152
https://doi.org/10.1007/7081_2020_44

Published online: 21 April 2021

Synthesis of Paralytic Shellfish Toxins: Saxitoxins

Hayate Ishizuka and Kazuo Nagasawa

Contents

Abstract Saxitoxin (STX), which is isolated as a paralytic shellfish toxin, is an inhibitor of voltage-gated sodium channels (Na_VCh) such as tetrodotoxin. STX has two guanidine functional groups in its tricyclic core structure, and every carbon except at C11 connects heteroatoms. So far, more than 50 analogs of STX have been discovered from nature, and they have attracted considerable interest of synthetic

---

H. Ishizuka
Department of Biotechnology and Life Science, Faculty of Technology, Tokyo University of Agriculture and Technology, Tokyo, Japan

K. Nagasawa (✉)
Tokyo University of Agriculture and Technology, Tokyo, Japan
e-mail: knaga@cc.tuat.ac.jp

chemists due to their complicated structure as well as their potential as drug lead. This chapter provides an overview of reported syntheses of STX (**1**) and its natural analogs. In these syntheses, construction of the characteristic tricyclic structure of STX without loss of the protective groups on guanidine is crucial. The developed methodologies have enabled the synthesis of various natural and artificial analogs of STX with more complex structures.

Graphical Abstract

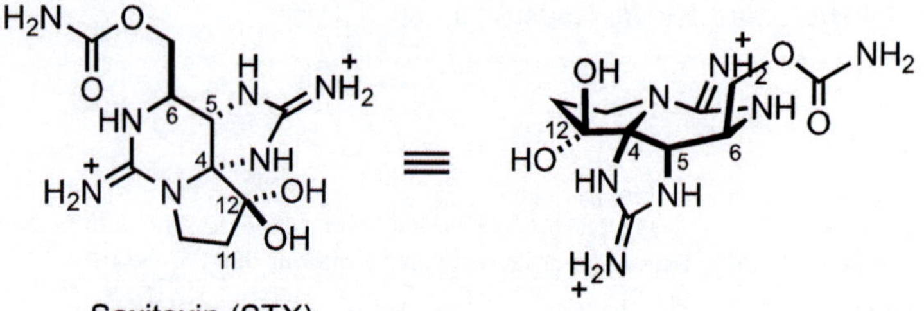

Keywords Guanidine alkaloid · Saxitoxin · Total synthesis · Voltage-gated sodium channel

1 Introduction: Saxitoxin and Its Derivatives

The paralytic shellfish toxin saxitoxin (STX; **1**, Fig. 1) is a guanidine alkaloid biosynthesized by dinoflagellates, and it potently inhibits voltage-gated sodium channels (Na_VCh), thereby blocking the influx of sodium ions into cells [1–3]. Na_VChs are distributed extensively in nerves, skeletal muscle, cardiac muscle, etc., and control neurotransmission through the generation and propagation of action potentials. So far, ten subtypes of Na_VCh, i.e., Na_V1.1–1.9 and Na_VX, have been identified, and these are classified into two categories on the basis of their sensitivity or resistance to tetrodotoxin (TTX), i.e. TTX-s (Na_V1.1–1.4, 1.6, and 1.7) and TTX-r (Na_V1.5, 1.8, and 1.9), respectively [3].

Groups led by Schantz and Rapoport independently identified the structure of STX (**1**) in 1975 by means of X-ray analysis. Schantz derivatized STX to the *p*-bromobenzenesulfonate salt, while Rapoport generated the ethyl hemiacetal at C12 [4, 5]. So far, over 50 STX analogs have been found in nature [6], and these commonly possess a tricyclic system with two cyclic guanidines, one five-membered and the other six-membered. The five-membered guanidine ring and the side chain at C6 take an anti-diaxial arrangement. Interestingly, the two cyclic guanidines show different pKa values of 8.8 (five-membered guanidine) and 11.2 (six-membered guanidine), and the difference is attributed to the non-planar structure of the five-membered guanidine ring [7]. STX (**1**) is highly oxidized, and every carbon except at

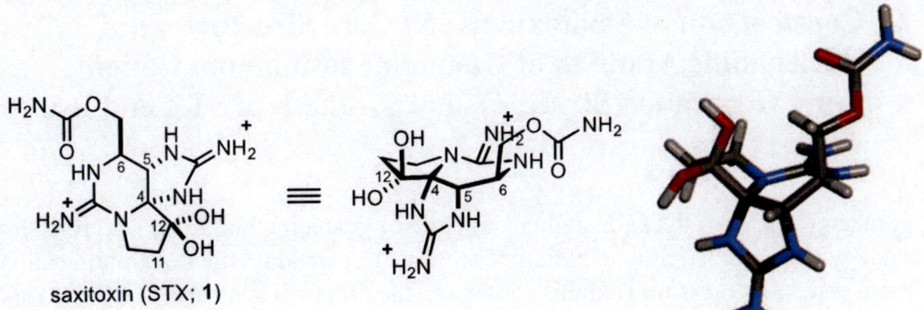

Fig. 1 Structure of saxitoxin (STX; **1**)

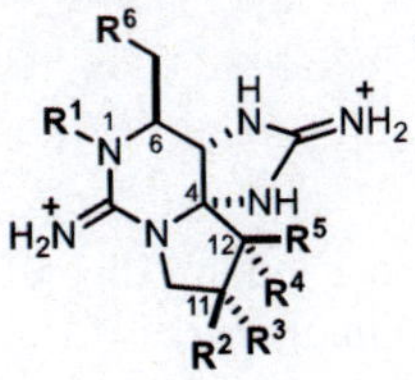

STX skelton

	R^1	R^2	R^3	R^4	R^5	R^6
STX (**1**)	H	H	H	OH	OH	$OCONH_2$
doSTX (**2**)	H	H	H	OH	OH	H
dcSTX (**3**)	H	H	H	OH	OH	OH
neoSTX (**4**)	OH	H	H	OH	OH	$OCONH_2$
GTX1 (**5**)	OH	H	OSO_3^-	OH	OH	$OCONH_2$
GTX2 (**6**)	H	H	OSO_3^-	OH	OH	$OCONH_2$
GTX3 (**7**)	H	OSO_3^-	H	OH	OH	$OCONH_2$
GTX4 (**8**)	OH	OSO_3^-	H	OH	OH	$OCONH_2$
β-STXol (**9**)	H	H	H	H	OH	$OCONH_2$
dc-α-STXol (**10**)	H	H	H	OH	H	OH

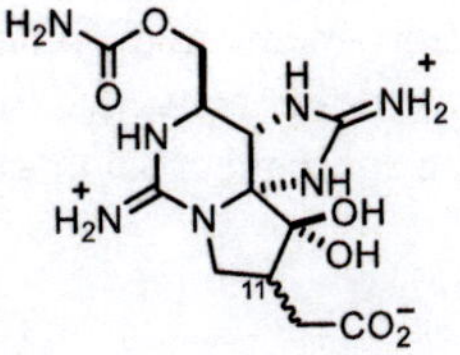

11-Saxitoxinethanoic acid (SEA; **11**)

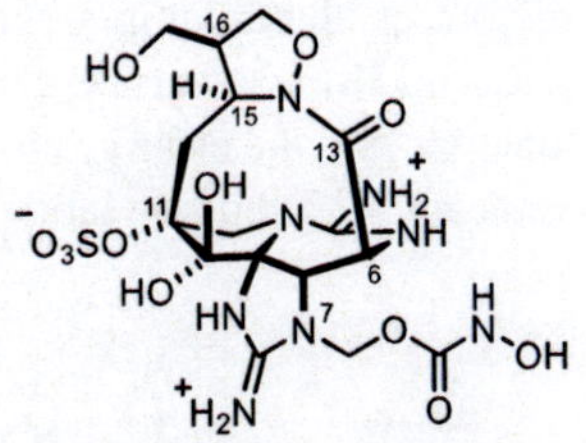

Zetekitoxin AB (ZTX; **12**)

Fig. 2 Structures of saxitoxin (STX; **1**) and representative analogs

C11 is connected to heteroatoms. The ketone group at C12 is stabilized as a hydrate form because of the neighboring effect of guanidine, which presumably lowers the LUMO level of the ketone group at C12. Most of the analogs are functionalized differently from STX (**1**) at C11, C13, and N5 (Fig. 2). Among them, zetekitoxin AB (ZTX; **12**) possesses an unusual macrocyclic structure, and shows extremely potent Na_VCh-inhibitory activity, with a picomolar IC_{50} value [8].

This chapter summarizes syntheses of STX (**1**) and its natural analogs that have been reported to date. We will focus particularly on construction strategies for the characteristic cyclic bis-guanidine structure, as well as the key *N, N*-aminal at C4 in the two cyclic guanidines.

2 Construction of Saxitoxin (STX) Core Structure via Nucleophilic Addition of Guanidine to Iminium Cation (First Generation Strategy), and Synthesis of STX and Its Analogs

For the synthesis of the *N, N*-aminal structure of guanidine in STX (**1**), a promising strategy is the nucleophilic addition of nitrogen in guanidine to the iminium cation. Kishi and co-workers utilized this strategy in the first total synthesis of STX (**1**) and dcSTX (**3**) (Scheme 1) [9, 10]. The groups led by Du Bois, Nishikawa, and Nagasawa also independently utilized similar iminium cation intermediates for the synthesis of STX (**1**) and its unnatural analog dc-α-STXol (**10**) [11–13].

2.1 *Synthesis of STX (1) by the Kishi Group*

The synthetic strategy for STX (**1**) adopted by Kishi's group is depicted in Scheme 2 [9, 10]. They planned to construct the STX skeleton from the iminium cation **13** via isomerization of the enamine **14** under acidic conditions. The generated cyclic thiourea and urea were further converted into the required cyclic guanidines at a later stage.

The details of the synthesis of STX (**1**) and dcSTX (**3**) are shown in Scheme 3. At the outset, three-component coupling reaction of enamine **16**, aldehyde **17**, and Si $(NCS)_4$ (**18**) was carried out to give six-membered cyclic thiourea **20** via intermediate **19**, and the ester group in **20** was further converted to urea to give **21**. Since the ketal at C12 would be labile under the subsequent acidic conditions, it was switched

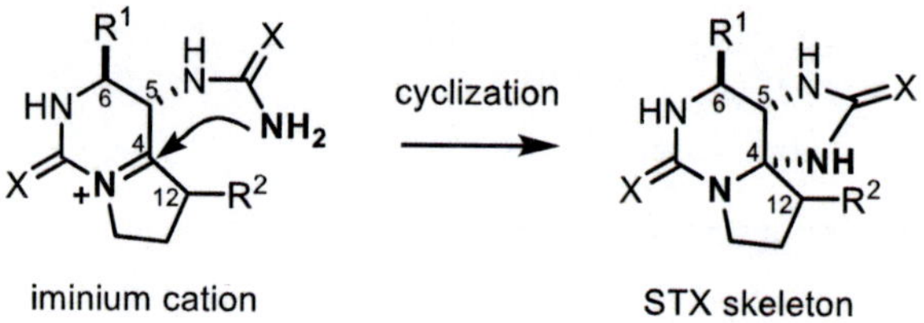

Scheme 1 Common strategy for the construction of the bis-cyclic guanidine structure in the STX skeleton: cyclization of the iminium cation with guanidine or its synthetic equivalent

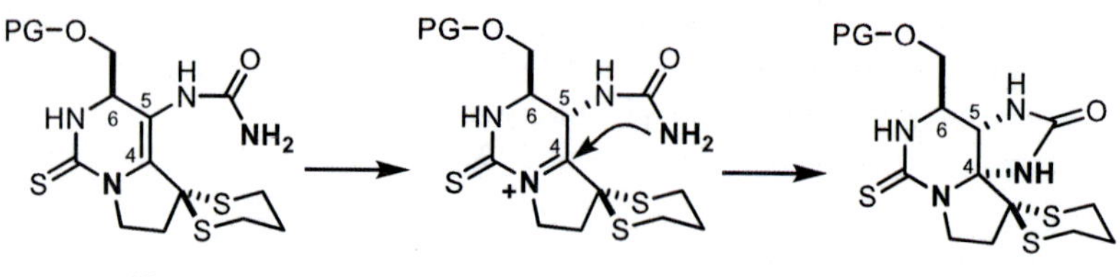

13 **14** **15**

Scheme 2 Synthetic strategy adopted by Kishi to obtain the saxitoxin skeleton

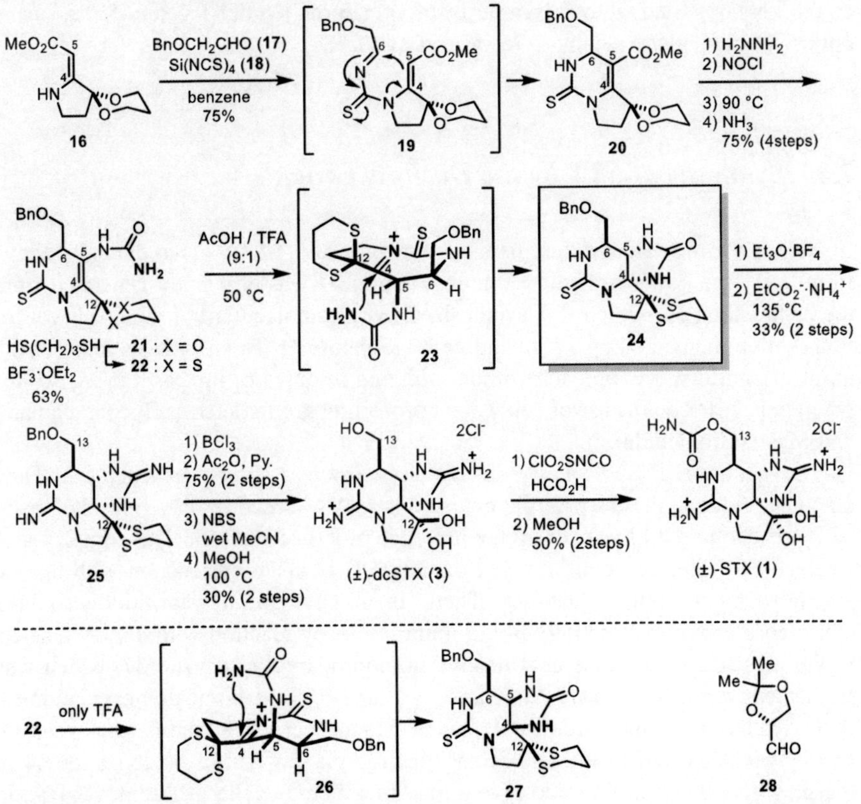

Scheme 3 Total synthesis of (±)-decarbamoylsaxitoxin (dcSTX; **3**) and (±)-saxitoxin (STX; **1**) by the Kishi group

to dithioketal at this stage to provide **22** as a precursor for the cyclization reaction. The key cyclization reaction was conducted in acetic acid-TFA (9:1) at 50°C. Under these conditions, the intramolecular cyclization took place via iminium cation intermediate **23** to afford bis-cyclic thiourea/urea **24** with an *N,N*-aminal moiety in 50% yield. In this cyclization reaction, the reaction rate was quite slow in acetic acid alone, presumably due to the steric effect of the thioacetal at C12. On the other hand, isomerization at C6 mainly proceeded in TFA alone to give **27**. Next, the bis-cyclic guanidine was generated by *S*- and *O*-alkylation of the cyclic thiourea and urea groups with Meerwein's reagent followed by reaction with ammonium propionate as an ammonia source under melting conditions. (±)-Decarbamoylsaxitoxin (dcSTX; **3**) was synthesized from **25** (Jacobi and co-workers have reported the synthesis of 25 via 1,3-dipolar cycloaddition strategy, see [14]) by deprotection of the dithioacetal and benzyl groups. The total synthesis of (±)-STX (**1**) was achieved by introducing a carbamoyl group at C13 with chlorosulfonyl isocyanate, followed by solvolysis of the resulting sulfonamide with methanol. In 1992, Kishi and co-workers reported a total synthesis of unnatural (–)-dcSTX (*ent*-**3**) by using optically active glyceraldehyde **28** instead of **17** [15].

2.2 Synthesis of STX by the Du Bois Group

In 2006, Du Bois reported the first total synthesis of (+)-STX (**1**) in optically active form [11]. The key intermediate iminium cation **31** was efficiently generated from the hemiaminal **30**, which was synthesized by intramolecular transannular cyclization of nine-membered cyclic guanidine **29** (Scheme 4). This synthesis represents a more straightforward approach without utilizing thiourea or urea as a precursor for guanidine, based on the use of convenient protecting groups such as Boc (carbamate) and Mbs (sulfonamide).

The synthesis of STX (**1**) by the Du Bois group is shown in Scheme 5. They obtained sulfonamide **34** with the characteristic *N,O*-acetal moiety from sulfamate **33** by reaction with $PhI(OAc)_2$ in the presence of a rhodium binuclear complex as a catalyst [16]. This reaction involved oxidative C-H insertion reaction with nitrene generated by the Rh (0) catalyst. Then, an alkynyl group was introduced in a diastereoselective manner into **34** via iminium **36** by reaction with alkynylzinc **35** in the presence of a Lewis acid to give homopropargylic tosylate **37**, which was further converted into azide **41** bearing (*Z*)-allylic- and homoallylic guanidines. After reduction of the azide in **41** under Staudinger's conditions, the resulting amine was subjected to intramolecular guanidinylation with *S*-methyl thiourea in the presence of silver nitrate to give a nine-membered cyclic guanidine **42** [17] in 65% yield (2 steps). At this stage, the carbamoyl group at C13 was introduced by reaction with trichloroacetyl isocyanate followed by solvolysis with methanol. Then, the double bond in **43** was oxidized with osmium (III) chloride-oxone to give α-hydroxy ketone **44**, which underwent transannular cyclization with N3 in guanidine to generate hemiaminal **45**. Upon treatment of **45** with $B(CF_3CO_2)_3$ in TFA, cyclization proceeded via iminium cation **46** to give (+)-β-saxitoxinol (β-STXol; **9**) in 82% yield. Finally, oxidation of the secondary alcohol at C12 with DCC and DMSO in Py-TFA [18] furnished (+)-STX (**1**).

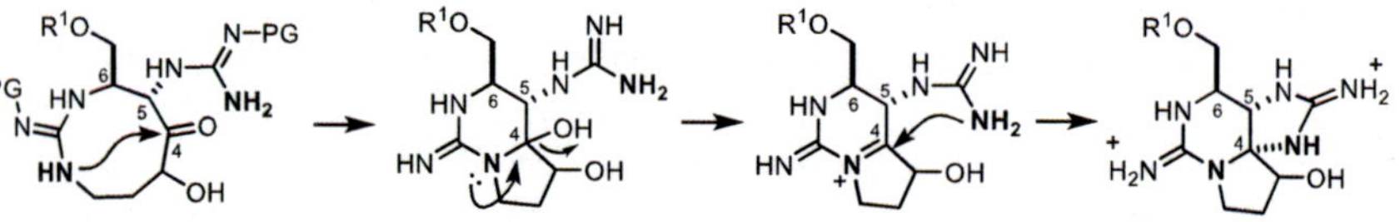

29 30 31 32

Scheme 4 Synthetic strategy for the saxitoxin skeleton by Du Bois

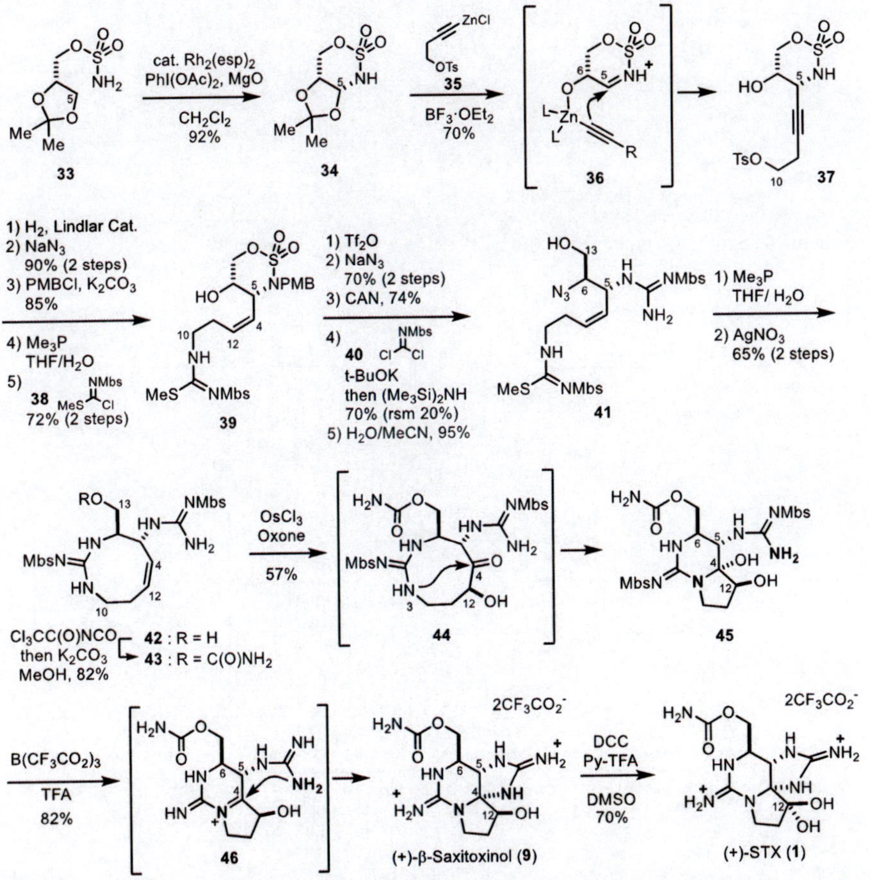

Scheme 5 Total synthesis of (+)-saxitoxin (STX; **1**) and (+)-β-saxitoxinol (β-STXol; **9**) by the Du Bois group

2.3 *Synthesis of STX Derivative dc-α-STXol (10) by the Nishikawa Group*

In 2011, Nishikawa's group used a similar key synthetic intermediate, hemiaminal **48**, for the synthesis of (+)-decarbamoyl-α-saxitoxinol (dc-α-STXol; **10**) [13], a natural analog of **1** [19]. The iminium cation **49** was generated by acid treatment of the tricyclic *N,O*-acetal compound **48**, which was obtained from the alkyne **47**, bearing guanidine and hydroxy groups, by bromine-mediated tandem-cyclization reaction (Scheme 6).

Their synthesis commenced with preparation of homopropagyl guanidine **57** (Scheme 7) [20]. Alkynylation was carried out with Garner's aldehyde (**50**) by reaction with alkyne **51** in the presence of *n*-BuLi to give alkynyl alcohol **52**. After installation of the guanidine group at C6 and the azide group at C5 via aziridine

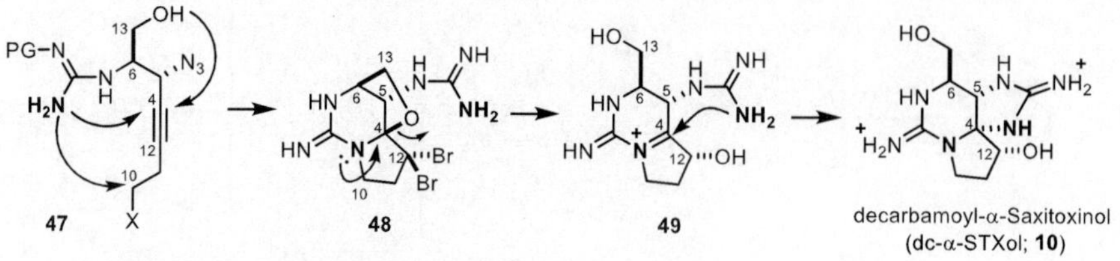

Scheme 6 Synthetic strategy of saxitoxin skeleton via iminium cation **49** by the Nishikawa group

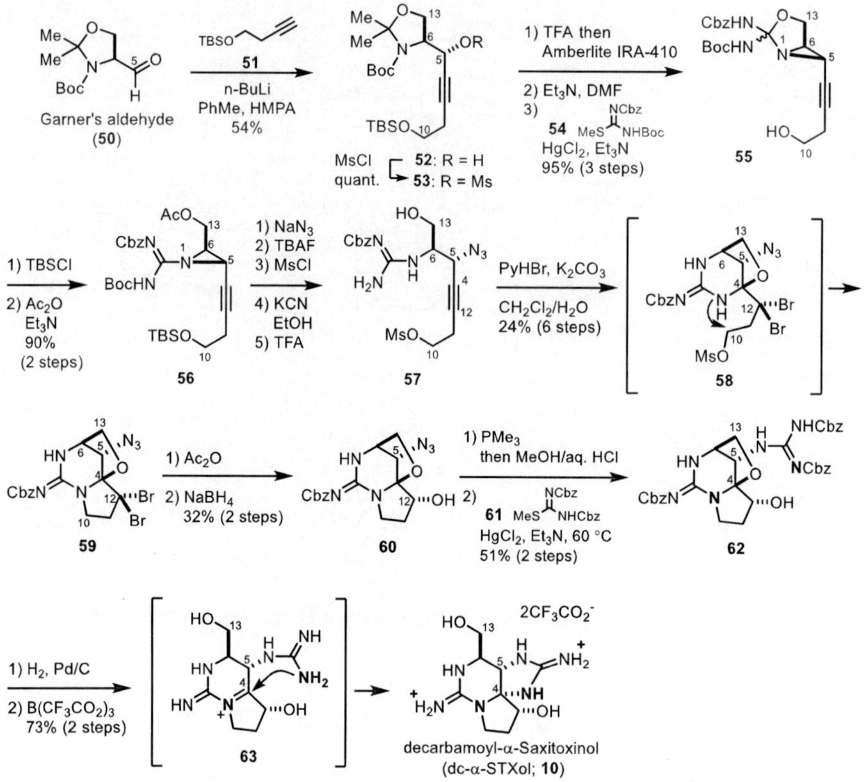

Scheme 7 Total synthesis of (+)-decarbamoyl- α-saxitoxinol (dc-α-STXol; **10**) via iminium cation **63** by the Nishikawa group

intermediate **56**, treatment of the resulting alcohol **57** with $PyHBr_3$ in the presence of potassium carbonate resulted in bromocyclization. The resulting bicyclic *N*,*O*-acetal **58** was cyclized at C10 with guanidine to afford cyclic guanidine **59**. Then, the azide at C5 and the geminal dibromide at C12 were converted into Cbz-protected guanidine and a hydroxy group, respectively, to give the key aminal intermediate **60**. After deprotection of the Cbz group in **60** by hydrogenation in the presence of Pd/C, the resulting bis-guanidine was subjected to acid treatment with $B(CF_3CO_2)_3$ in TFA to furnish dc-α-saxitoxinol (dc-α-STXol, **10**) in 73% yield through an iminium cation intermediate **63**.

2.4 Synthesis of STX by the Nagasawa Group

Nagasawa and co-workers employed hemiaminal **30** as a key synthetic intermediate, similarly to the Du Bois group, and accomplished a total synthesis of (+)-STX (**1**) via iminium cation **31** (Scheme 8) [12]. In their synthesis, the key hemiaminal **30** was synthesized by the IBX oxidation of α-amino ketone with cyclic guanidine **64**, which was obtained stereoselectively by 1,3-dipolar cycloaddition (1,3-DC) reaction of cyclic nitrone.

Details of the synthesis of (+)-STX (**1**) by Nagasawa's group are depicted in Scheme 9. A 1,3-DC reaction of nitroolefin **67** and chiral nitrone **66**, obtained from (-)-malic acid, gave isoxazolidine **68**, which was isomerized at C5 with DBU, followed by reduction of the nitro group to hydroxylamine with Zn in acetic acid to give isoxazolidine **69**. In the 1,3-DC reaction, three contiguous stereogenic centers at C4–6 were controlled by the chirality at C12 in **66**. The isomerization at C5 took place in equatorial orientation as the thermodynamically stable form. These three steps proceeded in one pot to give **68** in 70% yield. The next issue was differentiation of the two N-O bonds in **69**. The amine in the hydroxyamino group was selectively protected with Cbz, and then the two N-O groups were reduced with Zn and $TiCl_3$ in one pot to generate pyrrolidine **71**. In this process, the two nitrogens were differentiated, enabling further conversion to the cyclic guanidine **73** by 1) guanylation with *S*-methylthiourea **70**, 2) chloromethane sulfonate-mediated cyclization, and 3) deprotection of TIPS ether with TBAF. After oxidation of the hydroxy group in **73** under the Swern conditions, the resulting ketone **74** was subjected to IBX oxidation to give C4-hydroxylated ketone **76** in 94% yield. In the IBX reaction, generation of an intermediate complex of **75** with the enol of **74** and IBX was

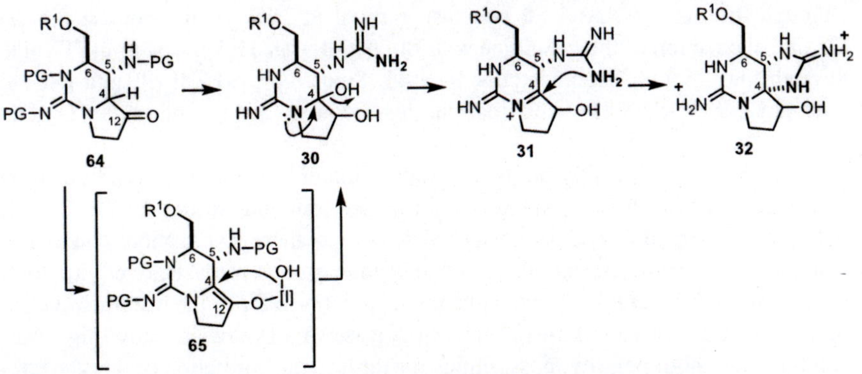

Scheme 8 Synthetic strategy for the saxitoxin skeleton by the Nagasawa group

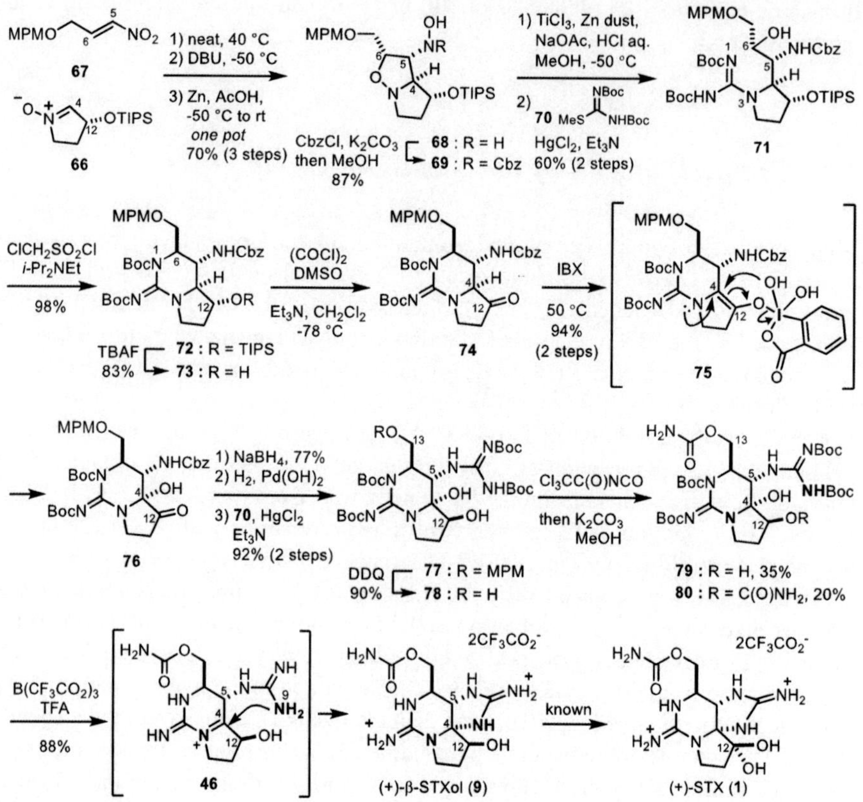

Scheme 9 Total synthesis of (+)-saxitoxin (STX; **1**) and (+)-β-saxitoxinol (β-STXol; **9**) by the Nagasawa group

proposed to take place [21]. Then, ketone **76** was converted to the key intermediate aminal **77** by installing guanidine at C5 and a carbamoyl group at C13 in five steps, although the bis-carbamoyl **80** was also formed in 20% yield from the diol **78**. Finally, cyclization of the guanidine with $B(CF_3CO_2)_3$ in TFA gave (+)-β-STXol (**9**) through the iminium cation intermediate **46**. Since (+)-β-STXol (**9**) is a one-step precursor for (+)-STX (**1**) via oxidation, this accomplished a synthesis of (+)-STX (**1**).

These STX syntheses by the four groups commonly employed cyclic guanidine construction at C4 through the key intermediate iminium cations **23**, **46**, and **63**. Although the approach to the STX skeleton via these cyclization reactions is straightforward and efficient, all of the protective groups were removed due to the need to use strong acids to generate the iminium cations. Since the resulting cyclized products, such as saxitoxinol (STXol) in Nagasawa's synthesis, show high water solubility and high polarity, possibilities for further transformation or derivatization are limited. Therefore, novel approaches are required to construct the STX skeleton without loss of the protective groups for the synthesis of other, more functionalized STX derivatives. In the next section, we will focus on the construction of the STX skeleton in protected form.

3 Synthesis of STX Skeletons in Protected Forms and Its Application to the Synthesis of STX (1) and Its Natural Analogs

3.1 Synthesis of STX Skeleton in Protected Form and Synthesis of STX Analogs by the Du Bois Group

For the synthesis of the STX skeleton in protected form, Du Bois and co-workers examined the intramolecular oxidative dearomative amination reaction of pyrrole **81** (Scheme 10) [22, 23]. In this reaction, amination should occur selectively at C4, and then regioselective nucleophilic addition of acetate to **82** should generate the STX skeleton as a protected form **83**.

Synthesis of a protected form of the STX skeleton **94** is depicted in Scheme 11. Urea **86** was synthesized by reaction with L-serine methyl ester hydrochloride (**84**) and pyrrole-1-carboxylic acid (**85**), followed by protection of the hydroxy group with TBDPS ether and reduction of the ester with DIBAH. The resulting aldehyde **86** was subjected to Pictet-Spengler-type reaction with allylamine in the presence of $BF_3 \cdot Et_2O$. In the reaction, the newly generated stereocenter at C5 is controlled by C6 through an allylic strain effect in the transition state of imine **87**, leading to the bicyclic urea **88** in 56% yield (2 steps) with >20:1 diastereoselectivity. After conversion of the cyclic urea and amine at C5 into guanidine by protection with Troc and Tces (trichloroethoxysulfonyl) groups, respectively, the resulting bis-guanidine was reacted with the $PhI(OAc)_2$ in the presence of rhodium catalyst, $Rh_2(esp)_2$ [24]. Oxidative amination proceeded at C4 selectively, followed by nucleophilic attack of the acetate in allyl-aziridine at C10 to give **94** with the STX skeleton in protected form.

With the synthesis of **94**, Du Bois and co-workers achieved a total synthesis of (+)-gonyautoxin 3 (GTX3; **7**) [22, 23], which was isolated from the Dinophyceae *Gonyaulax tamarensis* by Shimizu [25, 26], GTX3 shows high polarity due to its

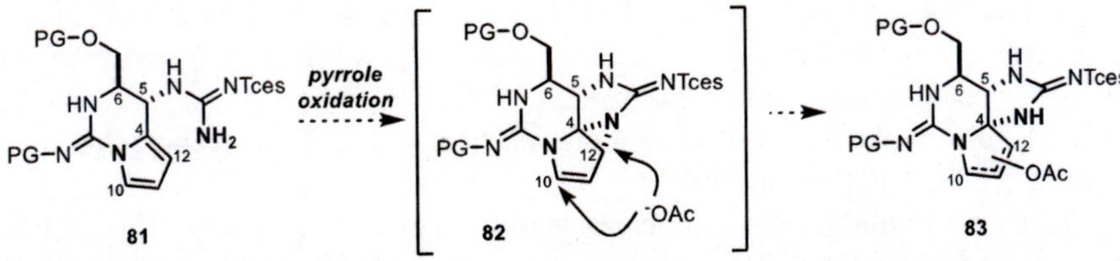

Scheme 10 Construction of protected STX skeleton **83** via C-H amination by Du Bois group

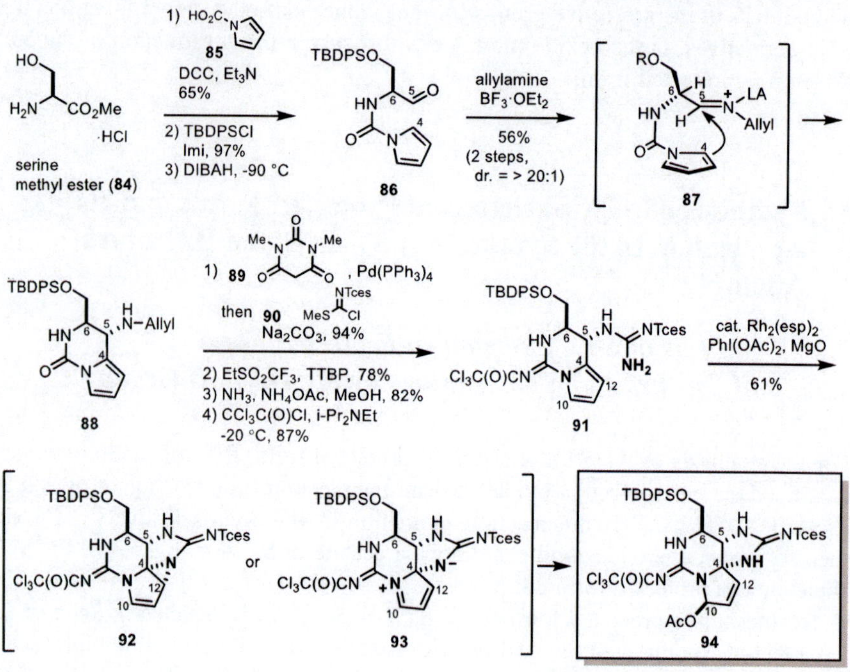

Scheme 11 Synthesis of protected STX skeleton **94** by the Du Bois group

characteristic sulfated secondary hydroxy group at C11 of STX (**1**), and the installation of this group is an important issue in the synthesis of (+)-GTX3 (**7**) (Scheme 12).

First, protected **94** was converted into the olefin **96** as follows: (i) reduction of the acetate at C10 with triethylsilyl hydride in the presence of a Lewis acid, (ii) deprotection of TPDPS ether, and (iii) introduction of the carbamoyl group at C13. Then, the double bond in **96** was reacted with osmium tetroxide to afford the diol **97**, whose hydroxy group at C11 was selectively protected with a benzoyl group to give **98** in 67% yield. After oxidation of the hydroxy group at C12 with Dess-Martin reagent, all of the protective groups, Troc, Tces, and benzoyl, in **99** were removed, and total synthesis of (+)-GTX3 (**7**) was achieved by sulfation of the hydroxy group at C11.

Du Bois and co-workers have also reported the synthesis of another STX analog, 11-saxitoxinethanoic acid (SEA; **11**) [27], which was isolated from xanthid crab *Atergatis floridus* by Arakawa and co-workers [28]. (+)-SEA (**11**) has a carbon–carbon bond connection of acetic acid at C11, which is unusual among STX analogs, and Du Bois applied the Stille coupling reaction to install the acetic acid unit in protected STX **101** (Scheme 13).

Firstly, they conducted allylic rearrangement of **101** to obtain allylic alcohol **104** by applying the Mislow-Evans conditions via sulfoxide **103**. After oxidation of the hydroxy group in **104**, the resulting enone **105** was reacted with iodine in the

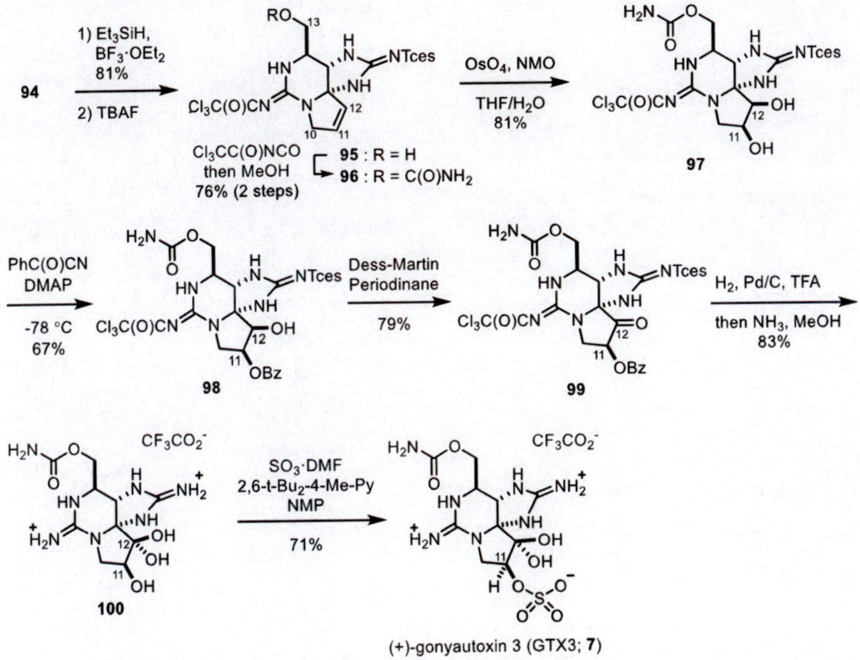

Scheme 12 Total synthesis of (+)-gonyautoxin 3 (GTX3; **7**) from the protected STX precursor **94** by the Du Bois group

presence of pyridine to generate vinyl iodide **107** via **106**. Then, the vinyl iodide **107** was subjected to the Stille coupling reaction with tributylvinyltin **108** in the presence of Pd(0) catalyst to install the acetic acid unit at C11, affording **109** in 60% yield. Total synthesis of (+)-SEA (**11**) was achieved from **109** by reduction of the double bond, selective deprotection of the TBDPS ether, installation of the carbamoyl group at C13, and removal of all the protecting groups.

3.2 *Synthesis of STX Skeleton in Protected Form and Synthesis of STX Analogs by the Nishikawa Group*

Nishikawa's group has developed a different type of protected STX skeleton **113**, focusing on construction of the aminal at C4, based upon the intramolecular bromocyclization of alkyne with guanidine-urea (**111** to **112**) followed by the introduction of a one-carbon unit at C6 via Strecker reaction (Scheme 14) [29].

The alkyne-containing guanidine-urea **116** was obtained from alkyne **114** in 5 steps. Then, cascade bromocyclization of **116** was carried out by treatment with PyHBr to give gem-bromide **118**. In this reaction, the stereochemistry at C4 was controlled by that at C5 in the TS **117**. Interestingly, no nucleophilic reaction

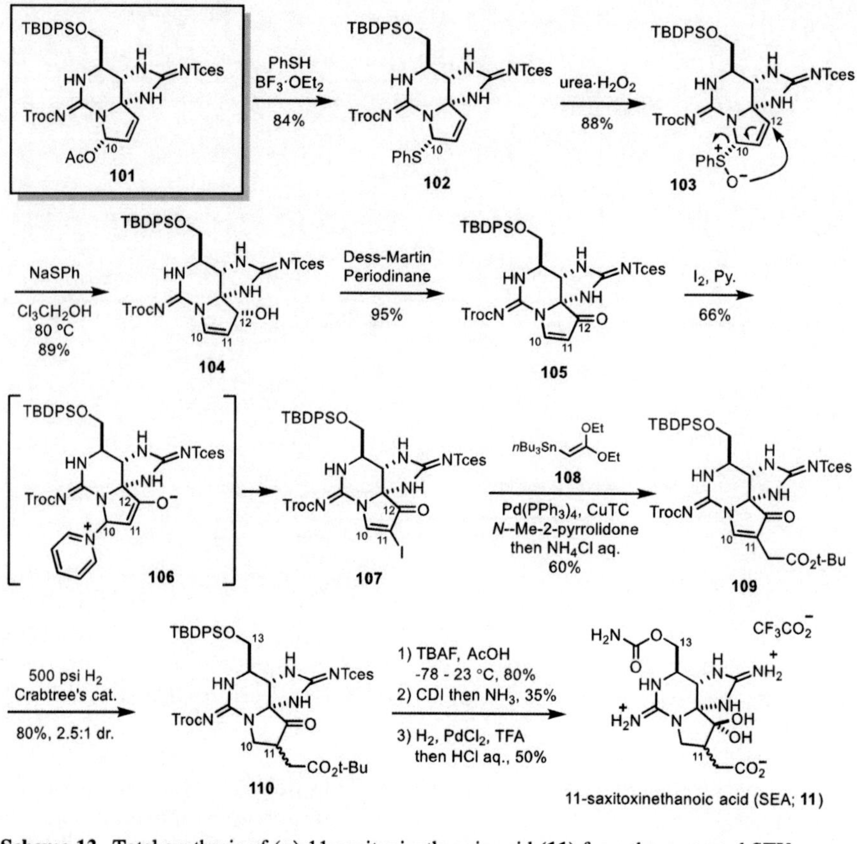

Scheme 13 Total synthesis of (+)-11-saxitoxinethanoic acid (**11**) from the protected STX precursor **101** by the Du Bois group

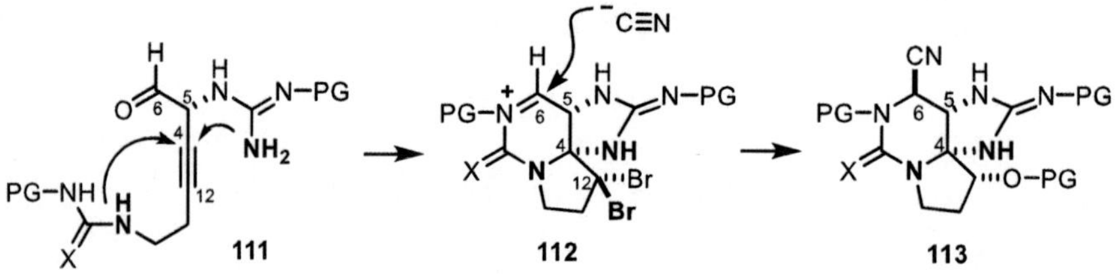

Scheme 14 Synthesis of a protected form of the STX skeleton **113** by the Nishikawa group

occurred in the case of guanidine instead of urea at N3. After conversion of the gem-bromide **119** into acetate at C12, deprotection of the acetonide and oxidative cleavage of the resulting diol with periodic acid gave the aldehyde, which cyclized with the urea to give hemiaminal **121** in 75% yield. Then, the Strecker type reaction was employed to install a cyano group at C6 by reaction with cyanohydrin, affording the protected STX skeleton **122**.

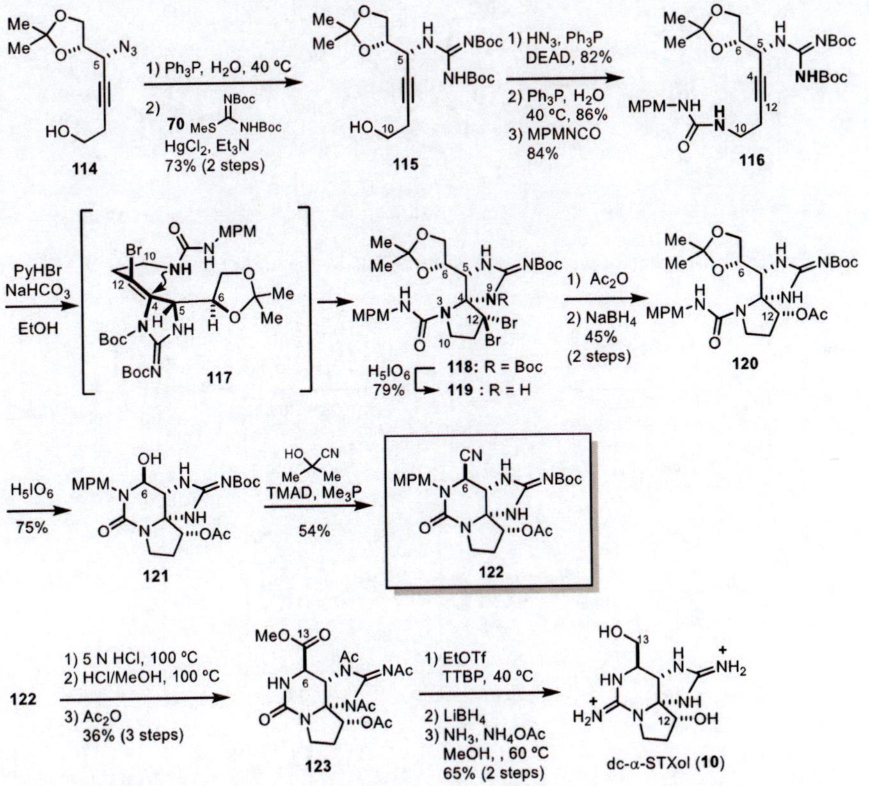

Scheme 15 Synthesis of a protected form of STX **122** and synthesis of dc-α-STXol (**10**) by the Nishikawa group

Nishikawa and co-workers achieved a total synthesis of dc-α-STXol (**10**) from **122**. Specifically, hydrolysis of the cyano group in **122** followed by reduction of the resulting carboxylic acid gave the diol, and conversion of the cyclic urea into guanidine followed by removal of all of the protective groups furnished dc–αSTXol (**10**) (Scheme 15).

3.3 Synthesis of STX Skeleton in Protected Form and Synthesis of STX Analogs by the Looper Group

Looper and co-workers have developed a synthesis of the protected STX skeleton **126** using the alkyne-bearing bis-guanidine **124** by applying a similar halocyclization to Nishikawa's, but with a different bond-connection pattern of the substrate [30]. In their synthesis, the cascade 5-exo-dig and 6-exo-trig cyclizations proceeded selectively to give the protected bicyclic guanidine **126** (Scheme 16).

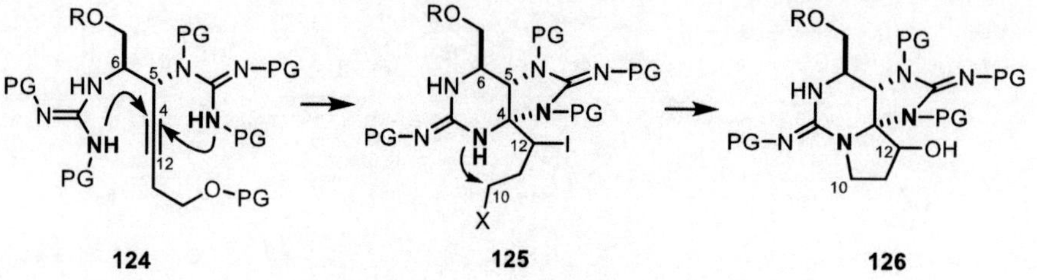

Scheme 16 Construction of protected STX skeleton **126** by the Looper group

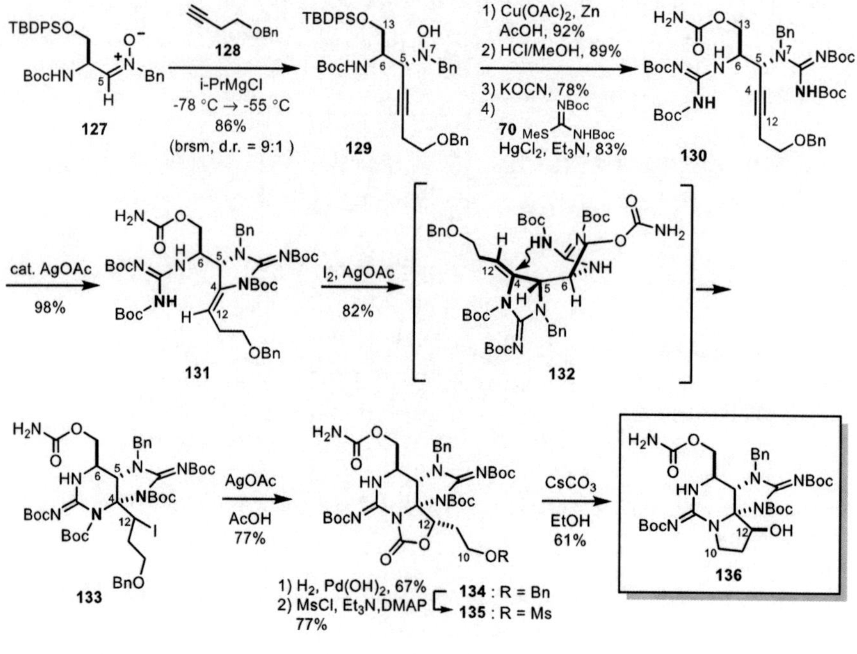

Scheme 17 Synthesis of a protected form of STXol **136** by the Looper group

The synthesis of protected STX skeleton **136** by Looper's group is depicted in Scheme 17. They synthesized the 4,5-diamino alkyne **129** by nucleophilic addition of alkyne **128** to nitrone **127** derived from L-serine. After the introduction of the carbamoyl group at C13, the guanidine groups at C5 and C6 were introduced to give **130**. Then, in the presence of a catalytic amount of silver acetate, the key 5-exo-dig type cyclization of bis-guanidine **130** occurred selectively to generate enamine **131**, which was subsequently reacted with iodine in the presence of a catalytic amount of silver acetate to furnish bis-cyclic guanidine **133** stereoselectively via intermediate **132** [31]. In the tandem-cyclization reaction, iodine was simultaneously introduced at C12 for further functionalization. Next, aminal **133** was treated with silver acetate in acetic acid to give the five-membered cyclic carbamate by cyclization of the Boc

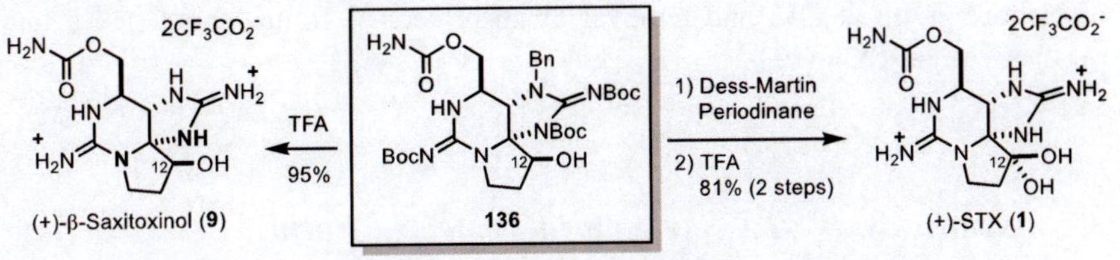

Scheme 18 Total synthesis of (+)-β-STXol (**9**) and (+)-STX (**1**) from **136** by the Looper group

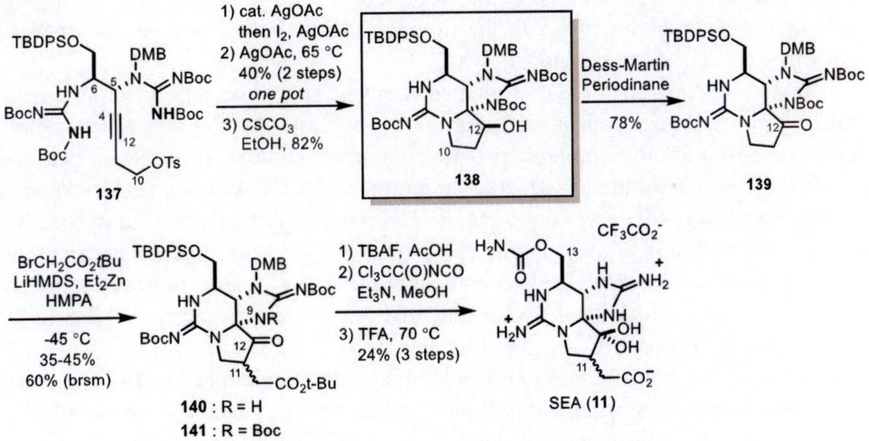

Scheme 19 Total synthesis of (+)-SEA (**11**) from **138** by the Looper group

group with iodine at C12. These processes from bis-guanidine **130** to cyclic guanidine **133** were carried out in one pot, improving the yield from 57% to 67%. After the conversion of the benzyl ether in cyclic carbamate **134** into mesylate, the resulting **135** was treated with cesium carbonate to give the protected STXol derivative **136** via hydrolysis of cyclic carbamate followed by cyclization to pyrrolidine.

With the protected form of STXol **136** in hand, Looper and co-workers synthesized (+)-β-STXol (**9**) by deprotection of the Boc group on the guanidine. Also, (+)-STX (**1**) was synthesized from **136** by oxidation of the hydroxy group at C12 followed by deprotection of the Boc group (Scheme 18) [30].

Looper and co-workers developed a total synthesis of (+)-SEA (**11**) from protected STXol **138**, in which the functional groups at C10 and C13 were changed to tosylate and TBDPS ether, respectively, from the intermediate **130** (Scheme 19) [32].

The protected form of STXol **138** was synthesized from **137** in a similar manner to that used for **136**. After oxidation of the hydroxy group at C12 with Dess-Matin reagent, the resulting ketone **139** was alkylated with *tert*-butyl 2-bromoacetate via zinc enolate to introduce the acetic acid unit at C11. Finally, installation of the carbamoyl group at C13 and removal of all protective groups completed a total synthesis of (+)-SEA (**11**).

3.4 Synthesis of STX Skeleton in Protected Form and Synthesis of STX Analogs by the Nagasawa Group

Nagasawa and co-workers synthesized a protected form of STXol by applying the cyclization of guanidine at C5 to the iminium cation, based on their earlier work illustrated in Scheme 9 [33, 34].

During construction of the five-membered cyclic guanidine, all protecting groups were removed and cyclization occurred to generate the highly water-soluble saxitoxinol. To avoid deprotection during the cyclization process, Nagasawa and co-workers examined the construction of a protected STX skeleton, i.e., conversion of **142** into **144**, under a variety of acidic conditions, but found that no cyclization reaction via **143** took place (Scheme 20). However, the cyclization proceeded under strongly acidic conditions after the removal of the Cbz guanidine protecting groups. Based on these results, they considered that two important processes should be involved in the cyclization reaction (Scheme 21).

For the cyclization reaction to proceed, firstly, the iminium cation **147** should be generated from **145**. Secondly, conformational change should take place from **147** to **148**, where the guanidine side chain is oriented axially, to construct the five-membered guanidine moiety. Thus, we planned to install a leaving group at C4 to accelerate iminium ion generation and an acyloxy group at C12 that would facilitate the conformational change by forming a bridged acyloxonium ion intermediate **148**.

Based upon the working hypothesis, substrates with various types of R group at C4 and C12 were explored for the cyclization reaction. Among the substrates tested, the bis-acetate **149**, derived from **77** by treatment with acetic anhydride in the presence of DMAP, cyclized smoothly in the presence of zinc(II) chloride at −20°C to afford **150**, a protected saxitoxinol, in 98% yield from **77** (Scheme 22). Since the acyl groups at C12 in **149** were mandatory for the cyclization reaction, the acyloxonium ion intermediate **151** presumably favors an appropriate conformation for the cyclization.

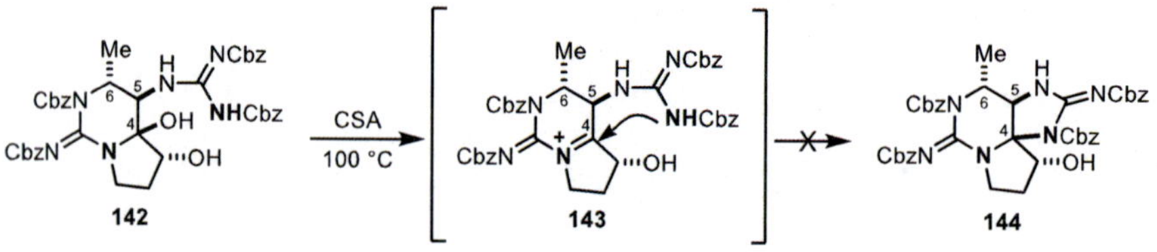

Scheme 20 Previous attempts to construct the five-membered guanidine under acidic conditions from **142**

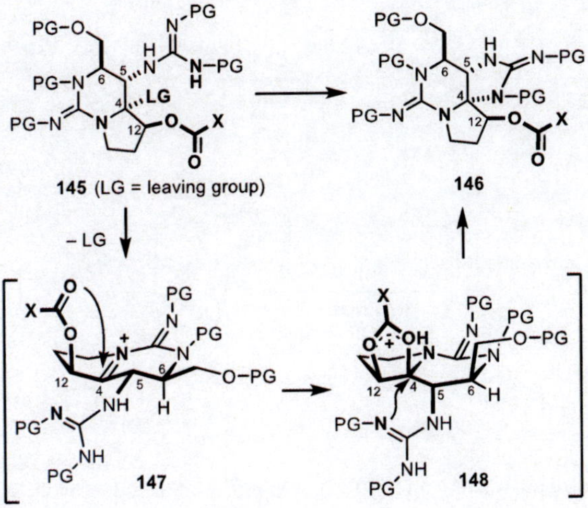

Scheme 21 Proposed transition state model for construction of the five-membered cyclic guanidine in the STX skeleton

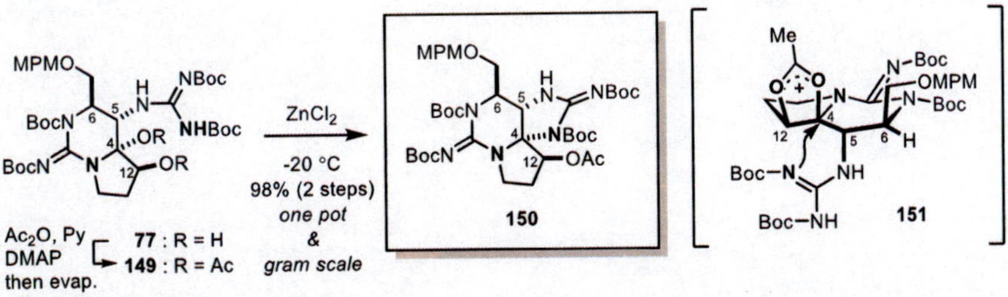

Scheme 22 Construction of the protected STX skeleton **150** by the Nagasawa group

With the protected STXol **150** in hand, Nagasawa and co-workers proceeded to synthesize (+)-GTX3 (**7**) as depicted in Scheme 23 [33, 34]. The reaction of **152** with TIPS chloride in the presence of NaHMDS followed by oxidation of the resulting silyl enol ether with *m*-CPBA to give the siloxy epoxide **154**. In this reaction, MPM ether was oxidatively cleaved simultaneously to generate the primary alcohol, which was further converted into carbamate **155**. After removal of the protective groups of Boc and TIPS ether, sulfation of the resulting hydroxyl group at C11 was conducted to complete the total synthesis of (+)-GTX3 (**7**).

Nagasawa also synthesized (+)-SEA (**11**) from protected STXol **150** (Scheme 24), employing the Mukaiyama-aldol condensation reaction to install acetic acid at C11 [35]. Thus, the Mukaiyama-aldol condensation reaction of silyl enol ether **156** with ethyl glyoxylate in the presence of $[Bu_4N][Ph_3SnF_2]$ gave the aldol condensation adduct **158** in 85% yield. In this reaction, the TBS ether at C13 remained intact. After reduction of the double bond with L-selectride, deprotection of the TBS group in **159** was carried out with HF·Et_3N complex to give **160** in 87% yield. Conversion of the hydroxy group of **160** to a carbamoyl group, hydrolysis of the ethyl ester in

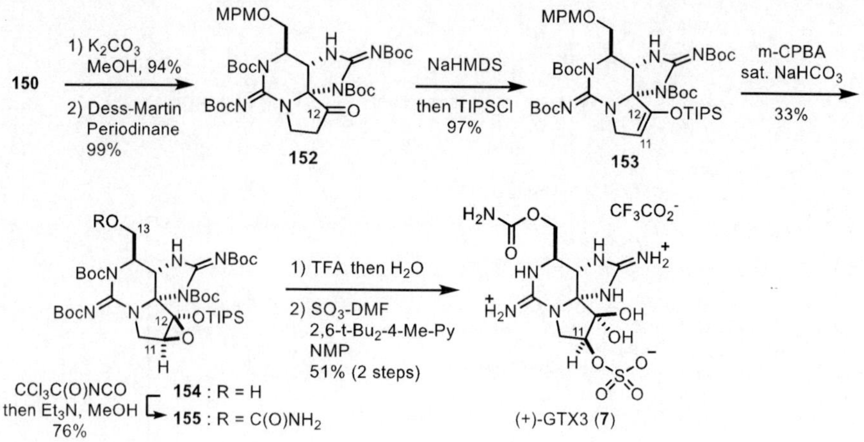

Scheme 23 Total synthesis of (+)-GTX3 (**7**) from protected STX **150** by the Nagasawa group

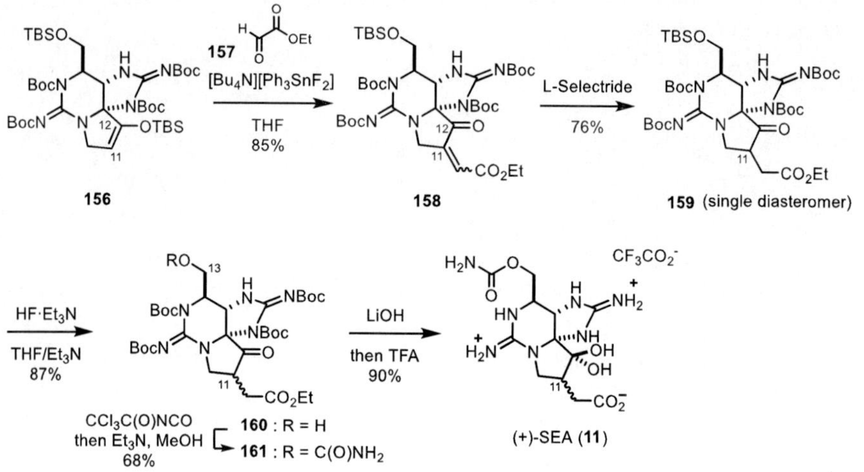

Scheme 24 Total synthesis of (+)-SEA (**11**) from protected STX **150** by the Nagasawa group

161 with lithium hydroxide, and deprotection of all four Boc groups with TFA completed a total synthesis of (+)-SEA (**11**).

4 Summary and Outlook

This chapter provides an overview of reported syntheses of STX (**1**) and its natural analogs. In these syntheses, construction of the characteristic tricyclic structure of STX without loss of the protective groups on guanidine as well as the other functional groups is crucial. The developed methodologies have enabled the synthesis of various natural analogs of STX with more complex structures, as well as artificial STX derivatives. In particular, SAR studies have led to ligands with interesting subtype-selectivity for Na_V [36–40]. Applications of the synthetic methods introduced in this chapter are expected to provide access to even more selective STX derivatives, which would be candidate drugs for the treatment of pain and cardiac disease.

References

1. Llewellyn LE (2006) Nat Prod Rep 23:200
2. Cusick KD, Sayler GS (2013) Mar Drugs 11:991
3. Thottumkara AP, Parsons WH, Du Bois J (2014) Angew Chem Int Ed 53:2
4. Schantz EJ, Mold JD, Stanger DW, Shavel J, Riel FJ, Bowden JP, Lynch JM, Wyler RS, Riegel B, Sommer H (1957) J Am Chem Soc 79:5230
5. Mold JD, Bowden JP, Stanger DW, Maurer JE, Lynch JM, Wyler RS, Schantz EJ, Riegel B (1957) J Am Chem Soc 79:5235
6. Wiese M, D'Agostino PM, Mihali TK, Moffitt MC, Neilan BA (2010) Mar Drugs 8:2185
7. Rogers PS, Rapoport H (1980) J Am Chem Soc 102:7335
8. Yotsu-Yamashita M, Kim YH, Dudley Jr SC, Choudhary G, Pfahnl A, Oshima Y, Daly JW (2004) Proc Natl Acad Sci U S A 101:4346
9. Tanino H, Nakata T, Kaneko T, Kishi Y (1976) J Am Chem Soc 99:2818
10. Kishi Y (1980) Heterocycles 14:1477
11. Fleming JJ, Du Bois J (2006) J Am Chem Soc 128:3926
12. Iwamoto O, Shinohara R, Nagasawa K Chem Asian J 4:277
13. Sawayama Y, Nishikawa T (2011) Angew Chem Int Ed 50:7176
14. Jacobi PA, Martinelli MJ, Polanc S (1984) J Am Chem Soc 106:5594
15. Hong CY, Kishi Y (1992) J Am Chem Soc 114:7001
16. Fleming JJ, Fiori KW, Du Bois J (2003) J Am Chem Soc 125:2028
17. Fleming JJ, McReynolds MD, Du Bois J (2007) J Am Chem Soc 129:9964
18. Koehn FE, Ghazarossian VE, Schantz EJ, Schnoes HK, Strong FM (1981) Bioorg Chem 10:412
19. Onodera H, Satake M, Oshima Y, Yasumoto T, Carmichael WW (1997) Nat Toxins 5:146
20. Sawayama Y, Nishikawa T (2010) Synlett 5:651
21. Iwamoto O, Koshino H, Hashizume D, Nagasawa K (2007) Angew Chem Int Ed 46:8625
22. Mulcahy JV, Du Bois J (2008) J Am Chem Soc 130:12630
23. Mulcahy JV, Walker JR, Merit JE, Whitehead A, Du Bois J (2016) J Am Chem Soc 138:5994
24. Kim M, Mulcahy JV, Espino CG, Du Bois J (2008) Org Lett 8:1073
25. Shimizu Y, Buckley LJ, Alam M, Oshima Y, Fallon WE, Kasai H, Miura I, Gullo VP, Nakanishi K (1976) J Am Chem Soc 98:5414
26. Boyer GL, Schantz EJ, Schnoes HK (1978) J Chem Soc Chem Commun:889
27. Walker RJ, Merit EJ, Thomas-Tran R, Tang DTY, Du Bois J (2018) Angew Chem Int Ed 58:1689
28. Arakawa O, Nishio S, Noguchi T, Shida Y, Onoue Y (1995) Toxicon 12:1577
29. Ueno S, Nakazaki A, Nishikawa T (2016) Org Lett 18:6368
30. Bhonde VR, Looper RE (2011) J Am Chem Soc 133:20172
31. Gainer MJ, Bennett NR, Takahashi Y, Looper RE (2011) Angew Chem Int Ed 50:684
32. Paladugu SR, James CK, Looper RE (2019) Org Lett 21:7999
33. Iwamoto O, Nagasawa K (2010) Org Lett 12:2150
34. Iwamoto O, Akimoto T, Nagasawa K (2012) Pure Appl Chem 84:1445

35. Wang C, Oki M, Nishikawa T, Harada D, Yotsu-Yamashita M, Nagasawa K (2016) Angew Chem Int Ed 55:11600
36. Andresen BA, Du Bois J (2009) J Am Chem Soc 131:12524
37. Shinohara R, Akimoto T, Iwamoto O, Hirokawa T, Yotsu-Yamashita M, Yamaoka K, Nagasawa K (2011) Chem Eur J 17:12144
38. Akimoto T, Masuda A, Yotsu-Yamashita M, Hirokawa T, Yamaoka K, Nagasawa K (2013) Org Biomol Chem 11:6642
39. Thomas-Tran R, Du Bois J (2016) Proc Natl Acad Sci U S A 113:5856
40. Adachi K, Yamada T, Ishizuka H, Oki M, Tsunogae S, Shimada N, Chiba O, Orihara T, Hidaka M, Hirokawa T, Odagi M, Konoki K, Yotsu-Yamashita M, Nagasawa K (2020) Chem Eur J 26:2025

Part III
Isolation and Synthesis of Low-Weight Molocular Compounds from Marine Bacteria Etc

Top Heterocycl Chem (2021) 58: 155–174
https://doi.org/10.1007/7081_2020_49

Published online: 12 January 2021

Natural Products from Marine Bacteria and Actinomycetes

Yasuhiro Igarashi

Contents

Abstract Instead of terrestrial microbes, much attention is being paid to marine microbes as an alternative source of new chemical entities for drug development. New compounds are steadily discovered from marine bacteria and actinomycetes. This short review summarizes structures, bioactivity, and biosynthesis of a limited but diverse class of natural products from marine bacteria and actinomycetes, including those discovered in our laboratory.

Keywords Marine actinomycetes · Marine bacteria · Natural products

Y. Igarashi (✉)
Biotechnology Research Center and Department of Biotechnology, Toyama Prefectural University, Toyama, Japan
e-mail: yas@pu-toyama.ac.jp

1 Introduction

Marine microbes are attracting much more attention than ever as a reservoir of new drug leads, alternative to terrestrial microbes [1]. This is mainly because the number of new bioactive compounds from terrestrial actinomycetes and fungi is likely reaching a plateau through intensive screening efforts in the past decades, while the number of new compounds from marine microbes is steadily increasing in recent years [2].

A large portion of marine microbes are associated with host organisms such as marine invertebrates, forming microbial communities unique to each host species [3]. Host-associated microbes are suggested to have a beneficial role in defensing the host from predators or pathogenic microbes through the production of toxic/antimicrobial substances [4]. Free-living microbes existing in sea water and sediment are also an important source of new natural products.

Despite the declined interest in drug development from natural products at pharmaceutical companies, natural products still keep providing opportunities to obtain novel scaffold structures for pharmacophore design [5]. In this chapter, several different classes of natural products from marine microorganisms are summarized regarding the structure, bioactivity, and biosynthesis.

2 Natural Products from Marine Bacteria

2.1 Dihydroisocoumarins

Dihydroisocoumarins, represented by amicoumacins, are the microbial natural products characterized by the common chromophore, 3,4-dihydro-8-hydroxyisocoumarin conjugating with an isopentylamine unit (Fig. 1). The right half of dihydroisocoumarins is basically a linear diacid which is modified by hydroxylation, amination, cyclization, or acylation. To date more than 20 related compounds are reported: baciphelacin [6], amicoumacins [7], Y-05460M-A [8], PM-94128 [9], AI-77s [10], and amicoumacin phosphates [11] from terrestrial *Bacillus* and lipoamicoumacins [12] from marine *Bacillus*. Zenocoumacins are the products of a Gram-negative bacterium *Xenorhabdus* [13]. Furthermore, amicoumacin-congeners possessing additional heterocyclic rings are reported: bacilosarcins from marine *Bacillus* [14], hetiamacins from terrestrial *Bacillus* [15–18], and Sg17-1-4 from a marine fungus *Alternaria tenuis* [19]. Dihydroisocoumarins exhibit a wide range of bioactivities including antibacterial [12], cytotoxic [12], antiulcer [20], anti-inflammatory [20], gastroprotective [10], and herbicidal activities [14]. Owing to these pharmaceutically attractive properties, the dihydroisocoumarin family has been a target of total synthesis [21].

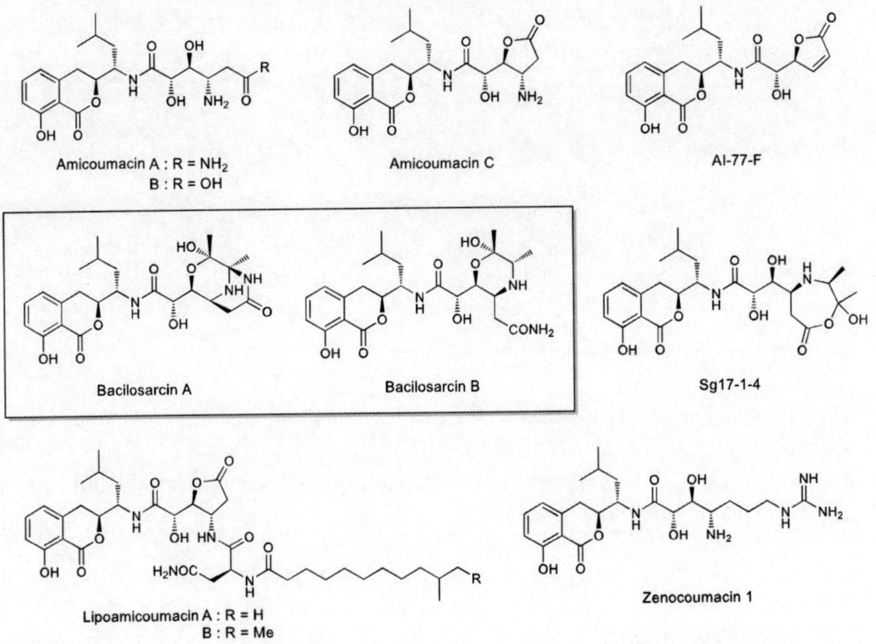

Fig. 1 Amicoumacins and related dihydroisocoumarins

In our screening of novel natural products from marine bacteria, a *Bacillus* strain was found to produce two dihydroisocoumarins, bacilosarcins A and B, decorated with unusual nitrogen- and oxygen-containing heterocyclic structure [14]. The producing strain *Bacillus* sp. TP-B0611 was isolated from the intestine content of a sardine collected in Toyama Bay, Japan. Bacilosarcin A has the 3-oxa-6,9-diazabicyclo[3.3.1]nonane ring system, which was unprecedented in natural and synthetic compounds at the time of discovery. The 2-hydroxymorpholine ring structure in bacilosarcin B is not unprecedented, yet very rare: convolutamine E and akashin C are the natural products bearing this ring system. Bacilosarcin A inhibited the growth of barnyard millet sprouts with 82% inhibition at 50 μM, more potent than herbimycin A (58% inhibition at the same concentration). Inspired by their intriguing structures as well as the potent herbicidal activity, total synthesis of bacilosarcins A and B was accomplished by Enomoto and Kuwahara [22].

Amicoumacin A is biosynthesized by the PKS-NRPS hybrid pathway (Scheme 1) [23]. The dihydroisocoumarin part with the isopentylamino side chain is assembled from four malonyl CoAs and one L-leucine. The diacid part is constructed from one hydroxymalonyl CoA and one L-asparagine. Interestingly, amicoumacin A is produced as an inactive form, preamicoumacins A and B, in which an *N*-acyl asparagine is attached onto the amino group of amicoumacin A. Biologically active amicoumacin A is released from the *N*-acyl asparagic acid by the action of peptidase. Bacilosarcins are presumably biosynthesized by the coupling of amicoumacin A and

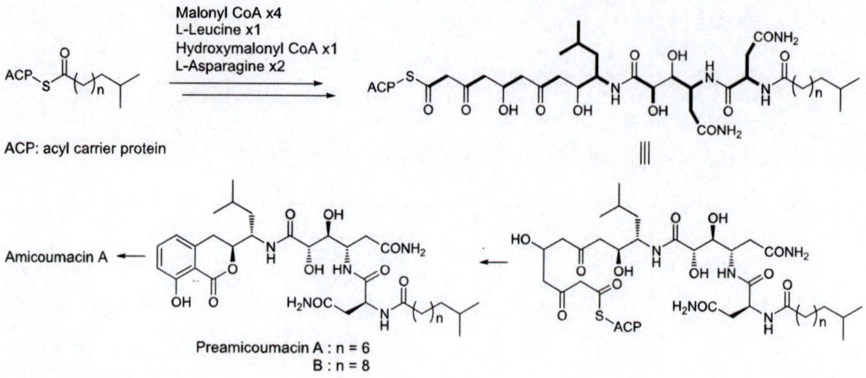

Scheme 1 Biosynthesis of amicoumacin A

diacetyl. Diacetyl is a byproduct of valine biosynthesis but is rarely utilized in secondary metabolite biosynthesis.

2.2 *Arylthiazolines*

Environmental microorganisms utilize non-proteinous, small molecules, termed "siderophore" to uptake ferric ions, which are essential to their growth, from surroundings [24]. One large group of microbial siderophores is the arylthiazoline, consisting of a hydroxyphenyl and a thiazoline unit (Fig. 2).

Aerugine, a metabolite of *Pseudomonas aeruginosa*, is the smallest member of this family biosynthesized from the condensation of salicylic acid and cysteine [25]. Pyochelin, a product of *Pseudomonas* species, is an extended derivative of aerugine in which a thiazoline and a thiazolidine are tandemly connected through the condensation of an additional cysteine [26]. Yersiniabactin is a further modified congener of pyochelin, bearing one more thiazoline ring. Biological function of

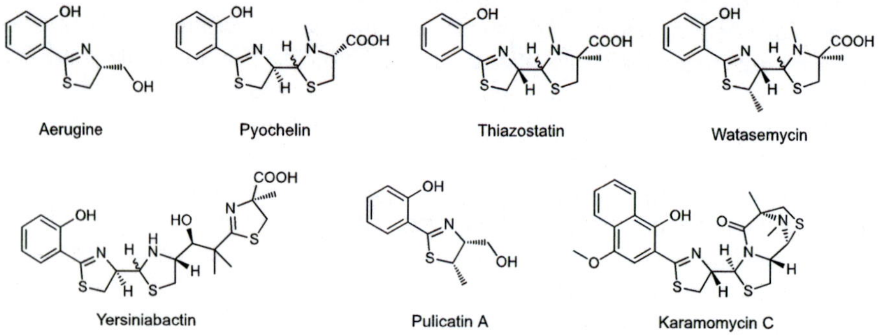

Fig. 2 Arylthiazolines from bacteria

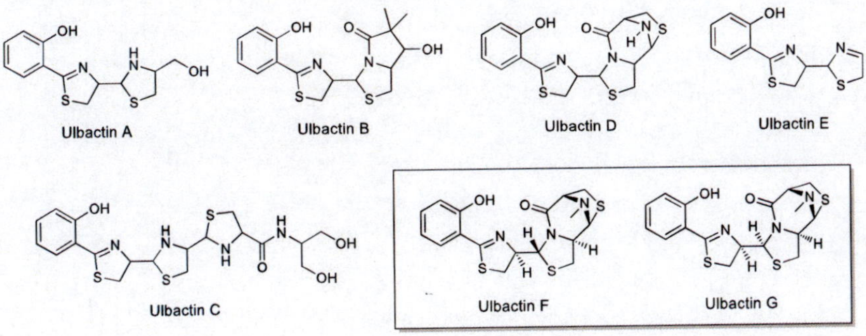

Fig. 3 Ulbactins A–G from marine bacteria

yersiniabactin is the acquisition of ferric ions for its producing pathogens such as *Yersinia pestis*, *Y. pseudotuberculosis*, and *Escherichia coli* to promote their growth and pathogenicity in host animals [27]. These siderophores are well known as the virulence factors of each producing-bacteria [28].

Watasemycin is a *C*-methylated congener of thiazostatin which is also a *C*-methylated congener of pyochelin. The producing strain of watasemycin, *Streptomyces* sp. TP-A0597, was isolated from a deep-sea water sample collected in Toyama Bay, Japan, in our laboratory [29], while thiazostatin was obtained from a soil-derived *Streptomyces* [30]. Watasemycin and thiazostatin exhibit anti-parasitic activities against malaria parasite, *Plasmodium*, with IC_{50} of 0.79–8.5 μg/mL. Both compounds are also active against *Trypanosoma brucei*, a causative agent of sleeping sickness in Africa, with IC_{50} of 0.24–2.5 μg/mL (unpublished data).

Pulicatins A–E are aerugine congeners produced by *Streptomyces* which was isolated from a cone snail collected in Philippines [31]. Pulicatin A binds to the human serotonin 5-HT_{2B} receptor at submicromolar concentrations. Recently, an unprecedented type of arylthiazoline, in which the hydroxyphenyl part is replaced by a hydroxynaphthyl unit, was discovered from a terrestrial rare actinomycete of the genus *Nonomuraea* [32]. Karamomycins A and B are the hydroxynaphthyl homologues of aerugine and aeruginol, respectively, and karamomycin C is the hydroxynaphthyl homologue of ulbactins F and G (see below).

Ulbactins are additional members of arylthiazolines produced by marine-derived bacteria (Fig. 3). Ulbactins A and B were found from *Vibrio* isolated from a sea grass and ulbactins C–E were from *Alteromonas* isolated from a seaweed [33, 34]. The genus *Vibrio* and *Alteromonas* are Gram-negative bacteria, belonging to the class *Gammaproteobacteria*. Ulbactins A–E are described in patents as UV absorbers useful for cosmetics. Ulbactin C possesses a tandemly connected thiazoline/thiazolidine/thiazolidine ring system.

In our screening for structurally unique bioactive compounds from marine bacteria, ulbactins F and G were discovered from the culture extract of a bacterial strain of the genus *Brevibacillus* [35]. The producing strain *Brevibacillus* sp. TP-B0800 was isolated from an unidentified orange-colored sponge collected at Otsuchi, Iwate, Japan. Ulbactins F and G are the *N*-methylated congeners of ulbactin D and their

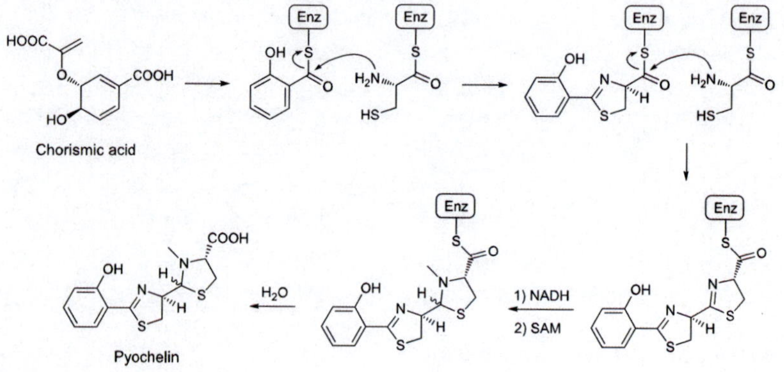

Scheme 2 Biosynthesis of pyochelin

absolute configurations were determined by X-ray crystallographic analysis. Ulbactins F and G exhibited potent inhibitory effects on the migration of epidermoid carcinoma A431 cells at non-cytotoxic concentrations with IC_{50} values of 6.4 and 6.1 μM, respectively. Further examination revealed that ulbactin F inhibited the cancer cell migration (IC_{50} 2.1 μM; esophageal squamous carcinoma EC109 cells) and tumor cell invasion (IC_{50} 1.7 μM; murine colon carcinoma 26-L5 cells).

Ulbactins F and G are featured by the tricyclic ring system, 6,9-imino-1*H*, 3*H*, 5*H*-thizaolo[4,3-c][1, 4]thiazepin-5-one. Intrigued by the structural and biological aspects, total synthesis of ulbactin F was accomplished by Wuest et al. [36]. This ring system was very recently found in karamomycin C, but quite surprisingly it is enantiomeric to the corresponding part of ulbactin F.

Ulbactins F and G are produced by a Gram-positive bacterium *Bacillus* in the phylum *Firmicutes*, whereas ulbactin D is produced by a Gram-negative bacterium *Alteromonas* in the phylum *Proteobacteria*. Karamomycin C is produced by an actinomycete *Nonomuraea* in the phylum *Actinobacteria*. In terms of the distribution of biosynthetic genes for secondary metabolites, it should be noted that these structurally rare and complex molecules are produced by microbial species phylogenetically apart.

Biosynthesis of arylthiazolines has been well studied for pyochelin and yersiniabactin [37]. The 2-hydroxyphenylthiazoline core is assembled by the condensation of L-cysteine and salicylate which is derived from chorismate. The second condensation of L-cysteine to the residual carboxyl group of the firstly-condensed L-cysteine gives rise to the second thiazoline ring which is reduced to the thiazolidine ring, followed by *N*-methylation, in pyochelin biosynthesis (Scheme 2).

Inahashi et al. disclosed the biosynthetic pathway for watasemycin and thiazostatin [38]. During the formation of the second thiazoline ring, *C*-methylation takes place at the carbon connecting to the carboxy group. After NADH-dependent reduction of the second thiazoline ring and *N*-methylation, watasemycin is released from the NRPS enzyme. Then, *C*-methylation again takes place in the first thiazoline ring to yield watasemycin (Scheme 3).

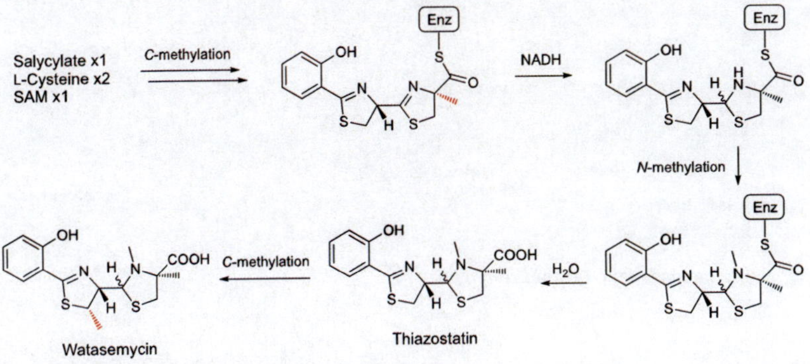

Scheme 3 Biosynthesis of thiazostatin and watasemycin

2.3 *Compounds from* Labrenzia

The genus *Labrenzia* is a Gram-negative, aerobic, motile, rod-shaped bacterium, categorized into the class *Alphaproteobacteria*. Sea salt-based medium is essential for its growth in laboratories and thus *Labrenzia* is defined as the marine obligate bacteria. Members of *Labrenzia* are widely found in marine habitats, often associated with marine organisms such as dinoflagellates, seashells, and corals [39, 40].

Labrenzia had been ignored among natural product chemists until the discovery of 18-*O*-demethylpederin (labrenzin) from a strain isolated from marine sediment [41]. Pederin is a polyketide characterized by two tetrahydropyran rings connected through the amide linkage with multiple methoxy groups. It was first discovered from *Paederus* beetles as the toxic substance [42]. Later, mycalamides [43] and onnamides [44], antitumor compounds bearing the common carbon skeleton, were isolated from sponges (Fig. 4). Isolation of these compounds from insects and sponges implied the production of pederin-class compounds by symbiotic microbes. Thereafter, Piel et al. identified the plausible biosynthetic gene clusters for pederin and onnamides through the metagenomic analysis of the beetles and sponges [45, 46]. The true producer of pederin-class compounds was unknown for many years, and it was believed that the producing microbe must be unculturable.

Quite recently, a culturable bacterium *Labrenzia* isolated from marine sediment was finally identified as a producer of pederin-class polyketides [41]. Biosynthetic

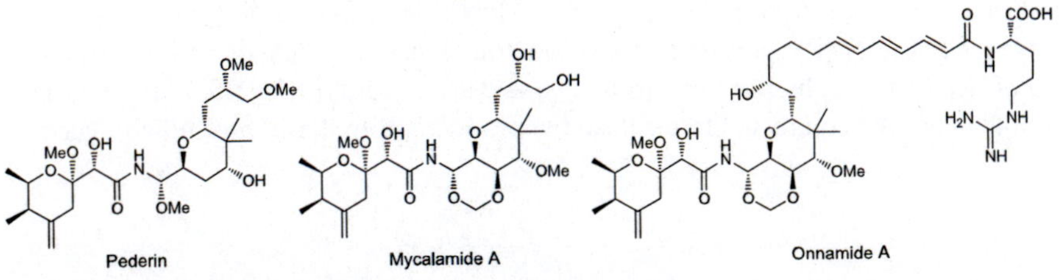

Fig. 4 Pederin and related marine natural products

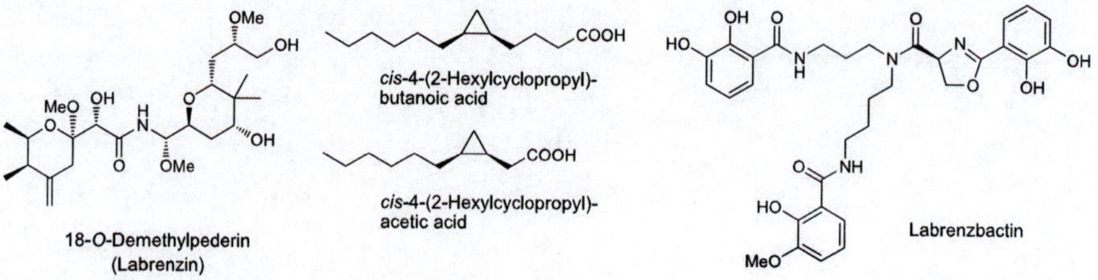

Fig. 5 Compounds isolated from *Labrenzia*

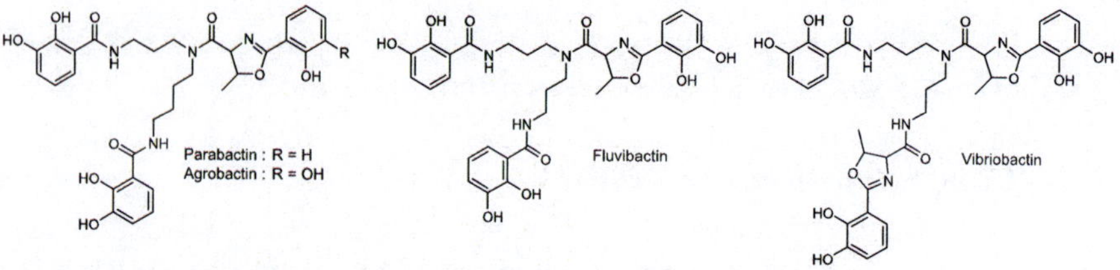

Fig. 6 Microbial siderophores related to labrenzbactin

genes for labrenzin and onnamide-like compounds were identified through the genome analysis of the producing-strain, *Labrenzia* sp. PHM005 [47]. In addition to labrenzin, two cyclopropane-containing fatty acids [48] are reported from *Labrenzia* (Fig. 5).

Labrenzia strain C1-1 was isolated from a Scleractinian (stony) coral *Montipora* sp. in our screening program on marine bacterial metabolites. Strain C1-1 showed 99.9% 16S rRNA gene sequence similarity to *Labrenzia aggregata* IAM 12614^T. This strain is marine obligate because it grows well in the sea water-based medium but not in fresh water-based medium. From the culture extract of strain C1-1, a new siderophoric metabolite, labrenzbactin was isolated [49]. This compound showed cytotoxicity against P388 murine leukemia cells with IC_{50} 13 μM and antibacterial activity against *Ralstonia solanacearum* and *Kocuria rhizophila* with MICs of 25 and 50 μg/mL, respectively.

Labrenzbactin is a new member of bacterial siderophores comprising a linear triamine moiety, spermidine or norspermidine, and aromatic acyl groups (Fig. 6). Several related compounds, parabactin from *Paracoccus*, agrobactin from a plant pathogen *Agrobacaterium*, and fluvibactin and vibriobactin from *Vibrio*, are known.

The genome size of *Labrenzia* is 5–8 Mbp in average and biosynthetic genes for type I PKS, NRPS, and terpenoids are found in the DNA database [50]. Some *Labrenzia* strains harbor three to five gene clusters for type I PKS. Further exploration should be conducted to disclose the products from these biosynthetic genes.

2.4 *Compounds from* Microbulbifer

The genus *Microbulbifer*, classified in the class *Gammaproteobacteria*, was first isolated from a black liquor sample of the pulp industry [51]. Members of *Microbulbifer* are Gram-negative, rod-shaped, non-motile, and halophilic. *Microbulbifer* species are marine obligate, requiring sodium salt for growth [51], and thus are frequently isolated from halophilic environments such as marine solar saltern and marine sediment [52–54]. *Microbulbifer* has been studied for its ability of degrading polysaccharides and lignins, but nothing was known about its metabolites when we started the metabolite analysis of this untapped genus.

The first natural products discovered from *Microbulbifer* are bulbiferates A and B [55] (Fig. 7). Bulbiferates have antibacterial activity against *Escherichia coli* and methicillin-sensitive *Staphylococcus aureus*.

In our metabolite investigation on marine bacteria, a unique fatty acid, (2*Z*,4*E*)-3-methyl-2,4-decadienoic acid, was identified as a major product of *Microbulbifer* isolated from a stony coral of the genus *Porites* [56]. The methyl group in this fatty acid is located at an uncommon position as a natural product. Usually in microbial products, methylation takes place at the carbon derived from the methyl carbon of acetate (Fig. 8), however this fatty acid has the methyl-branching at the carbon derived from the carbonyl carbon of acetate. Feeding experiments of ^{13}C-labeled precursors elucidated that the methyl group is derived from L-methionine, implying

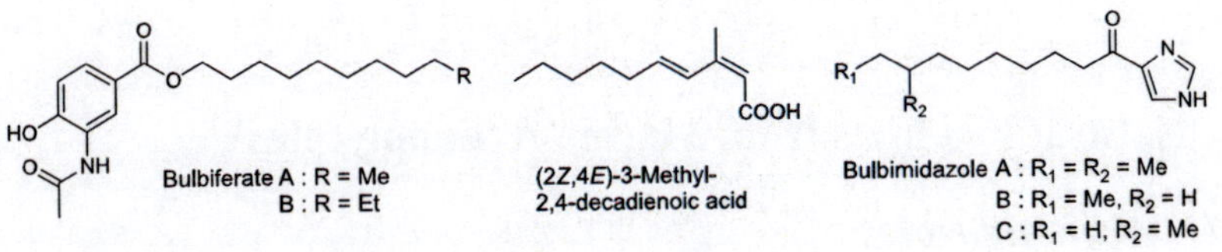

Fig. 7 Compounds isolated from *Microbulbifer*

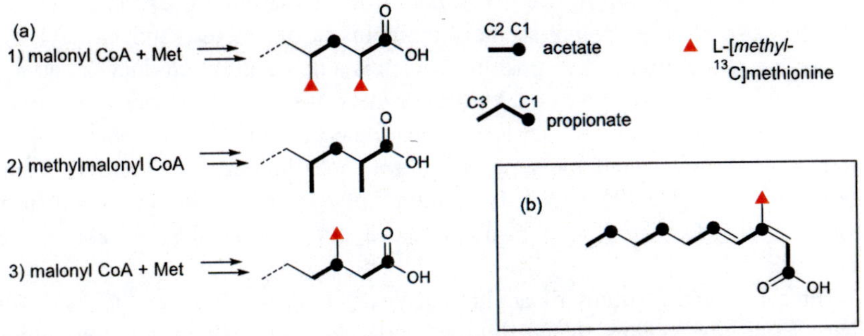

Fig. 8 Methylation pattern of fatty acids and polyketides (**a**) and ^{13}C-labeling pattern of (2*Z*,4*E*)-3-methyl-2,4-decadienoic acid (**b**)

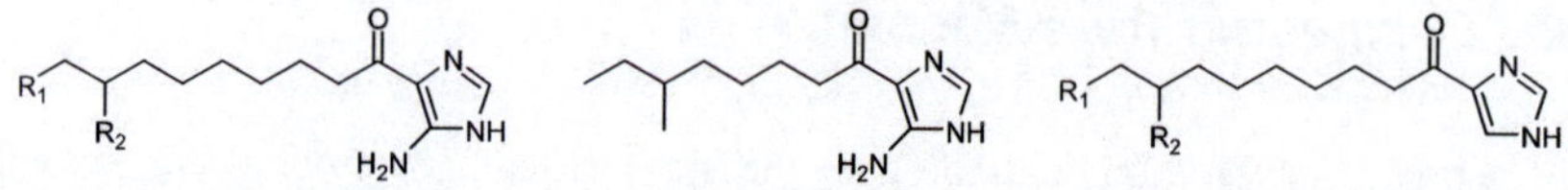

Fig. 9 Nocarimidazoles and bulbimidazoles

the involvement of unknown methylation mechanism using *S*-adenosylmethionine (SAM).

Subsequently, we found three new acylimidazoles, designated bulbimidazoles A–C (Fig. 9), from another *Microbulbifer* strain DC3-6 isolated from a stony coral of the genus *Tubastraea* [57]. The absolute configuration of the anteisomethyl group of bulbimidazole A was determined to be a mixture of (*R*)- and (*S*)-configurations in a ratio of 9:91 by applying the Ohrui-Akasaka method. Bulbimidazoles exhibit unique broad-spectrum antimicrobial profile against bacteria and fungi with MICs ranging from 0.78 to 12.5 μg/mL. They also showed cytotoxicity against P388 murine leukemia cells with IC_{50} in micromolar range.

According to the public DNA database, the genome size of *Microbulbifer* is around 3.6–5.0 Mbp, relatively small as a secondary metabolite-producer, but biosynthetic genes for NRPS and siderophore are present [50]. Discovery of bulbimidazole warrants further investigation of this genus because the production of bulbimidazoles was not predictable only from the gene analysis.

3 Natural Products from Marine Actinomycetes

3.1 *Acylimidazoles*

In advance of the discovery of bulbimidazoles, the first natural acylimidazoles, nocarimidazoles A and B, were reported from a marine *Nocardiopsis* (Fig. 9) [58]. Nocarimidazoles comprise the 4-aminoimidazole moiety and an acyl chain connecting at the imidazole 5-position, which is a new natural product skeleton. As described in the previous section, bulbimidazoles A–C are the new deamino congeners of nocarimidazoles, isolated from a marine bacterium *Microbulbifer* [57]. Furthermore, we discovered the new congeners, nocarimidazoles C and D, from a marine *Kocuria* [59]. The producing-strain *Kocuria* sp. T35-5, an isolate from a stony coral *Mycedium* sp., also produced nocarimidazoles A and B and bulbimidazole A.

Three producing-strains of acylimidazoles belong to phylogenetically distinct taxa. Both the genus *Nocardiopsis* and *Kocuria* belong to the phylum *Actinobacteria*, but they are separated at the family level. The most significant difference is their morphology: *Nocardiopsis* grows in a filamentous pattern whereas

Table 1 Production distribution of acylimidazoles

Compounds	Actinobacteria		Gammaproteobacteria
	Nocardiopsis sp. CNQ115	*Kocuria* sp. T35-5	*Microbulbifer* sp. DC3-6
Nocarimidazole A	○	○	–
B	○	○	–
C	–	○	–
D	–	○	–
Bulbimidazole A	–	○	○
B	–	–	○
C	–	–	○

Table 2 Antimicrobial activity of bulbimidazoles A–C and nocarimidazoles A–D

Microorganisms	MIC (μg/mL)						
	BA	BB	BC	NA	NB	NC	ND
Kocuria rhizophila[a]	0.78	0.78	1.56	6.25	12.5	6.25	6.25
Staphylococcus aureus[b]	1.56	1.56	3.12	12.5	25	12.5	25
Escherichia coli[c]	>100	>100	>100	>100	>100	>100	>100
Rhizobium radiobacter[d]	>100	>100	>100	>100	>100	>100	>100
Tenacibaculum maritimum[e]	3.12	6.25	12.5	NT	NT	NT	NT
Candida albicans[f]	6.25	12.5	6.25	25	25	12.5	12.5
Glomerella cingulata[g]	1.56	1.56	3.12	12.5	12.5	25	25
Trichophyton rubrum[h]	0.78	1.56	3.12	6.25	6.25	25	6.25

BA, BB, BC bulbimidazoles A, B, and C, *NA, NB, NC, ND* nocarimidazoles A, B, C, and D, *NT* not tested

[a]ATCC9341
[b]FDA209P JC-1
[c]NIHJ JC-2
[d]NBRC14554
[e]NBRC16015
[f]NBRC0197
[g]NBRC5907
[h]NBRC5467

Kocuria is single cellular cocci. The genus *Microbulbifer* belongs to the phylum *Proteobacteria*, phylogenetically apart from the phylum *Actinobacteria*. It is noteworthy that the common class of metabolites is produced by widely different genera. It should be also noted that all three producing-strains are derived from marine environment. The exclusive origin of these metabolites from marine bacteria, as well as the distribution among phylogenetically distinct taxa, implies their potential function in the adaptation of the producing organisms to the marine ecosystem. Production pattern of acylimidazoles is summarized in Table 1.

Bulbimidazoles and nocarimidazoles are active against Gram-positive bacteria, yeast, and fungi but inactive against Gram-negative bacteria (Table 2). Bulbimidazoles are more potent than nocarimidazoles in general, exhibiting a wide

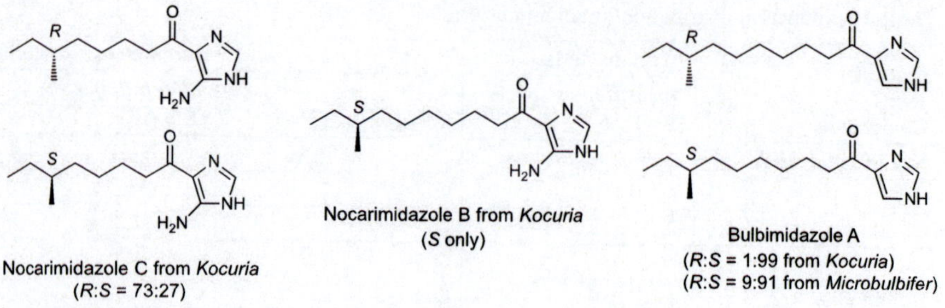

Fig. 10 Enantiomeric diversity of anteisomethyl chirality in acylimidazoles

antimicrobial spectrum against pathogens of humans (*S. aureus*, *C. albicans*, and *T. rubrum*), fish (*T. maritimum*), and plants (*G. cingulata*).

The absolute configurations of the anteisomethyl group in nocarimidazoles B and C and bulbimidazole A were analyzed by using low-temperature HPLC technique developed by Ohrui and Akasaka. Nocarimidazole B is a pure (*S*)-enantiomer, whereas nocarimidazole C is a mixture of (*R*)- and (*S*)-enantiomers with a ratio of 73:27. Bulbimidazole A from *Kocuria* is an (*S*)-enantiomer with 98% ee and that from *Microbulbifer* is an (*S*)-enantiomer with 82% ee (Fig. 10). Bacterial anteiso-fatty acids have been believed to be (*S*)-configured, because the starter unit of biosynthesis, 2-methylbutanoyl-CoA, is derived from L-isoleucine. However, this seems not always true in secondary metabolites. The reason of *R*/*S*-chirality fluctuation in bacterial anteisomethyl-branching is unknown. Elucidation of regulation mechanism in enzymatic reactions and biological differences between (*R*)- and (*S*)-enantiomers are the future subjects.

3.2 Rakicidins

Rakicidins are the 15-membered depsipeptides consisting of 4-amino-2,4-pentadienoate (APDA), glycine, and hydroxyasparagine/glutamine, and a fatty acid portion modified with hydroxy and methyl substitutions (Fig. 11). To date, six congeners, rakicidins A, B [60], and E [61] from *Micromonospora* and rakicidins C [62], D [63], and F [64] from *Streptomyces*, are known. Of these, producing-strains of rakicidins D, E, and F are marine origin.

Rakicidins A and B were originally isolated from soil-derived *Micromonospora* at Bristol-Myers Squibb. They were potently cytotoxic against murine lung carcinoma cells with IC_{50} of 40 to 200 ng/mL [60]. Rakicidin D, a shorter chain analog of rakicidin A, was isolated from *Streptomyces* collected from marine sediment of Thailand [63]. This compound inhibited invasion of murine colon carcinoma 26L5 cells with an IC_{50} of 6.7 μM.

After more than 20 years since the first discovery, rakicidin A was rediscovered during the screening for hypoxia selective cytotoxic agents from 20,000 microbial

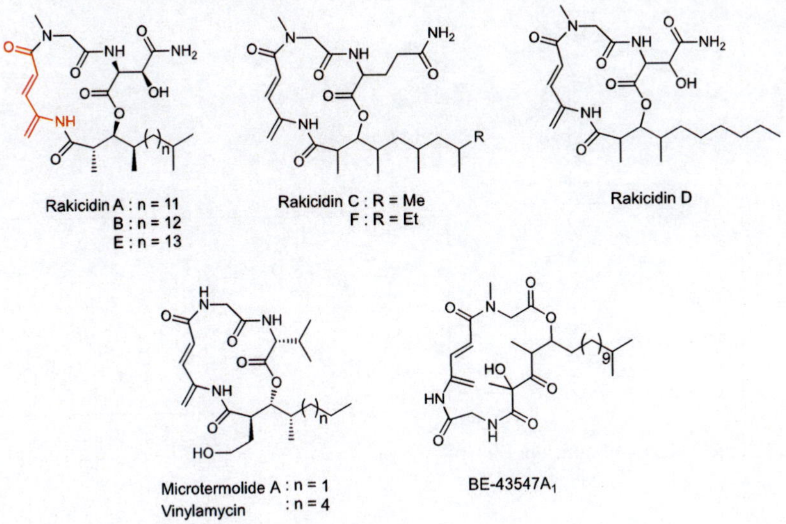

Fig. 11 Rakicidins and related cyclodepsipeptides

extracts [65]. Cancer cells adapted to hypoxic conditions inside the solid tumor become more resistant to antitumor drugs. Rakicidin A showed ca. 17 times more selective cytotoxicity to hypoxic cancer cells than to normoxic cancer cells. Furthermore, rakicidin A induced apoptosis in hypoxia-adapted leukemic cells with stem cell-like characteristics [66]. In addition, rakicidin A displayed synergistic anticancer activity in combination with Imanitib against hypoxia-adapted leukemic cells. These promising findings promoted rakicidin A to a potential lead for cancer chemotherapy, leading to extensive synthetic and structure–activity relationship studies [67–70].

4-Amino-2,4-pentadienoate (APDA), an unusual amino acid, present in rakicidins is found only in a limited range of secondary metabolites of actinomycetes. Microtermolides [71], vinylamycin [72], and BE-43547 [73], all of which are from *Streptomyces*, are additional members of APDA-containing cyclodepsipeptides. Despite the scarcity in nature, biosynthetic origin of APDA remained unknown. We carried out the draft genome analysis of *Streptomyces* sp. MWW064, the producing-strain of rakicidin D [74]. The 7.9 Mbp genome of strain MWW064 contained at least three type I polyketide synthase (PKS) gene clusters, seven nonribosomal peptide synthetase (NRPS) gene clusters, and four hybrid PKS/NRPS gene clusters, from which one hybrid PKS/NRPS gene cluster was identified responsible for rakicidin synthesis, based on the annotation results.

Starting from the *n*-hexanoyl unit, PKS/NRPS assembly line for rakicidin D connects the building blocks in the sequence of two methylmalonyl CoA, and one each of L-serine, malonyl CoA, glycine, and L-asparagine, to afford the linear intermediate, which is then cyclized to the depsipeptide structure. The final step is the hydroxylation of Asn residue to yield rakicidin D (Fig. 12). During the chain

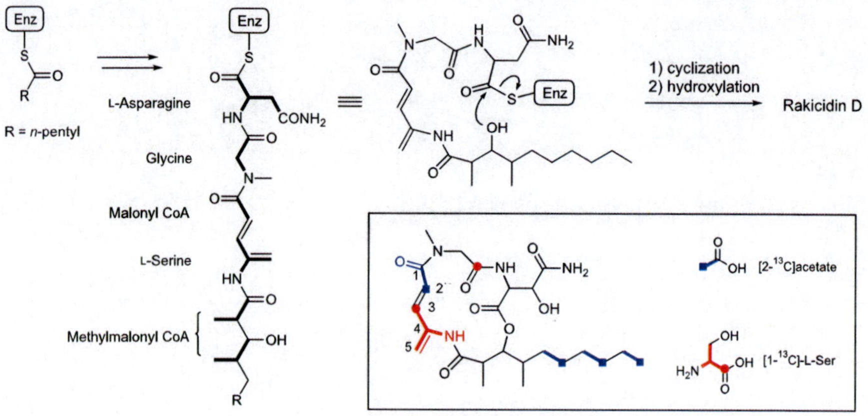

Fig. 12 Biosynthesis of rakicidin D

elongation step, the APDA unit is assembled from L-serine and malonate, followed by dehydration of the serine hydroxy group [74]. To verify this biosynthetic cascade predicted from the gene analysis, feeding experiments with ^{13}C-labeled precursors were performed. As expected, [2-^{13}C]acetate labeled the 2-position of APDA and [1-^{13}C]-L-serine labeled the 3-position of APDA, unambiguously establishing that the APDA unit is derived from an acetate and a serine (Fig. 12).

Furthermore, the draft genome sequence of a marine-derived *Micromonospora* sp. DSW705, the producing-strain of rakicidins A and B was analyzed [75]. The 6.8 Mbp genome of strain DSW705 harbored a hybrid PKS/NRPS gene cluster homologous to that found in strain MWW064. Accordingly, the biosynthetic gene cluster for rakicidins A and B was identified. These findings will make it possible to generate more potent rakicidin-based antitumor compounds by genetic engineering and to discover new APDA-containing natural products by genome mining.

3.3 *Nocapyrones*

Nocapyrones are a series of α- and γ-pyrones produced by marine-derived rare actinomycete of the genus *Nocardiopsis* (Fig. 13). Nocapyrones A–D were isolated from a *Nocardiopsis* strain collected from the marine sponge *Halichondria panicea* [76].

Three α-pyrones, nocapyrones E–G, were isolated from a marine-derived actinomycete *Nocardiopsis dassonvillei* [77]. Nocapyrones E–G were antibacterial against *Bacillus subtilis* with MIC at micromolar concentrations. Furthermore, 10 new congeners, nocapyrones H–Q, were identified from a symbiotic actinomycete of the genus *Nocardiopsis* isolated from a cone snail [78]. Almost in the same period, three α-pyrones were isolated from a marine *Nocardiopsis* and were

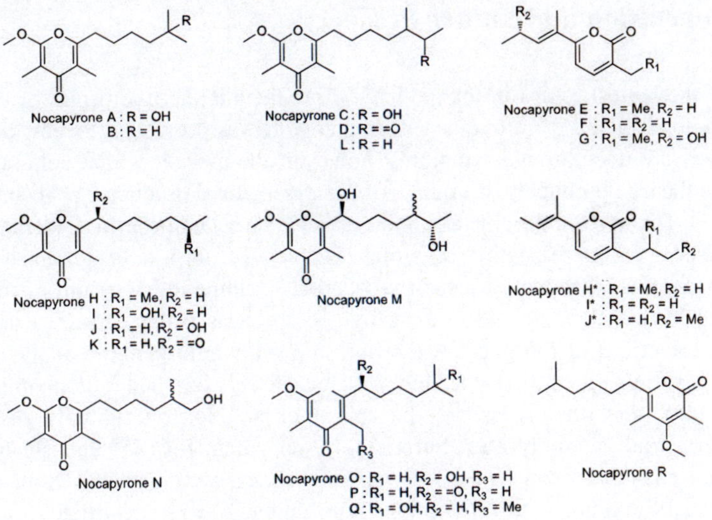

Fig. 13 Nocapyrones from marine actinomycetes

duplicately designated nocapyrones H–J [79] (these compounds are shown as nocapyrones H*, I*, and J* in Fig. 13).

In our screening for antidiabetic compounds from marine microbes, one new α-pyrone, nocapyrone R, along with three known γ-pyrones, nocapyrones B, H, and L, were isolated from a sediment-derived *Nocardiopsis* [80]. γ-Pyrones were found to induce adiponectin production in murine ST-13 preadipocyte cells while the α-pyrone, nocapyrone R, was inactive.

Furthermore, new nocapyrone congeners were isolated from marine *Nocardiopsis* [81–83]. Exclusively isolated from marine-derived *Nocardiopsis*, these pyrones are suggested to play an unknown role in the adaptation of the producing organisms to the marine environment. Meanwhile, some marine invertebrates are likely utilizing α- and γ-pyrones produced by symbiotic bacteria for chemical defense [78]. Nocapyrone-producing *Nocardiopsis* was isolated from the venom duct of cone snails. Astonishingly, the cone snail extracts contained a substantial amount (0.01–0.1% of snail dry weight) of nocapyrones B, H, and J, which were also produced by the *Nocardiopsis* strain isolated from this marine animal. These γ-pyrones displayed modulatory activity against nerve cell depolarization, implying a possible defensive role of nocapyrones in nature.

4 Conclusion and Future Prospect

One of the hottest topics in marine natural products is the true origin of bioactive compounds isolated from marine invertebrates. It was previously believed that the producers are unculturable symbionts; however, the evidences that culturable bacteria are the true producers of a number of marine natural products are accumulating (Fig. 14). The most noticeable examples are didemnin and pederin. Didemnin B, an anticancer cyclic depsipeptide, was first discovered from a tunicate, and was later rediscovered as a product of a sediment-derived culturable bacterium of the genus *Tistrella* [84]. Labrenzin (18-*O*-demethylpederin) was isolated from a culturable marine bacterium of the genus *Labrenzia* and the whole genome analysis of the producing-strain identified the biosynthetic genes for pederin/onnamide-class polyketides [47]. Enediynes are the antitumor antibiotics originally discovered from terrestrial actinomycetes. Surprisingly, namenamicin [85] and shishijimicin [86], the chalicheamicin-type enediyne polyketides, were isolated from colonial tunicates. In addition, the producer of manzamine A was identified as a marine *Micromonospora* [87]. These findings give hope to solve the supply problem associated with the natural products of marine invertebrates.

This short review deals with only a limited range of natural products from marine bacteria and actinomycetes including those discovered in our laboratory. For more information, readers should refer to other comprehensive, specific reviews on each topic. As a personal opinion, marine bacteria are still less explored than actinomycetes and fungi. This is partly because selection and isolation of diverse and unique bacteria are not easy, compared with actinomycetes and fungi which can be easily recognized from the colony morphology on isolation plates. Even from non-marine obligate bacteria, structurally novel compounds can be obtained, as exemplified by

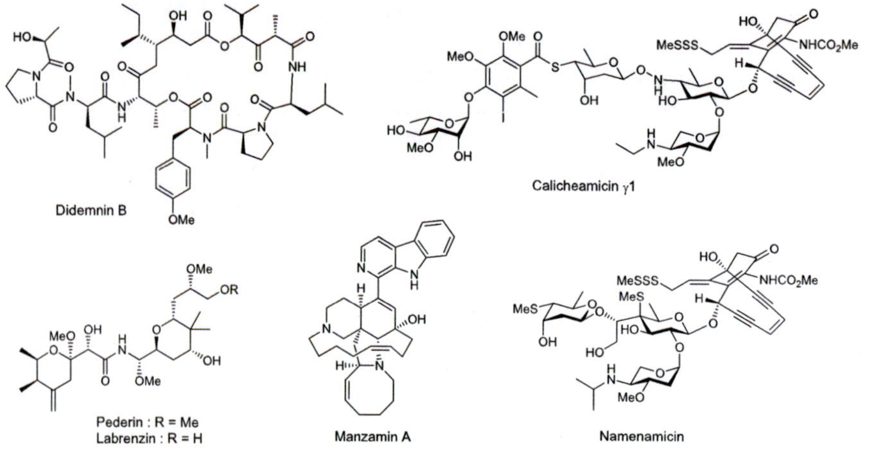

Fig. 14 Marine natural products of invertebrate-origin produced by culturable bacteria

bacilosarcins and ulbactins through our study. Continuous exploration of untapped marine bacteria will lead us to discover more examples of the "true producers."

References

1. Blunt JW, Copp BR, Keyzers RA, Munro MH, Prinsep MR (2016) Nat Prod Rep 33:382
2. Hu Y, Chen J, Hu G, Yu J, Zhu X, Lin Y, Chen S, Yuan J (2015) Mar Drugs 13:202
3. Pollock FJ, McMinds R, Smith S, Bourne DG, Willis BL, Medina M, Thurber RV, Zaneveld JR (2018) Nat Commun 9:4921
4. Rust M, Helfrich EJN, Freeman MF, Nanudorn P, Field CM, Rückert C, Kündig T, Page MJ, Webb VL, Kalinowski J, Sunagawa S, Piel J (2020) Proc Natl Acad Sci U S A 117:9508
5. Newman DJ, Cragg GM (2020) J Nat Prod 83:770
6. Okazaki H, Kishi T, Beppu T, Arima K (1975) J Antibiot 28:717
7. Itoh J, Omoto S, Nishizawa N, Kodama Y, Inouye S (1982) Agric Biol Chem 46:2659
8. Sato T, Nagai K, Suzuki K, Morioka M, Saito T, Nohara C, Susaki K, Takebayashi Y (1992) J Antibiot 45:1949
9. Canedo LM, Fernandez Puentes JL, Perez Baz J, Acebal C, de la Calle F, Garcia Gravalos D, Garcia de Quesada T (1997) J Antibiot 50:175
10. Shimojima Y, Hayashi H, Ooka T, Shibukawa M, Iitaka Y (1984) Tetrahedron 40:2519
11. Hashimoto M, Taguchi T, Nishida S, Ueno K, Koizumi K, Aburadab M, Ichinose K (2007) J Antibiot 60:752
12. Li Y, Xu Y, Liu L, Han Z, Lai PY, Guo X, Zhang X, Lin W, Qian P-Y (2012) Mar Drugs 10:319
13. Mclnerney BV, Taylor WC, Lacey MJ, Akhurst RJ, Gregson RP (1991) J Nat Prod 54:785
14. Azumi M, Ogawa K, Fujita T, Takeshita M, Yoshida R, Furumai T, Igarashi Y (2008) Tetrahedron 64:6420
15. Liu S-W, Jin J, Chen C, Liu J-M, Li J-Y, Wang F-F, Jiang Z-K, Hu J-H, Gao Z-X, Yao F, You X-F, Si S-Y, Sun C-H (2013) J Antibiot 66:281
16. Liu S, Han X, Jiang Z, Wu G, Hu X, You X, Jiang J, Zhang Y, Sun C (2016) J Antibiot 69:769
17. Wu G, Liu S, Wang T, Jiang Z, Lv K, Wang Y, Sun C (2018) Org Lett 20:3566
18. Wang T, Lu Q, Sun C, Lukianov D, Osterman IA, Sergiev PV, Dontsova OA, Hu X, You X, Liu S, Wu G (2020) Molecules 25:4446
19. Huang Y-F, Li L-H, Tian L, Qiao L, Hua H-M, Pei Y-H (2006) J Antibiot 59:355
20. Itoh J, Omoto S, Shomura T, Nishizawa N, Miyado S, Yuda Y, Shibata U, Inouye S (1981) J Antibiot 34:611
21. Tsukaguchi S, Enomoto M, Towada R, Ogura Y, Kuwahara S (2019) Eur J Org Chem 2019:6110. References for previous total syntheses of amicoumacins are cited therein
22. Enomoto M, Kuwahara S (2009) Angew Chem Int Ed Engl 48:1144
23. Li Y, Li Z, Yamanaka K, Xu Y, Zhang W, Vlamakis H, Kolter R, Moore BS, Qian PY (2015) Sci Rep 5:9383
24. Guerinot ML (1994) Annu Rev Microbiol 48:743
25. Zummundzhanov A, Bessonova IA, Abdullaev ND, Ogai DK (1988) Chem Nat Compd 23:461
26. Cox CD, Rinehart Jr KL, Moore ML, Carter Cook Jr J (1981) Proc Natl Acad Sci U S A 78:4256
27. Drechsel H, Stephan H, Lotz R, Haag H, Zähner H, Hantke K, Jung G (1995) Liebigs Ann 10:1727
28. Miethke M, Marahiel MA (2007) Microbiol Mol Biol Rev 71:413
29. Sasaki T, Igarashi Y, Saito N, Furumai T (2002) J Antibiot 55:249
30. Shindo K, Takenaka A, Noguchi T, Hayakawa Y, Seto H (1989) J Antibiot 42:1526
31. Lin Z, Antemano RR, Hughen RW, Tianero MDB, Peraud O, Haygood MG, Concepcion GP, Olivera BM, Light A, Schmidt EW (2010) J Nat Prod 73:1922

32. Shaaban KA, Shaaban M, Rahman H, Laatsch H, Shaaban M, Grun-Wollny I, Kampfer P, Kelter G, Fiebig H-H (2019) J Nat Prod 82:870
33. Kikuchi K, Chen CL, Adachi K, Nishishima M, Nishida F, Takadera T, Sano H (1998) Jpn Kokai Tokkyo Koho JP 10101657 A 19980421
34. Kikuchi K, Chen Y, Adachi K, Nishijima M, Nishida A, Takatera T, Sano H (1998) Jpn Kokai Tokkyo Koho JP 10245377 A 19980914
35. Igarashi Y, Asano D, Sawamura M, In Y, Ishida T, Imoto M (2016) Org Lett 18:1658
36. Shapiro JA, Morrison KR, Chodisetty SS, Musaev DG, Wuest WM (2018) Org Lett 20:5922
37. Crosa JH, Walsh CT (2002) Microbiol Mol Biol Rev 66:223
38. Inahashi Y, Zhou S, Bibb MJ, Lijiang S, Al-Bassam MM, Bibb MJ, Challis GL (2017) Chem Sci 8:2823
39. Biebl H, Pukall R, Lünsdorf H, Schulz S, Allgaier M, Tindall BJ, Wagner-Döbler I (2007) Int J Syst Evol Microbiol 57:1095
40. Camacho M, Redondo-Gómez S, Rodríguez-Llorente I, Rohde M, Spröer C, Schumann P, Klenk HP, Montero-Calasanz MDC (2016) Int J Syst Evol Microbiol 66:5173
41. Schleissner C, Cañedo LM, Rodríguez P, Crespo C, Zúñiga P, Peñalver A, de la Calle F, Cuevas C (2017) J Nat Prod 28:2170
42. Cardani C, Ghiringhelli D, Mondelli R, Quilico A (1965) Tetrahedron Lett 29:2537
43. Perry NB, Blunt JW, Munro MHG, Pannell LK (1988) J Am Chem Soc 110:4850
44. Sakemi S, Ichiba T, Kohmoto S, Saucy G, Higa T (1988) J Am Chem Soc 110:4851
45. Piel J (2002) Proc Natl Acad Sci U S A 99:14002
46. Piel J, Hui D, Wen G, Butzke D, Platzer M, Fusetani N, Matsunaga S (2004) Proc Natl Acad Sci U S A 101:16222
47. Kačar D, Schleissner C, Cañedo LM, Rodríguez P, de la Calle F, Galán B, García JL (2019) Front Microbiol 10:2561
48. Amiri Moghaddam J, Dávila-Céspedes A, Kehraus S, Crüsemann M, Köse M, Müller CE, König GM (2018) Mar Drugs 16:369
49. Raj Sharma A, Zhou T, Harunari E, Oku N, Trianto A, Igarashi Y (2019) J Antibiot 72:634
50. EZ BioCloud at https://www.ezbiocloud.net/
51. González JM, Mayer F, Moran MA, Hodson RE, Whitman WB (1997) Int J Syst Bacteriol 47:369
52. Yoon J-H, Jung S-Y, Kang S-J, Oh T-K (2007) Int J Syst Evol Microbiol 57:2365
53. Zhang DS, Huo Y-Y, Xu X-W, Wu Y-H, Wang C-S, Xu X-F, Wua M (2012) Int J Syst Evol Microbiol 62:505
54. Moh TH, Go Furusawa G, Al-Ashraf Amirul A (2017) Int J Syst Evol Microbiol 67:4089
55. Jayanetti DR, Braun DR, Barns KJ, Rajski SR, Bugni TS (2019) J Nat Prod 82:1930
56. Sharma AR, Harunari E, Zhou T, Trianto A, Igarashi Y (2019) Beilstein J Org Chem 2019 (15):2327
57. Karim MRU, Harunari E, Oku N, Akasaka K, Igarashi Y (2020) J Nat Prod 83:1295
58. Leutou AS, Yang I, Kang H, Seo EK, Nam S-J, Fenical W (2015) J Nat Prod 78:2846
59. Karim MRU, Harunari E, Raj Sharma A, Oku N, Akasaka K, Urabe D, Sibero MT, Igarashi Y (2020) Beilstein J Org Chem 16:2719
60. McBrien KD, Berry RL, Lowe SE, Neddermann KM, Bursuker I, Huang S, Klohr SE, Leet JE (1995) J Antibiot 48:1446
61. Oku N, Matoba S, Yamazaki YM, Shimasaki R, Miyanaga S, Igarashi Y (2014) J Nat Prod 77:2561
62. Hu J-F, Wunderlich D, Sattler I, Feng X-Z, Grabley S, Thiericke R (2000) Eur J Org Chem:3353
63. Igarashi Y, Shimasaki R, Miyanaga S, Oku N, Onaka H, Sakurai H, Saiki I, Kitani S, Nihira T, Wimonsiravude W, Panbangred W (2010) J Antibiot 63:563
64. Kitani S, Ueguchi T, Igarashi Y, Leetanasaksakul K, Thamchaipenet A, Nihira T (2017) J Antibiot 71:139
65. Yamazaki Y, Kunimoto S, Ikeda D (2007) Biol Pharm Bull 30:261

66. Takeuchi M, Ashihara E, Yamazaki Y, Kimura S, Nakagawa Y, Tanaka R, Yao H, Nagao R, Hayashi Y, Hirai H, Maekawa T (2011) Cancer Sci 102:591
67. Tsakos M, Clement LL, Schaffert ES, Olsen FN, Rupiani S, Djurhuus R, Yu W, Jacobsen KM, Villadsen NL, Poulsen TB (2016) Angew Chem Int Ed Engl 55:1030
68. Sang F, Ding Y, Wang J, Sun B, Sun J, Geng Y, Zhang Z, Ding K, Wu LL, Liu JW, Bai C, Yang G, Zhang Q, Li LY, Chen Y (2016) J Med Chem 59:1184
69. Villadsen NL, Jacobsen KM, Keiding UB, Weibel ET, Christiansen B, Vosegaard T, Bjerring M, Jensen F, Johannsen M, Tørring T, Poulsen TB (2017) Nat Chem 9:264
70. Chen J, Li J, Wu L, Geng Y, Yu J, Chong C, Wang M, Gao Y, Bai C, Ding Y, Chen Y, Zhang Q (2018) Eur J Med Chem 151:601
71. Carr G, Poulsen M, Klassen JL, Hou Y, Wyche TP, Bugni TS, Currie CR, Clardy J (2012) Org Lett 14:2822
72. Igarashi M, Shida T, Sasaki Y, Kinoshita N, Naganawa H, Hamada M, Takeuchi T (1999) J Antibiot 52:873
73. Nishioka H, Nakajima S, Nagashima M, Kojiri K, Suda H (1998) JP 10147594 A 19980602
74. Komaki H, Ishikawa A, Ichikawa N, Hosoyama A, Hamada M, Harunari E, Nihira T, Panbangred W, Igarashi Y (2016) Stand Genomic Sci 11:83
75. Komaki H, Ichikawa N, Hosoyama A, Hamada M, Harunari E, Ishikawa A, Igarashi Y (2016) Stand Genomic Sci 11:84
76. Schneemann I, Ohlendorf B, Zinecker H, Nagel K, Wiese J, Imhoff JF (2010) J Nat Prod 73:1444
77. Fu P, Liu P, Qu H, Wang Y, Chen D, Wang H, Li J, Zhu W (2011) J Nat Prod 74:2219
78. Lin Z, Torres JP, Ammon MA, Marett L, Teichert RW, Reilly CA, Kwan JC, Hughen RW, Flores M, Tianero MD, Peraud O, Cox JE, Light AR, Villaraza AJ, Haygood MG, Concepcion GP, Olivera BM, Schmidt EW (2013) Chem Biol 20:73
79. Kim MC, Kwon OW, Park JS, Kim SY, Kwon HC (2013) Chem Pharm Bull 61:511
80. Kim Y, Ogura H, Akasaka K, Oikawa T, Matsuura N, Imada C, Yasuda H, Igarashi Y (2014) Mar Drugs 12:4110
81. Zhang XM, Sun MW, Shi H, Lu CH (2017) Nat Prod Res 31:2245
82. Zou G, Liao XJ, Peng Q, Chen GD, Wei FY, Xu ZX, Zhao BX, Xu SH (2017) J Asian Nat Prod Res 19:1232
83. Wang JX, Sun CX, Shah M, Zhang GJ, Gu QQ, Zhu TJ, Che Q, Li DH (2020) J Asian Nat Prod Res 22:1031
84. Tsukimoto M, Nagaoka M, Shishido Y, Fujimoto J, Nishisaka F, Matsumoto S, Harunari E, Imada C, Matsuzaki T (2011) J Nat Prod 74:2329
85. McDonald LA, Capson TL, Krishnamurthy G, Ding W-D, Ellestad GA, Bernan VS, Maiese WM, Lassota P, Discafani C, Kramer RA, Ireland CM (1996) J Am Chem Soc 118:10898
86. Oku N, Matsunaga S, Fusetani N (2003) J Am Chem Soc 125:2044
87. Waters AL, Peraud O, Kasanah N, Sims JW, Kothalawala N, Anderson MA, Abbas SH, Rao KV, Jupally VR, Kelly M, Dass A, Hill RT, Hamann MT (2014) Front Mar Sci 1:54

Top Heterocycl Chem (2021) 58: 175–208
https://doi.org/10.1007/7081_2021_53

Published online: 14 February 2021

Siderophores from Fish Pathogenic Bacteria

Carlos Jiménez

Contents

Abstract Aquaculture is an important food source worldwide but one of its main problems is the spread of infectious diseases caused by Gram-negative pathogenic bacteria that results in considerable economic losses. Since siderophores, low molecular weight organic compounds involved in the iron-uptake mechanisms of the bacteria, are critical for the growth and virulence of the producer pathogens, the bacterial iron acquisition systems are promising targets for the design of new antimicrobial strategies. In this chapter, the current chemical knowledge of the siderophores involved in the iron-uptake mechanisms of Gram-negative pathogenic bacteria *Vibrio anguillarum*, *Photobacterium damselae* subsp. *piscicida*, and *Aeromonas salmonicida* subsp. *salmonicida,* responsible for the main fish infectious diseases Vibriosis, photobacteriosis, and furunculosis, respectively, is summarized. The isolation, structural elucidation, and the chemical synthesis of the siderophores biosynthesized from those bacteria are displayed. Their involvement in the iron-uptake mechanisms, virulence importance, and the application in the development of new strategies to fight against the infectious diseases will be also discussed.

C. Jiménez (✉)
Departmento de Química, Facultade de Ciencias e Centro de Investigacións Científicas Avanzadas (CICA), AE CICA-INIBIC, Universidade da Coruña, A Coruña, Spain
e-mail: carlos.jimenez@udc.es

Keywords *Aeromonas* · Fish diseases · Iron acquisition · Pathogenic bacteria · *Photobacterium* · Siderophore synthesis · Siderophores · *Vibrio*

1 Introduction

The need for a safe, reliable, and economic supply of food to feed the high increase of human population is transforming aquaculture as an important food source worldwide. According to the Food and Agriculture Organization (FAO) of the United Nations, global fish consumption (from wild fishing and aquaculture) grew at an average of 3.1% yearly from 1961 to 2017, which is a rate almost twice that of the 1.6% annual world population increase. While wild fishing production has stabilized in the last 20 years, global aquaculture production increased more than five times in the last three decades (from 14.9 million tonnes in the period 1986/1995 to 82.1 million tonnes in 2018). Right now, more than 50% of fish for human consumption comes from aquaculture [1].

However, the production intensification increases the likelihood of pathogen outbreaks and the emergence of several infectious diseases which are causing severe economic losses in aquaculture (Fig. 1a) [2]. The extensive use of antibiotics has contributed to the spread of antimicrobial resistance (AMR), which has become a global health emergency. Moreover, fish are reservoirs of zoonotic pathogens such as *Aeromonas hydrophila* and *Photobacterium damselae* that can infect not only the animal host but also humans who are in contact with the aquaculture facility and via foodborne infections [3]. For all these reasons, the search for new strategies to prevent and control diseases in aquatic species is urgently needed. It is one of the main research objectives within the field of health in the next decade.

For most bacteria, iron uptake ability during the naturally iron-limited conditions of an infection is a key virulence factor essential for multiplication within the host. Iron is a crucial nutrient in virtually all living organisms because it is needed in many metabolic processes, catalyzing a wide variety of indispensable enzymatic reactions. The wide range of properties of iron and its abundance, being the fourth most abundant element in the Earth's crust, could explain why early microorganisms, from the beginning of evolution, used this element to develop their metabolic processes in an oxygen-poor atmosphere. However, when the atmosphere became rich in oxygen, ferrous iron was oxidized to ferric ion which precipitates to form ferric hydroxide which is highly insoluble at physiological pH. Since then, the bioavailability of free ion became very low and the iron acquisition uptake mechanisms to combat the extremely low iron availability became an indispensable process for most living organisms [4]. Now, a high competition for this scarce iron is taking place and so, iron sequestration systems have been identified as important virulence mechanisms in several pathogenic bacteria. In fact, potential hosts keep the level of free iron inside the body to a minimum as a defense mechanism.

One of the major strategies used by bacteria and fungi entailed the use of siderophores, a Greek term that means iron carriers. These are low molecular weight organic compounds that have extremely high affinity for ferric iron [5]. Thus, bacteria and fungi excrete "endogenous" siderophores into the culture medium

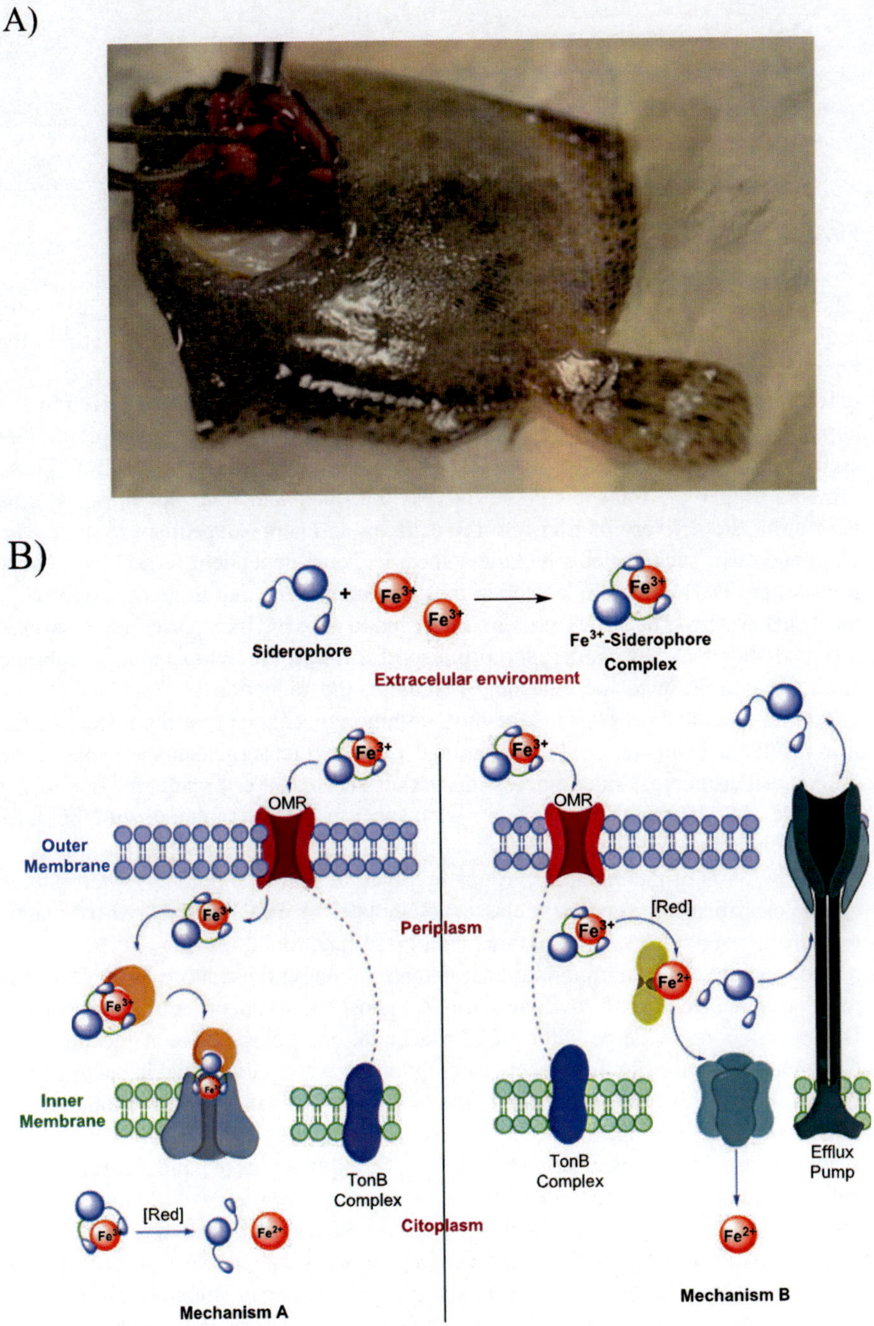

Fig. 1 (**a**) Turbot infected with photobacterioisis. (**b**) General iron uptake mechanisms in Gram-negative bacteria

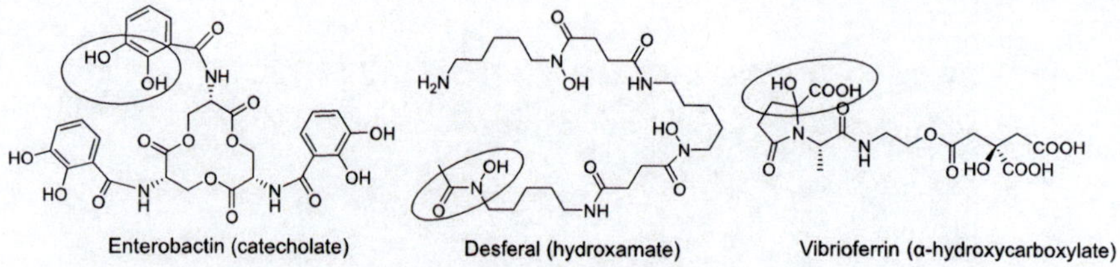

Fig. 2 Examples of the main types of microbial siderophores

where they are able to solubilize Fe (III) from its hydroxide or they can efficiently sequester the iron bound by transferrins and other iron-holding proteins within the host. Once siderophores are synthesized and secreted outside the bacterial cells, they bind Fe^{3+} ions to form a ferric-siderophore complex which is then internalized by the appropriate transport mechanism. In Gram-negative bacteria, ferric-siderophore complexes are recognized by specific outer membrane receptors (OMR). These proteins trigger the transport system across the outer and inner membranes, thus permitting the delivery of iron into the cells by an energy-dependent system. The components of such systems include a specific TonB-dependent ferric-siderophore transporter (TBDT) protein located in the outer membrane that is energized through the TonB system. The TBDT protein is coupled to an ABC transporter that catalyzes the final stages of ferric-siderophore transport through the cytoplasmic membrane from the periplasm to the cytosol [6]. Finally, the reduction of Fe^{3+} to Fe^{2+} by microbial-mediated processes is the most common mechanism for the release of the iron inside the cell (Fig. 1b) [7]. Additionally, some microorganisms are able to use the xenosiderophores, siderophores biosynthesized by other organisms, due to the presence of OMR that can recognize a wide spectrum of siderophores with the same type of chelating units.

On the basis of the bidentate groups involved in iron (III) binding, siderophores can be classified into four main classes: phenolates or catecholates, hydroxamates, α-hydroxycarboxylates, and mixtures with these functionalities (Fig. 2) [5].

Iron restriction is an important host defense strategy; thus, successful pathogens must possess mechanisms to acquire iron from host sources in order to cause disease. During infection, a fierce battle of iron acquisition occurs between the host and bacterial pathogens. For that reason, siderophores are not only crucial in bacteria iron metabolism but also are key factors in the virulence to enhance the infection disease. In this way, the iron-uptake mechanism mediated by siderophores along with their structures provides an ideal target for therapeutic applications [4]. Indeed, the natural siderophore "desferal," produced by *Streptomyces pilosus*, is used in clinics for the removal of excess iron in human blood, a disease known as thalassemia [8]. But one of the most promising applications of siderophores is for the development of new antibacterial treatments based on iron acquisition, such as novel antimicrobials and vaccines. Thus, iron-uptake mechanisms mediated by siderophores are providing the basis for the development of new antibacterial treatments: by the Trojan horse Strategy where the siderophore is coupled to a known antibiotic in order to overcome the permeability barrier of the Gram-negative outer membrane that uses a specific siderophore route of entry into the cell. The natural siderophore-antibiotic conjugates (sideromycins), such as albomycins and salmycins, inspired this methodology [9]. This approach seeks to avoid microbial resistance, thus enhancing sensitivity and specificity. In fact, the catechol based-siderophore cefiderocol (Fetcroja®) is the first siderophore-drug conjugate approved by the U.S. Federal Drug Administration (FDA) [10] and the European Medicines Agency (EMA) [11]. Other applications imply the inhibition of biosynthesis of siderophore and the development of vaccines [4].

Vibriosis, photobacteriosis, and furunculosis, caused by Gram-negative marine bacteria, constitute the main infectious diseases in aquaculture [12]. In the present chapter, studies of the siderophores from the fish Gram-negative pathogenic bacteria involved in those infectious diseases and its application in fish farming will be presented. Special attention will be paid to the isolation methodology used, the synthesis of natural siderophores and their derivatives, the establishment of some Structure Activity Relationship (SAR), the biosynthetic proposals and the OMR involved.

2 Vibriosis by *Vibrio anguillarum*

Vibriosis is a group of diseases that cause a serious hemorrhagic septicemia associated with both wild and farmed marine fishes. This leads to severe economic losses in the aquaculture industry worldwide. Different Gram-negative bacterial species, mainly classified into the *Vibrio* genus, belonging to the family of Vibrionaceae are responsible for this infection [13]. The bacillus *V. anguillarum* is considered the etiological agent of classical vibriosis, being the most common and also the most extensively studied Vibrionaceae species. Its virulence in marine settings was reported as early as 1718 along the coastline of continental Europe and was referenced on record as the red disease in eel (*Anguilla anguilla*) in 1893, serving as the root of the name *anguillarum*. By 1970, it was known to have devastating effects on several marine organisms (cod, eel, finnock, oyster, salmon, trout, etc.) and now has a wide geographical distribution and host range. *V. anguillarum* is pathogenic to a variety of crustaceans, bivalves and mainly to warm- and cold-water fish of economic importance, including Pacific and Atlantic salmon, rainbow trout, turbot, sea brass, sea bream, among others, accounting for more than 90 susceptible aquatic species in at least 28 different countries [14]. A total of 23 different O serotypes of *V. anguillarum* isolates have been recognized. Among them, O1 and O2 are the main ones implicated in the infections, although serotype O3, and to a lesser extent O4 and O5, have increasing importance in vibriosis outbreaks in specific geographic areas [15]. The remaining serotypes are considered to be environmental strains and only on rare occasions are isolated as responsible for vibriosis in fish [16].

For a long time, just two different siderophore-mediated systems were described in *V. anguillarum* strains belonging to serotypes O1 and O2, the anguibactin (Ang, **1**) and the vanchrobactin (Vcb, **2**). More recently, a third one was reported,

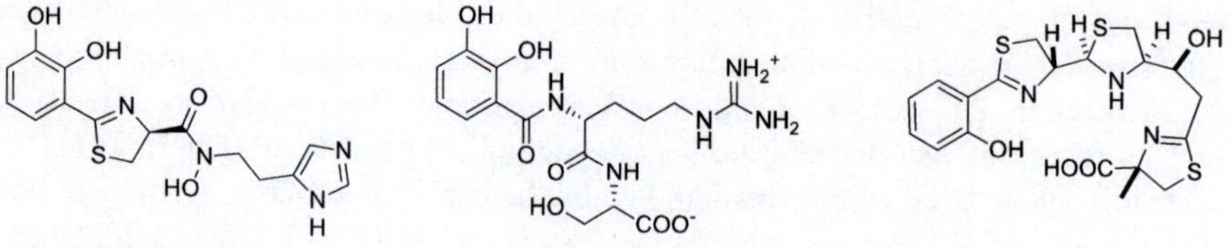

Anguibactin (Ang, **1**) Vanchrobactin (Vcb, **2**) Piscibactin (Pcb, **3**)

Fig. 3 Structure of the siderophores biosynthetized by *V. anguillarum*

piscibactin (Pcb, **3**), which turned out to be one of the most important virulence factors for this bacterium (Fig. 3). Additionally, *V. anguillarum* can also use xenosiderophores as iron sources, like enterobactin, ferrichrome or rhodotorulic acid.

2.1 *Anguibactin Iron-Uptake System*

The first siderophore-mediated system described in this bacterium, called anguibactin system, was found in pathogenic strains of serotype O1, which bears the 65 kb pJM1 plasmid that harbors the genes for the biosynthetic machinery and utilization of the siderophore anguibactin (Ang, **1**).

Ang (**1**) was purified from its supernatants of iron-deficient cultures by adsorption onto an XAD-7 resin and subsequent gel filtration on a Sephadex LH-20 column and reported in 1986 [17]. Three years later, its hybrid structure between a catechol and a hydroxamate type of siderophore was determined by ^{1}H- and ^{13}C-NMR spectral analysis of its Ga(III)-complex, FAB-MS, chemical degradation, and single-crystal X-ray diffraction studies of its anhydro derivative, anhydroanguibactin (**4**) [18]. The crystal structure of the Ga(III)-complex of anguibactin was reported in 1998 [19]. However, the absolute configuration of the cysteine moiety in Ang (**1**) was not determined at that time.

The structure of Ang (**1**) is similar to that of preacinetobactin (PreAcb, **5**), which contains instead an oxazoline ring. PreAcb (**5**) is the precursor of acinetobactin (Acb, **6**), the siderophore produced by the Gram-negative pathogen *Acinetobacter baumannii,* which constitutes one of the major health concerns because it is responsible for a large number of hospital-acquired and nosocomial infections and by its increasing development of antimicrobial resistance (Fig. 4) [20].

Since the content of siderophores from marine bacteria is very low and they have potentially valuable applications, the development of chemical synthesis is a crucial step to further study their biological activities, mechanisms, absolute configurations, and/or structure-activity relationships. Furthermore, the synthesis of analogues gives us very important information, for instance, for the preparation of conjugates to be used as biosensors to study these mechanisms [21], for the development of novel therapies against iron-overload related disease [8], for the preparation of new

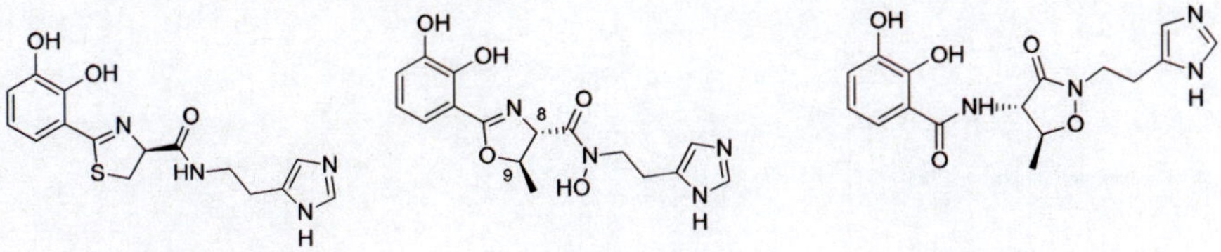

Anhydroanguibactin (**4**) Pre-acinetobactin (**5**) Acinetobactin (Acb, **6**)

Fig. 4 Similar structures to that of Ang (**1**)

antimicrobial agents that target bacterial pathways for the acquisition of iron [22], for a Trojan horse strategy to develop modified molecules with antimicrobial activity [9], or for the synthesis of bioconjugates of siderophores to be tested as antibacterial vaccines [23].

The total synthesis of Ang (**1**), published by H. J. Kim and co-workers in 2018, started with the preparation of N^{α}-benzyloxyhistamine (**9**) which was obtained from histamine dihydrochloride in three steps following the same procedure employed in the synthesis of Acb (**6**) [24]. The reaction of histamine dihydrochloride with $NaNO_2$ in acid media followed by treatment with $SOCl_2$ gave **7**. Coupling of **7** with *N*-*tert*-butoxycarbonyl(Boc)-*O*-benzyloxyamine afforded imidazole **9** after deprotection with TFA. In parallel, the thiazoline fragment **12** was obtained by the coupling of aryl nitrile **10**, prepared from *o*-vanillin, with L-cysteine.

Condensation of thiazoline acid **12** with N^{α}-benzyloxyhistamine (**9**) was one of the key steps of the synthesis in order to avoid the epimerization observed at position 8, because the coupling gives a near-racemic mixture. The lack of a substituent at position 9 in Ang (**1**), in relation to that in Pre-acb (**5**), seems to justify that racemization. The presence of an (*R*)-methyl group at C-9 in Pre-acb (**5**) could secure the sterically stable *trans* geometry with the (*S*)-configuration at C-8. The use of **H**exafluorophosphate **A**zabenzotriazole **T**etramethyl **U**ronium (HATU) in dichloromethane with a free amine form of **9** resolved this problem, giving the intermediate **14** with a stereoselective ratio of er = 98:2. The problematic removal of the two ether-type protecting groups present in **14** using boron trichloride (BCl_3) in the final step was solved by the addition of tetrabutylammonium iodide (TBAI), which allowed a clean removal of both benzyl and methyl groups. Using a similar procedure, the anguibactin analogue, 3-deoxyanguibactin (**16**), was also prepared (Scheme 1). The preparation of synthetic siderophore **1** made it possible to determine, by fluorescence titrations, a 2:1 binding stoichiometry between Ang (**1**) and Fe (III) in liquid phase. Furthermore, the iron delivery capability of Ang (**1**) and its thermal stability over PreAcb (**5**) led the authors to propose Ang (**1**) as a competent surrogate siderophore for *Acinetobacter baumannii* that can be useful for the future development of a siderophore-based antibiotic delivery system against that pathogen [25].

Evaluation of cytotoxic effects of Ang (**1**) on P388 murine leukemia cell lines displays an IC_{50} value of <15 μM [26]. Moreover, Ang (**1**) has been patented for performing deferration therapy due to its properties for removing ferric iron from

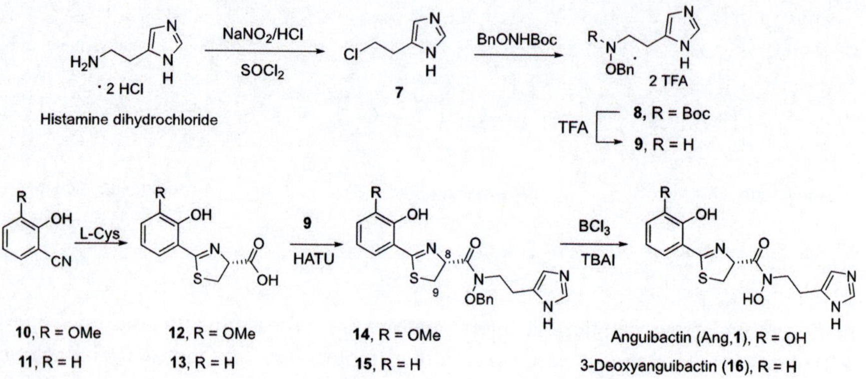

Scheme 1 Synthesis of Ang (**1**) and its analogue 3-deoxyanguibactin (**15**)

aqueous liquids. Among other properties, it was found that Ang (**1**) inhibits iron uptake by living cells, wrests iron from vertebrate tissues, removes iron from other siderophores and ferric hydroxide, and removes ferric iron from aqueous solutions, including cell-culture media [27].

Most biosynthetic genes of Ang (**1**) are located in the pJM1 plasmid and it is accomplished by a complex of non-ribosomal peptide synthetases (NRPS). However, its biosynthesis requires additional chromosomally encoded enzymes that are also involved in the production of Vcb (**2**) as the second siderophore described. The pJM1 plasmid series is present in most *V. anguillarum* O1 strains but is absent in serotype O2. The ferric iron-anguibactin complex is transported by the FatABCD proteins, all encoded by the pJM1 plasmid: the OM receptor FatA, the periplasmic binding protein FatB, and the inner membrane proteins FatC and FatD (ABC transporter) [28].

2.2 *Vanchrobactin Iron-Uptake System*

The second siderophore-mediated iron-uptake system reported in *V. anguillarum* was the vanchrobactin system. Vanchrobactin (Vcb, **2**) is encoded by a gene cluster located in the chromosome that was first found in all serotype O2 strains and some plasmid-less serotype O1 strains. The terms of *V. anguillarum* ("van"), chromosome ("chro") and the suffix "bactin" for chelating agent, were used to name this siderophore as vanchrobactin.

This compound was isolated for the first time from iron-deficient cultures of *V. anguillarum* serotype O2 strain RV22 by using a XAD-7 lipophilic resin, liquid–liquid fractionation, Sephadex LH-20, and RP-HPLC. Its planar structure was established by analysis of its 1D and 2D NMR, IR, UV, and mass spectral data as a dipeptide made up of the amino acids arginine and serine linked to a 2,3-dihydroxybenzoyl moiety [29]. A faster and a more convenient isolation procedure using Hydrophilic-Lipophilic Balance (HLB) resins followed by HPLC-HRMS analysis was developed and applied in the re-isolation of Vcb (**2**). This method, named as SPE-HLB/HPLC-HRMS, improved not only the efficiency of the isolation methodology but also the isolation time in relation to the existing methods [30].

The absolute configuration of Vcb (**2**) was established by synthesis of three out of four possible stereoisomers using 2,3-dihydroxybenzoic acid (DHBA), L/D-ornithine, and L/D-serine as starting materials and comparison to the natural product. Thus, 2,3-dibenzyloxybenzoic acid (**17**), prepared in three steps from DHBA, was coupled with N^{δ}-Cbz-D-ornithine-OMe (**16a**), obtained in four steps from D-ornithine, or with N^{δ}-Cbz-L-ornithine-OMe (**16b**), prepared from esterification of the commercial N^{δ}-Cbz- L-ornithine, to give acids **18a,b**, respectively, after saponification. Then, *tert*-butyl esters of L/D-serine, synthetized from commercial amino acids, were coupled to acids **18a,b** to afford **19a-c**. Deprotection of the benzyl groups in **19a-c**, followed by introduction of the guanidine functionality and deprotection of the *tert*-butyl group in acid media, yielded three out of the four possible stereoisomers of vanchrobactin. Comparison of the synthetic compounds with the natural sample by achiral HPLC, chiral capillary electrophoresis, along with NMR data, sign and value of the optical rotation allowed to establish the absolute structure of Vcb (**2**) as *N*-[*N'*-(2,3-dihydroxybenzoyl)-D-arginyl]-L-serine (Scheme 2) [31].

Two series of vanchrobactin analogues were synthetized using a similar strategy employed in the total synthesis of Vcb (**2**). The first one where the 2,3-dihydroxybenzoyl moiety is linked to Arg or Orn residues, compounds **20–24**, (Scheme 2) and the second series where it was the serine residue which was linked directly to the aromatic moiety, compounds **27–29** (Scheme 3).

The synthesized compounds were subjected to the chrome azurol-S (CAS) test and evaluated by growth promotion assay (siderophore activity test) with several bacteria that are well known for their ability to transport and use natural siderophores, specified in brackets: *V. anguillarum* RV22 (serotype O2) (Vcb (**2**) and Pcb (**3**)), *V. anguillarum* 775 (serotype O1) (Ang (**1**)), *Salmonella enterica* enb-1 (enterobactin), *Erwinia chrysanthemi* PPV20 (chrysobactin (**34**)), and *V. alginolyticus* TA15 (vibrioferrin). Most of analogues were positive in the CAS test and showed siderophore activity to both *V. anguillarum* serotypes tested. Although most of them displayed siderophore activity to *S. enterica* and *E. chrysanthemi*, none of them were able to restore the growth of *V. alginolyticus*. These results are in agreement with the catecholate structure of enterobactin and chrysobactin (**34**) and the non catecholate structure of vibrioferrin, an α-hydroxycarboxylate siderophore (Fig. 1). From the results, it was suggested that the aromatic ring in catecholate siderophores is crucial for the binding of the outer membrane receptors. Additionally, the results reveal the lack of stereoisomeric influence of the amino acid scaffold on the siderophore activity in both serotypes of *V. anguillarum*, although some differences can be observed in *E. chrysanthemi* PPV20 and *S. enterica*. This lack of specificity could be an evolutionary adaptation of bacterial pathogens to increase the number of possibilities to obtain iron from the

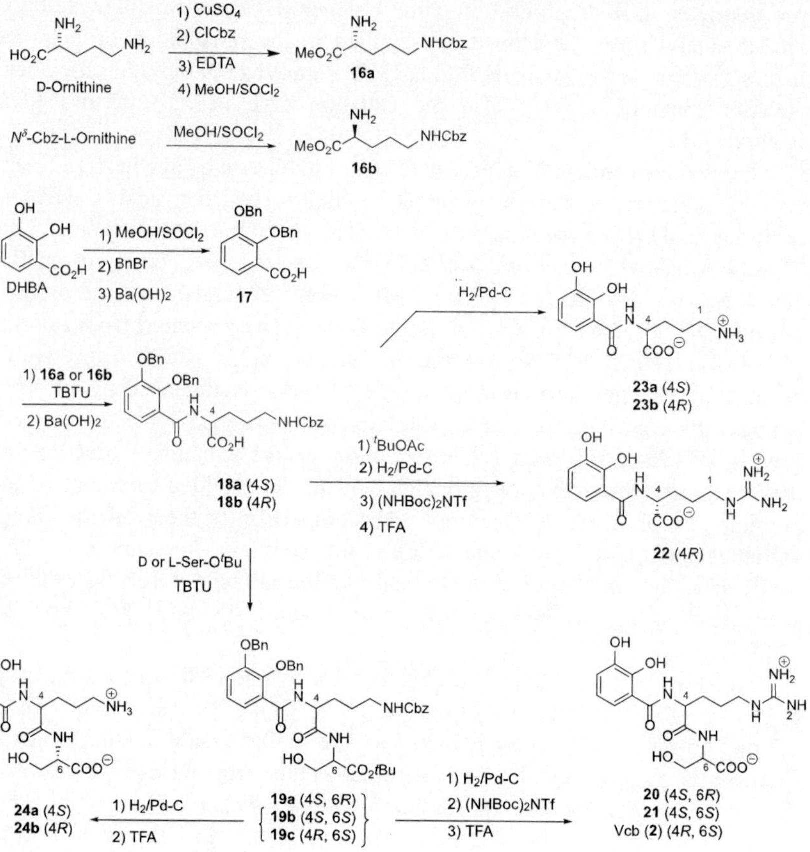

Scheme 2 Synthesis of Vcb (**2**) and its analogues DHBA-Arg/Orn **20–24**

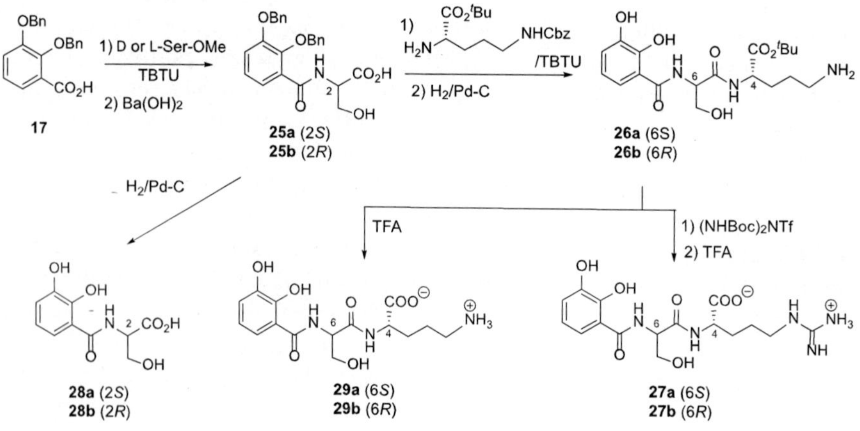

Scheme 3 Synthesis of Vcb analogues DHBA-Ser **27–29**

environment. The low specificity of molecular recognition of the recipients of these strains could facilitate the use of those conjugates to be prepared. Analogues having ornithine (**23–24**) instead of arginine (**20–22**) residue maintain the siderophore activity to *V. anguillarum*, as the Vb analogue DHBA- D-Orn- L-Ser (**24b**), and they are of special interest since these molecules possess an appropriate functionality, an amino group, that can be used as an anchor group to attach it to other bioactive agents. For these reasons, these compounds were considered as potential antibiotic vectors for the delivery of known antibacterial agents via the bacterial iron-uptake systems [32].

Vcb (**2**) is structurally related to chrysobactin (**30**), isolated from the phytopathogenic bacterium *E. chrysanthemi*, which bears a lysine residue instead of arginine. This compound also has the L configuration for the serine residue, while the lysine residue has a D configuration as the arginine in Vcb (**2**) [33]. From an identified species of *Vibrio*, DS40M4 strain, a bis- and triscatechol amide analogues of Vcb (**2**), named divanchrobactin (**31**) and trivanchrobactin (**32**), were isolated. Vcb (**2**) and Ang (**1**) were also isolated from this bacterium and represented the first case that the two siderophores were isolated from the same strain. Their isolation was carried out using Amberlite XAD-2 resins and HPLC. The isolation of **31** and **32** raised the question whether linear trivanchrobactin is the real siderophore or comes from a hypothetical cyclic ester, a cyclic trivanchrobactin product, which was not detected or isolated due to the instability of the cyclic triserine ester or the reactivity of a possible esterase [34]. Curiously, cyclic trichrysobactin, linear trichrysobactin (**33**), and dichrysobactin (**34**) along with the monomeric siderophore unit, chrysobactin (**30**) were isolated from plant pathogen *Dickeya chrysanthemi* EC16 (formerly known as *Pectobacterium chrysanthemi* EC16 and *E. chrysanthemi* EC16) (Fig. 5) [26].

On the basis of the results of the siderophore activity evaluation of Vcb analogues, three conjugates **35–37** between the known antibiotic norfloxacin and Vcb analogues **23b** and **24b** were synthesized using acetate as a spacer arm. A four conjugate **38** was also prepared between norfloxacin and aminochelin (**39**), one of the siderophores produced by the Gram-negative free-living soil bacterium *Azotobacter vinelandii* [35]. Although aminochelin (**39**) displayed siderophore activity in *V. anguillarum* serotype O2, in contrast to the other Vcb two analogues, aminochelin does not use the OMR of Vcb (**2**) FvtA as the route of entry.

The coupling of 2,3-diisopropyloxybenzoic acid (**40**), obtained in a similar way as **9**, with *tert*-butyl(4-aminobutyl)carbamate gave carbamate **41**. The free amine obtained from the deprotection of the Boc group in **41** was coupled with chloroacetyl chloride and then with norfloxacin to give the aminochelin–norfloxacin conjugate **38** after the removal of isopropyl protecting groups with BCl_3. For the synthesis of DHBA-D-ornithine-norfloxacin conjugates **36** and **37**, N^{δ}-Cbz-D-ornithine-O^{t}Bu, prepared from the commercially available N^{δ}-Cbz-D-ornithine, was coupled with 2,3-diisopropyloxybenzoic acid (**40**) to give carbamate **42**. Removal of the Cbz group in **42** by catalytic hydrogenation gave a free amine which was submitted to the same procedure as before to afford an intermediate that yielded conjugates **36** and **37**, after removal of isopropyl protecting groups with BCl_3 in MeOH or H_2O,

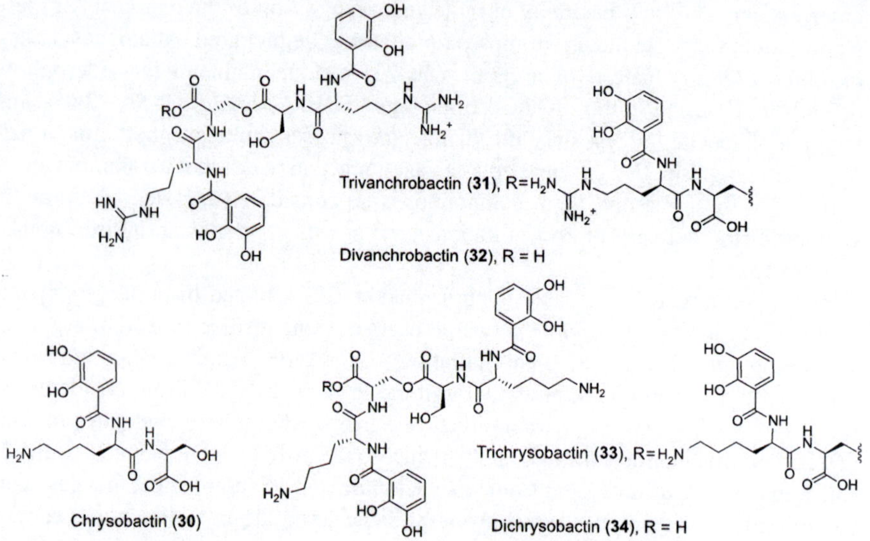

Fig. 5 Siderophores with similar structures to that of Vcb (**2**)

respectively. For the synthesis of DHBA-D-ornithine-norfloxacin conjugate **35**, N^{δ}-Cbz-D-ornithine-OMe (**16a**, Scheme 2) was coupled first with 2,3-diisopropyloxybenzoic acid (**40**) and then, after saponification, with O-*tert*-butyl-L-serine *tert*-butyl ester to give dipeptide **43**. Finally, the submission of the protected dipeptide **43** to the same sequence of reactions employed in the preparation of conjugate **38** from **41** gave conjugate **35** (Scheme 4). The four synthetic conjugates **35–38** gave a positive reaction in the CAS test indicating that the binding of the Vcb analogs to norfloxacin through the amino functionality did not affect the chelating ability of these compounds. However, the lower antibiotic activity displayed by the four conjugates **35–38** in relation to unconjugated norfloxacin indicates that they are not working as Trojan horse conjugates [36].

Vanchrobactin iron complexes studies using spectrophotometric and potentiometric methods were reported. Kinetic studies at different pHs showed that of free Vcb (**2**) is stable in the pH range from 4 to 10.4 while the presence of a single pK with a value of 6.79 was obtained by spectroscopic titration studies. The influence of pH on iron vanchrobactin complex formation, shown in its UV visible spectra, reflects the presence of three sequential complexation reactions, corresponding to the formation of the ferric mono-, bis-, and tris(Vcb) complexes. The color change from alkaline to acid media was a clear indication of a structural change under pH variation. The resulting potentiometric titration curve of ferric Vcb complexes indicates the stepwise formation of the ferric mono-, bis-, and tris(Vcb) complexes. A value of pFe(Vcb) of 20 (for enterobactin pFe for is 35.6) was obtained for Vcb (**2**) by modeling the Fe^{3+} uptake under physiological conditions [37].

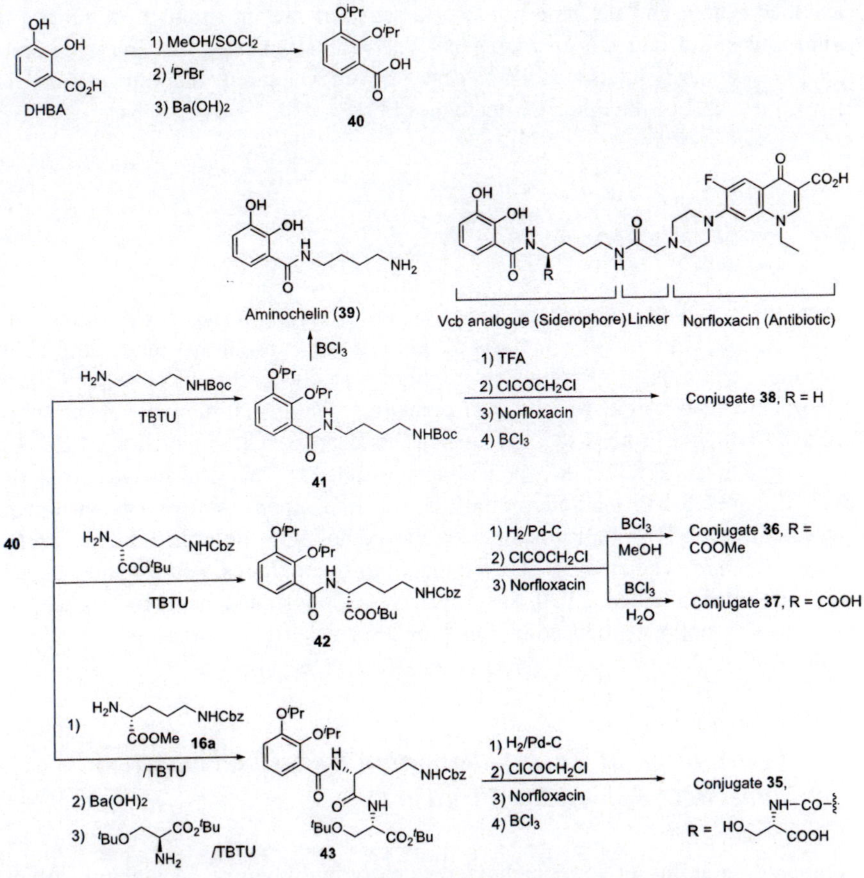

Scheme 4 Synthesis of the conjugates **35–38** between norfloxacin and Vcb analogues and aminochelin (**39**)

The biosynthesis and transport of Vcb (**2**) in strain RV22 is encoded by the *vab* gene cluster, a c. 26 kb chromosomal region, containing 13 genes. The *vabABCEF* genes are the main biosynthetic components which encode the enzymes necessary for DHBA biosynthesis and activation, and *vab*F, a gene encoding an NRPS involved in the assembly of the siderophore components [38]. The biosynthetic pathway proposal was very helpful in elucidating its structure. Indeed, the first proposed structure, 2,3-dihydroxybenzoyl- serinyl -arginine, was discarded for the corrected one, [(2,3-dihydroxybenzoyl)arginyl]serine, where the position of serine and arginine was interchanged because the domain responsible for incorporating the arginine residue is located first [29].

Vanchrobactin transport to the cell is achieved by means of the TonB-dependent outer membrane receptor (OMR) FvtA, a 78 kDa protein, encoded by the *fvt*A gene linked to the biosynthetic genes. The synthesis of Vcb (**2**) along with their analogues described before and the preparation of a series of mutant strains were crucial in confirming the actual role of *fvt*A gene. The results indicated that FvtA seems to recognize primarily the catechol-iron center since it is not only the route of entry for Vcb (**2**) but also of the other Vb analogues [39].

2.3 Piscibactin Iron-Uptake System

When the complete genome sequence of the *V. anguillarum* RV22 strain was described [40], the in silico analysis of this genome showed that the second chromosome II contains an additional siderophore gene cluster that encodes Pcb (**3**), the siderophore responsible for the iron uptake of *Photobacterium damselae* subsp *piscicida* which will be discussed later in this chapter. The production of Pcb (**3**) in *V. anguillarum* RV22 wild type was demonstrated by chemical analysis using the SPE-HLB/HPLC-MS methodology of the cell-free supernatants of this bacterium, detecting its Ga (III) complex using the same methodology described in *P. damselae* subsp. *piscicida*. The analysis showed that both siderophores, Pcb (**3**) and Vcb (**2**), are produced simultaneously. In this way, it was demonstrated that Pcb (**3**) is actually the third siderophore produced by this *Vibrio* species [41].

2.4 Distribution of Each Siderophore Iron-Uptake System and Their Role in the Virulence

Although the ability of some pathogens to carry multiple iron acquisition systems seems redundant in laboratory culture conditions, the co-expression of siderophores with different chemical properties can play specialized roles at the host-pathogen interface. Indeed, the ability to produce more than one siderophore can enhance niche flexibility and pathogenesis of the bacteria.

The vanchrobactin biosynthesis and transport gene clusters (*vabB* and *fvtA*) are ubiquitous in both vanchrobactin- and anguibactin-producing *V. anguillarum* strains. However, anguibactin strains do not produce Vcb (**2**) because the *vab*F gene, which encodes the NRPS that assembles the siderophore molecule, is inactivated by the insertion of the transposable element encoded on the plasmid pJM1. On the other hand, piscibactin genes (*irp*-HPI) are not only widespread in *V. anguillarum* strains but also many other species of *Vibrio* [42]. Moreover, the piscibactin biosynthesis gene cluster is never present in anguibactin-producing strains. Most pathogenic *V. anguillarum* strains lacking the anguibactin system produce simultaneously both siderophores, Vcb (**2**) and Pcb (**3**).

It is assumed that the vanchrobactin mediated system is the ancestral iron-uptake mechanism mediated by siderophores of *V. anguillarum*. The anguibactin-mediated system could have been acquired later by *V. anguillarum*, likely by horizontal transfer, during the evolution of this species, leading to the inactivation of the vanchrobactin system, since anguibactin-mediated iron transport seems to be more efficient due to the Ang (**1**) stronger affinity for iron. The Highly Pathogenically Island encoding piscibactin mediated system (*irp*-HPI) could also have been acquired by horizontal transfer [16].

On the other hand, *V. anguillarum* is able to modulate the synthesis of Vcb (**2**) and Pcb (**3**) according to the surrounding temperature. Expression of piscibactin biosynthetic genes is three times higher at 18 °C than at 25 °C. Thus, Pcb (**3**) is preferentially produced at low temperatures, which are the temperature that *V. anguillarum* finds during infection of fish cultivated cold-water temperatures. In contrast, when *V. anguillarum* grows at warm-water temperatures (approx. 25 °C), piscibactin genes are down-regulated (fourfold lower) in relation to Vcb (**2**) [41].

It has been demonstrated that the ability to produce Ang (**1**) is a prerequisite for the virulence of *V. anguillarum* serotype O1 plasmid bearing strains and so, Ang (**1**) is a key virulence factor for those strains. In relation to the efficiency or virulence of Vcb (**2**) versus Ang (**1**), it was found that when an anguibactin-producing *V. anguillarum* strain co-infects a fish with a vanchrobactin producing strain, the former is the only one recovered from diseased fish in the first days of infection [16].

On the other hand, Pcb (**3**) contributes to a greater extent than Vcb (**2**) to virulence in fish and also, the ability to produce Pcb (**3**) seems to be sufficient to confer maximum virulence to *V. anguillarum*. In fact, piscibactin genes are associated with a highly virulent phenotype for several fish species. Pcb (**3**) plays a relevant role in the cell fitness and contributes to a greater extent than Vcb (**2**) to the virulence for fish [41]. Experimental infections in turbot showed that although both siderophores are simultaneously produced, Pcb (**3**) is a key virulence factor to infect fish, while Vcb (**2**) seems to have a secondary role in virulence. A recent analysis of 44 *V. anguillarum* genomes confirmed this hypothesis [15]. Thus, it is likely that Vcb (**2**) biosynthesis could be more related to persistence in a marine environment than to pathogenesis [43].

3 Photobacteriosis by *Photobacterium damselae* subsp. *piscicida*

Photobacteriosis, formerly known as fish pasteurellosis, is a septicemia caused by the Gram-negative pathogenic halophilic bacterium *Photobacterium damselae* subsp. *piscicida* (Vibrionaceae). The disease was originally named after the classification of this aforesaid agent as *Pasteurella piscicida*. It acquired its present name following the reassignment to the genus *Photobacterium*. The presence of white nodules in the internal viscera, particularly, spleen and kidney in the fish is the reason why it is also known as pseudotuberculosis. Since 1969, photobacteriosis has been one of the most important diseases in Japan, affecting mainly to yellowtail (*Seriola quinqueradiata*), and from 1990 became the major pathological problem in

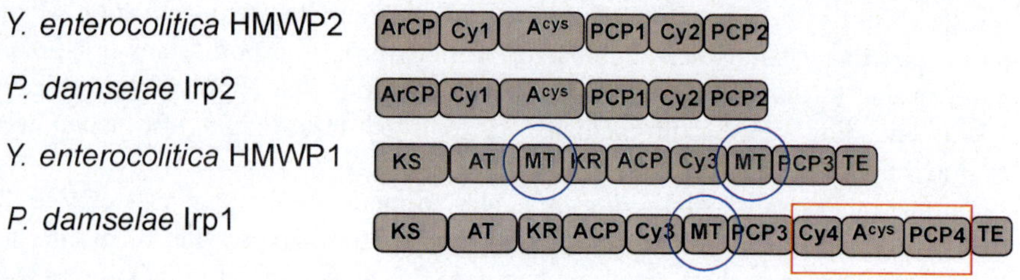

Fig. 6 Comparison of the domain composition of the multifunctional enzymes HMWP1 and HMWP2 of *Y. enterocolitica* and the corresponding Irp1 and Irp2 of strain DI21 of *P. damselae* subsp. *piscicida*

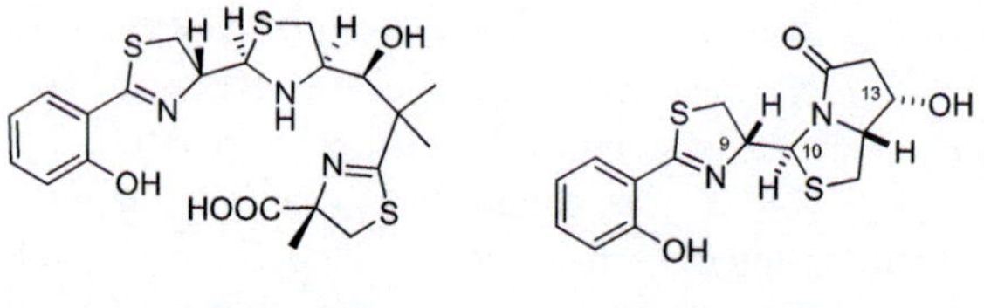

Fig. 7 Structures of yersiniabactin (**44**) and prepiscibactin (PrePcb, **45**)

the culture of gilthead sea bream (*Sparus aurata*), seabass (*Dicentrarchus labrax*), and sole (*Solea senegalensis* and *S. solea*) in the European Mediterranean countries, including Spain. Additionally, this pathogen is also able to infect a wide variety of marine fish such as striped bass (*Morone saxatilis*), white perch (*M. americana*), and hybrid striped bass (a hybrid between the striped bass (*M. saxatilis*) and the white bass (*M. chrysops*)) in the USA, cobia (*Rachycentron canadum*) in Taiwan, and golden pompano (*Trachinotus ovatus*) in China. Now, photobacteriosis is now considered one of the most dangerous bacterial diseases in aquaculture worldwide due to its wide host range, high mortality rate, and ubiquitous distribution [44].

The structure of the siderophore responsible for the iron uptake in *P. damselae* subsp. *piscicida*, named as piscibactin (**3**), was predicted from the study of genes in the strain DI21 involved in its biosynthesis which are organized in a typical structure of a Pathogenicity Island (PAI) [45]. Database searches revealed that this PAI shows a high degree of homology to the High Pathogenicity Island (HPI) described in *Yersinia* spp., such as *Yersinia pestis*, which encodes the synthesis of the siderophore yersiniabactin (**44**) [46]. More specifically, the structural prediction was deduced from the differences between the domain composition of the multifunctional enzymes HMWP1 and HMWP2 of *Y. enterocolitica* and the corresponding Irp1 and Irp2 of strain DI21 of *P. damselae* subsp. *piscicida*. Thus, the presence of just one methyltransferase domain (MT) in Irp1 of *P. damselae* subsp. *piscicida* DI21 strain instead of the two MT in HMWP1 of *Y. enterocolitica* (Fig. 6) suggested the lack of some methyl groups in the structure of Pcb (**3**) in relation to that of yersiniabactin (**44**) (Fig. 7). *Y. pestis* is the causative agent of the bubonic plague or Black Death that killed one third of the European population in the Middle Age, while *Y. enterocolitica* causes severe enteric disorders in humans. Moreover, *P. damselae* subsp. *piscicida* can use yersiniabactin (**44**) as xenosiderophore [45].

Due to its high instability, Pcb (**3**) was isolated as its Ga^{3+} and Fe^{3+} complexes. Its Ga^{3+} complex made an NMR study possible, while a bioguided fractionation could be performed due to its Fe^{3+} complex. Thus, filtered supernatants of cell-free culture broth of *P. damselae* subsp. *piscicida* DI21 were incubated with the corresponding Ga^{3+} and Fe^{3+} salts and submitted to the SPE- HLB/HPLC-HRMS methodology where the use of acidic conditions being avoided at any time. Fractions containing the piscibactin Ga^{3+} complex were easily identified in the MS by the presence of the *m/z* cluster with the distinctive isotopic ratio of gallium (Mr = 69 and 71, ratio 3:2). The structure of Pcb (**3**) was established from the 1D/2D NMR and HR-ESIMS data analysis of its piscibactin-Ga(III) complex. Furthermore, a possible intermediate of its biosynthesis, called prepiscibactin (prePcb, **45**), was also isolated in the process (Fig. 7) [47].

The stereoselective synthesis of prePcb (**45**), whose structure is characterized by the presence of four stereogenic centers and three heterocycle rings, was reported. The chiral sources for positions (9*R*) and (12*R*) were L- and D-cysteines while the secondary alcohol at position (13*S*) was generated by a stereoselective SmI_2-mediated Reformatsky reaction. The synthesis of prePcb (**45**) started with the conversion of L-cysteine to acid **46** which was converted to the thiazolidinic aldehyde **47** through the Weinreb amide and posterior reduction. The crucial SmI_2-induced Reformatsky reaction between the thiazolidinic aldehyde **47** and (*R*)-chloroacetyl-2-oxazolidinone **48** resulted in the formation of the desired secondary alcohol **49** as a single diastereoisomer with a very high yield. Removal of the chiral auxiliary, the acetonide and Boc protecting groups afforded the thiol amino alcohol ethyl ester **50**. In parallel, condensation of D-cysteine and 2-hydroxy benzonitrile gave a thiazolinic carboxylic acid which was transformed to the thiazolidinic aldehyde **51** via the corresponding Weinreb amide followed by reduction. Finally, the coupling between the thiol amino **50** and the thiazolidinic aldehyde **51** in the presence of potassium acetate followed by refluxing in toluene gave prePcb (**45**) along with its epimer at C10 in a 1:1 ratio. The stereoselectivity of the required isomer prepiscibactin from 1:1 to 4:1 ratio was improved by addition of $ZnCl_2$ (Scheme 5) [48].

In the recently reported stereoselective convergent synthesis of Pcb (**3**), D-cysteine was used as a chiral source for the (9*R*) and (17*S*)-positions, L-cysteine was employed to generate the (12*R*)-position while Meldrum's acid was utilized in the preparation of the secondary alcohol at the (13*S*)-position. One of the major challenges of this synthesis was the presence of a high sensitive β-hydroxy-2,4-disubstituted thiazoline moiety in Pcb (**3**) lacking the bulky *gem*-dimethyl groups at C-14 present in yersiniabactin (**44**) [49].

The synthesis started with the protection of the thiol and the primary amine of L-cys with Trt and Fmoc protecting groups, respectively, to give Fmoc-Cys(Trt)-OH (**52**) which was subsequently coupled with Meldrum's acid to afford **53**. Refluxing **53** in EtOAc followed by stereoselective reduction of the resulting tetramic acid with $NaBH_4$ gave lactam **54** where the stereogenic centers present in Pcb (**3**) at C-12 and C-13 positions were efficiently generated. Then, the selective basic cleavage of the lactam **54** in the presence of the Fmoc group to afford statine **55** was performed by

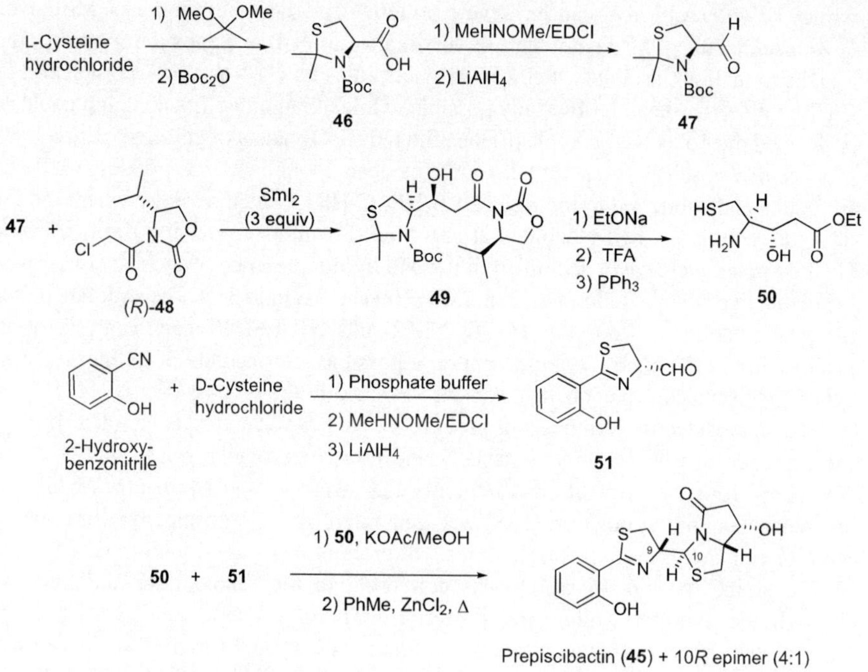

Scheme 5 Synthesis of PrePcb (**45**)

using LiOH with added $CaCl_2$. Posterior protection of the hydroxyl group as triethylsilyl (TES) ether in **55** yielded the conveniently protected carboxylic acid **56**. The next step implied the coupling of a freshly prepared solution of the methyl ester of (*S*)-α-azido-α-methylcysteine (**57**), obtained from D-cysteine following reported procedures, with statine **56**, previously activated with EDC, to furnish thioester **58**. The key step for the formation of the sensitive β-hydroxy-2,4-disubstituted thiazoline moiety implied the use of the mild and non-dehydrative conditions from the azide of thioester **58** via Staudinger reduction and the subsequent intramolecular aza-Wittig (S-AW process) with PPh_3 in 2,6-lutidine to give **59**. A very mild method for removal of protecting groups in **59** was carried out by deprotection of the Trt group using I_2 in CH_2Cl_2, followed by treatment of the resulting disulfide **60** with NaN_3 in the presence of the scavenger octanethiol and ending with deprotection of the TES ether with TBAF to give the methylester thiazoline **62**.

For the completion of the synthesis of **3**, thiazolinic aldehyde **51**, already described in the synthesis of prePcb (**45**), was coupled with methylester thiazoline **62**, both freshly prepared, to afford an epimeric mixture of methyl esters **63a,b** at C-9. Finally, basic hydrolysis of the ester mixture **63a,b** with LiOH gave the very unstable acid **3** and its C-9 epimer which were subjected without purification to complexation with gallium $Ga(acac)_3$. The resulting product was submitted to chromatographic separation using RP-C18 cartridges followed by HPLC purification to deliver Pcb-Ga^{3+}-complex **64** which resulted to be identical to the natural product. In this way, the absolute configuration of Pcb (**3**) could be determined as **(9*R*,10*R*,12*R*,13*S*,17*S*)-3**. Additionally, when the product obtained from the basic hydrolysis with LiOH of methyl esters **63a,b** was subjected to complexation with gallium in CD_3OD furnished the deuterated Pcb-d_2 (**65**), where the two hydrogens at C-14 position were interchanged by two deuterium atoms. Pcb-d_2 (**65**) could be used as a new molecular tool to carry out further studies on the iron-uptake mechanisms.

Pcb (**3**) is a siderophore encoded by *irp* genes in a high-pathogenic island named *irp*-HPI [45] which is widespread among *Vibrionaceae* [42]. In *P. damselae* subsp. *piscicida*, the *irp*-HPI is part of a plasmid pPHDP70 [50]. It was suggested that this plasmid could be acquired by horizontal transfer, with an evolutionary origin similar to that of *Yersinia* HPI. The proposal of an assembly line for the biosynthesis of Pcb (**3**), showing the presence of just one methyltransferase (MT) domain in the multifunctional enzymes PKS/NRPS Irp1, agrees with the lack of the two methyl groups at C-14 in relation to that of yersiniabactin (**44**). Moreover, a cryptic metabolite was also predicted, although not detected yet, from the presence of additional module in Irp1 [47].

The plasmid pPHDP70 that encodes Pcb (**3**) biosynthesis could be transferred by conjugation from *P. damselae* subsp. *piscicida* to a mollusk pathogenic *Vibrio alginolyticus* mutant strain, where it was abolished the production of its native siderophore, vibrioferrin [51]. Acquisition of plasmid pPHDP70 by conjugation restored the capacity of that mutant to grow under low-iron conditions since it could synthesize Pcb (**3**). Piscibactin production was demonstrated by chemical analysis of its culture supernatants using the same SPE-HLB/HPLC-MS methodology employed in the isolation of gallium (III) and iron (III) piscibactin complexes from *P. damselae* subsp. *piscicida*. These experiments confirmed not only that the plasmid pPHDP70 is responsible for the production of Pcb (**3**) but also showed that the horizontal transfer of that plasmid by conjugation can be very easy and common [50].

Piscibactin production was proven as one of the major virulence factors not only in *V. anguillarum* but also in *Photobacterium damselae* subsp. *piscicida*. The key role of Pcb (**3**) in the virulence of this bacterium was demonstrated by the fact that *P. damselae* subsp. *piscicida* strains cured of pPHDP70 not only were incapable of producing Pcb (**3**) and growing under iron-limited conditions but also exhibited markedly decreased virulence in fish [50].

The plasmid pPHDP70 encodes a hitherto uncharacterized TonB-dependent transporter named FrpA that would act as the Pcb (**3**) outer membrane receptor (OMR) [45]. In order to find a new vaccine formulation as alternative to the existence one based on the use of inactivated bacterial cells, FrpA was cloned, expressed in *E. coli* using an arabinose-inducible promoter, and the resulting rFrpA purified from its cultures. Evaluation of its effectiveness against photobacteriosis and immunogenicity as vaccine on sole (*Solea senegalensis*) showed that the rFrpA protein is immunogenic, displaying an immunity response that was equivalent to that observed in fish immunized with the corresponding bacterin, obtained from formalin-killed whole cells of *P. damselae* subsp. *piscicida*. Furthermore, experimental infection challenge trials showed that fish groups vaccinated with bacterin or rFrpA were protected against photobacteriosis with similar protection levels, reaching survival rates (RPS) of ca. 79% and 73%, respectively [52]. Indeed, rFrpA represents a promising antigen candidate for the development of novel more effective vaccines for the prevention of fish photobacteriosis. Although there are precedents for the use of OMR of siderophores from pathogenic bacteria as vaccines, this is one of the first applications against an infectious disease in fish [53].

4 Furunculosis by *Aeromonas salmonicida*

Furunculosis is a disease which causes economically devastating losses in cultivated salmonids in fresh and marine waters. It shows a widespread distribution. Although all salmonid species such as Atlantic salmon (*Salmo salar*), which is considered particularly susceptible, and rainbow trout (*Oncorhynchus mykiss*) can be affected, it also affects several non-salmonids species including turbot (*Scophthalmus maximus*) and halibut (*Hippoglossus hippoglossus*). The disease was first described by Lehmann and Neumann in 1896 and its name comes from the deep ulcerations that it causes on skin. Most Atlantic salmon infected by the disease presents a hemorrhagic septicemia with high mortality rates, in both juvenile and adult fish, within as little as 2 or 3 days [54].

The etiological agent of furunculosis is the non-motile rod Gram-negative γ-proteobacteria *Aeromonas salmonicida* subsp. *salmonicida* (*A. salmonicida*). This bacterium can be defined as biochemically, antigenically, and genetically homogeneous with no biotypes, serotypes, or genotypes. Three subspecies of *A. salmonicida*, *masoucida*, *achromogenes,* and *smithia*, named atypical strains, are also pathogenic [12].

The first studies of siderophore biosynthesis genes of *A. salmonicida* displayed the presence of a chromosomal gene cluster (*asb*GFDCBI) with high similarity to genes related to the synthesis of the siderophore Acb (**6**) from *A. baumannii* [55]. The completion of its genome sequence several years later revealed, by bioinformatic genomic analysis, the presence of a second gene cluster with high similarity to those of amonabactins (Fig. 8) from *Aeromonas hydrophila*. Finally, it was demonstrated that most *A. salmonicida* subsp. *salmonicida* strains produce two

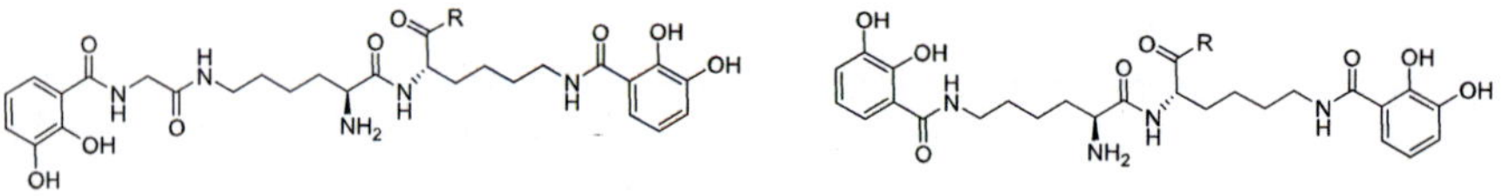

Amonabactin P750, R = D-Phe (**66**)
Amonabactin T789, R = D-Trp (**67**)

Amonabactin P693, R = D-Phe (**68**)
Amonabactin T732, R = D-Trp (**69**)

Fig. 8 Structure of the amonabactins biosynthetized (**66–69**) by *A. salmonicida* subsp. *salmonicida*

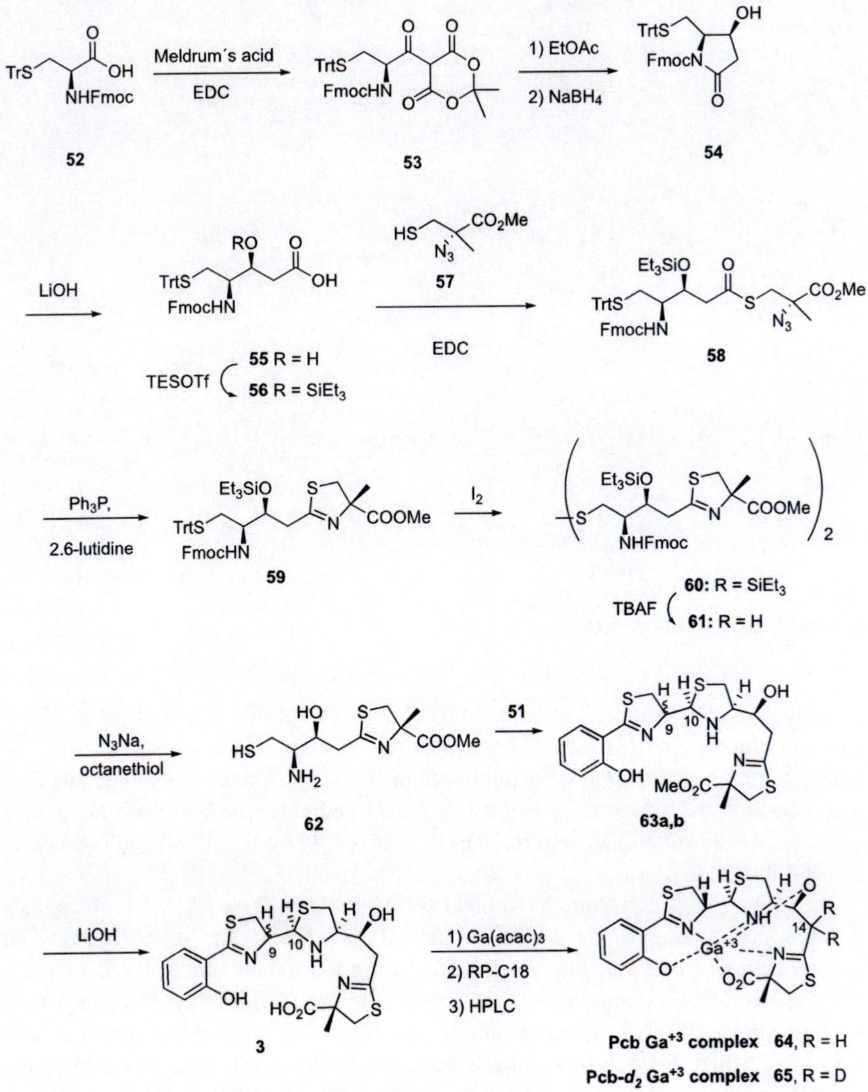

Scheme 6 Synthesis of Pcb (**3**) and its Ga^{+3} complex **64**

catechol siderophores simultaneously: Acb (**6**) and the four amonabactin variants that are denoted as P750, T789, P693, and T732 (**66–69**) according to the presence of either a Phe or Trp residue and their molecular weights (by the presence or absence of a Gly residue) [56].

The initial structure assigned for Acb as **5** when it was first described in 1994 from the human pathogen *A. baumannii* ATCC19606 [57] was corrected in 2009 as **6**. It was suggested that **5**, named as now preacinetobactin (preAcb, **5**), is an unstable

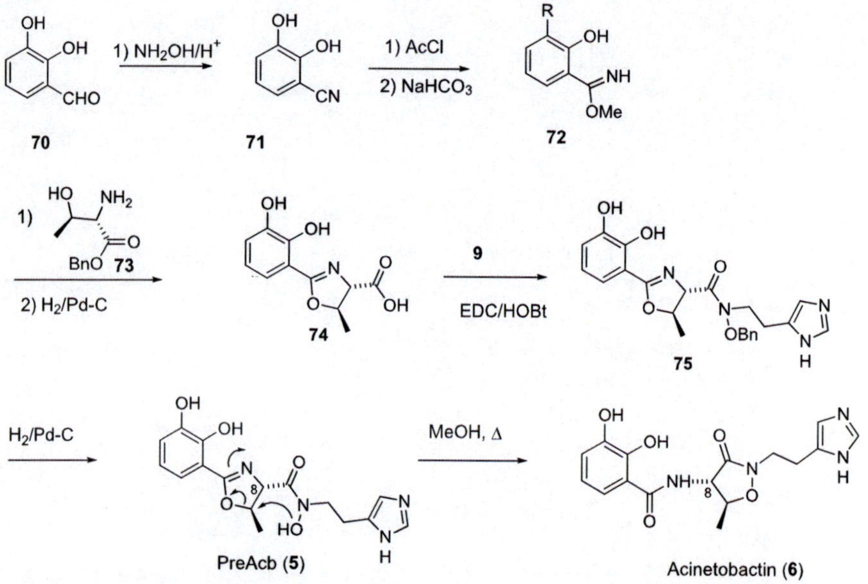

Scheme 7 Synthesis of Acb (**6**)

oxazoline which suffers an intramolecular nucleophilic substitution to give isoxazolidinone Acb (**6**) (see Scheme 7) [58]. A complete study of the Acb (**6**) isomerization at different pH was published in 2015 [59]. On the other hand, the four amonabactins (**52–55**) were previously isolated and characterized from the pathogenic bacterium of animals and humans *A. hydrophila* by Raymond and coworkers in 1994 [60].

Acb (**6**) was isolated from the virulent strain *A. salmonicida* RSP74.1 which does not produce amonabactins due to a natural deletion in the *amoG* gene. The amonabactins were isolated and/or detected from the mutant strain *A. salmonicida* VT45.1ΔasbD. This mutant was obtained from another highly virulent strain, *A. salmonicida* VT45.1, which produces simultaneously both siderophores, where it was abolished the Acb biosynthesis (deletion of the *asbD* gene) to facilitate the isolation of the amonabactins. The methodology SPE-HLB/HPLC-HRMS [30] was successfully applied in both cases confirming its high efficiency to accelerate the isolation process of siderophores containing catecholate/salicylate moieties [56].

The structure of Acb as **6**, including its absolute configuration, was confirmed by chemical synthesis developed by Takeuchi et al. [24]. This synthesis started from 2,3-dihydroxybenzaldehyde (**70**) which was transformed to the imidate **72** via the nitrile **71**. Condensation of L-threonine benzyl ester (**73**) with imidate **72** gave an oxazoline intermediate which was submitted to hydrogenolysis to afford the oxazoline acid **74**. Then, imidazole **9**, whose preparation was described in the synthesis of Ang (**1**) displayed in Scheme 1, was coupled with acid **74** to yield the oxazoline protected **75**. Finally, benzoyl deprotection by hydrogenolysis of **75** gave preAcb (**5**) which under reflux in MeOH afforded Acb (**6**) (Scheme 7).

Although the synthesis of the four natural amonabactins were first reported by Telford and Raymond in 1997 [61], an improved variation of that synthesis was published in 2019 [62]. The more recent synthesis of amonabactins started with the preparation of the activated building blocks **77** and **79**. Thus, activation with *N*-hydroxysuccinimide (NHS) of the isopropyl-protected 2,3-dihydroxybenzoic acid **40** (see Scheme 4) affording **76** and posterior coupling with commercially available N^{α}-Boc-L-lysine gave, after activation of the carboxylic acid with NHS, the first building block **77**. In parallel, activation of the carboxylic acid of commercially available N^{α}-Cbz-glycine with NHS and posterior coupling with N^{α}-Boc-L-lysine afforded, after hydrogenolysis of the Cbz group, the protected dipeptide **78**. This dipeptide was coupled with the NHS activated isopropyl-protected 2,3-dihydroxybenzoic acid **76** followed by activation with NHS to give the second building block **79**. Condensation of D-phenylalanine *tert*-butyl ester with the first building block **77** gave, after selective removal of the Boc protecting group, the protected dipeptide **80**. Coupling of dipeptide **80** with building blocks **79** or **77** yielded amonabactins P750 (**66**) and P6903 (**68**) (bearing a Phe residue), respectively, after the simultaneous removal of the isopropyl and *tert*-butyl protecting groups with BCl_3. In a similar way, the commercially available D-tryptophan methyl ester was coupled with the first building block **77** to give, after Boc deprotection, the dipeptide **81**. Dipeptide **81** was coupled with building blocks **79** or **77** to afford amonabactins T789 (**67**) and T732 (**69**) (bearing a Tyr residue), respectively, after basic hydrolysis of methyl ester with LiOH and removal of the isopropyl protecting groups with BCl_3 (Scheme 8).

The biosynthesis of both types of siderophores in *A. salmonicida* depends on *entABCE* genes, located within the amonabactin gene cluster, which encode the functions necessary for the synthesis of 2,3-dihydroxybenzoic acid (DHBA). Once DHBA is synthesized, *asbBCD* and *amoFGH* genes encode acinetobactin and amonabactins synthesis, respectively. Indeed, both siderophores share the first step of their biosynthesis, with the production of DHBA being the amonabactin gene cluster that supplies DHBA for both siderophores. Although both, acinetobactin and amonabactin, siderophore systems are present in all *A. salmonicida* strains, some isolates carry a mutation in the *amoG* gene that inactivates amonabactin synthesis. Interestingly, all *A. salmonicida* strains analyzed up to this point are able to use amonabactin as an iron source. The amonabactin biosynthesis genes are highly conserved in most *Aeromonas* species including the human pathogen *A. hydrophila*. On the contrary, Acb (**6**) is encoded by a gene cluster restricted to *A. salmonicida*. From the analysis of the genomic context of acinetobactin cluster, it seems clear that this gene cluster was likely acquired from other bacteria through horizontal gene transfer at some point during the speciation process of *A. salmonicida* while amonabactin must be the ancestral siderophore of the species [56].

Although FstB and FstC were previously identified as two outer membrane proteins specifically induced under iron deficiency and under in vivo conditions,

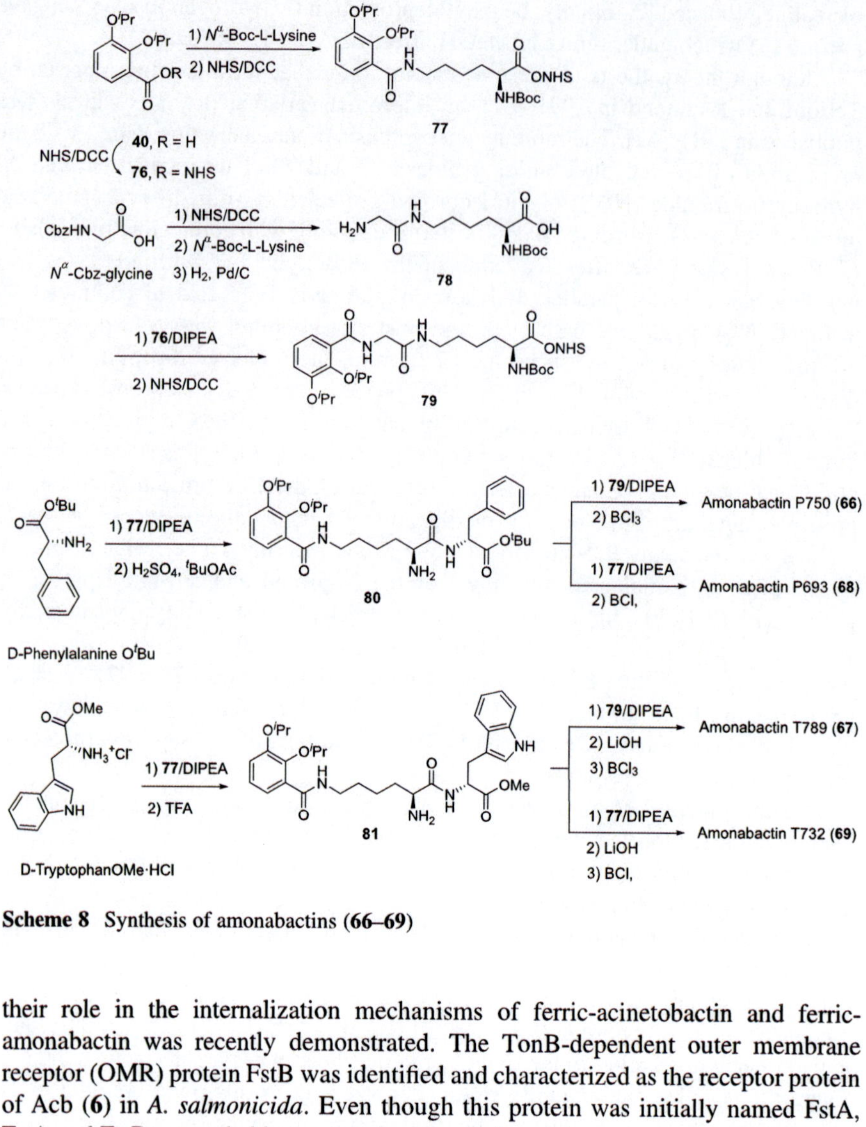

Scheme 8 Synthesis of amonabactins (**66–69**)

their role in the internalization mechanisms of ferric-acinetobactin and ferric-amonabactin was recently demonstrated. The TonB-dependent outer membrane receptor (OMR) protein FstB was identified and characterized as the receptor protein of Acb (**6**) in *A. salmonicida*. Even though this protein was initially named FstA, FstA and FstB are probably two versions or alleles of the same protein [63]. On the other hand, FstC is the OMR of all four amonabactin forms. The FstC receptor is widespread in the genus *Aeromonas*, not only among *A. salmonicida* strains but also in most *Aeromonas* species, including relevant human and animal pathogens as *A. hydrophila* [62].

Several analogues of Acb (**6**) and amonabactins (**66–69**) were prepared based on their simplicity in order to deduce some correlations between the structure and the siderophore activity to facilitate the preparation of conjugates for the development of novel applications of the bacterial iron-uptake mechanisms based on siderophores.

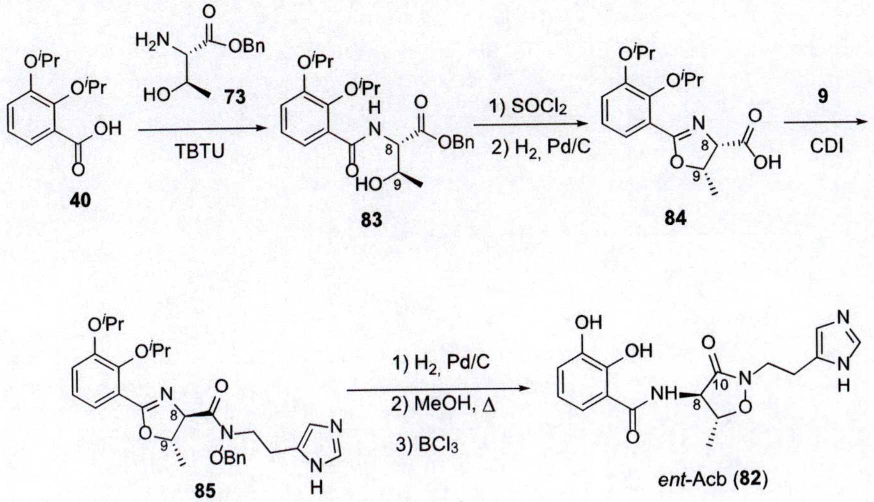

Scheme 9 Synthesis of *ent*-Acb (**68**)

In order to evaluate the influence of the stereogenic centers on the siderophore activity of Acb (**6**), its enantiomer, *ent*-Acb (**82**), was synthetized using a similar strategy employed in the synthesis of Acb (**6**) and Ang (**1**). Condensation of 2,3-diisopropyloxybenzoic acid (**40**) with L-threonine benzyl ester (**73**) gave amide **83** which was submitted to dehydrative-cyclization followed by debenzylation by hydrogenolysis to afford the oxazoline acid **84**. An inversion of configuration of the oxymethine carbon at C-9 took place in **84** during the cyclization process. Coupling of N^{α}-benzyloxyhistamine (**9**) (Scheme 1) with oxazoline acid **84** using CDI in the absence of base gave the sterically stable *trans*-isomer **85** rather than the *cis*-isomer, probably due to epimerization of the activated acyl intermediate. Debenzylation of oxazoline **85** by hydrogenolysis followed by reflux in MeOH gave the expected rearrangement of oxazoline to isoxazolidinone which was treated with BCl_3 to remove the isopropyl groups to yield *ent*-Acb (**82**) (Scheme 9).

With the aim of comparing the influence of the methyl group on the isoxazolidinone ring and the presence of similar heterocyclic rings, such as triazol instead of imidazole, on the siderophore activity of Acb analogues, four demethylacinetobactin derivatives (compounds **86–89**) were prepared. They were synthetized from a common intermediate **90** which was prepared by coupling of 2,3-dibenzyloxybenzoic acid (**17**) (Scheme 2) and the commercially available D-cycloserine. *N*-alkylation of intermediate **90** with mesylate **91**, prepared by mesylation of the commercial 3-benzyloxycarbonylaminopropanol, followed by subsequent removal of Cbz and Bn protecting groups by hydrogenolysis, gave the first demethylacinetobactin **86**. The second demethylacinetobactin **87** was synthetized by *N*-alkylation of intermediate **90** with the substituted imidazole

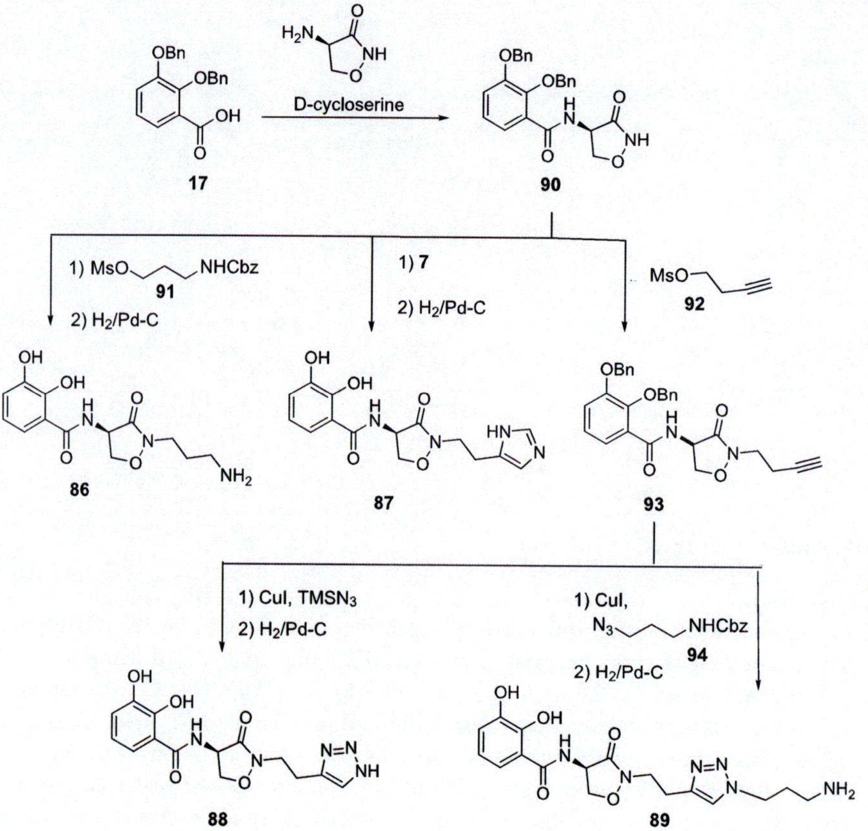

Scheme 10 Synthesis of demethylacinetobactin derivatives **86–89**

7 (see Scheme 1) and posterior debenzylation by hydrogenolysis. Demethylacinetobactin derivatives **88** and **89** were synthetized by using the "click-chemistry" reaction involving the efficient Cu(I)-catalyzed azide-alkyne couplings. Thus, *N*-alkylation of **90** with the mesylate **92**, obtained by mesylation of the commercial homopropargyl alcohol, gave the alkyne **93**. Coupling of trimethysilyl azide or azide derivative **94**, prepared by displacement of the mesylate group in **91** with sodium azide, with alkyne **93** afforded demethylacinetobactin analogues **88** and **89**, respectively, after removal of the protecting groups by hydrogenolysis (Scheme 10).

Several structure-activity relationships in relation to the OMR protein FstB implied in the internalization of ferric-Acb were deduced from the evaluation of the biological activity of synthetic *ent*-Acb (**82**) and four demethylacinetobactin analogues (**86–89**). All of them showed CAS values indistinguishable from those of Acb (**6**), indicating that they bind iron efficiently. The *ent*-Acb (**82**) displayed the same siderophore activity as the natural one, suggesting the lack of enantiomer preference on the molecular recognition by the protein receptor FstB. The recognition of Acb (**6**) by its cognate OMR protein FstB mostly resides in the presence of imidazole (**87**) or a similar heterocyclic ring (**88**) in its structure because the presence of a primary amino group (**86**) or a propylamino substituted triazole (**89**) was devoid of such activity. Moreover, removal of the methyl group at the isoxazolidinone, in combination with the presence of either an imidazole or a triazole ring (**87** and **88**), leads to higher levels of biological activity. Additionally, it was suggested an alternative route(s) of entry for analogues **87** and **88** since they displayed siderophore activity in the growth promotion assay of a mutant strain which is unable to produce or transport acinetobactin [63].

On the other hand, with the aim of evaluating the minimum structural requirements for amonabactin recognition by its OMR protein FstC, four simplified analogues of the amonabactins were synthetized, compounds **95–99** using a similar synthetic sequence as that described for the natural ones.

Coupling the activated NHS carboxylic acid intermediate **76** with the commercially available *N*-Boc cadaverine (**99**) gave amide **100** which was treated with BCl_3 for removal of isopropyl and Boc protecting groups to give monocatecholate amonabactin analogue **95**. Removal of the Boc protecting group in **100** with TFA gave an intermediate which was coupled with activated NHS carboxylic acid intermediates **77** or **79** (Scheme 8) to give biscatecholate amonabactin analogues **97** and **98**, respectively, after deprotection with BCl_3. Finally, activation of N^{α}-Cbz-glycine with NHS, posterior coupling with *N*-Boc cadaverine (**99**) followed by deprotection of the Cbz group by hydrogenolysis gave amide **101** which was coupled with activated NHS carboxylic acid **76** to yield monocatecholate amonabactin analogue **96** after treatment with BCl_3 (Scheme 11).

Important structure-activity correlations were deduced from the evaluation of the biological activity of each of the four natural amonabactins individually **66–69** and of the four synthetic amonabactin analogues **95–99**. The lack of siderophore activity for monocatecholate amonabactin analogues **95** and **96** indicated that the presence of at least two catechol moieties is necessary to display biological activity. A quite different growth promotion activity was observed for each natural amonabactin form. The results suggested that natural amonabactins in which the length of the linker between the two iron-binding catecholamide units is 15 atoms, P750 (**66**) and T789 (**67**), instead of 12 atoms, P693 (**68**) and T732 (**69**), recognize more efficiently the OMR protein FstC. This was confirmed from the results obtained from the siderophore activity evaluation of the synthetic amonabactin analogues **97** and **98**. Indeed, the higher biological activity displayed by the bis-catechol amonabactin analogue **98** with longer linker (15 atoms linker length) in relation to that of bis-catechol amonabactin analogue **97** with a shorter linker (12 atoms linker length) is in agreement with the hypothesis that the length of the linker backbone (presence or absence of glycine) between the two iron-binding catecholamide units is relevant to maximize ferric-siderophore acquisition. Even though the presence of Phe or Trp residues seems not be required for siderophore recognition, their presence could also participate in the recognition by the receptor since biscatecholate amonabactin analogues **97** and **98** show less biological activity than their natural counterparts.

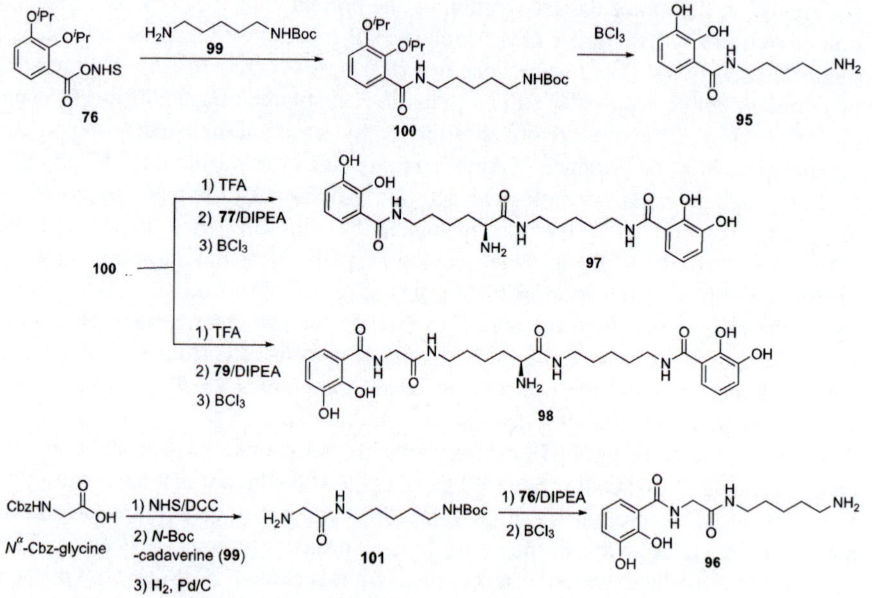

Scheme 11 Synthesis of amonabactin analogues **95–98**

In summary, three important conclusions were deduced from these studies. Firstly, the OMR protein FstC possesses a considerable functional plasticity that could be exploited for delivery of antimicrobial compounds into the cell. Second, the optimum linker length between the two catecholamide units for maximizing the siderophore activity in the amonabactin type-structures is 15 atoms, preferentially a Lys-Lys-Gly fragment which is present in the natural longer amonabactins **66** and **67** and analogue **98**. Finally, the synthetic simplified analogues **97** and **98** could be used as vectors for the preparation of conjugates in the Trojan horse strategy to develop new effective antimicrobials against many different pathogens of the genus *Aeromonas* including the human pathogen *A. hydrophila,* and for the synthesis of fluorescence probes to study the iron-uptake mechanism in this bacterium [62].

5 Concluding Remarks

The iron-uptake pathway is a key process for microbial survival and virulence. The knowledge of the mechanisms of iron acquisition mediated by siderophores can be of great help for the design of new strategies against microbial infections.

Fish disease outbreaks, vibriosis, photobacteriosis, and furunculosis, are considered as the major threats alarming the aquaculture industry and food security. Chemical and biological studies of the Gram-negative bacteria responsible for

Table 1 Siderophores from pathogenic bacteria involved in the main aquaculture infectious fish disease

Bacterium (fish disease)	Siderophore	OMR proteins	Biosynthetic gene cluster	Synthesis (Ref.)	Synthesis of analogues (Ref.)
V. anguillarum (vibriosis)	Anguibactin (**1**)	FatA	*AngBDEMNR* in the pJM1 plasmid	[25]	
	Vanchobactin (**2**)	FtvA	*VabABCEF* in the chromosome	[31]	[32]
	Piscibactin (**3**)			[49]	[48]
P. damselae subsp. *piscicida* (photobacteriosis)	Piscibactin (**3**)	FrpA	*Irp*-HPI in the pPHDP70 plasmid	[49]	[48]
A. salmonicida (furunculosis)	Acinetobactin (**6**)	FstB	*EntABCE asbBCD*	[24]	[63, 64]
	Amonabactins (**66–69**)	FstC	*EntABCE amoFGH*	[61, 62]	[62]

these infectious diseases allowed not only the isolation, structural elucidation and synthesis of the siderophores involved in their iron-siderophore uptake mechanisms but also the identification of their corresponding OMR proteins and the gene cluster involved in their biosynthesis (Table 1). All this information is being used in the development of new antimicrobial and vaccines based on the iron-siderophore uptake mechanism of these bacteria.

It is worth mentioning the great structural similarity of the siderophores biosynthesized by both human and fish virulent pathogenic bacteria. The structure of yersiniabactin (**44**), the siderophore produced by *Yersinia pestis*, responsible for epidemics of high mortality throughout the history of mankind (the plague of Justinian, the Black Death and the third pandemic), is very similar to that of Pcb (**3**) whose production confers maximum virulence in both *V. anguillarum* and *P. damselae* subsp. *piscicida*. Moreover, pyochelin, one of the siderophores produced by the human multidrug resistant bacteria pathogen *Pseudomonas aeruginosa*, is also structurally related to Pcb (**3**). One of the siderophores produced by *A. salmonicida*, Acb (**2**), is also biosynthesized by the human pathogen *A. baumannii*. This bacterium has been identified as one of the group of pathogens with a high rate of resistance to antibiotics and responsible for the majority of nosocomial infections. Moreover, the structure of Acb (**2**) is closely related to that of Ang (**1**) produced by some virulent strains of *V. anguillarum*. The fact that the anguibactin and piscibactin iron-uptake systems in *V. anguillarum* and the acinetobactin iron-uptake system to *A. salmonicida* had been probably acquired by horizontal transfer from other pathogenic bacteria seems to confirm the high efficiency of these iron absorption pathways.

Acknowledgements This work was supported by grants RTI2018-093634-B-C22 (AEI/FEDER, EU) from the State Agency for Research (AEI) of Spain, co-funded by the FEDER Programme from the European Union.

References

1. FAO. The State of World Fisheries and Aquaculture [Internet]. 2020 [cited 17 Oct 2020]. http://www.fao.org/state-of-fisheries-aquaculture/es/
2. Santos L, Ramos F (2018) Antimicrobial resistance in aquaculture: current knowledge and alternatives to tackle the problem. Int J Antimicrob Agents 52(2):135–143
3. Watts JEM, Schreier HJ, Lanska L, Hale MS (2017) The rising tide of antimicrobial resistance in aquaculture: sources, sinks and solutions. Mar Drugs 15(6):158
4. Wilson BR, Bogdan AR, Miyazawa M, Hashimoto K, Tsuji Y (2016) Siderophores in Iron metabolism: from mechanism to therapy potential. Trends Mol Med 22(12):1077–1090
5. Hider RC, Kong X (2010) Chemistry and biology of siderophores. Nat Prod Rep 27(5):637–657
6. Ji C, Juárez-Hernández RE, Miller MJ (2012) Exploiting bacterial iron acquisition: siderophore conjugates. Future Med Chem 4(3):297–313
7. Sandy M, Butler A (2009) Microbial iron acquisition: marine and terrestrial siderophores. Chem Rev 109(10):4580–4595
8. Bentur Y, McGuigan M, Koren G (1991) Deferoxamine (Desferrioxamine). Drug Saf 6 (1):37–46
9. Schalk IJ (2018) Siderophore–antibiotic conjugates: exploiting iron uptake to deliver drugs into bacteria. Clin Microbiol Infect 24(8):801–802
10. Cefiderocol Approval Letter (2020) [Internet]. https://www.accessdata.fda.gov/drugsatfda_docs/nda/2019/209445Orig1s000ltr.pdf. Accessed 17 Oct 2020
11. Fetcroja (cefiderocol) (2020) European Union [Internet]. https://www.ema.europa.eu/en/medicines/human/EPAR/fetcroja. Accessed 17 Oct 2020
12. Toranzo AE, Magariños B, Romalde JL (2005) A review of the main bacterial fish diseases in mariculture systems. Aquaculture 246(1–4):37–61
13. Mohamad N, Amal MNA, Yasin ISM, Zamri Saad M, Nasruddin NS, Al-saari N et al (2019) Vibriosis in cultured marine fishes: a review. Aquaculture 512:734289
14. Hickey ME, Lee JL (2018) A comprehensive review of *Vibrio* (*Listonella*) *anguillarum*: ecology, pathology and prevention. Rev Aquac 10(3):585–610
15. Hansen MJ, Kudirkiene E, Dalsgaard I (2020) Analysis of 44 *Vibrio anguillarum* genomes reveals high genetic diversity. PeerJ 8:e10451
16. Lemos ML, Balado M, Osorio CR (2010) Anguibactin- versus vanchrobactin-mediated iron uptake in *Vibrio anguillarum*: evolution and ecology of a fish pathogen. Environ Microbiol Rep 2(1):19–26
17. Actis LA, Fish W, Crosa JH, Kellerman K, Ellenberger SR, Hauser FM, Sanders-Loehr J (1986) Characterization of anguibactin, a novel siderophore from *Vibrio anguillarum* 775(pJM1). J Bacteriol 167(1):57–65
18. Jalal MAF, Hossain MB, Van der Helm D, Sanders-Loehr J, Actis LA, Crosas JH (1989) Structure of anguibactin, a unique plasmid-related bacterial siderophore from the fish pathogen *Vibrio anguillarum*. J Am Chem Soc 111(2):292–296
19. Hossain MB, Jalal MAF, Van Der Helm D (1998) Gallium-complex of anguibactin, a siderophore from fish pathogen *Vibrio anguillarum*. J Chem Crystallogr 28(1):57–60
20. Shrivastava SR, Shrivastava PS, Ramasamy J (2018) World Health Organization releases global priority list of antibiotic-resistant bacteria to guide research, discovery, and development of new antibiotics. J Med Soc 32:76–77

21. Noël S, Guillon L, Schalk IJ, Mislin GLA (2011) Synthesis of fluorescent probes based on the pyochelin siderophore scaffold. Org Lett 13(5):844–847
22. Lamb AL (2015) Breaking a pathogen's iron will: inhibiting siderophore production as an antimicrobial strategy. Biochim Biophys Acta 1854(8):1054–1070
23. Sassone-Corsi M, Chairatana P, Zheng T, Perez-Lopez A, Edwards RA, George MD, Nolan EM, Raffatellu M (2016) Siderophore-based immunization strategy to inhibit growth of enteric pathogens. Proc Natl Acad Sci 113(47):13462–13467. https://doi.org/10.1073/pnas.1606290113
24. Takeuchi Y, Ozaki S, Satoh M, Mimura K, Hara S, Abe H, Nishioka H, Harayama T (2010) Synthesis of acinetobactin. Chem Pharm Bull (Tokyo) 58(11):1552–1553
25. Lee H, Song WY, Kim M, Lee MW, Kim S, Park YS, Kwak K, Oh MH, Kim HJ (2018) Synthesis and characterization of anguibactin to reveal its competence to function as a thermally stable surrogate Siderophore for a gram-negative pathogen, *Acinetobacter baumannii*. Org Lett 20(20):6476–6479
26. Sandy M, Butler A (2011) Chrysobactin siderophores produced by *Dickeya chrysanthemi* ec16. J Nat Prod 74(5):1207–1212
27. Crosa JH. Deferration using anguibactin siderophore US5393777 A 1995-02-28; United States
28. Stork M, Di Lorenzo M, Welch TJ, Crosa LM, Crosa JH (2002) Plasmid-mediated iron uptake and virulence in *Vibrio anguillarum*. Plasmid 48(3):222–228
29. Soengas RG, Anta C, Espada A, Paz V, Ares IR, Balado M, Rodríguez J, Lemos ML, Jiménez C (2006) Structural characterization of vanchrobactin, a new catechol siderophore produced by the fish pathogen *Vibrio anguillarum* serotype O2. Tetrahedron Lett 47(39):7113–7116
30. Espada A, Anta C, Bragado A, Rodríguez J, Jiménez C (2011) An approach to speed up the isolation of hydrophilic metabolites from natural sources at semipreparative level by using a hydrophilic-lipophilic balance/mixed-mode strong cation exchange-high-performance liquid chromatography/mass spectrometry system. J Chromatogr A 1218(13):1790–1794
31. Soengas RG, Anta C, Espada A, Nieto RM, Larrosa M, Rodríguez J, Jiménez C (2007) Vanchrobactin: absolute configuration and total synthesis. Tetrahedron Lett 48(17):3021–3024
32. Soengas RG, Larrosa M, Balado M, Rodriguez J, Lemos ML, Jimenez C (2008) Synthesis and biological activity of analogues of vanchrobactin, a siderophore from *Vibrio anguillarum* serotype O2. Org Biomol Chem 6(7):1278–1287
33. Persmark M, Expert D, Neilands JB (1989) Isolation, characterization, and synthesis of chrysobactin, a compound with siderophore activity from *Erwinia chrysanthemi*. J Biol Chem 264(6):3187–3193
34. Sandy M, Han A, Blunt J, Munro M, Haygood M, Butler A (2010) Vanchrobactin and anguibactin siderophores produced by *Vibrio* sp. DS40M4. J Nat Prod 73(6):1038–1043
35. Page WJ, Tigerstrom MV (1988) Aminochelin, a catecholamine Siderophore produced by *Azotobacter vinelandii*. Microbiology 134(2):453–460
36. Souto A, Montaos MA, Balado M, Osorio CR, Rodríguez J, Lemos ML, Jiménez C (2013) Synthesis and antibacterial activity of conjugates between norfloxacin and analogues of the siderophore vanchrobactin. Bioorg Med Chem 21(1):295–302
37. Iglesias E, Brandariz I, Jiménez C, Soengas RG (2011) Iron(iii) complexation by Vanchrobactin, a siderophore of the bacterial fish pathogen *Vibrio anguillarum*. Metallomics 3(5):521–528
38. Balado M, Osorio CR, Lemos ML (2006) A gene cluster involved in the biosynthesis of vanchrobactin, a chromosome-encoded siderophore produced by *Vibrio anguillarum*. Microbiology 152(12):3517–3528
39. Balado M, Osorio CR, Lemos ML (2009) FvtA is the receptor for the siderophore vanchrobactin in *Vibrio anguillarum*: utility as a route of entry for vanchrobactin analogues. Appl Environ Microbiol 75(9):2775–2783
40. Naka H, Dias GM, Thompson CC, Dubay C, Thompson FL, Crosa JH (2011) Complete genome sequence of the marine fish pathogen *Vibrio anguillarum* harboring the pJM1 virulence

plasmid and genomic comparison with other virulent strains of *V. anguillarum* and *V. ordalii*. Infect Immun 79(7):2889–2900
41. Balado M, Lages MA, Fuentes-Monteverde JC, Martínez-Matamoros D, Rodríguez J, Jiménez C, Lemos ML (2018) The siderophore piscibactin is a relevant virulence factor for *Vibrio anguillarum* favored at low temperatures. Front Microbiol 9:1766
42. Thode SK, Rojek E, Kozlowski M, Ahmad R, Haugen P (2018) Distribution of siderophore gene systems on a Vibrionaceae phylogeny: database searches, phylogenetic analyses and evolutionary perspectives. PLoS One 13(2):e0191860
43. Ruiz P, Balado M, Fuentes-Monteverde JC, Toranzo AE, Rodríguez J, Jiménez C, Avendaño-Herrera R, Lemos ML (2019) The fish pathogen *Vibrio ordalii* under iron deprivation produces the siderophore piscibactin. Microorganisms 7(9):313
44. Andreoni F, Magnani M (2014) Photobacteriosis: prevention and diagnosis. J Immunol Res 793817
45. Osorio CR, Juiz-Río S, Lemos ML (2006) A siderophore biosynthesis gene cluster from the fish pathogen *Photobacterium damselae* subsp. *piscicida* is structurally and functionally related to the *Yersinia* high-pathogenicity island. Microbiology 152(11):3327–3341
46. Miller DA, Walsh CT, Luo L (2001) C-methyltransferase and cyclization domain activity at the intraprotein PK/NRP switch point of yersiniabactin synthetase [2]. J Am Chem Soc 123 (34):8434–8435
47. Souto A, Montaos MA, Rivas AJ, Balado M, Osorio CR, Rodríguez J, Lemos ML, Jiménez C (2012) Structure and biosynthetic assembly of piscibactin, a siderophore from *Photobacterium damselae* subsp. *piscicida*, predicted from genome analysis. Eur J Org Chem:5693–5700
48. Segade Y, Montaos MA, Rodríguez J, Jiménez C (2014) A short stereoselective synthesis of prepiscibactin using a SmI2-mediated reformatsky reaction and Zn^{2+}-induced asymmetric thiazolidine formation. Org Lett 16(21):5820–5823
49. De Fuente MC, Segade Y, Valderrama K, Rodríguez J, Jiménez C (2021) Convergent total synthesis of the siderophore piscibactin as its Ga^{3+} complex. Org Lett 23(2):340–345
50. Osorio CR, Rivas AJ, Balado M, Fuentes-Monteverde JC, Rodríguez J, Jiménez C, Lemos ML, Waldor MK (2015) A transmissible plasmid-borne pathogenicity island confers piscibactin biosynthesis in the fish pathogen *Photobacterium damselae* subsp. *Piscicida*. Appl Environ Microbiol 81(17):5867–5879
51. Wang Q, Liu Q, Ma Y, Rui H, Zhang Y (2007) LuxO controls extracellular protease, haemolytic activities and siderophore production in fish pathogen *Vibrio alginolyticus*. J Appl Microbiol 103(5):1525–1534
52. Valderrama K, Balado M, Rey-Varela D, Rodríguez J, Vila-Sanjurjo A, Jiménez C, Lemos ML (2019) Outer membrane protein FrpA, the siderophore piscibactin receptor of *Photobacterium damselae* subsp. *Piscicida*, as a subunit vaccine against photobacteriosis in sole (Solea senegalensis). Fish Shellfish Immunol 94:723–729
53. Lemos Ramos ML, Balado Dacosta M, Jiménez González C, Rodríguez González J, Vila Sanjurjo A, Valderrama Pereira AK. Recombinant ferri-piscibactin receptor protein for vaccination against pasteurelosis in fish. WO2017009511 A1 2017-01-19
54. Valderrama K, Soto-Dávila M, Segovia C, Vásquez I, Dang M, Santander J (2019) *Aeromonas salmonicida* infects Atlantic salmon (*Salmo salar*) erythrocytes. J Fish Dis 42(11):1601–1608
55. Najimi M, Lemos ML, Osorio CR (2008) Identification of siderophore biosynthesis genes essential for growth of *Aeromonas salmonicida* under iron limitation conditions. Appl Environ Microbiol 74(8):2341–2348
56. Balado M, Souto A, Vences A, Careaga VP, Valderrama K, Segade Y, Rodríguez J, Osorio CR, Jiménez C, Lemos ML (2015) Two catechol Siderophores, acinetobactin and amonabactin, are simultaneously produced by *Aeromonas salmonicida* subsp *salmonicida* sharing part of the biosynthetic pathway. ACS Chem Biol 10(12):2850–2860
57. Yamamoto S, Okujo N, Sakakibara Y (1994) Isolation and structure elucidation of acinetobactin., a novel siderophore from *Acinetobacter baumannii*. Arch Microbiol 162 (4):249–254

58. Wuest WM, Sattely ES, Walsh CT (2009) Three siderophores from one bacterial enzymatic assembly line. J Am Chem Soc 131(14):5056–5057
59. Shapiro JA, Wencewicz TA (2016) Acinetobactin isomerization enables adaptive Iron acquisition in *Acinetobacter baumannii* through pH-triggered siderophore swapping. ACS Infect Dis 2(2):157–168
60. Telford JR, Leary JA, Tunstad LMG, Byers BR, Raymond KN (1994) Amonabactin: characterization of a series of siderophores from *Aeromouas hydrophila*. J Am Chem Soc 116 (12):4499–4500
61. Telford JR, Raymond KN (1997) Amonabactin: a family of novel siderophores from a pathogenic bacterium. J Biol Inorg Chem 2(6):750–761
62. Rey-Varela D, Cisneros-Sureda J, Balado M, Rodríguez J, Lemos ML, Jiménez C (2019) The outer membrane protein FstC of *Aeromonas salmonicida* subsp *salmonicida* acts as receptor for amonabactin siderophores and displays a wide ligand plasticity. Structure-activity relationships of synthetic amonabactin analogues. ACS Infect Dis 5(11):1936–1951
63. Balado M, Segade Y, Rey D, Osorio CR, Rodríguez J, Lemos ML, Jiménez C (2017) Identification of the ferric-acinetobactin outer membrane receptor in *Aeromonas salmonicida* subsp *salmonicida* and structure-activity relationships of synthetic acinetobactin analogues. ACS Chem Biol 12(2):479–493
64. Shapiro JA, Wencewicz TA (2017) Structure-function studies of acinetobactin analogs. Metallomics 9(5):463–470

Top Heterocycl Chem (2021) 58: 209–252
https://doi.org/10.1007/7081_2021_56

Published online: 12 May 2021

Marine Natural Products with Bioactivity Against Neglected Tropical Diseases

Sofia Kokkaliari, Nicole E. Avalon, Kristin Herrera, Ryan M. Young, Joshua Welsch, Bingjie Yang, Sarah Dietrick, and Bill J. Baker

Contents

Sofia Kokkaliari and Nicole E. Avalon contributed equally to this work.

S. Kokkaliari, N. E. Avalon, K. Herrera, J. Welsch, B. Yang, S. Dietrick, and B. J. Baker (✉)
Department of Chemistry, University of South Florida, Tampa, FL, USA
e-mail: bjbaker@usf.edu

R. M. Young
School of Chemistry, National University of Ireland Galway, Galway, Ireland

Abstract Neglected tropical diseases (NTDs) are a group of disabling infectious diseases that disproportionately affect middle- and low-income people in Africa and Latin America. Although progress has been made in controlling and/or eradicating these devastating infections, there is still an urgent need for new and accessible drugs. For several neglected tropical diseases there are no available drugs or drug resistance has emerged to the existing ones, limiting overall efficacy of the current treatments. Natural products have been a source of active metabolites against various targets through the years, with over half of the approved pharmaceuticals either inspired by or directly representing natural products. In this review, we present secondary metabolites isolated from marine sources that have demonstrated activity against these diseases and are promising candidates for drug development. Herein, we discuss 114 marine derived metabolites active against the WHO's 2017 list of 20 NTDs. While not all these diseases have reported active marine natural products, we still describe these diseases and offer explanation as to why this rich source of bioactive metabolites has been omitted. With the development of new technologies in accessing new parts of the oceans, identifying novel chemistry or biological screening, the potential of marine natural products against NTDs has yet to be fully tapped into to.

Keywords Infectious disease · Marine natural products · Neglected tropical diseases · Parasites · Secondary metabolites

1 Introduction

1.1 Sources of Natural Products

Natural products, or secondary metabolites, are an invaluable source of compounds with pharmaceutical potential. These compounds are often used by the producing organism to enhance survival by providing a competitive advantage, serving as feeding deterrents, or promoting reproduction [1]. Nearly two-thirds of small-molecule pharmaceutical agents approved over the last four decades are natural products or inspired by natural product [2]. As a result of ease of access to terrestrial environments, the majority of natural products have been isolated terrestrial organisms; however, oceans have become more accessible due to SCUBA and submersible vehicle technology and marine organisms have become more attractive to the biomedical community. Over 35,000 marine natural products have been identified, with many demonstrating biological activity against human disease targets [3]. Most secondary metabolites from marine environments have been isolated from invertebrates, with Porifera and Cnidaria being the two phyla with the majority of bioactive metabolites. In recent years, marine microorganisms have also emerged as a prolific new source of active secondary metabolites [4–7].

This review focuses on natural products isolated from marine organisms that have demonstrated bioactivity against one or more of the neglected tropical diseases (NTDs) as defined by the World Health Organization (WHO). Metabolites isolated from a span of marine invertebrates are discussed, including from the phyla Porifera, Cnidaria, Bryozoa, Echinodermata, Mollusca, and Chordata. In addition, natural products isolated from algae, marine derived fungi, marine derived bacteria, and cyanobacteria are included. The growing contribution of marine microorganisms to the field of natural products is demonstrated in this review. To limit the scope of this review, only purified metabolites with an IC_{50} of 15 μM or less were included, though if a single metabolite was found to be effective against an NTD, this parameter was relaxed to highlight potential drug leads for a disease state that is otherwise lacking bioactivity from marine natural products. At times, members of compound families that lack potent bioactivity are discussed in order to highlight structural differences that may be responsible for the observed bioactivity of a metabolite.

1.2 Neglected Tropical Diseases

Neglected tropical diseases (NTD) are a group of largely infectious diseases that affect more than one billion people globally. NTDs are mostly found in low- and middle-income countries of Africa and Latin America, where vaccines and affordable treatment are still lacking or access to proper sanitation. Current therapeutic approaches for controlling and battling these diseases are inadequate, and often insufficient, due to limited local resources and the lack of interest from pharmaceutical companies. There is an urgent need for the development of new pharmaceutical agents and natural products are a great source for discovering new drug leads at lower cost [8, 9].

For the purposes of this chapter we used the list of NTDs from the WHO for 2017 which include Buruli ulcer, Chagas disease, dengue and severe dengue, chikungunya, echinococcosis, foodborne trematode infections, human African trypanosomiasis (HAT, sleeping sickness), leishmaniasis, leprosy (Hansen disease), lymphatic filariasis (elephantiasis), mycetoma, chromoblastomycosis and other deep mycoses, onchocerciasis, rabies, scabies and other ectoparasitoses, schistosomiasis, soil-transmitted helminthiases, snakebite envenoming, taeniasis and cysticercosis, trachoma, and yaws (endemic treponematoses). Due to the abundance of marine natural products with bioactivity against the causative agents of leishmaniasis, Chagas disease, and HAT, we have divided these sections by the marine organism from which the antiparasitic metabolites were isolated. The active metabolites against each remaining disease are grouped together by the NTD. While not all NTDs have active marine metabolites against the causative agent, we attempt to describe the limitations of marine drug discovery specific to these diseases and have grouped these specific descriptions toward the end of the review.

2 Marine Natural Products Active Against Leishmaniasis

Affecting over 1.5 million people per year and with a mortality rate of 70,000 cases per year, leishmaniasis is an endemic protozoan disease found in the tropics, subtropics, and southern Europe [10, 11]. There are three forms of leishmaniasis: cutaneous, mucocutaneous, and visceral. The severity can range from covert with minimal symptoms to infections associated with significant morbidity and mortality. Cutaneous leishmaniasis is most common and manifests as localized ulcerative skin lesions [12]. Mucocutaneous is the most disabling form of the disease, destroying the mucous membranes of the nasal and oropharyngeal cavities and causing significant disfigurement [13]. Visceral disease affects internal organs, primarily the liver and spleen, and is associated with a 95% fatality rate if left untreated [14].

There are over twenty *Leishmania* spp. that cause human infection, which include but are not limited to *Leishmania donovani* complex and its species, *L. mexicana* complex and its species, *L. tropica, L. major, L. aethiopica* and the subgenus *Viannia* with its species [12, 14]. Spread by about 30 species of sand flies, leishmaniasis is transmitted by infected female sand flies of the subfamily Phlebotominae [15]. There are two distinct phases of growth that occur in the development of the parasite, one within the sand fly (development of motile promastigotes) and the other in the infected mammal (development of intracellular amastigotes) [16]. The female sand fly is initially infected through the amastigotes in a blood-meal from an infected mammal [16–18]. The amastigotes transform into promastigotes within the gut of the insect and then migrate to the insect's proboscis. After the promastigote form is injected into humans, they are engulfed by macrophages and other phagocytic cells [16, 18]. This promotes the transformation of promastigotes into amastigotes, which then multiply by simple division and further infect other mononuclear phagocytic cells [16, 18]. The progression of the disease depends on the parasite strain and the host [18, 19]. This multistage lifestyle provides numerable opportunities for therapeutic intervention.

Five drugs are available for treatment of leishmaniasis, including amphotericin B, miltefosine, paromomycin, pentamidine, and pentavalent antimonials [20–23]. However, the efficacy of these agents is limited as they are associated with systemic toxicity and do not completely eliminate the parasite from the human body [24, 25]. In patients with concurrent immunosuppression, this often results in relapse [26]. Furthermore, there is an increase in drug resistant strains of the parasites, which has also reduced the overall impact of the current arsenal of leishmanicidal pharmaceuticals [20].

2.1 Antileishmanial Sponge Metabolites

Alkaloids represent a major chemical class for potent inhibition of leishmaniasis parasites from sponges. Extraction of the Japanese marine sponge *Neopetrosia* sp. led to the alkaloid renieramycin A (**1**) (Fig. 1) displaying sub-μM antileishmanial activity against recombinant *Leishmania amazonensis* promastigotes with an IC_{50} of 0.35 μM and greater than 10-fold selectivity over P388 murine leukemia cells [27]. A series of manzamine-type alkaloids isolated from *Acanthostrongylophora* sp. was shown to inhibit *L. donovani* promastigotes. Among this series, manzamine A (**2**) and its *N*-oxide (**3**) displayed the highest level of activity with respective IC_{50}s of 1.64 and 1.95 μM, while (+)-8-hydroxymanzamine A, manzamine E, 6-hydroxymanzamine E, manzamine F, 6-deoxymanzamine X, manzamine X, neo-kauluamine, ircinal A, ircinol A, manzamine Y, and 12,28-oxamanzamine A were all reported with IC_{50}s below 15 μM [28, 29]. Both **2** and **3** were found to have a selectivity index (SI, = mammalian cell line IC_{50}/parasite IC_{50}) below 4 relative to Vero cells. The pyrimidine alkaloid monalidine A (**4**) as well as four guanidine alkaloids batzelladine D, batzelladine F, batzelladine L (**5**), and norbatzelladine 1 (**6**) were tested for bioactivity against *L. infantum* strain MHOM/BR/1972/LD promastigotes following isolation from the Brazilian sponge *Monanchora arbuscula*. They displayed IC_{50}s of 2.46, 2.18, and 1.82 μM, and SI of 10, 11, and 47, respectively, relative to LLC-MK2 monkey kidney cells [30]. Related guanidine alkaloids ptilomycalin A (**7**), batzelladine M, batzelladine C, dehydrobatzelladine C, and crambescidine 800 from *Monanchora unguifera* were screened against *L. donovani* promastigotes. Ptilomycalin A was most potent, IC_{50} = 7.16 μM, with cytotoxicity >5.7 μM in Vero monkey kidney fibroblast cells [31]. Two bromopyrrole alkaloids, longamide B and dirbomopalau'amine (**8**), isolated from *Agelas dispar* and *Axinella verrucosa,* respectively, showed inhibition of *L. donovani* amastigotes of the strain MHOM/ET/67/L82 with **8** being the more

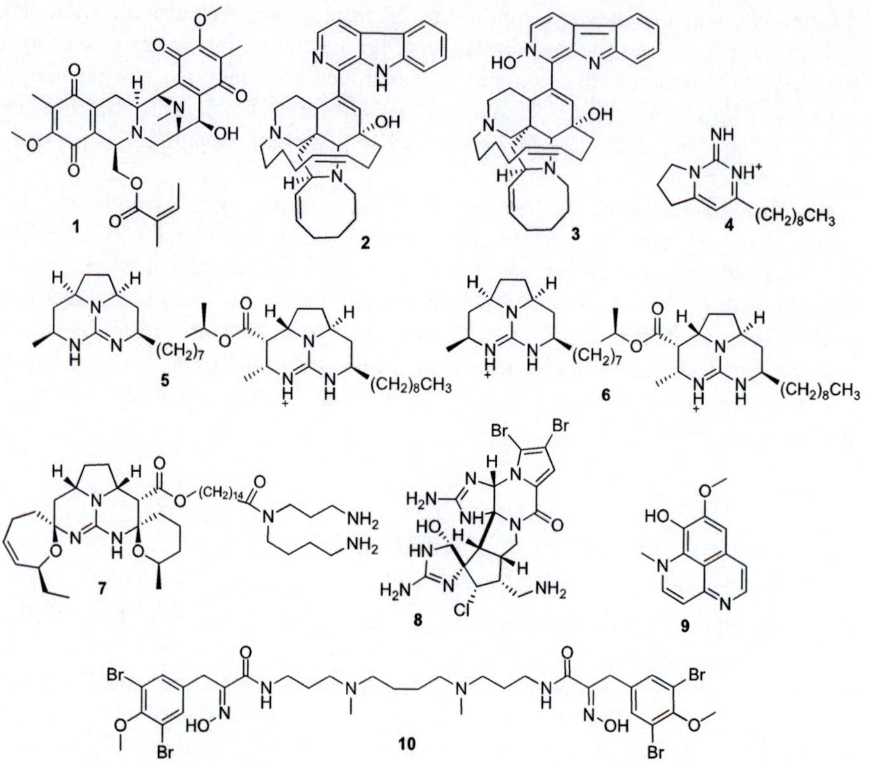

Fig. 1 Sponge alkaloids demonstrating activity against *Leishmania* spp. parasites

potent inhibitor with an IC_{50} of 1.88 μM representing fourfold selectivity over L6 rat skeletal myoblasts [32]. Isoaaptamine (**9**), an alkaloid isolated from *Aaptos* sp., exhibited an IC_{50} of 2.98 μM against *L. donovani* promastigotes, however with cytotoxicity against the murine leukemia cell line P388 at 1.23 μM it is not a selective antileishmanial metabolite [33]. A natural product screening library revealed the alkaloid spermatinamine (**10**) to inhibit *L. donovani* promastigotes and intracellular amastigotes with respective IC_{50}s of 11.87 and 6.5 μM [34].

Terpenes are another class of metabolites commonly isolated from marine organisms (Fig. 2). The *Agelas* sp. diterpene alkaloid agelasine D (**11**) has an SI of 4.5 for *Leishmania infantum* over MRC-5 fibroblast cells with respective IC_{50}s of 3.55 and 15.85 μM [35]. Spongian diterpenoids isolated from the Antarctic sponge *Dendrilla membranosa* demonstrate potent activity and high selectivity for axenic amastigotes of *L. donovani* [36]. Within the series of aplysulphurin, membranolide G (**12**), membranolide H, tetrahydroaplysulphurin, membranolide, and darwinolide, **12** was the most potent with a reported IC_{50} of 0.82 μM and >53.1 μM cytotoxicity against Vero cells.

In an effort to target leishmanicidal metabolites with low mammalian toxicity, a screen of three novel meroterpenoid disulfates from the Brazilian marine sponge

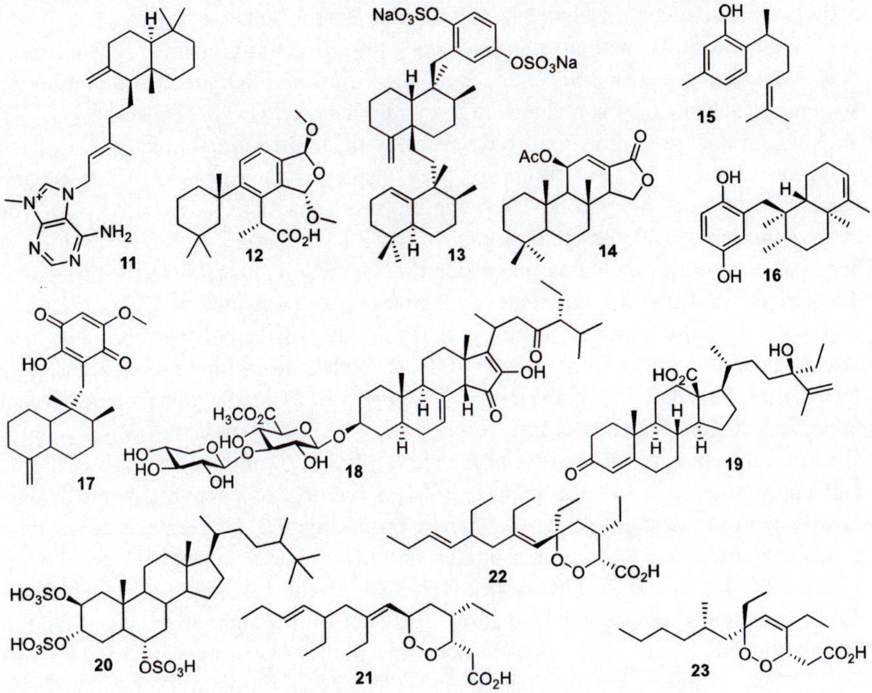

Fig. 2 Sponge terpenes displaying activity against *Leishmania* spp. parasites

Callyspongia sp. against *Leishmania* sp. adenosine phosphoribosyltransferase (L-APRT) returned isoakaterpin (**13**) as a 1.05 μM hit [37]. A linear meroterpene, 4-hydroxy-3-octraprenyl-benzoic acid, as well as the linear furanoterpene furospongin-1 and diterpene 11β-acetoxyspongi-12-en-16-one (**14**) were screened against *L. donovani* MHOM/ET/67/L82 amastigotes. The most potent inhibitor, **14**, displayed an IC_{50} of 2.07 μM with a fourfold selectivity over L6 rat skeletal myoblasts [38]. (*S*)-(+)-Curcuphenol (**15**), a sesquiterpene phenol from the Jamaican sponge *Myrmekioderma styx*, displayed an IC_{50} of 11 μM against *L. donovani* [39]. A SAR assessment at the C1 position achieved an 18-fold increase in the activity of **15** through an esterification with isonicotinic acid. Avarol (**16**) and avarone are sesquiterpenes isolated from *Dysidea avara*. Averol was found with IC_{50}s of 7.42, 7.08, and 3.19 μM, respectively, against *L. infantum* promastigotes, *L. tropica* promastigotes, and *L. infantum* amastigotes [40]. Assessment of averol cytotoxicity was performed using both human microvascular endothelial cell line HMEC-1 and THP-1 cells with associated IC_{50}s of 36.85 and 31.75 μM. *L. mexicana* strain NR promastigote growth was inhibited by illimaquinone (**17**), a mammalian Golgi apparatus disrupter, with an LD_{50} of 5.6 μM [41]. Chemical investigation into the composition of the Caribbean sponge *Pandaros acanthifolium* yielded a series of leishmanicidal steroidal glycosides including pandaroside F and pandaroside G as well as their methyl esters for which pandaroside G methyl ester (**18**) showed potent activity against *L. donovani* axenic amastigotes as represented by an IC_{50} of 51 nM [42]. Pandaroside G methyl ester displayed sub-μM activity against *Trypanosoma brucei rhodesiense, Trypanosoma cruzi,* and *Plasmodium falciparum* although screening against L6 cells revealed high cytotoxicity, $IC_{50} = 0.22$ μM.

Norselic acids are highly oxidized steroids with antipredation bioactivity, isolated from the Antarctic sponge *Crella* sp. They displayed activity against *Leishmania* parasites with norselic acid D (**19**) being the most potent (IC_{50} 2.0 μM) [43]. The sterol halistanol A (**20**) was active against L-APRT with an EC_{50} of 4.19 μM [44]. A series of polyketides including plakortide P (**21**), methyl (2*Z*,6*R*,8*R*,9*E*)-3,6-epoxy-4,6,8-triethyl-2,4,9-dodecatrienoate, spongosoritin A, and methyl (2*E*,6*R*,8*S*)-3,6-epoxy-4,6,8-triethyldodeca-2,4-dienoate, from the Brazilian sponge *Plakortis angulospiculatus*, displayed activity against both promastigote and amastigote forms of MHOM/BR/1972/LD strain *L. chagasi* with **21** displaying the most potent inhibition, $IC_{50} = 5.2$ and 1.4 μM, respectively [45]. Pachymatismin is a glycoprotein from the sponge *Pachymatisma johnstonii* that also inhibits both promastigote and amastigote forms of the parasite [46]. Screening of pachymatismin against *L. donovani* promastigotes, *L. braziliensis* promastigotes, *L. braziliensis* resistant promastigotes, *L. mexicana* promastigotes, and *L. mexicana* amastigotes found IC_{50}s of 13, 22, 54, 22, and 26 nM, respectively. The cyclic peroxide ent-3,6-epidoxy-4,6,8,10-tetraethyltetradeca-7,11-dienoic acid (**22**) from *Plakortis* sp. was found to inhibit the NR strain of *L. mexicana* promastigotes with an IC_{50} of 2.72 μM [47]. Two other cyclic peroxides (3*S*,6*R*,8*S*)-4,6-diethyl-3,6-epidioxy-8-methyldodeca-4-enoic acid (**23**) and 3,6-epidioxy-4,6,8,10-tetraethyltetradeca-7,11-dienoic acid as well as two furano-acids, methyl (2*Z*,6*R*,8*S*)-4,6-diethyl-3,6-epoxy-8-methyldodeca-2,4-dienoate and methyl (2*Z*,6*R*,8*S*)-3,6-epoxy-4,6,8-triethyldodeca-2,4-dienoate, were isolated from the Palauan sponge *P.* aff. *angulospiculatus* and assayed against *L. mexicana* promastigotes [48]. Metabolite **23** proved the most effective inhibitor with an IC_{50} of 0.97 μM. Two lectins, CaL from *Cinachyrella apion* and CvL from *Cliona varians*, have shown the ability to agglutinate *L. chagasi* promastigote cells up to 4 and 8 AU (agglutinating units) corresponding to the agglutination of 6.7×10^5 and 10^6 cells, respectively [49, 50]. The activity of CaL ceased with the addition of lactose, indicating the potential for lactose receptors in this parasitic stage.

2.2 *Antileishmanial Cnidaria Metabolites*

Shagenes A (**24**) (Fig. 3) and B were isolated from an undescribed octocoral collected in the South Georgia Islands of the southern Atlantic Ocean [51]. Shagene A was active against *Leishmania donovani* with an IC_{50} of 5 μM against the macrophage stage and 54 μM against the axenic amastigote form, with no notable cytotoxicity in J774.A1 macrophages. Both structures have a 3/6/5 tricyclic ring system and an isopropylidene on the three-membered ring, though **24** differs from shagene B by the presence of a methoxy group in place of the ketone found in

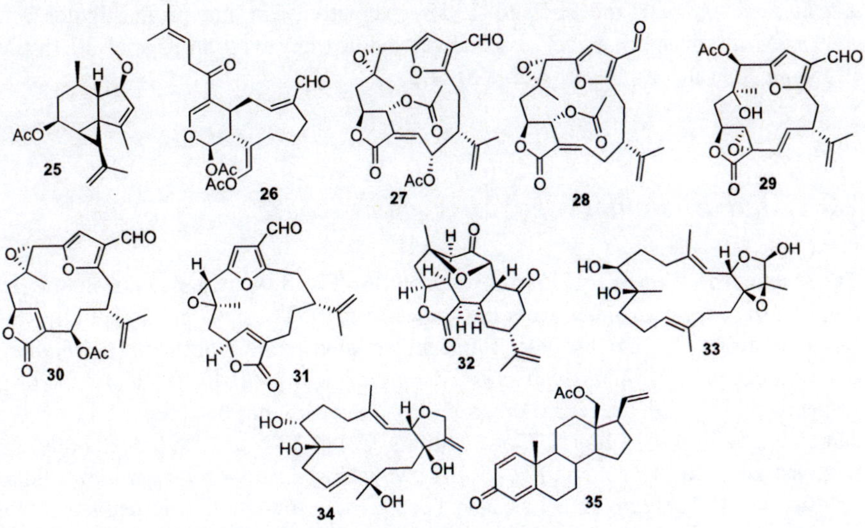

Fig. 3 Terpenes from Cnidaria with antileishmanial activity

shagene A. Shagene B had no activity against the parasite, highlighting the importance of the methoxy group.

Cristaxenicin A (**25**) was isolated from a Japanese deep-sea gorgonian *Acanthoprimnoa cristata* [52]. This xenicane diterpenoid was evaluated against *Leishmania amazonensis* (*La/egfp* promastigotes) which demonstrated potent activity with an IC_{50} of 88 nM and an SI of 23 to 53, with cytotoxicity tested in P388 and HeLa cells, respectively.

Keikipukalides B-E (**26-29**), pukalide aldehyde (**30**), and ineleganolide (**31**) are furanocembranoid diterpenes isolated from *Plumarella delictissima* collected at a depth of 800–950 m near the Falkland Islands in the Southern Ocean. These metabolites demonstrate activity against *Leishmania donovani* in the infected macrophage assay [53]. Of this group of briarane diterpenes and the norditerpenoid, the most potent of these is **30**, which had an IC_{50} of 1.9 μM. None demonstrated cytotoxicity in A549 cells at concentrations up to 50 μM.

Additional cembranoid diterpenes, lobocrasol A (**32**) and lobocrasol C (**33**) had very potent bioactivity against *Leishmania donovani* with IC_{50} of 0.18 μM and 0.17 μM, respectively, without significant cytotoxicity in L6 cells [54]. Interestingly, lobocrasol B had no activity against the parasite and represents a diastereomer of **33** at the position of the C-7 hydroxy group. These two cembranoid diterpenes were first isolated from *Lobophytum crassum*, a soft coral collected in Con Co, Quang Tri, Vietnam and initially found to have anti-inflammatory activity [55].

Initial screening of methanolic crude extracts from Brazilian cnidarians showed promising bioactivity that ultimately led to the rediscovery of the steroid 18-acetoxipregna-1,4,20-trien-3-one (**34**) from the octocoral *Carijoa riisei*, collected in the São Sebastião Channel, Brazil [56]. This steroid was tested for activity in

Leishmania chagasi and showed 15.53 μM activity in the promastigotes and 41.37 μM in amastigotes, but also showed cytotoxicity with an IC_{50} of 30.15 μM in mammalian macrophages, with a SI of 2.

2.3 Antileishmanial Algal Metabolites

The sesquiterpenoids elatol (**35**) and obtusol (**36**) (Fig. 4) were isolated from the red alga, *Laurencia dendroidea*, from the coast of Brazil. The most potent activity was reported from **35** against the intracellular amastigote form of *Leishmania amazonensis* and *Trypanosoma cruzi*, having an IC_{50} of 0.45 μM and 1.01 μM, respectively [57–60]. The enantiomer of (**35**), (−)-elatol, has been reported from the same alga collected within 300 km, with an IC_{50} of 13.5 μM [59]. Metabolite **36** demonstrates activity with an IC_{50} of 9.4 μM against *Le. amazonensis* intracellular amastigotes [59]. *Laurencia* has also been a rich source of triterpenoids, with fucosterol (**37**) isolated from *La. dendroidea* and iubol (**38**), saiyacenol A (**39**), saiyacenol B (**40**), and dehydrothyrsiferol (**41**) isolated from *La. viridis* [61, 62]. Fucosterol has reported activity against the intracellular amastigotes of both *Le. infantum* and *Le. amazonensis*, with an IC_{50} of 10.30 μM and 7.98 μM, respectively [61]. Iubol is active against *Le. amazonensis* and *T. cruzi* (7.00 μM and 9.20 μM). While saiyacenol A showed activity against both *Le. amazonensis* and *T. cruzi* (12.96 μM and 13.75 μM), saiyacenol B had a slightly more potent IC_{50} of 10.32 μM against *Le. amazonensis*. Dehydrothyrsiferol showed activity at similar concentrations against *Le. amazonensis* (8.36 μM) and *T. cruzi* (9.45 μM). Dehydrothyrsiferol (**41**) has also been manipulated synthetically to produce

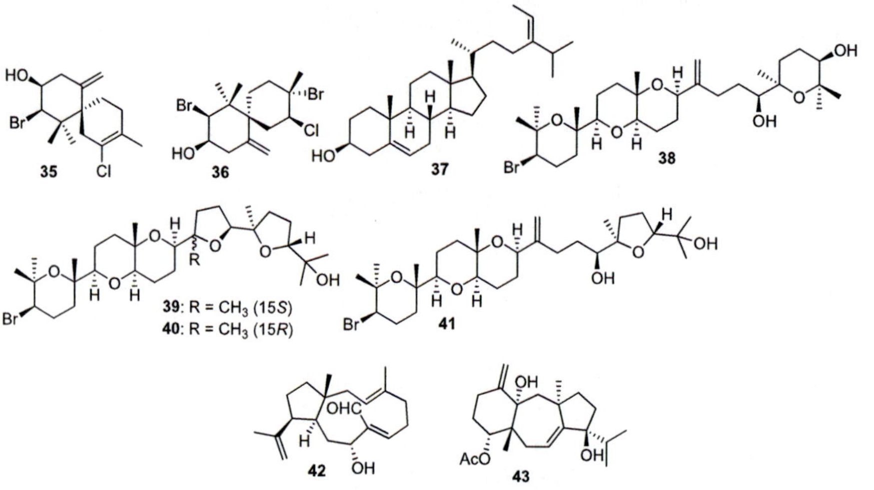

Fig. 4 Algal terpenes active against *Leishmania* spp. parasites

bioactive triterpenoids 18-ketodehydrothyrsiferol, 28-iodosaiyacenol A, and 28-iodosaiyacenol B, with only the latter showing an improved IC_{50} of 5.40 μM against *Le. amazonensis* and 16.18 μM against *Le. donovani* [62].

A diterpene, spiralyde A (**42**), isolated from brown algae, *Dictyota spiralis*, has bioactivity against *Leishmania amazonensis* and *Trypanosoma cruzi* with IC_{50}s of 15.47 μM and 5.62 μM, respectively [63]. Bioactive meroditerpenoid 4-acetoxydolastane (**43**), from brown alga *Cystoseira baccata*, has a similar activity against *L. amazonensis* with an IC_{50} of 5.5 μM [64].

2.4 Antileishmanial Echinoderm Metabolites

Isolation of a series of saponins, neothyosides A-C, from the Pacific sea cucumber *Neothyone gibbosa* has shown neothyoside C (**44**) (Fig. 5), a sulfated tetraglycoside chemically identical to previously isolated pervicoside B from *Holothuria pervicax*, to be active against MHOM/MX/84ISETGS and MHOM/MX/88HRCMC strains of *Leishmania mexicana* with IC_{100}'s of 4.20 and 8.41 μM, respectively [65, 66]. Metabolite **44** showed low cytotoxicity against *Artemia salina* shrimp with an LD_{50} of 52.98 μM. In contrast, neothyosides A and B, for which the terminal isobutylene substituent of C is substituted with an acetoxy group, displayed poor antileishmanial activity at or above 80 μM [65].

2.5 Antileishmanial Molluscan Metabolites

Crude extracts and bis-indole indorubin derivatives from the mollusc, *Hexaplex trunculus* [67] were screened against *Leishmania donovani*. The only purified mollusc metabolite reported to date with activity against leishmaniasis is from the sea hare *Dolabrifera dolabrifera*. The epidioxysterol, 5α,8α-epidioxycholest-6-en-3β-ol (**45**) (Fig. 6), was isolated from an aqueous extract and subject to biological screening against *L. donovani, Plasmodium falciparum,* and *Trypanosoma cruzi* [68]. The IC_{50} of *L. donovani* was most promising, 4.6 μM, while the other two IC_{50}s

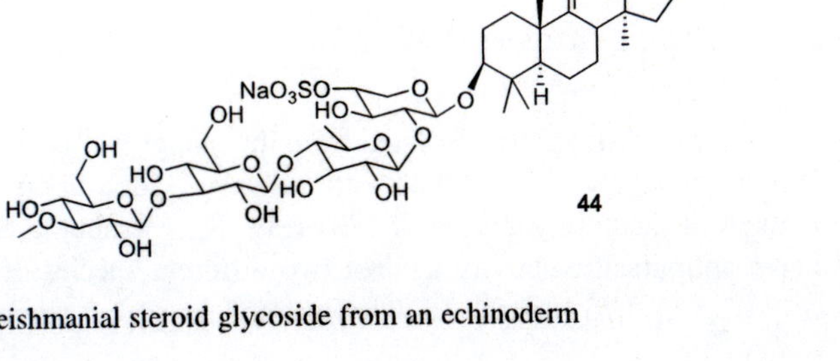

44

Fig. 5 Antileishmanial steroid glycoside from an echinoderm

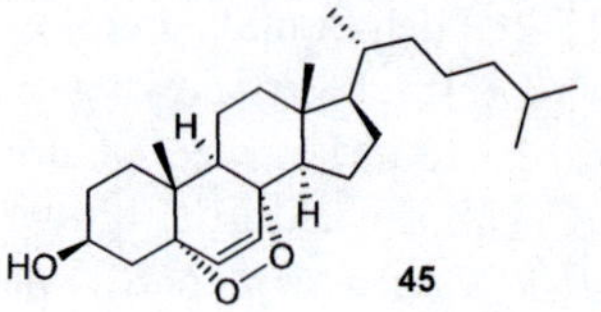

Fig. 6 Steroid from a sea hare with activity against the leishmaniasis parasite

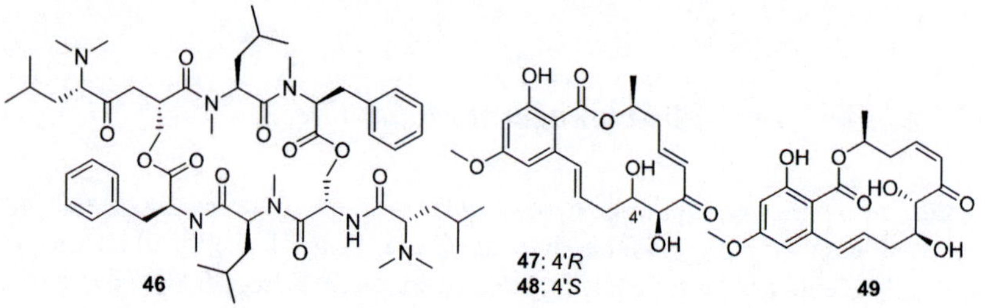

Fig. 7 Marine fungal antileishmanial metabolites

were reported to be >10 μM. Cytotoxicity of **45** against a mammalian Vero cell line was determined to be 281 μM, with an SI of 57.3 showing that this metabolite could be a promising drug lead.

2.6 Antileishmanial Marine Fungal Metabolites

IB-01212 (**46**) is a cyclodepsipeptide isolated from the marine fungus *Clonostachys* sp. (Fig. 7). It has shown activity against *Leishmania donovani* promastigotes with an IC_{50} of 10.5 μM and against *L. pifanoi* amastigotes with an IC_{50} of 7.1 μM [69, 70]. The fungus *Cochliobolus lunatus*, which was isolated from a zoanthid, upon fermentation under optimized conditions yielded a series of 14 membered resorcylic acid lactones. Of those, cochliomycin D (**47**), (7′*E*)-6′-oxozeanol (**48**), and LL-Z1640-2 (**49**) showed activity against *L. donovani* with an IC_{50} of 9.1, 1.9 and 2.2 μM, respectively [71, 72].

2.7 Antileishmanial Bacterial Metabolites

The bacterium *Paenibacillus* sp. was isolated from the mangrove sediment collected from the wetlands of the Digya National Park, Ghana, which upon fermentation yielded new alkaloid paenidigyamycin A (**50**) (Fig. 8). Metabolite **50** showed a 10-fold selective antiparasitic activity against two different species of *Leishmania* with IC_{50}s of 7.02 and 0.75 μM for *L. donovani* and *L. major*, respectively. The

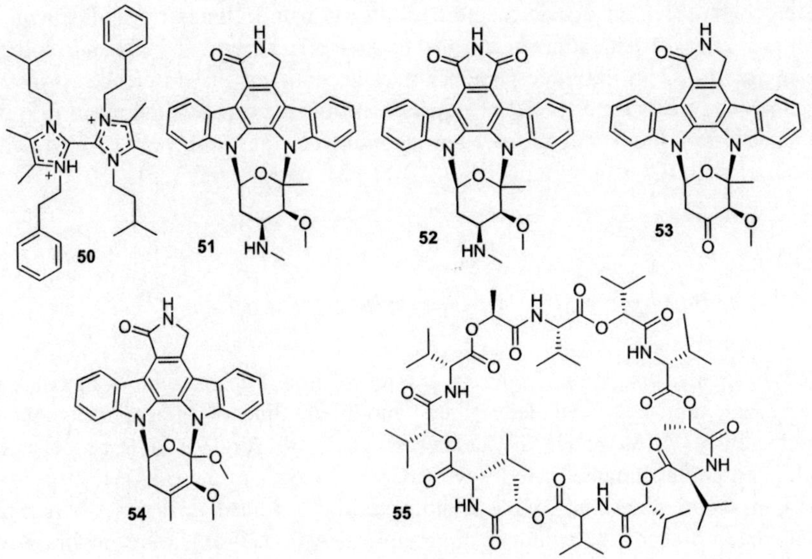

Fig. 8 Marine bacterial metabolites active against leishmaniasis

bacterial alkaloid also showed sub-μM activity against the trypanasomal parasite *Trypanosoma brucei brucei* with an IC_{50} of 0.78 μM. In addition, these authors tested **50** against *Schistosoma mansoni* cercariae and observed that at a concentration of 100 μM, cercariae viability decreased significantly within half an hour and complete mortality occurred after 1 h. The MLC of **50** on cercariae after 180 min was 25 μM, while the MEC after at the same time point was 6.25 μM [73]. Another mangrove-dwelling bacterium *Streptomyces sanyensis*, isolated from sediment collected from Jambelí mangroves in Ecuador, biosynthesized a series of indolocarbazoles (**51-54**). The researchers evaluated the antikinetoplastid activities of the indolocarbazole alkaloids against promastigotes of both *L. amazonensis* and *L. donovani*. Both **53** and **54** were deemed inactive against *L. donovani* at concentrations below 40 μM and showed moderate activity against *L. amazonensis* with IC_{50}s of 17.10 and 10.44 μM, respectively. Alkaloid **51** showed potent activity against *L. amazonensis* with an IC_{50}s of 80 nM, while less active against *L. donovani* ($IC_{50} = 2.07$ μM). Metabolite **52** showed the inverse, with a better activity against *L. donovani* ($IC_{50} = 0.56$ μM) when compared to *L. amazonensis* ($IC_{50} = 3.58$ μM). All four metabolites, **51-54**, exhibited activity against the epimastigotes of *Trypanosoma cruzi* IC_{50}s of 3.63, 1.58, 17.10 and 12.50 μM, respectively. Due to the promising results obtained for **52** the authors investigated a possible mechanism of action by evaluating the effects of **52** on the mitochondrial membrane potential in promastigotes of *L. amazonensis* and *L. donovani*, and epimastigotes *of T. cruzi*. When tested with **52** at the IC_{90} concentration (8.36 μM) the researchers observed an intense effect of the mitochondrial membrane potential in only *L. amazonensis* promastigotes. In a subsequent cytoplasmic membrane permeability study, researchers observed drastic membrane alteration in cultures of *L. donovani* and *T. cruzi* after 24 h treatment with the IC_{90} of **52**. These two different biological responses to **52** in the three parasites may account for the differences in activity against the three species [74]. Previously mentioned **51** and valinomycin (**55**) were isolated from the sponge-associated actinomyces and showed activity against *L. major* with IC_{50}s of 5.30 μM and < 0.11 μM, respectively [75].

2.8 Antileishmanial Cyanobacterial Metabolites

Activity-guided isolation from an extract of the Panamanian cyanobacterium *Lyngbya majuscula* yielded three modified linear lipopeptides, namely dragonamides A (**56**) and E (**57**) and herbamide B (**58**) (Fig. 9). All three metabolites exhibited antileishmanial activity with IC_{50}s of 6.5, 5.1, and 5.9 μM, respectively [76]. In a subsequent study of the cyanobacterium *L. majuscula*, a series of highly *N*-methylated linear lipopeptides, almiramides A-C (**59-61**) were reported with antileishmanial activity against *Leishmania donovani* with IC_{50}s of >13.5, 2.4 and 1.9 μM, respectively. While **60** and **61** exhibited antiparasitic activity they also showed cytotoxicity toward Vero cells with IC_{50}s of 52.3 and 33.1 μM, respectively. The semisynthesis of a series of derivatives highlighted structural features necessary for activity and provided one new metabolite with superior *in vitro* activity while

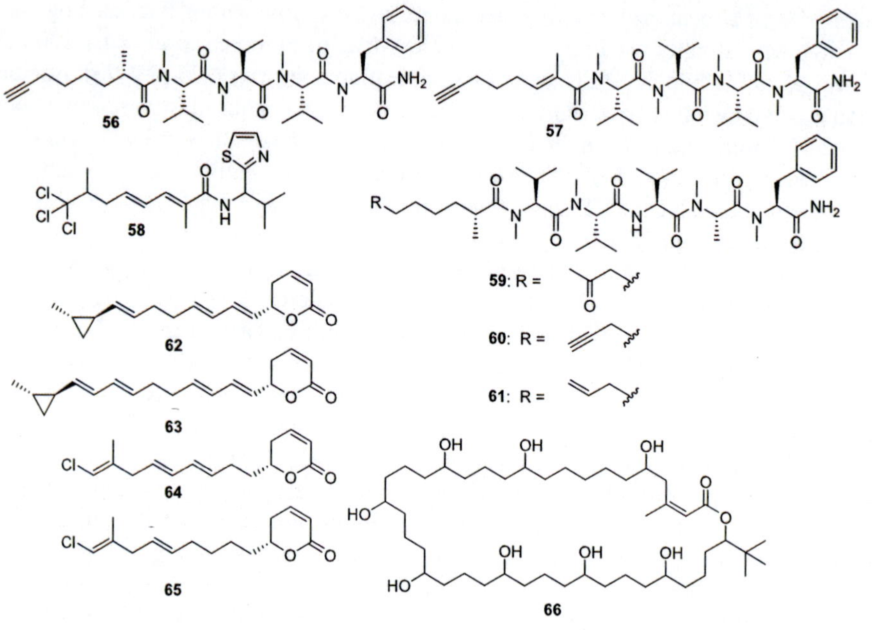

Fig. 9 Antileishmanial marine cyanobacterial metabolites

minimizing the cytotoxicity in the majority of the analogues [77]. Chemical analysis of another species of Panamanian cyanobacterium, cf. *Oscillatoria* sp. yielded four unsaturated polyketide lactones, coibacins A-D (**62–65**). These cyanobacterial metabolites **62–65** possess antileishmanial activity against *Le. donovani* axenic amastigotes with IC_{50}s of 2.4, 7.2, 18.7 and 7.8 μM, respectively [78]. Extracts of a tropical marine cyanobacterium, *Leptolyngbya* sp., collected at Palmyra Atoll were analyzed using NMR-guided fractionation and MS^2-based molecular networking to identify a complex polyhydroxy macrolide with a 40-membered ring, palstimolide A (**66**). Palstimolide A (**66**) was tested for activity against *Le. donovani* resulting in an IC_{50} of 4.67 μM [79].

3 Marine Natural Products Active Against Human African Trypanosomiasis

Human African trypanosomiasis (HAT), known as sleeping sickness, is a disease caused by two geographically distinct subspecies of the parasite *Trypanosoma brucei*; *T. b. gambiense*, found in western and central Africa, and *T. b. rhodesiense*, found in eastern and southern Africa [80]. Both subspecies lead to clinical manifestations in two phases in humans; a hemolymphatic phase which manifests clinically as chancre, irregular fever, headaches and adenopathy, and a meningoencephalitic phase, which occurs after the parasites cross the blood–brain barrier into the central nervous system and is associated with high mortality rates [80, 81]. Infections associated with *T. b. gambiense* tend to have a slower progression, while those due to *T. b. rhodesiense* cause severe disease even in the acute stage of the disease [82].

HAT is transmitted by the tsetse fly (*Glossina* spp.), which becomes infected by trypomastigotes when taking a blood-meal from an infected host [83, 84]. In the fly's midgut they transform into procyclic trypomastigotes, multiply and further transform into epimastigotes once they leave the midgut [83, 84]. The epimastigotes reach the salivary glands of the fly where they further multiply. The infected fly then injects metacyclic trypomastigotes into skin tissue during a blood-meal [83, 84]. The parasites enter the bloodstream through the lymphatic system and transform into bloodstream trypomastigotes [83, 84]. From there, they are transferred to other sites and continue to replicate [83, 84]. *T. b. gambiense* may be also transmitted from mother to child if infected during pregnancy.

Treatment of HAT is specific to the type of infection and disease stage, with four medications currently on the market and a fifth one approved for combination therapy. These include suramin and pentamidine for early-stage disease and melarsoprol, eflornithine, and nifurtimox combined with eflornithine for late-stage infections of *T. b. gambiense* [80, 85, 86]. Although control efforts have resulted in less than 2,000 cases annually, there are several challenges in treating HAT including the toxicity of the pharmaceutical agents, complicated administration, and emerging drug resistance [85, 86].

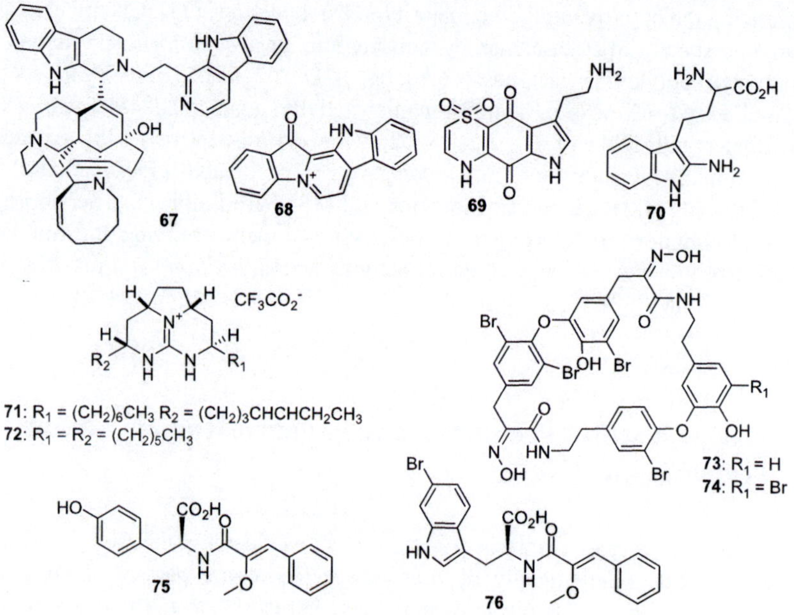

Fig. 10 Sponge alkaloids with activity toward the causative agent of HAT

3.1 Sponge Metabolites Active Against the HAT Parasites

Manzamine A (**2**) (Fig. 1) and related manzamine alkaloids zamamidines A-C and 3,4-dihydromanzamine J *N*-oxide isolated from *Amphimedon* sp. were shown to inhibit *Trypanosoma brucei* with **2** the most potent with (IC_{50} 0.07 μM) [87]. Zamamidine C (**67**) (Fig. 10) also possesses sub-μM activity at IC_{50} of 0.36 μM. The bromopyrrole alkaloids longamide B and dibromopalau'amine (**8**) from *Agelas dispar* and *Axinella verrucosa* also act as inhibitors of *Trypanosoma rhodesiense* with **8** more active, inhibiting 50% of the parasite at 0.79 μM [32]. Extraction of the Fijian sponge *Hyrtios* cf. *erecta* afforded the known metabolite fascaplysin (**68**) which was tested against *T. brucei* and found as another sub-μM inhibitor with an IC_{50} of 0.63 μM with greater than 10-fold selectivity against L6 cells [88]. Spermatinamine (**10**) and another alkaloid thiaplakortone A (**69**) were found to inhibit strain BS427 *T. brucei* parasites with an IC_{50}s of 1.0 and 3.94 μM, respectively [35]. Agelasine D (**11**) was found to inhibit Squib-427 strain *T. brucei* at an IC_{50} of 2.13 μM with an SI of 7.4 [35]. Extraction of the Red Sea marine sponge *Hyrtios* sp. has yielded a new antitrypanosomal alkaloid hyrtiodoline A (**70**) with an IC_{50} of 7.48 μM and no reported cytotoxicity against J774.1 macrophages [89]. Two tricylic guanadines, merobatzelladines A (**71**) and B (**72**), isolated from *Monanchora* sp. show similar potency in inhibiting *T. brucei* with respective IC_{50}s of 0.51 and 0.67 μM [90].

The bastadin class of bromotyrosine derivatives from the sponge *Ianthella* cf. *reticulata* revealed bastadins 13 (**73**) and 19 (**74**) (Fig. 10) to inhibit

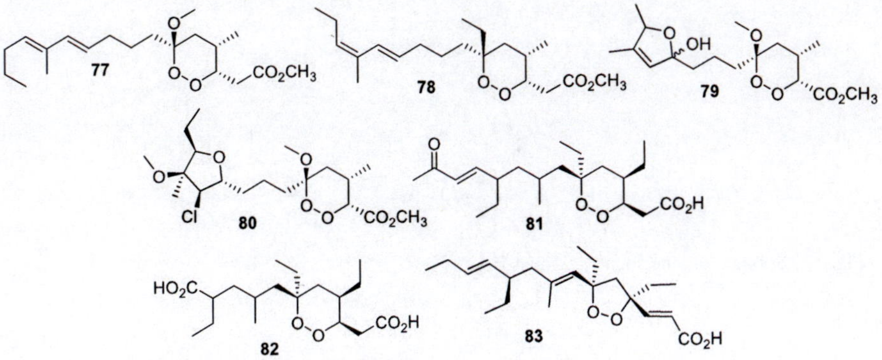

Fig. 11 Sponge polyketides active against HAT

Trypanosoma brucei parasites with IC_{50}s of 1.77 and 1.31 μM, respectively [91]. Two *N*-cinnamoyl-amino acids, Iotrochamides A (**75**) and B (**76**), isolated from the Australian sponge *Iotrochota* sp. inhibited 50% of *T. brucei* strain BS427 at concentrations of 4.7 and 3.4 μM, respectively [92].

A series of polyketides isolated from the Indonesian *Plakortis* cfr. *lita* including manadoperoxide B (**77**), 12-isomanadoperoxide B (**78**), manadoperoxidic acid, manadoperoxide C, manadoperoxide F, manadoperoxide G, manadoperoxide H, manadoperoxide I (**79**), and manadoperoxide K (**80**) as well as the peroxyplakortic ester B3. Metabolites **77-80** showed potent activity against *Trypanosoma brucei rhodesiense* with IC_{50}s of 8.8, 32, 170, and 210 nM, respectively (Fig. 11) [93, 94]. A 20-fold reduction in activity was noted between the activity of the *Plakortis* sp. cyclic polyketide peroxide 11,12-didehydro-13-oxo-plakortide Q (**81**) (IC_{50} 49 nM) when a carboxylic acid was present in the sidechain yielding 10-carboxy-11,12,13,14-tetranor-plakortide Q (**82**) (IC_{50} 0.94 μM) [95]. Plakortide E (**83**), an endoperoxide from the Caribbean sponge *P. halichondroides*, was found to be trypanocidal at a concentration of 5 μM, as well as an inhibitor of cathepsins and cathepsin-like parasitic proteases [96].

Terpene peroxide inhibitors of *Trypanosoma brucei* were discovered in *Diacarnus bismarckensis* including (+)-muqubilone B, (+)-sigmosceptrellin A, (+)-sigmosceptrellin A methyl ester, (−)-sigmosceptrellin B (**84**), (+)-*epi*-muqubillin A, and (−)-*epi*-nuapapuin B methyl ester with **84** demonstrating the strongest inhibition, $IC_{50} = 0.5$ μM (Fig. 12) [97]. The linear furanoterpene demethylfurospongin-4, linear monoterpenes 4-hydroxy-3-tetraprenylphenylacetic acid (**85**) and heptaprenyl-*p*-quinol, as well as diterpenes dorisenone D and **14** (Fig. 2) isolated from *Spongia* sp. and *Ircinia* sp. were all found to be inhibitors of *Trypanosoma rhodesiense* strain STIB 900 [38]. Metabolite **85** displayed an IC_{50} of 1.46 μM. The Caribbean sponge *Pandaros acanthifolium* has yielded a host of antitrypanosomal steroidal glycosides including padarosides A, E, F, G (**18**), I, and J as well as the methyl esters of E and G (**86**) and acanthifolioside F methyl ester [42, 98]. The IC_{50}s for **18** and **86** are 0.8 μM and 38 nM, respectively.

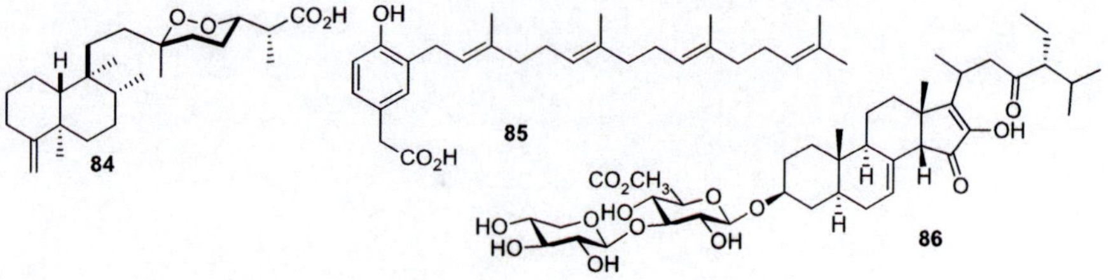

Fig. 12 Sponge terpenes active against HAT

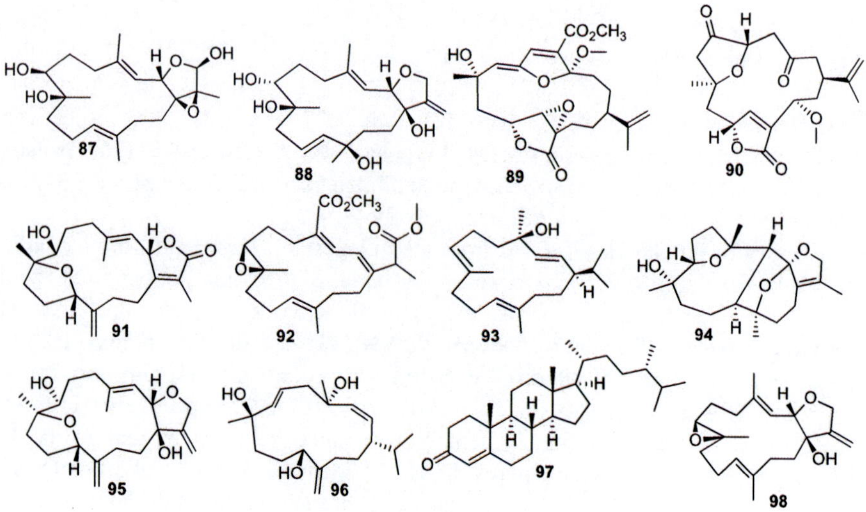

Fig. 13 Metabolites from Corals active against HAT

3.2 *Cnidarian Metabolites Active Against HAT*

A suite of thirty-four terpenoids derived from the Vietnamese soft corals *Lobophytum crassum*, *Lobophytum laevigatum*, and *Sinularia maxima*, were tested against *Trypanosoma brucei rhodesiense*, of which ten demonstrated significant bioactivity and favorable SI when considering cytotoxicity in rat skeletal myoblast cell line L6 [54]. Lobocrasol A (**87**) and C (**88**) (Fig. 13), active against *Leishmania donovani*, were also active against *T. b. rhodesiense* with IC_{50} of 9.97 μM and 9.37 μM, respectively, though the selectivity for both was modest with an SI of 5.6 and 4.3. Sinumaximol C (**89**), 13-*epi*-scabrolide C (**90**), and crassumol G (**91**) all demonstrated activity, though with slightly higher IC_{50}s (9.88, 11.76, and 12.20 μM) and less favorable selectivity. 7*S*,8*S*-Epoxy-1,3,11-cembratrien-16-oic acid methyl ester (**92**) had an IC_{50} of only 1.00 μM and no demonstratable cytotoxicity (SI > 54). (1*R*,4*R*,2*E*,7*E*,11*E*)-Cembra-2,7,11-trien-4-ol (**93**) had an IC_{50} of 1.14 μM and an SI of 43. Crassumol D (**94**), crassumol E (**95**), and (1*S*,2*E*,4*S*,6*E*,8*S*,11*S*)-2,6,12(20)-

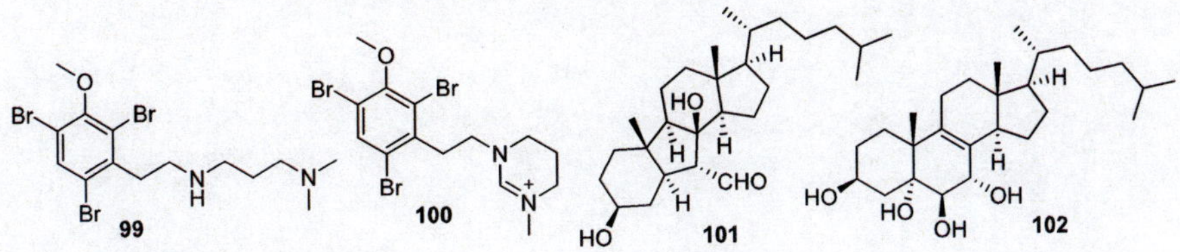

Fig. 14 Bryozoan and echinoderm metabolites active against HAT

cembrantriene-4,8,11-triol (**96**) all inhibited *T. b. rhodesiense* growth with an IC_{50} of less than 1 μM and SI greater than 59 [54]. Additionally, (24*S*)-ergost-4-ene-3-one (**97**) isolated from *Sinularia dissecta* with IC_{50} of 5.47 μM and no notable cytotoxicity [99]. Another promising metabolite is laevigatol B (**98**) isolated from *Lo. crassum* and *Lo. laevigatum*, a diterpene with activity against *T. brucei* with an IC_{50} of 5.34 μM and no notable cytotoxicity against HEK293T and HepG2 cells. The crude methanolic extract had greater activity than the purified metabolite, which suggested synergy with other metabolites that have yet to be discovered [99].

3.3 *Metabolites from Bryozoa Inhibiting Active Against HAT*

Two brominated alkaloids, convolutamines I (**99**) and J (**100**) (Fig. 14), isolated from the bryozoan *Amathia tortuosa* displayed promising bioactivity against *Trypanosoma brucei* parasites with respective IC_{50}s of 1.1 and 13.7 μM [100]. When tested for cytotoxicity against human embryonic kidney cell line HEK293 the pair showed IC_{50}s of 22 μM for convolutamine I and 41 μM for J, displaying a greater SI for the former. A lack of biological activity data exists for related convolutamines, and further evaluation could hold information regarding the structural components required for HAT inhibition [100].

3.4 *Echinoderm Metabolites Active Against HAT*

High throughput screening of 433 natural product extracts and 428 purified metabolites identified two echinoderm-derived steroids possessing activity against *Trypanosoma brucei* [99]. The sea star *Astropecten polyacanthus* produces astropectenol A (**101**) (Fig. 14) capable of inhibiting 50% of strain 427 bloodstream parasites at a concentration of 1.57 μM, while cholest-8-ene-3β,5α,6β,7α-tetraol (**102**) from the sea urchin *Diadema savignyi* showed an EC_{50} of 14.60 μM. Both metabolites showed CC_{50} cytotoxicity greater than 100 μM against both HEK293T and HepG2 cells.

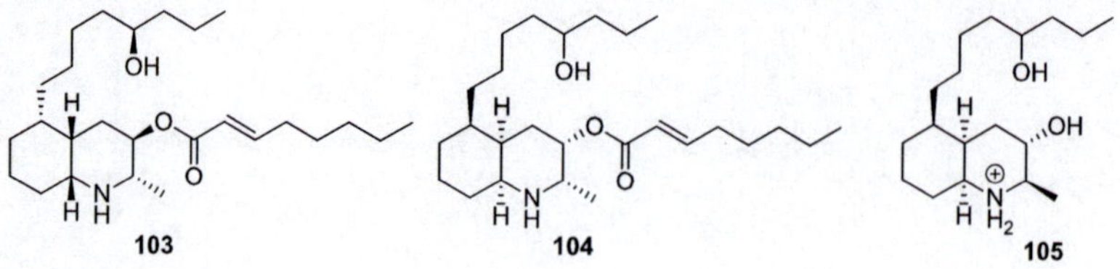

Fig. 15 Ascidian metabolites active against HAT

3.5 Ascidian Metabolites Active Against HAT

Lepadins E and F (**103** and **104**) are alkaloids containing a decahydroquinoline skeleton (Fig. 15). These metabolites, as well as the quaternary nitrogen derivative of lepadin D (**105**), were isolated from *Didemnum* sp. and screened for activity against *Trypanosoma rhodesiense* [101]. Both **103** and **104** had remarkable results with an IC_{50} of 0.90 and 0.55 μM, respectively, while **105** displayed an IC_{50} of 6.62 μM. The stereochemical configuration of **103** was later established through synthetic efforts and is reflected in the structure below [102].

3.6 Marine Bacterial Metabolites Active Against HAT

In a study of Croatian sponge-associated actinomyces, several extracts showed anti-infective activities. Most notably were two actinomyces isolates from two different sponge species, *Axinella polypoides* and *Aplysina aerophoba*, which both produced the cyclodepsipeptide valinomycin (**55**) (Fig. 8) which showed potent antiparasitic activity against *Trypanosoma brucei brucei* with IC_{50}s of 3.2 nM and 3.6 nM after a 48 h and 74 h incubation, respectively. Another active strain identified in this study was isolated from *Tedania* sp., which yielded the indolocarbazole alkaloid staurosporine (**51**) (Fig. 8), which exhibited significant inhibitory activities against *Tr. b. brucei* with IC_{50}s of 0.022 μM and 0.035 μM after 48 and 74 h incubation, respectively [75]. A comprehensive metabolomic analysis and dereplication strategy of the extract of a marine sponge-associated *Actinokineospora* sp. EG49 identified two new antitrypanosomal metabolites actinosporins A (**106**) and B (Fig. 16). These two new *O*-glycosylated angucyclines were tested for activity against *Tr. b. brucei*, where only the former exhibited activity with an IC_{50} of 15 μM suggesting the importance of the second sugar moiety to antitrypanosomal activity [103]. In a similar study of the extract of marine sponge-derived bacterium *Ac. spheciospongiae* sp. nov. isolated from the Red Sea sponge *Spheciospongia vagabunda*, two new fridamycins H (**107**) and I, and three known metabolites, actinosporin C, D, and G, were isolated. All metabolites were inactive against *Tr. b. brucei*, with the exception of **107** which showed significant antitrypanosomal activity after 48 and 72 h with IC_{50}s of 3.38 and 5.26 μM, respectively. No cytotoxicity against J774.1 macrophages was observed (IC_{50} of >200 μM) [104]. The rare actinomycete,

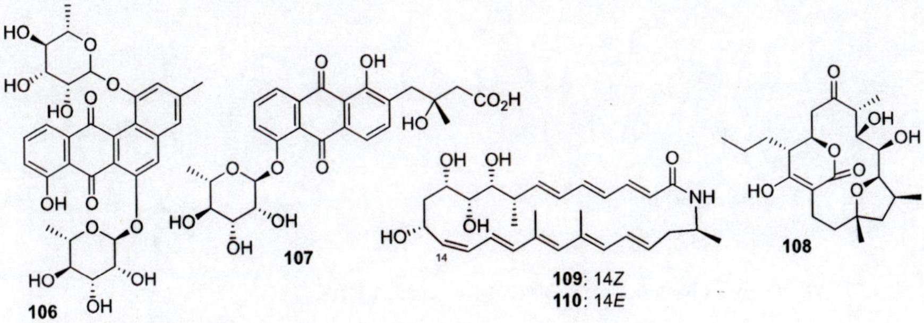

Fig. 16 Metabolites from marine microbes with anti-HAT bioactivity

Lechevalieria aerocolonigenes, was isolated from a mangrove sediment sample collected on Iriomote Island, Japan. A purified culture of *L. aerocolonigenes* yielded two new cyclopentadecane metabolites mangromicins A (**108**) and B, which exhibited *in vitro* antitrypanosomal activity against *Tr. b. brucei* with IC_{50}s of 5.85 and 110 μM, respectively [105]. Screening of a marine bacterially derived natural product library against axenic *Tr. b. brucei* identified a fraction originating from *Micromonospora* sp. RL09-050-HVF-A isolated from a marine sediment sample collected from Point Lobos in Monterey Bay, California. Further spectroscopic analysis identified three novel polyene macrolactams, lobosamides A–C. Lobosamides A (**109**) and B (**110**) exhibited activity against *Tr. b. brucei* parasites with IC_{50}s of 0.8 and 6.1 μM, respectively [106].

3.7 Marine Cyanobacterial Metabolites with Activity Against HAT

As highlighted throughout this chapter, cyanobacteria are a prolific source of bioactive metabolites. The marine cyanobacteria *Okeania* sp. collected near Janado, Okinawa Prefecture, Japan produced the cyclic polyketide–peptide hybrid, janadolide (**111**) (Fig. 17). This metabolite exhibited potent antitrypanosomal activity against *Trypanosoma brucei brucei* with an IC_{50} of 47 nM and no cytotoxicity against human cells at 10 μM [107]. In a subsequent study this same research group isolated and synthesized the antitrypanosomal lactam hoshinolactam (**112**), from a marine cyanobacteria from the genus *Oscillatoria*. This metabolite was active against *T. b. brucei* with an IC_{50} of 3.9 nM without cytotoxicity against human fetal lung fibroblast MRC-5 cells (IC_{50} > 25 μM) [108]. Chemical analysis of the marine *Dapis* sp. cyanobacterium yielded three linear peptides iheyamides A (**113**), B, and C. While, iheyamides B and C were deemed inactive, **113** showed moderate antitrypanosomal activities against *T. b. rhodesiense* and *T. b. brucei* (IC_{50} = 1.5 μM) [109]. An *in vitro* screening initiative against *T. b. brucei* identified two previously isolated Panamanian collected *Lyngbya majuscula* metabolites,

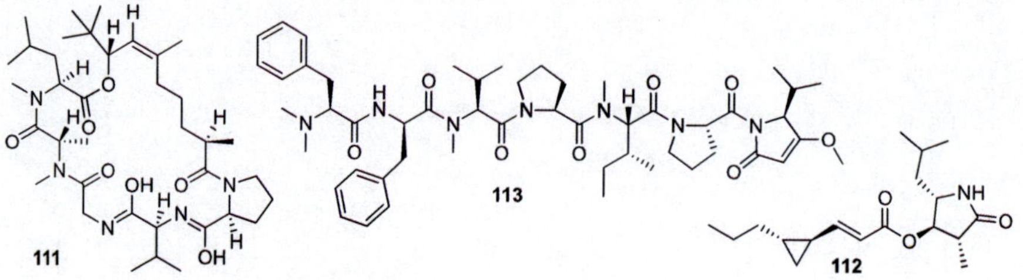

Fig. 17 Marine cyanobacterial metabolites active against HAT

almiramides B (**59**) and C (**61**) (Fig. 9) with IC_{50}s of 6.0 and 3.0 μM, respectively. The molecular target of these metabolites and their analogues were investigated in *T. brucei* lysates, and site localization was probed using epifluorescence microscopy. Compounds **59** and **61** likely interfere with the function of glycosomes, organelles both specific to kinetoplastid parasites and essential to their survival in the bloodstream stage. In addition, these metabolites were evaluated for toxicity in *in vivo* zebrafish models, which showed that they have little effect on organism development, even at high concentrations [77, 110].

4 Marine Natural Products Active Against Chagas Disease

Chagas is a chronic disease caused by the protozoan *Trypanosoma cruzi* and can be found in the Americas, predominantly in rural Latin America [111]. Transmitted by insect vectors to both humans and animals, the parasite affects around eight million people annually and leads to around 50,000 deaths a year [112]. The early clinical manifestations of the disease include inflammation at the site of entry, while systemic disease includes the onset of cardiac arrhythmias, splenomegaly, lymphadenopathy, tachycardia, and fever [113]. Chronic infection is associated with cardiac disease, resultant liver congestion, digestive tract abnormalities, and denervation of the nervous system [113].

Chagas disease is transmitted through the feces of an insect vector, primarily the triatomine bug, which can be infected by *Trypanosoma cruzi* after feeding on an infected human or animal [114]. Trypomastigotes transform in two stages within the insect; epimastigote asexual multiplication in the midgut and differentiation into infective metacyclic trypomastigotes in the hindgut [114]. The insect then releases the trypomastigotes in its feces near the bite wound allowing the parasites to enter the host through the wound [114]. Alternatively, they can traverse intact mucosal membranes [114]. Once inside the host, the trypomastigotes invade cells and differentiate into intracellular amastigotes which then multiply and differentiate into trypomastigotes, which are released into the vascular circulation and affect different organ systems, leading to the clinical symptoms [114]. The trypomastigotes transform into intracellular amastigotes in the new infection sites

[114]. *Trypanosoma cruzi* can also be transmitted through blood transfusions, laboratory accidents, organ transplant, and from mother to unborn baby [112].

There are currently only two drugs available in the affected regions: nifurtimox and benznidazole [115]. These treatments are effective if given immediately, but if the disease has reached its chronic phase, pharmaceutical agents can only slow the progression and manage the symptoms [116].

4.1 Antitrypanosomal Sponge Metabolites Active Against Chagas Disease

Alkaloids **4-6** (Fig. 1) as well as batzelladine F from *Monanchora arbuscula* are active against *Trypanosoma cruzi* with batzelladine L (**5**) the most potent inhibitor with an IC_{50} of 2.35 μM [30]. The thiazine-derived alkaloid **69** (Fig. 10) is also an effective Chagas inhibitor with an IC_{50} of 4.26 μM [34]. In addition to activity against both leishmaniasis and HAT, agelasine D (**11**) (Fig. 2) is capable of inhibiting *T. cruzi* with an IC_{50} of 2.12 μM [35]. Pandaroside G and its methyl ester (**18**) (Fig. 2) as well as the steroid glycoside acanthifolioside F from *Pandaros acanthifolium* have been shown to inhibit C2C4 strain with **18** the most potent, IC_{50} = 0.77 μM [42, 98]. Linear monoterpene heptaprenyl-*p*-quinol (**114**) (Fig. 18) and diterpene **14** (Fig. 2) also inhibit the C2C4 strain at IC_{50}s of 6.80 and 12.4 μM, respectively [38]. Polyketide **21** (Fig. 2) from *Plakortis angulospiculatus* has an IC_{50} of 6.27 μM against *T. cruzi* [45].

4.2 Antitrypanosomal Algal Metabolites Active Against Chagas Disease

Atomaric acid (**115**) (Fig. 18), a meroditerpene isolated from brown alga *Stypopodium zonale,* has demonstrated activity against *Trypanosoma cruzi* with an IC_{50} of 5.42 μM [117]. The only reported metabolite from a green alga to be active against these parasites is the sesquiterpene (+)-curcudiol (**116**). It was isolated from the green alga *Udotea orientalis*, and has shown activity against *T. cruzi* with an IC_{50} of 10 μM [118].

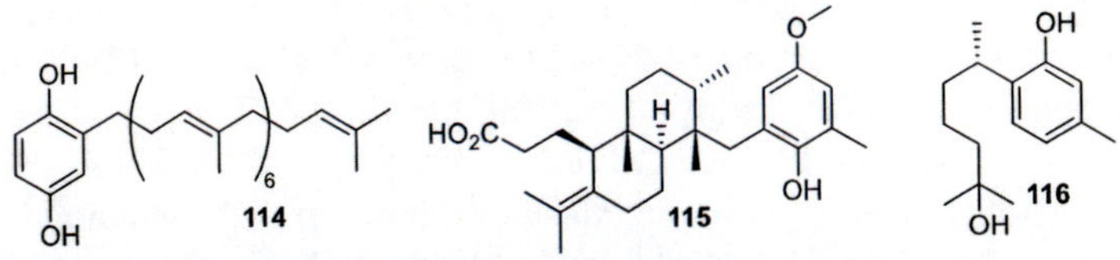

Fig. 18 Sponge and algal metabolites active against Chagas

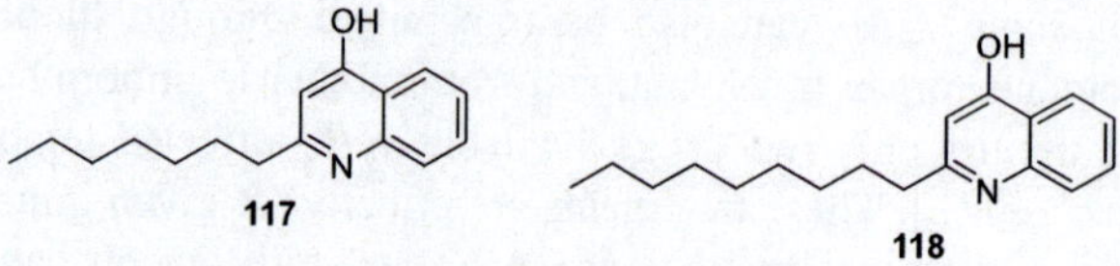

117 118

Fig. 19 Alkaloids from marine bacteria active against Chagas disease

4.3 Antitrypanosomal Ascidian Metabolites Active Against Chagas Disease

Lepadins E (**103**) and F (**104**) (Fig. 15), isolated from the ascidian *Didemnum* sp. showed activity against *Trypanosoma cruzi* in the same study, showing an IC_{50} of 5.22 μM and 6.17 μM.

4.4 Antitrypanosomal Marine Bacterial Metabolites Active Against Chagas Disease

The Panamanian marine bacterium *Pseudomonas aeruginosa* was found to produce five metabolites, with two (**117**, **118**) (Fig. 19) showing modest activity against *Trypanosoma cruzi*, $IC_{50} = 15.0$ and 14.7 μM [119].

4.5 Antitrypanosomal Marine Cyanobacterial Metabolites Active Against Chagas Disease

Two new PKS-NRPS-derived metabolites, viridamides A (**119**) and B, were isolated from the Panamanian marine cyanobacterium *Oscillatoria nigro-viridis* (Fig. 20) [120]. Only **119** displayed antitrypansomal activity with an IC_{50} of 1.1 μM against *Trypanosoma cruzi*. Another Panamanian cyanobacterium from the same genus, *Oscillatoria* sp., yielded two new modified cyclic hexapeptides, venturamides A (**120**) and B (**121**). Both cyclic peptides showed moderate activity against *T. cruzi* with IC_{50}s of 14.6 and 15.8 μM, respectively [121]. A series of 2,2-dimethyl-3-hydroxy-7-octynoic acid (Dhoya)-containing cyclic depsipeptides, named dudawalamides, were isolated from a Papua New Guinean field collection of the cyanobacterium *Moorea producens*. Dudawalamides A (**122**), B (**123**) and D (**124**) were evaluated for their ability to inhibit growth of *T. cruzi* at 10 μg/mL. Depsipeptides **122** and **123** showed minor activity with 12% and 7% growth inhibition, while **124** showed significantly more growth inhibition at 60% [122]. Gallinamide A (**125**) isolated from a Panamanian *Schizothrix* sp., also isolated from *Symploca* sp. in the Florida Keys and known as symplostatin 4, was found to

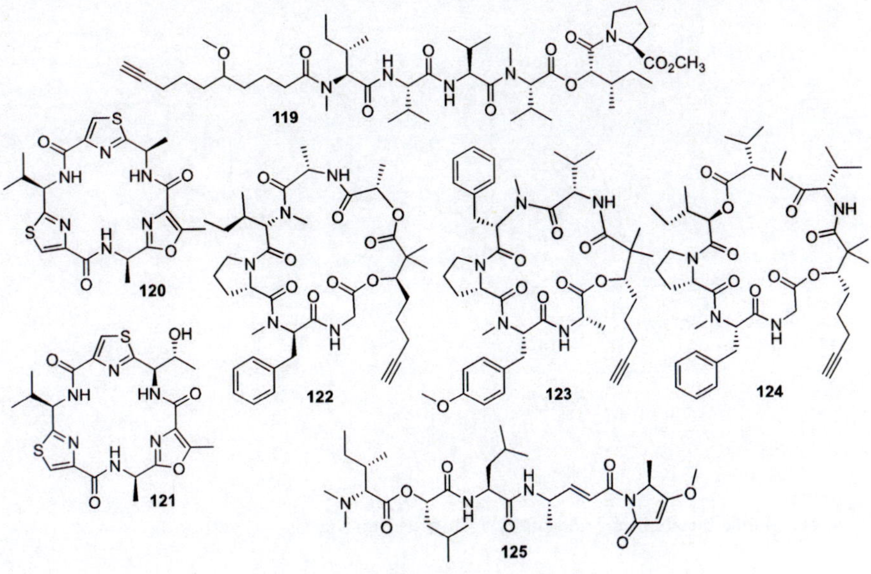

Fig. 20 Cyanobacterial metabolites active against Chagas disease

have potent bioactivity against the intracellular amastigote stage of *T. cruzi* with an IC_{50} of 0.26 nM against recombinant cruzain [123, 124].

5 Marine Natural Products Active Against Schistosomiasis

Schistosomiasis (commonly known as bilharzia) is a disease caused by trematodes (blood born fluke worms) of the genus *Schistosoma* and has been reported in 78 countries resulting in at least 229 million people requiring preventative treatment in 2018 [125]. The three primary species causing human infection are *Schistosoma haematobium, S. japonicum,* and *S. mansoni.* Three additional species that result in fewer cases due to being more localized geographically are *S. mekongi, S. intercalatum,* and *S. guineensis*. Infected humans release *Schistosoma* eggs with faces or urine depending on the species, which under the correct environmental conditions hatch releasing miracidia. A specific aquatic snail intermediate host is found and infected by miracidia. Once in the snail host, two generations of sporocysts are produced and ultimately form into cercariae which exit the snail into the surrounding water. The cercariae swim and come into contact with humans where they penetrate the skin, shed their forked tails, and develop into schistosomulae. The schistosomulae migrate to the lungs via venous circulation, followed by migration to the heart and ultimately to the liver, where they mature and exit the liver via the portal vein system. Adult worms primarily reside and copulate in the mesenteric venules, although this location may vary by species. The females deposit eggs into

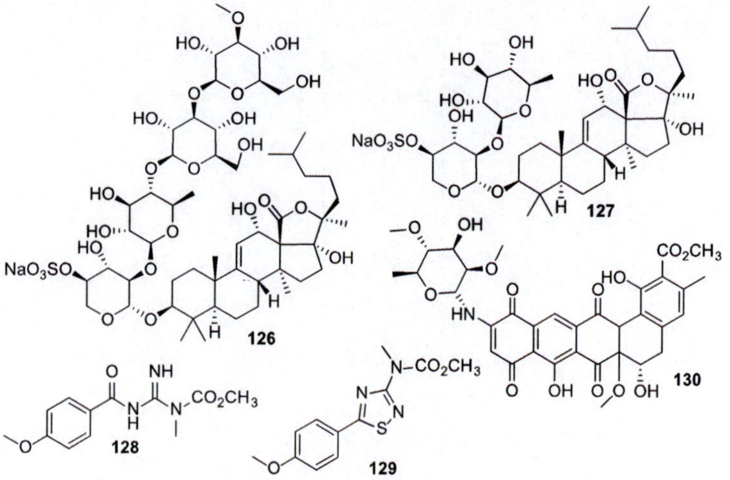

Fig. 21 Marine metabolites with activity against schistosomiasis

the small venules of the portal and perivesical systems which are moved toward the lumen of the intestine (in the cases *of S. mansoni, S. japonicum, S. mekongi, S. intercalatum/guineensis* infections) and of the bladder and ureters (in *S. haematobium* infections), and are eliminated with feces or urine, respectively resulting in the cycle beginning once again [126]. Schistosomiasis presents in two major forms: namely intestinal and urogenital. The only existing drug is praziquantel. It has several drawbacks including a high dosage and a lower effectiveness of total lack of activity against the younger parasite stages relative to the mature worms, resulting in immature parasites surviving and causing morbidity [127].

In an effort to identify new metabolites with schistosomicidal activity, the methanolic extracts of 78 marine organisms from the Red Sea were screened [128]. The most active crude extracts originated from two sea cucumbers *Actinopyga echinites* and *Holothuria polii*, which yielded the triterpene glycosides echinosides A (**126**) and B (**127**), respectively (Fig. 21). The purified metabolites were tested against adult *S. mansoni* showing an LD_{50} of 157.4 nM and 322.6 nM, respectively.

Chemical analysis of the Indonesian tunicate *Polycarpa aurata* yielded two novel alkaloids, polyaurines A (**128**) and B (**129**) (Fig. 21), along with six *p*-substituted benzoyl derivatives [129]. Both **128** and **129** showed no activity against schistosomula of *S. mansoni*, however **128** did impair egg production. Some eggs laid by **128**-treated parasites appeared deformed and several fragments of eggs were observed.

A sponge-associated actinomyces, *Streptomyces* sp. strain RV15, was isolated from the marine sponge *Dysidea tupha* offshore Rovinj, Croatia. Purification of the bacterial fermentation yielded the naphthacene glycoside, SF2446A2 (**130**) (Fig. 21). While no change was observed in *Sc. mansoni* couples treated with **130** below 50 μM relative to untreated control, at 50 μM the number of distorted eggs

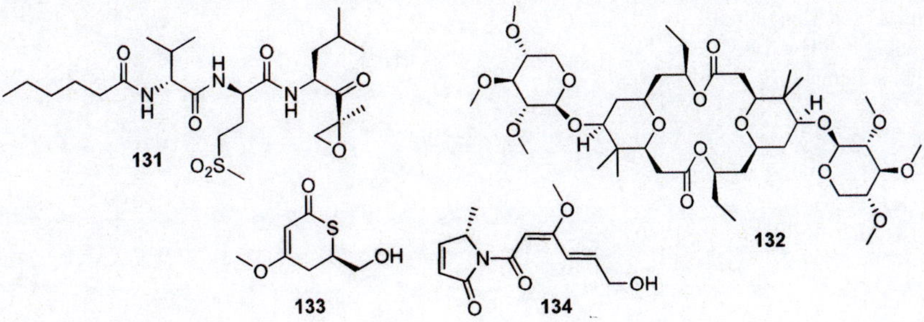

Fig. 22 Marine Cyanobacterial metabolites active against schistosomiasis

was more than double that found in the control. Over 48 h the egg deposition was dramatically reduced to five eggs in the treated samples, compared to 379 in the untreated worms. After 78 h egg production was halted completely, and this was also seen following treatment with 100 μM **130** after 24 h [130].

Schistosome proteolytic enzymes facilitate invasion of the human host and digestion of host proteins and therefore are fundamental to successful parasitism. The proteasome is a multi-subunit protein complex that regulates protein turn over and degradation of misfolded proteins. While the proteasomes of other infectious parasitic diseases such as *Plasmodium, Trypanosoma,* and *Leishmania* infections have been studied as potential drug targets for treatment, the proteasome in the schistosome has received little attention. However, a screening campaign was conducted using 11 peptide epoxyketone inhibitors derived from the marine natural product carmaphycin B (**131**) (Fig. 22) against the *Schistosoma mansoni* proteasome (Sm20S) [131]. Utilizing fluorogenic proteasome substrates found Sm20S contains caspase-type (β1), trypsin-type (β2), and chymotrypsin-type (β5) activities. Carmaphycin B (**131**) is a peptide derivative isolated from the marine cyanobacterium *Symploca* sp. from Curaçao [132]. Metabolite **131** inhibited Sm20S β2 and β5 activity with IC_{50}s of 9.8 and 0.6 nM, respectively, but showed cytotoxicity toward HepG2 cells (EC_{50} of 12.6 nM). Twenty **131** analogues were synthesized of which 11 were less cytotoxic. One analogue improved selectivity for the *Sc. mansoni* proteasome over the human proteasome while reducing toxicity. The antischistosomal activities of **131** and the analogue were evaluated at 1 μM using mixed-sex adult *Sc. mansoni* in culture. Both metabolites decreased movement in the worms by at least 80% and induced the same deleterious phenotypic changes in worms as the known proteasome inhibitor, bortezomib.

As previously mentioned, the intermediate host snails of the genera *Biomphalaria* play a pivotal role in the lifecycle of these parasitic trematodes. Over the past 50 years various molluscicides have been developed to kill the intermediate host in an effort to break disease transmission cycle. Chemical analysis of extracts *Lyngbya bouillonii* from Papua New Guinea produced the glycosidic macrolide, cyanolide A (**132**) (Fig. 22), which was identified as a potent molluscicidal agent against *Biomphalaria glabrata* with an LC_{50} of 1.2 μM [133]. The environmental assemblage of two cyanobacteria, cf. *Oscillatoria* and *Hormoscilla* spp., from

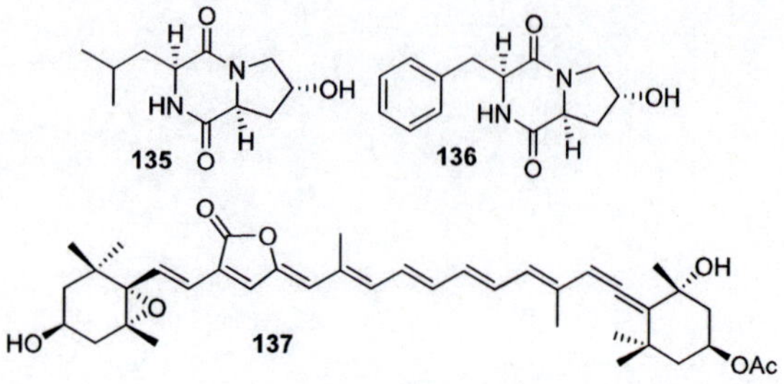

Fig. 23 Marine natural products with activity against dengue

Palmyra Atoll, yielded two potent molluscicidal chemotypes, namely thiopalmyrone (**133**) and palmyrrolinone (**134**). Assessing the lethality of the thioester **133** and amide **134** against *B. glabrata* showed LC_{50}'s of 8.3 and 6.0 μM, respectively. Interestingly, an equimolar binary mixture of **133** and **134** gave a slight enhancement of the molluscicidal effect of $LC_{50} = 5.0$ μM [134].

6 Marine Natural Products Active Against Dengue

Dengue fever is a mosquito transmitted virus affecting more than 100 countries around the world. The disease is spread through the bite of an infected *Aedes* spp. mosquito. Dengue is caused by one of the dengue viruses (1, 2, 3, and 4) and for this reason a person can be infected up to four times in their lifetime [135]. A dengue vaccine is currently available in some countries but can only be given post-infection. Cyclo(4-*trans*-hydroxy-L-proline-L-leucine) (**135**) and cyclo(4-*trans*-hydroxy-L-proline-L-phenylalanine) (**136**) (Fig. 23) were isolated from *Microbulbifer variabilis* and they were reported to have bioactivity against the dengue virus with IC_{50}s of 12.3 and 11.2 μM [136]. Peridinin (**137**), a common photopigment from dinoflagellates, was isolated from Formosan zoanthid *Palythoa mutuki*, displaying bioactivity against the dengue virus with IC_{50} of 7.1 μM [137, 138].

7 Marine Natural Products Active Against Chikungunya

Chikungunya virus is spread to humans by the same mosquitoes that transmit the dengue virus, *Aedes aegypti* and *A. albopictus*, and can also be transmitted through blood transfusions and very rarely from mother to child [139]. There is no vaccine; treatment is limited to treatment of the symptoms. Debromoaplysiatoxin (**138**) and 3-methoxydebromoaplysiatoxin (**139**) (Fig. 24) were isolated from the filamentous marine cyanobacterium *Trichodesmium erythraeum*. They showed bioactivity against Chikungunya virus with IC_{50}s of 1.3 and 2.7 mM, respectively [140].

Fig. 24 Cyanobacterial toxins with activity against Chikungunya

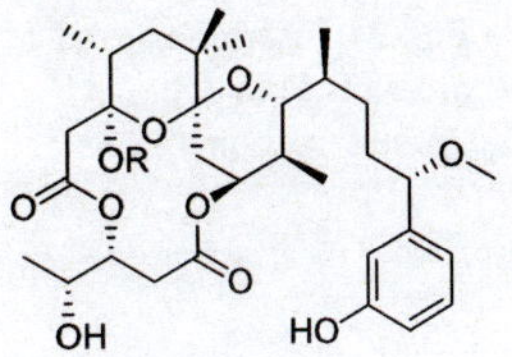

138: R = H
139: R = CH_3

Fig. 25 Algal metabolites with promising activity to treat snakebite envenomation

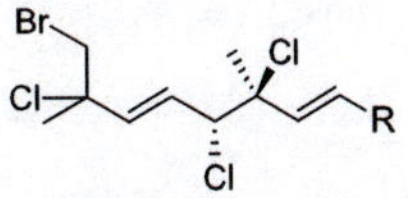

140: R = H
141: R = Br

8 Marine Natural Products Active Against Snakebite Envenoming

The only NTD that is not infectious, snakebite envenoming is traditionally treated by administration of an antivenom. Envenoming by different snakes require different antivenoms, and access to them can prove challenging. Different envenomation can also cause a variety of symptoms and be complicated by specific underlying conditions the victim may have, making the choice of antivenom difficult in life-threatening situations [141].

Two monoterpenes, 8-bromo-3,4,7-trichloro-3,7-dimethyl-1*E*,5*E*-octadiene (**140**) and 1,8-dibromo-3,4,7-trichloro-3,7-dimethyl-1*E*,5*E*-octadiene (**141**) (Fig. 25) isolated from the alga *Plocamium brasiliense* have shown promising antihemorrhagic and antiproteolytic effects against the venom of *Bothrops jararaca*, suggesting that more marine natural products should be investigated for the treatment of this disease [142].

9 Marine Natural Products Active Against Trachoma

Trachoma is a disease caused by the bacterium *Chlamydia trachomatis*. It is the leading cause of preventable blindness and is spread through direct or indirect contact. Although overcoming infection is possible for healthy individuals, re-infections are common resulting in complications [143]. It is the leading cause of ocular morbidity and preventable blindness in the world. Although natural products have proven to be good lead compounds for this disease, only a few of those are derived from marine sources. Among those gelliusterol E (**142**) (Fig. 26), a sterol from the Red Sea sponge *Callyspongia aff. Implexa,* is active, with an IC_{50} of 2.3 μM [144]. Additionally, SF2446A2 (**130**) (Fig. 21), a naphthacene glycoside

Fig. 26 Marine natural products with activity against trachoma

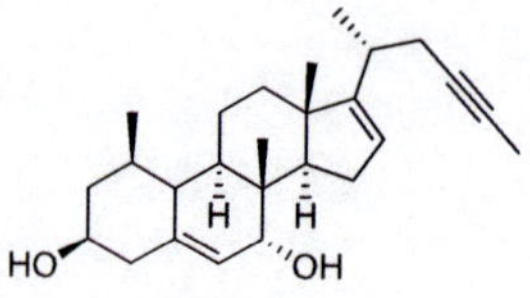

142

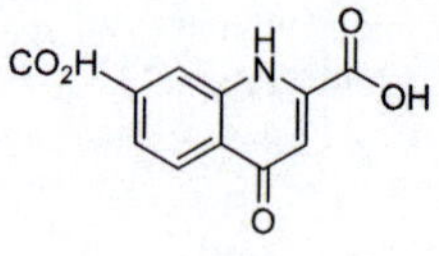

143

from a marine *Streptomyces* sp., has an IC_{50} of 4 μM [130]. Finally, ageloline A (**143**) is a quinolone isolated from a marine sponge-derived *Streptomyces* sp., which has an IC_{50} of 9.5 μM [145].

10 Marine Natural Products Active Against Soil-Transmitted Helminths (STH)

This group of infections is caused by intestinal worms coming in contact with humans through soil. The worms responsible are *Ascaris lumbricoides*, *As. suum, Trichuris trichiura* (whipworm), *Ancylostoma duodenale, An. ceylanicum,* and *Necator americanus* (hookworm*)* [146]. Together these helminths infect approximately 1.7 million people per year. The helminths live in the intestine of the patient from where their eggs are passed through the feces [147]. In the case of the first two, the eggs mature in the soil and then become infective. Humans ingest the eggs either through consuming fruits and vegetables that have not been properly cooked, washed, or peeled or by contact of dirt with their mouth, and become infected. Hookworm is contracted when eggs hatch in the soil and release larvae, which then mature into a form that can penetrate human skin and cause infection [148]. *An. duodenale* can also be transmitted by ingesting the larvae. Most infected people do not demonstrate symptoms, unless the infection has reached a severe stage, where multiple symptoms can be seen varying from abdominal pain to physical and cognitive growth retardation [149, 150]. Albendazole and mebendazole are two of the existing medications for the treatment of all STH infections and are not specific to the type of infection. These existing drugs appear to have few side effects and result in complete recovery of the patient [146].

The only marine natural product that has shown potential against a helminth, *Ancylostoma suum*, is nafuredin (**144**) (Fig. 27), isolated from the culture of *Aspergillus niger* FT-0554, which was derived from a marine sponge [151–153]. Nafuredin inhibits the NADH-fumarate reductase complex I and II with an IC_{50} of 12 nM for the adult *An. suum*, the NADH-ubiquinone reductase complex I with an IC_{50} of 8 nM and 8.9 nM for the adult and larvae stage and the NADH-rhodoquinone reductase complex I with an IC_{50} of 24 nM and 9 nM for the adult and larvae, respectively [153].

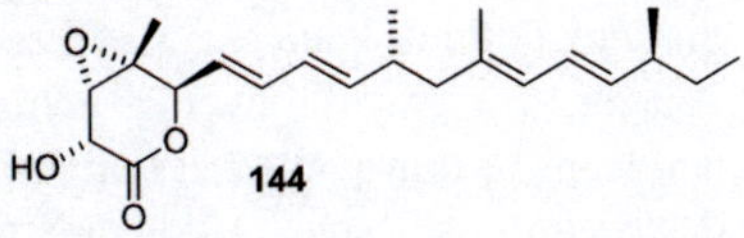
144

Fig. 27 A bacterial product active against soil-transmitted helminths

11 Buruli Ulcer

Mycobacterium ulcerans is the bacterium responsible for a severe skin disease known as Buruli ulcer (BU) [154]. Despite the severe effects on rural populations, there is a lack of investigation on isolated natural products' effect on this disease. Communities affected by BU tend to use traditional medicine techniques centered around treatment of the ulcers with plants and plant extracts [155, 156]. Main components have been identified, but the investigation of isolated metabolites from the active extracts is still needed. Current treatment consists of a combination of antibiotics and traditional medicine treatments. The most common combinations are rifampicin and streptomycin or rifampicin and clarithromycin [157]. The biggest limitation in regard to this NTD is the lack of knowledge regarding disease transmission.

12 Dracunculiasis

Dracunculiasis is caused by the parasite *Dracunculus medinensis* and also known as Guinea worm disease. Humans become infected when consuming the water containing copepods or tiny crustaceans who have ingested the worm larvae. Once swallowed by humans, the larvae are released, penetrate the intestinal walls and migrate through the body [158, 159]. The fertilized female worm will emerge from the skin once it has migrated to an exit point. A person can become infected multiple times throughout their life, as immunity cannot be developed [160]. Currently there is no drug to treat or prevent this disease. Patient care involves removing the worm and caring for the wound [161]. Dracunculiasis is one of the few NTDs that are close to being eradicated through a strategy developed by WHO and the Center for Disease Control [158, 160]. Although a few medicinal plants have been tested against this diseases, no marine natural products have shown activity thus far [162].

13 Echinococcosis

Infection with dog, fox, or coyote tapeworms leads to echinococcosis. There are two classifications of echinococcosis: cystic and alveolar, with the majority of worldwide cases being the former and the more deadly of the two being the latter [163, 164]. Cystic echinococcosis is caused by *Echinococcus granulosus* larvae that create thick-walled cysts in several internal organs and can remain asymptomatic for years, while alveolar echinococcosis is caused by *E. multilocularis* larvae and leads to thin-walled vesicles mainly in the liver, marked by abdominal pain and destruction of tissues [165]. Treatment for these diseases includes surgery, evacuation of the cyst, and chemotherapy with benzimidazoles [166]. While dogs, foxes, and coyotes have been found to be the definitive host of these parasites, there are several intermediate hosts as well including sheep, cattle, and goats, leading to increased risk for populations that raise these animals. *Echinococcus* eggs are shed from the infected animals through feces and can be transmitted to humans through accidental consumption. This vector is particularly difficult to deal with due to the fact that the eggs are viable in soil up to a year. Prevention lies in controlling stray dog populations, avoiding infected meats, and practicing good hygiene [166].

14 Foodborne Trematodiasis

Foodborne trematodiasis is caused by one of four parasites *Clonorchis, Opisthorchis, Fasciola,* and *Paragonimus*. The life cycle of these parasites is complex, and it involves two intermediate and one final host [167]. The life cycle of the parasites is similar but some differences can be observed. Clonorchiasis, also called Chinese liver fluke disease, is caused by *Clonorchis sinensis*. Opisthorchiasis, also called cat liver fluke disease, is another foodborne trematodiasis. It is caused by *Opisthorchis viverrini* and *O. felinea*. Fascioliasis is caused by *Fasciola hepatica* and *F. gigantica*. Both have similar symptoms and life cycles. Finally, paragonimiasis, also called lung fluke disease is caused by *Paragonimus westermani, P. heterotremus,* and *P. philippinensis* in Asia, *P. africanus* and *P. uterobilateralis* in Africa, and *P. kellicotti* and *P. mexicanus* in north and central America.

For the first three, the life cycle of the parasites includes the adult flukes laying eggs in the bile ducts, which then develop into miracidia when ingested by the carrier and develop into cercariae. For clonorchiasis and opisthorchiasis the carrier is aquatic snails, while for fascioliasis the carrier are leaves and stems of aquatic and semi-aquatic plants [168]. Cercariae leave the carrier and enter various freshwater fish, whereas metacercariae form cysts. The cysts hatch when ingested and the young worms can now repeat the cycle. Humans get infected when consuming raw or partially cooked fish.

For clonorchiasis the symptoms include as fever, pain and oftentimes intermittent colic pain [169, 170]. The symptoms of opisthorchiasis are similar to the ones mentioned above, but in chronic cases recurrent pyogenic cholangitis may be frequent by the former. The latter causes similar symptoms to clonorchiasis but there is no evidence of increased liver cancer risk [171–173]. The only drug available for clonorchiasis and opisthorchiasis is praziquantel.

Fascioliasis has an acute and chronic phase, with the worms destroying the liver cells causing fever, nausea, rashes, and swollen liver amongst others for the former, while the chronic symptoms often include jaundice, anemia, pain, pancreatitis, and liver fibrosis [174, 175].

Paragonimiasis can be found in crustacean-eating mammals and the cycle of the parasites is similar to the one mentioned above with the difference that the adult flukes inhabit the lungs and their eggs are dispersed through coughing. The rest of the life cycle is similar to those described above. Symptoms of paragonimiasis are similar to those of tuberculosis. Triclabendazole is the drug available for both paragonimiasis and fascioliasis [176].

Although the need for new drugs against this group of parasites is obvious, there are very few reports of natural products screened against them and most target the intermediate snail host. No marine natural products have been reported that are active against the worms or the snail [177].

15 Leprosy

Leprosy, or Hansen's disease, is an infection caused by *Mycobacterium leprae.* It affects the skin, eyes, nerves, and lining of the nose [178]. A combination of antibiotics, most commonly dapsone combined with rifampicin, and on some occasions clofazimine, is used as treatment. Permanent nerve damage is possible if left untreated [179]. A multidrug approach is needed to combat drug resistance due to the extended period of time required for treatment which usually lasts one to two years. Although natural product extracts and some isolated secondary metabolites have shown activity against this bacterium, there is no report of marine natural products that have activity [180, 181]. Although multiple attempts have been made to culture *M. leprae ex vivo,* it remains recalcitrant to cultivation. This combined with the slow growth rate *in vivo* has made it difficult to screen against this disease [182, 183].

16 Lymphatic Filariasis

Lymphatic filariasis, also referred to as elephantiasis, is caused by the filarial nematodes *Wuchereria bancrofti, Brugia malayi* and *B. timori. W. bancrofti* is the cause of most cases of this disease. The disease is transmitted after an infected mosquito introduces filarial larvae to the host through a bite [184]. The larvae develop into adults in the lymphatic system and produce microfilariae which migrate to the lymph and enter the bloodstream [185]. A mosquito can then ingest microfilariae, which then develop into larvae and can infect the next host during a blood-meal [186]. Current treatment consists of the use of diethylcarbamazine, but treatment is not effective once the patient has developed clinical symptoms, and patients require surgery or symptomatic treatment depending on the form and stage of the disease. The existing literature has many examples of active metabolites from plant sources, but there are very limited reports of active extracts from marine sources. There are no secondary metabolites that are active at 15 μM or less against the microfilariae or the adult worms [187–189].

17 Mycetoma and Chromoblastomycosis

Mycetoma and chromoblastomycosis are two diseases on which more research is needed to enhance the search for treatment. Mycetoma is caused by different fungal species or aerobic filamentous bacteria, while chromoblastomycosis is only caused by the former [190]. Biomarkers are being investigated to obtain the necessary information to expedite the drug discovery process [191]. Similar to other NTDs, a common treatment for these infections is with medicinal plants [192].

18 Onchocerciasis

Onchocerciasis, also known as River Blindness, is caused by the parasitic worm *Onchocerca volvulus.* Blackflies of the genus *Simulium* are the transmission vector. It is the second most important cause of infectious blindness after trachoma and it also results in chronic dermatitis [193]. A female fly introduces the larvae into the human host through a blood-meal which 6–12 months later develop into adult worms. The adult worms reside in fibrous sub-cutaneous nodules and can produce 1,000–1,500 baby worms daily, which require 1–2 years to develop into the adult stage [194]. Currently no vaccine or medication exists to prevent infection and attempts have been made to control the spread of the disease through vector control and treatment with ivermectin [194, 195]. Ivermectin needs to be administered once or twice a year for 10–15 years, because it does not kill the worm but only reduces the fertility of the female worm. This fact in combination with the adverse effects of the high doses needed is indicative of the need for new drugs. Although there are multiple secondary metabolites from terrestrial sources that have been screened against this disease, there are no active metabolites isolated from marine organisms [196].

19 Rabies

Rabies is a preventable viral disease that can become fatal if left untreated. The mortality associated with this disease is extremely high and depends on presenting symptoms and on available treatments in the area. The standard of care for aggressive treatment of the rabies-related viral encephalitis includes antivirals, immunotherapy, and administration of neuroprotective agents [197, 198]. Unfortunately, many patients are not candidates for aggressive therapy and in resource-limited settings do not have access to combination therapy; the alternative course of action is palliative care [197, 199]. No marine natural products have been developed to treat causative agent of rabies, *Rabies lyssaviru*; however, coelomic fluid from urchins has demonstrated some efficacy against the virus [200]. Further exploration as a treatment modality for the rabies virus is warranted.

20 Scabies

Natural products are used as insecticides in the treatment of the mite *Sarcoptes scabiei*, which causes scabies in humans and animals. These treatments, however, often arise from terrestrial sources. Ivermectin was isolated from fermentation of soil-derived *Streptomyces avermitilis* and is an oral treatment of scabies worldwide [201, 202]. Many other natural products from botanical sources have demonstrated efficacy as insecticides, such as the pyrethrins isolated from *Chrysanthemum* sp., indicating that marine natural product drug discovery may lead to novel therapies for this NTD [202–204].

21 Taeniasis/Cysticercosis

Taeniasis refers to a human infection with an adult *Taenia* sp. tapeworm in the gut, while cysticercosis refers to an infection with the *T. solium* larvae, and is marked by cysticerci forming in the muscles, brain, and/or spinal cord. The most serious form of the disease, neurocysticercosis, results from cysticerci in the brain and can cause epilepsy and sudden death. Transmission occurs through the oral-fecal route, by swallowing the eggs shed in feces, either through contaminated food or water or through poor hand hygiene, can develop cysticercosis. Treatment is currently limited to surgery or administration of albendazole or praziquantel, although anthelminthic treatment can lead to an unwanted and dangerous inflammatory response. Access to CT and MRI equipment is necessary for choosing a treatment, making treatment in developing countries a challenge. Prevention is key to controlling this disease by practicing good hygiene, washing raw foods before eating, and drinking either boiled or filtered and iodine-treated water [205, 206].

22 Yaws

Yaws is an infection caused by the bacterium *Treponema pertenue,* a subspecies of *T. pallidum*, the bacterium responsible for venereal syphilis. It can be found in tropical areas of Africa, Asia, and Latin America and is transmitted through skin-to-skin contact. Although the disease is not fatal, it can lead to chronic disfigurement and disability if left untreated [207]. Treatment of the disease consists of using the benzathine penicillin and is highly effective. There are no reports of marine natural products that have shown activity against this disease potentially due to the existing inexpensive and readily available treatment [208].

23 Conclusion

NTDs have attracted the interest of the biomedical community in the past decades and there is a plethora of bioactive metabolites from marine sources. The need for new drugs to combat drug resistance, side effects, and increase their efficiency is urgent and can be seen from the lack of drugs against some of the NTDs. Marine organisms have outstanding potential as drug leads against NTD targets, as judged by the 145 diverse metabolites described here, demonstrating an IC_{50} of 15 μM or below. Discovery of novel metabolites against these targets will increase as more species are screened, including organisms derived from lesser-studied geographic regions such as the extreme environments of the deep-sea and Southern Ocean. Highlighted in this review, many NTDs have multistage infective cycles, providing numerous routes for therapeutic intervention by marine natural products.

Porifera are the major source of active metabolites against NTDs found in this review, with bacteria and cyanobacteria emerging as a new source due to the recent advances in culturing, isolation techniques, and genome mining for silent gene sequences. Furthermore, marine natural products from biodiversity found in afflicted countries could be explored, as there are many reports of active metabolites against other targets from some of those regions, demonstrating the untapped potential.

The importance and potential of marine natural products in NTD drug discovery is highlighted in this review, as well as the need for more studies to address the biological activity of the secondary metabolites isolated from these sources. Finally, new techniques for screening against these diseases in a high throughput way are needed, as only a few marine natural products have been tested, limiting the potential of identifying drug leads.

References

1. Williams DH, Stone MJ, Hauck PR, Rahman SK (1989) J Nat Prod 52:1189
2. Newman DJ, Cragg GM (2020) J Nat Prod 83:770
3. Blunt JW, Munro MHG (2020) R Soc Chem. http://pubs.rsc.org/marinlit/. Accessed 28 Nov 2020
4. Carroll AR, Copp BR, Davis RA, Keyzers RA, Prinsep MR (2020) Nat Prod Rep 37:175
5. Blunt JW, Carroll AR, Copp BR, Davis RA, Keyzers RA, Prinsep MR (2018) Nat Prod Rep 35:8
6. Blunt JW, Copp BR, Keyzers RA, Munro MHG, Prinsep MR (2014) Nat Prod Rep 31:160
7. Blunt JW, Copp BR, Keyzers RA, Munro MHG, Prinsep MR (2016) Nat Prod Rep 33:382

8. Hotez PJ, Molyneux DH, Fenwick A, Kumaresan J, Sachs SE, Sachs JD, Savioli L (2007) N Engl J Med 357:1018
9. Centers for Disease Control and Prevention (2017). https://www.cdc.gov/globalhealth/ntd/index.html
10. Reithinger R, Dujardin J-C, Louzir H, Pirmez C, Alexander B, Brooker S (2007) Lancet Infect Dis 7:581
11. https://www.who.int/health-topics/leishmaniasis#tab=tab_1. Accessed 28 Nov 2020
12. Peacock CS, Seeger K, Harris D, Murphy L, Ruiz JC, Quail MA, Peters N, Adlem E, Tivey A, Aslett M, Kerhornou A, Ivens A, Fraser A, Rajandream MA, Carver T, Norbertczak H, Chillingworth T, Hance Z, Jagels K, Moule S, Ormond D, Rutter S, Squares R, Whitehead S, Rabbinowitsch E, Arrowsmith C, White B, Thurston S, Bringaud F, Baldauf SL, Faulconbridge A, Jeffares D, Depledge DP, Oyola SO, Hilley JD, Brito LO, Tosi LRO, Barrell B, Cruz AK, Mottram JC, Smith DF, Berriman M (2007) Nat Genet 39:839
13. Desjeux P (1996) Clin Dermatol 14:417
14. Arenas R, Torres-Guerrero E, Quintanilla-Cedillo MR, Ruiz-Esmenjaud J (2017) F1000 Res 6
15. Akhoundi M, Kuhls K, Cannet A, Votýpka J, Marty P, Delaunay P, Sereno D (2016) PLoS Negl Trop Dis 10:e0004349
16. Gossage SM, Rogers ME, Bates PA (2003) Int J Parasitol 33:1027
17. Schlein Y (1993) Parasitol Today 9:255
18. Killick-Kendrick R (1990) Ann Parasitol Hum Comp 65:37
19. Bates PA (2018) Nat Microbiol 3:529
20. Alves F, Bilbe G, Blesson S, Goyal V, Monnerat S, Mowbray C, Ouattara GM, Pécoul B, Rijal S, Rode J, Solomos A, Strub-Wourgaft N, Wasunna M, Wells S, Zijlstra EE, Arana B, Alvar J (2018) Clin Microbiol Rev 31:1
21. Eiras DP, Kirkman LA, Murray HW (2015) Curr Treat Options Infect Dis 7:52
22. Bekhit AA, El-Agroudy E, Helmy A, Ibrahim TM, Shavandi A, Bekhit AEDA (2018) Eur J Med Chem 160:229
23. Singh N, Kumar M, Singh RK (2012) Asian Pac J Trop Med 5:485
24. Lockwood D, Moore E (2010) J Glob Infect Dis 2:151
25. Gupta G, Oghumu S, Satoskar AR (2013) Adv Appl Microbiol 82:155. https://doi.org/10.1016/B978-0-12-407679-2.00005-3
26. Desjeux P (1999) Clin Dermatol 17:317
27. Nakao Y, Shiroiwa T, Murayama S, Matsunaga S, Goto Y, Matsumoto Y, Fusetani N (2004) Mar Drugs 2:55
28. Rao KV, Kasanah N, Wahyuono S, Tekwani BL, Schinazi RF, Hamann MT (2004) J Nat Prod 67:1314
29. Rao KV, Donia MS, Peng J, Garcia-Palomero E, Alonso D, Martinez A, Medina M, Franzblau SG, Tekwani BL, Khan SI, Wahyuono S, Willett KL, Hamann MT (2006) J Nat Prod 69:1034
30. Santos MFC, Harper PM, Williams DE, Mesquita JT, Pinto ÉG, Da Costa-Silva TA, Hajdu E, Ferreira AG, Santos RA, Murphy PJ, Andersen RJ, Tempone AG, Berlinck RGS (2015) J Nat Prod 78:1101
31. Hua HM, Peng J, Dunbar DC, Schinazi RF, de Castro Andrews AG, Cuevas C, Garcia-Fernandez LF, Kelly M, Hamann MT (2007) Tetrahedron 63:11179
32. Scala F, Fattorusso E, Menna M, Taglialatela-Scafati O, Tierney M, Kaiser M, Tasdemir D (2010) Mar Drugs 8:2162
33. Gul W, Hammond NL, Yousaf M, Bowling JJ, Schinazi RF, Wirtz SS, de Castro AG, Cuevas C, Hamann MT (2006) Bioorg Med Chem 14:8495
34. Zulfiqar B, Jones AJ, Sykes ML, Shelper TB, Davis RA, Avery VM (2017) Molecules 22:1
35. Vik A, Proszenyák Á, Vermeersch M, Cos P, Maes L, Gundersen L-LL (2009) Molecules 14:279
36. Shilling AJ, Witowski CG, Maschek JA, Azhari A, Vesely BA, Kyle DE, Amsler CD, McClintock JB, Baker BJ (2020) J Nat Prod 83:1553

37. Gray CA, De Lira SP, Silva M, Pimenta EF, Thiemann OH, Oliva G, Hajdu E, Andersen RJ, Berlinck RGS (2006) J Org Chem 71:8685
38. Orhan I, Şener B, Kaiser M, Brun R, Tasdemir D (2010) Mar Drugs 8:47
39. Gul W, Hammond NL, Yousaf M, Peng J, Holley A, Hamann MT (2007) Biochim Biophys Acta Gen Subj 1770:1513
40. Imperatore C, Gimmelli R, Persico M, Casertano M, Guidi A, Saccoccia F, Ruberti G, Luciano P, Aiello A, Parapini S, Avunduk S, Basilico N, Fattorusso C, Menna M (2020) Mar Drugs 18:112
41. Rangel H (1997) Cell Biol Int 21:337
42. Regalado EL, Tasdemir D, Kaiser M, Cachet N, Amade P, Thomas OP (2010) J Nat Prod 73:1404
43. Ma WS, Mutka T, Vesley B, Amsler MO, McClintock JB, Amsler CD, Perman JA, Singh MP, Maiese WM, Zaworotko MJ, Kyle DE, Baker BJ (2009) J Nat Prod 72:1842
44. Kossuga MH, De Lira SP, Nascimento AM, Gambardella MTP, Berlinck RGS, Torres YR, Nascimento GGF, Pimenta EF, Silva M, Thiemann OH, Oliva G, Tempone AG, Melhem MSC, De Souza AO, Galetti FCS, Silva CL, Cavalcanti B, Pessoa CO, Moraes MO, Hajdu E, Peixinho S, Rocha RM (2007) Quim Nova 30:1194
45. Kossuga MH, Nascimento AM, Reimão JQ, Tempone AG, Taniwaki NN, Veloso K, Ferreira AG, Cavalcanti BC, Pessoa C, Moraes MO, Mayer AMS, Hajdu E, Berlinck RGS (2008) J Nat Prod 71:334
46. Le Pape P, Zidane M, Abdala H, Moré MT (2000) Cell Biol Int 24:51
47. Lim CW, Kim Y-K, Youn HD, Park H-Y (2006) Agric Chem Biotechnol 49:21
48. Compagnone RS, Piña IC, Rangel HR, Dagger F, Suárez AI, Reddy MVR, Faulkner DJ (1998) Tetrahedron 54:3057
49. Medeiros DS, Medeiros TL, Ribeiro JKC, Monteiro NKV, Migliolo L, Uchoa AF, Vasconcelos IM, Oliveira AS, de Sales MP, Santos EA (2010) Comp Biochem Physiol B Biochem Mol Biol 155:211
50. Moura RM, Queiroz AFS, Fook JMSLL, Dias ASF, Monteiro NKV, Ribeiro JKC, Moura GEDD, Macedo LLP, Santos EA, Sales MP (2006) Comp Biochem Physiol A Mol Integr Physiol 145:517
51. Von Salm JL, Wilson NG, Vesely BA, Kyle DE, Cuce J, Baker BJ (2014) Org Lett 16:2630
52. Ishigami S-T, Goto Y, Inoue N, Kawazu S-I, Matsumoto Y, Imahara Y, Tarumi M, Nakai H, Fusetani N, Nakao Y (2012) J Org Chem 77:10962
53. Thomas SAL, Von Salm JL, Clark S, Ferlita S, Nemani P, Azhari A, Rice CA, Wilson NG, Kyle DE, Baker BJ (2018) J Nat Prod 81:117
54. Thao NP, Luyen BTT, Brun R, Kaiser M, Van Kiem P, Van Minh C, Schmidt TJ, Kang JS, Kim YH (2015) Molecules 20:12459
55. Thao NP, Luyen BTT, Ngan NTT, Song SB, Cuong NX, Nam NH, Van Kiem P, Kim YH, Van Minh C (2014) Bioorg Med Chem Lett 24:228
56. Reimão JQ, Migotto AE, Kossuga MH, Berlinck RGS, Tempone AG (2008) Parasitol Res 103:1445
57. Brennan MR, Erickson KL, Minott DA, Pascoe KO (1987) Phytochemistry 26:1053
58. Dos Santos AO, Veiga-Santos P, Ueda-Nakamura T, Dias Filho BP, Sudatti DB, Bianco ÉM, Pereira RC, Nakamura CV (2010) Mar Drugs 8:2733
59. Da Silva MacHado FL, Pacienza-Lima W, Rossi-Bergmann B, De Souza Gestinari LM, Fujii MT, Campos De Paula J, Costa SS, Lopes NP, Kaiser CR, Soares AR (2011) Planta Med 77:733
60. Veiga-Santos P, Pelizzaro-Rocha KJ, Santos AO, Ueda-Nakamura T, Filho BPD, Silva SO, Sudatti DB, Bianco EM, Pereira RC, Nakamura CV (2010) Parasitology 137:1661
61. Becerra M, Boutefnouchet S, Córdoba O, Vitorino GP, Brehu L, Lamour I, Laimay F, Efstathiou A, Smirlis D, Michel S, Kritsanida M, Flores ML, Grougnet R (2015) Phytochem Lett 11:418

62. Díaz-Marrero AR, López-Arencibia A, Bethencout-Estrella CJ, Cen-Pacheco F, Sifaoui I, Hernández Creus A, Duque-Ramírez MC, Souto ML, Hernández Daranas A, Lorenzo-Morales J, Piñero JE, Fernández JJ (2019) Bioorg Chem 92:103276
63. Chiboub O, Sifaoui I, Lorenzo-Morales J, Abderrabba M, Mejri M, Fernández JJ, Piñero JE, Díaz-Marrero AR (2019) Mar Drugs 17:192
64. Dos Santos AO, Britta EA, Bianco EM, Ueda-Nakamura T, Dias Filho BP, Pereira RC, Nakamura CV (2011) Mar Drugs 9:2369
65. Encarnacion-Dimayuga R, Iván Murillo-Álvarez J, Christophersen C, Chan-Bacab M, Reiriz MLG, Zacchino S (2006) Nat Prod Commun 1:541
66. Mayer AMS, Rodríguez AD, Taglialatela-Scafati O, Fusetani N (2013) Mar Drugs 11:2510
67. Efstathiou A, Gaboriaud-Kolar N, Smirlis D, Myrianthopoulos V, Vougogiannopoulou K, Alexandratos A, Kritsanida M, Mikros E, Soteriadou K, Skaltsounis AL (2014) Parasites Vectors 7:1
68. Clark KE, Capper A, Della TG, Paul VJ, Romero LI, Johns T, Cubilla-Rios L, Capson TL (2013) Nat Prod Commun 8:1537
69. Cruz LJ, Martínez Insua M, Pérez Baz J, Trujillo M, Rodriguez-Mias RA, Oliveira E, Giralt E, Albericio F, Cañedo LM (2006) J Org Chem 71:3335
70. Luque-Ortega JR, Cruz LJ, Albericio F, Rivas L (2010) Mol Pharm 7:1608
71. Zhang XQ, Spadafora C, Pineda LM, Ng MG, Sun JH, Wang W, Wang CY, Gu YC, Shao CL (2017) Sci Rep 7:1
72. Trepos R, Cervin G, Hellio C, Pavia H, Stensen W, Stensvåg K, Svendsen J-S, Haug T, Svenson J (2014) J Nat Prod 77:2105
73. Osei E, Kwain S, Mawuli G, Anang A, Owusu K, Camas M, Camas A, Ohashi M, Alexandru-Crivac C-N, Deng H, Jaspars M, Kyeremeh K (2018) Mar Drugs 17:9
74. Cartuche L, Sifaoui I, López-Arencibia A, Bethencourt-Estrella CJ, San Nicolás-Hernández D, Lorenzo-Morales J, Piñero JE, Díaz-Marrero AR, Fernández JJ (2020) Biomolecules 10:657
75. Pimentel-Elardo SM, Kozytska S, Bugni TS, Ireland CM, Moll H, Hentschel U (2010) Mar Drugs 8:373
76. Balunas MJ, Linington RG, Tidgewell K, Fenner AM, Ureña LD, Della TG, Kyle DE, Gerwick WH, Ureña L-D, Della TG, Kyle DE, Gerwick WH (2010) J Nat Prod 73:60
77. Sanchez LM, Lopez D, Vesely BA, Della Togna G, Gerwick WH, Kyle DE, Linington RG (2010) J Med Chem 53:4187
78. Balunas MJ, Grosso MF, Villa FA, Engene N, McPhail KL, Tidgewell K, Pineda LM, Gerwick L, Spadafora C, Kyle DE, Gerwick WH (2012) Org Lett 14:3878
79. Keller L, Siqueira-Neto JL, Souza JM, Eribez K, LaMonte GM, Smith JE, Gerwick WH (2020) Molecules 25:1604
80. Simarro PP, Cecchi G, Paone M, Franco JR, Diarra A, Ruiz JA, Fèvre EM, Courtin F, Mattioli RC, Jannin JG (2010) Int J Health Geogr 9:57
81. Stich A (2002) Br Med J 325:203
82. Baker CH, Welburn SC (2018) Trends Parasitol 34:818
83. Brun R, Blum J, Chappuis F, Burri C (2010) Lancet 375:148
84. Franco JR, Simarro PP, Diarra A, Jannin JG (2014) Clin Epidemiol 6:257
85. Fairlamb AH (2003) Trends Parasitol 19:488
86. Franco JR, Cecchi G, Priotto G, Paone M, Diarra A, Grout L, Simarro PP, Zhao W, Argaw D (2020) PLoS Negl Trop Dis 14:e0008261
87. Yamada M, Takahashi Y, Kubota T, Fromont J, Ishiyama A, Otoguro K, Yamada H, Omura S, Kobayashi J (2009) Tetrahedron 65:2313
88. Kirsch G, Köng GM, Wright AD, Kaminsky R (2000) J Nat Prod 63:825
89. Shady NH, Fouad MA, Ahmed S, Pimentel-Elardo SM, Nodwell JR, Kamel MS, Abdelmohsen UR (2018) J Antibiot (Tokyo) 71:1036
90. Takishima S, Ishiyama A, Iwatsuki M, Otoguro K, Yamada H, Õmura S, Kobayashi H, Van Soest RWM, Matsunaga S (2009) Org Lett 11:2655

91. Calcul L, Inman WD, Morris AA, Tenney K, Ratnam J, McKerrow JH, Valeriote FA, Crews P (2010) J Nat Prod 73:365
92. Feng Y, Davis RA, Sykes ML, Avery VM, Quinn RJ (2012) Bioorg Med Chem Lett 22:4873
93. Chianese G, Scala F, Calcinai B, Cerrano C, Dien HA, Kaiser M, Tasdemir D, Taglialatela-Scafati O (2013) Mar Drugs 11:3297
94. Chianese G, Fattorusso E, Scala F, Teta R, Calcinai B, Bavestrello G, Dien HA, Kaiser M, Tasdemir D, Taglialatela-Scafati O (2012) Org Biomol Chem 10:7197
95. Feng Y, Davis RA, Sykes M, Avery VM, Camp D, Quinn RJ (2010) J Nat Prod 73:716
96. Oli S, Abdelmohsen UR, Hentschel U, Schirmeister T (2014) Mar Drugs 12:2614
97. Rubio BK, Tenney K, Ang KH, Abdulla M, Arkin M, McKerrow JH, Crews P (2009) J Nat Prod 72:218
98. Regalado EL, Jiménez-Romero C, Genta-Jouve G, Tasdemir D, Amade P, Nogueiras C, Thomas OP (2011) Tetrahedron 67:1011
99. Thao NP, No JH, Luyen BTT, Yang G, Byun SY, Goo J, Kim KT, Cuong NX, Nam NH, Van Minh C, Schmidt TJ, Kang JS, Kim YH (2014) Molecules 19:7869
100. Davis RA, Sykes M, Avery VM, Camp D, Quinn RJ (2011) Bioorg Med Chem 19:6615
101. Wright AD, Goclik E, König GM, Kaminsky R (2002) J Med Chem 45:3067
102. Pu X, Ma D (2004) Angew Chemie 116:4318
103. Abdelmohsen UR, Cheng C, Viegelmann C, Zhang T, Grkovic T, Ahmed S, Quinn RJ, Hentschel U, Edrada-Ebel RA (2014) Mar Drugs 12:1220
104. Tawfike A, Attia EZ, Desoukey SY, Hajjar D, Makki AA, Schupp PJ, Edrada-Ebel R, Abdelmohsen UR (2019) AMB Express 9:12
105. Nakashima T, Iwatsuki M, Ochiai J, Kamiya Y, Nagai K, Matsumoto A, Ishiyama A, Otoguro K, Shiomi K, Takahashi Y, Ōmura S (2014) J Antibiot (Tokyo) 67:253
106. Schulze CJ, Donia MS, Siqueira-Neto JL, Ray D, Raskatov JA, Green RE, McKerrow JH, Fischbach MA, Linington RG (2015) ACS Chem Biol 10:2373
107. Ogawa H, Iwasaki A, Sumimoto S, Kanamori Y, Ohno O, Iwatsuki M, Ishiyama A, Hokari R, Otoguro K, Ōmura S, Suenaga K (2016) J Nat Prod 79:1862
108. Ogawa H, Iwasaki A, Sumimoto S, Iwatsuki M, Ishiyama A, Hokari R, Otoguro K, Ōmura S, Suenaga K (2017) Org Lett 19:890
109. Kurisawa N, Iwasaki A, Jeelani G, Nozaki T, Suenaga K (2020) J Nat Prod 83:1684
110. Sanchez LM, Knudsen GM, Helbig C, De Muylder G, Mascuch SM, MacKey ZB, Gerwick L, Clayton C, McKerrow JH, Linington RG (2013) J Nat Prod 76:630
111. Pérez-Molina JA, Molina I (2018) Lancet 391:82
112. Rassi A, Rassi A, Marin-Neto JA (2010) Lancet 375:1388
113. Prata A (2001) Lancet Infect Dis 1:92
114. Gourbière S, Dorn P, Tripet F, Dumonteil E (2012) Heredity (Edinb) 108:190
115. Apt W (2017) American trypanosomiasis chagas disease. Elsevier, Amsterdam, p 751
116. Sales PA, Molina I, Murta SMF, Sánchez-Montalvá A, Salvador F, Corrêa-Oliveira R, Carneiro CM (2017) Am J Trop Med Hyg 97:1289
117. Teixeira VL, Lima JCR, Lechuga GC, Ramos CJB, de Pereira MCS, Calvet CM, Bourguignon SC (2019) Brazilian J Pharmacogn 29:735
118. Sabry OMM, Goeger DE, Gerwick WH (2017) Nat Prod Res 31:1245
119. Martínez-Luis S, Cherigo L, Spadafora C, Gutiérrez M (2019) Nat Prod Commun 14:1934578X1901400
120. Simmons TL, Engene N, Ureña LD, Romero LI, Ortega-Barría E, Gerwick L, Gerwick WH (2008) J Nat Prod 71:1544
121. Linington RG, González J, Ureña L-D, Romero LI, Ortega-Barría E, Gerwick WH (2007) J Nat Prod 70:397
122. Almaliti J, Malloy KL, Glukhov E, Spadafora C, Gutiérrez M, Gerwick WH (2017) J Nat Prod 80:1827
123. Boudreau PD, Miller BW, McCall L-II, Almaliti J, Reher R, Hirata K, Le T, Siqueira-Neto JL, Hook V, Gerwick WH (2019) J Med Chem 62:9026

124. Taori K, Liu Y, Paul VJ, Luesch H (2009) ChemBioChem 10:1634
125. Caffrey CR, El-Sakkary N, Mäder P, Krieg R, Becker K, Schlitzer M, Drewry DH, Vennerstrom JL, Grevelding CG (2019) Drug discovery and development for schistosomiasis. Wiley, Hoboken p 187. https://doi.org/10.1002/9783527808656.ch8
126. Palmer PES (1998) Semin Roentgenol 33:6
127. Caffrey CR (2015) Future Med Chem 7:675
128. Melek FR, Tadros MM, Yousif F, Selim MA, Hassan MH (2012) Pharm Biol 50:490
129. Casertano M, Imperatore C, Luciano P, Aiello A, Putra MY, Gimmelli R, Ruberti G, Menna M (2019) Mar Drugs 17:278
130. Reimer A, Blohm A, Quack T, Grevelding CG, Kozjak-Pavlovic V, Rudel T, Hentschel U, Abdelmohsen UR (2015) J Antibiot (Tokyo) 68:674
131. Bibo-Verdugo B, Wang SC, Almaliti J, Ta AP, Jiang Z, Wong DA, Lietz CB, Suzuki BM, El-Sakkary N, Hook V, Salvesen GS, Gerwick WH, Caffrey CR, O'Donoghue AJ, O'Donoghue AJ (2019) ACS Infect Dis 5:1802
132. Pereira AR, Kale AJ, Fenley AT, Byrum T, Debonsi HM, Gilson MK, Valeriote FA, Moore BS, Gerwick WH (2012) ChemBioChem 13:810
133. Pereira AR, McCue CF, Gerwick WH (2010) J Nat Prod 73:217
134. Pereira AR, Etzbach L, Engene N, Müller R, Gerwick WH (2011) J Nat Prod 74:1175
135. Malavige GN, Fernando S, Fernando DJ, Seneviratne SL (2004) Postgrad Med J 80:588
136. Lin CK, Wang YT, Hung EM, Yang YL, Lee JC, Sheu JH, Liaw CC (2017) Planta Med 83:158
137. Lee J-C, Chang F-R, Chen S-R, Wu Y-H, Hu H-C, Wu Y-C, Backlund A, Cheng Y-B (2016) Mar Drugs 14:151
138. Waller RF, Kořený L (2017) Advances in botanical research. Academic Press Inc., Oxford, p 105
139. Sudeep AB, Parashar D (2008) J Biosci 33:443
140. Gupta D, Kaur P, Leong S, Tan L, Prinsep M, Chu J (2014) Mar Drugs 12:115
141. Williams DJ, Faiz MA, Abela-Ridder B, Ainsworth S, Bulfone TC, Nickerson AD, Habib AG, Junghanss T, Fan HW, Turner M, Harrison RA, Warrell DA (2019) PLoS Negl Trop Dis 13: e0007059
142. da Silva G, Domingos T, Fonseca R, Sanchez E, Teixeira V, Fuly A (2015) J Venom Anim Toxins Incl Trop Dis 21:2
143. Burton MJ, Mabey DCW (2009) PLoS Negl Trop Dis 3:e460
144. Abdelmohsen UR, Cheng C, Reimer A, Kozjak-Pavlovic V, Ibrahim AK, Rudel T, Hentschel U, Edrada-Ebel R, Ahmed SA (2015) Planta Med 81:382
145. Cheng C, Othman EM, Reimer A, Grüne M, Kozjak-Pavlovic V, Stopper H, Hentschel U, Abdelmohsen UR (2016) Tetrahedron Lett 57:2786
146. Starr MC, Montgomery SP (2011) Am J Trop Med Hyg 85:680
147. Meltzer E (2006) Lancet 368:283
148. Savioli L, Albonico M (2004) Nat Rev Microbiol 2:618
149. Hagel I, Giusti T (2010) Infect Disord Drug Targets 10:349
150. Bethony J, Brooker S, Albonico M, Geiger SM, Loukas A, Diemert D, Hotez PJ (2006) Lancet 367:1521
151. Spiegler V, Liebau E, Hensel A (2017) Nat Prod Rep 34:627
152. Ui H, Shiomi K, Yamaguchi Y, Masuma R, Nagamitsu T, Takano D, Sunazuka T, Namikoshi M, Omura S (2001) J Antibiot (Tokyo) 54:234
153. Omura S, Miyadera H, Ui H, Shiomi K, Yamaguchi Y, Masuma R, Nagamitsu T, Takano D, Sunazuka T, Harder A, Kölbl H, Namikoshi M, Miyoshi H, Sakamoto K, Kita K (2001) Proc Natl Acad Sci U S A 98:60
154. Yemoa A, Gbenou J, Affolabi D, Moudachirou M, Bigot A, Anagonou S, Portaels F, Quetin-Leclercq J, Martin A (2011) Planta Med 77:641
155. Brahmachari G (2019) Discovery and development of therapeutics from natural products against neglected tropical diseases. Elsevier, Amsterdam, p 373

156. Tsouh Fokou PV, Nyarko AK, Appiah-Opong R, Tchokouaha Yamthe LR, Ofosuhene M, Boyom FF (2015) Biomed Res Int 2015:1
157. Van Der Werf TS, Barogui YT, Converse PJ, Phillips RO, Stienstra Y (2020) Expert Rev Clin Pharmacol 13:391
158. Parenti DM (2009) Medical parasitology. CRC Press, Boca Raton, pp 58–62
159. Nithiuthai S, Anantaphruti MT, Waikagul J, Gajadhar A (2004) Vet Parasitol 126:167
160. Biswas G, Sankara DP, Agua-Agum J, Maiga A (2013) Philos Trans R Soc B Biol Sci 368:20120146
161. Ruiz-Tiben E, Hopkins DR (2006) Adv Parasitol 61:275
162. Wink M (2012) Molecules 17:12771
163. Hemphill A, Mller J (2009) J Helminthol 83:99
164. Higuita NIA, Brunetti E, McCloskey C (2016) J Clin Microbiol 54:518–523
165. Wen H, Vuitton L, Tuxun T, Li J, Vuitton DA, Zhang W, McManus DP (2019) Clin Microbiol Rev 32:e00075
166. McManus DP, Gray DJ, Zhang W, Yang Y (2012) Br Med J 344:39
167. Keiser J, Utzinger J (2004) Expert Opin Pharmacother 5:1711
168. Stürchler D (2018) Travel Med Infect Dis 21:85
169. Prueksapanich P, Piyachaturawat P, Aumpansub P, Ridtitid W, Chaiteerakij R, Rerknimitr R (2018) Gut Liver 12:236
170. Qian MB, Utzinger J, Keiser J, Zhou XN (2016) The lancet. Lancet Publishing Group, London, p 800
171. Schuster RK (2010) Infect Disord Drug Targets 10:402
172. Pakharukova MY, Mordvinov VA (2016) Trans R Soc Trop Med Hyg 110:28
173. Pengput A, Schwartz DG (2020) J Infect Public Health 13:1265
174. Keiser J, Utzinger J (2009) Clin Microbiol Rev 22:466
175. Mas-Coma S, Adela Valero M, Dolores Bargues M (2014) Adv Exp Med Biol 766:77
176. Liu Q, Wei F, Liu W, Yang S, Zhang X (2008) Trends Parasitol 24:318
177. Singh SK, Yadav RP, Singh A (2010) J Appl Toxicol 30:1
178. Mungroo MR, Khan NA, Siddiqui R (2020) Microb Pathog 149:104475
179. Gomes Lima Santos N, Perez Pereira Ramos K, Shanmugam S, Oliveira De Carvalho F, Garcez Barreto Teixeira L, Ramos Silva É, De Vasconcelos Cerqueira-Braz J, Santos Nunes P, Antunes De Souza Araújo A (2018) Epidemiol Infect 146:1746
180. Copp BR (2003) Nat Prod Rep 20:535
181. Barbosa-Filho JM, Do Nascimento Júnior FA, De Andrade Tomaz AC, De Athayde-Filho PF, Da Silva MS, Da Cunha EVL, De De Souza MFV, Batista LM, Diniz MFFM (2007) Braz J Pharmacogn 17:141
182. Davis GL, Ray NA, Lahiri R, Gillis TP, Krahenbuhl JL, Williams DL, Adams LB (2013) PLoS Negl Trop Dis 7:e2404
183. Pattyn SR (1973) Bull World Health Organ 49:403
184. Lourens GB, Ferrell DK (2019) Nurs Clin North Am 54:181
185. Paily KP, Hoti SL, Das PK (2009) J Parasit Dis 33:3
186. Davis BR (1989) Dermatol Clin 7:313
187. Gupta J, Misra S, Mishra SK, Srivastava S, Srivastava MN, Lakshmi V, Misra-Bhattacharya S (2012) Exp Parasitol 130:449
188. Lakshmi V, Srivastava S, Kumar Mishra S, Misra S, Verma M, Misra-Bhattacharya S (2009) Parasitol Res 105:1295
189. Cheuka PM, Mayoka G, Mutai P, Chibale K (2017) Molecules 22:58
190. Queiroz-Telles F, Fahal AH, Falci DR, Caceres DH, Chiller T, Pasqualotto AC (2017) Lancet Infect Dis 17:e367
191. Bienvenu AL, Picot S (2020) Molecules 25:2
192. Asong JA, Ndhlovu PT, Khosana NS, Aremu AO, Otang-mbeng W (2019) S Afr J Bot 126:11
193. Nelson GS (1991) Ann Trop Med Parasitol 85:83

194. Gyasi ME, Okonkwo ON, Tripathy K (2020) Onchocerciasis. StatPearls Publishing, Treasure Island. Available from: https://www.ncbi.nlm.nih.gov/books/NBK559027/
195. Ashour DS (2019) Int J Antimicrob Agents 54:134
196. Ndjonka D, Djafsia B, Liebau E (2018) Parasitol Res 117:2697
197. Hemachudha T, Ugolini G, Wacharapluesadee S, Sungkarat W, Shuangshoti S, Laothamatas J (2013) Lancet Neurol 12:498
198. Dacheux L, Delmas O, Bourhy H (2011) Infect Disord Drug Targets 11:251
199. Tarantola A, Crabol Y, Mahendra BJ, In S, Barennes H, Bourhy H, Peng Y, Ly S, Buchy P (2016) Trop Med Int Health 21:564
200. Salas-Rojas M, Galvez-Romero G, Anton-Palma B, Acevedo R, Blanco-Favela F, Aguilar-Setién A (2014) Fish Shellfish Immunol 36:158
201. Dunne CL, Malone CJ, Whitworth JAG (1991) Trans R Soc Trop Med Hyg 85:550
202. Chandler DJ, Fuller LC (2019) Dermatology 235:79
203. Haustein UF (1991) Hautarzt 42:9
204. Tabassum N, Hamdani M (2014) Pharmacogn Rev 8:52
205. Okello A, Thomas L (2017) Risk Manag Healthc Pol 10:107
206. Willingham AL, Engels D (2006) Adv Parasitol 61:509
207. Maxfield L, Crane JS (2018) Yaws (Frambesia tropica, Thymosis, Polypapilloma tropicum, Parangi, Bouba, Frambosie, Pian). StatPearls Publishing, Treasure Island
208. Marks M (2018) Trop Med Infect Dis 3:92

Top Heterocycl Chem (2021) 58: 253–274
https://doi.org/10.1007/7081_2021_55

Published online: 14 April 2021

Total Synthesis of the Amicoumacin Family of Natural Products

Masaru Enomoto and Shigefumi Kuwahara

Contents

Abstract The amicoumacin family of natural products, as represented by AI-77-B and bacilosarcin A, is reported to exhibit various pharmacologically and agriculturally important biological profiles, such as antiulcer, herbicidal, cytotoxic, and antimicrobial activities. Owing to their promising biological properties as well as attractive molecular architectures, characterized by a structurally diverse carboxylic acid moiety bonded via an amide linkage to a common dihydroisocoumarin unit, a number of total and partial synthetic studies on this family of natural products have been reported thus far. This review describes the total syntheses of several amicoumacins isolated from marine organisms.

Keywords Amicoumacins · Bacilosarcins · Isocoumarin · Total synthesis

M. Enomoto (✉) and S. Kuwahara
Graduate School of Agricultural Science, Tohoku University, Sendai, Japan
e-mail: masaru.enomoto.a2@tohoku.ac.jp; shigefumi.kuwahara.e1@tohoku.ac.jp

Abbreviations

aq	Aqueous
Boc	*Tert*-butoxycarbonyl
CAN	Cerium(IV) ammonium nitrate
Cbz	Benzyloxycarbonyl
CSA	10-Camphorsulfonic acid
DBU	1,8-Diazabicyclo[5.4.0]undec-7-ene
DCC	Dicyclohexylcarbodiimide
DEAD	Diethyl azodicarboxylate
DEPC	Diethylphosphoryl cyanide
DMAP	(4-Dimethylamino)pyridine
DMF	*N,N*-dimethylforamide
DPPA	Diphenylphosphoryl azide
EDCI	1-Ethyl-3-[3-(dimethylamino)propyl]carbodiimide hydrochloride
HBTU	1-[Bis(dimethylamino)methylene]-1*H*-benzotriazolium 3-oxide hexafluorophosphate
HOBt	1-Hydroxybenzotriazole
KHMDS	Potassium hexamethyldisilazanide
LDA	Lithium diisopropylamide
liq	Liquid
MCPBA	*m*-chloroperbenzoic acid
MOM	Methoxymethyl
Ms	Methanesulfonyl
NMO	*N*-methylmorpholine *N*-oxide
PMB	4-Methoxybenzyl
Py	Pyridine
quant.	Quantitative
SAR	Structure–activity relationship
TBAF	Tetrabutylammonium fluoride
TBAI	Tetrabutylammonium iodide
TBS	*Tert*-butyldimethylsilyl
TES	Triethylsilyl
Tf	Trifluoromethanesulfonyl
TFA	Trifluoroacetic acid
THF	Tetrahydrofuran
TMEDA	*N,N,N′,N′*-tetramethylethylenediamine
TMS	Trimethylsilyl
Ts	Tosyl, [(4-methylphenyl)sulfonyl]

1 Introduction

The amicoumacin family of natural products, mostly produced by bacteria of the genus *Bacillus*, is structurally characterized by a dihydroisocoumarin unit linked via an amide bond to an unusual amino acid segment, as exemplified by AI-77-B (**1**), bacilosarcins A–C (**2–4**), PM-94128 (**5**), and Y-05460M-A (**6**) (Fig. 1) [1, 2]. While the left-hand dihydroisocoumarin-containing amine segment is a common structure among this family of natural products, the right-hand acid portion displays considerable structural diversity and presumably originates from various α-amino acids, such as aspartic acid, asparagine, leucine, and valine. This family of secondary metabolites is known to exhibit various therapeutically and agriculturally important biological profiles such as antibacterial, antiulcer, cytotoxic, and herbicidal activities. Prompted by their unique molecular architectures and promising biological activities, numerous research groups, including us, have conducted synthetic studies on this class of natural products as they are considered as new lead compounds for the development of therapeutic agents as well as agrochemicals. This review describes the total syntheses of AI-77-B (**1**), bacilosarcins A–C (**2–4**), PM-94128 (**5**), and Y-05460M-A (**6**).

2 AI-77-B

AI-77-B (**1**) (also known as amicoumacin B), a representative member of the amicoumacin family of natural products, was first isolated from the fermentation broth of *Bacillus pumilus* as a potent antiulcer agent [3–6]. Since the first isolation of **1** in 1982, several research groups have reported the isolation of **1** mainly from bacteria of the genus *Bacillus*, including a marine-derived *B. subtilis*. This structurally unique pseudopeptide exhibits potent antiulcerogenic activity toward stress ulcers without anticholinergic, antihistaminergic, or central suppressive side effects [5–7]. Thus, several SAR studies for **1** have been reported so far [8]. Additionally, efforts to develop orally active prodrug analogs derived from **1** have also been made [8, 9]. Owing to these therapeutically important biological profiles as well as its intriguing molecular architecture, AI-77-B (**1**) has attracted much attention from many synthetic organic chemists, which culminated in the total synthesis of **1** by eight research groups including our own. This section describes seven reported total

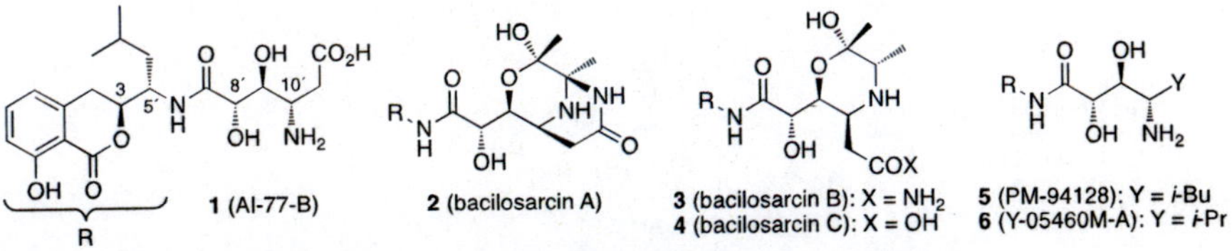

Fig. 1 Representative members of the amicoumacin family of natural products

syntheses of AI-77-B (**1**), omitting our synthesis which will be discussed in Section 3.

2.1 Hamada–Shioiri's Total Synthesis of AI-77-B

The first total synthesis of AI-77-B (**1**) was accomplished by Hamada–Shioiri's group in 1989 (Scheme 1) [10, 11].

They concisely prepared the full carbon skeleton of the amine segment via diastereoselective addition of a lithiated 6-methylsalicylate derivative, generated by treating **7** (1.4 equiv) with excess LDA (2.6 equiv), to *N*-Boc-(*S*)-leucinal (**8**) (1.0 equiv) accompanied by spontaneous lactonization to afford a separable 81:19 mixture of the desired product **9** and 3-*epi*-**9** in 32% yield (61% based on recovered starting material). Removal of the Boc and MOM protecting groups with HCl/MeOH furnished the amine segment **10** as its hydrochloride salt.

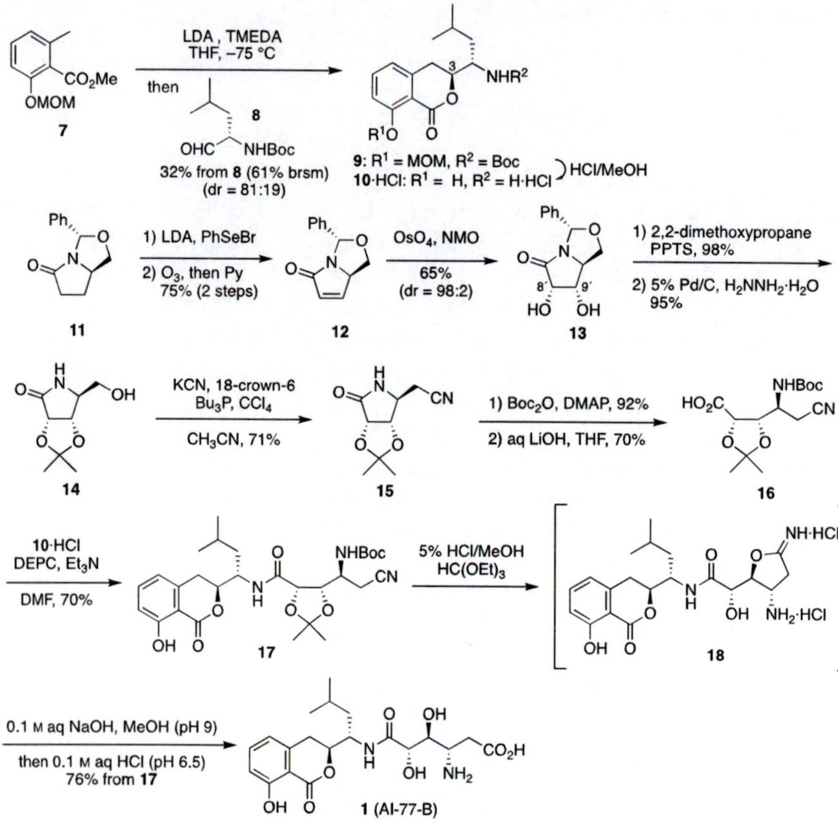

Scheme 1 Hamada–Shioiri's total synthesis of AI-77-B

On the other hand, their synthesis of the acid segment commenced with α-selenenylation of bicyclic lactam **11**, prepared in four steps from D-glutamic acid [12]. Oxidation of the resulting selenide followed by concomitant *syn*-elimination of the intermediary selenoxide gave **12**. Dihydroxylation of **12** with OsO_4 proceeded with excellent diastereoselectivity (98:2), furnishing **13** after separation of the minor diastereomer (8′*R*,9′*R*)-**13** by SiO_2 column chromatography. Protection of **13** as its acetonide followed by removal of the benzylidene acetal gave **14**, which was subjected to a one-carbon homologation reaction (KCN, 18-crown-6, Bu_3P, and CCl_4 in CH_3CN) developed by their group [13], thus providing nitrile **15** in 71% yield. Hydrolysis of the lactam ring after *N*-protection of **15** with Boc_2O/DMAP afforded acid segment **16**.

Condensation between **10**·HCl (1.2 equiv) and **16** (1.0 equiv) using DEPC as a coupling reagent in the presence of Et_3N gave **17** in 70% yield. Acidic methanolysis of the acetonide group of **17** induced an intramolecular Pinner reaction to afford **18**, which was carefully treated with 0.1 M aqueous NaOH (pH 9) and then with 0.1 M HCl (pH 6.5) to complete the first total synthesis of AI-77-B (**1**).

2.2 *Thomas's Total Synthesis of AI-77-B*

The second synthesis of AI-77-B (**1**) was reported by Thomas and coworkers in 1991 (Scheme 2) [14, 15].

For the preparation of the amine segment **10**, they utilized a phenyl oxazoline derivative as a salicylic acid synthon. The synthesis of the amine segment commenced with the conversion of 2-methoxy-6-methylbenzoic acid (**19**) into amide **20** by condensation of **19** with **21** via an acid chloride intermediate and subsequent hydrolysis of the resulting acyl oxazolidinone derivative. Treatment of **20** with $SOCl_2$ followed by NaOH afforded oxazoline **22**, which was deprotonated with *n*-BuLi and then allowed to react with **23**, prepared by treating *N*-Cbz-leucinal with *t*-BuMgCl at −78°C, to provide a mixture of **24** and **25**. Upon exposure to SiO_2, the mixture was transformed to a separable 87:13 mixture of **26** and its C3-epimer in 44% yield from *N*-Cbz-leucinal. Removal of the methoxy group with BBr_3 followed by hydrogenolysis in EtOH/aqueous HCl furnished **10**·HCl.

The acid segment was synthesized by employing substrate-controlled diastereoselective dihydroxylation as the key step. Aldehyde **27**, prepared in five steps from L-aspartic acid [16, 17], was subjected to the Horner–Wadsworth–Emmons reaction under Still–Gennari's conditions to afford a separable mixture of **28** and its *E*-isomer in a ratio of 85:12. Dihydroxylation of the *Z*-olefin **28** using OsO_4 and NMO proceeded in a diastereoselective manner, albeit with modest diastereoselectivity, to give **29** in 58% yield along with its (8′*R*,9′*R*)-diastereomer (15%). Protection of the diol **29** as its acetonide followed by hydrogenolysis of the benzyl ester afforded the acid segment **30**.

Condensation of **30** with **10**·HCl (1.1 equiv) using DCC and DMAP provided **31** in 54% yield. The amide **31** was treated with aqueous HCl (3 M)/THF (1:1) with the

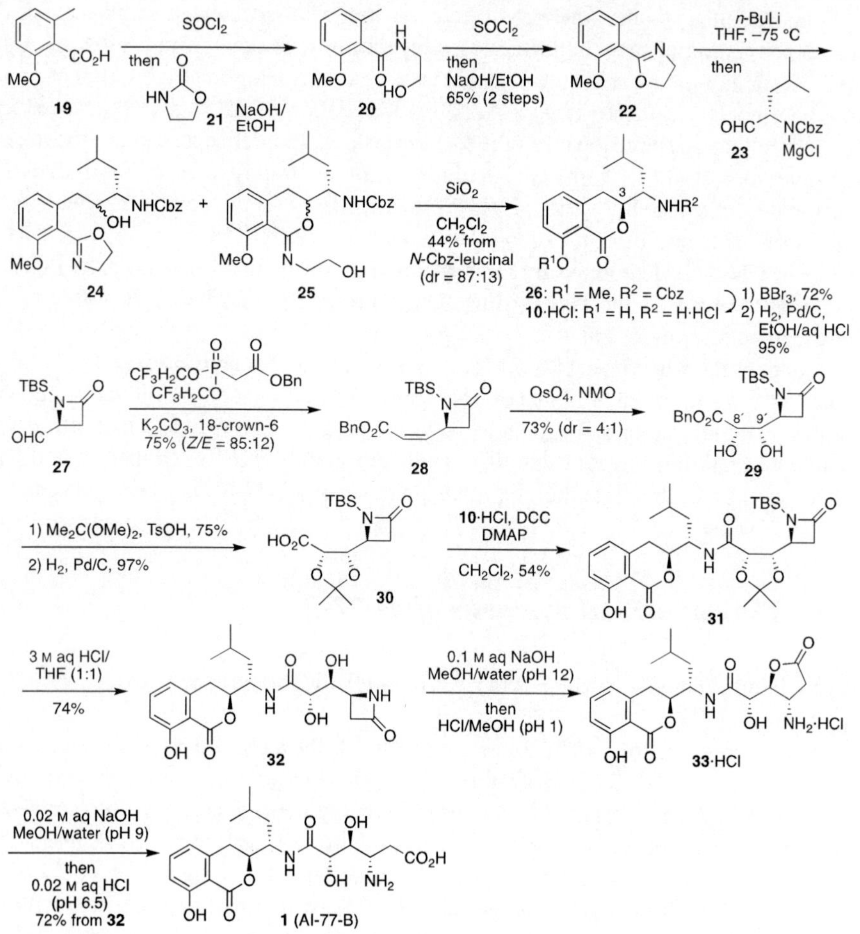

Scheme 2 Thomas's total synthesis of AI-77-B

intention of inducing removal of the TBS and acetonide protecting groups to afford azetidinone **32** together with its spontaneous lactonization to di-lactone **33** (amicoumacin C). Contrary to their expectation, however, the product was not **33**, but instead **32**. Thus, the five-membered lactone ring was constructed by exposure of **32** to aqueous NaOH (pH 9) followed by acidification with methanolic HCl to give **33**·HCl; in the first operation, both the six-membered lactone and azetidinone rings were presumed to be cleaved. Finally, selective opening of the five-membered lactone ring afforded the target compound AI-77-B (**1**).

2.3 Procter's Total Synthesis of AI-77-B

Procter and coworkers reported the total synthesis of AI-77-B in 1992 (Scheme 3) [18, 19].

Similar to the first synthesis by Hamada and Shioiri, they also utilized the nucleophilic addition of a metallated 6-methylsalicylate derivative to *N*-Boc-leucinal (**8**) to obtain the amine segment. Interestingly, however, they found that the undesired diastereomer 3-*epi*-**35** could be quantitatively transformed to the desired diastereomer **35** by saponification of 3-*epi*-**35** followed by mesylation of the resulting hydroxy carboxylate in the presence of Et_3N. Mild deprotection of the methoxy group of **35** with MgI_2 and subsequent removal of the Boc group with methanolic HCl furnished **10**·HCl in 87% yield from **35**.

Synthesis of acid segment **40** was achieved from commercially available L-aspartic acid derivative **36** in four steps. First, the protected amino acid **36** was converted to aldehyde **37** by a conventional two-step sequence. Chain elongation of **37** by the Wittig reaction using **38** in MeOH in the presence of benzoic acid afforded an *E*/*Z* mixture of desired *Z*-ester **39** (38% isolated yield) and its *E*-isomer (37%). The acid segment **40** derived by acidic treatment of **39** was condensed with the amine segment **10** prior to the installation of the 8′,9′-*erythro* diol functionality of

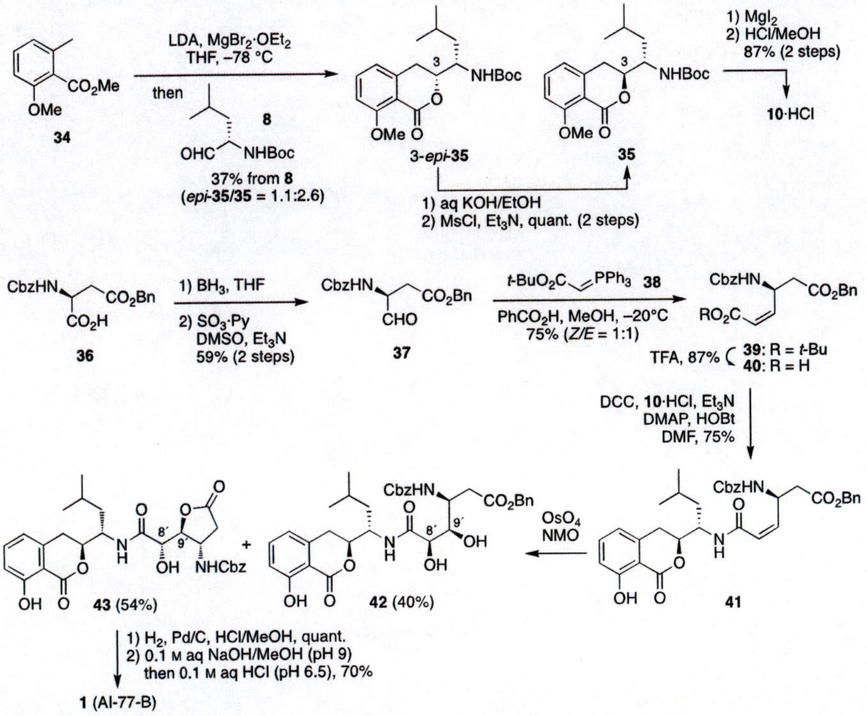

Scheme 3 Procter's total synthesis of AI-77-B

AI-77-B. Dihydroxylation of the resulting condensation product **41** proceeded with low diastereoselectivity, providing a separable mixture of undesired diol **42** (40%) and desired lactone **43** (54%), the latter of which was considered to be generated via spontaneous cyclization of the corresponding dihydroxy ester [namely (8′*S*,9′*S*)-**42**]. Deprotection of the Cbz protecting group in **43** followed by selective hydrolysis of the γ-lactone ring furnished AI-77-B (**1**).

2.4 Vogel's Total Synthesis of AI-77-B

Vogel and coworkers disclosed the total synthesis of AI-77-B using a unique strategy involving regio- and stereoselective functionalization of 7-oxobicyclo [2.2.1]heptane derivatives (Scheme 4) [20].

Their synthesis of AI-77-B began with the conversion of (±)-2-cyano-7-oxabicyclo[2.2.1]hept-5-en-2-yl acetate [(±)-**44**], obtained from the Diels–Alder reaction of furan with 1-cyanovinyl acetate, into **45** by a four-step sequence consisting of (1) regio- and stereoselective addition of PhSeBr to **44**, (2) subsequent

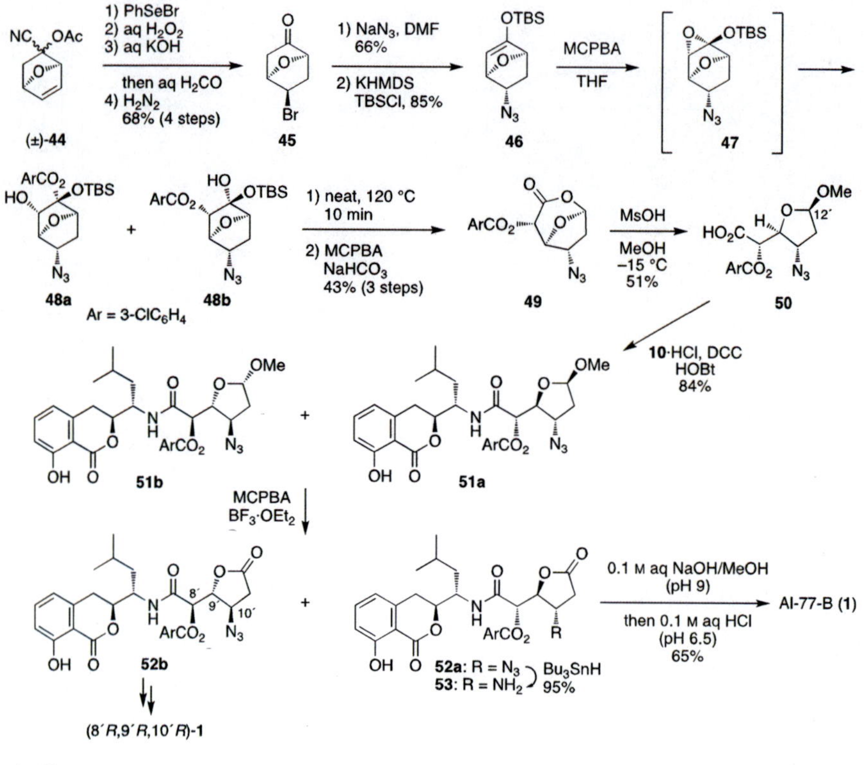

Scheme 4 Vogel's total synthesis of AI-77-B

oxidative elimination of the resulting selenide, (3) saponification of the acetyl group followed by treatment of the resultant cyanohydrin with formalin to form the corresponding bromoketone, and (4) stereoselective hydrogenation using diimide [21–23]. After nucleophilic substitution of the bromo group in **45** with NaN_3, the resulting product was converted to silyl enol ether **46**. The Rubottom oxidation of **46** with MCPBA gave intermediate silyloxy epoxide **47**, which underwent concomitant nucleophilic addition of *m*-chlorobenzoic acid to furnish a mixture of **48a** and **48b** in a ratio of 3.8:1. After heating the mixture at 120°C for 10 min, the resulting ketone was subjected to the Baeyer–Villiger oxidation to afford lactone **49** in 43% yield from **46**. Since **49** was unstable at room temperature, its methanolysis was conducted at −15°C in the presence of methanesulfonic acid, thus providing a 9:1 mixture of acid segment **50** and its 12′-epimer in 63% yield, from which **50** was isolated in 51% yield by crystallization.

Condensation of the carboxylic acid **50** prepared as a racemate with an equimolar amount of optically active **10**·HCl, obtained according to the Hamada–Shioiri procedure [10, 11], gave an inseparable 1:1 mixture of **51a** and **51b** in 84% yield. The mixture was then oxidized with MCPBA in the presence of a catalytic amount of $BF_3 \cdot OEt_2$, providing lactones **52a** and **52b** in isolated yields of 44% and 52%, respectively. Reduction of the azido group in **52a** with Bu_3SnH followed by saponification of the resulting amino ester **53** according to the literature procedure completed the total synthesis of AI-77-B (**1**) [10, 11]. On the other hand, subjection of **52b** to the same procedure produced (8′*R*,9′*R*,10′*R*)-**1**.

2.5 Kotsuki's Total Synthesis of AI-77-B

Kotsuki and coworkers reported an efficient total synthesis of AI-77-B utilizing D-ribose as the common chiral source for the construction of both the acid and amine segments (Scheme 5) [24].

The key step for the synthesis of amine segment **10**·HCl involved a copper-catalyzed Grignard coupling of **54a** with triflate **55**. The Grignard reagent **54a**, a salicylic acid synthon, was prepared as a mixture with its regioisomer **54b** by CsF-mediated rearrangement of *m*-bromophenyl propargyl ether (**56**) followed by treatment of the resulting inseparable mixture of benzofurans (**57a/57b**) with Mg. The triflate **55**, on the other hand, was prepared from D-ribose-derived lactol **58** in three steps [25], and allowed to react in the presence of a catalytic amount of CuBr with the mixture of **54a** and **54b** to afford an inseparable mixture of **59a** and **59b** in 93% yield. Ozonolysis of the mixture followed by methanolysis of the resulting *O*-acetyl salicylaldehyde derivatives gave **60a** in 64% isolated yield along with **60b** (28%). After protection of the phenolic hydroxy group of **60a** as its benzyl ether and removal of the acetonide group, the resulting lactol **61** was transformed into the amine segment **10**·HCl by a conventional four-step sequence.

The preparation of the acid segment **62** commenced with the conversion of lactone **64**, readily available from D-ribose in two steps [26], into lactam **63** in two

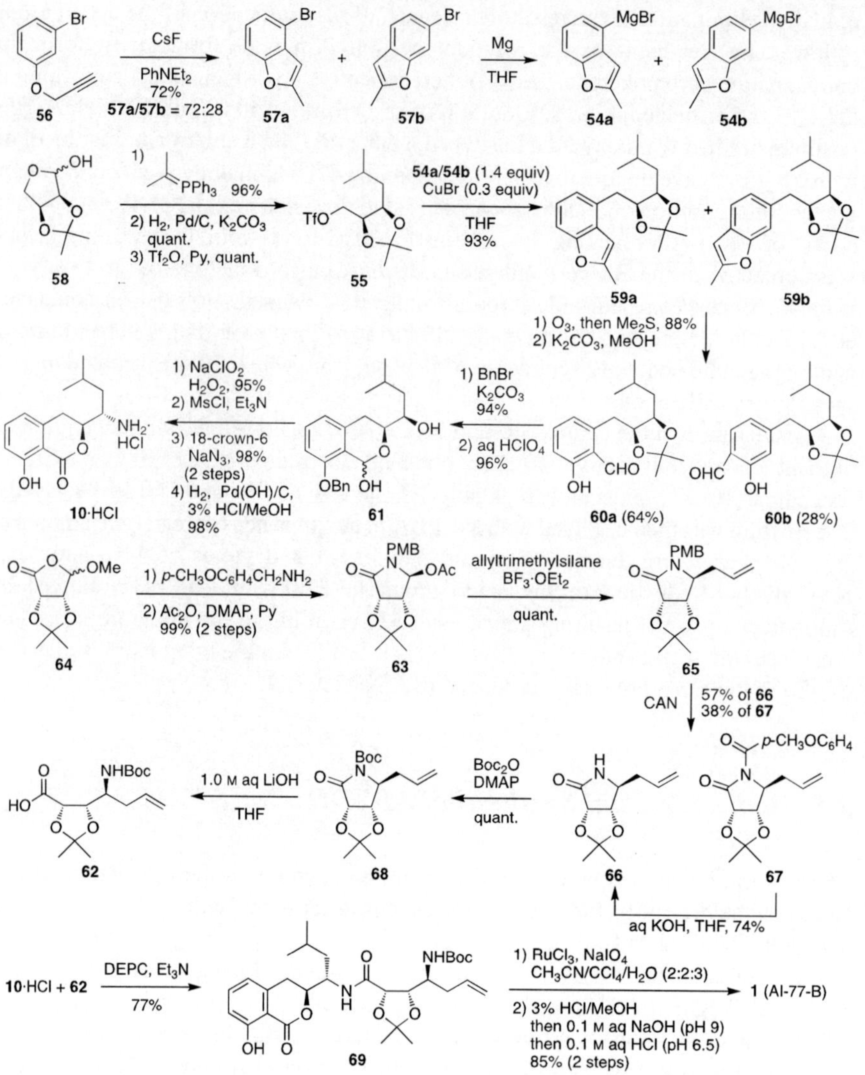

Scheme 5 Kotsuki's total synthesis of AI-77-B

steps via an *N*-PMB-protected aminal intermediate. Nucleophilic addition of allyltrimethylsilane to **63** in the presence of $BF_3 \cdot OEt_2$ proceeded stereoselectively to afford **65** in quantitative yield. Removal of the PMB protecting group with CAN afforded **66** (57% isolated yield) along with benzylic oxidation product **67** (38% isolated yield), the latter of which could be hydrolyzed to the desire lactam **66** in 74% yield.

After protection of **66** with Boc_2O in the presence of DMAP, the resulting lactam **68** was hydrolyzed with aqueous LiOH to furnish acid segment **62**, which was employed in the next step without purification.

Condensation of **10**·HCl with an equimolar amount of **62** using DEPC delivered amide **69** in 77% yield. Finally, oxidative cleavage of the terminal double bond with $RuCl_3/NaIO_4$, removal of the acetonide and Boc protecting groups, and a γ-lactone opening/neutralization sequence furnished AI-77-B (**1**).

2.6 *Ghosh's Total Synthesis of AI-77-B*

The Ghosh group disclosed an efficient total synthesis of AI-77-B featuring a titanium enolate-mediated asymmetric aldol reaction of an optically active indanol ester with an aldehyde (Scheme 6) [27, 28].

The preparation of amine segment **70** began with esterification of γ-methylvaleric acid (**71**) with chiral indanol **72** to give **73** [29]. The aldol reaction of **73** with 4,4,4-trichlorobutanal (**74**) in the presence of $TiCl_4$ and *i*-Pr_2NEt proceeded *anti*-selectively to furnish a 19:1 mixture of *anti*-aldol **75** and its *syn*-diastereomer [30, 31], from which the desired product **75** was isolated in 90% yield by SiO_2 chromatography. After mild saponification of **75**, the resulting acid was subjected to Curtius rearrangement using DPPA to afford oxazolidinone **76** as a single isomer. Protection of **76** with Boc_2O in the presence of DMAP followed by hydrolysis of the resulting *N*-Boc-oxazolidinone intermediate yielded the corresponding amino alcohol derivative, which was then protected with 2,2-dimethoxypropane to provide *N,O*-acetal **77**. Conversion of **77** to **78** was realized in 85% yield by treatment of **77** with *n*-BuLi (4.5 equiv) and TMEDA (3.0 equiv) and in situ trapping of the resulting acetylide with $ClCO_2Me$. The alkynyl ester **78** was also prepared in an alternative manner by utilizing diastereoselective addition of allyltributyltin to *N*-Boc-(*S*)-leucinal (**8**) as the key step [32]. The ester **78** thus obtained was subjected to the Diels–Alder reaction with **79** to give the corresponding cycloadduct, which upon prolonged heating underwent extrusion of ethylene to afford **80** in 76% yield. Acidic hydrolysis of the acetonide moiety of **80** accompanied by concomitant lactonization provided **81**. Deprotection of the Boc group of **81** with TFA was conducted just before the use of the resulting amine segment **70** in the next condensation step to prevent its change into the corresponding seven-membered lactam.

The preparation of acid segment **82** began with condensation of 4-pentenoic acid (**83**) with chiral indanol **72**. The asymmetric aldol reaction of the product **84** with (benzyloxy)acetaldehyde (**85**) furnished *syn*-aldol **86** as a single diastereomer in 97% yield [31]. After conversion of **86** into **88** via carbamate **87** using essentially the same procedure as that employed for the preparation of **77** from **75**, the *N,O*-acetonide **88** was derivatized into aldehyde **89** by reductive removal of the Bn group followed by the Swern oxidation. One-carbon homologation of **89** was performed by following Dondoni's procedure [33]. Treatment of **89** with 2-(trimethylsilyl)thiazole (**90**) in CH_2Cl_2 followed by exposure of the resulting

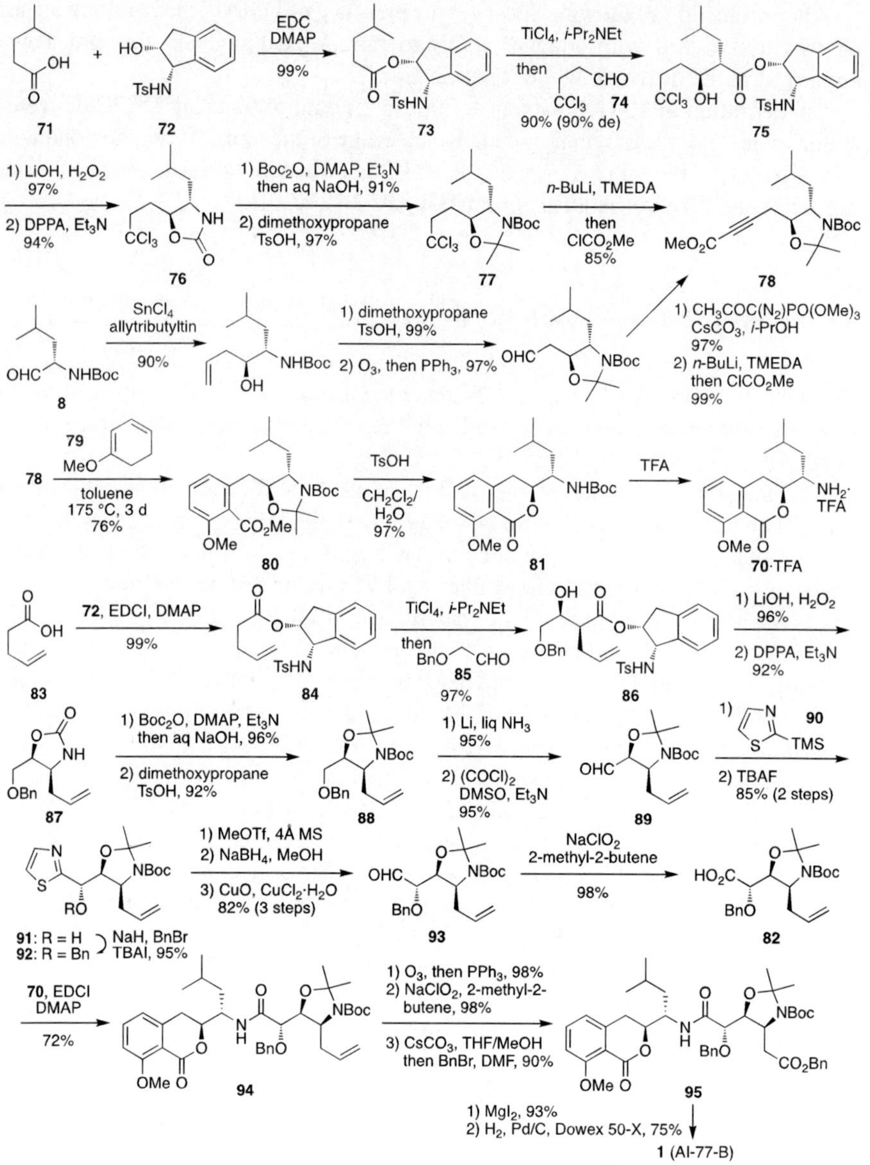

Scheme 6 Ghosh's total synthesis of AI-77-B

crude mixture containing **91** and its TMS ether (ratio 1:1) to TBAF delivered pure **91** in 85% yield. After protection of the hydroxy group as its Bn ether, the resultant thiazole **92** was transformed to aldehyde **93** by the following one-pot three-step procedure: (1) *N*-methylation of the thiazole ring with MeOTf; (2) reduction of the intermediary thiazolinium C=N bond; and (3) hydrolysis of the resulting thiazolidine in the presence of CuO and $CuCl_2$. Oxidation of the aldehyde **93** under the Pinnick–Kraus conditions yielded the acid segment **82**.

Condensation of the acid segment **82** with the amine segment **70** was performed using EDCI, providing amide **94** in 72% yield. Ozonolytic cleavage of the terminal double bond of **94** followed by the Pinnick–Kraus oxidation gave a carboxylic acid intermediate, which was temporarily protected as its Bn ester for clean *O*-demethylation in the subsequent step. Unmasking of the methyl-protected phenolic hydroxy group of **95** was conducted with MgI_2 and subsequent removal of the remaining protecting groups furnished AI-77-B (**1**).

2.7 *Rao's Total Synthesis of AI-77-B*

Rao and coworkers achieved the total synthesis of AI-77-B by using a chiral pool strategy (Scheme 7) [34].

They prepared the amine segment **10**·TFA by a concise three-step sequence that involved conversion of known triflate **96** into *o*-toluic acid derivative **97**, nucleophilic addition of the benzyl anion, generated by treating **97** with LDA, to

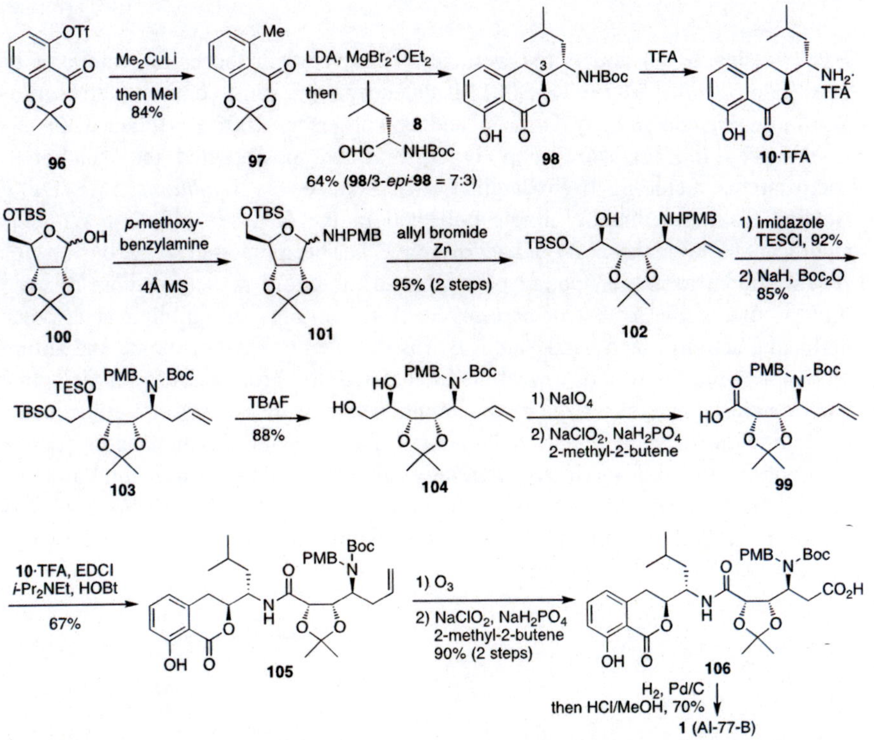

Scheme 7 Rao's Total Synthesis of AI-77-B

N-Boc-(*S*)-leucinal (**8**) to form **98**, and removal of the Boc protecting group with TFA [35, 36].

The preparation of the acid segment **99** commenced with the transformation of **100**, readily preparable in two steps from D-ribose [37], into aminal **101** by refluxing a solution of **100** and *p*-methoxybenzylamine in MeOH in the presence of 4 Å MS. The resulting crude product **101** was subjected to the Barbier reaction using allyl bromide to afford **102** in 95% yield as a single diastereomer [38]. After protection of the hydroxy and amino groups of **102** as its TES ether and *tert*-butyl carbamate, respectively, the TBS- and TES-protecting groups of the resulting product **103** were removed with TBAF. Oxidative cleavage of the resulting vicinal diol moiety in **104** followed by the Pinnick–Kraus oxidation of the resultant aldehyde provided **99**.

The acid segment **99** was condensed with an equimolar amount of freshly prepared **10**·TFA using EDCI, *i*-Pr_2NEt, and HOBt to give amide **105** in 67% yield. The olefinic amide **105** was converted, in a conventional manner, to carboxylic acid **106**, the global deprotection of which provided AI-77-B (**1**).

3 Bacilosarcins A–C

Bacilosarcins A (**2**) and B (**3**) were first discovered in the culture broth of the bacterium *Bacillus subtilis* TP-B0611 isolated from intestinal contents of the sardine *Sardinops melanostica* by Igarashi and coworkers in 2008 [39]. Since the first isolation by the Igarashi group, other three groups reported the isolation of bacirosarcins including bacirosarcin C (**4**) produced by *Bacillus subtilis* B1779 isolated from a sediment sample collected in the Red sea [40–42]. While no significant biological activity of bacilosarcin C has been reported so far, bacilosarcin A was shown to exhibit potent plant growth inhibitory activity against barnyard millets comparable to that of herbimycin A (an ansamycin antibiotic with potent herbicidal activity), and bacilosarcin B was revealed to have cytotoxic and antimicrobial activities as well as a weak herbicidal activity. From a structural viewpoint, bacilosarcins are characterized by the unique heterocyclic ring systems embedded in the acid segment. In particular, the 3-oxa-6,9-diazabicyclo[3.3.1]nonane ring system incorporated in bacilosarcin A was totally unprecedented both in natural and synthetic compounds. In this section, our first total synthesis of bacilosarcins A–C (**2**–**4**) exploiting thermodynamic control strategies is presented together with a new synthesis of AI-77-B (**1**) [43].

3.1 Enomoto–Kuwahara's Total Synthesis of Bacilosarcins A–C

As shown in Scheme 8, our total synthesis of bacilosarcins A (**2**) and B (**3**) commenced with the conversion of known epoxy alcohol **107**, obtainable by the

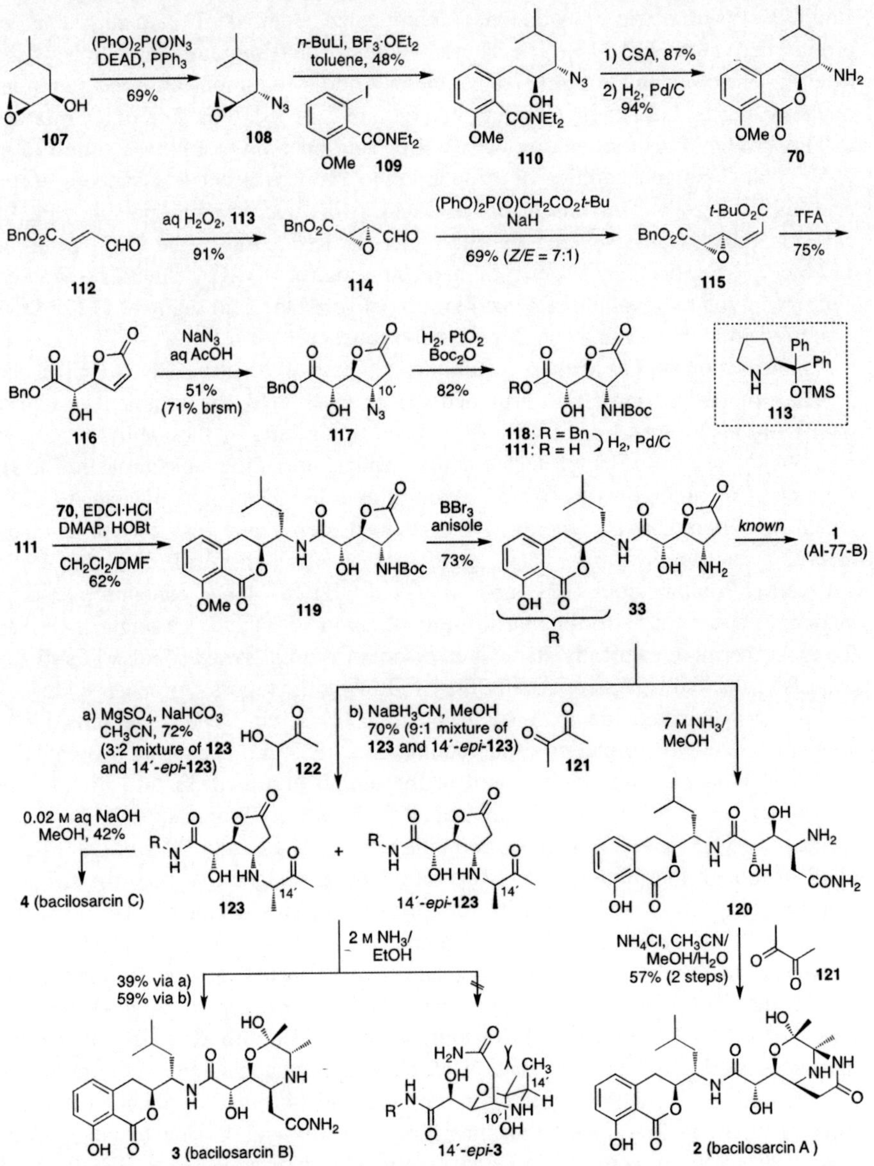

Scheme 8 Enomoto–Kuwahara's total synthesis of bacilosarcins A and B

Sharpless kinetic resolution of 5-methyl-1-hexen-3-ol [43], into azido epoxide **108** with inversion of configuration. The epoxide **108** was allowed to react in the presence of $BF_3{\cdot}OEt_2$ with the organolithium species, generated by treating **109** with *n*-BuLi, to give **110** [44–46]. Lactonization of **110** with CSA and subsequent catalytic hydrogenation of the azido functionality provided the amine segment **70** of 96% enantiomeric excess (ee) in 27% overall yield from **107** through four steps.

Our acid segment **111** was synthesized by a six-step sequence beginning with Córdova's asymmetric epoxidation of conjugated aldehyde **112** mediated by L-proline derivative **113** [47]. The Horner–Wadsworth–Emmons olefination of the resulting epoxy aldehyde **114** (97% ee) under Ando's conditions afforded a mixture of epoxy diester **115** and its *E*-isomer in a ratio of 7:1 [48]. Exposure of the mixture to TFA brought about intramolecular epoxide ring-opening to afford lactone **116** in 75% yield. Conjugate addition of azido anion to **116** was effected in aqueous acetic acid, furnishing a 5:1 mixture of **117** and its C10′ epimer in an acceptable yield of 51% (71% based on recovered starting material). Reduction of the azido group of **117** by PtO_2-catalyzed hydrogenation in the presence of Boc_2O yielded **118**. Finally, hydrogenolytic removal of the benzyl group delivered the acid segment **111** in 20% overall yield from **112** in a concise six-step sequence.

Condensation of **111** with an equimolar amount of **70** using EDCI·HCl in the presence of DMAP and HOBt provided **119** in 62% yield, the treatment of which with BBr_3 in the presence of anisole induced unmasking of the methyl-protected phenolic hydroxy and Boc-protected amino groups, providing amicoumacin C (**33**). Ammonolysis of the five-membered lactone ring of **33** gave amicoumacin A (**120**), which was immediately employed in the next step due to its instability. The heterobicyclic ring system in bacilosarcin A (**2**) was best installed by mixing crude **120** with 2,3-butanedione (**121**) and NH_4Cl in CH_3CN/MeOH containing a small amount of water, thus affording bacilosarcin A (**2**) in 57% yield as a single as well as the most thermodynamically stable stereoisomer. A total synthesis of AI-77-B (**1**) was also achieved by hydrolysis of **33** by following the reported procedure [15].

Having completed the total synthesis of bacilosarcin A (**2**), we turned our attention to the transformation of amicoumacin C (**33**) into bacilosarcin B (**3**). For the installation of a four-carbon unit at the amino group of **33**, we utilized the Amadori reaction. Thus, **33** was treated with 3-hydroxy-2-butanone (**122**) in the presence of $MgSO_4$ to furnish a 3:2 epimeric mixture of **123** and 14′-*epi*-**123** in 72% yield. Although the mixture was inseparable, we envisaged that both the epimers could be stereoconvergently transformed to **3** via epimerization at C14′ since the formation of 14′-*epi*-**3** seemed to be difficult due to steric repulsion between C10′ carbamoyl methyl and C14′ methyl groups. As expected, **3** was obtained in an acceptable yield of 39% by ammonolysis of the mixture, while 14′-*epi*-**3** was not observed at all. An improved transformation of **33** into **3** was realized by diastereoselective reductive amination of **33** with 2,3-butanedione (**121**) followed by ammonolysis of the resulting mixture of **123** and 14′-*epi*-**123** (obtained as a 9:1 mixture in 70% yield from **33**), thus furnishing **3** in 59% yield [7]. On the other hand, hydrolysis of the mixture of **123** and 14′-*epi*-**123** with aqueous NaOH (0.02 M) afforded bacilosarcin C (**4**) in 42% yield [49].

4 PM-94148 and Y-05460M-A

PM-94128 (**5**), an antitumor aminoisocoumarin, was isolated from the culture broth of the marine bacterium *Bacillus* sp. PhM-PHD in 1996 by Cañedo and coworkers [50]. On the other hand, Y-05460M-A (**6**), a one-carbon lower homologue of PM-94128 exhibiting antimicrobial activity and cytotoxicity, was first isolated in 1992 by Sato and coworkers from the culture broth *Bacillus* sp. Y-05460M of soil origin, and in 2013 by Kalinovskaya's group from *Paenibacillus profundus* sp. isolated from a deep sea sediment [51, 52]. The structures of these natural products were originally proposed as their planar structures, and neither the absolute nor the relative configurations of PM-94128 and Y-05460M-A have been elucidated until Py–Valleé's synthesis of **5** and our synthesis of **6**, respectively [53, 54].

4.1 Py–Valleé's Total Synthesis of PM-94128

The first synthesis and stereochemical elucidation of PM-94128 was accomplished in 2003 by the Py–Valleé group through their highly diastereoselective approach (Scheme 9) [53]. Since PM-94128 and its related compound AI-77-B were isolated from bacteria of the same genus, they assumed that its absolute stereochemistry would be shown as **5**.

Their synthesis of **5** began with the preparation of chiral auxiliary **124** from (*R*)-valinol (**125**) by a four-step conversion involving oxidation of the imine generated by treating **126** with *p*-methoxybenzaldehyde followed by treatment of the resultant

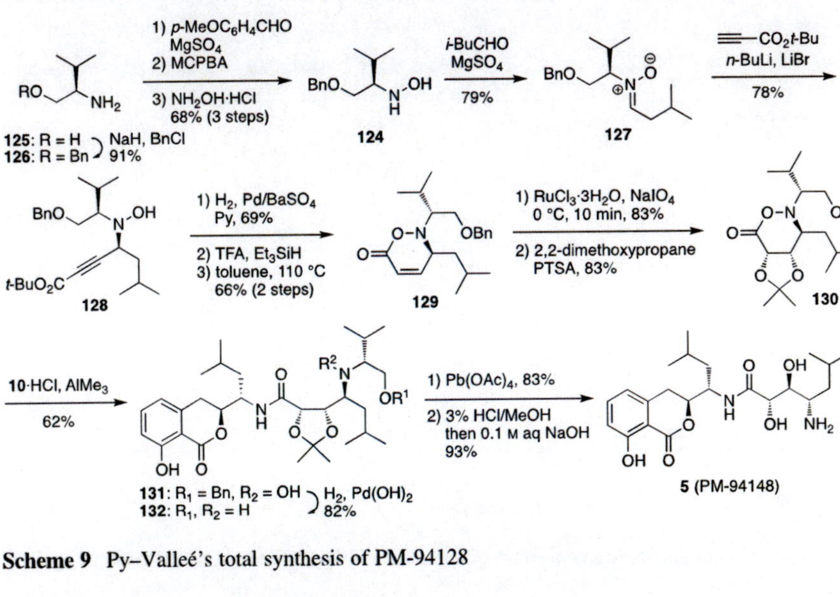

Scheme 9 Py–Valleé's total synthesis of PM-94128

oxime with NH_2OH. After condensation of **124** with isovaleraldehyde, the resulting nitrone **127** was allowed to react with *tert*-butyl lithiopropiolate to give **128** in 78% yield as a single diastereomer. In order to circumvent formation of a nitrone in a subsequent oxidation step, the hydroxylamine **128** was transformed by a three-step sequence to oxazinone **129**. Diastereoselective dihydroxylation of **129** was realized by its exposure to $RuCl_3{\cdot}3H_2O$ in the presence of $NaIO_4$ for a very short time, furnishing the corresponding diol in 83% yield as a single diastereomer, whose diol moiety was then protected as its acetonide. Coupling of the resultant **130** with 3.5 equivalent of **10**·HCl, prepared according to Kotsuki's procedure [24], was conducted under Ikegami's conditions to afford **131** in 62% yield [55]. After deprotection of the Bn group of **131** accompanied by concomitant reduction of the hydroxylamine moiety, the chiral auxiliary of the resulting amide **132** was removed by oxidative cleavage using $Pb(OAc)_4$ to yield the corresponding primary amine. Finally, deprotection of the acetonide moiety with methanolic HCl followed by adjustment of pH provided **5**, whose spectral data and specific rotation were identical with those of natural PM-94148.

4.2 *Enomoto–Kuwahara's Total Synthesis of PM-94128 and Y-05460M-A*

In 2009, we achieved a concise synthesis of PM-94128 and Y-05460M-A (Scheme 10) [54].

The synthesis commenced with conversion of known β-hydroxy lactam **133**, which was readily available from *N*-Boc-L-leucine in ca. 70% yield by a three-step sequence [56, 57], into **134** via a corresponding mesylate intermediate.

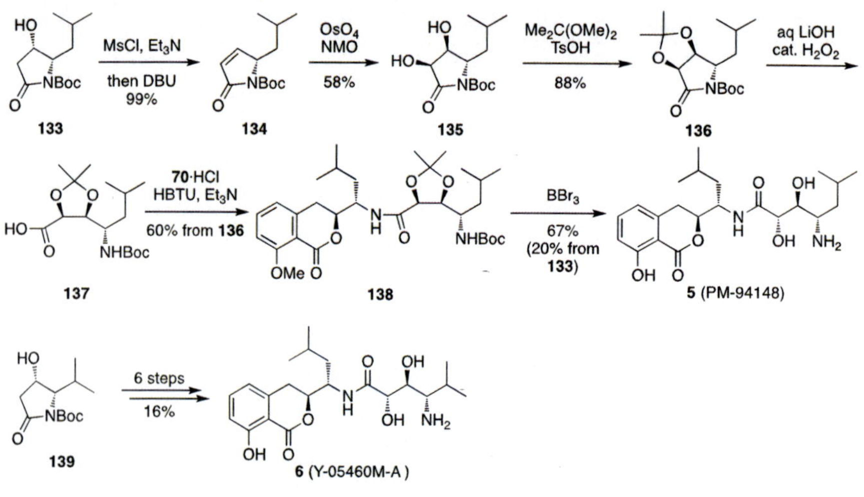

Scheme 10 Enomoto–Kuwahara's total synthesis of PM-94128 and Y-05460M-A

Dihydroxylation of **134** with OsO_4 and NMO provided **135** as a single diastereomer, whose diol moiety was then protected as its acetonide. The resulting lactam **136** was hydrolyzed with aqueous LiOH in the presence of a catalytic amount of H_2O_2 to give the acid segment **137**. Condensation of **137** with the known amine segment **70**·HCl, prepared by our procedure [43], was effected by employing HBTU and Et_3N to afford **138** in 60% yield from **136**. Treatment of **138** with BBr_3 induced deprotection of all the protecting groups to furnish PM-94128 (**5**). Thus, the enantioselective total synthesis of **5** was accomplished in 20% overall yield from **133** in the concise six-step sequence.

Having completed the total synthesis of **5**, we turned our attention to the stereochemical elucidation of Y-05460M-A (**6**) through synthesis. From the structural similarity between PM-94128 (**5**) and Y-05460M-A (**6**) as well as the fact that both the natural products were produced by the same genus of bacteria, we presumed that the absolute stereochemistry of Y-05460M-A would be represented by **6**. Thus, L-valine-derived hydroxy lactam **139** was transformed to **6** by following essentially the same set of the reactions as employed for the synthesis of **5**. While the NMR spectra of **6** disagreed with those of natural Y-05460M-A, the NMR spectra of its hydrochloride salt were identical with those of the natural product. Although we could not verify that the NMR spectra of the natural product had actually been recorded as its hydrochloride salt due to the unavailability of the natural sample, the complete agreement in the NMR spectra between **6**·HCl and natural Y-05460M-A, coupled with the stereochemical uniformity of this class of natural product, led us to the conclusion that the stereochemistry of natural Y-05460M-A should be depicted as **6**.

5 Conclusion

Herein, the total syntheses of AI-77-B (**1**), bacilosarcins A–C (**2**–**4**), PM-94128 (**5**), and Y-05460M-A (**6**) are summarized. Since the discovery of amicoumacin A (**120**), the first reported amicoumacin, in 1981 [58], new congeners of the amicoumacin family of natural products have continually been isolated to date. Hetiamacins A–F and damxungmacins A and B are the representative members of the recently isolated amicoumacins [59–62]. Synthetic studies on this family of natural products, such as our improved synthesis of amicoumacin C [63], Sun's total synthesis of hetiamacin A [64], and our unified total synthesis of hetiamacins A–D [65], have also been reported. Furthermore, some of amicoumacins have been revealed to exhibit additional biological activities such as antiplasmodial activity of bacilosarcin A (**2**) and quorum-sensing inhibitory activity of AI-77-B (**1**) [41, 66]. We hope that the synthetic studies of the amicoumacin family of natural products will contribute to the elucidation of their SARs and ultimately to the development of novel medicines and agrochemicals.

References

1. Kaspar F, Neubauer P, Gimple M (2019) J Nat Prod 82:2038
2. Saeed A (2016) Eur J Med Chem 116:290
3. Shimojima Y, Hayashi H, Ooka T, Shibukawa M (1982) Tetrahedron Lett 23:5435
4. Itoh J, Omoto S, Nishizawa N, Kodama Y, Inouye S (1982) Agric Biol Chem 46:2659
5. Shimojima Y, Hayashi H, Ooka T, Shibukawa M (1982) Agric Biol Chem 46:1823
6. Itoh J, Shomura T, Omoto S, Miyado S, Yuda Y, Shibata U, Inouye S (1982) Agric Biol Chem 46:1255
7. Shimojima Y, Hayashi H, Ooka T, Shibukawa M (1984) Tetrahedron 40:2519
8. Shimojima Y, Hayashi H (1983) J Med Chem 26:1370
9. Shimojima Y, Shirai T, Baba T, Hayashi H (1985) J Med Chem 28:3
10. Hamada Y, Kawai A, Kohno Y, Hara O, Shioiri T (1989) J Am Chem Soc 111:1524
11. Hamada Y, Hara O, Kawai A, Kohno Y, Shioiri T (1991) Tetrahedron 47:8635
12. Thottathil JK, Moniot JL, Mueller RH, Wong MKY, Kissick TP (1986) J Org Chem 51:3140
13. Mizuno A, Hamada Y, Shioiri T (1980) Synthesis:1007
14. Broady SD, Rexhausen JE, Thomas EJ (1991) J Chem Soc Chem Commun:708
15. Broady SD, Rexhausen JE, Thomas EJ (1999) J Chem Soc Perkin Trans 1:1083
16. Salzmann TN, Ratcliffe RW, Christensen BG, Bouffard FA (1980) J Am Chem Soc 102:6161
17. Labia R, Morin C (1984) Chem Lett:1007
18. Ward RA, Procter G (1992) Tetrahedron Lett 33:3359
19. Ward RA, Procter G (1995) Tetrahedron 51:12301
20. Durgnat J-M, Vogel P (1993) Helv Chim Acta 76:222
21. Warm A, Vogel P (1987) Helv Chim Acta 70:690
22. Warm A, Vogel P (1985) Tetrahedron Lett 26:5127
23. Warm A, Vogel P (1986) J Org Chem 51:5348
24. Kotsuki H, Araki T, Miyazaki A, Iwasaki M, Datta PK (1999) Org Lett 1:499
25. Kotsuki H, Miyazaki A, Ochi M (1991) Tetrahedron Lett 32:4503
26. Ali SM, Ramesh K, Borchardt RT (1990) Tetrahedron Lett 31:1509
27. Ghosh AK, Bischoff A, Cappiello J (2001) Org Lett 3:2677
28. Ghosh AK, Bischoff A, Cappiello J (2003) Eur J Org Chem:821
29. Ghosh AK, Fidanze C, Senanayake CH (1998) Synthesis:937
30. Ghosh AK, Onishi M (1996) J Am Chem Soc 118:2527
31. Ghosh AK, Fidanze S, Onishi M, Hussain KA (1997) Tetrahedron Lett 38:7171
32. Ghosh AK, Cappiello J (1998) Tetrahedron Lett 39:8803
33. Dondoni A, Perrone D, Semola MT (1995) J Org Chem 60:7927
34. Rao MV, Rao BV, Ramesh B (2014) Tetrahedron Lett 55:5921
35. Fürstner A, Thiel OR, Blanda G (2000) Org Lett 2:3731
36. Fürstner A, Konetzki I (1996) Tetrahedron 52:15071
37. Kumar DN, Rao BV, Ramanjaneyulu GS (2005) Tetrahedron Asymm 16:1611
38. Rao JP, Rao BV, Swarnalatha JL (2010) Tetrahedron Lett 51:3083
39. Azumi M, Ogawa K, Fujita T, Takeshita M, Yoshida R, Furumai T, Igarashi Y (2008) Tetrahedron 64:6420
40. Yongxin L, Xu Y, Liu L, Han Z, Lai PY, Guo X, Zhang X, Lin W, Yuan Qian P-Y (2021) Mar Drugs 10:319
41. Boya CA, Herrera L, Guzman HM, Gutierrez M (2021) J Pharm Bioallied Sci 4:66
42. Liu S, Xia M, Gao Q, Jiang Z, Hu X, He Q, You X, Zhang Y, Sun C (2016) Nat Prod Res Dev 28:5
43. Enomoto M, Kuwahara S (2009) Angew Chem Int Ed 48:1144
44. Casas R, Cavá C, d'Angelo J (1995) Tetrahedron Lett 36:1039
45. Kalinin AV, Bower JF, Riebel P, Snieckus V (1999) J Org Chem 64:2986
46. Eis MJ, Wrobel JE, Ganem B (1984) J Am Chem Soc 106:3693
47. Sundén H, Ibrahem I, Córdova A (2006) Tetrahedron Lett 49:99

48. Ando K (1999) J Org Chem 64:8406
49. Kurasawa K, Kuwahara S, Enomoto M (2016) Tetrahedron Lett 57:4997
50. Cañedo LM, Fernández Puentes JL, Baz JP (1997) J Antibiot 50:175
51. Sato T, Nagai K, Suzuki K, Morioka M, Saito T, Nohara C, Susaki K, Takebayashi Y (1992) J Antibiot 45:1949
52. Kalinovskaya NI, Romanenko LA, Kalinovsky AI, Dmitrenko PS, Dyshlovoy SA (2013) Nat Prod Commun 8:381
53. Patel SK, Murat K, Py S, Vallée Y (2003) Org Lett 5:4081
54. Enomoto M, Kuwahara S (2009) J Org Chem 74:7566
55. Takahashi H, Hitomi Y, Iwai Y, Ikegami S (2000) J Am Chem Soc 122:2995
56. Jouin P, Castro B (1987) J Chem Soc Perkin Trans 1:1177
57. Ma D, Ma J, Ding W, Dai L (1996) Tetrahedron Asymmetry 7:2365
58. Itoh J, Omoto S, Shomura T, Nishizawa N, Miyado S, Yuda Y, Shibata U, Inouye S (1981) J Antibiot 34:611
59. Liu S, Jin J, Chen C, Liu JM, Li JY, Wang FF, Jiang Z, Hu JH, Gao ZX, Yao F, You XF, Si SY, Sun C (2013) J Antibiot 66:281
60. Liu S, Han X, Jiang Z, Wu G, Hu X, You X, Jiang J, Zhang Y, Sun C (2016) J Antibiot 69:769
61. Wang T, Lu Q, Sun C, Lukianov D, Osterman IA, Sergiev PV, Dontsova OA, Hu X, You X, Liu S, Wu G (2020) Molecules 25:4446
62. Tang HL, Sun CH, Hu XX, You XF, Wang M, Liu SW (2016) Molecules 21:1601
63. Suzuki T, Nagasawa T, Enomoto M, Kuwahara S (2015) Tetrahedron 71:1992
64. Wu G, Liu S, Wang T, Jiang Z, Lv K, Wang Y, Sun C (2018) Org Lett 20:3566
65. Tsukaguchi S, Enomoto M, Towada R, Ogura Y, Kuwahara S (2019) Eur J Org Chem:6110
66. Shi WP, Zeng H, Wan CX (2020) Nat Prod Res. https://doi.org/10.1080/14786419.2020.1788554

Part IV
Isolation and Synthesis of Marine Macrolides, Cyclic Peptides, Depsipeptides Etc

Top Heterocycl Chem (2021) 58: 277–296
https://doi.org/10.1007/7081_2020_45

Published online: 20 November 2020

Bioactive Substances from Marine Cyanobacteria

Kiyotake Suenaga and Arihiro Iwasaki

Contents

K. Suenaga (✉) and A. Iwasaki
Department of Chemistry, Faculty of Science and Technology, Keio University, Yokohama, Japan
e-mail: suenga@chem.keio.ac.jp

Abstract Marine cyanobacteria have been known as prolific producers of secondary metabolites that have intriguing structures and remarkable biological activities. In fact, some of them have been regarded as promising candidates for drugs and have attracted attention from pharmaceutical companies. For the last 14 years, our group has investigated the secondary metabolites of marine cyanobacteria collected in Japanese coastal areas and has discovered more than twenty types of new natural products from them. In addition, we have evaluated the biological activities of these compounds and have clarified the mode of action of some of them. Also, we have achieved the total syntheses of several of these compounds. In this review, we summarize our efforts regarding the novel natural products that we have discovered so far.

Keywords Biological activity · Marine cyanobacteria · Marine natural products · Structure elucidation · Total synthesis

1 Introduction

The chemical studies of bioactive natural products are important not only in the field of chemistry but also in a wide range of related fields such as medicinal and pharmaceutical sciences. Marine organisms such as microorganisms, algae, sponges, tunicates, and mollusks have attracted considerable attention as a rich source of new bioactive substances, and a large number of bioactive substances have been found from them [1]. Among marine organisms, marine cyanobacteria have been considered as a treasure trove of bioactive substances. Cyanobacteria (sometimes called blue-green algae) are a group of bacteria that perform oxygen-producing photosynthesis. They are very old organisms that have existed on Earth for about 3 billion years. They have been greatly involved in the formation of the Earth's atmosphere, the evolution of life, and the supply of organic substances on a global scale. The secondary metabolites produced by cyanobacteria are often peptides or peptide-polyketide hybrids, and many of the peptides are highly *N*-methylated. In this review, we describe that our studies on isolation, structure determination, total syntheses, and biological activities of natural products isolated from marine cyanobacteria collected mainly at Okinawa and Amami islands, semitropical area of southern end of Japan. Several excellent reviews of bioactive compounds from marine cyanobacteria from other areas have been published [2–4].

2 Polyketides and Macrolides

2.1 Biselyngbyasides

2.1.1 Isolation and Structure

Biselyngbyaside (BLS, **1**) (Fig. 1) is an 18-membered macrolide glycoside with skipped and conjugated dienes isolated from a marine cyanobacterium *Lyngbya* sp. collected at Bise in main island of Okinawa [5]. The gross structure of BLS (**1**)

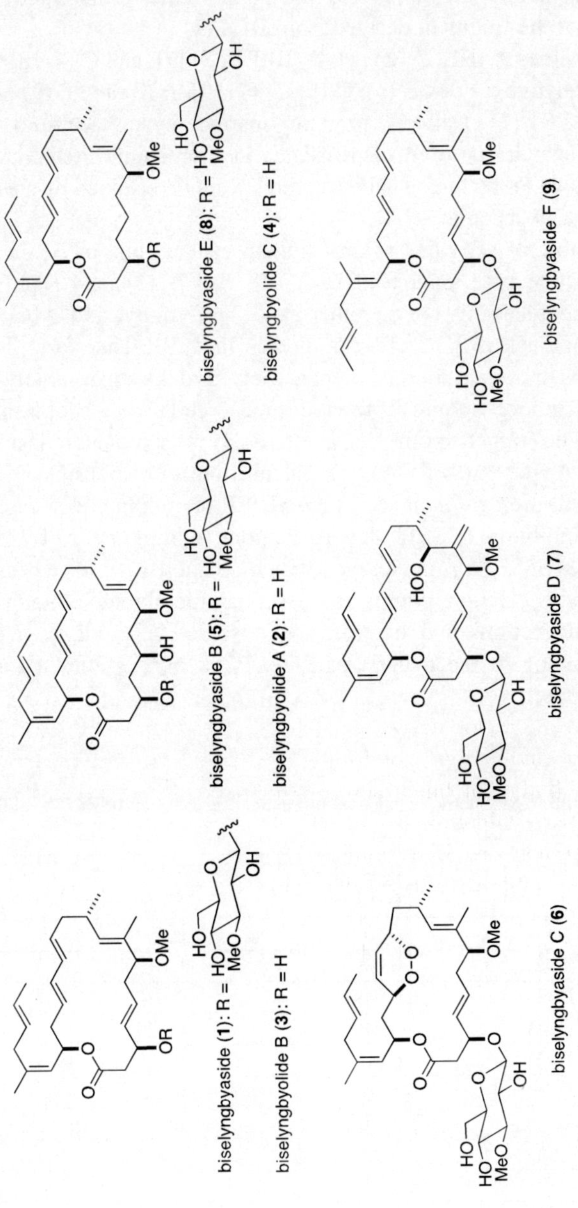

Fig. 1 The structures of biselyngbyasides

was elucidated on the basis of detailed spectral (2D NMR and HRMS) analyses. The absolute stereostructure of BLS (**1**) was determined by NMR analyses (NOESY experiments and coupling constants), application of a modified Mosher method to the derived secondary alcohols, and the asymmetric synthesis of the possible diastereomers of the fragment derived from BLS (**1**).

Biselyngbyolides A (BLLA, **2**) [6], B (BLLB, **3**) [7], and C (**4**) [8] without sugar moiety and biselyngbyasides B (**5**), C (**6**), D (**7**) [9], E (**8**) and F (**9**) [8] (Fig. 1) are analogs of BLS (**1**) isolated from a marine cyanobacterium collected at Tokunoshima Island, one of Amami islands in Kagoshima Prefecture, or Ishigaki Island in Okinawa Prefecture. Their structures were determined by spectral analyses including their CD spectra.

Growth inhibitory activities against human cancer cells of BLSs are shown in Table 1. Aglycone type analogs, BLLA (**2**), BLLB (**3**), and BLLC (**4**), exhibit particularly potent activity. On the other hand, biselyngbyaside C (**6**) whose conjugated diene has been oxidatively modified is inactive. Therefore, the conjugated diene of BLSs is important for the potent activity, and the sugar moiety decreases the activity. BLSs induce apoptosis toward cancer cells via endoplasmic reticulum (ER) stress. The target biomolecule of BLS is sarco/endoplasmic reticulum (SR/ER) Ca^{2+}-ATPases (SERCA), the calcium pump responsible for establishing the Ca^{2+} concentration gradient across the SR/ER membrane. BLSs are high affinity (*K*i ca 10 nM) inhibitors of SERCA with a unique binding mode (Table 1) [10]. The crystal structures of SERCA in complex with BLS and BLLB have been determined. The SERCA-type calcium pumps are essential membrane proteins that control various cellular functions, and therefore are conserved in a wide range of organisms. A SERCA ortholog of the malaria parasite, PfATP6, has attracted attention as a malarial drug target.

Table 1 Growth inhibitory activities against human cancer cells and SERCA inhibitory activities of BLSs

	IC_{50} (μM)		Ki (nM)
Compounds	HeLa cells	HL60 cells	SERCA1a
Biselyngbyaside (**1**)	0.30[a]	0.057[a]	19[b]
Biselyngbyolide A (**2**)	0.039[c]	0.012[c]	9[b]
Biselyngbyolide B (**3**)	0.049[a]	0.030[a]	17[b]
Biselyngbyolide C (**4**)	0.046[a]	0.024[a]	18,000[b]
Biselyngbyaside B (**5**)	6.5[c]	0.040[c]	nd[d]
Biselyngbyaside C (**6**)	>10[c]	nd[d]	nd[d]
Biselyngbyaside D (**7**)	>1.0[c]	nd[d]	nd[d]
Biselyngbyaside E (**8**)	0.19[a]	0.071[a]	nd[d]
Biselyngbyaside F (**9**)	3.1[a]	0.66[a]	nd[d]

[a]Ref. [8]
[b]Ref. [10]
[c]Ref. [9]
[d]Not determined

Biselyngbyaside inhibited both early and late stage osteoclast differentiation in RAW264 cells and decreased the cell viability of mature osteoclasts, but had no effect on the viability of undifferentiated RAW264 cells [11]. Therefore, BLS may be useful for the prevention of bone lytic diseases such as osteoporosis.

The author's group achieved total syntheses of BLLA (**2**) [12], BLLB (**3**) [13], and BLS (**1**) [14]. Their 18-membered rings were constructed by intramolecular Stille coupling reactions. Total synthesis of BLLB (**3**) was also achieved by Goswami [15] and Maier [16] groups. Goswami group recently achieved total syntheses of BLLA and BLLC [17]. Based on the co-crystal structures of SERCA with BLS and BLLB, several artificial analogs of BLLB were designed and synthesized, and their biological activities against malarial parasites were evaluated [18].

2.2 *Koshikalide*

Koshikalide (**10**) (Fig. 2) is a 14-membered macrolide isolated from a marine cyanobacterium *Lyngbya* sp. collected at Koshika, Shima City in Mie Prefecture [19]. Koshikalide (**10**) exhibited weak cytotoxicity against HeLa S_3 cells. The structural features of **10** are two trisubstituted olefins sequenced as a skipped-form. The gross structure of koshikalide **10** was elucidated by spectral (2D NMR and HRMS) analyses. The relative stereostructure of **10** was determined by NOESY experiments and by analyses of vicinal ^1H–^1H coupling constants. Total synthesis of koshikalide (**10**) was achieved by the author's group, and the absolute stereochemistry of **10** was determined [20]. Synthetic study of **10** has been carried out by Babu and Rao [21].

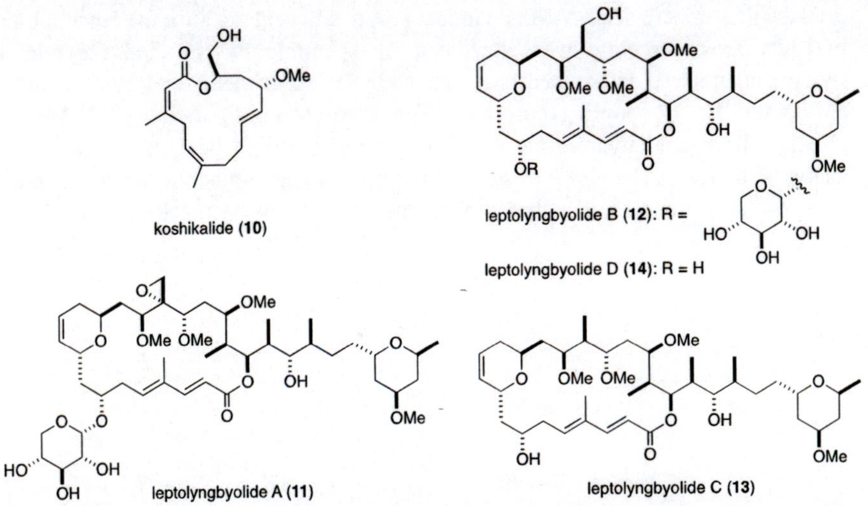

Fig. 2 The structures of koshikalide and leptolyngbyolides (highly probable stereochemistry is shown for **a**, **b**, and **d**)

2.3 Leptolyngbyolides

Leptolyngbyolides A-D (**11–14**) (Fig. 2) are 22-membered macrolides isolated from a marine cyanobacterium *Leptolyngbya* sp. collected on the coast of Itoman City, Okinawa Prefecture [22]. Leptolyngbyolides exhibit potent growth inhibition activities against HeLa S_3 cells and induce apoptosis to the cancer cells. The target molecule of leptolyngbyolides is acitin, a cytoskeleton protein: leptolyngbyolides induce depolymerization of filamentous actin (F-actin). The gross structures of leptolyngbyolides were determined based on spectral (2D NMR and HRMS) analyses. The relative configurations of leptolyngbyolides were partially determined by NOESY experiments. The authors assumed that leptolyngbyolides had a relative stereochemistry identical to that of lobophorolide [23], a structurally closely related compound isolated from the sea alga *Lobophora variegata*: their NOESY data did not contradict this assumption. To verify the predicted stereochemistry, asymmetric total synthesis of leptolyngbyolide C was achieved by Shibasaki and co-workers. A key reaction to construct the 22-membered lactone was Yamaguchi lactonization reaction. All the spectral data of the synthetic leptolyngbyolide C were identical to those of the natural product to determine the absolute configuration of leptolyngbyolide C as shown in formula **13**.

2.4 Kanamienamide

Kanamienamide (**15**) (Fig. 3) is a hybrid of polyketide and amino acid with an 11-membered ring isolated from the marine cyanobacterium *Moorea bouillonii* collected at Kanami on Tokunoshima Island and is the first discovered natural product that contains an *N*-Me-enamide group adjacent to an enol ether moiety [24]. Its gross structure was elucidated by spectral analyses and the relative stereostructure was determined based on NOESY experiments and by an analysis of vicinal ^1H–^1H coupling constants. The absolute configuration of *N*-Me-Leu obtained from acid hydrolysate of **15** was determined to be L by chiral HPLC analyses, therefore the absolute stereostructure of kanamienamide was determined as shown in formula **15**. The first total synthesis of **15** has been achieved by He and

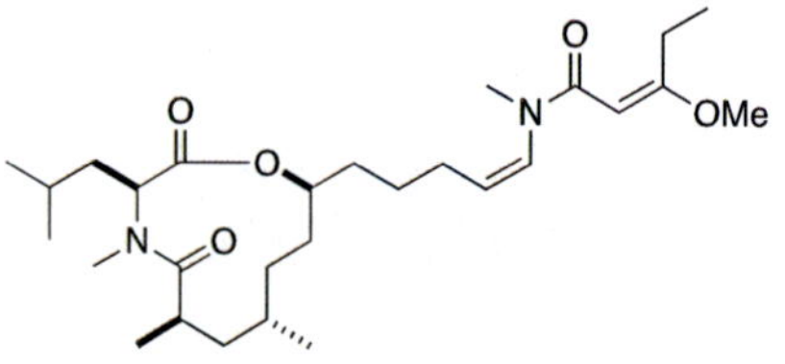

kanamienamide (**15**)

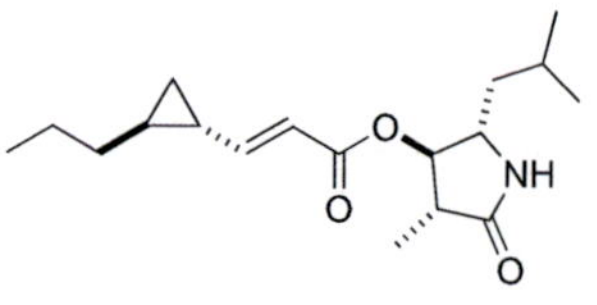

hoshinolactam (**16**)

Fig. 3 The structures of kanamienamide and hoshinolactam co-workers [25]. The author's group has also achieved total synthesis of **15** and evaluated biological activities [26], this clarified that kanamienamide was not an apoptosis-like cell death inducer, as reported in the isolation paper [24], and revealed its real biological activity as a necrosis-like cell death inducer. A formal total synthesis of **15** has been accomplished by Ye and co-workers [27].

2.5 *Hoshinolactam*

Hoshinolactam (**16**) (Fig. 3) is an antitrypanosomal lactam isolated from a marine cyanobacterium morphologically classified into the genus *Oscillatoria*, collected at the coast near Hoshino on Ishigaki Island, Okinawa [28]. Structural feature of **16** is possession of both a cyclopropane ring and a γ-lactam ring. The gross structure of **16** was elucidated by spectral analyses and the relative configuration of cyclopropane ring was determined based on analyses of the coupling constants and NOESY correlations. The absolute configurations of three stereocenters on the γ-lactam ring were determined based on a combination of derivatization reaction, a modified Mosher method, and synthetic means. To clarify the absolute configuration of the cyclopropane moiety, the authors synthesized two possible diastereomers of **16** and compared their ^{1}H NMR spectra and specific rotations with those of natural **16** [28]. Thus, the absolute stereochemistry was determined as shown in formula **16**. The total synthesis of **16** has also been achieved by Colobert and co-workers [29].

2.6 *Yoshinone A*

Yoshinone A (**17**) and related analogs yoshinones B1 (**18**) and B2 (**19**) (Fig. 4) were isolated from the marine cyanobacterium *Leptolyngbya* sp. collected at Ishigaki island, Okinawa [30]. Yoshinones are polyketides with a γ-pyrone ring and are structurally similar to kalkipyrone (**20**). The gross structures of yoshinones were

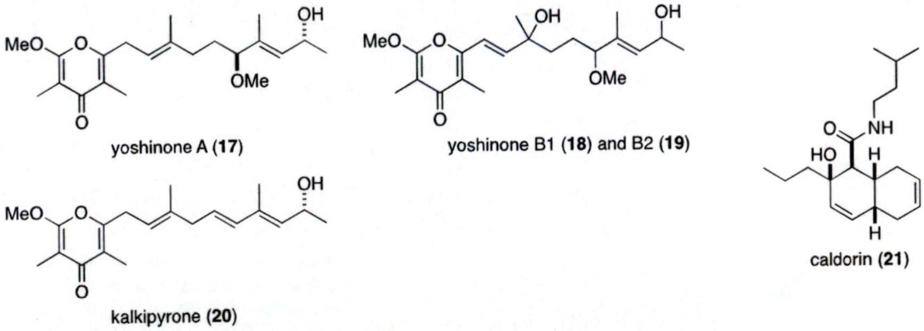

Fig. 4 The structures of yoshinones, kalkipyrone, and caldorin

determined based on spectral analyses. Yoshinone A (**17**) inhibited adipogenic differentiation of 3T3-L1 cells, but **18** and **19** did not show such activity. Kalkipyrone (**20**) exhibited anti-obesity effects for in vivo animal experiments using mice fed with a high-fat diet [31]. The author's group has achieved total synthesis of **17** and its diastereomer to determine absolute stereostructure of yoshinone A [32].

2.7 Caldorin

Caldorin (**21**) (Fig. 4) is a polyketide with a cis-fused decalin ring scaffold, isolated from the marine cyanobacterium *Caldora penicillate* [33]. The gross structure and relative configuration of **21** were elucidated by spectroscopic analyses. Caldorin (**21**) inhibited both sterol *O*-acyltransferase (SOAT) activity and osteoblast differentiation without cytotoxicity against HeLa or HL60 cells, but the activities are moderate or weak.

3 Linear Peptides with Heterocycles

3.1 Bisebromoamides

Bisebromoamide (BBA, **22**) (Fig. 5) was isolated from the marine cyanobacterium *Lyngbya* sp. collected at Bise in Okinawa Prefecture [34]. Structural feature of BBA (**22**) is possession of a 2-substituted 4-methylthiazoline (Tzn) fused to a 4-methylproline, *N*-methyl-3-bromotyrosine, and 2-(1-oxopropyl)pyrrolidine (Opp) residues, and a high degree of D-amino acids and *N*-methylated amino acids. The gross structure was elucidated by spectral analyses and the absolute stereostructure of **22** was determined by chemical degradation followed by chiral HPLC analyses. The absolute structure of **22** has been revised based on the first total synthesis of **22** and the other diastereomer concerning Tzn moiety by Ye and co-workers [35].

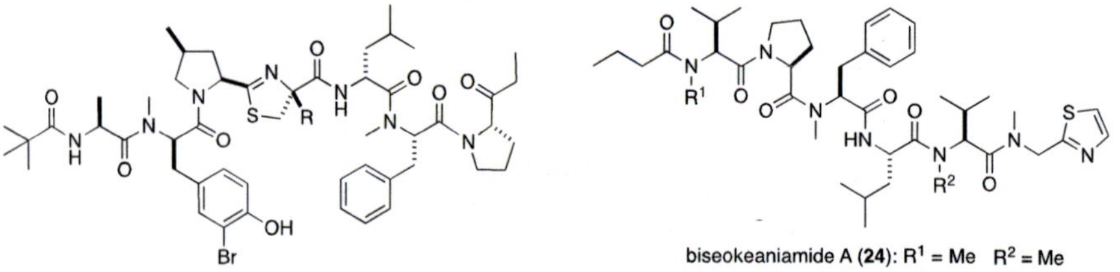

bisebromoamide (**22**): R = Me
norbisebromoamide (**23**): R = H

biseokeaniamide A (**24**): R^1 = Me R^2 = Me
biseokeaniamide B (**25**): R^1 = Me R^2 = H
biseokeaniamide C (**26**): R^1 = H R^2 = Me

Fig. 5 The structures of bisebromoamides and biseokeaniamides

Bisebromoamide (**22**) exhibited potent cytotoxicity against HeLa S_3 cells with an IC_{50} value of 0.04 μg/mL and potent protein kinase inhibition: the phosphorylation of extracellular signal regulated protein kinase (ERK) in NRK cells by platelet-derived growth factor (PDGF) stimulation was selectively inhibited by treatment with 10–0.1 μM of BBA (**22**) [34]. Bisebromoamide (**22**) suppresses renal cell carcinoma (RCC) proliferation and potentiates apoptosis by inhibiting both the Raf/MEK/ERK and the PI3K/Akt/mTOR pathways [36]. The target biomolecule of BBA (**22**) is actin: Bisebromoamide stabilized actin filaments (F-actin), and inhibited depolymerization of F-actin. This is the first example of a linear peptide actin-targeting molecule [37].

Norbisebromoamide (**23**) is an analog of BBA, isolated from the same marine cyanobacteria *Lyngbya* sp. [38]. The structure–activity relationships (SARs) of BBA (**22**) were investigated with the use of natural and synthetic analogs [38]. Ma and co-workers achieved the total synthesis of BBA and related analogs and investigated SARs: the configuration of alanine residue is important for potent cytotoxicity of BBA (**22**) [39]. Hulme and co-workers synthesized thiazole analogs of BBA (**22**) and evaluated their cytotoxicities: *N*-methyl-3-bromotyrosine and *N*-terminal pivalamide are required for potent cytotoxicity of BBA (**22**) [40].

3.2 *Biseokeaniamides*

Biseokeaniamides A (**24**), B (**25**), and C (**26**) (Fig. 5) are linear peptides possessing a terminal thiazole, isolated from an *Okeania* sp. marine cyanobacterium collected at Bise, Okinawa, Japan [41]. Their structures were established using a combination of spectroscopic analyses, degradation reactions, and HPLC analyses. Biseokeaniamide B (**25**) showed moderate cytotoxicity against HeLa cells, but the activities of biseokeaniamides A (**24**) and C (**26**) were much lower than that of biseokeaniamide B (**25**). Biseokeaniamides A (**24**), B (**25**), and C (**26**) inhibited sterol *O*-acyltransferase (SOAT), not only at an enzyme level but also at a cellular level. Ohno, Matsuno, and co-workers identified biseokeanidemide A (**24**) as a selective inhibitor of lipopolysaccharides (LPS) signal transduction, a potential anti-inflammatory agent [42].

4 Linear Peptides and Depsipeptides

4.1 *Maedamide*

Maedamide (**27**) (Fig. 6) is a linear depsipeptide with an unusual γ-amino acid, 4-amino-3-hydroxy-5-phenylpentanoic acid, isolated from a marine cyanobacterial assemblage that mostly consisted of *Lyngbya* sp. collected at Cape Maeda, Okinawa [43]. Its structure was elucidated by spectroscopic analyses and chiral HPLC

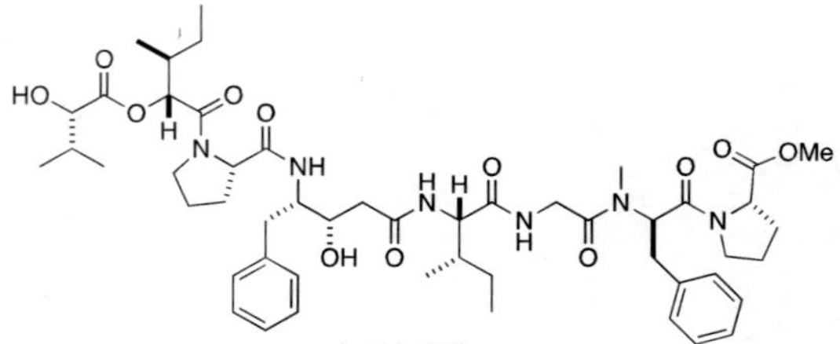

maedamide (**27**)

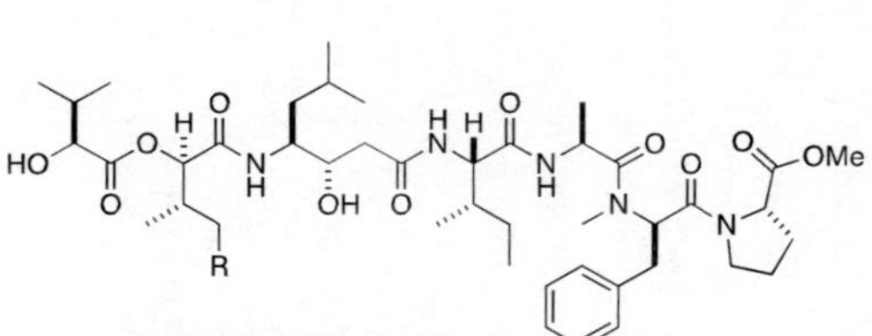

izenamide A (**28**): R = H
izenamide B (**29**): R = Me

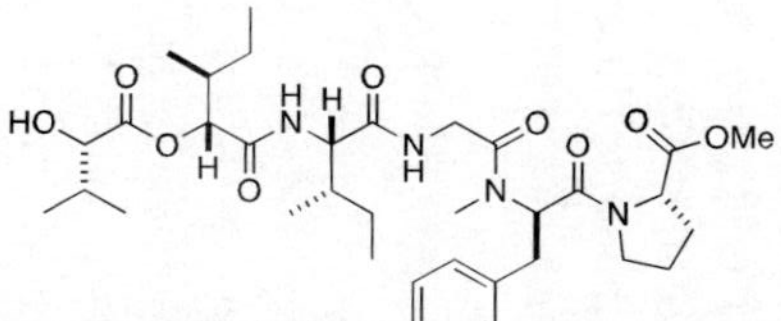

izenamide C (**30**)

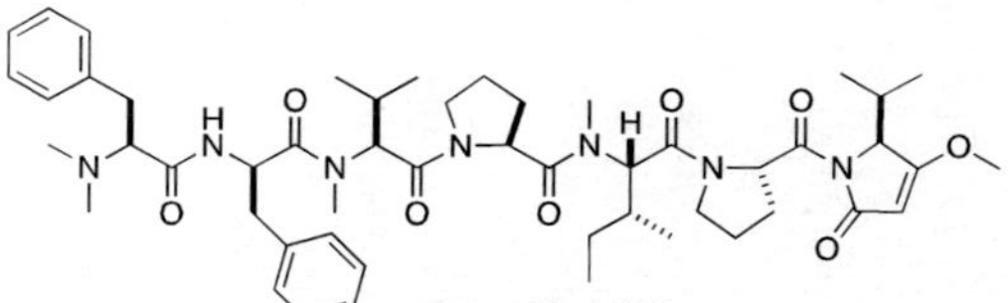

iheyamide A (**31**)

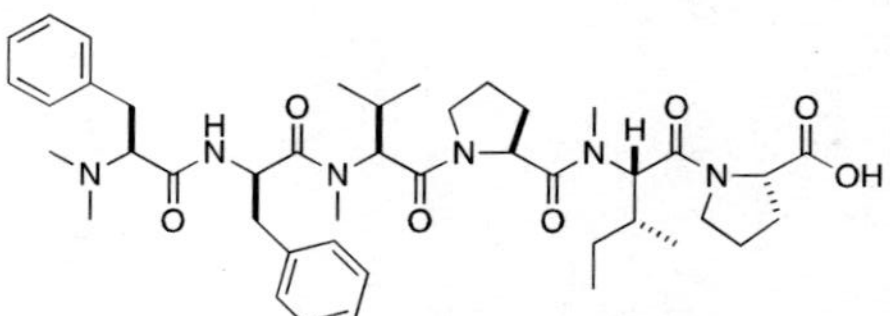

iheyamide B (**32**)

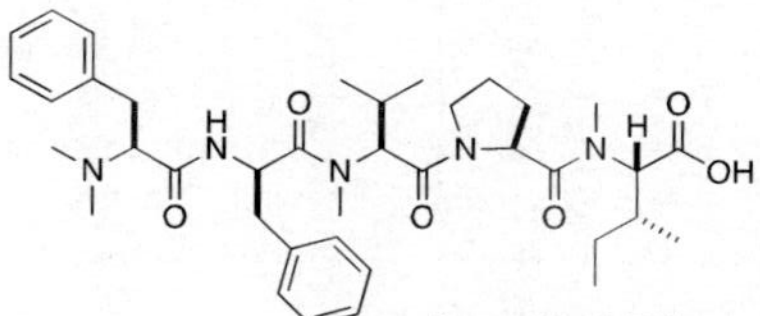

iheyamide C (**33**)

Fig. 6 The structures of medamide, izenamides, and iheyamides analyses of hydrolysis products of **27**. The first total synthesis of **27** was achieved by the author's group, and the absolute configuration of **27** has been revised [44]. Maedamide (**27**) exhibited strong protease-inhibitory activity selectively against chymotrypsin and inhibited the growth of HeLa cells and HL60 cells [43].

4.2 Izenamides

Izenamides A (**28**), B (**29**), and C (**30**) (Fig. 6) were isolated from a taxonomically independent cyanobacterium collected at Izena Island, Okinawa [45]. Izenamides A and B contain 4-amino-3-hydroxy-6-methylheptanoic acid called statin, whereas izenamide C does not. The structures of **28–30** were elucidated by a combination of spectroscopic analyses and degradation/derivatization reactions. Izenamides A and B inhibited the activity of cathepsin D, an aspartic peptidase, without exhibiting cytotoxicity against HeLa, HL60, or MCF-7 cells. Total syntheses of izenamides A, B, and C were achieved by Lim [46].

4.3 Iheyamides

Iheyamide A (**31**) (Fig. 6) is a linear peptide isolated from a marine *Dapis* sp. cyanobacterium collected at Iheya Island, Okinawa [47]. Iheyamide A (**31**) is the first natural product that has both an *N,N*-dimethylphenylalanine and an unusual isopropyl-*O*-Me-pyrrolinone (iPr-*O*-Me-pyr) moieties at *N*- and *C*- terminus, respectively. Iheyamides B (**32**) and C (**33**), related amino acid analogs of **31**, were isolated from the same cyanobacterium together with **31** [47]. Their structures were elucidated by spectroscopic analyses and degradation reactions. Iheyamide A (**31**) showed moderate antitrypanosomal activities against *Trypanosoma brucei rhodesiense* and *Trypanosoma brucei brucei* without significant cytotoxicity against normal human cells, but the other two analogs, iheyamides B (**32**) and C (**33**), which lack iPr-*O*-Me-pyr, did not. Therefore, the iPr-*O*-Me-pyr moiety of **31** was necessary for the antitrypanosomal activity.

5 Lipopeptides

5.1 Kurahynes

Kurahyne (**34**) and kurahyne B (**35**) (Fig. 7) are terminal acetylene-containing lipopeptides with a ketone moiety isolated from a cyanobacterial assemblage that mostly consisted of *Lyngbya* sp. collected at Kuraha, Okinawa [48, 49]. Their gross structures were elucidated based on spectroscopic analyses, and the absolute

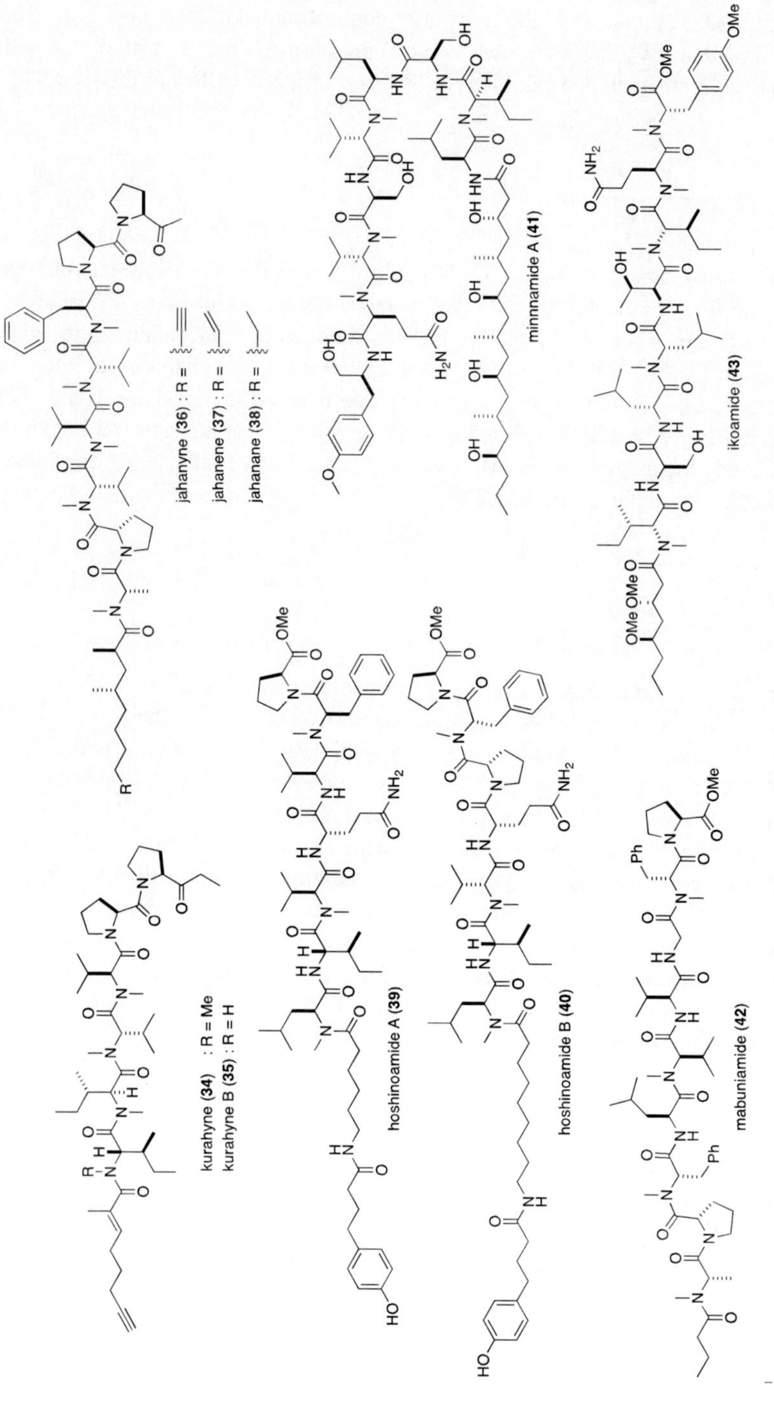

Fig. 7 The structures of kurahynes, jahanynes, hoshinoamides, minnamide A, mabuniamide, and ikoamide

configuration of **34** was determined by chiral HPLC analyses of acid hydrolysate of **34**. The author's group achieved total syntheses of kurahyne (**34**) and kurahyne B (**35**) and confirmed their absolute stereostructures [49]. Kurahynes exhibited growth inhibitory activities against HeLa and HL60 cells and induced apoptosis and ER-stress in HeLa cells. The intracellular target molecule of **34** was identified as SERCA [50].

5.2 *Jahanyne Family*

Jahanyne (**36**) (Fig. 7) was isolated from the marine cyanobacterium *Lyngbya* sp. collected at the coast near Jahana, Okinawa [51]. Structural features are a terminal acetylene, a 2-(1-oxoethyl)pyrrolidine, and five *N*-methyl amides. The absolute stereostructure of **36** was determined based on a combination of chiral HPLC analyses, spectroscopic analyses, and derivatization reactions. Jahanyne (**36**) inhibited the growth of both HeLa cells and HL60 cells and induced apoptosis in HeLa cells. Total synthesis of **36** was achieved by Brimble and co-workers for the first time [52], and also by the author's group [53] and Wang, Chen, and co-workers [54]. Chandrasekhar and co-workers synthesized desmethyl jahanyne and related analogs and deduced from biological experiments using the synthetic analogs that the target molecule of **36** had been Bcl-2, antiapoptotic protein [55]. Wang, Chen, and co-workers designed and synthesized artificial analogs of **36** and showed that jahanyne and its analogs could induce G0/G1 cell cycle arrest in H820 cells [56]. Two natural jahanyne analogs, jahanene (**37**) and jahanane (**38**), were isolated from a marine cyanobacterium *Okeania* sp. [53]. Their structures were determined by spectral analyses and confirmed by total syntheses of **37** and **38**. It has been clarified that the degree of unsaturation at the terminus of the fatty acid moiety affects the growth inhibitory activity against human cancer cells.

5.3 *Hoshinoamides*

Hoshinoamides A (**39**) and B (**40**) (Fig. 7) are acyclic lipopeptides isolated from the marine cyanobacterium *Caldora penicillate*, and possess an unusual long-chain amino acid moiety and a hydroxyphenylbutanoic acid moiety [57]. Their structures were elucidated by spectroscopic analyses and degradation reactions. Hoshinoamides A (**39**) and B (**40**) did not inhibit the growth of HeLa cells, but inhibited the in vitro growth of the malarial parasite *Plasmodium falciparum*.

5.4 *Minnamide A*

Minnamide A (**41**) (Fig. 7) is a lipopeptide that contains a fatty acid with a repeating structure consisting of hydroxy and proposed β-branched methyl groups isolated from the marine cyanobacterium *Okeania hirsuta* collected at Minna Island, Okinawa [58]. The gross structure of **41** was elucidated by spectral analyses, and the absolute stereochemistry was determined based on a combination of chiral HPLC analyses, a modified Mosher method, analyses of vicinal coupling constants and NOE experiments for the derived cyclic ethers, chemical degradation reactions, and synthetic methods. Minnamide A (**41**) inhibited the growth of HeLa cells and induced necrosis-like regulated cell death, which may involve the copper-mediated accumulation of reactive oxygen species (ROS) [58].

5.5 *Mabuniamide*

Mabuniamide (**42**) (Fig. 7) was isolated from an *Okeania* sp. marine cyanobacterium collected from the coast of Odo, in Okinawa [59]. The absolute configuration of **42** was determined by spectral analyses and Marfey's analyses of the acid hydrolysate of **42** and confirmed by total synthesis. Mabuniamide (**42**) stimulated glucose uptake in cultured rat L6 myotubes in a dose-dependent and insulin-independent manner and exhibited moderate antimalarial activity.

5.6 *Ikoamide*

Ikoamide (**43**) (Fig. 7) is a highly *N*-methylated lipopeptide possessing a 3,5-dimethoxyoctanoic acid moiety isolated from an *Okeania* sp. [60] marine cyanobacterium collected at Iko-pier, Kuroshima Island, Okinawa. Its absolute configuration was clarified based on a combination of spectroscopic analyses, chiral- phase HPLC analyses, derivatization reactions, and synthetic means. Ikoamide (**43**) showed potent and selective antimalarial activity against the malarial parasite *Plasmodium falciparum* 3D7 clone without cytotoxicity.

6 Cyclic Peptides and Depsipeptides

6.1 *Kurahamide*

Kurahamide (**44**) (Fig. 8), an analog of dolastatin 13, was isolated from a marine cyanobacterial assemblage, consisting mostly of *Lyngbya* sp. collected at Kuraha,

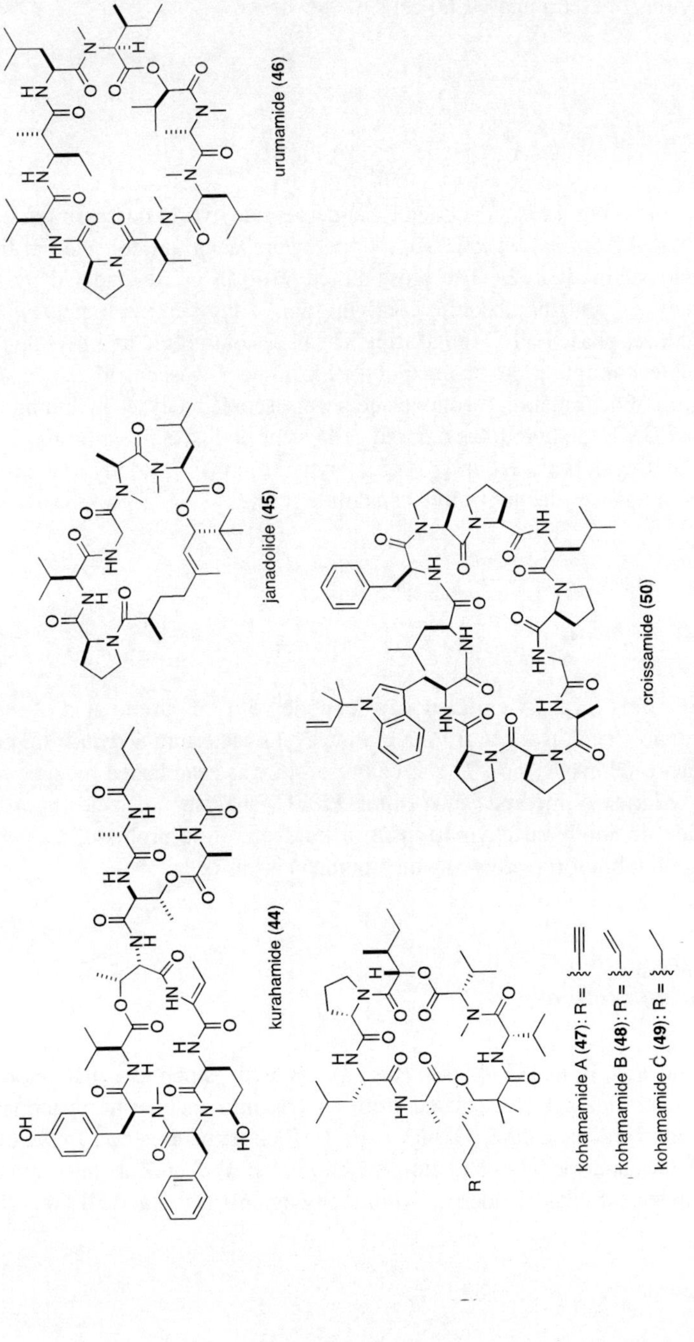

Fig. 8 The structures of kurahamide, janadolide, urumamide, kohamamides, and croissamide

Okinawa [61]. The absolute configuration of **44** was established by spectroscopic and chiral HPLC analyses of acid hydrolysates. Kurahamide (**44**) exhibited strong protease-inhibitory activity against elastase and chymotrypsin and the growth inhibitory activities of HeLa and HL60 cells.

6.2 *Janadolide*

Janadolide (**45**) (Fig. 8) is a cyclic polyketide-peptide hybrid possessing a tert-butyl group, isolated from an *Okeania* sp. marine cyanobacterium collected at the coast near Janado, Okinawa [62]. The gross structure of **45** was elucidated by spectroscopic analyses, and the absolute configuration of the peptide moiety was determined by chiral-phase HPLC and Marfey's analyses of the acid hydrolysate from **45**. The absolute configuration of the polyketide moiety was elucidated based on a combination of degradation reactions and spectroscopic analyses including application of the PGME method to the derived carboxylic acid. Total syntheses of **45** were achieved by the author's group [63] and Payne group [64]. Reddy and co-workers reported synthesis of des-tert-butyl Janadolide [65].

6.3 *Urumamide*

Urumamide (**46**) (Fig. 8) is a cyclodepsipeptide with a β-amino acid, 2-methyl-3-aminopentanoic acid, isolated from a marine cyanobacterium *Okeania* sp. collected at Ikei Island, Okinawa [66]. The structure of **46** was established by spectroscopic analyses, Marfey's method and chiral HPLC analyses of acid hydrolysates. Urumamide **46** inhibited chymotrypsin, a kind of serine protease, and exhibited weak growth inhibitory activity against human cancer cells.

6.4 *Kohamamides*

Kohamamides A, B, and C (**47–49**) (Fig. 8), cyclic depsipeptides that belong to the kulolide superfamily, were isolated from an *Okeania* sp. marine cyanobacterium collected at Kohama Island, Okinawa [67]. Their structures were elucidated by spectroscopic analyses, degradation reactions, and chemical interconversions. Kohamamides exhibited moderate cytotoxicity against HeLa and HL60 cells.

6.5 *Croissamide*

Croissamide (**50**) (Fig. 8) is a proline-rich cyclic peptide that contains an *N*-prenylated tryptophan, isolated from a marine cyanobacterium *Symploca* sp. collected at Minna Island (called "croissant island" due to its crescent shape), Okinawa [68]. The structure of **50** was determined by spectroscopic analyses and chiral HPLC analyses of acid hydrolysates.

7 Conclusion

This review has summarized new bioactive substances found by the author's group. We isolated a variety of new bioactive substances from the marine cyanobacteria collected at Okinawa and Amami islands, and determined their chemical structures. Some compounds such as biselyngbyaside (**1**) and its analogs, bisebromoamide (**22**), jahanyne (**36**), and minnamide A (**41**) exhibit potent biological activities and possess unusual chemical structures. In most cases, these natural products were obtained in trace amounts from the marine cyanobacteria. We achieved total syntheses of 16 natural products and provided ample amounts for elucidation of biological activities. Target biomolecules and/or mechanism of biological activities have been elucidated for some compounds.

References

1. Carroll AR, Copp BR, Davis RA, Keyzers RA, Prinsep MR (2020) Nat Prod Rep 37:175
2. Tan LT, Phyo MY (2020) Molecules 25:2197
3. Demay J, Bernard C, Reinhardt A, Marie B (2019) Mar Drugs 17:320
4. Ali Shah SA, Akhter N, Auckloo BN, Khan I, Lu Y, Wang K, Wu B, Guo Y-W (2017) Mar Drugs 15:354
5. Teruya T, Sasaki H, Kitamura K, Nakayama T, Suenaga K (2009) Org Lett 11:2421
6. Morita M, Ohno O, Suenaga K (2012) Chem Lett 41:165
7. Ohno O, Watanabe A, Morita M, Suenaga K (2014) Chem Lett 43:287
8. Watanabe A, Ohno O, Morita M, Inuzuka T, Suenaga K (2015) Bull Chem Soc Jpn 88:1256
9. Morita M, Ohno O, Teruya T, Yamori T, Inuzuka T, Suenaga K (2012) Tetrahedron 68:5984
10. Morita M, Ogawa H, Ohno O, Yamori T, Suenaga K, Toyoshima C (2015) FEBS Lett 589:1406
11. Yonezawa T, Mase N, Sasaki H, Teruya T, Hasegawa S, Cha B-Y, Yagasaki K, Suenaga K, Nagai K, Woo J-T (2012) J Cell Biochem 113:440
12. Tanabe Y, Sato E, Nakajima N, Ohkubo A, Ohno O, Suenaga K (2014) Org Lett 16:2858
13. Sato E, Tanabe Y, Nakajima N, Ohkubo A, Suenaga K (2016) Org Lett 18:2047
14. Sato E, Sato M, Tanabe Y, Nakajima N, Ohkubo A, Suenaga K (2017) J Org Chem 82:6770
15. Das S, Paul D, Goswami RK (2016) Org Lett 18:1908
16. Kämmler L, Maier ME (2018) J Org Chem 83:4554
17. Paul D, Sahana MH, Mandal P, Chakrabarti P, Goswami RK (2020) Org Bio Chem 18:7151. https://doi.org/10.1039/D0OB00576B

18. Sato E, Morita M, Ogawa H, Iwatsuki M, Hokari R, Ishiyama A, Ōmura S, Iwasaki A, Suenaga K (2018) Bioorg Med Chem Lett 28:298
19. Iwasaki A, Teruya T, Suenaga K (2010) Tetrahedron Lett 51:959
20. Kunifuda K, Iwasaki A, Nagamoto M, Suenaga K (2016) Tetrahedron Lett 57:3121
21. Venkanna A, Sreedhar E, Siva B, Babu KS, Kothakonda Prasad KR, Rao JM (2013) Tetrahedron Asymmetry 24:1010
22. Cui J, Morita M, Ohno O, Kimura T, Teruya T, Watanabe T, Suenaga K, Shibasaki M (2017) Chem Eur J 23:8500
23. Kubanek J, Jensen PR, Keifer PA, Sullards MC, Collins DO, Fenical W (2003) Proc Natl Acad Sci U S A 100:6916
24. Sumimoto S, Iwasaki A, Ohno O, Sueyoshi K, Teruya T, Suenaga K (2016) Org Lett 18:4884
25. Reddy DS, Zhang N, Yu Z, Wang Z, He Y (2017) J Org Chem 82:11262
26. Ojima D, Iwasaki A, Suenaga K (2017) J Org Chem 82:12503
27. Li Y, Guo Y, Xu Z, Ye T (2018) Syntlett 29:964
28. Ogawa H, Iwasaki A, Sumimoto S, Iwatsuki M, Ishiyama A, Hokari R, Otoguro K, Ōmura S, Suenaga K (2017) Org Lett 19:890
29. Jerhaoui S, Poutrel P, Djukic J-P, Wencel-Delord J, Colobert F (2018) Org Chem Front 5:409
30. Inuzuka T, Yamamoto K, Iwasaki A, Ohno O, Suenaga K, Kawazoe Y, Uemura D (2014) Tetrahedron Lett 55:6711
31. Koyama T, Kawazoe Y, Iwasaki A, Ohno O, Suenaga K, Uemura D (2016) J Antibiot 69:348
32. Shinomiya S, Iwasaki A, Ohno O, Suenaga K (2016) Phytochemistry 132:109
33. Shiota I, Iwasaki A, Sumimoto S, Tomoda H, Suenaga K (2018) Tetrahedron Lett 59:1261
34. Teruya T, Sasaki H, Fukazawa H, Suenaga K (2009) Org Lett 11:5062
35. Gao X, Liu Y, Kwong S, Xu Z, Ye T (2010) Org Lett 12:3018
36. Suzuki K, Mizuno R, Suenaga K, Teruya T, Tanaka N, Kosaka T, Oya M (2013) Cancer Med 2:32
37. Sumiya E, Shimogawa H, Sasaki H, Tsutsumi M, Yoshita K, Ojika M, Suenaga K, Uesugi M (2011) ACS Chem Biol 6:425
38. Sasaki H, Teruya T, Fukazawa H, Suenaga K (2011) Tetrahedron 67:990
39. Li W, Yu S, Jin M, Xia H, Ma D (2011) Tetrahedron Lett 52:2124
40. Johnston HJ, Boys SK, Makda A, Carragher NO, Hulme AN (2016) Chembiochem 17:1621
41. Iwasaki A, Tadenuma T, Sumimoto S, Ohshiro T, Ozaki K, Kobayashi K, Teruya T, Tomoda H, Suenaga K (2017) J Nat Prod 80:1161
42. Ohno O, Terasaki T, Sano T, Hitomi Y, Miyamoto J, Matsuno M (2020) Bioorg Med Chem Lett 30:127069
43. Iwasaki A, Ohno O, Sumimoto S, Suda S, Suenaga K (2014) Tetrahedron Lett 55:4126
44. Takayanagi A, Iwasaki A, Suenaga K (2015) Tetrahedron Lett 56:4947
45. Kanamori Y, Iwasaki A, Sumimoto S, Matsubara T, Sato T, Suenaga K (2018) J Nat Prod 81:1673
46. Lim C (2019) Molecules 24:3424
47. Kurisawa N, Iwasaki A, Jeelani G, Nozaki T, Suenaga K (2020) J Nat Prod 83:1684
48. Iwasaki A, Ohno O, Sumimoto S, Suda S, Suenaga K (2014) RSC Adv 4:12840
49. Okamoto S, Iwasaki A, Ohno O, Suenaga K (2015) J Nat Prod 78:2719
50. Iwasaki A, Ohno O, Katsuyama S, Morita M, Sasazawa Y, Dan S, Simizu S, Yamori T, Suenaga K (2015) Bioorg Med Chem Lett 25:5295
51. Iwasaki A, Ohno O, Sumimoto S, Ogawa H, Nguyen KA, Suenaga K (2015) Org Lett 17:652
52. Siow A, Opiyo G, Kavianinia I, Li FF, Furkert DP, Harris PWR, Brimble MA (2018) Org Lett 20:788
53. Iwasaki A, Fujimura H, Okamoto S, Kudo T, Hoshina S, Sumimoto S, Teruya T, Suenaga K (2018) J Org Chem 83:9592
54. Ye B, Jiang P, Zhang T, Ding Y, Sun Y, Hao X, Li L, Wang L, Chen Y (2018) J Org Chem 83:6741

55. Kallepu S, Kavitha M, Yeeravalli R, Manupati K, Jadav SS, Das A, Mainkar PS, Chandrasekhar S (2018) ACS Omega 3:63
56. Ye B, Gong J, Li Q, Bao S, Zhang X, Chen J, Meng Q, Chen B, Jiang P, Wang L, Chen Y (2020) Mar Drugs 18:176
57. Iwasaki A, Tadenuma T, Sumimoto S, Shiota I, Matsubara T, Saito-Nakano Y, Nozaki T, Sato T, Suenaga K (2018) J Nat Prod 81:2545
58. Sumimoto S, Kobayashi M, Sato R, Shinomiya S, Iwasaki A, Suda S, Teruya T, Inuzuka T, Ohno O, Suenaga K (2019) Org Lett 21:1187
59. Ozaki K, Iwasaki A, Sezawa D, Fujimura H, Nozaki T, Saito-Nakano Y, Suenaga K, Teruya T (2019) J Nat Prod 82:2907
60. Iwasaki K, Iwasaki A, Sumimoto S, Matsubara T, Sato T, Nozaki T, Saito-Nakano Y, Suenaga K (2020) J Nat Prod 83:481
61. Iwasaki A, Sumimoto S, Ohno O, Suda S, Suenaga K (2014) Bull Chem Soc Jpn 87:609
62. Ogawa H, Iwasaki A, Sumimoto S, Kanamori Y, Ohno O, Iwatsuki M, Ishiyama A, Hokari R, Otoguro K, Ōmura S, Suenaga K (2016) J Nat Prod 79:1862
63. Ojima D, Mine H, Iwasaki A, Suenaga K (2018) Tetrahedron Lett 59:1360
64. Chung JH, Tang AH, Geraghty K, Corcilius L, Kaiser M, Payne RJ (2020) Org Lett 22:3089
65. Athawale PR, Jachak GR, Shukla A, Shanmugam D, Reddy DS (2018) ACS Omega 3:2383
66. Kanamori Y, Iwasaki A, Sumimoto S, Suenaga K (2016) Tetrahedron Lett 57:4213
67. Iwasaki A, Shiota I, Sumimoto S, Matsubara T, Sato T, Suenaga K (2017) J Nat Prod 80:1948
68. Iwasaki K, Iwasaki A, Sumimoto S, Sano T, Hitomi Y, Ohno O, Suenaga K (2018) Tetrahedron Lett 59:3806

Top Heterocycl Chem (2021) 58: 297–316
https://doi.org/10.1007/7081_2020_47

Published online: 14 November 2020

Cyclic Depsipeptides, Callipeltins

Reda A. Abdelhamid and Hiroyuki Konno

Contents

Abstract Marine sponges are considered as an untapped reservoir of biologically active natural compounds exhibiting numerous interesting and potentially useful pharmacological activities, including antimicrobial, antifungal, cytotoxic, anti-inflammatory, and immunostimulatory activities. One class of such compounds is the cyclic depsipeptides, which are characterized by the incorporation of at least one ester linkage into their structures. Callipeltin A and its 14 analogs, neamphamide A, papuamides A and B, pipecolidepsin A and 10 homophymines are cyclic depsipeptides isolated from marine sponges, have structurally unique features incorporating several modified amino acid residues, and are well known for their potent

R. A. Abdelhamid
Department of Pharmacognosy, Faculty of Pharmacy, Al-Azhar University, Assuit, Egypt

Graduate School of Science and Engineering, Yamagata University, Yonezawa, Japan
e-mail: reda.ahmed@azhar.edu.eg

H. Konno (✉)
Graduate School of Science and Engineering, Yamagata University, Yonezawa, Japan
e-mail: konno@yz.yamagata-u.ac.jp

HIV-inhibitory activity. A variety of synthetic studies have been undertaken, to elucidate the motifs essential for biological activity and meet the challenge posted by their synthesis. In this chapter, recent progress in the isolation and synthesis of the callipeltins will be reviewed.

Graphical Abstract

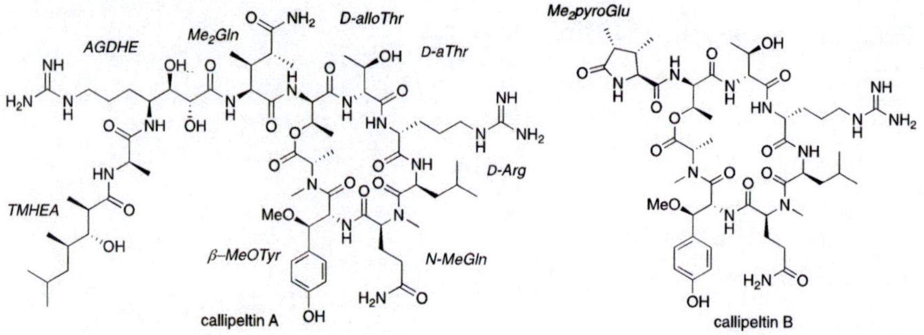

callipeltin A callipeltin B

Keywords Anti-HIV activity · Cyclic depsipeptide · Cytotoxicity · Isolation · Marine sponge · Unusual amino acid

1 Introduction

Depsipeptides are a class of peptides incorporating both amide and ester bonds, and are found in many marine organisms including bacteria, mollusks, and tunicates. There are two types of depsipeptides: linear and cyclic depsipeptides [1]. *Cyclic depsipeptides (CDPs),* also known as "cyclodepsipeptides or peptolides" are cyclooligomers comprising a core ring constructed from amino acids and at least one hydroxy acid such that its constituent residues are joined by amide and at least one ester (lactone) bonds. CDPs are thus an extraordinarily diverse class of natural products [2] differing in the structures of both their ring and side chains, and incorporating both natural and unnatural amino acid building blocks [3, 4]. Cyclic tri-, tetra-, penta-, hexa-, hepta-, octa-, nona-, deca-, and trideca CDPs are all known, so classified based on the total number of amino and hydroxy acids [4].

Marine organism-derived cyclic depsipeptides are potential sources of biologically active lead compounds, exhibiting anti-HIV, antifungal, anti-inflammatory, and cytotoxic activities [5]. Natural product researchers have focused on the isolation and identification of bioactive cyclic peptides from sponges [6], and synthetic chemists have undertaken stereostructural determination and synthetic studies.

2 Isolation of Cyclic Depsipeptides

2.1 *Distribution of Cyclic Depsipeptides in Marine Sponges*

Marine sponges are considered as abundant source of bioactive peptides, particularly cyclic peptides that are distributed in many sponge genera as *Dysidia*, *Callipelta*, *Discodermia*, *Phakellia*, *Stylotella*, *Stylissa*, and *Theonella*. Table 1 listed a selection of bioactive cyclic depsipeptide and their corresponding natural source [6–8].

2.2 *Isolation and Structural Analysis of Callipeltins*

Callipeltins are a large group of cyclodepsipeptides characterized by the presence of the unusual Me_2Gln residue, initially isolated from the marine sponge *Callipelta* sp., and subsequently found in other species such as *Latrunculia* sp. [32]. There are linear and cyclic callipeltins: examples of the latter are jaspamide, geodiamolide H, neamphamide, callipeltins, papuamides, mirabamides, stellettapentins, and stellatolides; and all have demonstrated a wide range of biological activities including antifungal and antiviral activity, cytotoxicity against several human tumor cell lines, and the ability to regulate the contraction of myocardial cells [1, 5, 21, 24]. Many of these compounds have been isolated, characterized, synthesized, and used for the synthesis of analogs, for the optimization of their activities. Callipeltins B (**3**) – M (**4**) are believed to be degradative products of callipeltin A (**1**) [1].

Callipeltin A (1) (Fig. 1) ($C_{68}H_{116}N_{18}O_{20}$) was the first callipeltin to be isolated from the sponges *Callipelta* and *Latrunculia* sp. [1, 9, 35–37]. It is a macrocyclic depsipeptide, incorporating numerous nonribosomal amino acids, a unique *N*-terminal aliphatic hydroxy acid moiety and three unusual amino acid residues: (2*R*,3*R*,4*S*)-4-amino-7-guanidino-2,3-dihydroxyheptanoic acid (AGDHE, **5**), (3*S*,4*R*)-3,4-dimethyl-L-glutamine (Me_2Gln, **6**), and β-methoxytyrosine (-β-MeOTyr, **7**) in addition to four amino acids present in their L configuration (Ala, Leu, and two Thr residues); one D-Arg; two *N*-methyl amino acids, *N*-MeAla and *N*-MeGln [1, 38]. Callipeltin A (**1**) has anti-HIV and antifungal activity, is cytotoxic to several human carcinoma cell lines, and is also a potent selective inhibitor of Na^+/Ca^{2+} exchange, exhibiting a positive inotropic-effect in the left atria of the Guinea pig [1, 32, 39].

The neamphamides B-D and callipeltins N (**8**) and O (**9**) are closely related to callipeltin A (**1**), having identically sized lactone rings, and incorporating both β-MeOTyr (**7**) and AGDHE (**5**) residues. The papuamides, mirabamides, stellatolides, and stellettapeptins are also related to callipeltin A (**1**) but lack an AGDHE (**5**) residue and incorporate the dipeptide Thr-dAbu also found in homophymines, theopapuamides, and some other related peptides [24] (Fig. 1).

Callipeltin B (3) was isolated from *Callipelta* sp. and *Latrunculia* sp. Its structure incorporates one residue each of *N*-methylalanine (MeAla), β-MeOTyr (**7**),

Table 1 Sources of sponge cyclic depsipeptides and their biological activities

Compound	Source organism	Bioactivity	Ref.
Arenstatin A	*Dysidea arenaria*	Anticancer	
Callipeltin A	*Callipelta* sp.	Antiviral	[9]
Celebesides	*Siliquariaspongia mirabilis*	Antiviral	[10]
Discodermins A-H	*Discodermia* sp.	Anticancer	[6]
Geodiamolide H	*Geodia corticostylifera*	Anticancer	[11]
Gunungamide A	*Discodermia* sp.	Anticancer	[12]
Homophymines B-E	*Homophymia* sp.	Anticancer	[13]
Homophymines A1-E1	*Homophymia* sp.	Anticancer	[13]
Halileptins	*Haliclona* sp., *Leiosella* cf. *arenifibrosa*	Anti-inflammatory	[14]
Jaspamide	*Jaspis and Hemiastrella genus*	Insecticidal, antimicrobial, antiplasmodial, anticancer, antiparasitic	[15]
Microspinosamide	*Geodia (Sidonops) microspinosa*	Antiviral	[16]
Mirabamides A-D	*Siliquariaspongia mirabilis*	Antiviral	[17]
Mirabamides E-H	*Stelletta clavosa*	Antiviral	[17]
Neamphamide A	*Neamphius huxleyi*	Antiviral	[15]
Papuamides A and B	*Theonella mirabilis* and *Theonella swinhoei*	Antiviral	[18]
Theopapuamide B	*Siliquariaspongia mirabilis*	Antiviral	[19]
Neamphamides	*Neamphius huxleyi*	Anticancer	[20, 21]
Pipecolidepsins A and B	*Homophymia lamellosa*	Anticancer	[22]
Stylissatins A-D	*Stylissa massa*	Anticancer	[7, 23]
Stellatolide H	*Discodermia sp.*	Anticancer	[24]
Stellettapeptin B,	*Discodermia sp.*	Antiviral	[24]
Swinhopeptolides A and B	*Theonella swinhoei*	Anticancer	[5]
Nagahamide A	*Theonella swinhoei*	Antibacterial	[25]
Theonellapeptolides	*Theonella sp.*	Immunosuppressive, antimicrobial	[26, 27]
Keramamides A-N	*Theonella sp.*	Anticancer	[28–31]

N-MeGln, Leu, D-Arg, and two residues of D-alloThr, in addition to one residue of 3,4-dimethylpyroglutamic acid (Me_2pyroGlu, **10**), which replaces the AGDHE (**5**), Ala and Me_2Gln (**6**) residues in callipeltin A (**1**) (Scheme 1) [33]. Callipeltin B (**3**) was cytotoxic to human bronchopulmonary non-small cell lung carcinoma NSCLC-N6 (IC_{50} = 1.3 μg/mL), murine leukemia P388 (IC_{50} < 3.3 μg/mL), and human melanoma M96 cells (IC_{50} < 3.3 μg/mL) [32, 35, 37, 40], and inhibited Na^+/Ca^{2+}

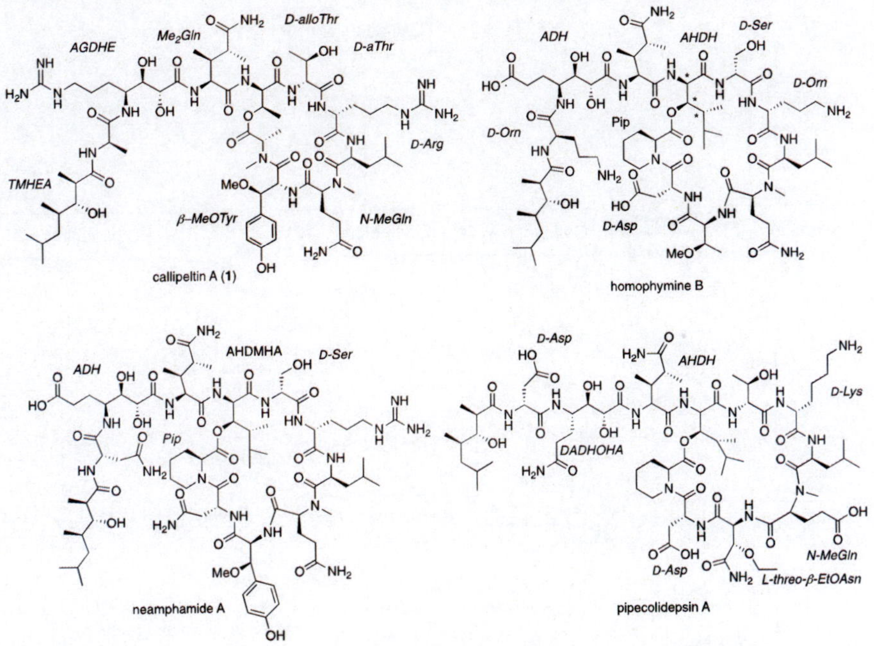

Fig. 1 Chemical structure of callipeltin A (**1**), homophymine B, neamphamide A, and pipecolidepsin A

exchange. However, callipeltin B (**3**) did not exhibit significant anti-HIV activity. Comparison of the structures of callipeltin A (**1**) (which does exhibit anti-HIV activity) and callipeltin B (**3**) led to the conclusion that the *N*-terminal side chain of callipeltin A (**1**) (which is not present in callipeltin B (**3**)) is important for this antiviral activity [41].

Callipeltin C (11) is an acyclic callipeltin (Fig. 2), obtained from *Latrunculia* sp. and having the same amino acid configurations as callipeltin A (**1**) [42]. Callipeltin C (**11**) showed antifungal activity against *Fusarium oxysporum*, *Helminthosporium sativum*, *Phytophthora hevea*, and *Candida albicans* [37, 40, 43]. Callipeltin C (**11**) was significantly less cytotoxic to human cancer cell lines than callipeltin A (**1**) and callipeltin B (**3**), suggesting the presence of the macrocyclic ring to be critical for this bioactivity [33].

Callipeltin D (12) ($C_{32}H_{60}N_8O_{11}$) is an acyclic peptide (Fig. 2) obtained from *Latrunculia* sp. [37]. By comparison of its ^{1}H NMR spectra with that of callipeltin A (**1**), it was shown to be a truncated open-chain derivative incorporating one residue each of Ala, D-alloThr; two nonproteinogenic amino acids, AGDHE (**5**) and Me_2Gln (**6**); and the fatty acid (2*R*,3*R*,4*R*)-3-hydroxy-2,4,6-trimethylheptanoic acid (TMHEA, **13**) [37, 41].

Callipeltin E (14) ($C_{36}H_{61}N_{10}O_{11}$) (Fig. 2) is a truncated, acyclic hexapeptide derivative of callipeltin A (**1**) isolated from *Latrunculia* sp., and incorporating unique nonproteinogenic amino acids such as D-alloThr, D-Arg, Leu, MeGln,

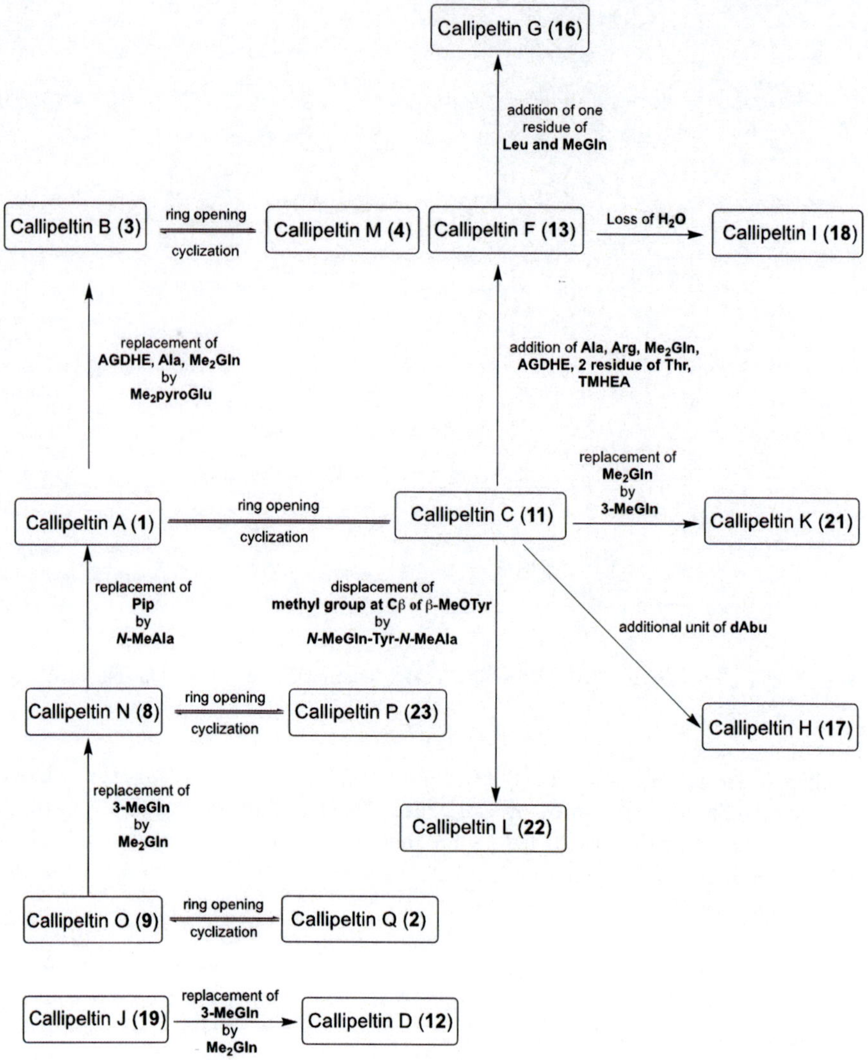

Scheme 1 Relationship between different types of Callipeltins

MeAla, and β-MeOTyr (**7**) in addition to AGDHE (**5**) and TMHEA (**13**) [1, 32, 39]. 1D and 2D NMR analyses revealed the presence of six amino acid residues which are also present in callipeltin A (**1**) [37] (Fig. 2).

Callipeltin F (15) ($C_{42}H_{79}N_{13}O_{14}$) was isolated as colorless amorphous solid. Structurally it is similar to callipeltin C (**11**), though incorporating one residue each of Ala, D-Arg, Me_2Gln (**6**), AGDHE (**5**), TMHEA (**13**), and of two residues of D-alloThr [43] (Fig. 3).

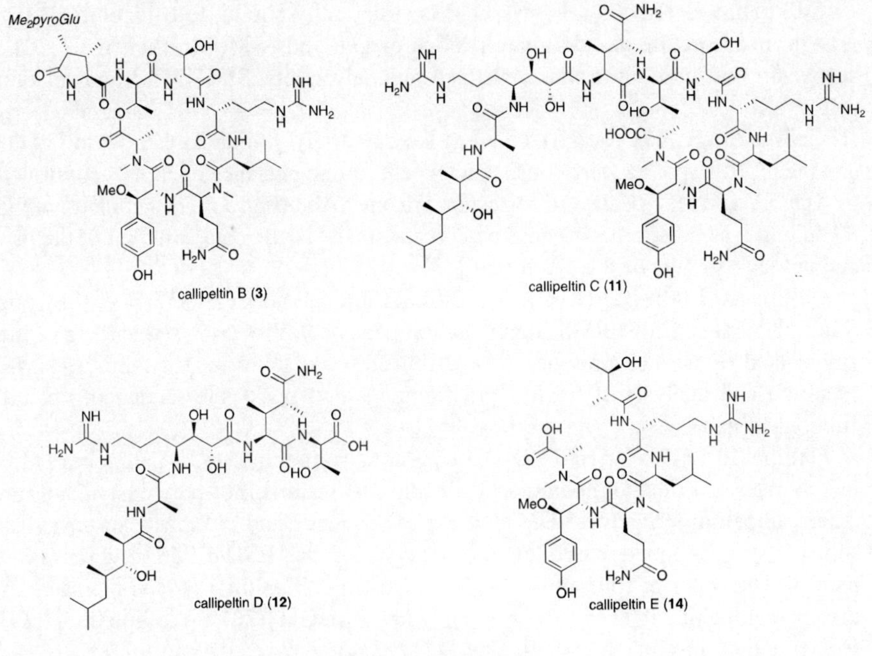

Fig. 2 Chemical structure of callipeltins B (**3**)-E (**14**)

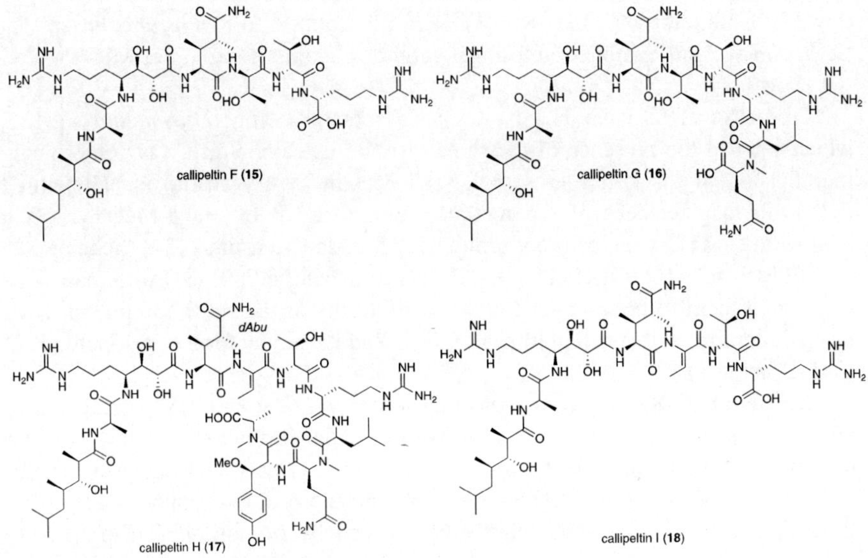

Fig. 3 Chemical structure of callipeltins F (**15**) – I (**18**)

Callipeltin G (16) ($C_{54}H_{100}N_{16}O_{17}$) is structurally similar to callipeltin F (**15**), though incorporating an additional residue of Leu and *N*-MeGln (Scheme 1). The sequencing of these units was confirmed by analysis of ESIMS/MS fragmentation peaks [43].

Callipeltin H (17) ($C_{68}H_{116}N_{18}O_{20}$) is structurally similar to callipeltin C (**11**) and their NMR spectra were consistent, except for the presence of signals confirmed by TOCSY and HMBC data to correspond to one dAbu (Fig. 3). The geometry of the dAbu unit was determined to be *Z* by comparison of NMR data with that of the two geometrical isomers of the dAbu unit [43].

Callipeltin I (18) ($C_{42}H_{77}N_{13}O_{13}$) differs from callipeltin F (**15**) (Fig. 3) by one water molecules. ESIMS/MS fragmentation peaks confirmed the sequencing of the amino acid units. The presence of a dAbu unit was confirmed by the ^{1}H NMR spectrum and analysis of ESIMS/MS fragmentation peaks the sequencing of the amino acid units [43].

Callipeltin J (19) ($C_{31}H_{58}N_8O_{11}$) is similar to callipeltin D (**12**). Callipeltin J (**19**) incorporates a 3-methylglutamine (3-MeGln, **20**) residue, not previously found in other callipeltin derivatives. The amino acid sequence and placement of acyl substituent were assigned based on the analysis of the ESIMS/MS fragmentation pattern. The order of four amino acid residues in callipeltin J (**19**) is the same as that of callipeltin D (**12**) (Scheme 1), with 3-MeGln (**20**) in callipeltin J (**19**) replacing the Me_2Gln (**6**) in callipeltin D (**12**) [34].

Callipeltin K (21) ($C_{67}H_{116}N_{18}O_{21}$) is similar to callipeltin C (**11**) except for the presence of 3-MeGln (**20**) in the place of the Me_2Gln (**6**) in callipeltin C (**11**) [34] (Fig. 4). Callipeltins F (**15**) -K (**21**) were all isolated from *Latrunculia* sp. and *Callipelta* sp. and exhibited antifungal activities against *Candida* at MIC value of ~10^{-4} M [32].

Callipeltin L (22) ($C_{66}H_{112}N_{18}O_{20}$) was obtained from *Latrunculia* sp. [32], which showed the presence of the same residues as callipeltin C (**11**) other than a modification of the *C*-terminus moiety (*N*-MeGln-Tyr-*N*-MeAla) in callipeltin L (**22**) with displacement of the methoxy group at Cβ of the β-MeOTyr (**7**) in callipeltin C (**11**) by the carboxy group of *N*-MeAla C-terminus [34] (Scheme 1).

Callipeltin M (4) ($C_{47}H_{77}N_{12}O_{15}$) differs from callipeltin B (**3**) by 18 mass units, suggesting them to be the cyclic and acyclic forms of the same compound. NMR analysis confirmed that callipeltin M (**4**) is indeed the acylic form of callipeltin B (**3**) [34] (Scheme 1).

Callipeltin N (8) showed a molecular ion at *m/z* 766.4563 $[M + 2H]^{2+}$ in the HRESIMS spectrum, corresponding to the molecular formula $C_{70}H_{120}N_{18}O_{20}$. The sequence of amino acids in callipeltin N (**8**) (Fig. 5) was determined by HMBC correlations and was found to be consistent with those in callipeltin A (**1**) with the exception of *N*-MeAla which has been replaced by homoproline (Pip) [32, 42] (Scheme 1).

Callipeltin O (9) showed the doubly charged molecular ion at *m/z* 759.4392 $[M + 2H]^{2+}$ in the HRESIMS spectrum corresponding to the molecular formula $C_{69}H_{118}N_{18}O_{20}$. The initial structure (Fig. 5) was determined by extensive analysis of both 1D and 2D NMR data and was consistent with that of callipeltin N (**8**) except

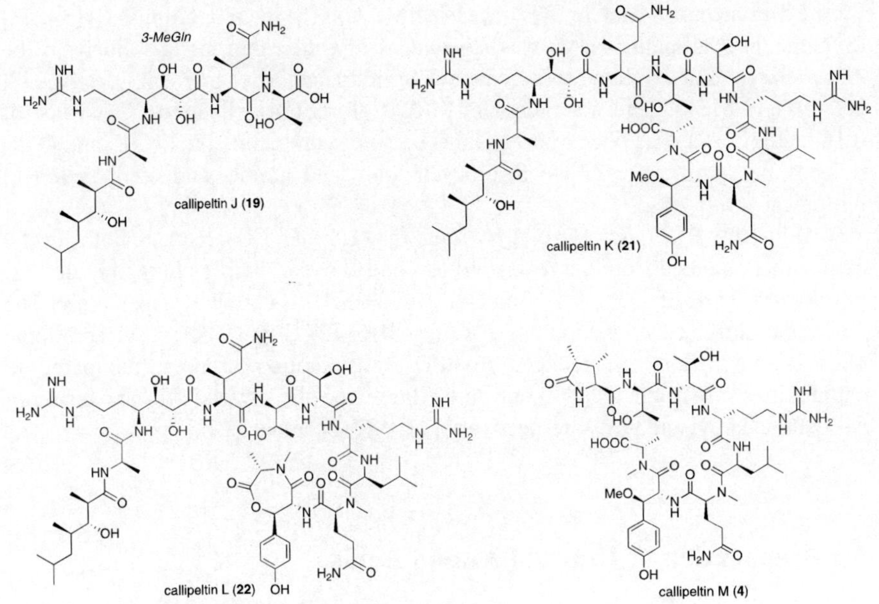

Fig. 4 Chemical structures of callipeltins J (**19**) – M (**4**)

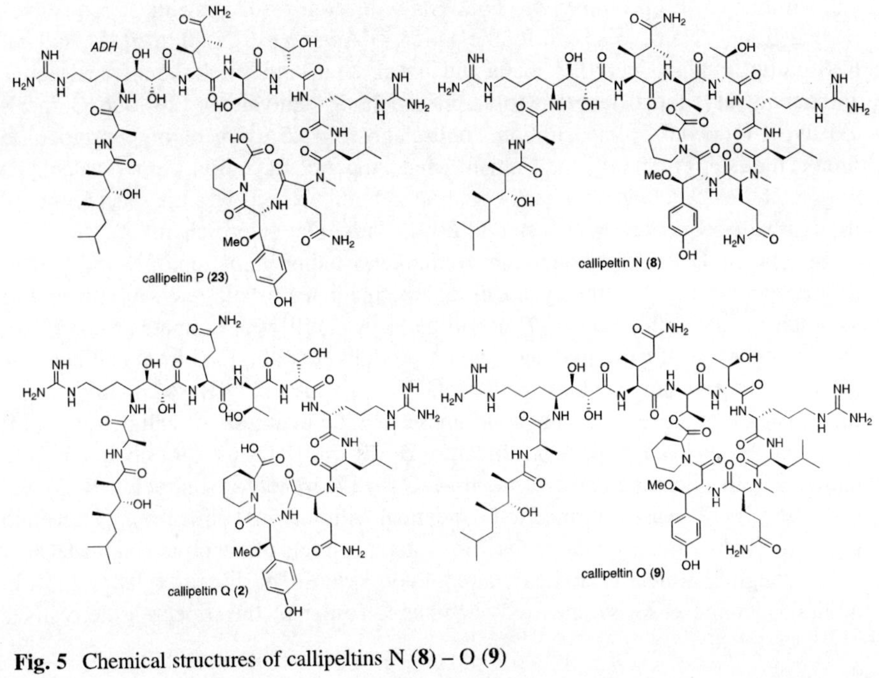

Fig. 5 Chemical structures of callipeltins N (**8**) – O (**9**)

for the presence of 3-MeGln (**20**) instead of Me_2Gln (**6**) in callipeltin N (**8**) [32, 42] (Scheme 1). Callipeltin N (**8**) was isolated as a white amorphous solid from the *Asteropus* sp. and exhibited significant cytotoxicity against cancer cell lines A2058, HT-29, and MCF-7 and nonmalignant MRC-5 fibroblast cells at an IC_{50} value of 0.16 μM [32, 42]. The potent biological activity of both callipeltins N (**8**) and O (**9**) reflects the significance of the macrocyclization and amino acid composition in biological activity [32, 42].

Callipeltins P (23) ($C_{70}H_{122}O_{21}N_{18}$) and **Q (2)** ($C_{69}H_{120}O_{21}N_{18}$) are both acyclic callipeltins obtained from the *Asteropus* sp., and showed doubly charged molecular ion at *m/z* 775.4505 $[M + 2H]^{2+}$ and *m/z* 768.4437 $[M + 2H]^{2+}$, respectively. The molecular formulae of callipeltins P (**23**) and Q (**2**) (Fig. 5) show an additional 18 mass units and one reduced (by one) unsaturation number, compared to callipeltins N (**8**) and Q (**2**), confirming these are the acyclic (linear) forms of callipeltins N (**8**) and O (**9**), respectively [32, 42] (Scheme 1).

3 Preparation of Unusual Amino Acids

3.1 Preparation of β-MeOTyr and Determination of Absolute Structure

The unspecified stereostructure of the β-MeOTyr (**7**) moiety in the initial studies of callipeltin A (**1**) [9] prompted the synthesis of all four of its possible stereoisomers by five groups (Hamada's [44], Joullie's [45], D'Auria's [46], Lipton's [47], and our [48] group) independently. Hamada and Joullie both undertook the stereoselective addition of aryl metal reagents to serine aldehyde equivalents (**24**) and Garner's aldehyde, followed by methylation. Joullie accessed **25** using Lajoie's protocol to protect the ester group [49] and Hamada used the popular synthon Garner's aldehyde to give **26** [50–52]. One drawback common to both methods was the sheer length of the synthetic sequences, which started from L- and D-serine (Scheme 2).

In contrast, D'Auria's group employed hν-assisted bromination of D-and L-tyrosine derivative (**27**) for the synthesis of intermediates as diastereo-mixtures. The stereochemistry of β-MeOTyr (**7**) in callipeltin A (**1**) [9] was estimated to be 2*R*,3*R* by a comparison of oxidative degradation product (**28**) obtained from callipeltin A (**1**) and four separately synthesized β-MeOTyr derivative (**29**). Based on this assignment, Lipton et al. reported the solid phase peptide synthesis of callipeltin E (**14**) [37] and a cyclized peptide, callipeltin B (**3**) [33]. Using **30**, obtained from cinnamate, the stereostructure of the β-MeOTyr (**7**) moiety in callipeltin B (**3**) and callipeltin E (**14**) was confirmed to be identical with that in callipeltin A (**1**). Konno et al. reported a simple synthesis of all four stereoisomers of the protected β-MeOTyr (**31**) through Sharpless asymmetric aminohydroxylation or dihydroxylation [53]. In addition, Konno et al. synthesized each stereoisomer of the tripeptide derivatives

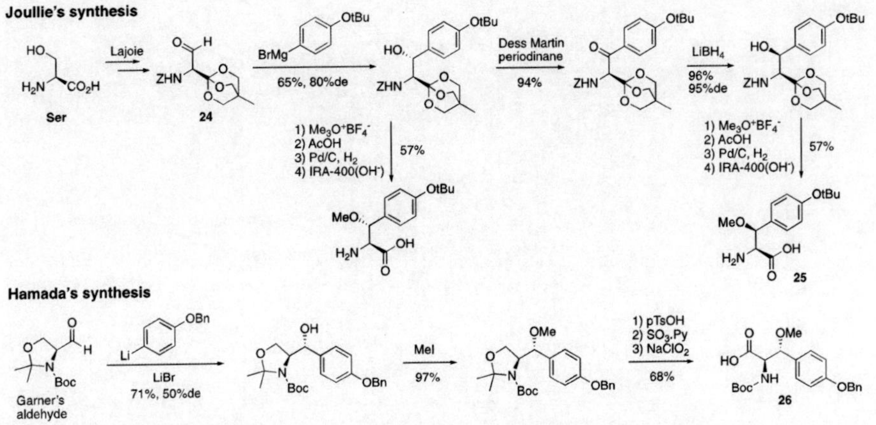

Scheme 2 Synthesis of β-MeOTyr derivatives (**25**) and (**26**) by Joullie and Hamada

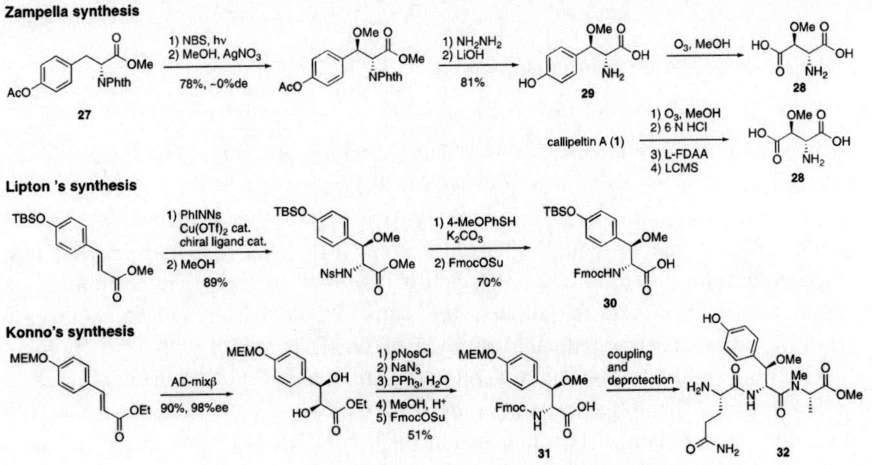

Scheme 3 Synthesis of β-MeOTyr derivatives (**29**) (**30**) (**31**) by Zampella, Lipton, and Konno groups

(**32**) and determined the absolute configuration of β-MeOTyr (**7**) in the callipeltins to be 2*R*,3*R* by NMR analysis (Scheme 3).

3.2 *Preparation of 3-MeGln and Me₂pyroGln*

Preparation of 3-MeGln was independently reported by Inoue and Konno. Inoue started from 2-hydroxypinan-3-one (**33**), the Gly-Schiff base of which underwent a Michael addition reaction with ethyl crotonate to give (2*S*,3*S*)-Fmoc-3-MeGln (**34**)

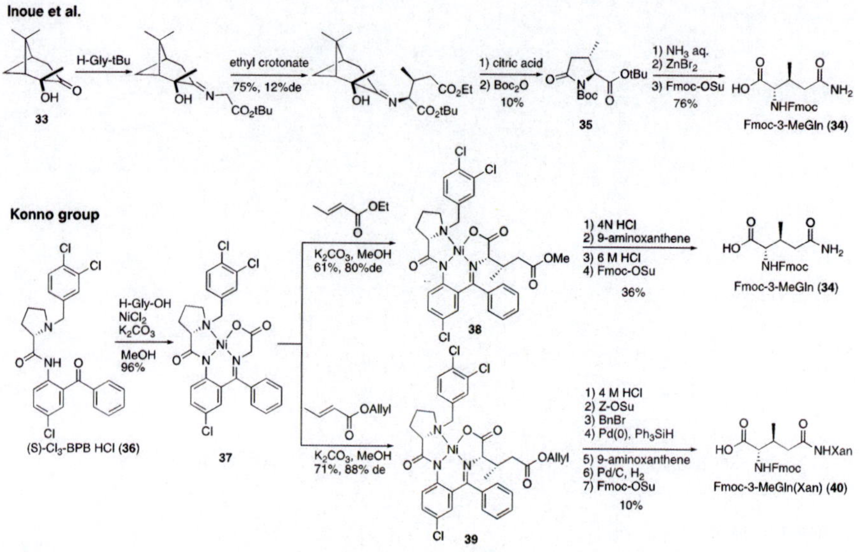

Scheme 4 Synthesis of Fmoc-3-MeGln (**34**)

via the pyroGlu derivative (**35**) [54]. The diastereomeric ratio associated with the Michael addition reaction was low (12%, 75% chemical yield). Konno applied Schiff base Ni(II)-complex methodology [55–57] to the synthesis of Fmoc-3-MeGln (**34**). After the coupling of Gly with Ni(II) with (*S*)-Cl_3-BPB.HCl (**36**), Michael addition of chiral Gly-Ni(II) Schiff base with two different crotonates was conducted to obtain desired adducts (**38**) and (**39**) in diastereomeric excesses of 80–88%, respectively. After the removal of Ni(II) complex with acid, Fmoc-3-MeGln (**34**) and Fmoc-3-MeGln(Xan) (**40**) were obtained [58] (Scheme 4).

Syntheses of the protected Me_2Gln derivative (**43**) and Fmoc-3-MeGln(Xan) (**40**) were first reported by the Joullie group [59]. Me_2Gln (**43**) was obtained via an asymmetric Michael addition with the chiral auxiliary (+)-camphorsultam (**41**) developed by Capet [60]. Installation of the α-amino group was accomplished by electrophilic azidation, and **42** was obtained after reduction with $SnCl_2$ and Boc protection. Hydrolysis of the chiral auxiliary with LiOH provided the protected Me_2Gln (**43**). In contrast, Hamada's group chose Thottathil's synthon (**44**) [61, 62] as the starting material for Me_2pyroGlu (**45**) in 6 steps [63], including a diastereo-controlled dimethylation reaction.

Lipton et al. obtained Me_2pyroGlu (**46**) in 9 steps from L-pyroGlu via cuprate addition and alkylation of the unsaturated lactam [64, 65], followed by the kinetic epimerization of the methyl substituent. Me_2Gln (**47**) was obtained from **46** by a pyrrolidone ring opening reaction in the presence of $Eu(OTf)_3$ without epimerization (Scheme 5).

Due to discrepancies between peaks corresponding to the Me_2pyroGlu (**10**) moiety in the NMR spectra of natural callipeltin B (**3**) and samples of

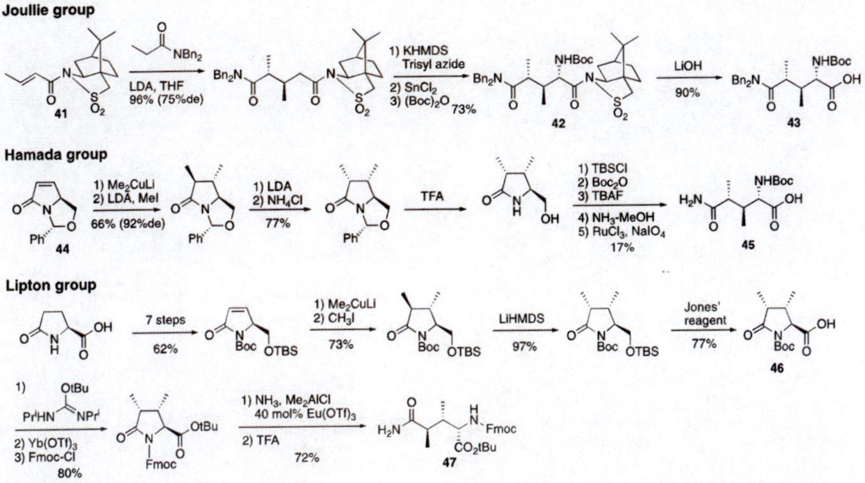

Scheme 5 Synthesis of Me_2Gln derivatives (**43**) (**45**) (**47**) and Me_2pyroGlu derivative (**46**) by Joullie, Hamada and Lipton

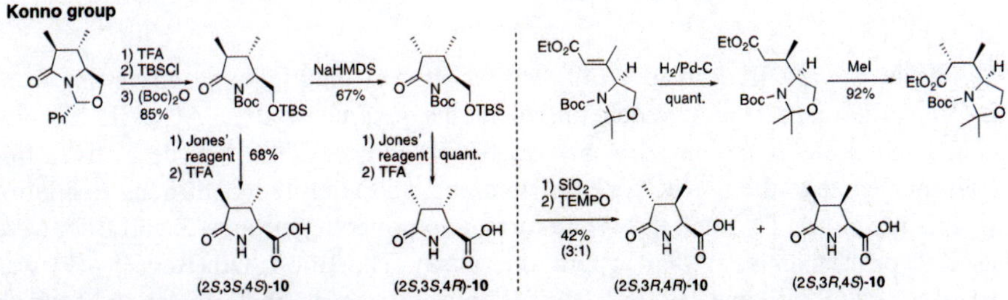

Scheme 6 Synthesis of four diastereoisomers of Me_2pyroGlu (**10**) by Konno group

(2*S*,3*S*,4*R*)-Me_2pyroGlu (**10**) synthesized by the Lipton and Hamada groups, the Konno group synthesized all four diastereoisomers of Me_2pyroGlu (**10**). Thottathil's synthon (**44**) was selected as the starting material for the synthesis (2*S*,3*S*,4*S*)- and (2*S*,3*S*,4*R*)-**10**, the other diastereomers of **10** were prepared from Garner's aldehyde. Diastereoselective hydrogenation of the *E*-olefin proceeded via the *N*-inside conformer to give desired compound. The ^{1}H NMR spectra of (2*S*,3*S*,4*R*)-**10** exhibited only a small coupling constant between H-2 and H-3-protons at $J_{H2,H3} = 3.2$ Hz and other three isomers had similar coupling constants of $J_{H2,H3} = 7.8$–8.7 Hz. Comparison of this data with that of natural Me_2Gln (**6**) ($J_{H2,H3} = 3.0$ Hz) confirmed its stereochemistry [66, 67] (Scheme 6).

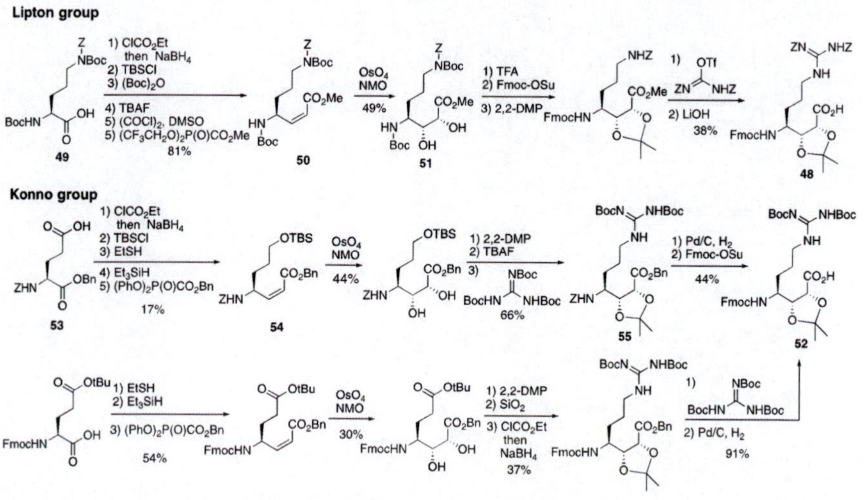

Scheme 7 Synthesis of AGDHEs (**48**) and (**52**) by Lipton and Konno

3.3 *Preparation of AGDHE*

Synthesis of AGDHE derivative (**48**) was reported by the Lipton group starting from L-Orn; as a result, the stereochemistry of the natural form of AGDHE (**5**) was conducted to be 2*R*,3*R*,4*S*. Orn is a reasonable intermediate from derivatives frequently cyclize to the unreactive cyclic aminal and therefore it is difficult to control several reactions. To block the cyclic form, two protecting groups, Z and Boc, were used to protect amino group in Orn side chain. The Boc-L-Orn(Boc,Z) (**49**) was transformed to Z-alkene (**50**) in 81% yield over six steps, dihydroxylation of which proceeded smoothly with catalytic OsO_4 to give diol (**51**) in 98% yield as a single diastereomer. The desired compound **48** was obtained in five steps from **51** [68].

The Konno group's synthesis of AGDHE derivative (**52**) started from Z-L-Glu-OBn (**53**), which was converted into Z-alkane (**54**) in a five-step sequence in 17% overall yield. After dihydroxylation with catalytic OsO_4, acetonide formation, and deprotection of the TBS group, guanidination under Mitsunobu conditions yielded **55**. Finally, sequential hydrogenolysis of the Z and Bn groups and Fmoc formation of the corresponding amino functionality gave the Fmoc-AGDHE derivative (**52**) [69]. In a second generation synthesis, Fmoc-L-Glu(tBu) was used as a starting material, and Fmoc-AGDHE derivative (**52**) was obtained via similar protocols [70, 71] (Scheme 7).

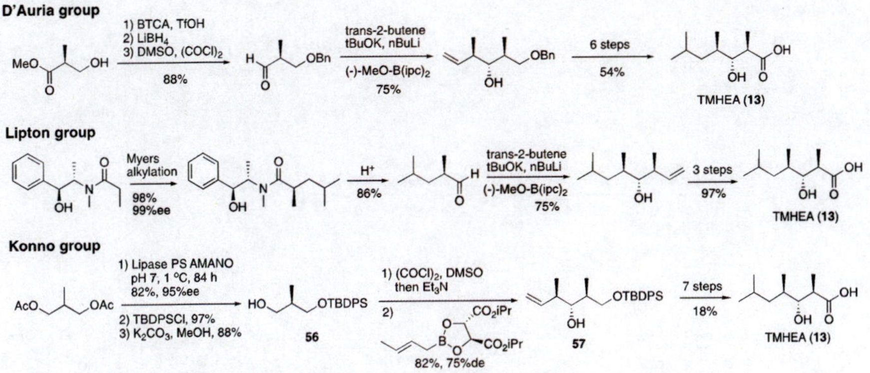

Scheme 8 Synthesis of TMHEA (**13**) and determination of its stereochemistry

3.4 *Synthesis of TMHEA and Determination of Absolute Structure*

The absolute configuration of TMHEA (**13**) was initially concluded to be 2*R*,3*R*,4*S* based on its ^{1}H NMR data collected after its isolation. However, the D'Auria group revised it to be 2*R*,3*R*,4*R* by comparison with the NMR spectra of isolated TMHEA (**13**) and those of a synthetic sample, obtained using Brown's crotylboration [72]. This stereochemical assignment was confirmed by the Lipton group, who prepared all four diastereoisomers of TMHEA (**13**) using a Myers alkylation of ephedrine, and Brown's crotylboration [73].

Tokairin and Konno synthesized a selection of analogues of TMHEA (**13**) using a 2-methyl-1,3-propanediol derivative as the symmetric building block. (*R*)-**56** was prepared by an enantioselective enzymatic approach. The two adjacent stereocenters at C-3 and C-4 of the homoallylic alcohol were established by crotylboration of the aldehyde derived from alcohol (**56**) with a chiral borate derived from diisopropyl tartrate to give the homoallylic alcohol (2*R*,3*R*,4*R*)-**57** in a diastereomeric excess of 75%. The isopropyl unit was introduced by the oxidative cleavage and Wittig reaction [74] (Scheme 8).

3.5 *Synthesis of D-alloThr*

D-alloThr derivative (**60**) was prepared from alcohol (**58**), obtained from L-serine in 24% overall yield (6 steps) according to a published procedure [52]. Protection of the hydroxy group of **58** with MEMCl/iPr$_2$NEt followed by deprotection of the acetonide afforded the corresponding alcohol (**59**) in 59% yield. Oxidation of the alcohol with Jones' reagent followed by deprotection of the Boc group and the reprotection with Fmoc afforded the desired Fmoc-D-alloThr (**60**) [66].

Konno group

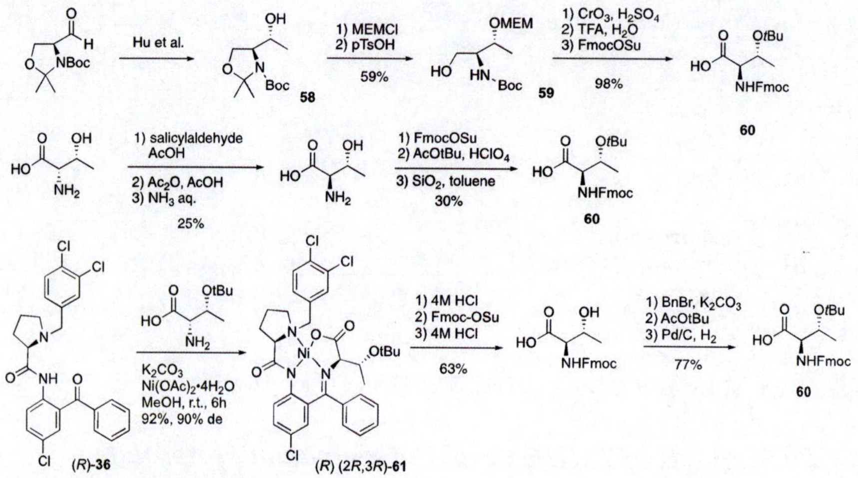

Scheme 9 Synthesis of D-alloThr (**60**)

An improved synthesis of the protected D-alloThr derivative was subsequently described [75]. Epimerization of cheap L-Thr with catalytic salicylaldehyde afforded a mixture of L-Thr and D-alloThr, separation of which gave D-alloThr in 96% de. After the chemoselective deprotection of the t-butyl ester of Fmoc-D-alloThr (tBu)-OtBu, the desired product Fmoc-D-alloThr(tBu) (**60**) was easily obtained [75].

In a third generation synthesis, (*R*)-**36** and H-(*S*)-Thr(tBu)-OH were treated with $Ni(OAc)_2 \cdot 4H_2O$ and K_2CO_3 in MeOH at 50°C to afford D-alloThr complex **61** with excellent yield and diastereoselectivity (92% yield, 90% de). Disassembly of major diastereomer **61** with 4 M HCl in THF for 30 min gave a 1:1 mixture of D-alloThr and D-alloThr(tBu) which was subjected to Fmoc protection followed by deprotection of the residual tBu group, affording Fmoc-D-alloThr in 63% yield over three steps [58] (Scheme 9).

4 Total Synthesis of Cyclic Depsipeptides

4.1 *Total Synthesis of Callipeltin B and Its Analogues*

Lipton et al. reported the solid phase total synthesis of callipeltin B (**3**) and revised its chemical structure as a result [76]. The amino and hydroxy acid sequence was built up starting from Tenta gel-based TG-Sieber amide resin, anchored using the side chain of MeGln. Benzyl and nitro groups were used for the side chain protections of β-MeOTyr and D-Arg, given the lability of the β-MeOTyr residue under acidic conditions. Cleavage from the resin was accomplished under mild acidic conditions. Lipton's group synthesized callipeltin E (**14**) starting from the MeAla at the

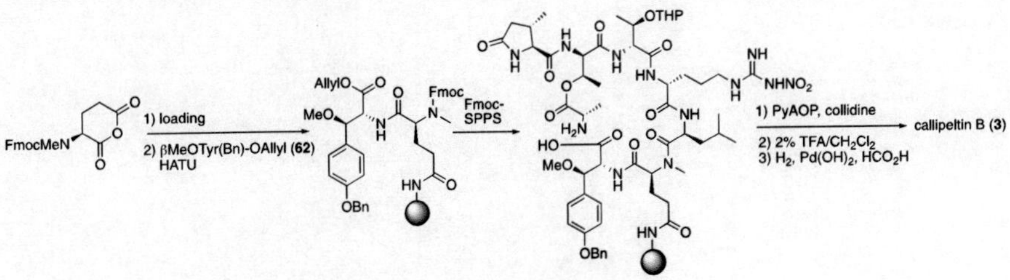

Scheme 10 Lipton's Total synthesis of callipeltin B (3)

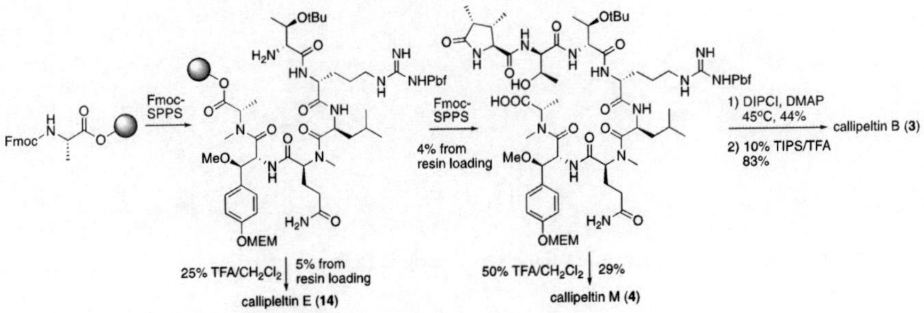

Scheme 11 Konno's Total synthesis of callipeltin B (3)

C-terminus [77]; in contrast, the synthesis of callipeltin B (**3**) started from β-MeOTyr derivative (**62**) to avoid macrolactonization during the deprotection process (Scheme 10).

Kikuchi and Konno report the total syntheses of the cyclodepsipeptide callipeltin B (**3**) and the linear octapeptide, callipeltin M (**4**). Both these syntheses commenced with conventional Fmoc-SPPS starting from the *C*-terminus, using MEM, Pbf and tBu protecting groups. Macrolactonization was accomplished between the D-alloThr and MeAla residues. Although the synthesis of cyclic depsipeptide by macrolactonization of the corresponding linear precursor peptide is generally difficult, it was assumed that the conformation of callipeltin M (**4**) in nature would be similar to that of callipeltin B (**3**) and favoring the direct conversion of callipeltin M (**4**) to callipeltin B (**3**). These approaches were used to prepare a variety of callipeltin analogues for structure activity relationship studies pursuant to the development of antitumor and/or anti-HIV agents [78] (Scheme 11).

In addition, the dimethoxy analogue of the cytotoxic, cyclic depsipeptide callipeltin B (**63**) was synthesized to evaluate the role of its β-MeOTyr residue by Lipton group [79]. The cytotoxicities of **63** and **3** were similar and methoxy group of β-position of callipeltin B (**3**) does not affect the cytotoxicity. Kikuchi and Konno synthesized the callipeltin B analogues containing D/L-Tyr residues using Fmoc-SPPS. An alternative route of intramolecular amide bond formation with Tyr-Ala was needed to avoid diketopiperadine formation. However, this strategy was not effective to give natural callipeltin B (**3**). Methoxy group of β-MeOTyr residue was

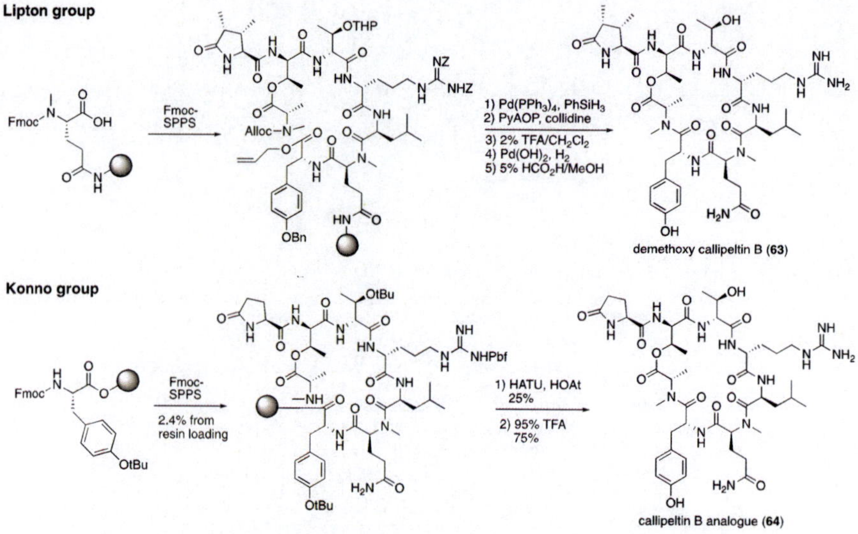

Scheme 12 Konno's synthesis of callipeltin B analogue (**63**)

left to obtain α,β-unsaturated product [80] (Scheme 12). In the assay of synthetic peptides, **64** had hardly exhibited the cytotoxicity.

5 Summary and Perspective

Marine organisms have yielded numerous cyclic depsipeptides (CDPs) with variable chemical structures and incorporating unusual amino acid building blocks. Callipeltins, one of the most intensively studied and synthesized classes of cyclic depsipeptides, exhibit fascinating biological effects and have served as lead compounds in drug discovery programs. The synthetic approaches to the callipeltins described herein are expected to support these efforts.

References

1. Vitali A (2018) Am J Clin Microbiol Antimicrob 1:1
2. Kitagaki J, Shi G, Miyauchi S, Murakami S, Yang Y (2015) Anti-Cancer Drug 2:259
3. Taevernier L, Wynendaele E, Gevaert B, De Spiegeleer B (2017) Curr Protein Peptide Sci 18:425
4. Wang X, Gong X, Li P, Lai D, Zhou L (2018) Molecules 23:169
5. Kim CK, Wang D, Bokesch HR, Fuller RW, Smith E, Henrich CJ, Durrant DE, Morrison DK, Bewley CA, Gustafson KR (2020) J Nat Prod
6. Jo C, Khan FF, Khan MI, Iqbal J (2017) Food Rev Int 33:44

7. Sun J, Cheng W, de Voogd NJ, Proksch P, Lin W (2016) Tetrahedron Lett 57:4288
8. Rangel M, José Correia de Santana C, Pinheiro A, dos Anjos L, Barth T, Rodrigues Pires Júnior O, Fontes W, Castro SM (2017) Curr Protein Peptide Sci 18:72
9. Zampella A, D'Auria MV, Paloma LG, Casapullo A, Minale L, Debitus C, Henin Y (1996) J Am Chem Soc 118:6202
10. Plaza A, Bifulco G, Keffer JL, Lloyd JR, Baker HL, Bewley CA (2009) J Org Chem 74:504
11. Freitas VM, Rangel M, Bisson LF, Jaeger RG, Machado-Santelli GM (2008) J Cell Physiol 216:583
12. Tarazona G, Fernandez R, Cruz PG, Perez M, Rodriguez J, Jimenez C, Cuevas C (2019) Org Chem Front 6:15
13. Zampella A, Sepe V, Luciano P, Bellotta F, Monti MC, D'Auria MV, Jepsen T, Petek S, Adeline MT, Laprevote O, Aubertin AM, Debitus C, Poupat C, Ahond A (2008) J Org Chem 73:5319
14. Randazzo A, Bifulco G, Giannini C, Bucci M, Debitus C, Cirino G, Gomez-Paloma L (2003) Chemtracts 16:688
15. Braekman JC, Daloze D, Moussiaux B, Riccio R (1987) J Nat Prod 50:994
16. Rashid MA, Gustafson KR, Cartner LK, Shigematsu N, Pannell LK, Boyd MR (2001) J Nat Prod 64:117
17. Plaza A, Gustchina E, Baker HL, Kelly M, Bewley CA (2007) J Nat Prod 70:1753
18. Ford PW, Gustafson KR, McKee TC, Shigematsu N, Maurizi LK, Pannell LK, Williams DE, Dilip de Silva E, Lassota P, Allen TM (1999) J Am Chem Soc 121:5899
19. Ratnayake AS, Bugni TS, Feng X, Harper MK, Skalicky JJ, Mohammed KA, Andjelic CD, Barrows LR, Ireland CM (2006) J Nat Prod 69:1582
20. Oku N, Gustafson KR, Cartner LK, Wilson JA, Shigematsu N, Hess S, Pannell LK, Boyd MR, McMahon JB (2004) J Nat Prod 67:1407
21. Tran TD, Pham NB, Fechner G, Zencak D, Vu HT, Hooper JNA, Quinn RJ (2012) J Nat Prod 75:2200
22. Coello L, Reyes F, Martín MAJS, Cuevas C, Fernández R (2014) J Nat Prod 77:298
23. Kita M, Gise B, Kawamura A, Kigoshi H (2013) Tetrahedron Lett 54:6826
24. Nakamukai S, Takada K, Furihata K, Ise Y, Okada S, Morii Y, Yamawaki N, Takatani T, Arakawa O, Gustafson KR (2018) Tetrahedron Lett 59:2532
25. Okada Y, Matsunaga S, van Soest RW, Fusetani N (2002) Org Lett 4:3039
26. Roy MC, Ohtani I, Ichiba T, Tanaka J, Satari R, Higa T (2000) Tetrahedron 56:9079
27. Tsuda M, Shimbo K, Kubota T, Mikami Y, Kobayashi J (1999) Tetrahedron 55:10305
28. Kobayashi J, Itagaki F, Shigemori I, Takao T, Shimonishi Y (1995) Tetrahedron 51:2525
29. Kobayashi J, Sato M, Ishibashi M, Shigemori H, Nakamura T, Ohizumi Y (1991) J Chem Soc Perkin Trans 1:2609
30. Uemoto H, Yahiro Y, Shigemori H, Tsuda M, Takao T, Shimonishi Y, Kobayashi J (1998) Tetrahedron 54:6719
31. Tsuda M, Ishiyama H, Masuko K, Takao T, Shimonishi Y, Kobayashi J (1999) Tetrahedron 55:12543
32. Gogineni V, Hamann MT (2018) Biochim Biophys Acta 1862:81
33. D'Auria MV, Zampella A, Paloma LG, Minale L, Debitus C, Roussakis C, Le Bert V (1996) Tetrahedron 52:9589
34. D'Auria MV, Sepe V, D'Orsi R, Bellotta F, Debitus C, Zampella A (2007) Tetrahedron 63:131
35. Kikuchi M, Konno H (2016) Biosci Biotechnol Biochem 80:1066
36. Ngo DH, Vo TS, Ngo DN, Wijesekara I, Kim SK (2012) Int J Biol Mcromol 51:378
37. Zampella A, Randazzo A, Borbone N, Luciani S, Trevisi L, Debitus C, D'Auria MV (2002) Tetrahedron Lett 43:6163
38. Andavan GSB, Lemmens-Gruber R (2010) Mar Drugs 8:810
39. Kikuchi M, Nosaka K, Akaji K, Konno H (2011) Tetrahedron Lett 52:3872
40. Zampella A, D'Auria MV, Minale L, Debitus C (1997) Tetrahedron 53:3243
41. Cranfill DC, Morales-Ramos ÁI, Lipton MA (2005) Org Lett 7:5881

42. Stierhof M, Hansen KØ, Sharma M, Feussner K, Subko K, Díaz-Rullo FF, Isaksson J, Pérez-Victoria I, Clarke D, Hansen E (2016) Tetrahedron 72:6929
43. Sepe V, D'Orsi R, Borbone N, D'Auria MV, Bifulco G, Monti MC, Catania A, Zampella A (2006) Tetrahedron 62:833
44. Okamoto N, Hara O, Makinio K, Hamada Y (2002) J Org Chem 67:9210
45. Hansen DB, Wan X, Carroll PJ, Joullie MM (2005) J Org Chem 70:3120
46. Zampella A, D'Orsi R, Sepe V, Casapullo A, Monti MC, D'Auria MV (2005) Org Lett 7:3585
47. Cranfill DC, Lipton MA (2007) Org Lett 9:3511
48. Konno H, Aoyama S, Nosaka K, Akaji K (2007) Synthesis 2007:3666
49. Blaskovich MA, Lajoie GA (1993) J Am Chem Soc 115:5021
50. Garner P, Ramakanth S (1986) J Org Chem 51:2609
51. Liang X, Andersch J, Bols M (2001) J Chem Soc Perkin Trans 1:2136
52. Hu XE, Kim NK, Ledoussai B (2002) Org Lett 4:4499
53. Kolb HC, VanNieuwhenze M, Sharpless KB (1994) Chem Rev 94:2483
54. Inoue M, Shinohara N, Tanabe S, Takahashi T, Okura K, Itoh H, Mizoguchi Y, Iida M, Lee N, Matsuoka M (2010) Nat Chem 2:280
55. Romoff TT, Ignacio BG, Mansour N, Palmer AB, Creighton CJ, Abe H, Moriwaki H, Han J, Konno H, Soloshonok VA (2020) Org Process Res Dev 24:294
56. Tokairin Y, Soloshonok VA, Konno H, Moriwaki H, Röschenthaler G-V (2019) J Fluor Chem 227:109376
57. Han J, Romoff TT, Moriwaki H, Mkrtchyan AF, Saghyan AS, Konno H, Soloshonok VA (2020) Molecules 25:2739
58. Tokairin Y, Soloshonok VA, Moriwaki H, Konno H (2019) Amino Acids 51:419
59. Liang B, Carroll PJ, Joullie MM (2000) Org Lett 2:4157
60. Capet M, David F, Bertin L, Hardy JC (1995) Synth Commun 25:3323
61. Thottathil JK, Przybyla C, Malley M, Gougoutas JZ (1986) Tetrahedron Lett 27:1533
62. Thottathil JK, Moniot JL, Mueller RH, Wong MK, Kissick TP (1986) J Org Chem 51:3140
63. Okamoto N, Hara O, Makino K, Hamada Y (2001) Tetrahedron Asymmetry 12:1353
64. Acevedo CM, Kogut EF, Lipton MA (2001) Tetrahedron 57:6353
65. Calimsiz C, Lipton MA (2005) J Org Chem 70:6218
66. Konno H, Takebayashi Y, Nosaka K, Akaji K (2010) Heterocycles 81:79
67. Kikuchi M, Konno H (2014) Heterocycles 89:1620
68. Thoen JC, Morales-Ramon ÁI, Lipton MA (2002) Org Lett 4:4455
69. Yoshino R, Tokairin Y, Konno H (2017) Tehrahedron Lett 58:1604
70. Tokairin Y, Takeda S, Kikuchi M, Konno H (2015) Tetrahedron Lett 56:2809
71. Tokairin Y, Maita K, Takeda S, Konno H (2015) Synthesis 47:351
72. D'Auria MV, Zampella A (2002) Tetrahedron Asymmetry 13:1237
73. Turk JA, Visbal GS, Lipton MA (2003) J Org Chem 68:7811
74. Tokairin Y, Konno H (2017) Tetrahedron 73:39
75. Kikuchi M, Konno H (2013) Tetrahedron 69:7098
76. Krishnamoorthy R, Vazquez-Serrano LD, Tunk JA, Kowalski JA, Benson AG, Breaux NT, Lipton MA (2006) J Am Chem Soc 128:15392
77. Calimisiiz S, Romos AIM, Lipton MA (2006) J Org Chem 71:6351
78. Kikuchi M, Konno H (2014) Org Lett 16:4324
79. Krishnamoorthy R, Richardson BL, Lipton MA (2007) Bioorg Med Chem Lett 17:5136
80. Kikuchi M, Watanabe Y, Tanaka M, Akaji K, Konno H (2011) Bioorg Med Chem Lett 21:4865

Top Heterocycl Chem (2021) 58: 317–360
https://doi.org/10.1007/7081_2020_50

Published online: 21 January 2021

Synthesis of Marine C_2-Symmetrical Macrodiolide Natural Products

Jun Ishihara

Contents

J. Ishihara (✉)
Graduate School of Biomedical Sciences, Nagasaki University, Nagasaki, Japan
e-mail: jishi@nagasaki-u.ac.jp

Abstract Marine natural products play a pivotal role in the discovery of structurally novel and diverse compounds with the unique pharmacological profiles. The marine C_2-symmetrical macrodiolides belong to a relatively small group in marine natural products, but most of them exhibit various different biological activities and possess a fascinating structure. In this review, the total syntheses and the synthetic studies of C_2-asymmetrical macrodiolides, swinholide A, clavosolide A, cyanolide A, marinomycin A will be discussed. All marine natural products shown here are potential candidates for drug discovery and drug lead.

Keywords Macrodiolide · Marine natural product · Polyketide · Total synthesis

Abbreviations

15-C-5	15-Crown-5
Ac	Acetyl
BAIB	(Diacetoxyiodo)benzene
Bn	Benzyl
Bpin	Boronic acid pinacol ester
Bz	Benzoyl
CDI	Carbonyldiimidazole
c-Hex	Cyclohexyl
cod	1,5-Cyclooctadienyl
Cp	Cyclopentadienyl
CSA	Camphorsulfonic acid
Cy	Cyclohexyl
DACH-Phenyl Trost	1,2-Diaminocyclohexane-*N,N′*-bis (2-diphenylphosphinobenzoyl)
DBU	1,8-Diazabicyclo[5.4.0]undec-7-ene
DDQ	2,3-Dichloro-5,6-dicyano-1,4-benzoquinone
DEAD	Diethyl azodicarboxylate
DIAD	Diisopropyl azodicarboxylate
DIBAL-H	Diisobutylaluminum hydride
diop	2,3-*O*-isopropylidene-2,3-dihydroxy-1,4-bis (diphenylphosphino)butane
DIPT	Diisopropyl tartrate
DMAP	4-(Dimethylamino)pyridine
DMF	*N,N*-dimethylformamide
DMPI	Dess–Martin periodinane
DMSO	Dimethyl sulfoxide
DtBMP	2,6-Di-*t*-butyl-4-methylpyridine
Et	Ethyl
HMDS	Hexamethyldisilazane
HMPA	Hexamethylphosphoramide
IBX	2-Iodoxybenzoic acid
imid	Imidazol

Ipc	Isopinocampheyl
LHMDS	Lithium hexamethyldisilazide
Me	Methyl
MNBA	2-Methyl-6-nitrobenzoic anhydride
MOM	Methoxymethyl
Ms	Methanesulfonyl
MS	Molecular sieves
MVK	Methyl vinyl ketone
NBS	*N*-bromosuccinimide
NIS	*N*-iodosuccinimide
NMO	*N*-methylmorpholine oxide
NMP	*N*-methyl-2-pyrrolidone
Ph	Phenyl
PMB	*p*-methoxybenzyl
PMP	*p*-methoxyphenyl
PPTS	Pyridinium *p*-toluenesulfonate
Pr	Propyl
PTS	*p*-toluenesulfonic acid
Py	Pyridine
Red-Al	Sodium bis(2-methoxyethoxy)aluminum hydride
SpiroSAP	7-[*N*-(1,3-dithian-2-yl)methylamino]-7′-[bis(3,5-di-*t*-butylphenyl)phosphino]-2,2′,3,3′-tetrahydro-1,1′-spirobindane
TBAF	Tetra(*n*-butyl)ammonium fluoride
TBDPS	*t*-butyldiphenylsilyl
TBHP	*t*-butyl hydroperoxide
TBS	*t*-butyldimethylsilyl
TEMPO	2,2,6,6-Tetramethylpiperidine *N*-oxyl
TES	Triethylsilyl
Tf	Trifluoromethanesulfonyl
TFA	Trifluoromethanesulfonic acid
THF	Tetrahydrofuran
TIB	2,4,6-Triisopropylbenzoyl
TIPS	Triisopropylsilyl
TMDS	1,1,3,3-Tetramethyldisilioxane
TMEDA	Tetramethylethylenediamine
TMS	Trimethylsilyl
TolBINAP	2,2′-Bis(di-*p*-tolylphosphino)-1,1′-binaphthyl
TPAP	*Tetra*(*n*-propyl)ammonium perruthenate

1 Introduction

Sponges, along with coelenterates, are one of the most abundant phylums of marine organisms, with more than half a million species in the coral reefs. Most sponges have a long life span and are rarely preyed upon by other organisms, thus they have been thought to produce some defense materials as protection against predators [1]. In this context, the biologically active substances from sponges have been widely sought throughout the world and various substances of interest from the perspective of pharmacological activity and high tumor activity have been discovered as a consequence of broad monitoring. One of prominent representatives isolated from sponge is a C_2-symmetric macrodiolide family, which exhibits remarkable biological activities.

The C_2-symmetric macrodiolides belong to a relatively small group in natural products, but most of them possess fascinating structures including a polyketide backbone and exhibit various biological activities. In the 1960s, unusual C_2-symmetric dimeric macrolides such as elaiophylin **1** [2], pyrenophorin **2** [3, 4], and vermiculine **3** [5] were discovered from terrestrial sources and later it was revealed that the artistic dimeric macrodiolides such as aplasmomycin **4** [6, 7] also appeared in marine organisms. In this review, the total syntheses and the synthetic studies of marine C_2-symmetric macrodiolides, which are potential candidates for drug discovery and drug lead, will be discussed (Fig. 1).

2 Biogenetic Formation of Macrodiolide

The biochemical polyketide synthesis resembles the synthesis of fatty acid [8–10]. Concerning to the biosynthesis of a polyketide chain in C_2-symmetric macrodiolides, the polyketide synthases (PKSs) catalyze the iterative decarboxylative condensation of small carboxylic acids, concomitant specifically programmed reduction and dehydration steps [11]. The fully processed linear polyketide chain is bound to the acyl carrier protein (ACP) by a thioester through a thioesterase-domain-mediated transesterification and concomitant cyclization with a specific hydroxy group present in the polyketide backbone through the action of a thioesterase and cyclase (TE) domain. The C_2-symmetric dimer TE must catalyze totally two acylation and two deacylation reactions to form the macrodiolide. The dimerization process of either biosynthesis or chemical synthesis is intriguing for the formation of C_2-symmetric macrodiolides, and the biogenetic research of macrodiolide formation has been flourished so far.

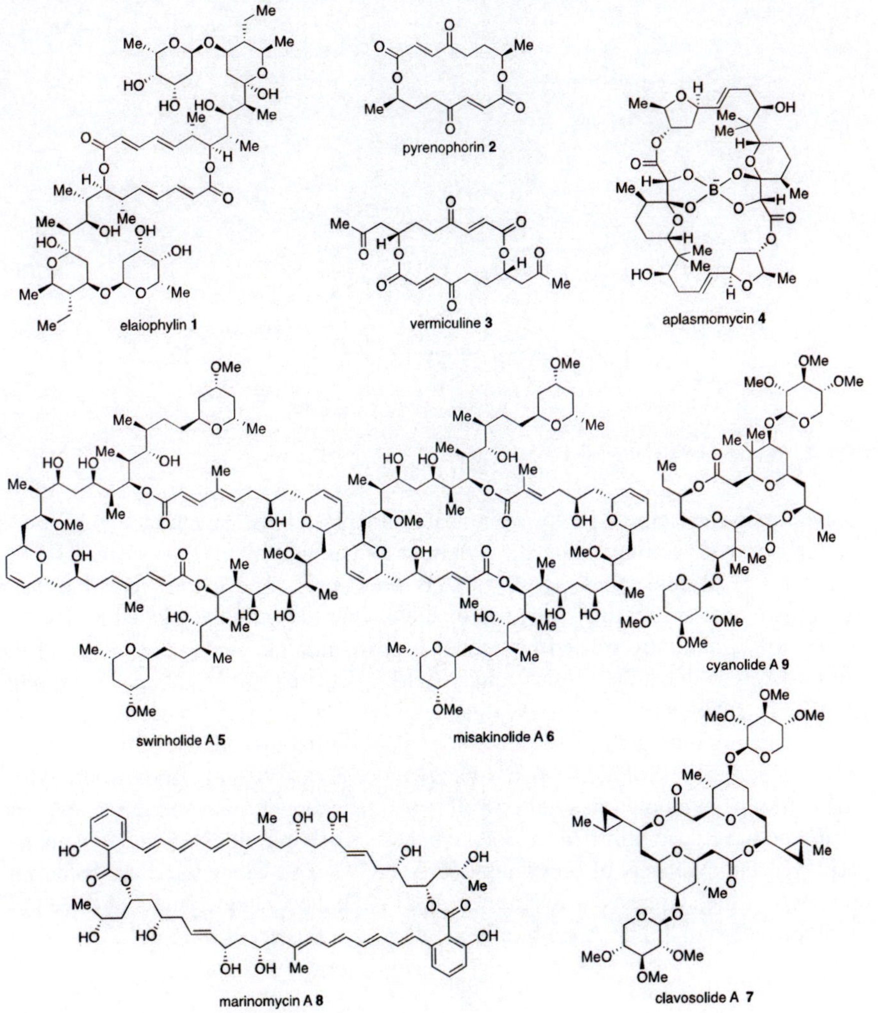

Fig. 1 C_2-symmetric macrodiolide natural products

3 Swinholide A

Swinholide A **5** is a natural product first reported in 1985 by Carnely and Kashman as one of several antifungal isolates obtained from *Theonella swinhoei*, a sponge which is yellowish-brown in color and reside in the Red Sea [12]. Their original structural assignment for the intriguing agent was 22-membered macrolide **10**, but later Kitagawa and co-workers elucidated that the correct structure was a C_2-symmetrical dimeric compound of the original [13–16]. Eventually, the established architecture of this natural product is a 44-membered macrodiolide with 15 stereocenters, a trisubstituted tetrahydropyran ring, a disubstituted

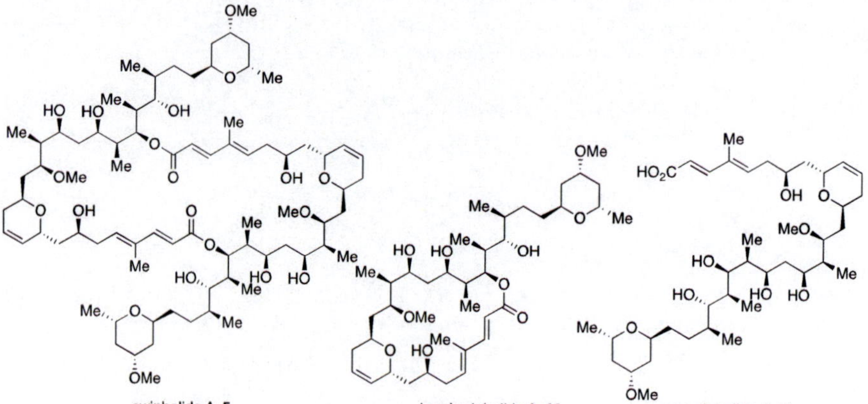

Fig. 2 Swinholide A and related compounds

dihydropyran system and two isomerizable double bonds. Swinholide A displays strong cytotoxic activity against a variety of human tumor cell lines (IC_{50} 0.03 μg/mL for L1210 cells, 0.04 μg/μL for KB cells) [17]. The cytotoxicity has been attributed to its ability to dimerize actin and disrupt the actin cytoskeleton. Interestingly, the structurally related compounds, preswinholide A **11**, [18, 19] and the 40-membered dilactone misakinolide **6** [20] isolated from Okinawan sponge *Theonella swinhoei*, do not serve filamentous actin like swinholide A. Based on the fascinating biological action and a formidable structure and a need for additional material supplies, swinholide A has been one of attractive target molecules. While many groups attempted to synthesize this natural product, Paterson, Nicolaou, and Krishche have accomplished the total synthesis of swinholide A **5** and Nakata has achieved the synthesis of preswinholide A **11** [21–24]. Since there are some fine reviews for the synthesis of swinholide A [25–27], the strategic points of three total syntheses of swinholide A are briefly mentioned in this section (Fig. 2).

3.1 Paterson's Total Synthesis of Swinholide A

In 1994, Paterson and the co-workers reported the first total synthesis of swinholide A [28–39]. Their synthetic strategy for swinholide A is filled with the very essence of aldol reaction. The Paterson's total synthesis was based on the regioselective dimerization of fully protected monomeric compound **12**, which was prepared with a convergent synthetic strategy from the two building blocks representing C19-C32 segment (**13**) and C1-C15 segment (**15**) (Scheme 1). The segments were efficiently prepared by various types of aldol reactions.

The aldol reaction of *E*-boron enolate **17**, derived from ketone **16** with Cy_2BCl and Et_3N, and aldehyde **18**, provided 1,2-*anti*-2,4-*anti* aldol adduct **20** in 84% yield through chair-like transition structure **19** (Scheme 2) [40, 41]. The reduction of

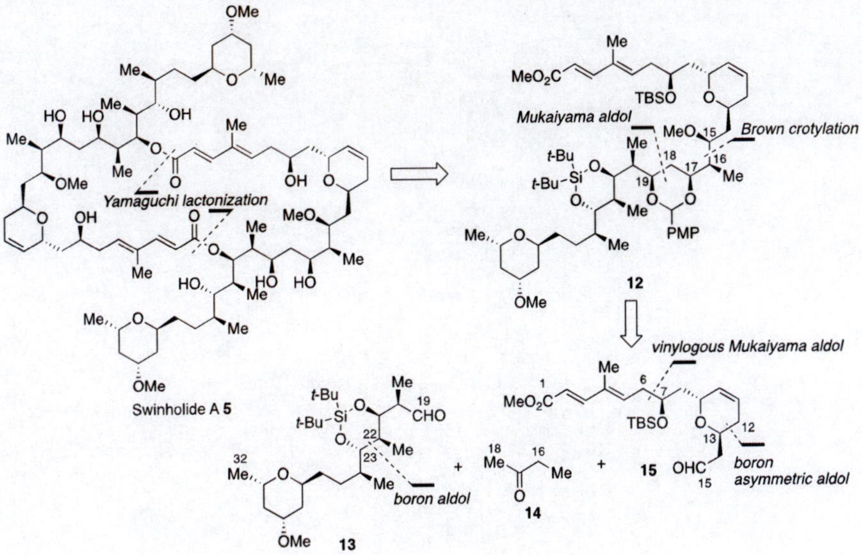

Scheme 1 Strategy of Paterson's total synthesis of swinholide A

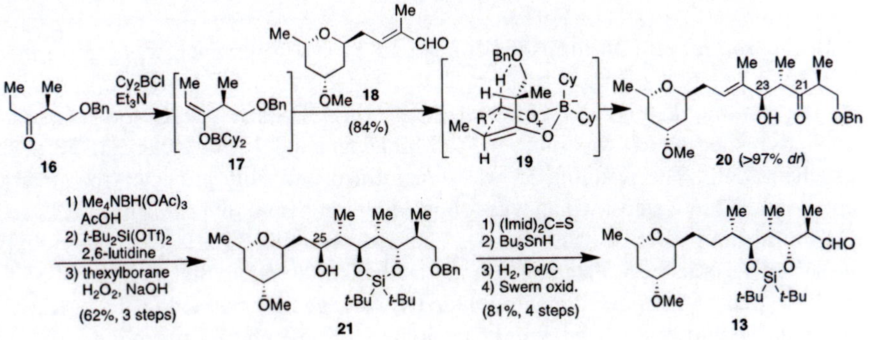

Scheme 2 Synthesis of segment **13**

ketone **20** at C21 with $Me_4NBH(OAc)_3$ afforded the 1,3-*anti* diol [42, 43], which was protected with a silylene and hydroborated to give compound **21**. A further 4 step-sequence led to segment **13**.

On the other hand, the preparation of segment **15** commenced with the asymmetric aldol reaction of aldehyde **24** with boron enolate **23**, derived from **22** with (+)-Ipc_2BCl, to provide chiral adduct **26** in 80% ee (Scheme 3) [44]. Compound **26** was transformed into aldehyde **27** in 4 steps including Ferrier rearrangement [45, 46]. Resulting aldehyde **27** was subjected to Mukaiyama aldol reaction with **28**, affording a 4:1 ratio of the desired product **29** and C7 epimer [47]. Compound **29** was subjected to Horner–Wadsworth–Emmons reaction and further 3 steps transformation yielded aldehyde **15**, C1-C15 segment. The crotylboration of **15**

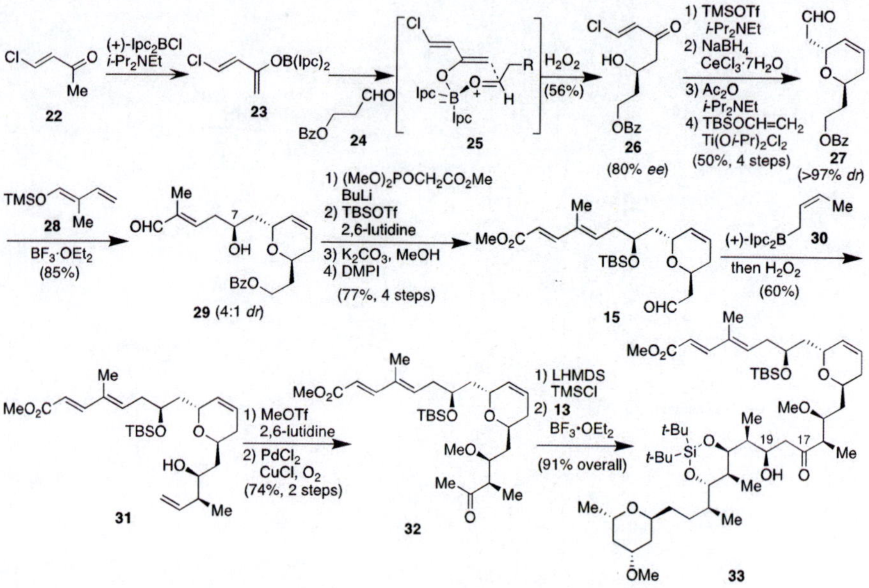

Scheme 3 Synthesis of segment **33**

with *Z*-boron reagent **30** furnished compound **31** [48, 49], which was transformed into **32** in 2 steps.

The resulting ketone was then treated with LiHMDS and TMSCl to generate an enol silyl ether, which was subjected to Mukaiyama aldol with **13** leading **33** in an excellent yield. The resulting **33** was transformed into fully protected monomeric compound **12** in 2 steps, which was followed by removal of the siloxane at C21 and C23 furnishing ester **34** (Scheme 4). The hydrolysis of **34** afforded the suitably protected seco-acid **35**. Paterson's group attempted the direct dimerization of **35** to form desired 44-membered macrodiolide by Yamaguchi's method [50] and Keck's method [51]; however, a mixture of monomeric 22-membered macrolide linked at C21 and 24-membered macrolide at C23 was generated and neither dimer nor trimer product was observed. Thus the stepwise esterification reactions were explored. The intermolecular esterification of **12** and **34** furnished a 2:1 mixture of dimeric products acylated at C21 and C23. After further 3 steps, Yamaguchi esterification led the desired macrodiolide and removal of the protective group provided swinholide A.

3.2 *Nicolaou's Total Synthesis of Swinholide A*

Nicolaou and co-workers achieved the second total synthesis of swinholide A based on the convergent strategy involving the union of protected monomeric compounds

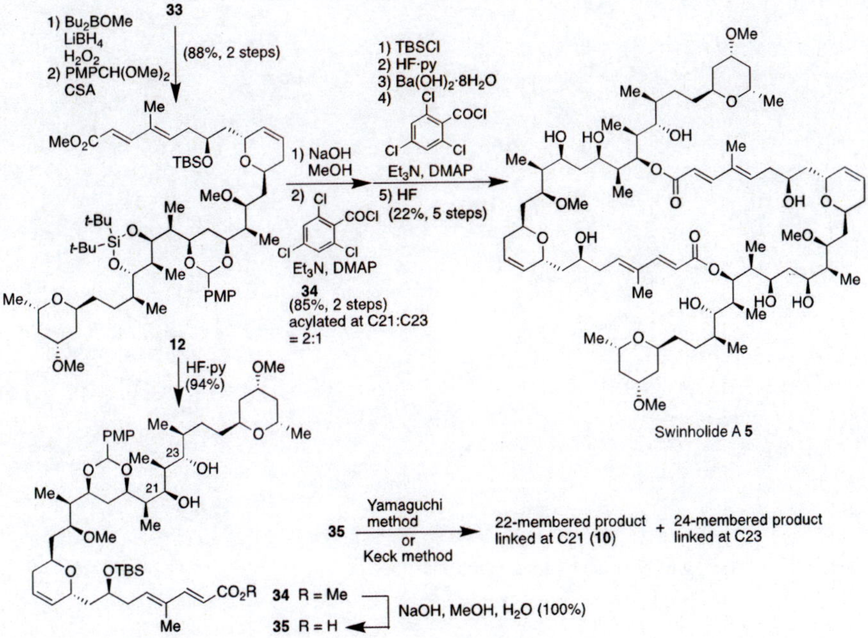

Scheme 4 Accomplishment of total synthesis of swinholide A

derived from **36** (Scheme 5). Compound **36** was prepared from two segments **37** and **38**, which were synthesized by taking advantage of ingenious aldol type couplings [52–55].

Nicolaou's total synthesis commenced with Ghosez cyclization [56] in which the lithiation of sulfone **39** was coupled with epoxide **40**, followed by acidic treatment to lead desired product **41** in excellent yield (Scheme 6). DIBAL-H reduction of **41** and Hosomi–Sakurai reaction of the resulting hemiacetal furnished **42** [57, 58], which was transformed into aldehyde **43** in 2 steps. The vinylogous Mukaiyama reaction of **43** and the silyl enol ether furnished a 5:1 ratio of the desired α-product **44** and its epimer [59–62]. Conversion of **44** to compound **38** was carried out in 6 steps including installation of dithiane.

On the other hand, the synthesis of another segment **37** began with tetrahydropyran **45**, which was derived from L-rhamnose (Scheme 7). The titanium-mediated aldol reaction of ketone **45** and aldehyde **46** provided a 3:1 mixture of the desired **48** and its C21-epimer via titanium-mediated transition structure **47**. The resulting **48** was transformed into cyclic sulfonate **37** in 5 steps. The coupling of **37** and dithiane fragment **38** was accomplished in 72% in 2 steps, leading **49** [63–65]. Compound **49** was transformed into segment **36** in further 7 steps and hydrolysis of **36** and protection of the C23 alcohol furnished another building block **50**. The coupling of the two monomeric compounds, **36** and **50**

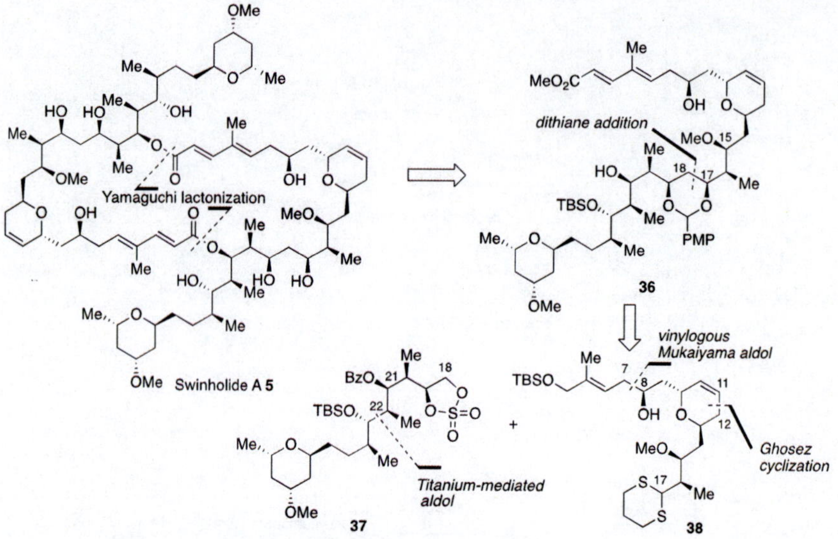

Scheme 5 Strategy of Nicolaou's total synthesis

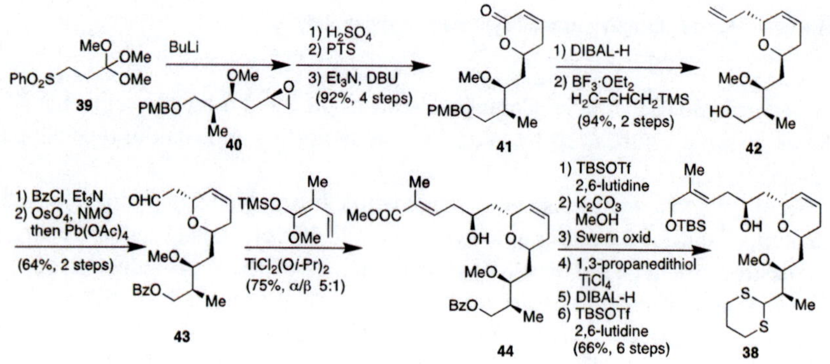

Scheme 6 Synthesis of segment **38**

bearing differentiable protective groups on the alcohol functions was performed by Yamaguchi method and removal of the protective groups led swinholide A.

3.3 *Krische's Total Synthesis of Swinholide A*

Krishche reported the total synthesis of swinholide A in only 15 steps for the longest linear sequence, which involves diverse hydrogen-mediated C–C bond formation [66–69]. They have developed catalytic carbonyl reductive couplings induced via hydrogenation or hydrogen auto-transfer. By virtue of the highly chemo- and

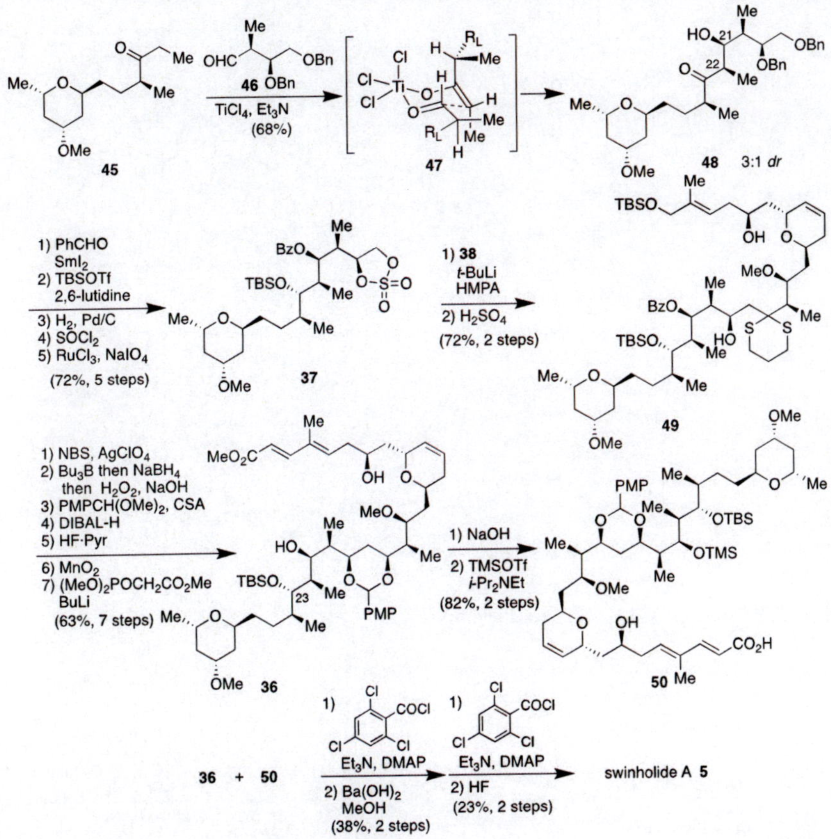

Scheme 7 Accomplishment of synthesis of swinholide A

stereoselectivities of these reactions, enantioselective C–C bond formation can be achieved in the absence of protecting group. In their strategy of total synthesis of **5**, swinholide A would be accessed through a direct macrodiolide formation by cross-coupling of monomeric compound **51**, which would be synthesized via Mukaiyama aldol reaction of **52** and **53** (Scheme 8). Each polyketide **52** and **53** would be prepared in high selectivity through their catalytic carbonyl reductive couplings.

Krische's synthesis of segment **52** began with enantioselective preparation of **56** (Scheme 9). When alcohol **54** was exposed upon iridium-catalyst **55** and allyl acetate, *C*-allylation proceeded to furnished homoallylic alcohol **55** in 93% ee [70–72]. Cross metathesis of **56** with acrolein, followed by treatment of the resulting enal with allyl trimethylsilane, resulted in allylation leading *trans*-2,6-disubstituted pyran **57** with good diastereoselectivities [73–75]. Selective dihydroxylation of terminal olefin in **57** and the cleavage of the resulting diol furnished aldehyde **58**. When aldehyde **58** was subjected to vinylogous Mukaiyama aldol with **28**, a 5:1 mixture of desired aldehyde **59** and its epimer was generated, which is similar to Paterson's result. Resulting **59** was then transformed into aldehyde **61** in 3 steps. The

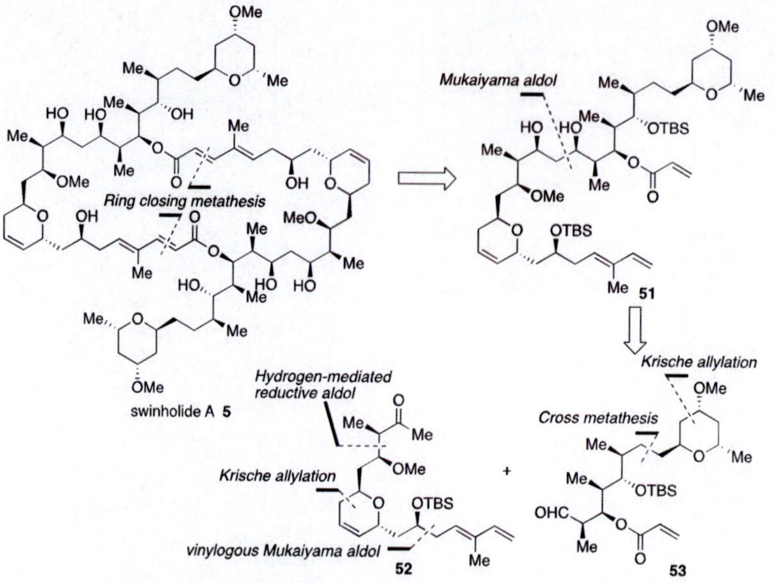

Scheme 8 Strategy of Krische's total synthesis

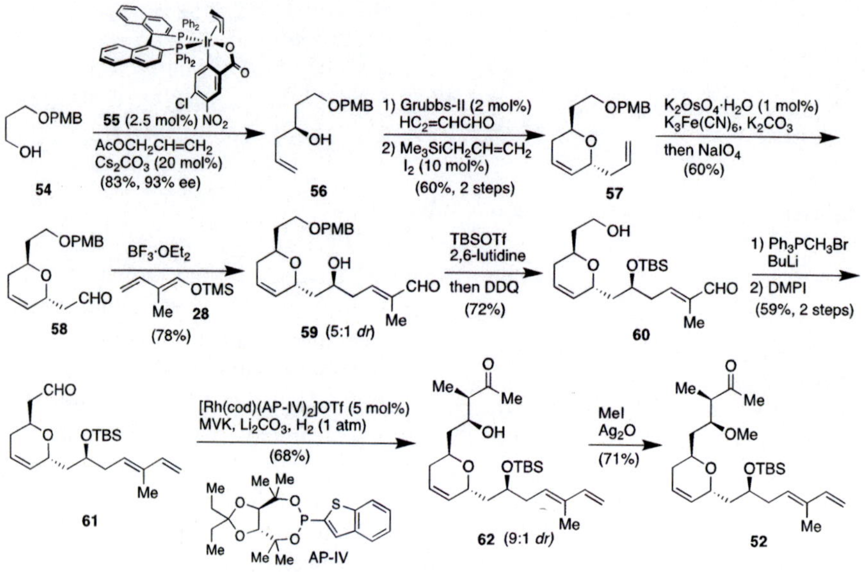

Scheme 9 Preparation of segment **52**

hydrogen-mediated reductive aldol coupling of **61** and methyl vinyl ketone with rhodium catalyst furnished **62** in 68% without hydrogenation of the other alkene and subsequent methylation of **62** provided segment **52** [76, 77].

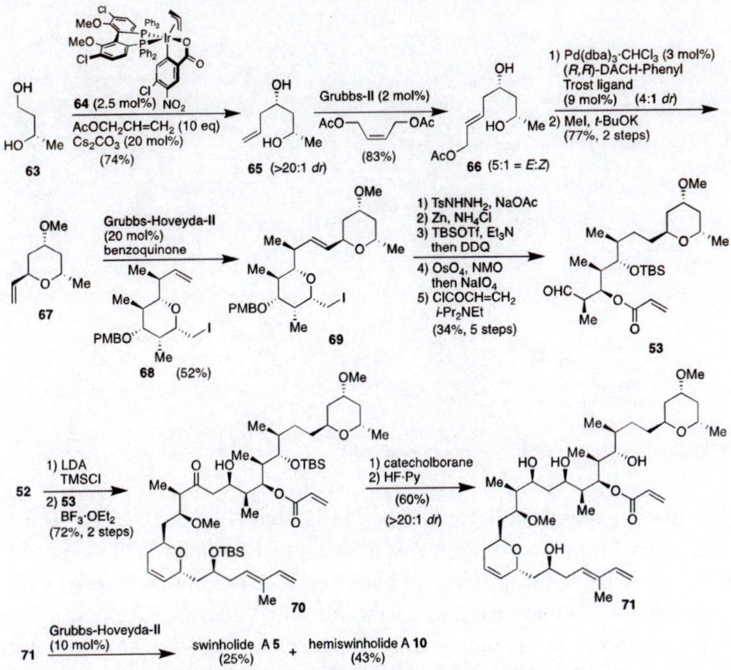

Scheme 10 Krische's total synthesis of swinholide A

On the other hand, exposure of (*S*)-1,3-butanediol **63** to iridium-catalyst **64** and allyl acetate results in diastereo- and site-selective allylation to form **56** (Scheme 10). The resulting homoallyl alcohol **65** was subjected to cross metathesis, followed by palladium-catalyzed Tsuji–Trost cyclization and methylation afforded **67**. The cross-coupling of 2,6-*trans*-substituted pyran **67** and alkenylpyran **68** provided **69** [78]. Diimide reduction of the alkene in **69** [78], and Bernet–Vasella cleavage of iodoether [79, 80], followed by functional group conversion in 3 steps to afford segment **53**. To merge segments **52** and **53**, Mukaiyama aldol reaction was again employed. Thus, segment **52** was transformed to the enol silyl ether and treated with **53** with $BF_3{\cdot}OEt_2$ to give **70**. The *syn* selective reduction and removal of the TBS group gave **71**. Exposure of **71** to the second generation Grubbs–Hoveyda catalyst provided swinholide A in 25% yield along with hemiswinholide A in 43% yield.

4 Clavosolide A

In 1998, Fu and co-workers isolated clavosines A-C from the extract of the sponge *Myriastra clavosa* collected from Palau [81]. Clavosine A is known to be a potent cytotoxin and inhibitor of protein phosphatase 1 and 2A. Furthermore, Faulkner and Rao isolated clavosolide A (**7**) and B-D (**72–74**) in trace quantities from the same

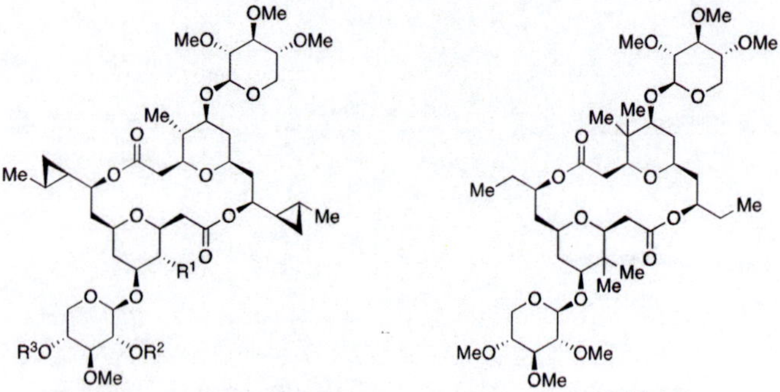

clavosolide A **7** $R^1 = R^2 = R^3 = Me$
clavosolide B **72** $R^1 = Me, R^2 = H, R^3 = Me$
clavosolide C **73** $R^1 = Me, R^2 = Me, R^3 = H$
clavosolide D **74** $R^1 = H, R^2 = R^3 = Me$

cyanolide A **9**

Fig. 3 Clavosolide A and related compounds

species of marine sponge collected in the Philippines (Fig. 3) [82, 83]. The structure initially proposed for clavosolide A was misassigned and later Will [84] and Lee [85] independently revised the structure of clavosolide A by total synthesis. Clavosolide A is C_2-symmetric 16-membered macrodiolide with two *trans*-cyclopropyl units and permethylated **D**-xylose moiety. On the other hand, Gerwick and co-workers isolated C_2-symmetric 16-membered macrodiolide, cyanolide A, from a collection of *Lyngbya bouillonii* in Papua New Guinea in 2010 [86]. Cyanolide is structurally closely related to clavosolides A-D. Clavosolides and cyanolide share a flat dimeric macrocyclic skeleton that is decorated with xylose appendages, and possess the same configurations for their stereocenters throughout the entire molecule. In this section, ten total and formal syntheses of clavosolide A will be discussed [87].

4.1 Lee's Total Synthesis of Clavosolide A

In 2006, Lee's group reported the enantioselective synthesis of clavosolide A and the determination of the absolute configuration of (+)-clavosolide A (Scheme 11) [88]. The key steps of their synthesis of **7** were the formation of the tetrahydropyran by intramolecular oxa-Michael reaction of hydroxyl unsaturated ester **83** and stereoselective aldol of cyclopropylketone **78** and aldehyde **79**. The starting material **75** was prepared from **D**-mannitol according to the literature [89]. Brown's asymmetric methallylation [90, 91] of **75** and protection of the alcohol with PMB group afforded **76**, which was transformed into ketone **78** in 4 steps. The aldol reaction of the boron enolate of **78** and aldehyde **79** proceeded with 1,5-anti-selectivity to afford **80** in excellent yield. The reduction of **80** with $Me_4NB(OAc)_3H$, followed by protection of the resulting diol, furnished **81**. Aldehyde **82**, derived from **81** in 2 steps, was subjected to Horner–Wadsworth–Emmons reaction to lead unsaturated **83**. Treatment of **83** with NaH raised to oxa-Michael reaction to furnish

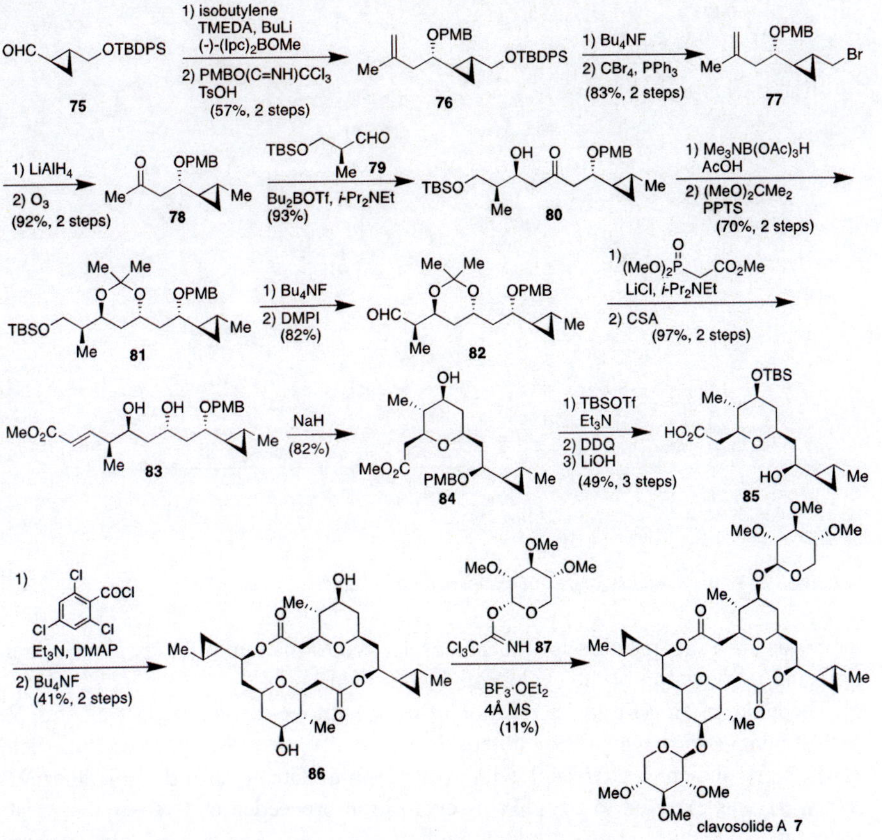

Scheme 11 Lee's total synthesis of clavosolide A

tetrahydropyran **84** in 82% yield [92–94]. Compound **84** was transformed into protected seco-acid **85** in 3 steps. When **85** was exposed to 2,4,6-trichlorobenzoyl chloride and Et_3N with DMAP, the macrodiolide formation successfully proceeded and subsequent removal of the TBS group provided **86**. Schmidt glycosylation [95, 96] of **86** with permethylated **D**-xylose derivative **87** [97] afforded clavosolide A in 11% yield.

4.2 *Willis's Total Synthesis of Clavosolide A*

At the almost same time of Lee's works, Willis's group independently completed the synthesis of the proposed structure of clavosolide A and confirmed the proposed structure was incorrect. In 2006, they also accomplished the total synthesis of the revised structure of (−)-clavosolide A (Scheme 12) [98–100]. The key steps of their

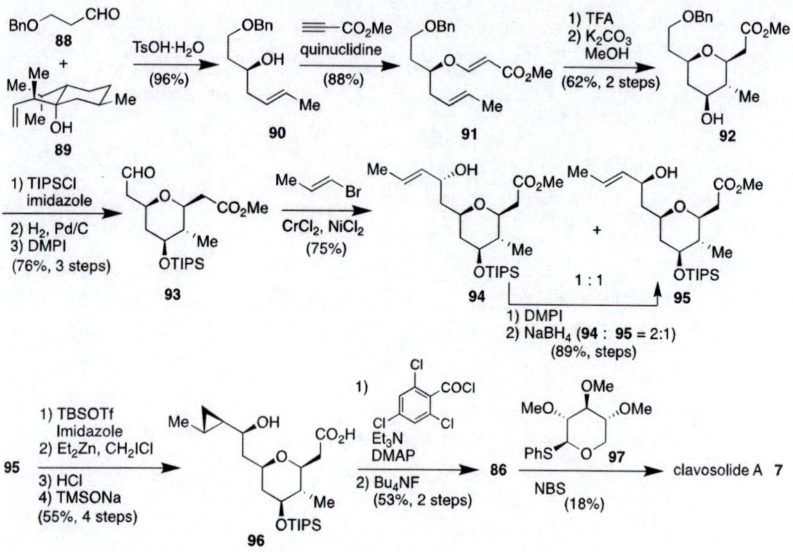

Scheme 12 Willis's total synthesis of clavosolide A

synthesis are Prins reaction for the stereoselective formation of the tetrahydrofuran core and introduction of the cyclopropyl side chain.

The Nokami crotyl-transfer reaction of the menthone-derived tertiary alcohol **89** and 3-benzyloxypropanal **88** stereoselectively afforded **90** in excellent yield [101, 102], which was treated with methyl propiolate to afford enol ether **91**. When **91** was exposed to TFA, Prins cyclization proceeded and subsequent treatment of the cyclization product with K_2CO_3 in MeOH led the desired tetrahydropyran **92** [103, 104]. Compound **92** was transformed into aldehyde **93** in 3 steps, followed by Nozaki–Hiyama–Takai–Kishi coupling [105–110] with bromopropene to afford a 1:1 mixture of secondary allylic alcohols **94** and **95**. Undesired **94** could be converted to **95** via Dess–Martin oxidation and $NaBH_4$ reduction. The secondary alcohol of **95** was protected with a TBS group and the subsequent Simmons–Smith cyclopropanation using Et_2Zn and CH_2ICl [111] provided a 4:1 mixture of diastereomers in favor of the required anti product due to the bulkiness of the TBS group. Removal of the TBS group and hydrolysis of the methyl ester gave protected monomer **96**. Compound **96** was exposed to Yamaguchi macrolactonization, followed by removal of the TIPS to furnish the desired dimeric product **86**. The glycosylation with thioglucoside using a modification of Nicolaou's glycosylation protocol [112, 113] provided clavosolide A.

4.3 Smith's Total Synthesis of Clavosolide A

Smith and co-workers reported the total synthesis of clavosolide A, which involves the Petasis–Ferrier union/rearrangement and one-pot esterification/lactonization (Scheme 13) [114]. Their synthesis began with the construction of cyclopropyl aldehyde **101**. $TiCl_4$-mediated Nagao acetate aldol reaction employing **98** and crotonaldehyde gave allyl alcohol **99** [115]. Conversion of **99** into the corresponding Weinreb amide and subsequent Simmons–Smith reaction [116] furnished *syn*-cylclopropane as a single diastereoisomer, which was subjected to Mitsunobu inversion leading product **100** [117]. Silylation of secondary alcohol of **100** and DIBAL-H reduction afforded aldehyde **101**.

On the other hand, Abiko–Masamune boron-mediated *anti*-aldol of **102** and **103** delivered alcohol **104** with excellent selectivity [118] and the removal of the chiral auxiliary furnished hydroxy acid **105**. Bissilylation of **105** followed by TMSOTf-promoted condensation with aldehyde **101** furnished dioxanone **106** as a 7:1 mixture. The Petasis–Tebbe methylenation of **106** afforded unstable enol product **107**, which was exposed to Me_2AlCl and 4Å-MS to raise rearrangement to furnish tetrahydropyranone **108** in 65% yield for 2 steps [119–122]. Compound **108** was transformed into protected seco-acid **109** in 4 steps and Yamaguchi macrolactonization of **109** and subsequent removal of benzyl ether furnished macrodiolide **86**. Same as Lee's synthesis, the glycosylation of **86** with **87** provided clavosolide A.

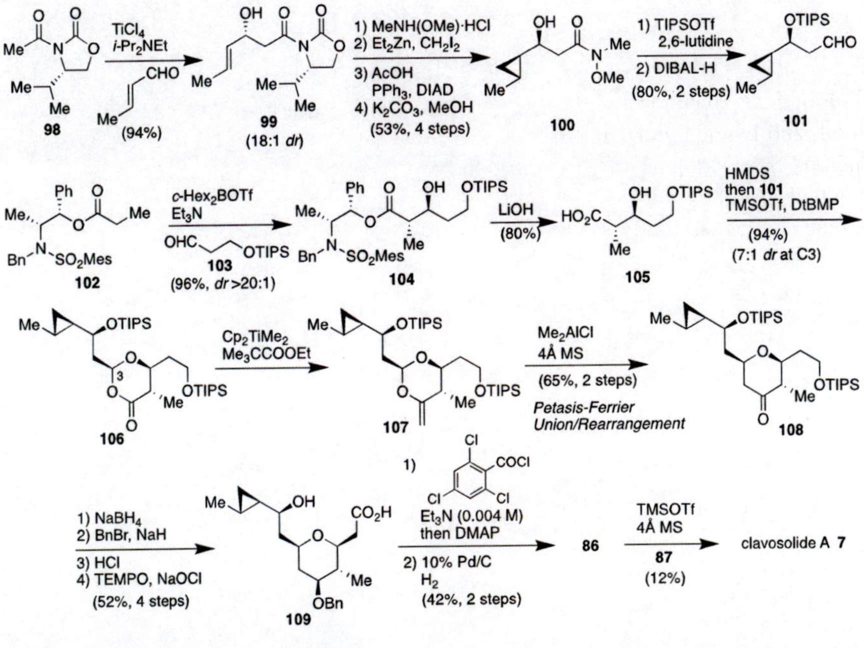

Scheme 13 Smith's total synthesis of **7**

4.4 Chakraborty's Total Synthesis of Clavosolide A

Chakraborty's group also reported total synthesis of clavosolide A (Scheme 14) [123–125]. The key points of their synthesis are the Schmidt glycosylation at an early stage in the synthesis and the direct coupling of monomer bearing glycoside **121**. Chakraborty's synthesis began with **110**, which was prepared by the Ti(III)-mediated opening reaction of a trisubstituted 2,3-epoxy alcohol [126]. Oxidation of **110**, followed by Wittig reaction and DIBAL-H reduction, gave allyl alcohol **111**.

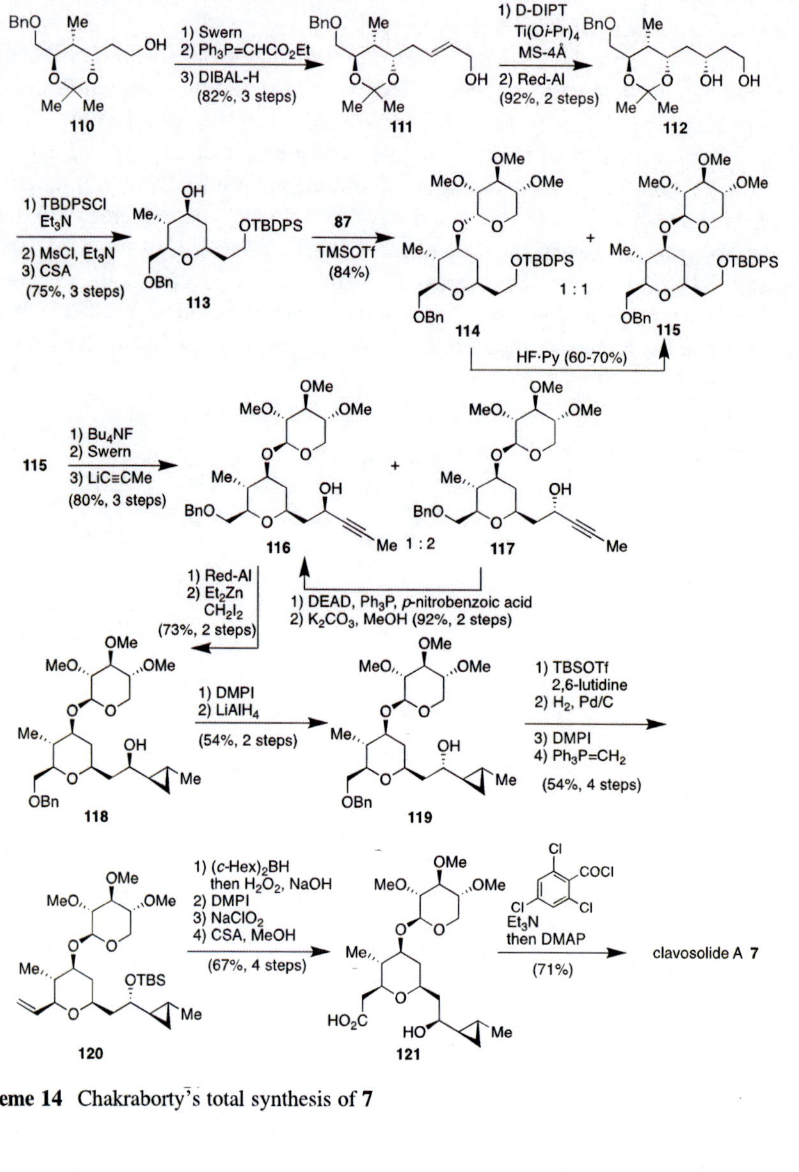

Scheme 14 Chakraborty's total synthesis of **7**

Sharpless epoxidation of **111** and Red-Al reduction afforded *syn*-diol **112**, which was then subjected to silyl protection of the primary alcohol, mesylation of the secondary alcohol, and acidic treatment to lead tetrahydropyran **113**. The glycosylation of **113** with **87** was performed under Schmidt condition, which delivered a 1:1 mixture of **114** and **115**. The undesired compound **114** could be converted to anomer **115** by exposure to HF·Py. Removal of TBPDS group and Swern oxidation gave aldehyde, followed by nucleophilic addition of acetylide to furnish a 1:2 mixture of **116** and its epimer **117**. Epimer **117** was convertible to **116** via Mitsunobu inversion. Red-Al reduction of **116** and subsequent cyclopropanation furnished *syn*-product **118** due to the directing effect of the secondary alcohol. The secondary alcohol was then inverted to **119** via DMPI oxidation and $LiAlH_4$ reduction. Compound **119** was transformed to monomeric seco-acid **121** in further 8 steps. Finally, treatment of **121** with Yamaguchi protocol raised to macrodiolide formation to furnish clavosolide A in 71% yield.

4.5 *Jennings's Formal Synthesis of Clavosolide A*

Jennings and co-workers reported the formal synthesis of clavosolide A, which is highlighted by one-pot oxocarbenium cation formation/reduction sequence (Scheme 16) [127]. Their synthesis commenced with aldol reaction of **122** and crotonaldehyde developed by Phillips to afford product **123** [128]. Compound **123** was transformed into Weinreb amide **124** and subsequent cyclopropanation furnished cyclopropane **125** with high selectivity [129, 130]. Mitsunobu inversion of the secondary alcohol in **125** and treatment with Grignard reagent afforded **126**. The reduction of **126** under Prasad protocol generated the 1,3-syn-diol [131], which, upon protection as its acetonide followed by ozonolysis, afforded aldehyde **127**. The boron-mediated Evans aldol of **127** with **128** furnished *syn*-aldol **129** [132, 133]. After protection of the secondary alcohol with a MOM group, exposure to BuOLi, followed by acidic treatment, furnished lactone **130**. The allylation of **130** with Grignard reagent gave hemiacetal **131**, which underwent tandem stereoselective oxocarbenium cation formation and reduction to afford **133**. Ozonolysis and oxidation generated **134**, followed by Yamaguchi lactonization and cleavage of MOM group provided macrodiolide aglycon **86**.

4.6 *Floreancig's Total Synthesis of Clavosolide A*

In the total synthesis by Floreancig and co-workers, the key steps involve an early-stage Feringa asymmetric cyclopropanation reaction and tetrahydropyran formation by oxidative C-H cleavage and subsequent electrophilic cyclization (Scheme 15) [134]. They assumed that the oxidative C-H cleavage of cyclopropane-containing molecules **136** would form oxocarbenium ion **137**, which could be trapped to form

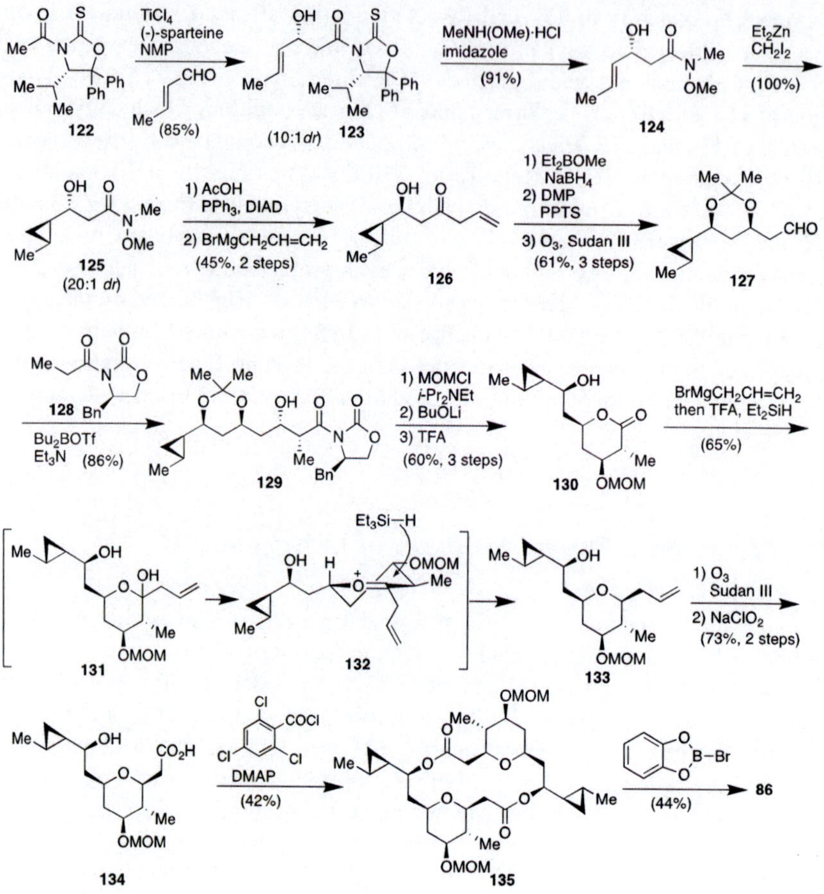

Scheme 15 Jennings's synthesis of 7

tetrahydropyrans **138** under non-acidic conditions [135, 136]. Cyclopropane compound **140** was prepared from by Feringa's asymmetric 1,4-addition [137]. The methyl ketone of **140** was converted to the terminal alkyne by Negishi's method [138], which, upon addition of paraformaldehyde and mesylation, furnished compound **141**. On the other hand, addition of **142** to aldehyde **143** under Marshall's conditions provided homopropargylic alcohol **144** with high diastereoselectivity [139]. Etherification of **141** and **144** and ruthenium-mediated addition of AcOH furnished compound **145** [140]. When **145** was exposed to DDQ, the oxidation of an acetoxy-substituted allylic ether and simultaneous cyclization proceeded stereoselectively to provide **146** as a single isomer. The reduction of **146**, followed by methanolysis of the enol acetate and glycosylation furnished **147**. The sequential 3 step-conversion afforded **121**, which underwent Yamaguchi lactonization to furnish clavosolide A (Scheme 16).

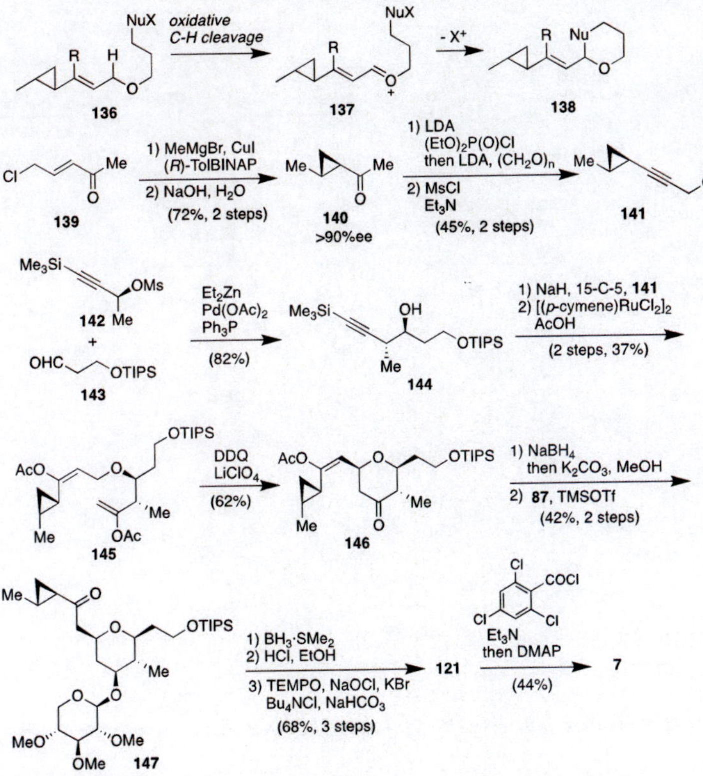

Scheme 16 Floreancig's total synthesis of **7**

4.7 *Hong's Total Synthesis of Clavosolide A*

Hong and co-workers achieved the total synthesis of clavosolide A utilizing tandem allylic oxidation/oxa-conjugate addition reaction as the key step (Scheme 17) [141]. They prepared epoxide **150** from known allylic alcohol **148** by Smith's procedure [142]. The coupling of **150** and dithiane **151** afforded **152**, which was exposed to MnO_2 and subsequent dimethyltriazolium iodide **153**, DBU, and MnO_2 to lead the desired tetrahydropyran **154** [143]. In this reaction, tandem allylic oxidation and oxa-conjugate addition/oxidation proceeded in one-pot. Cleavage of the dithiane, reduction of the resulting ketone, and glycosylation of **97** furnished a 3.8:1 mixture of **156** and its anomer. Simons–Smith reaction of **156** and removal of the TIPS group afforded **157** as a single isomer, which was subjected to hydrolysis and Shiina macrolactonization [144, 145] to provide clavosolide A.

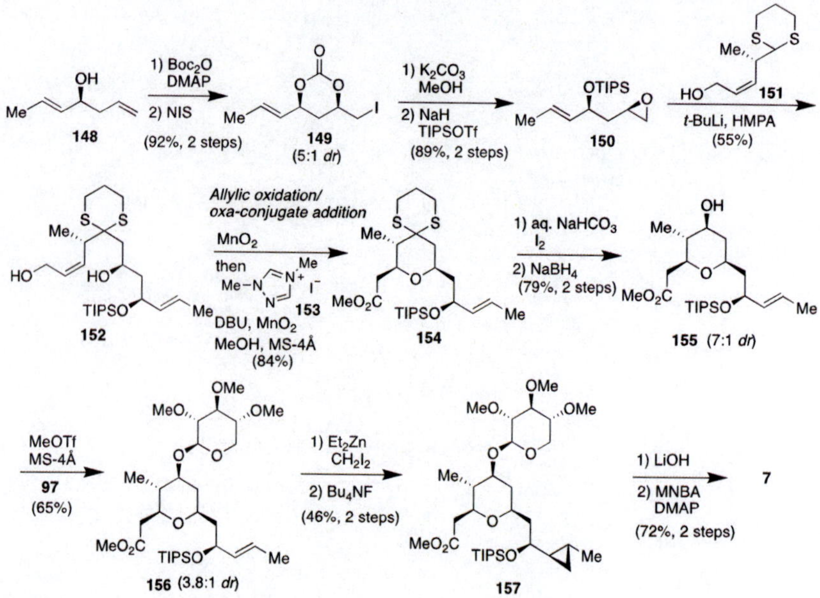

Scheme 17 Hong's total synthesis of **7**

4.8 *Breit's Total Synthesis of Clavosolide A*

Standard construction of macrodiolide requires stoichiometric amount of a reagent for the activation of the monomers. Breit's group has developed a rhodium-catalyzed atom-economical and regioselective addition of a carboxylic acid to an allene (Scheme 18) [146, 147]. They applied this rhodium-catalyzed reaction to synthesis of clavosolide A [148]. Treatment of known homoallenyl aldehyde **159** with Nokami crotyl-transfer reagent **158** furnished chiral allenyl alcohol **160**. Compound **160** was converted to the β-alkoxyacyclic acid ester, which was subjected to Prins cyclization followed by hydrolysis to afford tetrahydropyran **161**. The glycosylation of **161** with **97** generated a 77:23 mixture of **162** and its anomer and subsequent hydrolysis of **162** afforded allenyl acid **163**. Exposure of **163** to $[Rh(cod)Cl]_2$ and (*R*, *R*)-diop furnished head-to-tail dimer **164** in 92:8 *dr* [149]. The cross metathesis of **164** with (*Z*)-2-butene, followed by Simmons–Smith reaction [150] provided clavosolide A.

4.9 *Aggarwal's Total Synthesis of Clavosolide A*

Lithiation–borylation has emerged as a powerful tool for the synthesis of chiral boronic esters. Aggarwal's group achieved the total synthesis of clavololide A [151],

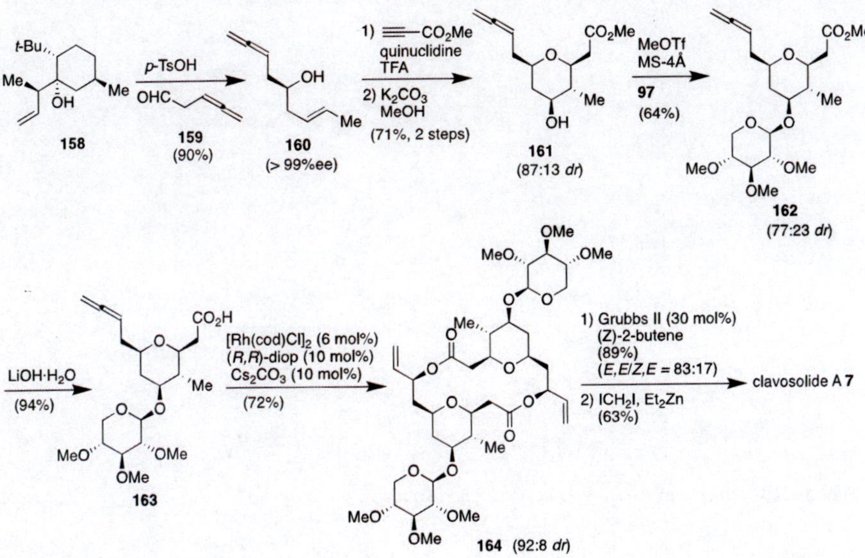

Scheme 18 Breit's total synthesis of **7**

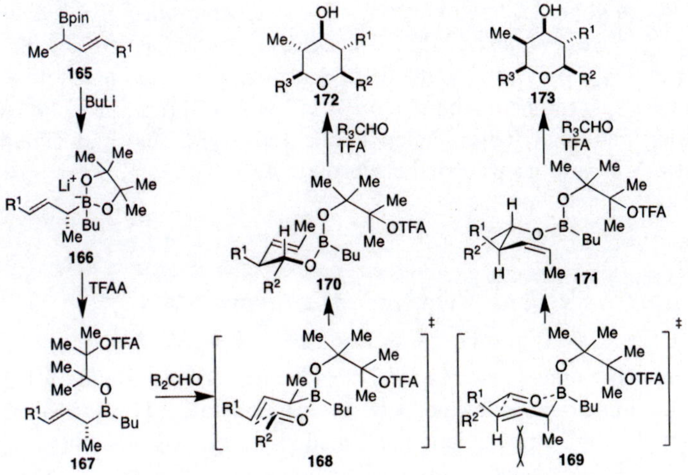

Scheme 19 Aggarwal's three-component allylboration/Prins cyclization

employing their three-component allylboration–Prins reaction [152, 153]. Treatment of allyl boronic ester with BuLi and TFAA generated boronic ester **165**, which was reacted with the first aldehyde to give intermediate **170** (Scheme 19). Following solvent exchange to CH_2Cl_2, subsequent reaction with the second aldehyde in the presence of TFA followed by base mediated hydrolysis furnished the hydroxy tetrahydropyran **172**. The allyl boration reaction would occur through more preferential Zimmerman–Traxler chair transition state **168** than **169** [154].

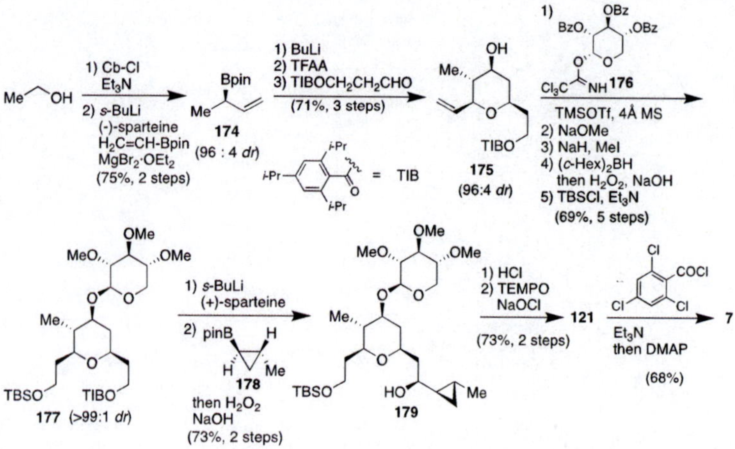

Scheme 20 Aggarwal's total synthesis of clavosolide A

Based on the key reaction, (*R*)-boronic ester **174** was prepared from ethanol in two steps (Scheme 20). The three-component allylboration–Prins reaction of **174** with 3-(2,4,6-triisopropylbenzoyloxy)propanal furnished tetrahydropyran **175** in high yield and good diastereoselectivity. For stereocontrolled glycosylation, xylose derivative **176** was employed [155]. Thus, the glycosylation of **175** with **176** in the presence of TMSOTf furnished the desired adduct, which was transformed into **177** in further 4 steps. Lithiation–borylation of **177** with **178** furnished **179**. Removal of TBS group and selective oxidation to the carboxylic acid and dimerization of Yamaguchi's conditions provided clavosolide A.

4.10 Krische's Total Synthesis of Clavosolide A

Krische and co-workers reported the 7 step-synthesis of clavosolide A [156], which involves asymmetric alcohol-mediated carbonyl allylation [157] (Scheme 21). The bidirectional asymmetric allylation of 2-methyl-1,3-propanediol **180** [158] followed by Fenton–Semmelhack alkoxypalladation–carbonylation of the resulting C_2-symmetric diol **181** furnished tetrahydropyran **182** [159]. The glycosylation of **182** with **87** catalyzed by BINOL phosphoric acid derivative (PA) displayed good β-diastereoselectivity to afford product **183** [160]. DIBAL-H reduction followed by nucleophilic addition of the chiral cyclopropyl lithium generated from **184** furnished a 4:1 mixture of desired **185** and its epimer. Exposure of **185** to $KMnO_4/NaIO_4$ effected chemoselective alkene oxidative cleavage [161] to form carboxylic acid and subsequent Yamaguchi macrolactonization furnished clavosolide A.

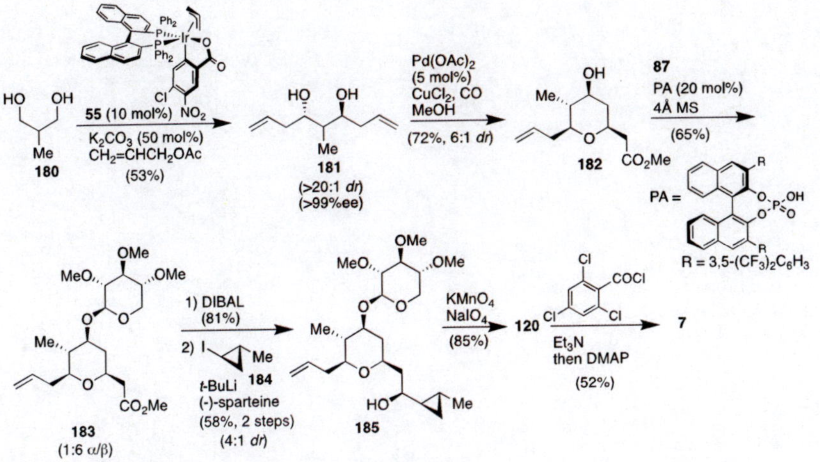

Scheme 21 Krische's total synthesis of clavosolide A

5 Cyanolide A

Cyanolide A was isolated in 2010 from a collection of *Lyngbya bouillonii* in Papua New Guinea by Gerwick and co-workers [86]. Cyanolide A (**9**) exhibits highly potent molluscicidal activity against the snail vector *Biomphalaria glabrata* (LC_{50} = 1.2 μM) but shows only modest toxicity toward brine shrimp (LC_{50} = 10.8 μM). The structure of cyanolide A was elucidated through extensive NMR spectroscopic analyses, showing it is closely related in structure to the clavosolide family. Cyanolide A is a C_2-symmetric macrodiolide consisting of a central 20-membered macrocycle fused to two tetrahydropyran moieties, each bearing a pendant xylose residue. Selected five syntheses will be discussed in this section for want of space; however, several other syntheses have been reported [162–165].

5.1 *Hong's Total Synthesis of Cyanolide A*

In 2010, Hong and co-workers reported the first total synthesis of cyanolide A [166]. The Hong's total synthesis was based on a tandem allylic oxidation/oxa-Michael reaction promoted by the *gem*-disubstituent effect and dimerization by Shiina method. This methodology was also employed for their total synthesis of clavosolide A. (cf. Sect. 4.7).

The coupling of known epoxide **186** and dithiane **187** furnished **188**, which was subjected to the tandem allylic oxidation/oxa-Michael reaction in the presence of **153** to furnish tetrahydropyran **190** stereoselectively. The oxa-Michael reaction proceeded through chair transition state **189**, leading 1,3-*syn*-tetrahydropyran **190**. Compound **190** was transformed into protected seco-acid **193** in 4 steps, which was

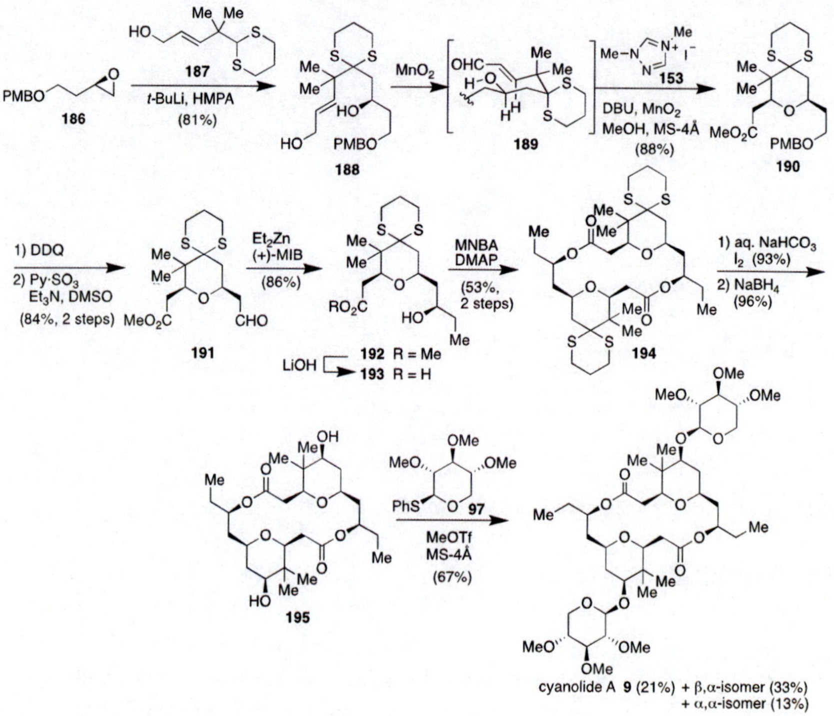

Scheme 22 Hong's total synthesis of cyanolide A

exposed to MNBA and DMAP, leading macrodiolide **194**. Cleavage of dithiane, reduction of the resulting ketone, and glycosylation of **97** provided a mixture of cyanolide A and its anomers (Scheme 22).

5.2 *Reddy's Synthesis of Cyanolide A Aglycon*

Reddy and co-workers accomplished the formal synthesis of (−)-clavosolide A [167], involving polyketide construction by Mukaiyama aldol reaction and tetrahydropyran formation by oxa-Michael reaction. (−)-Pantolactone was transformed into **196** in 6 steps and the following Mukaiyama aldol reaction of **196** and silyl enol ether **197** furnished **198** with high selectivity (>95% *dr*). Compound **198** was then treated with catecholborane under Evans conditions to raise 1,3-*syn*-reduction [168] and subsequent protection of the resulting diol afforded **199**. Compound **199** was subjected to removal of the TBS group, Swern oxidation, and Wittig reaction to furnish product **200**. When **200** was exposed to PTS·H_2O, removal of the acetonide and subsequent oxa-Michael cyclization proceeded in one-pot to afford tetrahydropyran **201** in high stereoselectivity. Hydrolysis of the

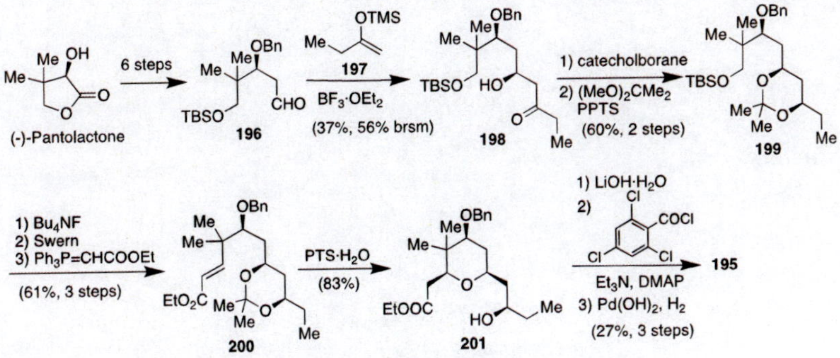

Scheme 23 Reddy's synthesis of cyanolide A aglycon

ethyl ester of **201**, followed by Yamaguchi protocol and cleavage of the benzyl group furnished cyanolide A aglycon **195** (Scheme 23).

5.3 Rychnovsky's Synthesis of Cyanolide A Aglycon

Rychnovsky's group has developed Prins dimerization/macrocyclization strategy to form tetrahydropyran-containing macrodiolides [169]. Based on this method, they completed the synthesis of cyanolide A aglycon utilizing Sakurai macrocyclization/dimerization reaction without the use of protecting groups [170, 171]. Readily available **202** was converted to the methyl ketone [172], from which enol triflate **203** was obtained in good yield. Kumada coupling and hydrolysis of the acetal furnished aldehyde **204**. The aldol reaction of **204** using Nagao's auxiliary under Sammakia's conditions provided the aldol adduct as a single diastereomer [173], which was then hydrolyzed to **205**. Esterification of **205** with **206** under Yamaguchi protocol afforded ester **207**. When **207** was treated with TMSOTf, Sakurai macrocyclization/dimerization reaction smoothly proceeded to furnish desired **208** in 78%. Oxidative cleavage of alkenes and subsequent diastereoselective reduction afforded cyanolide A aglycon **195** (Scheme 24).

5.4 Bates's Total Synthesis of Cyanolide A

Bates and co-workers reported synthesis of cyanolide A [174], which involves a diastereoselective Barbier reaction, early-stage glycosylation, and intramolecular oxa-Michael addition to form the THP ring. Nucleophilic addition of lithium acetylide to epoxide **209**, followed by Lindlar hydrogenation, afforded **210**. Cross metathesis of **210** with methyl acrylate furnished **211**, which was transformed into aldehyde **212** in 3 steps. Barbier reaction of **212** with prenyl bromide in the presence

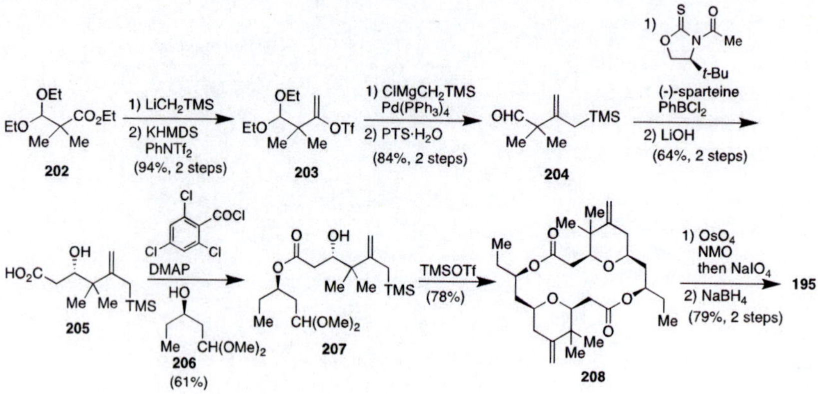

Scheme 24 Rychnovsky's synthesis of cyanolide A aglycon

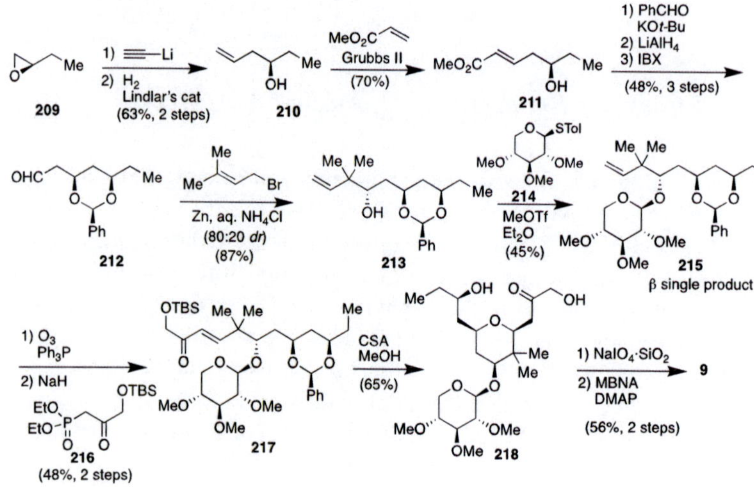

Scheme 25 Bates's total synthesis of cyanolide A

of zinc [175] provided an 80:20 mixture of **213** and its epimer. Glycosylation of **213** with **214** afforded **215**, which was transformed into **217** in 2 steps. Removal of the benzylidene acetal and subsequent oxa-Michael reaction furnished **218** (Scheme 25).

Treatment of compound **218** with $NaIO_4{\cdot}SiO_2$ yielded the seco-acid, which was subjected to Shiina macrolactonization to provide cyanolide A.

5.5 *Zhou's Total Synthesis of Cyanolide A*

Zhou's group developed the highly diastereo- and enantioselective construction of 1,3-polyols via iridium-catalyzed asymmetric hydrogenation of β-alkyl-β-keto esters [176]. They completed a formal total synthesis of cyanolide A based on this protocol [177]. The enantioselective hydrogenation of β-keto ester **220** in the presence of Ir–SpiroSAP ((*R*)-**219**) and *t*-BuOK proceeded under 100 atm of H_2 delivered alcohol **221** in 98% yield with 98.5% ee. The protection of alcohol **221** with a MOM group, followed by hydrolysis and Knoevenagel reaction furnished β-keto ester **222**. The iridium-catalyzed hydrogenation of compound **222** delivered alcohol **223** with high diastereoselectivity. The resulting compound **223** was transformed to alcohol **226** through the iterative asymmetric hydrogenation. After installation of two methyl groups, reduction/oxidation and Wittig reaction furnished unsaturated ester **227**. Methylation of alcohol of **227** and subsequent acidic treatment raised to oxa-Michael reaction to afford tetrahydropyran **228**. Hydrolysis and Shiina macrolactonization, followed by removal of methyl group afforded cyanolide A aglycon **195** (Scheme 26).

6 Marinomycin A

In 2006, Fenical and co-workers disclosed the isolation of marinomycins A–C from the saline culture broth of a marine actinomycete, *Marinospora* strain CNQ-140, which was cultured from a sediment sample collected deep off the coast of La Jolla

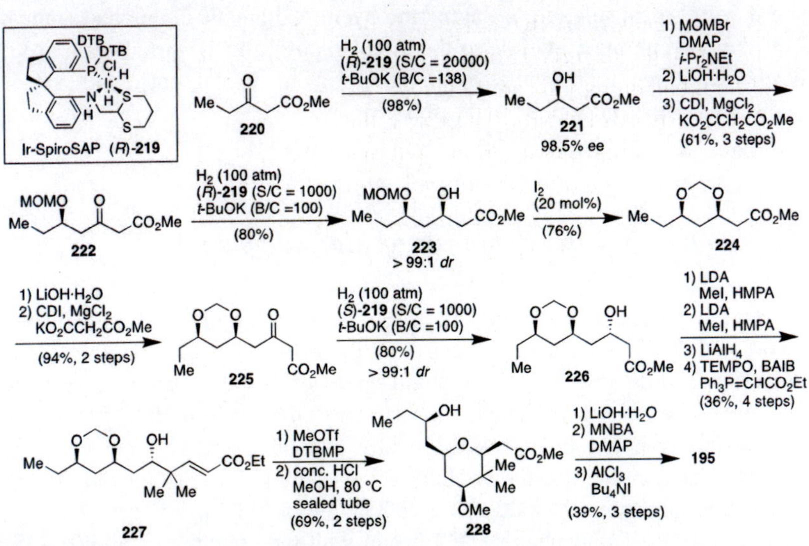

Scheme 26 Zhou's total synthesis of cyanolide A

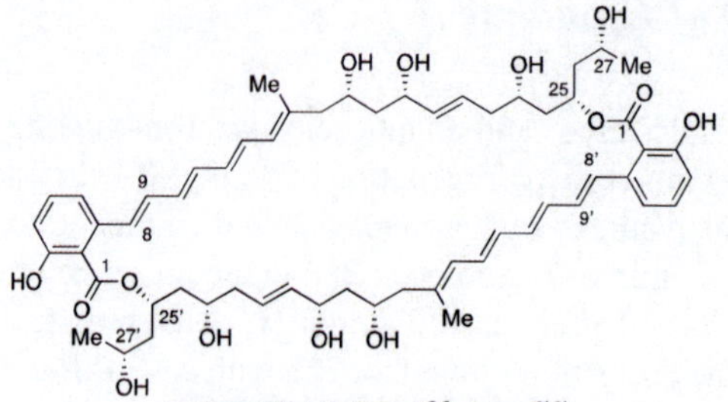

marinomycin A **8** (*trans* $\Delta^{8,9}$, *trans* $\Delta^{8',9'}$)
marinomycin B **229** (*cis* $\Delta^{8,9}$, *cis* $\Delta^{8',9'}$)
marinomycin C **230** (*trans* $\Delta^{8,9}$, *cis* $\Delta^{8',9'}$)

Photoinduced isomerization of marinomycins

marinomycin A ⇌ marinomycin C ⇌ marinomycin B

entry	time (min)	Ratio **8 / 229 / 230**
1	0	19 : 0 : 1
2	30	12.3 : 1 : 5.8
3	60	3.6 : 1 : 3.0
4	90	1.7 : 1 : 2.0
5	120	1.1 : 1 : 1.5

Fig. 4 Marinomycin family

in California [178]. These natural products possess novel 44-membered C_2-symmetrical polyene–polyol macrodiolide architectures. Marinomycins exhibit potent antibiotic activity (MIC_{90} = 0.1–0.6 μM) against methicillin-resistant *Staphylococcus aureus* (MRSA) and vancomycin-resistant *Enterococcus faecium* (VREF), and inhibit cancer cell proliferation in the National Cancer Institutes 60 cell-line panel (LC_{50} = 0.2–2.7 μM) with excellent selectivity against six of the eight melanoma cell lines. The significant biological properties and structural challenges have made the marinomycins attractive targets for synthesis.

It has been shown that marinomycin A photochemically isomerizes to marinomycins B **229** and C **230**. The isomerization ratio changed in favor of marinomycin C **229**, reaching a value of ca. 1.1:1:1.5 (**8/229/230**) after 2 h [179]. Despite the promising bioactivity of many agents, the photo-instability of the polyene motif would hamper their development as drug candidates. Evans found that the SpEC encapsulation of the marinomycin A dramatically increases the half-life of the polyene macrodiolide to the direct exposure to UV irradiation [180]. In this chapter, three total synthesis of marinomycin A and one synthesis of its monomeric compound will be discussed (Fig. 4).

6.1 *Nicolaou's Total Synthesis of Marinomycin A*

First total synthesis of marinomycin A has been accomplished by Nicolaou and co-workers [179, 181]. Their synthetic strategy involves Suzuki coupling of monomeric esters and the construction of chiral induction based on Brown's asymmetric allylation (Scheme 27). Based on this strategy, the macrodiolide framework of marinomycin A could be retrosynthetically constructed by Suzuki coupling of ester **231**. The monomeric ester would be accessed by HWE reaction and Mitsunobu reaction from phosphonate **232**, aldehyde **233**, and acid **234**.

The synthesis of phosphonate **232** began with the coupling of alkyne **235** and epoxide **236** (Scheme 28). The protection of alcohol **237** with a TBS group, removal

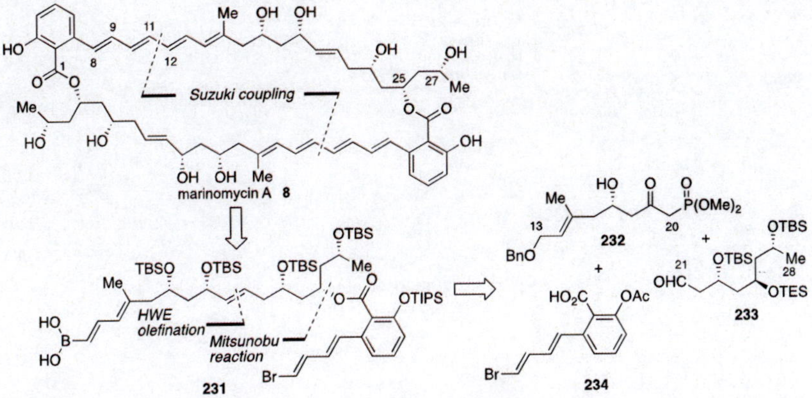

Scheme 27 Strategy of Nicolaou's total synthesis of marinomycin A

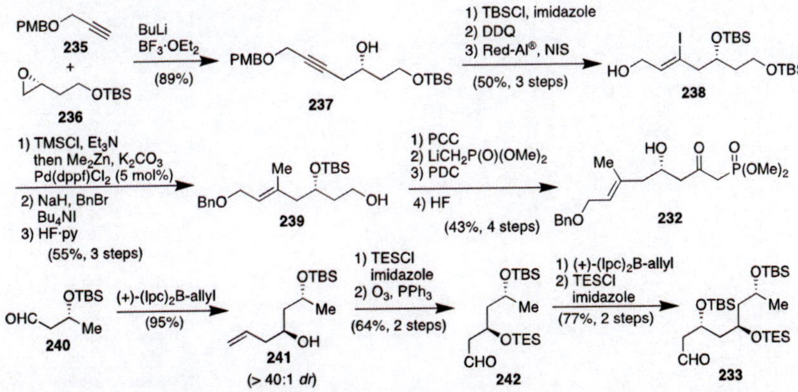

Scheme 28 Nicolaou's synthesis of marinomycin segment

of the PMB group, and hydroalumination with Red-Al followed by iodination furnished **238**. The primary alcohol of **238** was protected as its TMS ether, and Negishi coupling [182–184] of the vinyl iodide with Me_2Zn in the presence of palladium catalyst, followed by further 2 steps afforded product **239**.

Compound **239** was transformed in 4 steps into the desired phosphonate **232**. On the other hand, treatment of **240** with (+)-$(Ipc)_2$B-allyl raised Brown allylation to produce **241** with excellent selectivity. The protection of the secondary alcohol with a TES group and subsequent ozonolysis of the terminal alkene afforded aldehyde **242**. Brown's allylation of **242** and silyl protection furnished aldehyde **233**.

The resulting aldehyde **233** was then subjected to HWE reaction with **232**, followed by Evans reduction and removal of the benzyl group to afford **243** (Scheme 29). The allyl alcohol was transformed into the terminal alkyne in 3 steps to lead **244**. The coupling reaction of **243** and **244** was performed under Mitsunobu reaction,

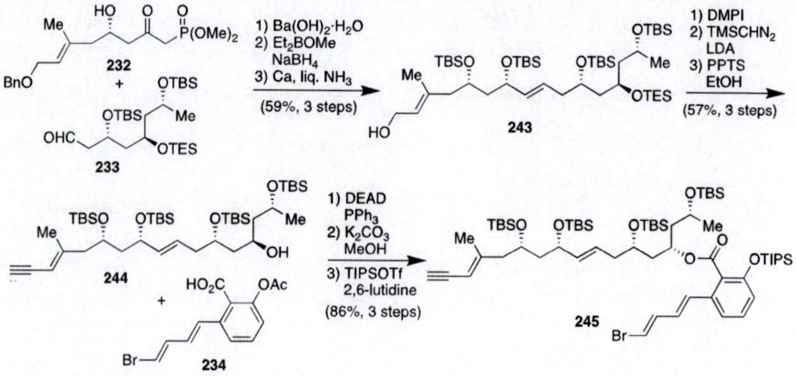

Scheme 29 Nicolaou's synthesis of marinomycin segment

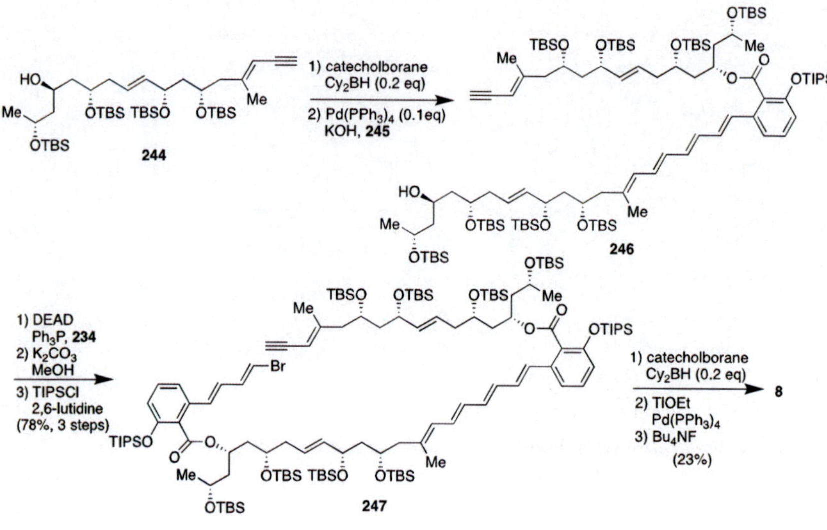

Scheme 30 Nicolaou's total synthesis of marinomycin

subsequent hydrolysis of the acetyl group, silylation of phenol by TIPS delivered monomeric compound **245**.

Nicolaou's group initially attempted to directly merge monomers by Suzuki coupling after hydroboration of the terminal alkyne of **245**; however, only intramolecular coupling proceeded to afford the corresponding 22-membered products. Next, the stepwise approach was explored (Scheme 30). Hydroboration of **244** and subsequent Suzuki coupling with **245** furnished **246**. Mitsunobu reaction of **246** with **234** and replacement of the acetyl group in the phenol by a TIPS group afforded **247**. Finally, hydroboration of **246**, Suzuki coupling using a catalytic amount of Pd $(PPh_3)_4$ and TlOEt [185], and removal of the silyl group provided marinomycin A.

6.2 *Evans's Total Synthesis of Marinomycin A*

As mentioned in Sect. 2, the macrolide framework is constructed from the acyclic polyketide bound to the acyl carrier protein (ACP) by a thioester, through a thioesterase-domain-mediated transesterification and concomitant cyclization with a specific hydroxy group present in the polyketide. Evans proposed that the salicylate serving as a molecular switch would play an important role for macrolactonization involving an ACP thioester and thioesterase. Based on this salicylate switch to modulate the reactivity of the ester, Evans's group completed the total synthesis of marinomycin A [186] (Scheme 31). In the synthetic strategy, marinomycin A would be accessed from monomeric compound **248**, *syn*-1,3-dioxane of which would provide a conformational lock to enforce the linear chain conformation to suppress the intramolecular cyclization. Compound **248** would be merged from three components, **249**, **250**, and **251** through cross-coupling and HWE reaction.

Ring opening reaction of **252** with the lithium anion of 1,3-dithiane and subsequent protection of the resulting alkoxide with MOMCl furnished the mono-substituted dithiane [187] (Scheme 32). The mono-substituted dithiane was lithiated with BuLi and reacted with **253** to raise ring opening reaction/closure to lead the corresponding epoxide. The epoxide thus obtained was reacted with vinylmagnesium bromide in the presence of CuI to furnish **254** in good yield. Compound **254** was transformed into product **249** in 4 steps through SmI_2-mediated stereoselective reduction under Keck condition [188]. On the other hand, Appel iodination of 1-*O*-methyl-2-deoxy-D-ribose **256** and protection of the secondary alcohol with a TES group afforded **257**. Vasella fragmentation [79] of the primary alkyl iodide with zinc and subsequent Wittig reaction afforded **258**.

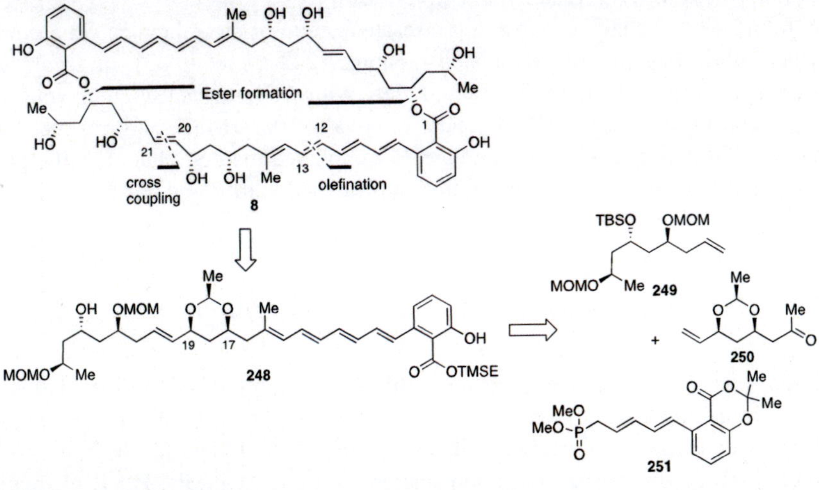

Scheme 31 Strategy of Evans's total synthesis of marinomycin A

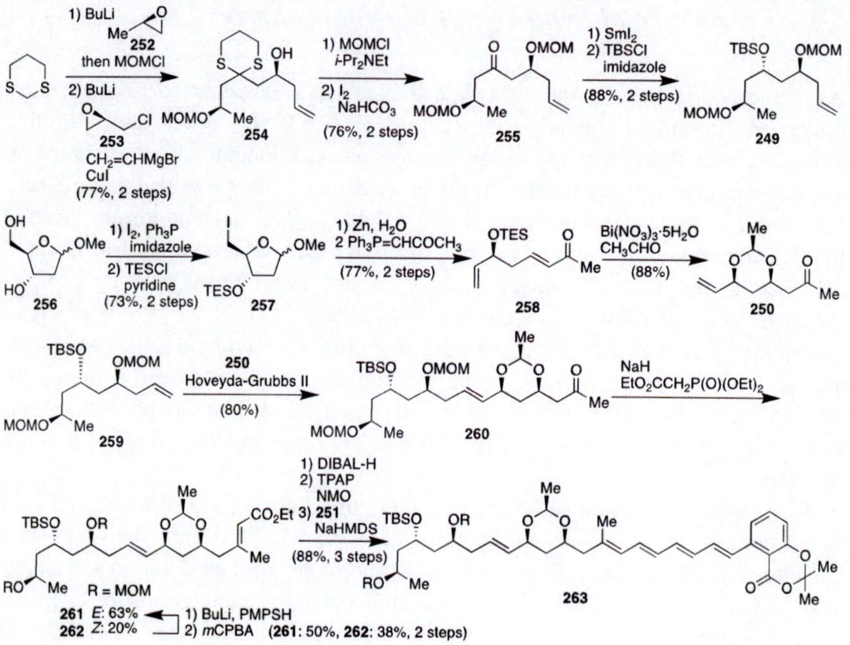

Scheme 32 Evans's total synthesis of marinomycin A

Bismuth-mediated hemiacetal/oxa-conjugate addition of acetaldehyde to **258** furnished *syn*-1,3-dioxane **250** [189]. Cross metathesis of **250** and **259**, followed by HWE reaction afforded a mixture of *E*-product **261** and *Z*-product **262**. Undesired **262** could be isomerized to **261** under 4-methoxybenzenethiol followed by the oxidation. After DIBAL reduction of **261** and oxidation, the resulting aldehyde was subjected to HWE reaction to deliver **263**.

Removal of the silyl group and transesterification of the salicylic acid acetonide with TMSEOH provided deactivated monomer **248** (Scheme 33). In accord with their hypothesis, transesterification of **248** with **263** furnished compound **264**. Protection of **264** with a MOM group, removal of the TMSE, and treatment with modified Mukaiyama salt furnished the 44-membered macrodiolide. Finally, global deprotection of the resulting macrodiolide furnished marinomycin A.

6.3 Cossy's Synthesis of Marinomycin A Monomer

Cossy's group reported an efficient and highly convergent synthesis of the monomeric counterpart of marinomycin A [190, 191] (Scheme 34). The strategy involves the chiral titanium complexes-mediated asymmetric allylation which they have developed, highly stereo- and regio-selective cross metathesis, HWE olefination, and Stille cross-coupling. In order to synthesize this molecule by dimerization of

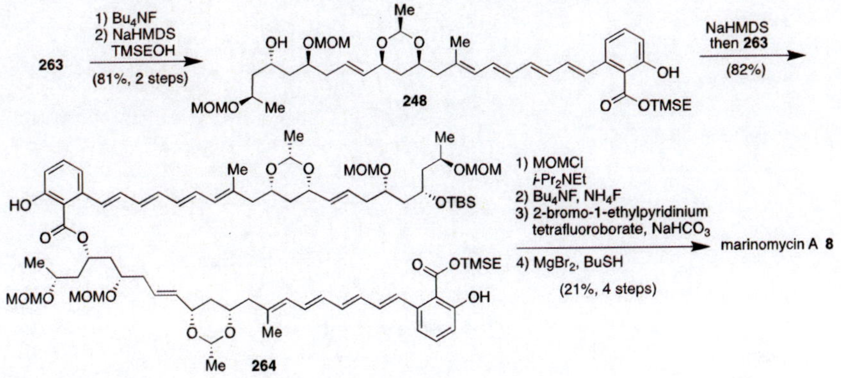

Scheme 33 Evans's total synthesis of marinomycin

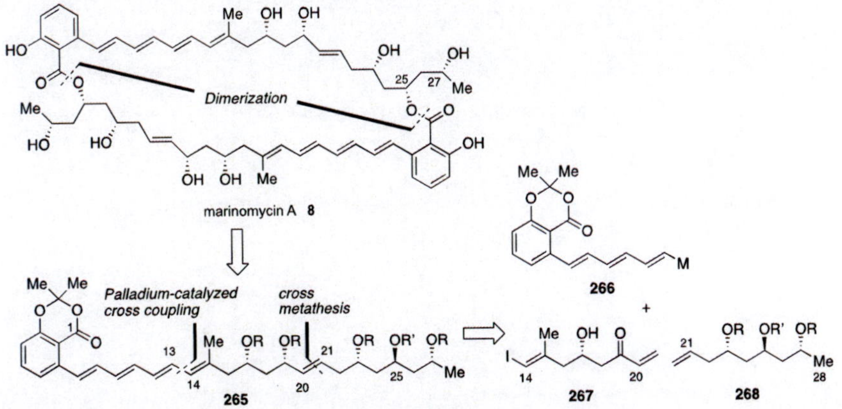

Scheme 34 Strategy of Cossy's synthesis of marinomycin A monomer

esters, their efforts were concentrated toward the synthesis of monomeric counterpart **265**. Monomer **265** could be accessed by performing a palladium-catalyzed cross-coupling of trienic vinyl metal **266** and vinyl iodide **267** for the formation of the C13-C14 bond, and cross metathesis of alkene **268** with α,β-unsaturated **267** for the construction of the C19-C20 bond.

For the preparation of trienic vinyl metal **266**, phosphonate **271** was synthesized by a 3-step sequence involving Stille coupling between aryl triflate **269** and vinylstannane **270**, bromination, and Arbuzov reaction (Scheme 35). HWE olefination of **271** and **272** afforded trienic stannane **273**. On the other hand, 3-butyn-1-ol was converted to **274** by regio- and stereo-selective carboalumination employing Cp_2ZrCl_2/Me_3Al [192, 193], followed by iodination and oxidation. Treatment of **274** with Hafner–Duthaler complex (*S*,*S*)-**275** [194–196] raised highly enantioselective allylation and compound **276** was produced after protection of the alcohol as a PMB-ether. Oxidative cleavage of the terminal alkene **276** gave

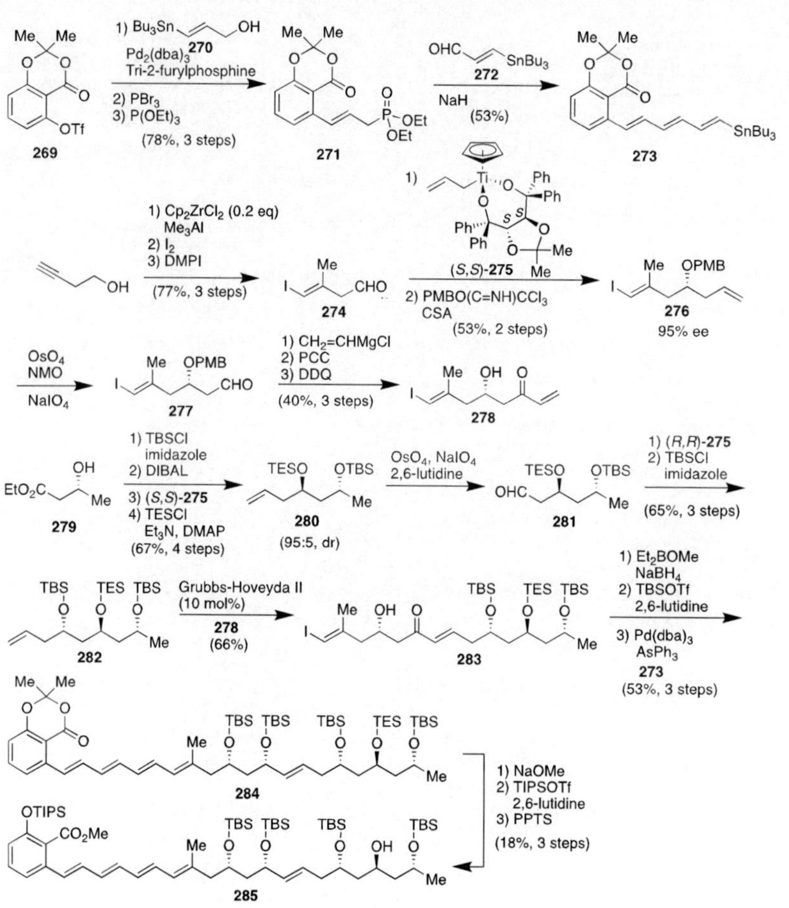

Scheme 35 Cossy's synthesis of marinomycin A monomer

aldehyde **277**, which was transformed to **278** in 3 steps. On the other hand, (*R*)-β-hydroxybutyrate **279** was converted to **280** in good yield by a 5-step sequence involving protection as a TBS ether, DIBAL reduction allyl titanation, and TES protection. Oxidative cleavage of the terminal alkene afforded aldehyde **281**, which was treated again with (*R*,*R*)-**275** to furnish **282** after silyl protection. Cross metathesis of **282** with **278** delivered **283**. Stereoselective reduction of **283**, silyl protection, and Pd-mediated coupling with **273** provided **284**. Compound **284** was subjected to methanolysis and subsequent protection of the resulting phenol as a TIPS ether, which was followed by removal of the TES group to provide the monomeric counterpart of marinomycin A **285**.

6.4 Hatakeyama's Total Synthesis of Marinomycin A

Hatakeyama also completed the total synthesis of marinomycin A [197] (Scheme 36). Based on their conformational studies, the protected monomeric compound **286** was selected for dimerization. The molecular mechanical calculations suggested that the intramolecular cyclization of **286** to the 22-membered macrolide was unlikely because of the remote distance between the salicylic ester carbonyl and hydroxy group (7.562 Å) in this molecule, whereas the formation of the corresponding 44-membered dimeric product looked favorable owing to the proximity of the reaction sites (3.489 Å) between C1 and the oxygen atom at C25' of dimeric compound **287**. Monomeric compound **286** could be accessed from the merge of dienyl bromide **288**, alkyne **290**, and epoxide **289** through Suzuki–Miyaura coupling and alkyne–epoxide coupling. Both alkyne **290** and epoxide **289** would be delivered from C_7 building block **291**.

Hatakeyama and co-workers have developed the preparation of enantiomerically pure C_7 building blocks with 1,2- and 1,3-*anti*-diol and 1,2- and 1,3-*syn*-diol functionalities from σ-symmetrical dialkenyl carbinol **293** [198–202]. Dialkenyl carbinol **293** was readily prepared by nucleophilic addition of the acetylyde

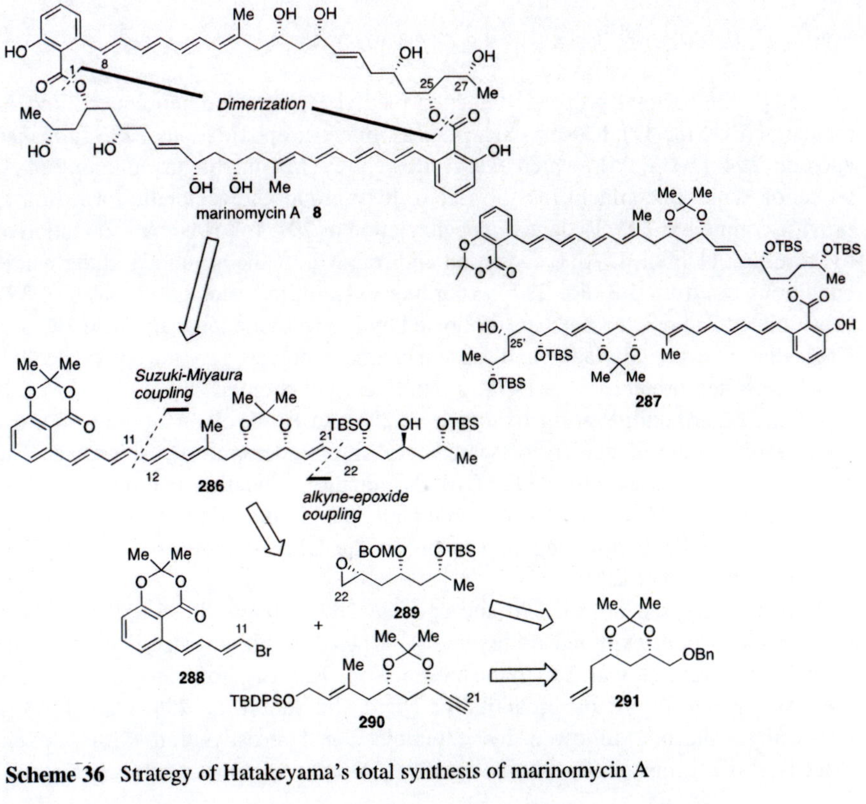

Scheme 36 Strategy of Hatakeyama's total synthesis of marinomycin A

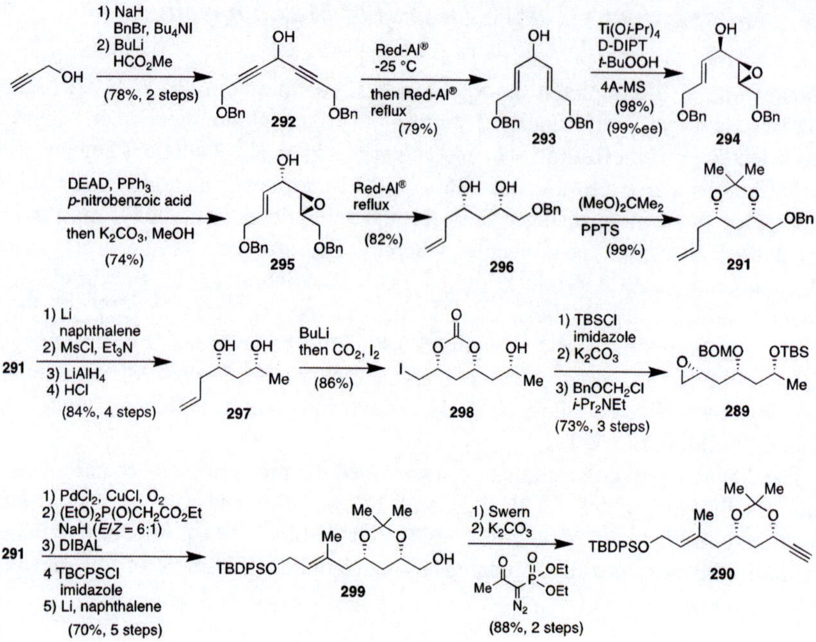

Scheme 37 Hatakeyama's total synthesis of marinomycin A

generated from benzyl propargyl ether to methyl formate and subsequent Red-Al reduction (Scheme 37). Katsuki–Sharpless asymmetric epoxidation of **293** furnished epoxide **294** [203–205], which was followed by Mitsunobu inversion, Red-Al reduction with concomitant loss of a benzyloxy group and acetonide formation to provide compound **291**. Reductive debenzylation of **291**, followed by mesylation of the resulting alcohol, $LiAlH_4$ reduction and removal of the acetonide under acidic conditions to afford 1,3-diol **297**. According to Cardillo's method [206, 207], **297** was lithiated and reacted with CO_2 followed by I_2 to provide iodocarbonate **298** as a single diastereomer. Silylation and subsequent methanolysis gave the epoxy alcohol, which was then protected as a BOM group to deliver epoxide **289**.

On the other hand, Wacker oxidation of **291** and HWE olefination afford a 6:1 mixture of *E*- and *Z*-isomers of which was chromatographically separated. The resulting *E*-isomer was subjected to DIBAL reduction, silylation, and debenzylation with lithium naphthalenide to lead compound **299** in good yield. Swern oxidation and Seyferth–Gilbert homologation by using the Ohira–Bestmann reagent [208–210] provided alkyne **290**.

With the required alkyne **290** and epoxide **289** in hand, the coupling of both segments was next explored (Scheme 38). When the lithium acetylide generated from **290** was reacted with **289** in the presence of $BF_3{\cdot}OEt_2$, homopropargyl alcohol **300** was given in almost quantitative yield. Silylation of **299** with 1,1,3,3-tetramethyldisilazane followed by intramolecular hydrosilylation with [Cp*Ru $(MeCN)_3]PF_6$ furnished the dihydrooxasiline, which was directly subjected to AgF

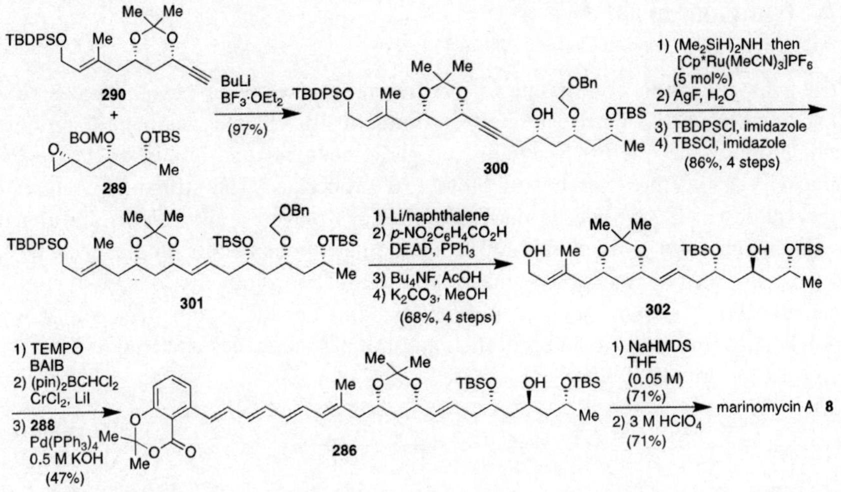

Scheme 38 Hatakeyama's total synthesis of marinomycin A

to afford the *E*-alkene [211]. In this reaction, TBDPS ether was partially cleaved, and thus the resulting primary alcohol was protected as a TBPDS group again. Protection of the secondary alcohol as the corresponding TBS ether to furnish product **301**. Removal of the BOM group of **301** and Mitsunobu inversion of the C25 hydroxy group afforded the 4-nitrobenzoate, which, upon selective desilylation and methanolytic removal of the 4-nitrobenzoyl group, delivered diol **302**. After TEMPO oxidation, Takai olefination [212, 213] of the resulting aldehyde with pinacol (dichloromethyl)boronate stereoselectively produced the *E*-boronate, which was then coupled with segment **288** under Suzuki–Miyaura reaction conditions to provide the monomeric hydroxy salicylate **286**.

To complete the total synthesis, Hatakeyama's group attempted the direct dimerization of monomer **286** under various transesterification conditions. When the reaction was conducted by adding NaHMDS, KHMDS, LHMDS, or LDA to a THF solution of **286** at −10 °C, the desired 44-membered macrodiolide was observed within 2 min but the desired product was yielded merely in 30%. However, reverse slow addition of **286** to NaHMDS in THF at −10 °C dramatically improved the dimerization reaction to furnish the desired macrodiolide in 71% yield. Treatment of the resulting macrodiolide with aqueous $HClO_4$ raised the removal of all of the protecting groups to cleanly furnish marinomycin A (**8**) in 71% yield.

7 Conclusion

The synthetic studies on marine C_2-symmetric macrodiolides have been described. These molecules comprise numerous synthetically challenging natural products, which attract the interest of synthetic organic chemists and biologists due to their unusual architecture and potent biological properties. The structural features of polyketides in C_2-symmetric macrodiolide have played a key role in drafting the synthetic chemistry and numerous outstanding synthetic methodologies were developed for the synthesis of macrodiolides. The synthesis of C_2-symmetric macrodiolides shown here would shed light on the synthetic chemistry of polyketides, macrolides, and biofunctional middle molecules and lead to the development for drug discovery and drug lead.

References

1. Faulkner DJ (2001) Nat Prod Rep 18:1R
2. Arai M (1960) J Antibiot 13:46
3. Ishibashi KJ (1961) Agric Chem Soc Jpn 35:257
4. Nozoe S, Hirai K, Tsuda K, Ishibashi K, Shirasaka M, Grove JF (1965) Tetrahedron Lett 6:4675
5. Fuska J, Nemec P, Kuhr I (1972) J Antibiot 25:208
6. Okami Y, Okazaki T, Kitahara T, Umezawa H (1976) J Antibiot 29:1019
7. Nakamura H, Iitaka Y, Kitahara T, Okazaki T, Okami Y (1977) J Antibiot 30:714
8. The biosynthesis of macrolide has been already discussed in some reviews; Ikeda H, Ōmura S (2002) Biosynthesis, regulation, and genetics of macrolide production. In: Ōmura S (ed) Macrolide antibiotics (second edition) chemistry, biology, and practice. Elsevier Science, San Diego, p 285
9. Masamune S, Bates GS, Corcoran JW (1977) Angew Chem Int Ed Engl 16:585
10. O'Hagan D (1989) Nat Prod Rep 6:205
11. Zhou Y, Prediger P, Dias LC, Murphy AC, Leadlay PF (2015) Angew Chem Int Ed Engl 64:5232
12. Carmely S, Kashman Y (1985) Tetrahedron Lett 26:511
13. Kobayashi M, Tanaka J, Katori T, Matsuura M, Kitagawa I (1989) Tetrahedron Lett 30:2963
14. Kobayashi M, Tanaka J, Katori T, Matsuura M, Yamashita M, Kitagawa I (1990) Chem Pharm Bull 38:2409
15. Kobayashi M, Tanaka J, Katori T, Kitagawa I (1990) Chem Pharm Bull 38:2960
16. Kitagawa I, Kobayashi M, Katori T, Yamashita M, Tanaka J, Doi M, Ishida TJ (1990) J Am Chem Soc 112:3710
17. Doi M, Ishida T, Kobayashi M, Kitagawa I (1991) J Org Chem 56:3629
18. Tsukamoto S, Ishibashi M, Sasaki T, Kobayashi J (1991) J Chem Soc Perkin Trans 1:3185
19. Todd JS, Alvi KA, Crews P (1992) Tetrahedron Lett 33:441
20. Tanaka J, Higa T, Kobayashi M, Kitagawa I (1990) Chem Pharm Bull 38:2967
21. Nakata T, Komatsu T, Nagasawa K (1994) Chem Pharm Bull 42:2403
22. Nakata T, Komatsu T, Nagasawa K, Yamada H, Takahashi T (1994) Tetrahedron Lett 35:8225
23. Nagasawa K, Shimizu I, Nakata T (1996) Tetrahedron Lett 37:6881
24. Nagasawa K, Shimizu I, Nakata T (1996) Tetrahedron Lett 37:6885
25. Nicolaou KC, Snyder SA (2003) In: Swinholide A (ed) Classic in total synthesis II more targets, strategies, methods, vol 3. Wiley-VCH GmbH & Co. KGaA, Weinheim, p 31

26. Norcross RD, Paterson I (1995) Chem Rev 95:2041
27. Kang EJ, Lee E (2005) Chem Rev 105:4348
28. Paterson I (1992) Pure Appl Chem 64:1821
29. Paterson I, Cumming JG (1992) Tetrahedron Lett 33:2847
30. Paterson I, Smith JD (1992) J Org Chem 57:3261
31. Paterson I, Smith JD (1993) Tetrahedron Lett 34:5351
32. Paterson I, Cumming JG, Smith JD, Ward RA (1994) Tetrahedron Lett 35:441
33. Paterson I, Cumming JG, Smith JD, Ward RA, Yeung KS (1994) Tetrahedron Lett 35:3405
34. Paterson I, Smith JD, Ward RA, Cumming JG (1994) J Am Chem Soc 116:2615
35. Paterson I, Yeung K-S, Ward RA, Cumming JG, Smith JD (1994) J Am Chem Soc 116:9391
36. Paterson I, Cumming JG, Ward RA, Lamboley S (1995) Tetrahedron 51:9393
37. Paterson I, Smith JD, Ward RA (1995) Tetrahedron 51:9413
38. Paterson I, Ward RA, Smith JD, Cumming JG, Yeung K-S (1995) Tetrahedron 51:9437
39. Paterson I, Yeung K-S, Ward RA, Smith JD, Cuimming JG, Lamboley S (1995) Tetrahedron 51:9467
40. Brown HC, Dhar RK, Bakshi RK, Pandiarajan PK, Singaram B (1989) J Am Chem Soc 111:3441
41. Brown HC, Dhar RK, Ganesan K, Singaram B (1992) J Org Chem 57:499
42. Saksena AK, Mangiaracina P (1983) Tetrahedron Lett 24:273
43. Evans DA, Chapman KT, Carreira EM (1988) J Am Chem Soc 110:3560
44. Paterson I, Osborne SA (1990) Tetrahedron Lett 31:2213
45. Ferrier RJ (1964) J Chem Soc:5443
46. Ferrier RJ, Prasad N (1969) J Chem Soc C:570
47. Krohn K, Tolkiehn K, Lehne V, Schmalle HW, Grützmacher HF (1985) Liebigs Ann Chem:1311
48. Brown HC, Bhat KS (1986) J Am Chem Soc 108:5919
49. Brown HC, Bhat KS, Randad RS (1989) J Org Chem 54:1570
50. Inanaga J, Hirata K, Saeki H, Katsuki T, Yamaguchi M (1979) Bull Chem Soc Jpn 52:1989
51. Boden EP, Keck GE (1985) J Org Chem 50:2394
52. Patron AP, Richter PK, Tomaszewski MJ, Miller RA, Nicolaou KC (1994) J Chem Soc Chem Commun:1147
53. Richter PK, Tomaszewski MJ, Miller RA, Patron AP, Nicolaou KC (1994) J Chem Soc Chem Commun:1151
54. Nicolaou KC, Patron AP, Ajito K, Richter PK, Khatuya H, Bertinato P, Miller RA, Tomaszewski MJ (1996) Chem A Eur J 2:847
55. Nicolaou KC, Ajito K, Patron AP, Khatuya H, Richter PK, Bertinato P (1996) J Am Chem Soc 118:3059
56. Carretero JC, Ghosez L (1988) Tetrahedron Lett 29:2059
57. Lewis MD, Cha JK, Kishi Y (1982) J Am Chem Soc 104:4976
58. Nicolaou KC, Hwang CK, Duggan M (1989) J Am Chem Soc 111:6682
59. Fleming I, Lee TV (1981) Tetrahedron Lett:705
60. Fleming I, Goldhill J, Paterson I (1979) Tetrahedron Lett:3205
61. Mukaiyama T, Ishida A (1975) Chem Lett:319
62. Mukaiyama T, Ishida A (1975) Chem Lett:1201
63. Gao Y, Sharpless KB (1988) J Am Chem Soc 110:7538
64. Kim B, Sharpless KB (1989) Tetrahedron Lett 30:655
65. Lohray B, Gao Y, Sharpless KB (1989) Tetrahedron Lett 30:2623
66. Shin I, Hong S, Krische MJ (2016) J Am Chem Soc 138:14246
67. Hassan A, Krische MJ (2011) Org Process Res Dev 15:1236
68. Ketcham JM, Shin I, Montgomery TP, Krische MJ (2014) Angew Chem Int Ed 53:9142
69. Feng J, Kasun ZA, Krische MJ (2016) J Am Chem Soc 138:5467
70. Kim IS, Ngai M-Y, Krische MJ (2008) J Am Chem Soc 130:6340
71. Kim IS, Ngai M-Y, Krische MJ (2008) J Am Chem Soc 130:14891

72. Hassan A, Lu Y, Krische MJ (2009) Org Lett 11:3112
73. Mohapatra DK, Das PP, Pattanayak MR, Yadav JS (2010) Chem A Eur J 16:2072
74. Shin I, Wang G, Krische MJ (2014) Chem A Eur J 20:13382
75. Shin I, Krische MJ (2015) Top Curr Chem 372:85
76. Gao X, Han H, Krische MJ (2011) J Am Chem Soc 133:12795
77. Gao X, Woo SK, Krische MJ (2013) J Am Chem Soc 135:4223
78. Shin I, Krische MJ (2015) Org Lett 17:4686
79. Bernet B, Vasella A (1979) Helv Chim Acta 62:1990
80. Bernet B, Vasella A (1979) Helv Chim Acta 62:2400
81. Fu X, Schmitz FJ, Kelly-Borges M, McCready TL, Holmes CFB (1998) J Org Chem 63:7957
82. Rao MR, Faulkner DJ (2002) J Nat Prod 65:386
83. Erickson KL, Gustafson KR, Pannell LK, Beutler JA, Boyd MR (2002) J Nat Prod 65:1303
84. Barry CS, Bushby N, Charmant JPH, Elsworth JD, Harding JR, Willis CL (2005) Chem Commun 40:5097
85. Lee K, Lanier ML, Kwak J-H, Kim H, Hong J (2016) Nat Prod Rep 33:1393
86. Pereira AR, McCue CF, Gerwick WH (2010) J Nat Prod 73:217
87. Other synthetic studies; Yakambram P, Puranik V, Gurjar M (2006) Tetrahedron Lett 22:3781
88. Son JB, Kim SN, Kim NY, Lee DH (2006) Org Lett 8:661
89. Morikawa T, Sasaki T, Hanai R, Shibuya A, Taguchi T (1994) J Org Chem 59:97
90. Jadhav PK, Bhat KS, Perumal PT, Brown HC (1986) J Org Chem 51:432
91. Brown HC, Jadhav PK, Perumal PT (1984) Tetrahedron Lett 25:5111
92. Evans DA, Ripin DH, Halstead DP, Campos KR (1999) J Am Chem Soc 121:6816
93. Vakalopoulos A, Hoffmann HMR (2001) Org Lett 3:177
94. Micalizio GC, Pinchuk AN, Roush WR (2000) J Org Chem 65:8730
95. Schmidt RR, Michel J (1980) Angew Chem Int Ed Engl 19:731
96. Schmidt RR (1986) Angew Chem Int Ed Engl 25:212
97. Furstner A, Albert M, Mlynarski J, Matheu M, DeClercq E (2003) J Am Chem Soc 125:13132
98. Barry CS, Elsworth JD, Seden PT, Bushby N, Harding JR, Alder RW, Willis CL (2006) Org Lett 8:3319
99. Barry CS, Bushby N, Harding JR, Willis CL (2005) Org Lett 7:2683
100. Barry CS, Bushby N, Charmant JPH, Elsworth JD, Harding JR, Willis CL (2005) Chem Commun:5097
101. Nokami J, Nomiyama K, Shafi S, Kataoka K (2004) Org Lett 6:1261
102. Nokami J, Ohga M, Nakamoto H, Matsubara T, Hussain I, Kataoka K (1991) J Am Chem Soc 123:9168
103. Cloninger MJ, Overman LE (1999) J Am Chem Soc 121:1092
104. Al-Mutairi EH, Crosby SR, Darzi J, Harding JR, Hughes RA, King CD, Simpson TJ, Smith RW, Willis CL (2001) Chem Commun:835
105. Okude O, Hirano S, Hiyama T, Nozaki H (1977) J Am Chem Soc 99:3179
106. Hiyama T, Kimura K, Nozaki H (1981) Tetrahedron Lett 22:1037
107. Hiyama T, Okude O, Kimura K, Nozaki H (1982) Bull Chem Soc Jpn 55:561
108. Takai K, Kimura K, Kuroda T, Hiyama T, Nozaki H (1983) Tetrahedron Lett 24:5281
109. Jin H, Uenishi J, Christ WJ, Kishi Y J Am Chem Soc 108:5644
110. Takai K, Tagashira M, Kuroda T, Oshima K, Utimoto K, Nozaki H (1986) J Am Chem Soc 108:6048
111. Denmark SE, Edwards JP (1991) J Org Chem 56:6974
112. Corey EJ, Nicolaou KC (1974) J Am Chem Soc 96:5614
113. Nicolaou KC, Seitz SP, Papahatjis DP (1983) J Am Chem Soc 105:2430
114. Smith III AB, Simov V (2006) Org Lett 8:3315
115. Nagao Y, Hagiwara Y, Kumagai T, Ochiai M (1986) J Org Chem 51:2391
116. Charette A, Lebel H (1995) J Org Chem 60:2966
117. Mitsunobu O (1981) Synthesis:1
118. Abiko A, Buske DC, Liu J, Inoue T (2002) J Org Chem 67:5250

119. Petasis NA, Bzowej EI (1990) J Am Chem Soc 112:6392
120. Petasis NA, Lu SP (1996) Tetrahedron Lett 37:141
121. Smith III AB, Minbiole KP, Verhoest PR, Schelhaas M (2001) J Am Chem Soc 123:10942
122. Smith III AB, Fox RJ, Razler TM (2008) Acc Chem Res 41:675
123. Chakraborty TK, Reddy VR (2006) Tetrahedron Lett 47:2099
124. Chakraborty TK, Reddy VR, Chattopadhyay AK (2006) Tetrahedron Lett 47:7435
125. Chakraborty TK, Reddy VR, Gajula PK (2008) Tetrahedron 64:5162
126. Chakraborty TK, Das S (1999) J Indian Chem Soc 76:611
127. Carrick JD, Jennings MP (2009) Org Lett 11:769
128. Guz NR, Phillips AJ (2002) Org Lett 4:2253
129. Charette AB, Label H (1995) J Org Chem 60:2966
130. Charette AB, Beauchemin A (2001) Org React 58:1
131. Chen K-M, Hardtmann GE, Prasad K, Repic O, Shapiro MJ (1987) Tetrahedron Lett 28:155
132. Evans DA, Glorius F, Burch JD (2005) Org Lett 7:3331
133. Evans DA, Trenkle WC, Zhang J, Burch JD (2005) Org Lett 7:3335
134. Peh G, Floreancig PE (2012) Org Lett 14:5614
135. Tu W, Liu L, Floreancig PE (2008) Angew Chem Int Ed 47:4184
136. Tu W, Floreancig PE (2009) Angew Chem Int Ed 48:4567
137. den Hartog T, Rudolph A, Maciá B, Minnaard AJ, Feringa BL (2010) J Am Chem Soc 132:14349
138. Negishi E, King AO, Klima WL, Patterson W, Silveira A Jr (1980) J Org Chem 45:2526
139. Marshall JA, Adams ND (1998) J Org Chem 63:3812
140. Goossen LJ, Paetzold J, Koley D (2003) Chem Commun:706
141. Baker JB, Kim H, Hong J (2015) Tetrahedron Lett 56:3120
142. Duan JJW, Smith III AB (1993) J Org Chem 58:3703
143. Kim H, Park Y, Hong J (2009) Angew Chem Int Ed 48:7577
144. Shiina I, Kubota M, Oshiumi H, Hashizume M (2004) J Org Chem 69:1822
145. Shiina I, Fukui H, Sasaki A (2007) Nat Protoc 2:2312
146. Lumbroso A, Koschker P, Vautravers NR, Breit B (2011) J Am Chem Soc 133:2386
147. Lumbroso A, Abermil N, Breit B (2012) Chem Sci 3:789
148. Haydl AM, Breit B (2015) Angew Chem Int Ed 54:15530
149. Koschker P, Lumbroso A, Breit B (2011) J Am Chem Soc 133:20746
150. Davies HML, Beckwith REJ (2003) Chem Rev 103:2861
151. Millán A, Smith JR, Chen JL-Y, Aggarwal VK (2016) Angew Chem Int Ed 55:2498
152. Althaus M, Mahmood A, Suárez JR, Thomas SP, Aggarwal VK (2010) J Am Chem Soc 132:4025
153. Chen JL-Y, Scott HK, Hesse MJ, Willis CL, Aggarwal VK (2013) J Am Chem Soc 135:5316
154. Mahmood A, Suárez JR, Thomas SP, Aggarwal VK (2013) Tetrahedron Lett 54:49
155. Chen L, Kong F (2002) Carbohydr Res 337:2335
156. Cabrera JM, Krische MJ (2019) Angew Chem Int Ed 58:10718
157. Lu Y, Kim IS, Hassan A, Del Valle DJ, Krische MJ (2009) Angew Chem Int Ed 48:5018
158. Fenton DM, Steinwand PJ (1972) J Org Chem 37:2034
159. Semmelhack MF, Bodurow C (1984) J Am Chem Soc 106:1496
160. Cox DJ, Smith MD, Fairbanks AJ (2010) Org Lett 12:1452
161. Lemieux RU, Von Rudloff E (1955) Can J Chem 33:1701
162. Yang Z, Xie X, Jing P, Zhao G, Zheng J, Zhao C, She X (2011) Org Biomol Chem 9:984
163. Pabbaraja S, Satyanarayana K, Ganganna B, Yadav JS (2011) J Org Chem 76:1922
164. Sharpe RJ, Jennings MP (2011) J Org Chem 76:8027
165. Waldeck AR, Krische MJ (2013) Angew Chem Int Ed 52:4470
166. Kim H, Hong J (2010) Org Lett 12:2880
167. Hajare AK, Ravikumar V, Khaleel S, Bhuniya D, Reddy DS (2011) J Org Chem 76:963
168. Evans DA, Hoveyda AH (1990) J Org Chem 55:5190
169. Gesinski MR, Tadpetch K, Rychnovsky SD (2009) Org Lett 11:5342

170. Tay GC, Gesinski MR, Rychnovsky SD (2013) Org Lett 15:4536
171. Gesinski MR, Rychnovsky SD (2011) J Am Chem Soc 133:9727
172. Nicolaou KC, Li A, Edmonds DJ, Tria S, Ellery SP (2009) J Am Chem Soc 131:16905
173. Zhang Y, Phillips AJ, Sammakia T (2004) Org Lett 6:23
174. Bates RW, Lek TG (2014) Synthesis:1731
175. Petier C, Luche JL (1985) J Org Chem 50:910
176. Che W, Li Y-Z, Liu J-C, Zhu S-F, Xie J-H, Zhou Q-L (2019) Org Lett 21:2369
177. Bao D-H, Wu H-L, Liu C-L, Xie J-H, Zhou Q-L (2015) Angew Chem Int Ed 54:8791
178. Kwon HC, Kauffman CA, Jensen PR, Fenical W (2006) J Am Chem Soc 128:1622
179. Nicolaou KC, Nold AL, Milburn RR, Schindler CS, Cole KP, Yamaguchi J (2007) J Am Chem Soc 129:1760
180. Bailey CS, Zarins-Tutt JS, Agbo M, Gao H, Diego-Taboada A, Gan M, Hamed RB, Abraham ER, Mackenzie G, Evans PA, Goss RJM (2019) Chem Sci 10:7549
181. Nicolaou KC, Nold AL, Milburn RR, Schindler CS (2006) Angew Chem Int Ed 45:6527
182. Negishi EI, Luo FT, Frisbee R, Matsushita H (1982) Heterocycles 18:117
183. Wipf P, Soth MJ (2002) Org Lett 4:1787
184. Williams DR, Nold AL, Mullins RJ (2004) J Org Chem 69:5374
185. Frank SA, Chen H, Kunz RK, Schnaderbeck MJ, Roush WR (2000) Org Lett 2:2691
186. Evans PA, Huang MH, Michael JL, Marcto S (2012) Nat Chem 4:680
187. Smith III AB, Xian M (2006) J Am Chem Soc 128:66
188. Keck GE, Wager CA (2000) Org Lett 2:2307
189. Evans PA, Grisin A, Lawler MJ (2012) J Am Chem Soc 134:2856
190. Amans D, Bellosta V, Cossy J (2007) Org Lett 9:1453
191. Amans D, Bareille L, Bellosta V, Cossy J (2009) J Org Chem 74:7665
192. Van Horn DE, Negishi E (1978) J Am Chem Soc 100:2252
193. Negishi E, Van Horn DE, Yoshida T (1985) J Am Chem Soc 107:6639
194. Hafner A, Duthaler RO, Marti R, Rihs J, Rothe-Streit P, Schwarzenbach F (1992) J Am Chem Soc 114:2321
195. Duthaler RO, Hafner A (1992) Chem Rev 92:807
196. Cossy J, BouzBouz S, Pradaux F, Willis C, Bellosta V (2002) Synlett:1595
197. Nishimaru T, Kondo M, Takeshita K, Takahashi K, Ishihara J, Hatakeyama S (2014) Angew Chem Int Ed 53:8459
198. Hatakeyama S, Satoh K, Takano S (1993) Tetrahedron Lett 34:7425
199. Esumi T, Hukuyama H, Oribe R, Kawazoe K, Iwabuchi Y, Irie H, Hatakeyama S (1997) Tetrahedron Lett 38:4823
200. Nishioka T, Iwabuchi Y, Irie H, Hatakeyama S (1998) Tetrahedron Lett 39:5597
201. Esumi T, Kimura R, Mori M, Iwabuchi Y, Irie H, Hatakeyama S (2000) Heterocycles 52:525
202. Masaki H, Maeyama J, Kamada K, Esumi T, Iwabuchi Y, Hatakeyama S (2000) J Am Chem Soc 122:5216
203. Hatakeyama S, Sakurai K, Takano S (1985) J Chem Soc Chem Commun:1759
204. Jäger V, Häfele B (1986) Angew Chem Int Ed Engl 25:87
205. Schreiber SL, Schreibers TS, Smith DB (1987) J Am Chem Soc 109:1525
206. Cardillo G, Orena M, Poni G, Sandri S (1981) J Chem Soc Chem Commun:465
207. Bongini A, Cardillo G, Orena M, Poni G, Sandri S (1982) J Org Chem 47:4626
208. Ohira S (1989) Synth Commun 19:561
209. Müller S, Liepold B, Roth GJ, Bestmann HJ (1996) Synlett:521
210. Roth GJ, Liepold B, Müller SG, Bestmann HJ (2004) Synthesis:59
211. Trost BM, Ball ZT (2005) J Am Chem Soc 127:17644
212. Takai K, Shinomiya N, Kaihara H, Yoshida N, Moriwake T, Utimoto K (1995) Synlett:963
213. Takai K, Kunisada Y, Tachibana Y, Yamaji N, Nakatani E (2004) Bull Chem Soc Jpn 77:1581

Top Heterocycl Chem (2021) 58: 361–386
https://doi.org/10.1007/7081_2020_43

Published online: 11 November 2020

Total Synthesis of (−)-Enigmazole A

Check for updates

Tomoya Sugai and Haruhiko Fuwa

Contents

Abstract (−)-Enigmazole A and its natural congeners, isolated from a Papua New Guinean sponge *Cinachyrella enigmatica*, constitute a family of cytotoxic phosphomacrolides. Due to the structural complexity and biological significance, (−)-enigmazole A has been an intriguing target among the synthetic community. This chapter summarizes total and formal syntheses of (−)-enigmazole A, independently completed by five groups, with emphasis on strategies for constructing the complex macrocyclic skeleton of the target.

Keywords Macrolides · Oxacycles · Synthetic strategies · Tandem reactions · Total synthesis

Abbreviations

acac	Acetylacetonate
aq.	Aqueous
Boc	*t*-Butoxycarbonyl
brsm	Based on recovery of starting material

T. Sugai and H. Fuwa (✉)
Department of Applied Chemistry, Faculty of Science and Engineering, Chuo University, Tokyo, Japan
e-mail: hfuwa.50m@g.chuo-u.ac.jp

CD	Circular dichroism
CSA	10-Camphorsulfonic acid
d.r.	Diastereomer ratio
DABCO	1,4-Diazabicyclo[2.2.2]octane
dba	Dibenzylideneacetone
DCC	*N,N′*-dicyclohexylcarbodiimide
DCE	1,2-Dichloroethane
DIBALH	Diisobutylaluminum hydride
DMAP	4-Dimethylaminopyridine
DMP	Dess–Martin periodinane
DMPU	*N,N′*-dimethylpropyleneurea
e.e.	Enantiomeric excess
Fm	9-Fluorenylmethyl
HMDS	1,1,1,3,3,3-Hexamethyldisilazane
HMPA	Hexamethylphosphoramide
Ipc	Isopinocampheyl
IPr	1,3-Bis(2,6-diisopropylphenyl)imidazol-2-ylidene
LDA	Lithium diisopropylamide
LHMDS	Lithium bis(trimethylsilyl)amide
MIB	3-*exo*-Morpholinoisoborneol
MNBA	*m*-Nitrobenzoic anhydride
MOM	Methoxymethyl
MS	Molecular sieves
NaHMDS	Sodium bis(trimethylsilyl)amide
NMO	*N*-Methylmorpholine *N*-oxide
NOE	Nuclear Overhauser effect
NOESY	Nuclear Overhauser effect spectroscopy
PMB	*p*-Methoxybenzyl
py	Pyridine
quant.	Quantitative
TBAF	Tetra-*n*-butylammonium fluoride
TBDPS	*t*-Butyldiphenylsilyl
TBS	*t*-Butyldimethylsilyl
TCB	2,4,6-Trichlorobenzoyl
TEMPO	(2,2,6,6-Tetramethylpiperidin-1-yl)oxyl
Tf	Trifluoromethanesulfonyl
TIPS	Triisopropylsilyl
TMS	Trimethylsilyl
TPAP	Tetra-*n*-propylammonium perruthenate
Troc	2,2,2-Trichloroethoxycarbonyl
Ts	*p*-Toluenesulfonyl

1 Introduction

Macrolides are the members of a large family of naturally occurring polyketides, which have a macrolactone skeleton comprised of more than 12 atoms [1]. Historically, the macrolide class of natural products were first discovered in the 1950s as the secondary metabolites of soil-derived actinomycete bacteria. As exemplified by erythromycin and its structurally relevant 14-membered macrolides [2–4], macrolide antibiotics are known to inhibit bacterial protein biosynthesis to exert their antibacterial activity, and are widely used for the treatment of infectious diseases. Moreover, there are known macrolide antibiotics that display additional biological activities, including antimalarial, anticancer, and immunosuppressive activities. From the 1980s, macrolide natural products were also identified from marine organisms such as invertebrates and algae. Represented by bryostatins [5] and halichondrins, [6] marine macrolides are structurally different from soil-derived macrolides, and many of them show promising anticancer activity.

Despite their common structural characteristic as having a macrolactone skeleton as the core structure, macrolide natural products have tremendous structural diversity, owing to their ring size variation and stereochemical complexity. Furthermore, the proven efficacy of classical macrolide antibiotics as chemotherapeutics renders macrolide natural products a promising scaffold for drug development. As such, the isolation, structure elucidation, total synthesis, and biological characterization of macrolide natural products are continuously important subjects in organic chemistry and related disciplines.

In 2010, Gustafson and co-workers from the National Cancer Institute (NCI) reported the isolation, structural characterization, and biological activity of (–)-enigmazole A (**1**, Fig. 1) and its natural congeners **2** and **3** [7]. These new macrolides were isolated from a Papua New Guinean sponge *Cinachyrella enigmatica*. The structures of **1–3** share a common structural motif characterized by an 18-membered macrolactone embedded with a 2,6-*cis*-substituted tetrahydropyran ring and substituted with a phosphate group at C5 position and an oxazole ring at C17 position. The planar structure of **1** was determined mainly by extensive 2D NMR experiments. The configuration of the stereogenic centers in **1** was assigned on the basis of careful derivatization/degradation experiments and spectroscopic analyses

Fig. 1 Structures of (–)-enigmazole A (**1**) and its natural congeners **2** and **3**

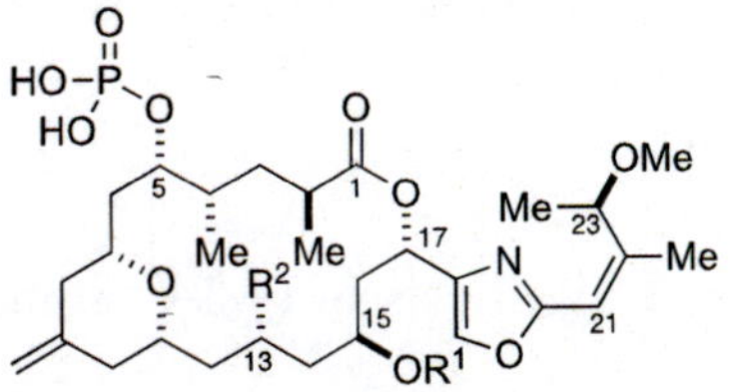

(–)-enigmazole A (**1**: R^1 = H, R^2 = H)
(–)-15-*O*-methylenigmazole A (**2**: R^1 = Me, R^2 = H)
(–)-13-hydroxy-15-*O*-methylenigmazole A (**3**: R^1 = Me, R^2 = OH)

of the resultant products. The absolute configuration of the C2 position was determined by a phenylglycine methyl ester (PGME) analysis [8], and the relative configuration of C2/C4 and C4/C5 was assigned through NOE experiments on a δ-lactone derivative of **1**, which was obtained via a dephosphorylation and a subsequent intramolecular transesterification. The absolute configuration of C15 position was established by means of a modified Mosher analysis [9], and correlated with C17 position via NOE experiments on a C15/C17 acetonide derivative. The configuration of C7 and C9 stereogenic centers was assigned through conformation analysis on the macrolactone skeleton of **1**, based on $^{3}J_{H,H}$ values and NOE enhancements observed in CD_3OD and in DMSO-d_6. The configurational assignment of the macrolactone of **1** was supported by molecular modeling. The configuration of the C21/C22 double bond was shown to be *Z* by a NOESY experiment. Finally, the absolute configuration of the C23 stereogenic center was determined by cleavage of the C21/C22 double bond of **1** with $RuCl_3/NaIO_4$ followed by trapping of the resultant 3-methoxy-2-butanone fragment with 2,4-dinitrophenylhydrazine, and comparison of the CD spectrum of the derived hydrazone with that of authentic reference. The structure assignment made by Gustafson et al. was finally confirmed by the first total synthesis achieved by Molinski and co-workers (vide infra). (−)-Enigmazole A (**1**) was shown to exhibit significant cytotoxic activity in the NCI-60 human cancer cell line panel, whilst no significant differential activity was observed among the cell lines tested. Meanwhile, some side fractions of the extracts of *C. enigmatica* showed selective cytotoxicity against cells expressing mutant c-Kit. Activating mutations of the receptor tyrosine kinase c-Kit is frequently observed for acute myelogenous leukemia and gastrointestinal stromal tumors, and is known to be responsible for the proliferation and invasion of cancer cells [7]. It is expected that (−)-enigmazole A (**1**) and its analogues provide a potential scaffold for the development of new chemotherapeutics targeting the c-Kit signaling pathway.

Given the structural complexity and potent cytotoxic activity, (−)-enigmazole A (**1**) has been an intriguing target of total synthesis. In 2010, Molinski and his colleagues described the first total synthesis of **1** and confirmation of its absolute configuration [10]. Later, the Smith [11, 12], Fürstner [13], and our groups [14] disclosed their total syntheses of **1** based on completely different synthetic strategies. Kadota et al. reported a formal synthesis of **1** by intercepting the Smith synthesis [15]. This chapter will attempt to illustrate the creativity and intellectual challenges of synthetic organic chemists toward the structural complexity of **1** and at the same time the current limitation of synthetic efficiency in total synthesis of polyketide macrolides.

2 Total Synthesis of (−)-Enigmazole A by Molinski

The Molinski group has accomplished the first total synthesis of (−)-enigmazole A (**1**). As summarized in Scheme 1, their synthesis plan was to construct the 18-membered macrocycle of **1** via a site-selective macrolactonization of dihydroxy

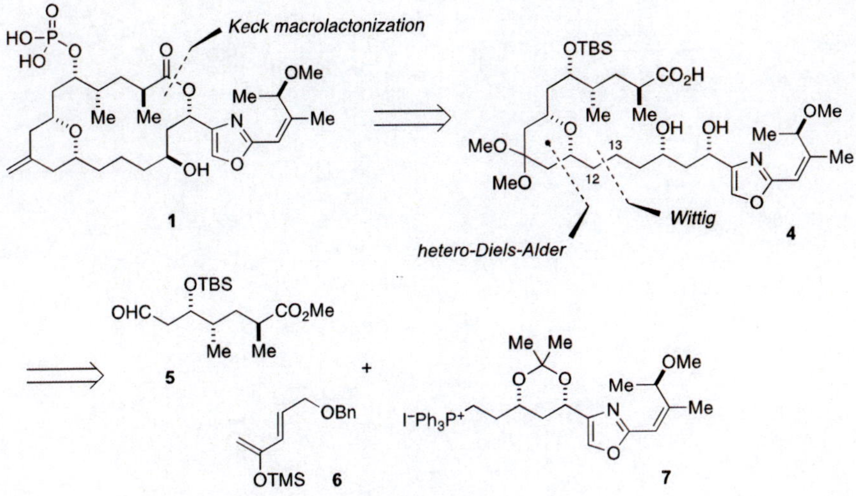

Scheme 1 Synthesis plan toward **1** by Molinski et al

acid **4**, which was expected to be available from aldehyde **5**, diene **6**, and phosphonium salt **7**, by considering a hetero-Diels–Alder cycloaddition for elaborating the 2,6-*cis*-substituted tetrahydropyran ring and a Wittig olefination for forging the C12/C13 bond.

The synthesis of phosphonium salt **7** commenced with oxidation of allylic alcohol **8**, available in one step from propargyl alcohol [16], with MnO_2 (82%) followed by (+)-MIB-catalyzed enantioselective addition of dimethylzinc [17, 18] to give optically active allylic alcohol **9** in 60% yield with 93% e.e. (Scheme 2). After methylation, Negishi reaction [19, 20] with an oxazol-2-ylzincate derived from 2-iodooxazole derivative **10** [21] provided cross-coupled product **11** in 86% yield. DIBALH reduction (89%) and asymmetric allylation with allylstannane **12** under Corey's conditions (**13**, BBr_3, CH_2Cl_2, −78°C) [22, 23] led to homoallylic alcohol **14** in 88% yield with greater than 20:1 diastereoselectivity. Selective cleavage of the exomethylene, Narasaka–Prasad 1,3-*syn* reduction [24] of the derived hydroxy ketone, and subsequent acetalization gave rise to acetonide **15** with correct configuration at the C15 position. Additional routine three-step sequence provided phosphonium salt **7**.

The synthesis of aldehyde **16**, the coupling partner of **7**, was achieved through a hetero-Diels–Alder reaction of aldehyde **5** and diene **6**, as illustrated in Scheme 3. The synthesis started with dehydration of dicarboxylic acid **17** (prepared in three steps from diethyl methylmalonate and methyl methacrylate) [25], methanolysis of the resultant anhydride, and chemoselective borane reduction to deliver alcohol **18** in 71% overall yield. Swern oxidation of **18** (85%) followed by Roush asymmetric allylation [26, 27] using chiral boronate **19** led to homoallylic alcohol **20** in 85% yield with 9:1 diastereoselectivity. Silylation and ozonolysis led to aldehyde **5**. Because hetero-Diels–Alder cycloaddition of **5** with diene **6** under chiral catalysis

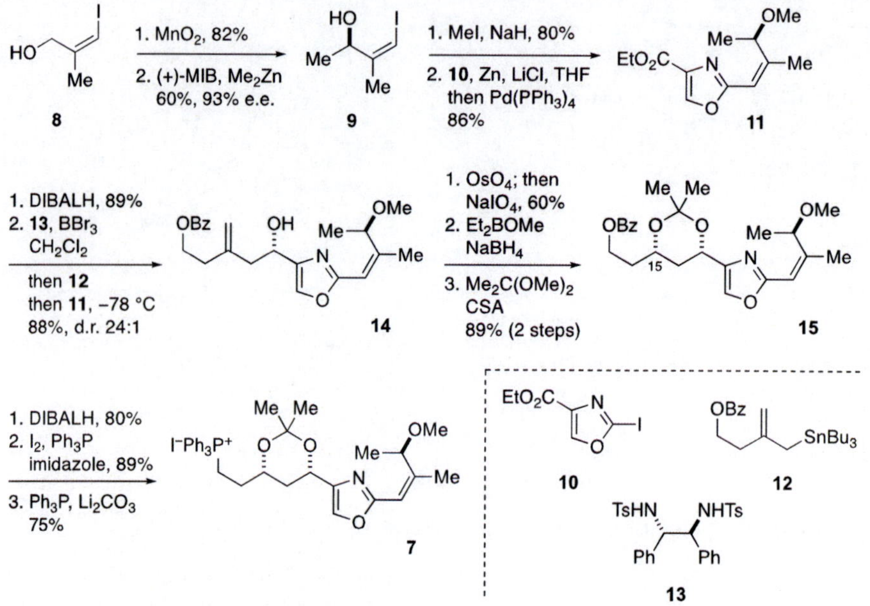

Scheme 2 Synthesis of phosphonium salt **7**

[28] was unsuccessful, the Molinski group reacted **5** with **6** in the presence of $BF_3{\bullet}OEt_2$ to provide, after in situ protection of the derived ketone, methyl acetal **21a–21c** in 81% yield as a mixture of three diastereomers. The desired 2,6-*cis*-substituted isomer **21a** was favored over 2,6-*trans* isomers **21b** and **21c**, although the diastereoselectivity remained moderate. After separation of these isomers by preparative HPLC, hydrogenolysis of the benzyl group and Swern oxidation of the resultant alcohol afforded aldehyde **16**.

Wittig reaction of **7** and **16** delivered *Z*-olefin **22** in 71% yield (Scheme 4). Subsequent three-step sequence, including hydrogenation of the *Z*-olefin, gave dihydroxy acid **4**. However, site-selective macrolactonization of **4** under Yamaguchi [29] or Shiina [30] conditions gave only complex, intractable mixtures of products, whereas that under Keck conditions [31] turned out to favor 16-membered macrolactone **23** as the major product. Intramolecular transesterification of **23** to the corresponding 18-membered macrolactone with $Ti(Oi\text{-}Pr)_4$ [32, 33] also failed. This site-selectivity problem was eventually overcome by using dihydroxy carboxylic acid **24** with *Z*-configured C12/C13 double bond. Thus, exposure of **24** to DCC, DMAP, and DMAP•HCl in refluxing $CHCl_3$, followed by in situ acetylation of the unreacted hydroxy group, gave rise to 18-membered macrolactone **25** in 35% overall yield from **22**. Presumably, the conformational constraint arising from the *Z*-configured C12/C13 double bond brought the C1 carboxylic acid and the C17 hydroxy group into proximity to facilitate the site-selective macrolactonization.

The final stage of the total synthesis required the introduction of the C9 exomethylene group and the C5 phosphate group. Thus, site-selective hydrogenation

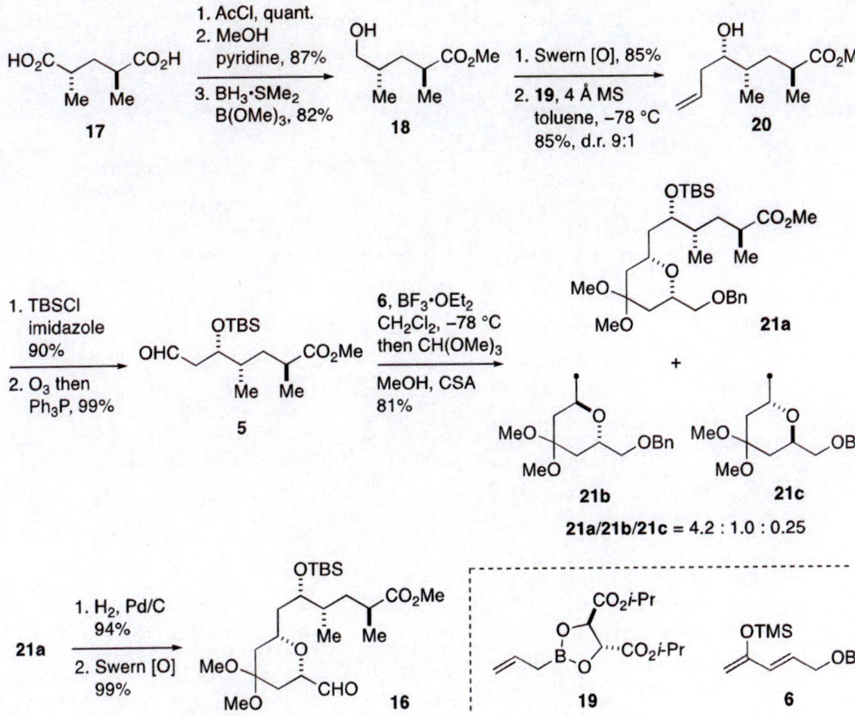

Scheme 3 Synthesis of aldehyde **16**

of the C12/C13 double bond and deprotection of the dimethyl acetal gave ketone **26**, whose exposure to Takai–Lombardo reagent (Zn, $TiCl_4$, CH_2Br_2) [34] resulted in exomethylene **27**. Removal of the TBS group, phosphorylation (i-$Pr_2NP(OFm)_2$, 1*H*-tetrazole; then aq. H_2O_2) [35], and global deprotection (K_2CO_3, aq. MeOH) furnished synthetic (−)-enigmazole A (**1**).

The Molinski synthesis of (−)-enigmazole A (**1**) proceeded in 23 longest linear steps (0.14% overall yield) from propargyl alcohol. The present synthesis involved a diastereoselective hetero-Diels–Alder cycloaddition for stereoselective synthesis of the 2,6-*cis*-substituted tetrahydropyran moiety and a site-selective Keck macrolactonization for crafting the macrocyclic skeleton as key transformations.

3 Total Synthesis of (−)-Enigmazole A by Smith

In 2015, the Smith group described the second total synthesis of (−)-enigmazole A (**1**). Their synthetic strategy, summarized in Scheme 5, depended on a three-stage Petasis–Ferrier union/rearrangement [36, 37] of aldehyde **28** and hydroxy acid **29** for the construction of the tetrahydropyran ring of **1**, and a Yamaguchi

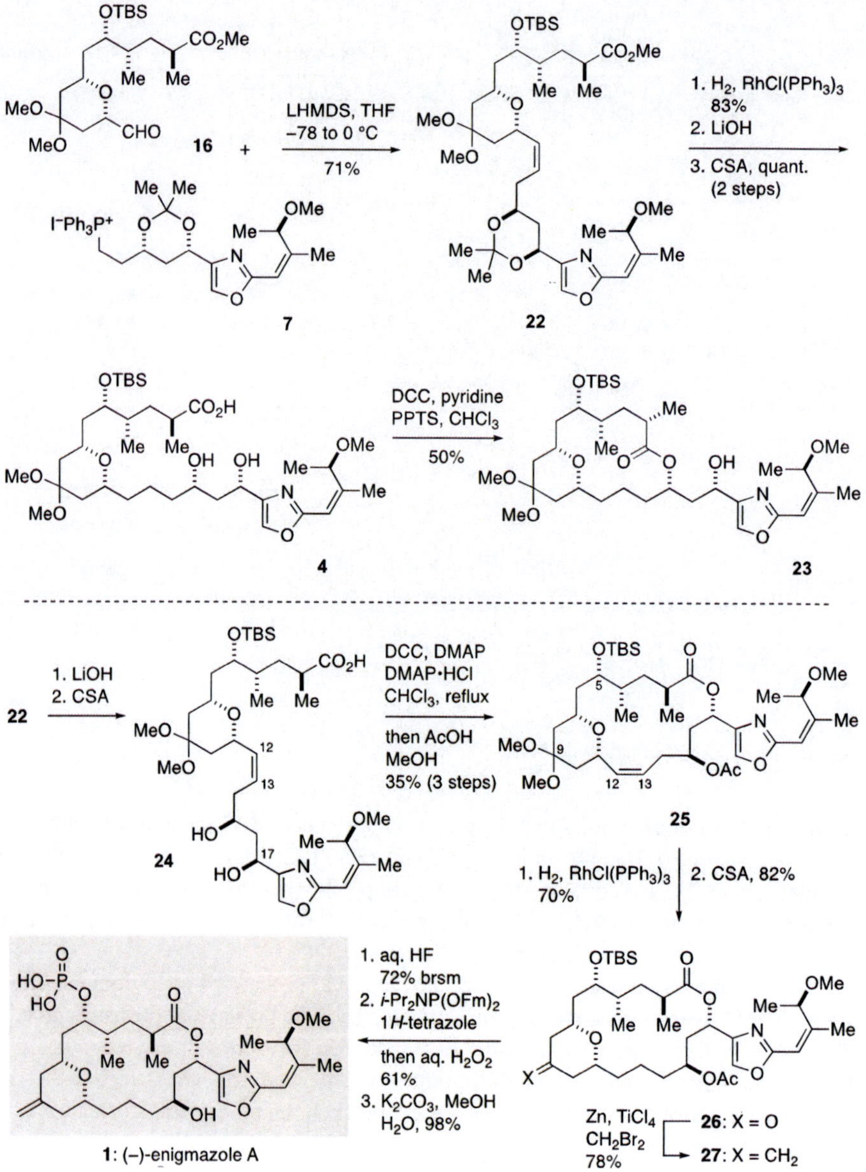

Scheme 4 Total synthesis of (−)-enigmazole A (**1**) by Molinski et al

macrolactonization [29] for the formation of the macrocyclic skeleton. Carboxylic acid **29** should be available from dithiane **30** and epoxide **31**. Dithiane **30** would be obtainable from 2-chlorooxazole derivative **32** and iodoolefin **33** by means of a Negishi reaction [19, 20]. The epoxide **31**, in turn, would arise via a

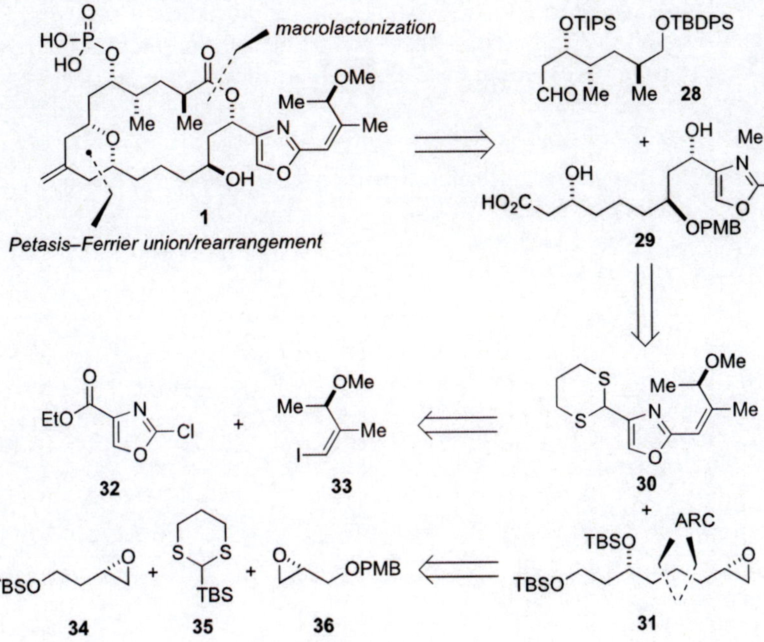

Scheme 5 Synthesis plan toward **1** by Smith et al

three-component "Anion Relay Chemistry" (ARC) union [38–40] of known fragments **34**, **35**, and **36**.

The synthesis of epoxide **31** started from ARC union of known dithiane **35** [39] and enantiopure epoxides **34** [40] and **36** [40] (Scheme 6). Deprotonation of **35** with *n*-BuLi, coupling of the generated lithiated dithiane with epoxide **34**, 1,4-Brook rearrangement, and subsequent addition of the resultant dithiane anion to epoxide **36** furnished the tri-component adduct **37** in 90% yield. Raney nickel reduction of the dithiane moiety and removal of the PMB group afforded 1,2-diol **39**. Epoxide **31** was obtained from **39** by a Fraser–Reid protocol [41] in a quantitative yield. Meanwhile, the synthesis of dithiane **30** commenced with commercially available (*R*)-3-butyn-2-ol (**40**). Copper-catalyzed carbometallation/iodination [42], followed by *O*-methylation, gave iodoolefin **33** in 77% yield for the two steps. Negishi cross-coupling [19, 20] of **32** and **33** provided ester **11** in 80% yield. Reduction of the ester with DIBALH, followed by dithioacetalization of the derived aldehyde provided dithiane **30**. Coupling of dithiane **30** and epoxide **31** required extensive optimization experiments because competitive lithiation at the C23 position of **30** occurred easily. The best result was obtained by deprotonation of **30** with *n*-BuLi (Et_2O, −78°C) followed by addition of epoxide **31** (−78 to 0°C, over 2 h), affording alcohol **41** in 77% yield. Removal of the dithioacetal [43] gave a β-hydroxy ketone (75%) and Narasaka–Prasad reduction [24] of the liberated ketone gave 1,3-*syn*-diol **42** in 95%

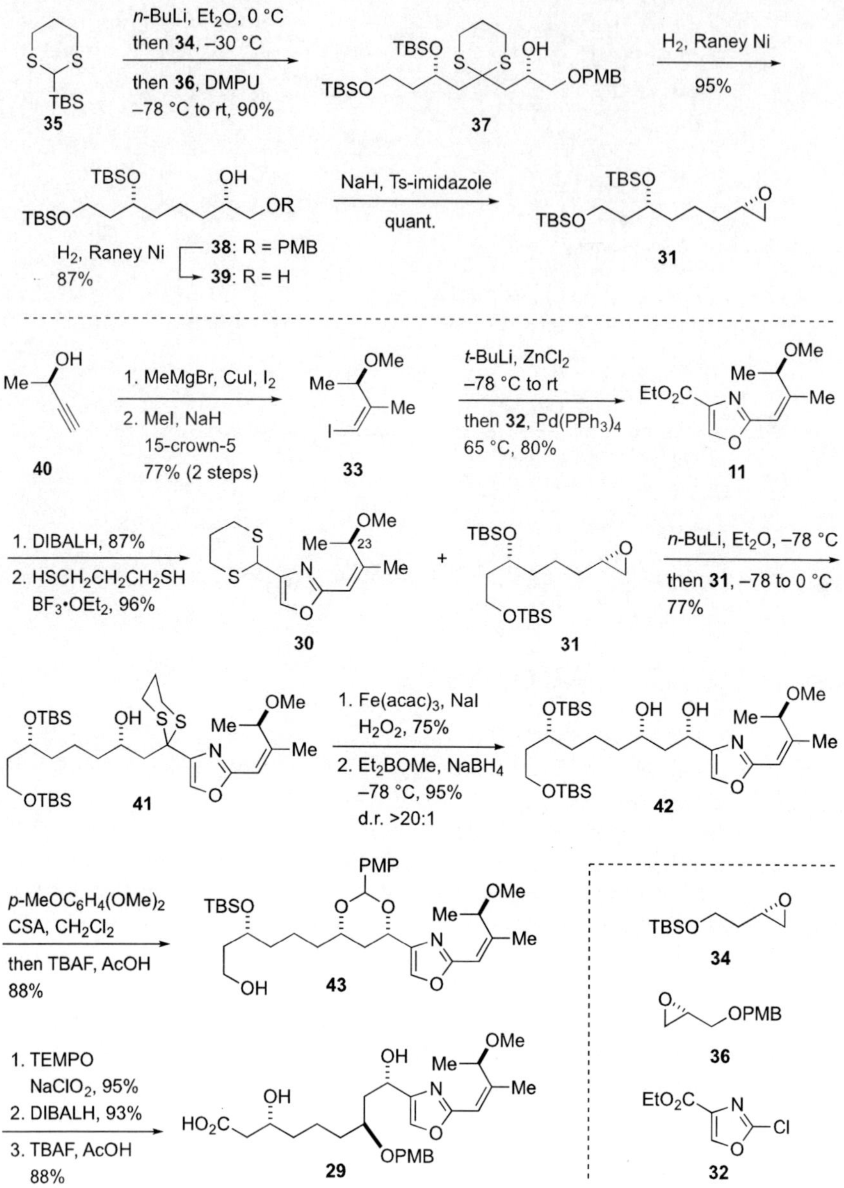

Scheme 6 Synthesis of carboxylic acid **29**

yield as a single diastereomer. Acetalization of **42** with *p*-methoxybenzylidene dimethyl acetal and removal of the primary TBS group in a single flask afforded **43** in 88% yield. TEMPO/NaClO$_2$-mediated oxidation, reductive opening of the *p*-

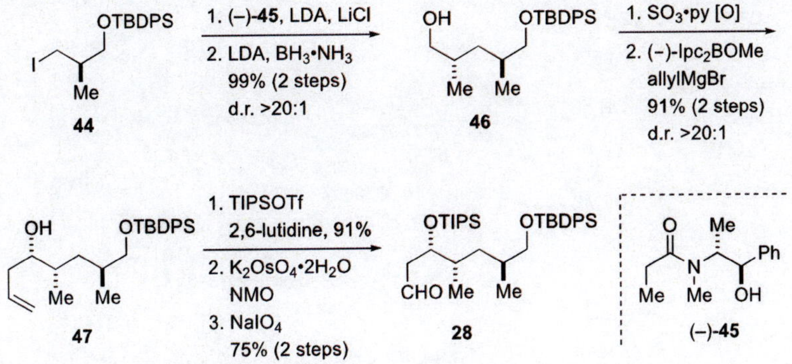

Scheme 7 Synthesis of aldehyde **28**

methoxybenzylidene acetal, and removal of the secondary TBS group provided carboxylic acid **29**.

The synthesis of aldehyde **28** started from known iodide **44** [44], prepared in three steps from commercially available (*R*)-Roche ester (Scheme 7). Myers asymmetric alkylation/reduction sequence [45] of **44** led to alcohol **46** in 99% yield (two steps) as a single diastereomer. Parikh–Doering oxidation and Brown asymmetric allylation [46] gave homoallylic alcohol **47** in 91% yield (two steps) with greater than 20:1 diastereoselectivity. After protection of **47** with TIPSOTf/2,6-lutidine, sequential dihydroxylation of the terminal olefin and cleavage of the derived diol provided aldehyde **28**.

With aldehyde **28** and carboxylic acid **29** in hand, their three-stage Petasis–Ferrier union/rearrangement was examined (Scheme 8). The union of **28** and **29** using HMDS/TMSOTf [37] gave dioxanone **48** in less than 10% yield under anhydrous conditions, and it was eventually found that traces of H_2O were necessary to obtain **48** in a satisfactory yield. Petasis olefination under microwave irradiation conditions provided enol ether **49** in 87% yield (cf. 40–50% yield under conventional heating conditions). Unfortunately, Ferrier rearrangement of **49** using Me_2AlCl did not deliver the corresponding rearrangement product, tetrahydropyranone **50**, in spite of attempting various additives and quenching methods. This was presumably because of Lewis acid-promoted intramolecular Michael fragmentation of the tetrahydropyranone ring. Eventually, in situ Wittig methylenation of **50** was found beneficial for overcoming this problem: after treatment of **49** with Me_2AlCl in CH_2Cl_2 at −78°C for 30 s, excess $Ph_3P{=}CH_2$ was introduced to the reaction mixture to transform **50** into 4-methylene tetrahydropyran **51**. Under these conditions, **51** was isolated in 84% yield with >20:1 diastereoselectivity. Selective deprotection of the TBDPS and TMS groups was achieved using KOH (84%). TEMPO-catalyzed oxidation of the liberated primary alcohol provided carboxylic acid **52** (74%). Macrolactonization of **52** under Yamaguchi conditions [29] afforded macrolactone **53** in 89% yield. Removal of the TIPS group was problematic due to competitive intramolecular transesterification and required carefully controlled conditions using HF•pyridine buffered with pyridine to deliver alcohol **54** in 70% yield. Installation

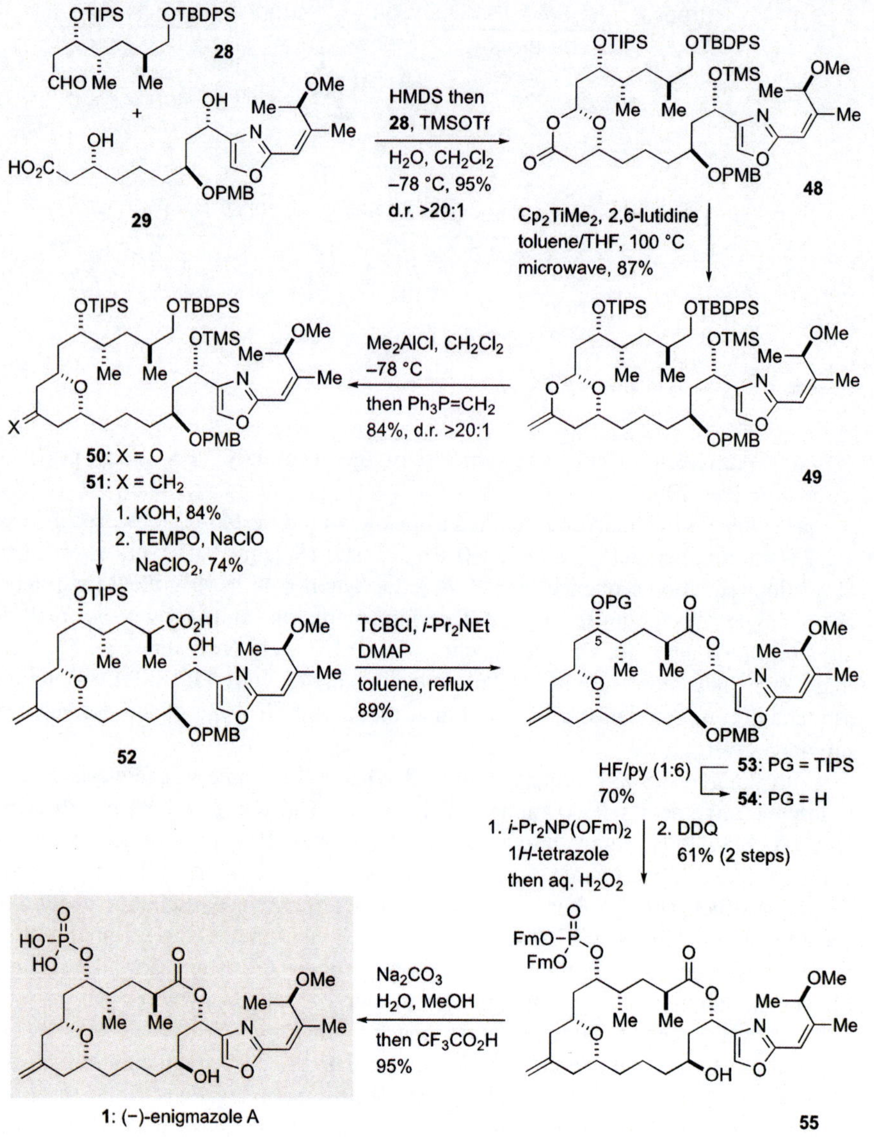

Scheme 8 Total synthesis of (−)-enigmazole A (**1**) by Smith et al

of the C5 phosphate [35] and deprotection of the PMB group gave bis (9-fluorenylmethyl) phosphate **55** (61%, two steps). Finally, exposure of **55** to Na_2CO_3/MeOH followed by acidification led to enigmazole A (**1**).

The Smith synthesis of (−)-enigmazole A (**1**) was achieved in 22 linear steps (4.4% overall yield) from (*R*)-3-butyn-2-ol (**40**). This synthesis featured a three-stage Petasis–Ferrier union/rearrangement of aldehyde **28** and carboxylic acid **29** for assembly of advanced fragments with concomitant construction of the 4-methylene tetrahydropyran ring of **1**.

4 Total Synthesis of (−)-Enigmazole A by Fürstner

In 2016, the Fürstner group described the third total synthesis of (−)-enigmazole A (**1**). The Fürstners synthetic strategy, summarized in Scheme 9, centered around a gold-catalyzed [3,3]-sigmatropic rearrangement of propargylic acetate **58** and a transannular hydroalkoxylation of the derived allenyl acetate **57** for a late-stage construction the 4-methylene tetrahydropyran ring of **1**. Propargylic acetate **58** should be available from diyne **59** via a ring-closing alkyne metathesis (RCAM) [47, 48]. Diyne **59** was retrosynthetically divided into three fragments, carboxylic acid **60**, allylstannane **61**, and aldehyde **62**.

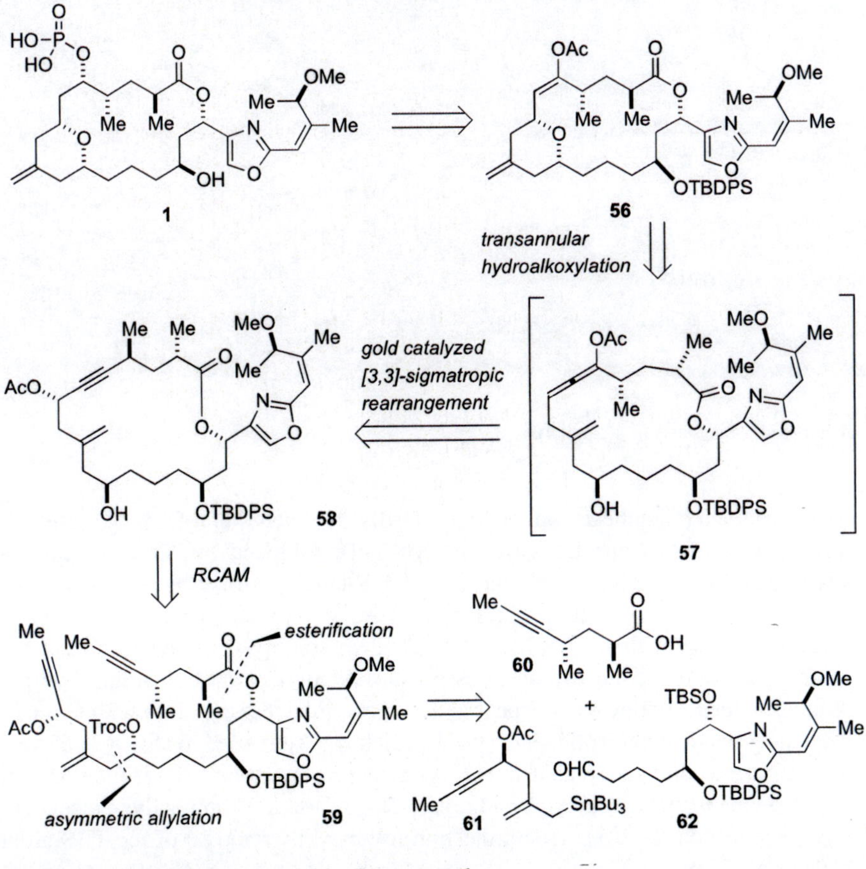

Scheme 9 Synthesis plan toward **1** by Fürstner et al

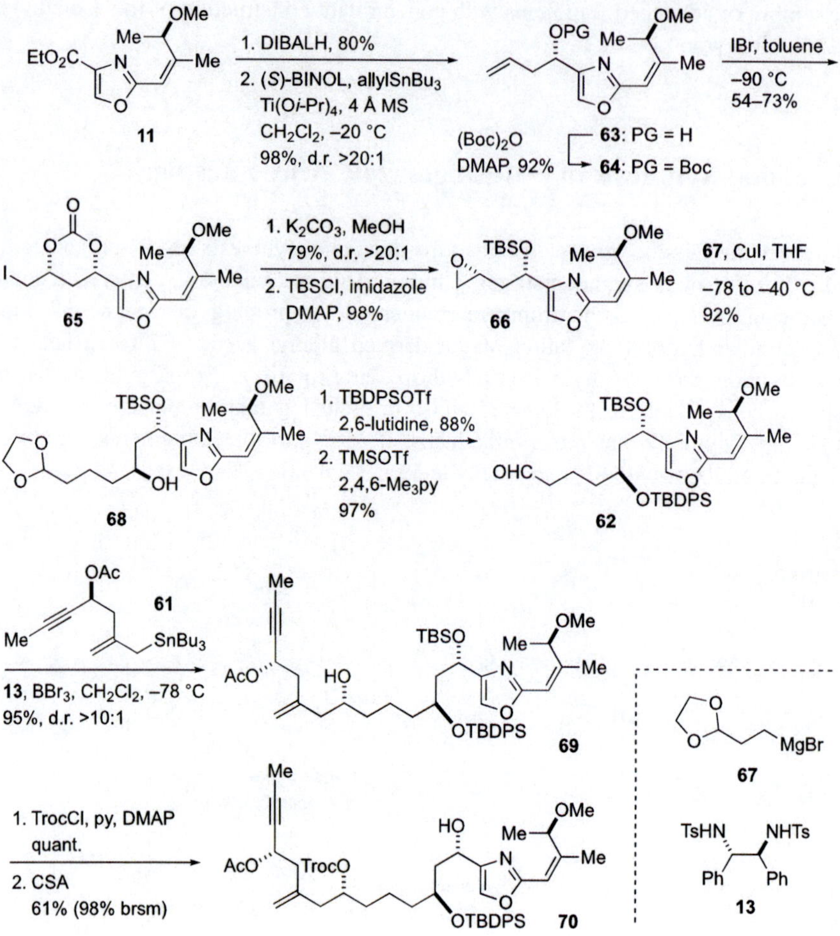

Scheme 10 Synthesis of alcohol **70**

The Fürstner synthesis started with DIBALH reduction of known ester **11** (prepared in three steps from (*R*)-3-butyn-2-ol), followed by Keck asymmetric allylation [49], to give allylic alcohol **63** (Scheme 10). Boc protection and IBr-mediated intramolecular cyclization [50] led to carbonate **65** in 54–73% yield. Treatment of **65** with K_2CO_3/MeOH resulted in solvolysis of the carbonate and spontaneous epoxide formation. Subsequent silylation of the remaining hydroxy group provided epoxide **66**. Copper-catalyzed epoxide opening of **66** with Grignard reagent **67** led to alcohol **68** in 92% yield, which was converted to aldehyde **62** via a two-step protecting group adjustment. Asymmetric allylation of aldehyde **62** with allylstannane **61** under Corey's conditions [22, 23] provided homoallylic alcohol **69** in 95% yield (d.r. >10:1). Troc protection followed by removal of the TBS group gave alcohol **70**.

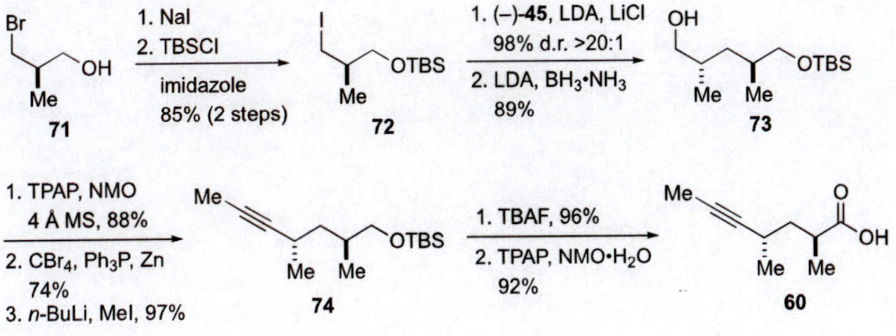

Scheme 11 Synthesis of carboxylic acid **60**

The synthesis of carboxylic acid **60** commenced with a two-step conversion of commercially available (*R*)-3-bromo-2-methyl-1-propanol (**71**) to iodide **72** (Scheme 11). Myers asymmetric alkylation/reduction [45] of **72** led to alcohol **73** as a single diastereomer. TPAP oxidation and Corey–Fuchs alkyne synthesis [51] led to alkyne **74**. After removal of the TBS group, TPAP/NMO•H_2O oxidation [52] provided carboxylic acid **60**.

Completion of the Fürstner synthesis of (−)-enigmazole A (**1**) is shown in Scheme 12. Yamaguchi esterification [29] of alcohol **70** and carboxylic acid **60** led to diyne **59** in a quantitative yield. RCAM [47, 48] of diyne **59** using molybdenum ate complex **75** in the presence of 4 Å MS and 5 Å MS [53, 54] in toluene provided macrolactone **76** in 79% yield. Zn/AcOH-mediated deprotection of the Troc group gave alcohol **58** (93%). [3,3]-Sigmatropic rearrangement of **58** and concomitant transannular hydroalkoxylation of the derived allenyl acetate was achieved by the action of (*R*)-**77**/$AgSbF_6$ [55] in CH_2Cl_2 to afford 4-methylene tetrahydropyran **56** in 91% yield. The success of this sequential process was found to be dependent on the configuration of the C7 stereogenic center and the axial chirality of the gold complex. After cleavage of the acetyl group (95%), the derived ketone was reduced with $NaBH_4$ to deliver alcohol **78** with approximately 2:1 diastereoselectivity (61% isolated yield). Phosphorylation of the resultant alcohol via phosphoramidite coupling/oxidation [35] and removal of the TBDPS group furnished enigmazole A (**1**).

The Fürstner synthesis of (−)-enigmazole A (**1**) was achieved in 23 linear steps (3.3% overall yield) from (*R*)-3-butyn-2-ol. Highlights of this synthesis include a ring-closing alkyne metathesis to construct the 18-membered macrocycle and a gold-catalyzed [3,3]-sigmatropic rearrangement/transannular hydroalkoxylation for a late-stage construction of the 4-methylene tetrahydropyran ring.

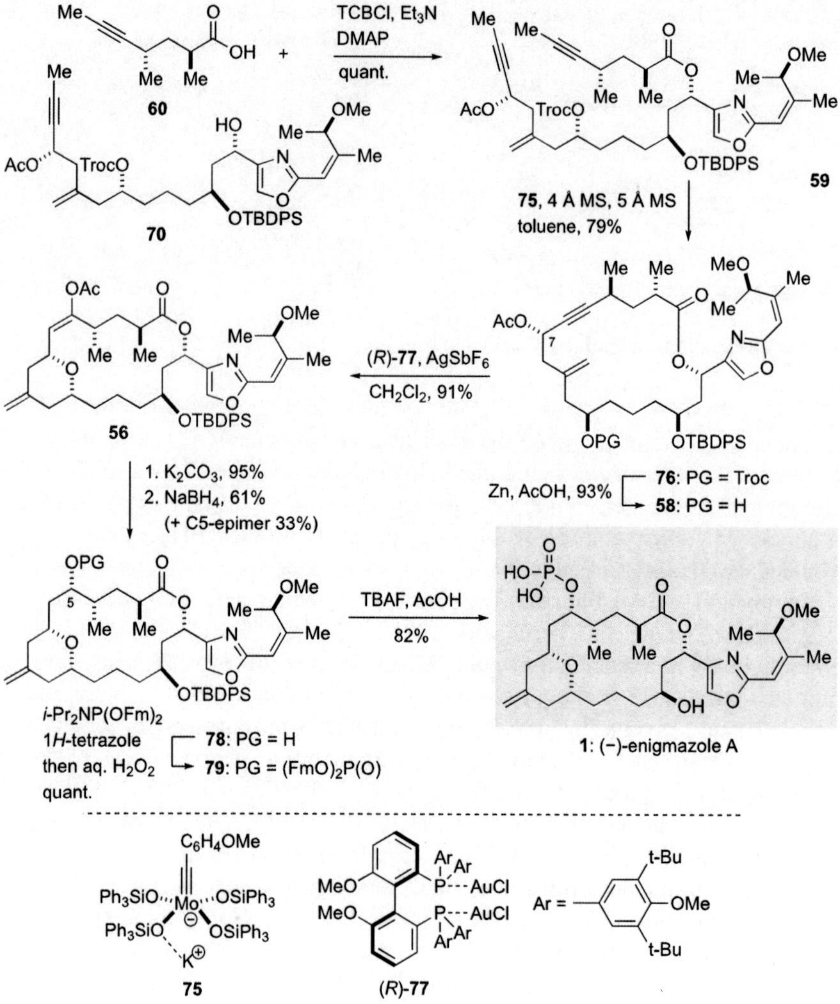

Scheme 12 Total synthesis of (−)-enigmazole A (**1**) by Fürstner et al

5 Total Synthesis of (−)-Enigmazole A by Fuwa

In 2018, the Fuwa group reported the fourth total synthesis of (−)-enigmazole A (**1**). The synthetic blueprint of the Fuwa synthesis of **1** is illustrated in Scheme 13. They considered macrocyclic α,β-unsaturated ketone **80** as an advanced intermediate, which has differentially protected C5 and C9 hydroxy groups as well as versatile α,β-unsaturated ketone moiety that is amenable to selective functionalization at the C13, C14, and C15 positions. Thus, structural diversification of the enigmazole macrocycle should be possible from advanced intermediate **80**. The macrocycle of

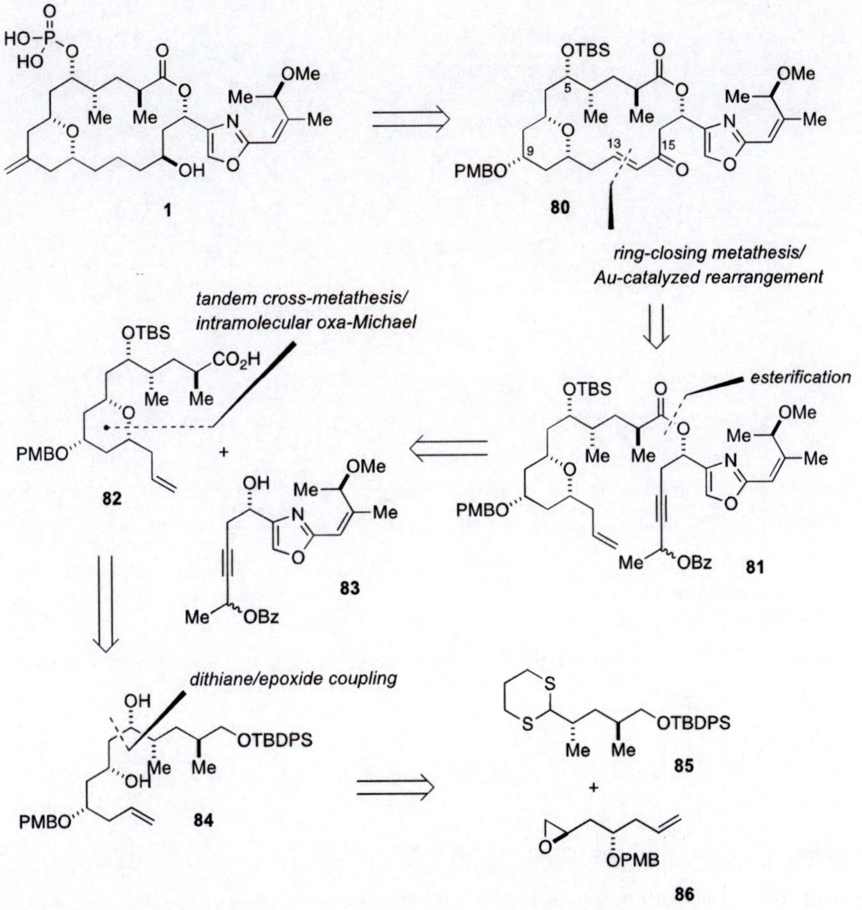

Scheme 13 Synthetic blueprint toward (−)-enigmazole A (**1**) by Fuwa et al

80 would be crafted through a gold-catalyzed propargylic benzoate rearrangement of benzoate **81** [56] and a subsequent ring-closing metathesis [48]. Ester **81** should be available from carboxylic acid **82** and alcohol **83**. Carboxylic acid **82**, in turn, was traced back to dihydroxy olefin **84** by considering a tandem olefin cross-metathesis/ intramolecular oxa-Michael addition [57, 58]. Compound **84** should be obtainable from dithiane **85** and epoxide **86**.

The synthesis of carboxylic acid **82** started from oxidation of known alcohol **46** [11] (prepared in the Smith synthesis of **1** from (*R*)-Roche ester over five steps), followed by dithioacetalization to give dithiane **85** (Scheme 14). Deprotonation of **85** with *t*-BuLi and coupling of the generated lithiated dithiane with epoxide **86** (HMPA, THF, −78 to −50°C) afforded alcohol **87** in 76% yield (88% based on recovered **86**). In this anion coupling, the stoichiometry of the reactants was critically important for clean reaction. Removal of the dithioacetal with $NaClO_2$ [59],

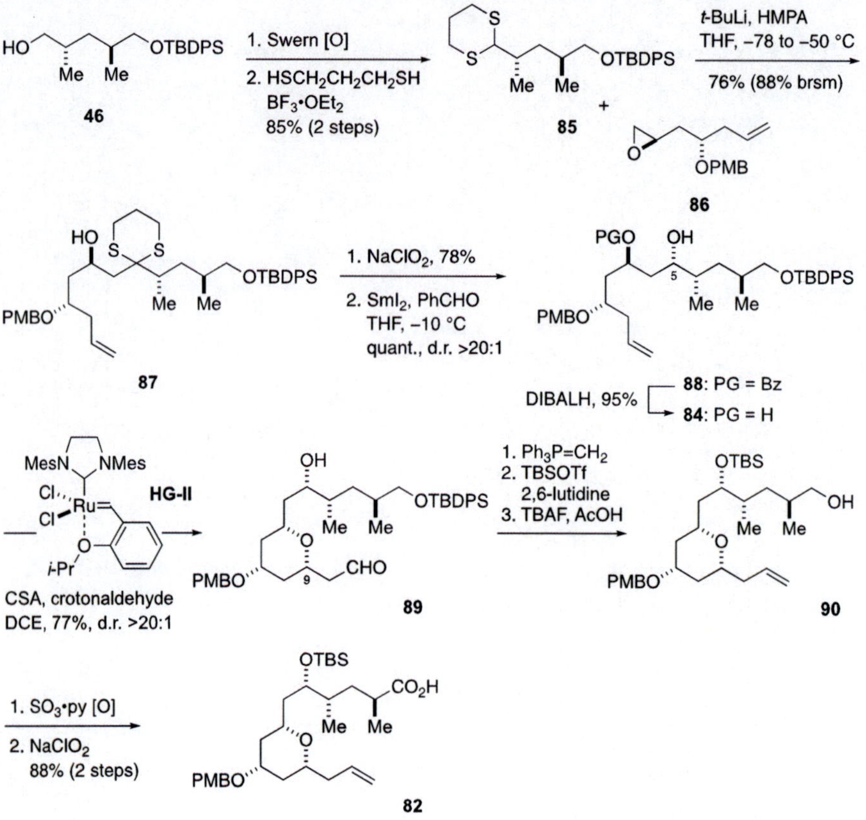

Scheme 14 Synthesis of carboxylic acid **82**

followed by Evans–Tishchenko reduction [60] (SmI_2, PhCHO, THF, −10°C), provided benzoate **88** quantitatively as a single stereoisomer with correct configuration at C5. Reductive removal of the benzoate led to dihydroxy olefin **84**. Olefin cross-metathesis/intramolecular oxa-Michael addition [57, 58] of **84** was achieved under co-catalysis of the second-generation Hoveyda–Grubbs (**HG-II**) complex [61] and CSA in the presence of crotonaldehyde in DCE at room temperature to furnish 2,6-*cis*-substituted tetrahydropyran **89** in 77% yield with greater than 20:1 diastereoselectivity. This tandem reaction, originally described by Fuwa et al., enables a step-economical, easy access to complex tetrahydropyran derivatives. In fact, it has been utilized in a number of total syntheses of complex natural products [62]. Wittig methylenation of **89** followed by silylation gave a TBS ether, from which the TBDPS group was selectively cleaved with TBAF/AcOH [63] to deliver alcohol **90**. A two-stage oxidation of **90** then afforded carboxylic acid **82**.

Alcohol **83** was synthesized from known homoallylic alcohol **62** [13] (prepared in the Fürstner synthesis of **1** over five steps from (*R*)-3-butyn-2-ol), as shown in Scheme 15. Silylation of **62** followed by selective double bond cleavage gave

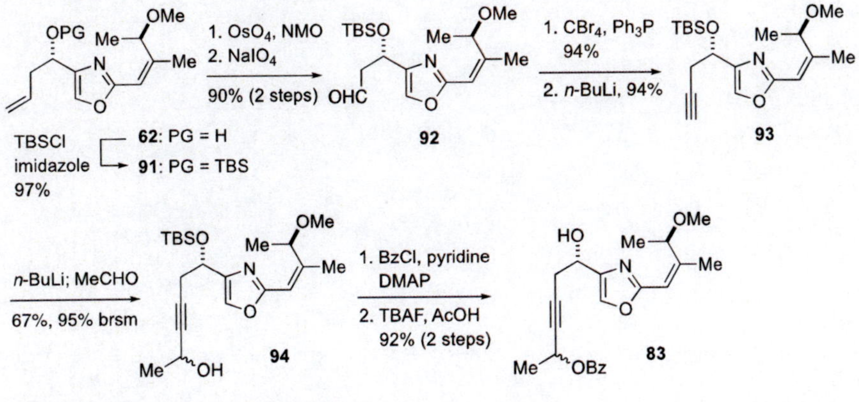

Scheme 15 Synthesis of alcohol **83**

aldehyde **92**. Corey–Fuchs alkyne synthesis [51] via a dibromoolefin delivered alkyne **93**, which was deprotonated and trapped with CH_3CHO to give propargylic alcohol **94**. Acylation of **94** with BzCl/pyridine and removal of the silyl group afforded alcohol **83**.

Completion of the total synthesis of (−)-enigmazole A (**1**) by the Fuwa group is illustrated in Scheme 16. Esterification of carboxylic acid **82** and alcohol **83** under Yamaguchi conditions [29] provided ester **81**. Propargylic benzoate rearrangement of **81** was achieved by the action of Nolan's catalytic system, IPrAuCl/$AgSbF_6$ [64], in THF/H_2O at 80°C under microwave irradiation to afford α,β-unsaturated ketone **95** in 82% yield as a single *E*-configured isomer. The success of this transformation depended on proper choice of the carboxylate moiety. While the original publication by Nolan et al. used propargylic acetates as substrates, model experiments by Fuwa et al. showed that propargylic benzoates are superior alternative to the acetate counterparts with respect to the reactivity and product yield. Ring-closing metathesis of **95** was easily accomplished by its exposure to the second-generation Grubbs (**G-II**) complex [65] in toluene at 40°C to deliver macrocyclic α,β-unsaturated ketone **80** in 81% yield as a single *E* isomer. Site-selective hydrogenation of **80** under the influence of Wilkinson's catalyst (91%) and stereoselective reduction of the resultant ketone gave rise to alcohol **96** in 86% yield with 3:1 diastereoselectivity. The major diastereomer with correct configuration at C15 was separated by flash column chromatography on silica gel. A three-step sequence including acetylation, PMB deprotection, and oxidation led to ketone **26**, which was previously synthesized as an intermediate in the Molinski synthesis of **1** [10]. The remaining four-step sequence was performed in accordance with the Molinski synthesis to furnish synthetic (−)-enigmazole A (**1**).

The Fuwa synthesis of (−)-enigmazole A (**1**) was achieved in 29 linear steps (2.5% overall yield) from (*R*)-Roche ester. The salient features of the present synthesis are a tandem olefin cross-metathesis/intramolecular oxa-Michael addition for stereoselective synthesis of the 2,6-*cis*-substituted tetrahydropyran ring and a gold-catalyzed propargylic benzoate rearrangement/macrocyclic ring-closing

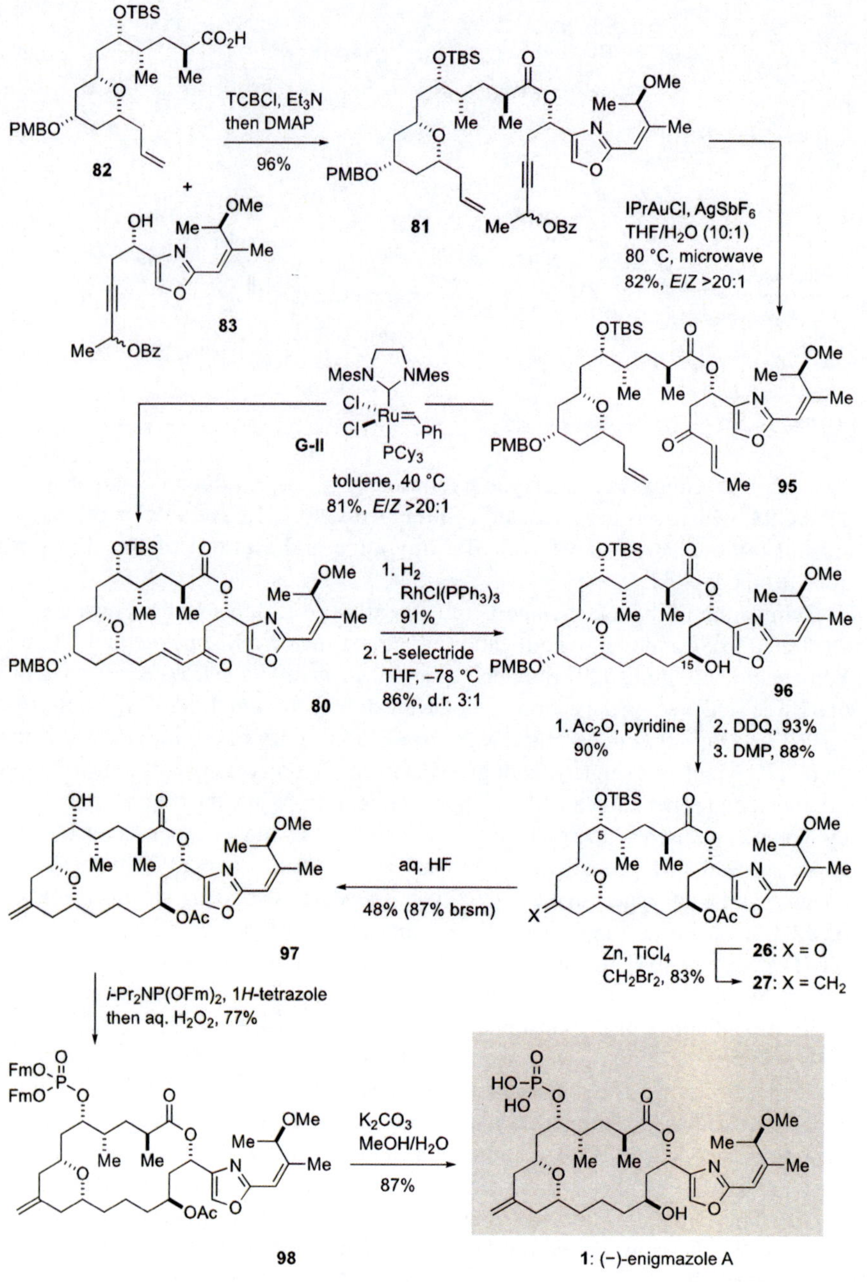

Scheme 16 Completion of the total synthesis of (−)-enigmazole A (**1**) by Fuwa et al

metathesis for forging the 18-membered macrocycle. Although the synthetic path of the Fuwa synthesis is longer than others, it involves versatile intermediates at late stage of the synthesis and therefore appears easily adaptable to structural diversification for analogue synthesis. The Fuwa group also completed the total synthesis of (−)-15-*O*-methylenigmazole A (**2**) from advanced intermediate **80** [66].

6 Formal Synthesis of (−)-Enigmazole A by Kadota

The Kadota group described a formal synthesis of (−)-enigmazole A (**1**) in 2018. Their synthetic strategy, summarized in Scheme 17, depended on a Yamaguchi macrolactonization [29] of hydroxy acid **99** for the construction of the 18-membered macrocyclic skeleton of **1**. The 2,6-*cis*-substituted tetrahydropyran ring of **99** would be crafted via a Lewis acid-mediated intramolecular allylation of α-monochloroacetoxy ether **100** [67], which in turn should be available from carboxylic acid **101** and alcohol **102**.

The synthesis of carboxylic acid **101**, as shown in Scheme 18, commenced with anion coupling of known dithiane **30** [11] (previously synthesized by Smith et al. from (*R*)-3-butyn-2-ol over five steps) and epoxide **103** (available in three steps from 5-hexen-1-ol) to give alcohol **104** (76%). After deprotection of the dithioacetal, the resultant ketone was reduced with DIBALH. The derived 1,3-diol **105** was acetalized with *p*-methoxybenzylidene dimethyl acetal and then reduced with

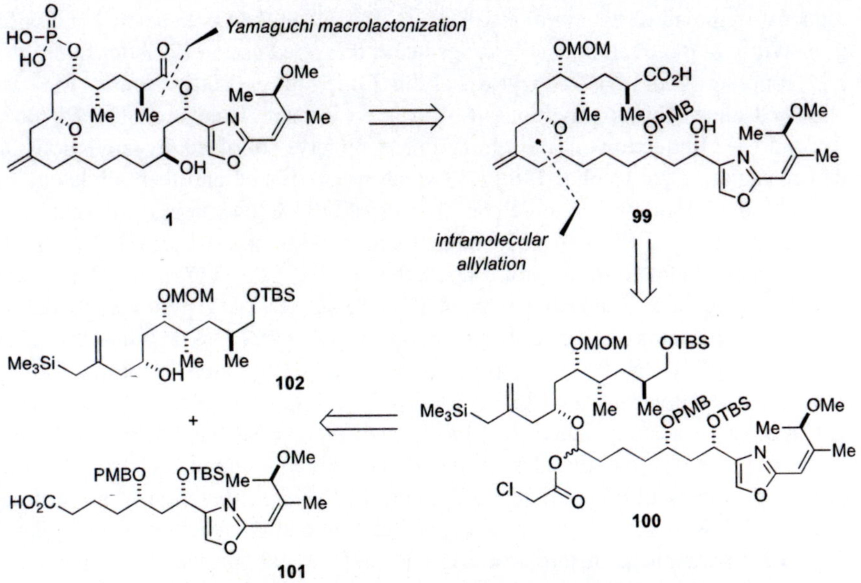

Scheme 17 Synthetic strategy toward (−)-enigmazole A (1) by Kadota et al

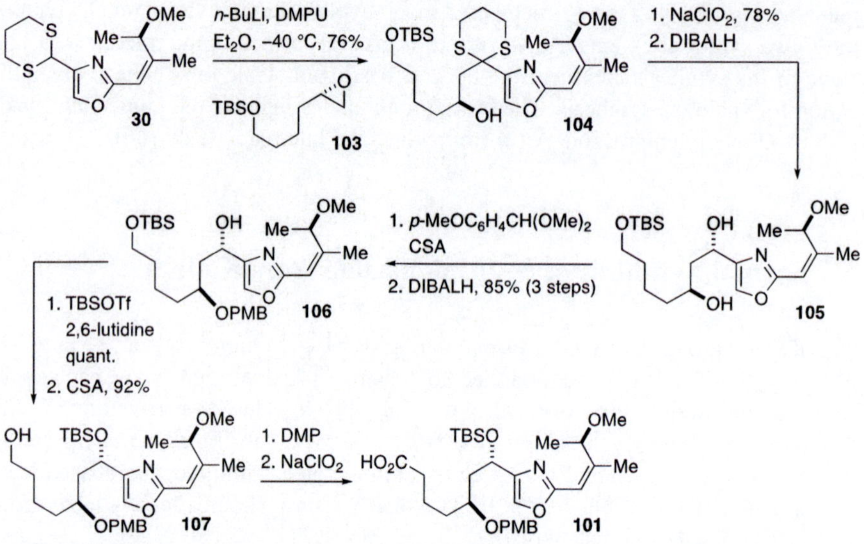

Scheme 18 Synthesis of carboxylic acid **101**

DIBALH to deliver alcohol **106**. Additional four-step sequence including protecting group manipulations and two-stage oxidation led to carboxylic acid **101**.

The synthesis of alcohol **102** started with Evans *syn*-aldol reaction [68] of 3-benzyloxypropanal (**108**) with *N*-propionyl (*R*)-4-benzyloxazolidin-2-one (**109**) (97%, d.r. >20:1), followed by MOM protection of the resultant alcohol **110** and reductive removal of the chiral auxiliary, to give alcohol **111** (Scheme 19). Oxidation, Wittig homologation, and hydrogenation delivered ester **112**. After hydrolysis and coupling with (*S*)-4-benzyloxazolidin-2-one, the resultant imide **113** was alkylated under Evans conditions [69] to provide methylated product **114** (86%, d.r. >20:1). Reduction of the chiral imide moiety, followed by silylation and debenzylation, gave alcohol **115** [70], which was oxidized and then allylated [46] with chiral allylborane **116** to deliver alcohol **102** (88%, two steps, d.r. >20:1).

Esterification of **101** and **102** under Shiina conditions [30] afforded ester **117** (83% overall yield from alcohol **107**) (Scheme 20). Ester **117** was reduced with DIBALH and in situ acylated with $(ClCH_2CO)_2O$ (DMAP, pyridine) to deliver α-monochloroacetoxy ether **100**, which was then exposed to $MgBr_2•OEt_2$ (5 Å MS, CH_2Cl_2/CH_3CN, 0°C) to induce intramolecular allylation, affording 2,6-*cis*-substituted tetrahydropyran **118** in 87% yield for the two steps. Intramolecular allylation of α-acetoxy ethers has been extensively investigated by the Kadota group over the past two decades and its utility has been demonstrated in a number of total syntheses of complex natural products [71, 72]. Liberation of the primary alcohol of **118**, followed by a two-stage oxidation and desilylation, gave hydroxy acid **99**. Yamaguchi macrolactonization [29] of **99** provided 18-membered macrolactone **119** (75%, four steps). Finally, deprotection of the MOM group afforded alcohol **54**, which had been transformed into (−)-enigmazole A (**1**) in

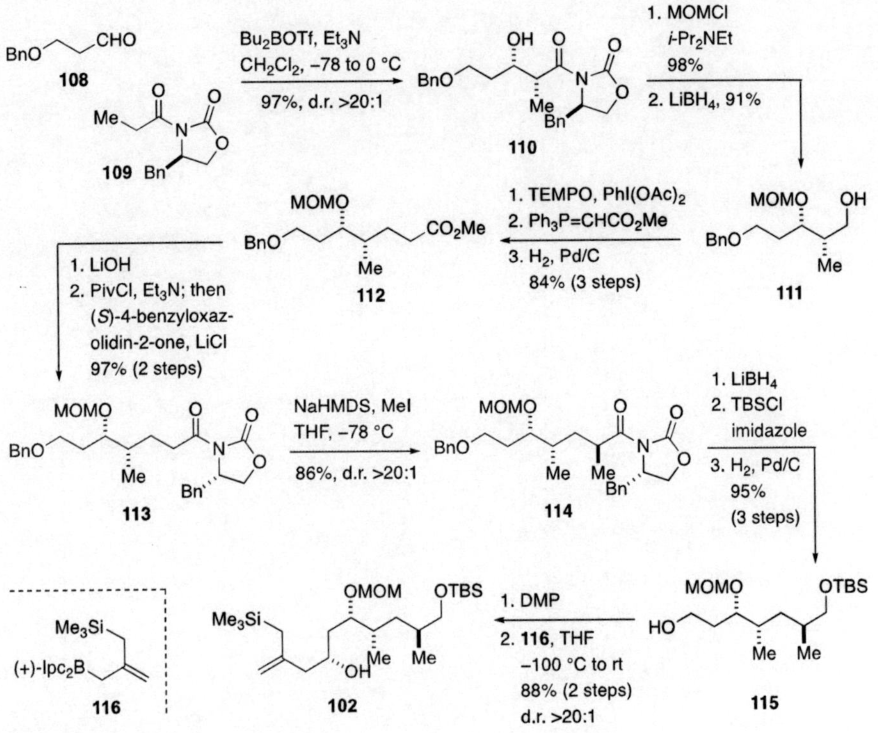

Scheme 19 Synthesis of alcohol **102**

three steps by Smith et al. [11, 12]. Thus, a formal synthesis of **1** was successfully completed.

The Kadota synthesis of (−)-enigmazole A (**1**) was achieved in 23 linear steps from 3-benzyloxypropanal (8.4% overall yield) to intercept the Smith intermediate **54**. Highlights of the present synthesis include a Lewis acid-mediated intramolecular allylation of an α-monochloroacetoxy ether for stereoselective synthesis of the 2,6-*cis*-substituted tetrahydropyran ring and a Yamaguchi macrolactonization to construct the 18-membered macrocycle. Noticeably, the entire synthetic path was fully stereocontrolled and the overall yield was remarkable.

7 Conclusions

As summarized in this chapter, five total and formal syntheses of (−)-enigmazole A (**1**) have been reported since the disclosure of the isolation paper by Gustafson. Each work is based on its own unique synthetic planning that reflects the creativity of individual groups. Specifically, five distinct ways to forge the 2,6-*cis*-substituted tetrahydropyran ring of **1**, i.e., hetero-Diels–Alder cycloaddition, Petasis–Ferrier

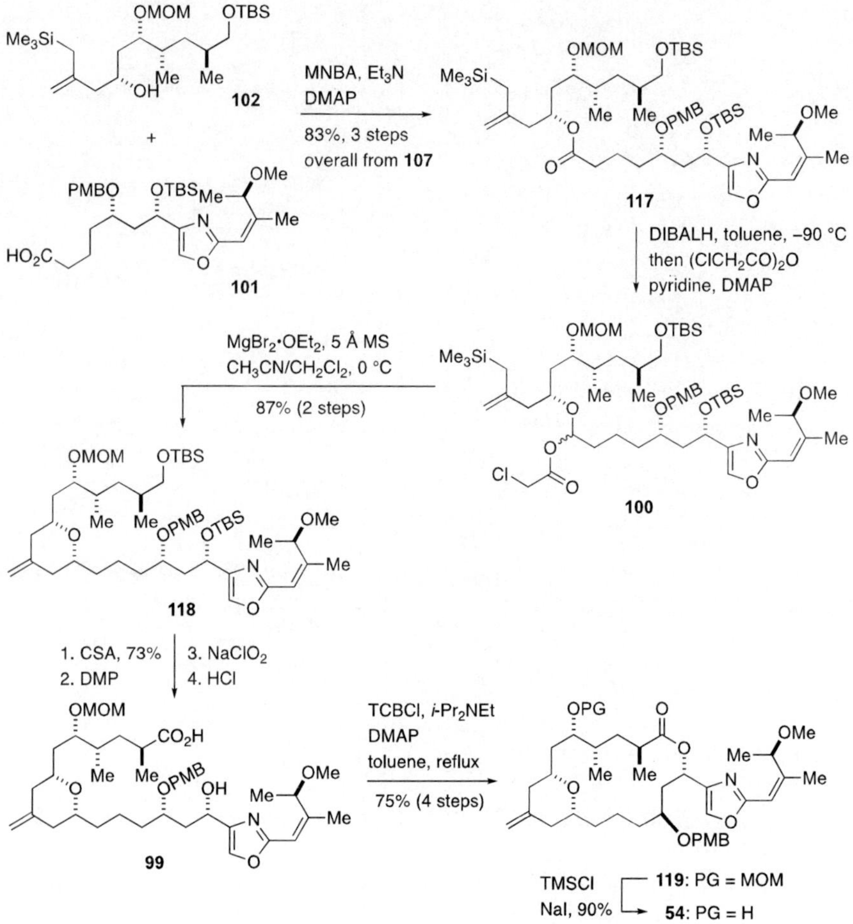

Scheme 20 Formal synthesis of (−)-enigmazole A (**1**) by Kadota et al

union/rearrangement, tandem [3,3]-sigmatropic rearrangement/transannular hydroalkoxylation, tandem olefin cross-metathesis/intramolecular oxa-Michael addition, and intramolecular allylation, are noticeable. Meanwhile, only limited methods are currently available for constructing the 18-membered macrocycle of **1**. Molinski, Smith, and Kadota demonstrated that macrolactonization of the 18-membered lactone skeleton of **1** required appropriately protected precursor *seco*-acids. As shown by Fürstner and Fuwa, ring-closing alkyne/alkene metathesis appears to be more versatile method for closing the macrocycle because of the flexibility of the bond-forming position and the functional group tolerance. All the five syntheses of **1** required 22–29 steps from commercially available materials, and these numbers may indicate the current limit of efficiency in total synthesis of complex polyketide macrolides. Future directions in this area will involve the development of new chemo-, regio- and stereoselective methods that enable rapid construction of polyketide fragments and complex ring systems.

References

1. Moss GP, Smith PAS, Tavernier D (1995) Pure Appl Chem 67:1307
2. McGuire JM et al (1952) Antibiot Chemother 2:281
3. Wiley PF, Gerzon K, Flynn EH, Sigal Jr MV, Weaver O, Quarck UC, Chauvette RR, Monahan R (1957) J Am Chem Soc 79:6062
4. Mills HH et al (1965) Tetrahedron Lett:679
5. Pettit GR, Herald CL, Doubek DL, Helard DL, Arnold E, Clardy J (1982) J Am Chem Soc 104:6846
6. Hirata Y, Uemura D (1986) Pure Appl Chem 58:701
7. Gustafson KR et al (2010) J Am Chem Soc 132:10278
8. Nagai Y, Kusumi T (1995) Tetrahedron Lett 36:1853
9. Ohtani I, Kusumi T, Kashman Y, Kakisawa H (1991) J Am Chem Soc 113:4092
10. Molinski TF et al (2010) J Am Chem Soc 132:10286
11. Smith III AB et al (2015) J Am Chem Soc 137:15426
12. Smith III AB et al (2018) J Org Chem 83:6110
13. Fürstner A et al (2016) Angew Chem Int Ed 55:1406
14. Fuwa H et al (2018) Angew Chem Int Ed 57:5143
15. Kadota I et al (2018) Tetrahedron Lett 59:4492
16. Duboudin JG, Jousseaume B, Bonakdar A, Saux A (1979) J Organomet Chem 168:227
17. Nugent WA (1999) Chem Commun:1369
18. Walsh PJ (2005) J Am Chem Soc 127:13138
19. Negishi E et al (1977) J Chem Soc Chem Commun:683
20. Negishi E et al (1977) J Org Chem 42:1821
21. Krasnokutskaya EA et al (2007) Synthesis:81
22. Corey EJ et al (1989) J Am Chem Soc 111:5495
23. Corey EJ et al (1993) Organic Synth 71:22
24. Narasaska K, Pai F-C (1984) Tetrahedron 40:2233
25. Huebner CF et al (1983) J Med Chem 26:1277
26. Roush WR et al (1985) J Am Chem Soc 107:8186
27. Roush WR et al (1990) J Am Chem Soc 112:6348
28. Jacobsen EN et al (1999) Angew Chem Int Ed 38:2398
29. Yamaguchi M et al (1979) Bull Chem Soc Jpn 52:1989
30. Shiina I (2004) J Org Chem 69:1822
31. Boden EP, Keck GE (1985) J Org Chem 50:2394
32. Paterson I et al (1998) Tetrahedron 54:11955
33. Paterson I et al (1998) Tetrahedron Lett 39:6041
34. Lombardo L (1987) Organic Synth 65:81
35. Watanabe Y et al (1997) Tetrahedron Lett 38:7407
36. Petasis NA et al (1996) Tetrahedron Lett 37:141
37. Smith III AB et al (2008) Acc Chem Res 41:675
38. Smith III AB et al (1997) J Am Chem Soc 119:6925
39. Smith III AB et al (2003) J Am Chem Soc 125:14435
40. Smith III AB et al (2006) J Am Chem Soc 128:66
41. Hicks DR, Fraser-Reid B (1974) Synthesis:203
42. Lu Z, Ma S (2006) J Org Chem 71:2655
43. Kirihara M et al (2013) Tetrahedron Lett 54:5477

44. Dai W-M (2010) Chem A Eur J 16:11530–11534
45. Myers AG et al (1997) J Am Chem Soc 119:6496
46. Brown HC et al (1986) J Org Chem 51:432
47. Fürstner A (2013) Angew Chem Int Ed 52:2794
48. Fürstner A et al (2013) Science 341:1229713
49. Keck GE et al (1993) J Am Chem Soc 115:8467
50. Smith III AB et al (1993) J Org Chem 58:3703
51. Corey EJ, Fuchs PL (1972) Tetrahedron Lett:3769
52. Stark CBW et al (2011) Org Lett 13:4164
53. Fürstner A et al (2010) J Am Chem Soc 132:11045
54. Fürstner A et al (2012) Chem A Eur J 18:10281
55. Toste FD et al (2014) Acc Chem Res 47:889
56. Dudley GB et al (2009) Org Biomol Chem 7:4149
57. Fuwa H et al (2010) Org Lett 12:1636
58. Fuwa H et al (2012) Org Biomol Chem 10:8108
59. Nakata M et al (2004) Synlett:1686
60. Evans DA, Hoveyda AH (1990) J Am Chem Soc 112:6447
61. Hoveyda AH et al (2000) J Am Chem Soc 122:8168
62. Fuwa H, Sasaki M (2016) Bull Chem Soc Jpn 89:1403
63. Nakata M et al (2000) Synlett:1306
64. Nolan SP et al (2007) Chem A Eur J 13:6437
65. Grubbs RH et al (1999) Org Lett 1:953
66. Fuwa H et al (2020) Chem Asian J. https://doi.org/10.1002/asia.202001015
67. Kadota I, Yamamoto Y et al (2007) J Org Chem 72:8371
68. Evans DA et al (1981) J Am Chem Soc 103:2127
69. Evans DA et al (1982) J Am Chem Soc 104:1737
70. Kadota I et al (2014) Heterocycles 89:515
71. Kadota I, Yamamoto Y et al (2006) Tetrahedron Lett 48:219
72. Kadota I, Yamamoto Y et al (2009) Org Lett 11:2531

Part V
Isolation and Synthesis of Middle-Weight Molecular Compounds with a Long Carbon Chain

Top Heterocycl Chem (2021) 58: 389–414
https://doi.org/10.1007/7081_2020_48

Published online: 17 January 2021

Polyketides Biosynthesis in Marine Sponges of the Family Theonellidae

Agustinus Robert Uria and Toshiyuki Wakimoto

Contents

Abstract Marine sponges often contain middle-weight molecular compounds with a long carbon chain, which are mainly polyketides, or polyketides partially fused with peptides. Some of them show highly potent cytotoxicity and promising anti-cancer activity. They have been the target molecules of total synthesis, because of their unique chemical structures and challenging scaffolds. Thus, the sponge-derived middle-weight molecules are hallmarks of natural products in terms of chemical structures and potent biological activities. However, the biosynthesis of these molecules remains largely unknown. The real producers have been proposed to be microbial symbionts, but only a few cases have been experimentally validated. This section introduces the recent advance of biosynthetic studies on sponge-derived compounds with the special emphasis on polyketides or polyketide-peptide hybrid molecules from marine sponges that belong to Theonellidae family.

Keywords Calyculin · Discodermolide · Marine sponges · Misakinolide · Onnamide · Polyketide biosynthesis

A. R. Uria and T. Wakimoto (✉)
Faculty of Pharmaceutical Sciences, Hokkaido University, Sapporo, Japan
e-mail: wakimoto@pharm.hokudai.ac.jp

1 Introduction

Marine sponges have been recognized as a prolific source of structurally diverse compounds with highly potent cytotoxic activity [1]. A notable example is halichondrin B from the marine sponges *Halichondria okadai* [2] and *Lissodendoryx* sp. [3, 4], which inspired the chemical synthesis of eribulin mesylate [5, 6] as a new chemotherapy agent for the treatment of metastatic breast cancer [7]. The unique structures and prominent biological activities of sponge-derived compounds [8] have attracted considerable attention of scientists to investigate their biosynthesis. However, so far only a few sponge-derived metabolites have been evaluated in terms of the biosynthetic pathways. One of the main reasons is that feeding experiment with labeled precursors, which is commonly used in the biosynthetic studies of terrestrial plants and microorganisms, has been difficult to apply in marine sponges, although there were a few successful examples reported [9].

Moreover, "symbiont hypothesis" has long been proposed that many sponge-derived compounds are produced by as-yet-unidentified symbiotic microorganisms, as they shared structural similarities of chemical scaffolds with typical bacterial compounds [10, 11]. This arises important questions about the identity of the actual microbial producers and how sponge-derived bioactive molecules are biosynthesized inside microbial cells. The limitation to cultivate the majority of microbial symbionts [12, 13] suggests the difficulty in identifying and growing the true producers for compound production. A potential problem is that laboratory cultivation may destroy any cell-to-cell communication between a target symbiont and the host or among symbionts in the natural environments that might be important for growth. Symbiotic bacteria may coexist with their host or other bacteria in natural habitats to get essential nutrients and substrates that are not present in standard media [14, 15]. Although cultivation conditions could be found, metabolites of interest might not be released due to the lack of required environmental signals [16].

Cultivation-independent experimental support for the microbial origin of sponge-derived natural products was initially provided by Faulkner and coworkers who conducted microbial cell separation from the Palauan *Theonella swinhoei* followed with chemical analysis of the cell fractions using ^{1}H- NMR spectroscopy and Transmission Electron Microscopy (TEM) observation. They found the highest concentration of the cytotoxic polyketide swinholide A in the mixed unicellular bacterial fraction, but not in sponge-associated cyanobacteria, while the antifungal peptide theopalauamide in the filamentous bacterial fraction [10, 17]. Further 16S-rRNA gene analysis showed the identity of such as-yet uncultivated filamentous bacteria as "*Candidatus* Entotheonella palauensis" [18]. However, chemical localization alone is not enough to provide convincing evidence for the bacterial origin due to the possibility that natural products transport across different cell types and whole tissues [11, 16].

The extreme complexity of sponge-associated microbiome containing numerous homologous genes from diverse nontarget pathways [19–21] suggests that genetic analysis for identifying a target pathway in a marine sponge is very challenging without an effective strategy. Using a metagenomics-based strategy, Piel and colleagues isolated onnamide biosynthetic gene cluster from the Japanese sponge *Theonella swinhoei* [22]. The typical bacterial architecture of the isolated cluster marked the first genetic proof for the microbial origin of sponge-derived natural products [22]. Further development of metagenomic approaches has led to the isolation of the biosynthetic gene clusters (BGCs) for other sponge-derived complex compounds, such as calyculin A [23] and misakinolide A [24], which were both attributed to as-yet uncultured symbionts *Entotheonella* as the producers [23, 24]. Recent development of metagenomics and single-cell analysis supported by new DNA sequencing technologies and bioinformatic tools have opened completely new opportunities for the rapid discovery of biosynthetic pathways from as-yet uncultivated bacteria [25–27]. This offers new exciting challenges to produce rare sponge-derived bioactive compounds in sustainable ways by heterologous production in easily culturable bacteria [16, 26]. Further understanding of the underlying biochemistry and enzymatic mechanisms would enable synthetic biology application for the rational diversification of natural products to produce structurally novel analogs with improved pharmacological profiles [28].

Among marine sponges, Theonellidae family (order Tetractinellida, previously known as order Lithistide) [29] has been well-studied for natural product chemistry. Particularly members of the two genera *Theonella* and *Discodermia* are known as an important source of unique biologically active polyketides and peptides [8]. In this review, we focus on the biosynthetic aspects of four types of Theonellidae sponge polyketides: (1) onnamides, (2) misakinolides from *Theonella*, (3) calyculin, and (4) discodermolide from *Discodermia*. The medium-sized structures of these polyketides with long carbon chains suggest their biosynthesis is catalyzed by the modular megaenzymes called type I polyketide synthases (PKSs-I). Polyketides, such as onnamides and calyculin, harbor amino acid residues or peptide segments, suggesting that hybrid PKS-I and non-ribosomal peptide synthetase (NRPS) assembly lines are responsible for their biosynthesis.

2 Type I Polyketide Synthase and Non-ribosomal Peptide Synthetase

Type I PKSs are multifunctional enzymes organized into modules. Each module contains catalytic domains responsible for one round of polyketide chain elongation. One module consists of a core domain set: an acyl carrier protein (ACP) domain that plays a role as an anchor for the building block and a ketosynthase (KS) domain that elongates the polyketide chain. The additional domain acyltransferase (AT), either integrated into each module or located externally outside the PKS assembly line, plays a role in selecting an appropriate acyl-CoA building block. During biosynthesis, various functional modifications occur on the β-position of the polyketide chain through the catalytic activity of optional tailoring domains, such as ketoreductase

(KR), dehydratase (DH), and enoylreductase (ER) domains. KR domain is responsible for reducing the β-ketoacyl moiety of an intermediate resulting from the KS-mediated condensation into a β-hydroxyacyl group. Then DH dehydrates a β-hydroxyacyl group to an α,β-enoyl that can subsequently be reduced by ER to generate a saturated acyl [30, 31].

The linear correlation between the resulting intermediates, the PKS domain organization and domain function allows the prediction of a polyketide structure. This correlation is called colinearity rule, which is usually applied for the PKS-I subclass "*cis*-AT PKSs" characterized by the presence of an AT domain integrated into each module [32, 33]. For the other subclass "*trans*-AT PKSs," no integrated AT domain is present in each module. Instead, an alone-standing AT domain usually located outside the assembly line acts in *trans* to load acyl building blocks into the ACP of each module [34]. This subclass contains unusual enzymatic features to generate diverse complex polyketides, such as modules split between two distinct proteins, some presumed inactive domains such as non-elongating KSs, multiple MTs (methyltransferases) and O-MTs (*O*-methyl transferases) that introduce methyl groups, and a PS (pyran synthase) domain generating a pyran ring, and a domain set that introduces a β-branching on the polyketide chain [19, 35].

Non-ribosomal peptide synthetase (NRPS) is composed of a series of modules. Each NRPS module contains minimally three domains: an adenylation (A) for amino acid activation and incorporation, a thiolation (T) for amino acid residue thioesterification, and a condensation (C) for peptide chain elongation via transpeptidation. During peptide biosynthesis, each module acts through the activity of such catalytic domains to incorporate and modify an amino acid unit into the growing peptide chain. Since chain elongation and post-assembly involve the existing 20 proteinogenic amino acids and a larger variety of nonproteinogenic amino and other acids as the building blocks, a remarkable structural and functional diversity of peptides is generated [36]. The biosynthetic mechanism of NRPS assembly line is similar to that of PKS-I assembly line; and therefore, PKS-catalyzed biosynthesis is capable of encompassing NRPS modules to offer the stage for highly modified flamework of the medium-sized compounds [37]. The PKS and NRPS hybrid module can generate nonproteinogenic amino acids, which features mixed polyketide-peptide structures of sponge origin. Some cyclic peptides (e.g., orbiculamide A) isolated from *Theonella* contain unique amino acids, theoalanine and theoleucine, that represent moieties formed by PKS-NRPS hybrid assembly lines [27, 38]. Many type-1 PKS-NRPS hybrid genes have recently been cloned from free-living bacteria such as actinomycetes, but only a very few of these types of biosynthetic gene clusters have been identified in sponge-microbe association.

3 Onnamide-Type Compounds

Onnamide A (**1**) was initially reported from Okinawan marine sponge *Theonella* sp. [39]. It exhibited potent antiviral activity against herpes simplex virus type-1, vesicular stomatitis virus, and coronavirus A-59 [39]. Structurally close analogs included dihydroonnamide A, onnamides B-E [40], and theopederins A-E [41] isolated from *T. swinhoei* chemotype yellow collected in Hachijo Island [42]. Theopederins A and B exhibited promising antitumor activity against P388 cells [43]. Onnamide and theopederin series structurally resemble the antitumor pederin (**2**) from the rove beetles *Paederus* and *Paederidus* [44], mycalamide A (**3**) from the New Zealand sponge *Mycale hentscheli* [45], and psymberin (**4**) from the sponge *Psammocinia* aff. *bulbosa* [46]. Psymberin was also known as irciniastatin A from the Indo-Pacific sponge *Ircinia ramosa* [47]. The occurrence of these structurally related compounds in different macroorganisms suggested the microbial origin of their biosynthesis [19, 42].

A large part of pederin biosynthetic gene cluster was successfully cloned by Piel from the beetle *Paederus fuscipus* [48]. Additional upstream region of the cluster was subsequently isolated [49]. The isolated pederin biosynthetic genes exhibited typical bacterial features, which were found to be located in a symbiotic bacterium closely related to *Pseudomonas aeruginosa* [50]. The predicted biosynthetic enzymes of pederin resemble hybrid *trans*-AT PKS-NRPS systems, consisting of three large proteins (PedI, PedF, and PedH) [48, 49] (Fig. 1a). The typical modular *trans*-AT PKS features of the assembly line are the lack of an integrated AT in each module. Instead, stand-alone ATs were identified separately from the gene clusters. Other unusual characteristics include a non-elongating KS (module 4), and MTs (module 1 and 6) that introduce methyl groups [22, 48]. The module architecture and domain organization of PedI and PedF perfectly matched the pederin structure. The additional PedH corresponding to a variable polyketide side chain and a terminal arginine residue matched to the extended congeners onnamide series [48]. It was proposed that the presence of oxygenase gene (*ped*G) inserted into the cluster was responsible for terminating further chain elongation through oxidative cleavage, leading to the release of pederin molecule instead of onnamide [48]. This pioneering work paved the way for identifying the biosynthetic pathways of onnamide-type compounds in more complex microbial communities of marine sponges known to contain onnamide and theopederin series [22, 51].

By developing metagenomic strategies that involved construction of highly complex 3D-metagenomic libraries and subsequent rapid library screening [52, 53], Piel and colleagues isolated the main part of onnamide biosynthetic genes from *T. swinhoei* chemotype yellow [22, 54], thereby providing the first genetic evidence of the bacterial origin of marine sponge-derived natural products. Further metagenomic sequencing of the enriched fraction of filamentous bacteria associated with this sponge combined with single-cell analysis revealed that onnamide biosynthetic genes were located on a plasmid in "*Candidatus* Entotheonella factor" that belongs to a newly proposed Phylum "Tectomicrobia"

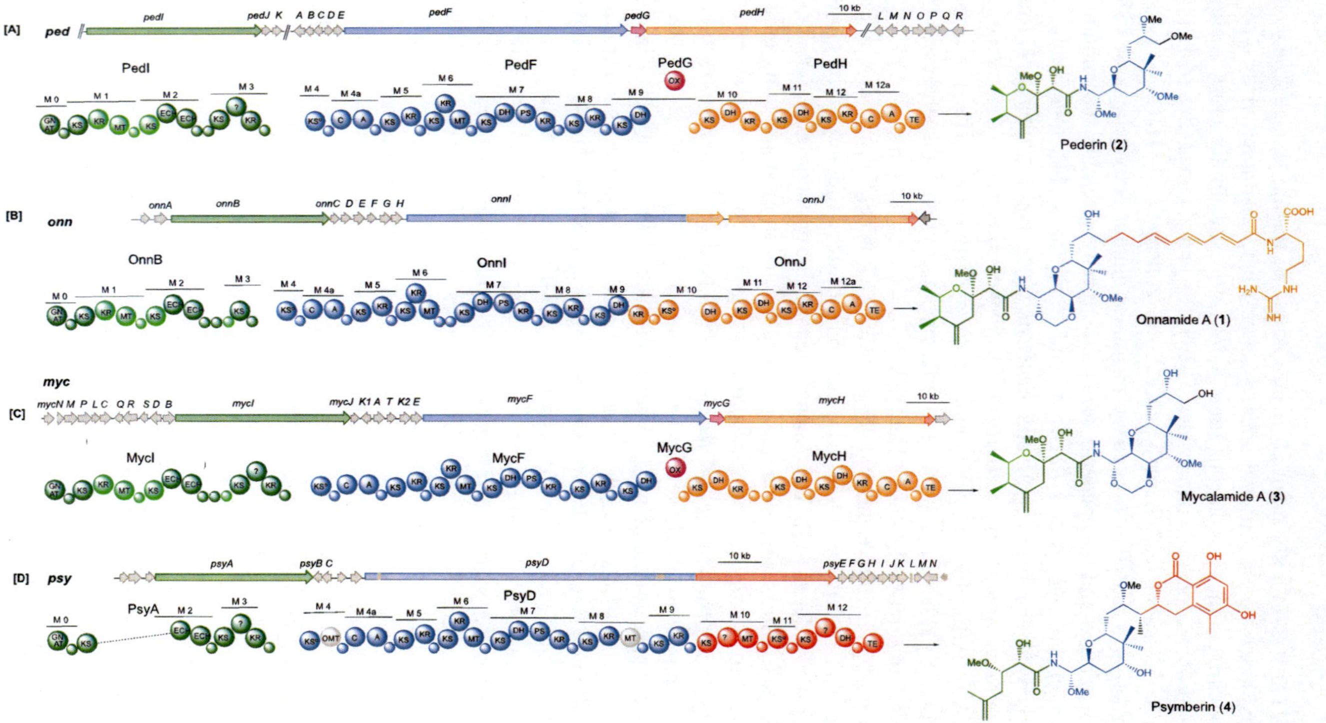

Fig. 1 Comparison of the BGC architecture and assembly line organization among the biosynthetic pathways of four onnamide-type compounds. *GNAT* GCN5-related *N*-acetyltransferase domain, *KS* ketosynthase domain, *KR* ketoreductase domain, *MT* methyltransferase domain, *CR* crotonase domain, *OXY* oxygenase, *CR* crotonase, *PS* pyran synthase, *A* adenylation, *C* condensation, *TE* thioesterase. The small and color circle bullet (•) is ACP, acyl carrier protein domain or T, thiolation domain

[27]. This approach enabled the identification of the missing downstream part of onnamide genes along with other gene clusters coding for the biosynthesis of almost all peptide natural products isolated from the sponge, which were all attributed to such uncultivated symbiont [27]. Piel Lab developed a targeting strategy that correlates between substrate specificity of *trans*-AT KS domain sequences and certain structure moieties [54], enabling the isolation of psymberin biosynthetic gene cluster from the metagenome of *Psammocinia* aff. *bulbosa* [51]. The isolated psymberin biosynthetic genes showed typical bacterial features, but the taxonomic identity of a bacterial producer was as-yet unknown [51]. More recently, the biosynthetic pathway of mycalamide A was described in the microbiome of *M. hentscheli*, which was specifically attributed to an uncultivated gammaproteobacterial symbiont called "*Candidatus* Entomycale ignis" as the producer [55].

The BGC architecture and the encoded biosynthetic enzymes of pederin, onnamide A, mycalamide A, and psymberin share remarkable similarity (Fig. 1). The core structure bearing a tetrahydropyran ring is shared among four compounds, and its formation is hypothetically catalyzed by the PedF/OnnI/McyF proteins and the corresponding upstream part of PsyD protein. In vitro functional analysis provided a convincing evidence that the pyran ring formation is catalyzed by pyran synthase (PS), a unique domain resembling DH domain with a deletion in the HxxxGxxxxP active-site motif [56]. Interestingly, cyclization to form the pyran ring undergoes in a stereoselective manner at the ring closing carbon irrespective of the configuration at the carbon having hydroxy group [56] (Fig. 2).

The PedI/OnnB/MyC proteins containing GCN5-related *N*-acyltransferase (GNAT) is predicted to incorporate acetate as the starter unit for chain initiation. In addition, these proteins contain two crotonase (CR) superfamily domains that might correspond to the formation of the exomethylene group as "Western region" [35, 49]. The hydroxymethylglutaryl-CoA synthase (HMGS)-like protein encoded on *OnnA*/*pedP* genes was proposed to be involved in exomethylene group formation [19, 49]. The corresponding PsyA in psymberin assembly line lacks four domains (KR-MT-ACP-KS) that correlate with a missing hydroxylated and methylated building block in the psymberin Western region (Fig. 3). It was proposed that a PKS module acquisition of PsyA generated the extended versions PedI/OnnB/MyC in pederin/onnamide/mycalamide assembly lines [51]. The downstream part of PsyD harboring the N-terminal five modules was predicted to contribute to the formation of a unique dihydroisocoumarin moiety as the "Eastern region." Compared with the onnamide "Eastern" part, the core structure is extended with three shifted double bonds and an arginine moiety, which is predicted to be formed by the activity of catalytic domains in OnnJ protein. Exchange of downstream modules was proposed as the factor that might covert the isocoumarin-like moiety into pederin/onnamide/mycalamide termini [35, 51].

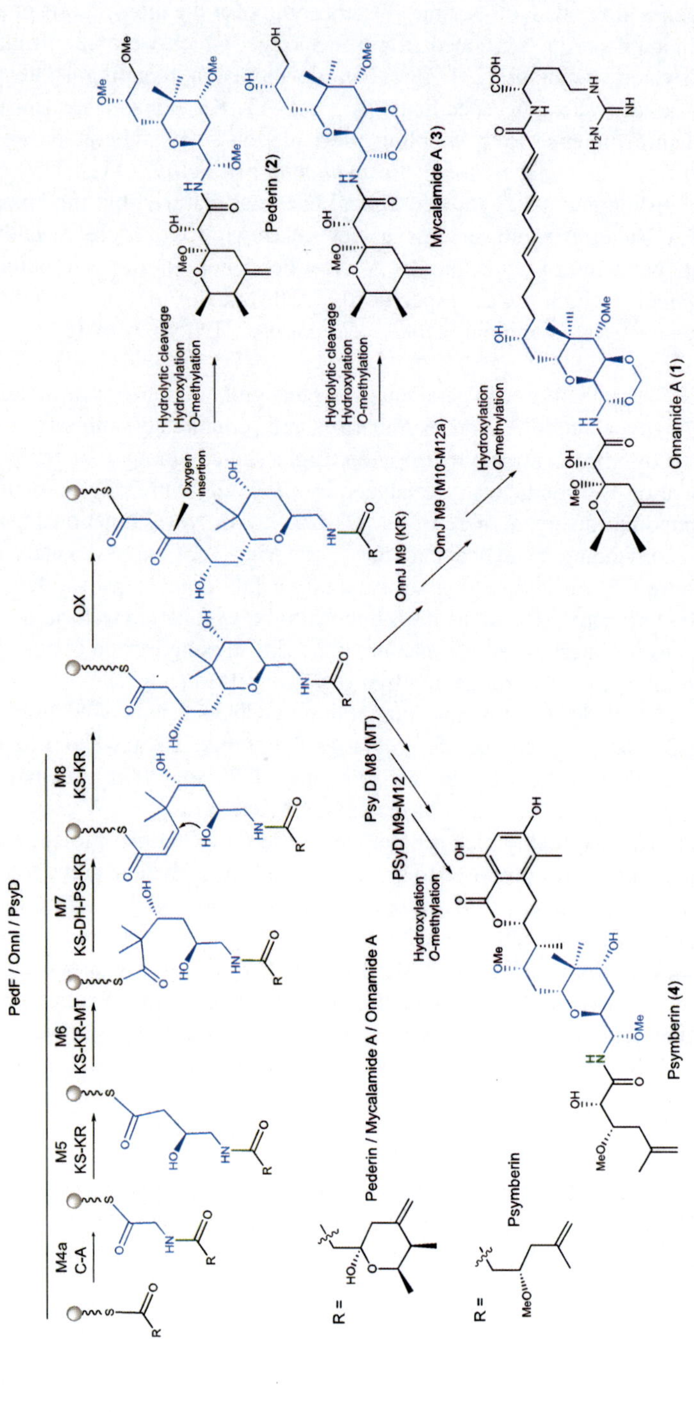

Fig. 2 The biosynthetic steps in the formation of the core structure bearing a tetrahydropyran ring (indicated by blue color) shared among four onnamide-type compounds

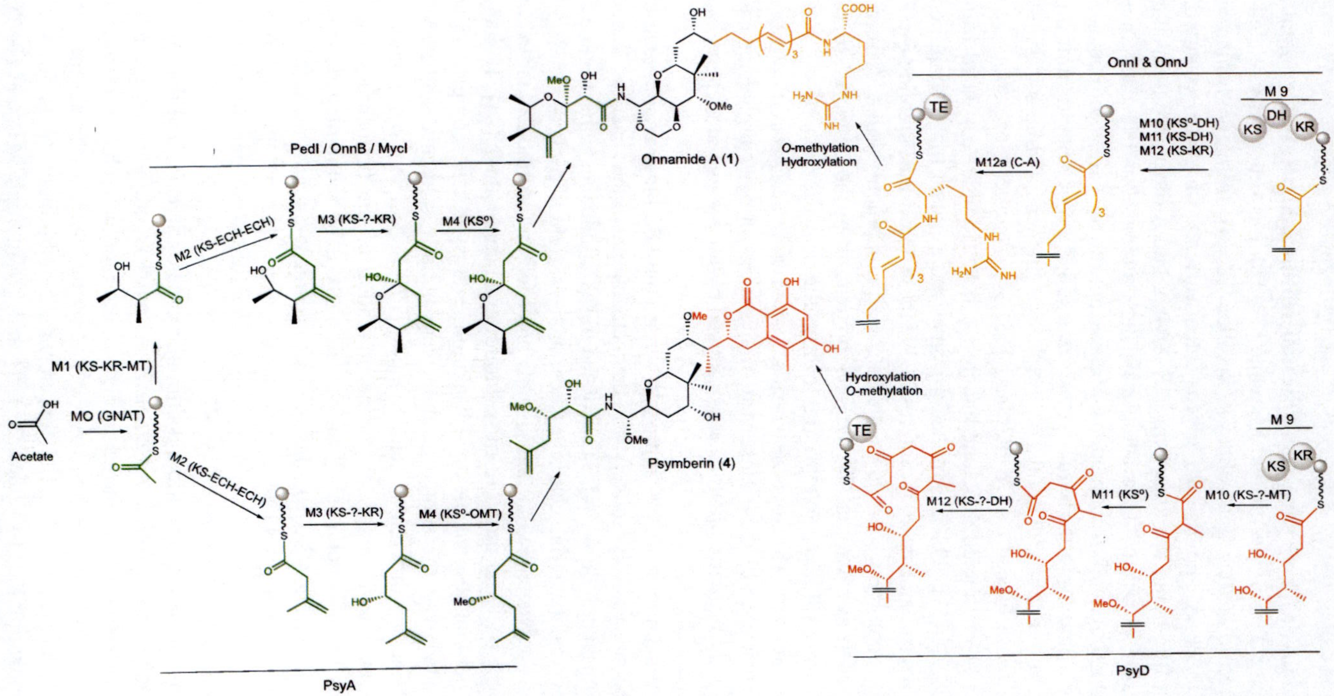

Fig. 3 Comparison between onnamide and psymberin assembly lines in the formation of the "Western" exomethylene group and the "Eastern" isocoumarin-like moiety

4 Misakinolide-Type Compounds

Misakinolide A (**5**) is a symmetric 40-membered dimeric polyketide found in the marine sponge *T. swinhoei* chemotype white [57, 58]. It is structurally related to the 42-membered dilactone macrolide swinholide A (**6**) from *T. swinhoei* [59–61]. Swinholide A and its two glycosylated variants were also reported from members of free-living cyanobacteria [62]. Swinholide A differs from misakinolide A only by the presence of an additional single double bond in each monomer structure. Although both compounds possess antifungal and cytotoxic activities through their binding to the actin cytoskeleton [58, 59, 63], they exhibit different mechanisms of action: swinholide A severs actin filaments, while misakinolide A caps the growing barbed end of filaments [63, 64]. Another close structure is the asymmetric 41-membered dilactone marcrolide hurghadolide A isolated from Red Sea *T. swinhoei* [65], which exhibited 10 times more potent than swinholide A at disrupting microfilaments [65]. Interestingly, these three misakinolide-type compounds share common scaffolds with cytotoxic compounds isolated from other organisms, such as tolytoxin (**7**) and scytophycins from the terrestrial blue-green algae *Scytonema* [66, 67], lobophorolide from the seaweed *Lobophora variegata* [68], and luminaolide (**8**) from the crustose coralline algae *Hydrolithon reinboldii* [69]. The occurrence of these close structural analogs in other organisms suggests that misakinolide-type compounds are produced by the sponge-associated bacterial symbionts [24, 26].

Using metagenomics-based approach, a misakinolide biosynthetic gene cluster (~90 kb) was successfully isolated from the metagenome of the Japanese *T. swinhoei* containing misakinolide A [24]. The isolated misakinolide biosynthesis gene cluster exhibited typical bacterial characteristics such as tightly packed genes, free of introns and polyadenylation sites, and preceded by Shine-Dalgarno motifs, suggesting the bacterial origin of misakinolide-type compounds [24]. To identify a bacterial symbiont responsible for misakinolide biosynthesis, complex microbial cells associated with the sponge sample were initially separated and sorted into several fractions by differential centrifugation [17]. Subsequent pyrosequencing of 16S rDNA amplicons suggested the accumulation of the filamentous bacterium "*Candidatus* Entotheonella" [24]. Single-cell analysis by combining fluorescence-activated cell sorting (FACS), multiple displacement amplification (MDA), and diagnostic PCR techniques [70–72] was proven as an efficient approach to connect misakinolide biosynthetic genes to *Entotheonella* 16S rRNA gene in the amplified DNA of some single filaments [24].

Bioinformatic analysis of the misakinolide gene cluster showed that it encoded four large polypeptides (*misC*, *misD*, *misE*, and *misF*) along with a stand-alone AT enzyme (*misG*) (Fig. 4). As shown in Fig. 4a, the module architecture and domain organization showed typical *trans*-AT PKS features, such as the absence of integrated AT domains, modules split between two distinct proteins (modules 5, 10, and 14), non-elongating KSs (modules 3, 11, 14 and 19), multiple MTs and *O*-MTs that introduce methyl groups (modules 3, 5, 6, 7, 9, 11, and 17), and a PS domain

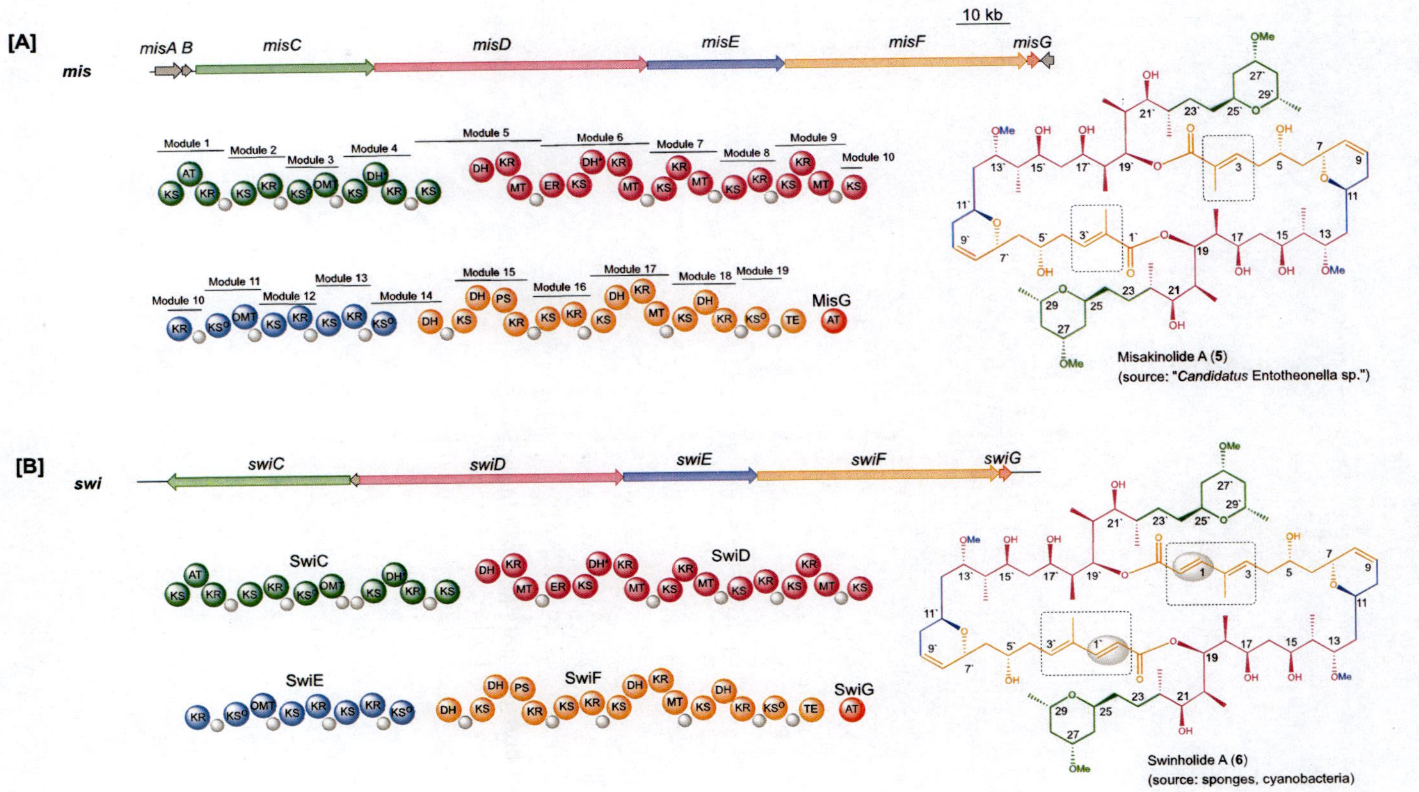

Fig. 4 Comparation of the module architecture and domain organization between the assembly lines of misakinolide A, swinholide A, and the two closely related structures tolytoxin and luminaolide. Biosynthetic gene clusters of swinholide (*swi*), misakinolide (*mis*), tolytoxin (*tto*), and luminaolide (*lum*) are shown in scale. Note: *AT* acyltransferase, *DH* dehydratase, *ER* enoyl reductase, *KR* ketoreductase, *KS* ketosynthase, *KS*° non-elongating KS, *MT* methyltransferase, *OMT O*-methyltransferase, *PS* pyran synthase, *TE* thioesterase, and (•) acyl carrier protein

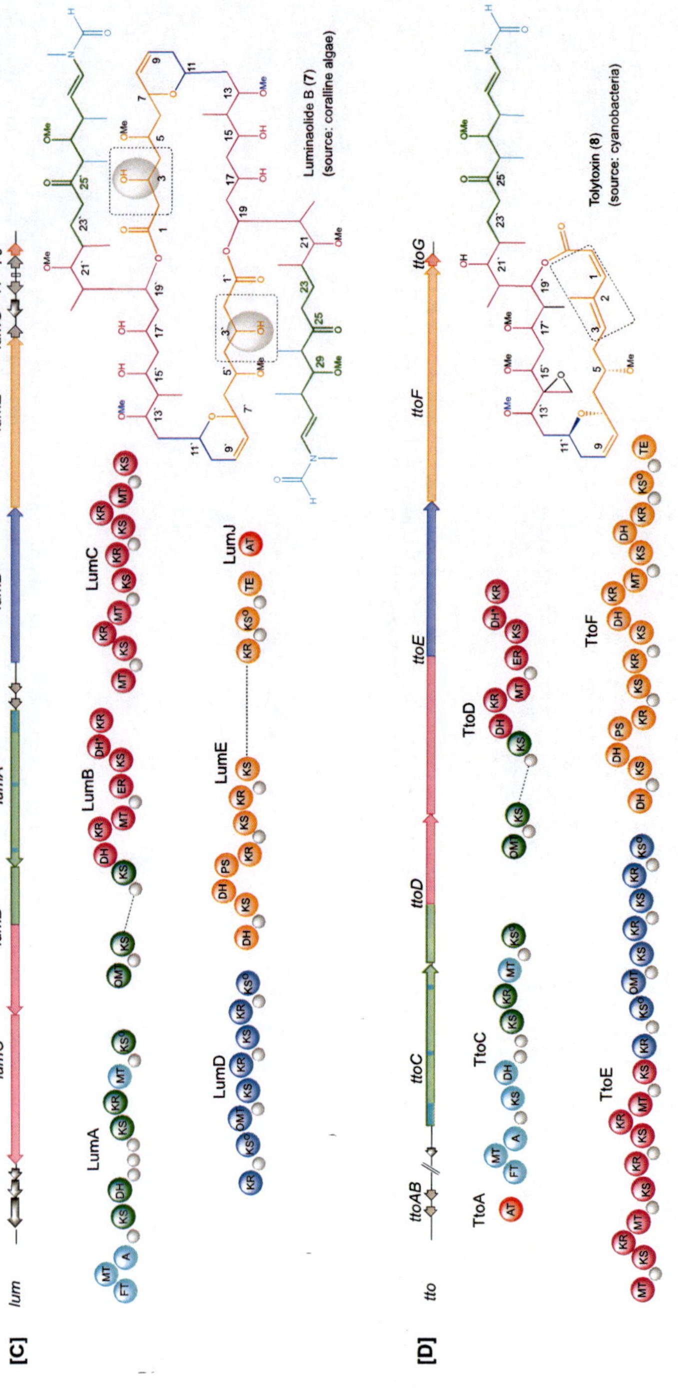

Fig. 4 (continued)

generating a pyran ring (module 15). The intermediate specificity and stereochemical configuration of the assembly line predicted by KS and KR sequence analyses indicated that the identified pathway matched to the structure and stereochemistry of misakinolide A [24, 26]. Interestingly, MisF protein of the misakinolide assembly line contains a penultimate module (module 18) that match to the extended congeners swinholide A and hurghadolide A. Since misakinolide A instead of swinholide A and hurghadolide was detected in the sponge extract, it was proposed that module skipping may occur during the biosynthesis probably due to the nonfunctional module 18 of MisF protein [24], leading to the direct transfer of the miskinolide monomer from ACP17 to ACP18 without elongation. Subsequent TE-catalyzed ligation of two misakinolide monomers would result in macrocyclization that produces the dimeric polyketide misakinolide A [26].

More recently, swinholide biosynthetic gene cluster was reported from the terrestrial cyanobacterium *Nostoc* sp. UHCC 0450 [73]. The module architecture and domain organization of the biosynthetic enzymes encoded by this swinholide gene cluster was very similar to those encoded by misakinolide gene cluster [24]. The only small difference is the presence of a single ACP in the module 3 of MisC protein while two ACPs in the corresponding module of the SwiC protein [24, 73] (Fig. 4a, b). Interestingly, the biosynthetic enzymes corresponding to the formation of cyclic core skeleton in misakinolide A (**5**) and swinholide A (**6**) shared similar domain organization with those in other structurally related macrolides, such as tolytoxin (**8**) and luminaolide B (**7**), suggesting the common ancestor of their biosynthetic genes [24].

Based on the bioinformatic analyses, it was proposed that the cyclic core skeleton formation starts from an intermediate bearing α-methyl moiety that undergoes a series of chain elongation and modification catalyzed by MisD/SwiD to MisF/SwiF in the misakinolide-type compound pathways (Fig. 5). The pyran rings as a part of the cyclic core skeleton are biosynthetically formed by the action of the first two modules in MisF/SwiF/TtoF/LumE containing DH and PS domains [24]. This was supported by in vitro enzymatic assay demonstrating that this PS generated a dihydropyran ring [24, 56]. Therefore, it was proposed that the pyran ring formation starts with generating an α, β-*E*-olefinic moiety (C6–C7) through DH-catalyzed dehydration of the C7 β-OH group. Subsequent reduction of the C11 hydroxy group to the olefinic moiety catalyzed by PS domain would give an intermediate bearing the C7–C11 pyran ring [26] (Fig. 5).

Five domains in the middle of MisF/SwiF/TtoF are missing in the corresponding protein (LumE) of the luminaolide assembly line [24] (Fig. 6). These domains correspond to the double bonds and methyl moieties next to an ester carbonyl group of the misakinolide/swinholide cyclic core skeleton (Fig. 6a). The lack of these domains in LumE corresponds to the presence of two hydroxy groups near the ester carbonyl of the luminaolide pathway. One of the OH groups would subsequently undergo *O*-methylation to form the methoxy group that might occur after

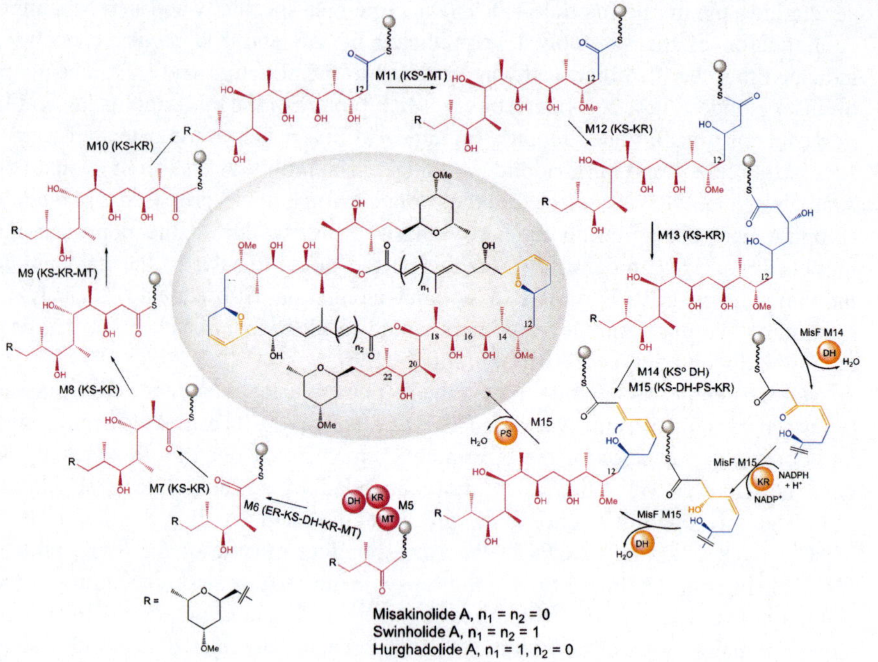

Fig. 5 Biosynthetic origin of the cyclic core skeleton of misakinolide-type compounds, which is shared with that of other structurally related macrolides tolytoxin and luminaolide B

polyketide assembly (Fig. 6b). The presence of this OH group instead of double bond(s) in the cyclic core skeleton may explain the inability of luminaolide B to bind to the actin-binding site surface [74, 75].

The domain organization of the misakinolide assembly line that corresponds to a part of the "tail" portion appeared different from that of tolytoxin and luminaolide B assembly lines. The 'tail' portion of each misaknolide monomer harbors a pyran ring. It was hypothesized that the formation of this pyran ring at the "tail" portion was catalyzed by DH domain in MisC/SwiC via a series of chain elongation and modification starting from acetate as the starter unit [24] (Fig. 6a). Although no PS was present in MisC/SwiC, the absence of glycine in the HxxxGxxxxP motif of the DH domains from MisC/SwiC may indicate the important role of this domain in pyran ring formation [73]. The "tail" portion of luminaolide monomer was proposed to biosynthetically be derived from *N*-methyl-*N*-vinylformamide group that involves three steps of chain extension, keto-reduction into a double bond, *C*-methylation, and *O*-methylation [24] (Fig. 6b). This suggests that the biosynthetic pathways of misaknolide, tolytoxin, and luminaolide emerged from a common ancestor, which were diversified by deletion or acquisition of some domains corresponding to the "tail" portion [24].

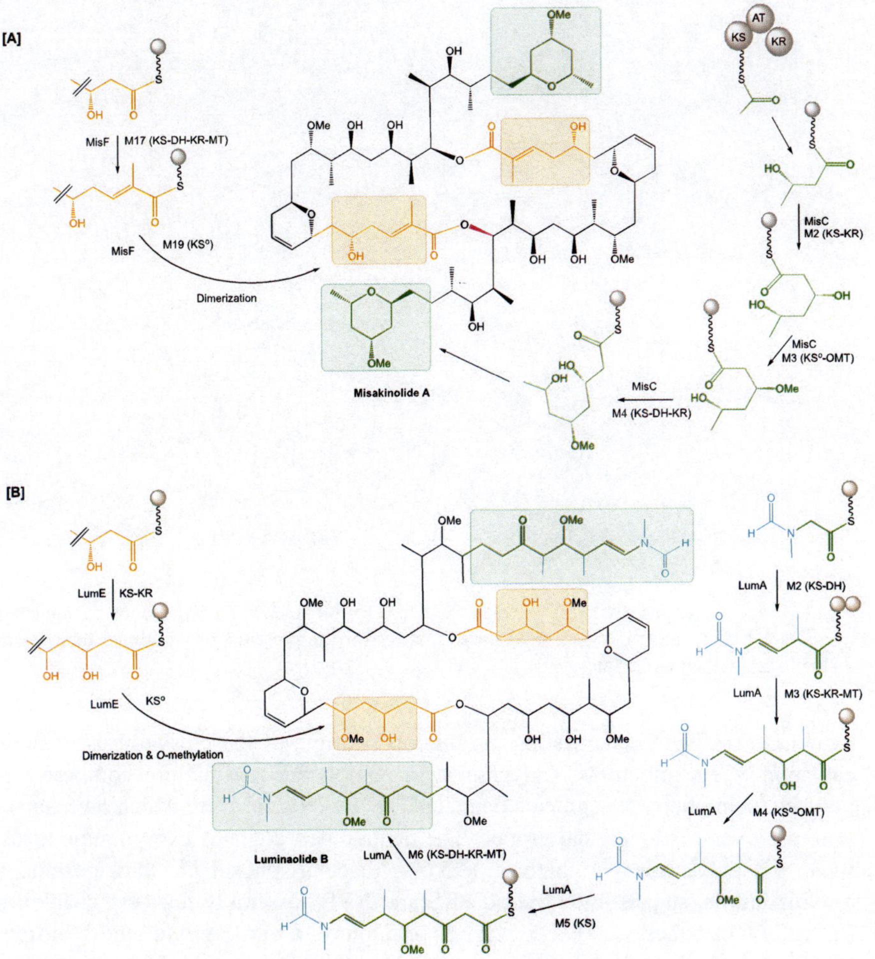

Fig. 6 Comparative biosynthetic origin of the core structure part and the "tail" portion between misakinolide A and luminaolide B

5 Calyculin A

The Japanese marine sponge *Discodermia calyx* was found to contain the highly cytotoxic compound, calyculin A (**9**) [76]. Calyculin A shows potent and specific inhibition against protein phosphatases 1 and 2A, as observed for microcystin and okadaic acid [77–79]. Structural analogs were reported from other marine sponges, such as calyculinamide A (**10**) in the New Zealand deep-water sponge *Lamellomorpha strongylata* [80], geometricin A (11) the Australian sponge *Luffariella geometrica* [81], swinhoeiamide A (12) in the the Papua New Guinean sponge *T. swinhoei* [82], and clavosine A (13) in the Palauan sponge *Myriastra*

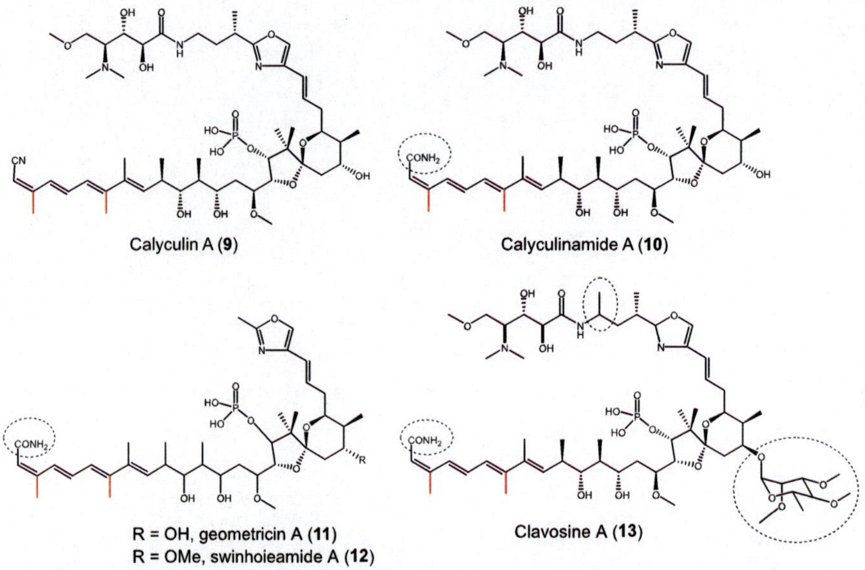

Fig. 7 Calyculin and its structurally related analogs. Two β-branched methyl groups are indicated by red color. Calyculin A differs from close analogs for the presence of additional moieties as marked with dashed circle lines

clavosa [83] (Fig. 7), suggesting the microbial origin of the biosynthesis of these calyculin-type compounds. Calyculin-type compounds are mainly composed of polyketide and peptide segments connected by an oxazole ring, which are characterized by various functional groups. The peptide part contains two γ-amino acids, while the polyketide part harbors a 5,6-spiroacetal, phosphate, tetraene, and a terminal nitrile, suggesting a hybrid PKS and NRPS assembly line responsible for calyculin biosynthesis [84]. The western terminus is a nitrile group whose biosynthetic origin is also of interest. The odd carbon chain number of this compound is apparently contrary to the two-carbon extension cycle in the polyketide assembly line, suggesting the requirement of a modification step to generate this 25-carbon polyketide portion. It was proposed that the tetraene moiety is photoisomerized to afford geometrical isomers, calyculin B, E, and F [84, 85]. The presence of two unusual β-branched methyl groups in this moiety suggests the involvement of *trans*-AT PKS system in calyculin biosynthesis [26].

In attempts to isolate calyculin biosynthetic genes, Wakimoto and colleagues initially identified several KS sequences that belong to *trans*-AT PKS from the sponge metagenome, which subsequently became a basis for designing probes to screen a metagenomic library of 250,000 fosmid clones. Screening of the fosmid library yielded 12 clones harboring calyculin biosynthetic gene clusters ranging from over 150 kb. The single-cell analysis employing laser microdissection revealed that the gene cluster was originated from the sponge-associated symbiont "*Candidatus* Entotheonella sp." [23]. The isolated gene cluster encodes multifunctional

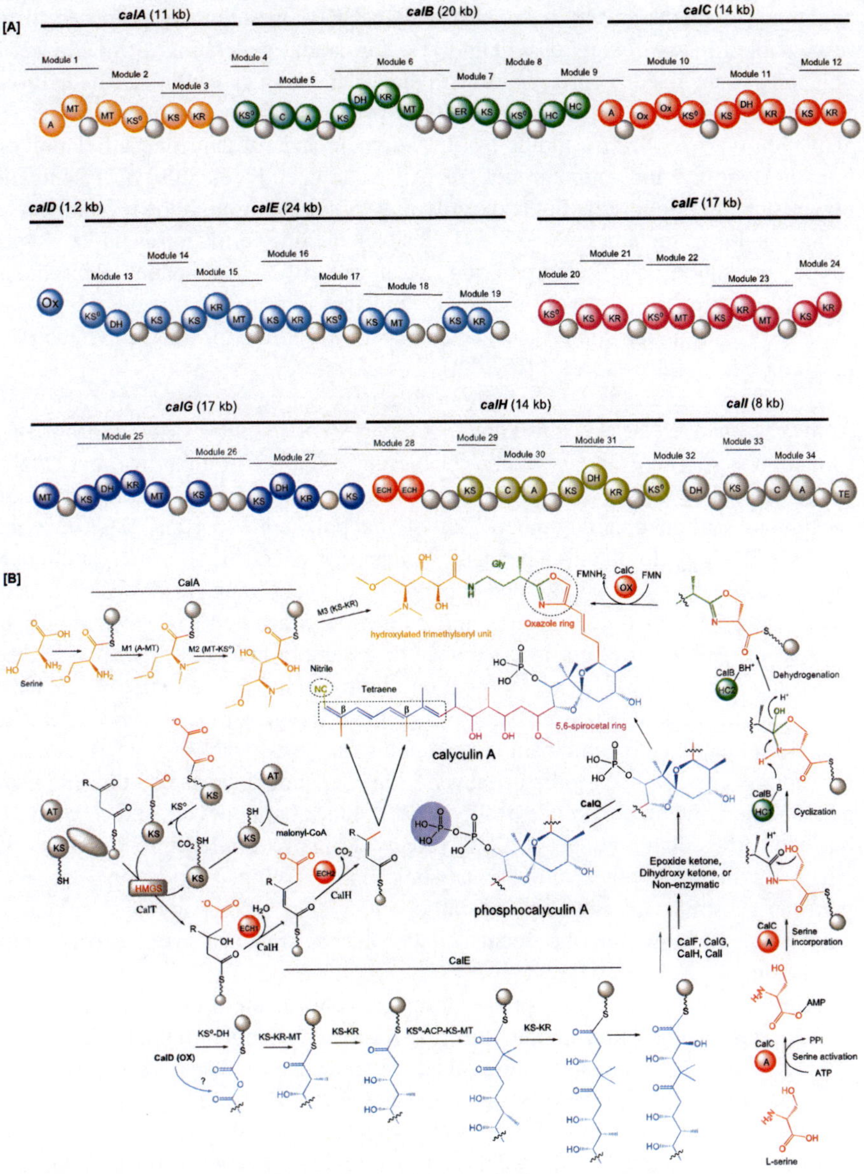

Fig. 8 The domain organization in the calyculin A assembly line (**a**) and biosynthetic origin of important moieties in calyculin A (**b**). Note: *HMGS* 3-hydroxy-3-methylglutaryl synthase (HMGS), *ECH* enoyl-CoA hydratase, *HC* heterocyclisation domain, *Ox* oxidation domain, *A* adenylation, *C* condensation, *TE* thioesterase

PKS-NRPS enzymes, designated as CalA to CalI (Fig. 8a). The upstream cluster region outside of the gene cluster encodes AT (CalY) that might introduce malonate for chain initiation and elongation. Three NRPS modules are present at the beginning of the assembly line. The first two NRPS modules are proposed to be responsible for two rounds of γ-amino acid formation, generating a trimethylseryl unit and glycine residue corresponding to the distal end of calyculin A. The hydroxylated trimethylseryl unit is biosynthetically originated from serine that undergoes activation onto an NRPS module by an A domain and subsequent methylation on both the hydroxyl and amino groups [23, 26]. Subsequently, the third NRPS module may play an important role in the oxazole ring formation through the recruitment of serine residue as the precursor [23, 84]. The oxazole ring formation is initiated with serine activation and incorporation catalyzed by A domain. Subsequent cyclodehydration generates an oxazoline ring that is further oxidized to form an oxazole ring through a similar mechanism recently reported for the C2-symmetrical macrodiolide conglobatin [86] (Fig. 8b).

As expected from the odd carbon chain number, a flavin-dependent oxygenase (CalD) is encoded at the downstream of two PKS modules after oxazole ring formation step. The exact enzymatic function of CalD is still unknown, but carbon chain shortening could take place by CalD-catalyzed reaction according to the contiguous carbon chain number C25 of the polyketide portion. This enzyme shows homology to flavin-dependent oxygenases encoded on *pedG* and *oocK* incorporated in the pederin [48] and oocydin biosynthetic gene clusters [87], respectively, suggesting that they share a similar function. Since PedG is encoded only in the pederin pathway but not in the onnamide assembly line [22], it was suggested that PedG is responsible for the termination of carbon chain of pederin [48]. Piel and coworkers conducted functional analysis of these oxygeneases. It was difficult to obtain the functional recombinant PedG, but instead OocK was obtained as a soluble protein and characterized to be a Baeyer-Villiger monooxygenase that inserts oxygen into the growing polyketide backbone [88]. Since both oocydin and calyculin A has odd carbon chain number, CalD is proposed to have similar function as OocK, which might be responsible for catalyzing Baeyer-Villiger oxidation of α-keto thioester substrate, followed by one-carbon elongation accompanying decarboxylation (Fig. 8). However, the details of the mechanism must await experimental validation.

After the chain shortening process proposed above, the 5,6-spiroacetal ring is formed, raising a question whether this ring formation occurs through cyclization of epoxide/ketone intermediates as described for monensin A [89], hydroxy ketone as reported for reveromycin A [90], or nonenzymatic route. The involvement of enzymes for the formation of spiroacetal moieties in biologically active natural products is best exemplified by RevJ, a spiroacetal synthase responsible for the 6,6-spiroacetal ring formation in reveromycin biosynthesis [90]. However, there is no homolog to RevJ encoded in the calyculin biosynthetic gene cluster. Considering that the 5,6-spiroacetal ring of calyculin A is a thermodynamically stable stereoisomer, it could be constructed in nonenzymatic manner from a keto-diol substrate. The calyculin assembly line is followed by the formation of tetraene moiety containing two β-branching methyl groups at C3 and C7. The domain set required for β-branching methylation is encoded in the corresponding region, which consists of a hydroxymethylglutaryl synthase (HMGS)-like enzyme and two enoyl-CoA hydratases (ECHs) [91]. Surprisingly, several modules are encoded in the downstream region of the domains corresponding to the tetraene part. The function of these extra modules in the nitrile formation is still unknown.

Further functional studies are required not only to verify whether the isolated gene cluster is specific to calyculin biosynthetic pathway but also to understand the mechanism of how calyculin is made inside the producer cell. Even though the entire pathway reconstitution is the best approach, it is still highly challenging to transfer and activate a large gene cluster of more than 150 kb in microbial hosts with different genetic backgrounds. Therefore, the function of key modification enzymes, such as the phosphotransferase CalQ, encoded on the cluster upstream region was investigated. Recombinant CalQ was initially produced by heterologous expression in *E. coli*. Functional analysis in vitro with some calyculin derivatives as putative substrates revealed that the phosphorylation of calyculin A occurred, generating a new derivative phosphocalyculin A, whose cytotoxicity was significantly lower compared to calyculin A [23]. This enzymatic phosphorylation process seems to be essential for the self-resistance of the sponge, since the accumulated potent phosphatase inhibitor is potentially harmful to the host sponge itself. It is therefore reasonable that the less cytotoxic protoxin is released as the end-product of the calyculin biosynthetic pathway. Interestingly, a phosphatase that converts the protoxin to a much more toxic form was observed in the crude enzyme fraction prepared from the sponge. The phosphatase activity is turned on in response to the sponge tissue disruption [23, 84]. The detailed mechanism of this wound-activated bioconversion process is currently under investigation.

6 Discodermolide

Discodermolide (**14**) is a 24-member polyketide carbon skeleton initially isolated in 1990 from the deep-water Caribbean sponge *Discodermia dissoluta* [92]. The gross structure of discodermolide was elucidated by extensive NMR studies, and the relative stereochemistry was defined by single-crystal X-ray crystallography [93]. The first total synthesis along with the absolute stereochemistry of this polyhydroxylated polyketide was reported in 1993 by Schreiber and his coworkers [94, 95]. Although (+)-discodermalide was initially found to exhibit immunosuppressive properties both in vitro [96] and in vivo [96], further bioactivity studies revealed its potent antiproliferative/antimitotic features [97–99] that target microtubule system in a mechanism of action similar to that of the clinically important anticancer drug taxol [98, 99]. Discodermolide belongs to prominent members of microtubule-stabilizing natural product agents and mitotic spindle poisons in the same group as taxol (**15**), epothilones A (**16**) and B (**17**) [100], and eleutherobin (**18**) [101] (Fig. 9). Compared with taxol, (+)-discodermolide was more potent but less cytotoxic [102] with the ability to inhibit taxol-resistant cell lines. More recently, it was found that (+)-discodermolide combined with taxol at low concentrations exhibited a synergetic toxicity by 20-fold towards human carcinoma cell lines

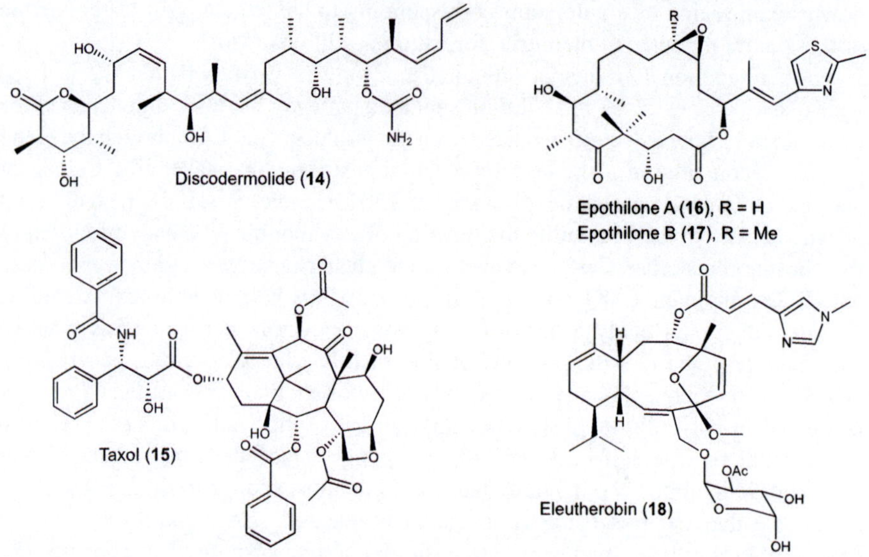

Fig. 9 Prominent members of microtubule-stabilizing natural products

[103, 104], a feature that was not found with both epothilones and eleutherobin [105].

The supply problem for discodermolide is chronic due to the extremely low quantity of this compound in the wild sponges. Obtaining this promising compound for preclinical and clinical trials have been pursued through initially sponge harvesting from natural sources and subsequently chemical synthesis of up to 60 g [93, 105], but the cost for the drug production is very high. Cultivation of the producer microorganisms or heterologous expression of the biosynthetic genes is considered as one of the promising approaches to provide discodermolide sustainably for drug production. In attempts to find discodermolide biosynthetic genes, researchers at Kosan Bioscience conducted metagenomic analysis of microbiome in *D. dissoluta* collected from the coast of Curacao, the Netherlands Antilles [106]. The symbiotic bacterial cells were initially separated from sponge cells as previously described by Bewley et al. (1996) [17]. The resulting separation consisted of (1) fraction containing highly enriched filamentous bacteria, (2) fraction containing a small number of unicellular bacteria and sponge cells, and (3) supernatant containing a mixture of sponge cells and filamentous and unicellular bacteria [106]. Highly abundant PKS and NRPS genes were found in the enriched filamentous bacterial fraction containing *Entotheonella*. Although discodermolide biosynthetic gene clusters have not been identified until now, this comprehensive study revealed the presence of highly diverse symbionts in the sponge *Discodermia* [106], as has been described in *Theonella* [27]. TEM-observation and Fluorescent In Situ Hybridization (FISH) analysis using targeting probes clearly indicated a highly

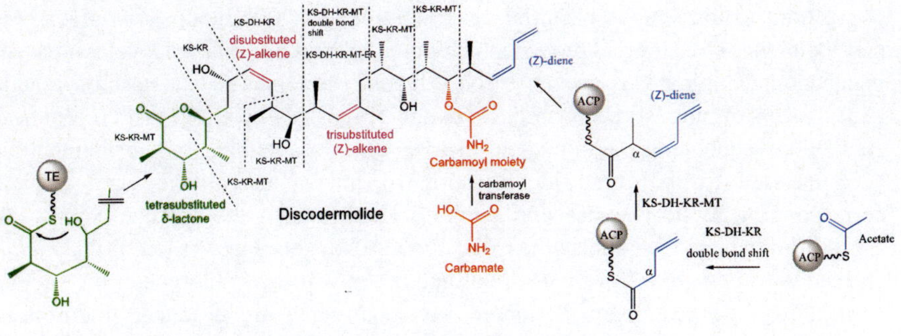

Fig. 10 Predictive model for the biosynthetic origin of discodermolide

abundant filamentous cells *Entotheonella* [106, 107]. The symbiont "Entotheonella sp." was also observed in *D. dissoluta* from the Caribbean Sea using FISH [107].

Discodermolide has 13 stereogenic centers, a tetrasubstituted δ-lactone, one di- and one trisubstituted (*Z*)-alkene, a carbamate moiety, and a terminal diene [105]. The structure of discodermolide suggests that its biosynthesis is catalyzed by a bacterial type I modular polyketide synthase (PKS) [31]. The biosynthetic origin of some structural moieties can be predicted, as described in Fig. 10, based on previous biosynthetic studies reported for polyketides. For example, carbamate/ carbamoyl group is present in some natural products, in which their biosynthetic pathways were identified, exemplified by corallopyronin A [108] and kalimatacin/ batunin [109]. It is proposed that carbamoyl moiety in discodermolide structure is derived from bicarbonate [108], and it is transferred onto the polyketide chain probably by the activity of carbamoyltransferase [109] (Fig. 10). A tetrasubstituted δ-lactone in discodermolide is likely resulted from polyketide chain termination/ release catalyzed by TE domain [31] (Fig. 10).

7 Conclusion

Theonellidae is considered as a gifted sponge family for having highly diverse structures of bioactive secondary metabolites, including polyketides and peptides. Due to the structural similarities of compound scaffolds from this sponge family with typical bacterial compounds, it has long been proposed that sponge-derived natural products are actually produced by microorganisms that live symbiotically or obtained from seawater through filter-feeding. This bacterial origin theory has been well-discussed for this sponge family, particularly the genus *Theonella*. Experimental support to provide convincing proof for the theory has been pursued through various cultivation-independent approaches that involve (1) bacterial cell separation from the sponge hosts, (2) microscopy observation, (3) chemical localization, (4) metagenome mining, (5) single-cell analysis, and (6) metagenomic sequencing. Mechanical cell separation and subsequent microscopy observation show a high population of filamentous bacterial cells identified as "Candidatus *Entotheonella*" that belongs to a newly proposed Phylum "Tectomicrobia." Development of metagenome mining was proven to be effective strategies to isolate biosynthetic gene clusters of interest from complex sponge-bacteria association. With recent next generation sequencing technologies, metagenomic sequencing of the filamentous cells has led to the rapid identification of many biosynthetic gene clusters corresponding to polyketides and peptides known from the sponge. Single-cell analysis has enabled to connect the isolated/identified gene clusters to *Entotheonella*. Of the two *Entotheonella* variants identified in the yellow chemotype of the sponge *T. swinhoei*, "*Ca.* Entotheonella factor" was identified as the producer of almost all secondary metabolites isolated from the sponge. The white chemotype of *T. swinhoei* harbors another variant called "*Ca.* Entotheonella serta," which was shown as the producer of misakinolide and theonellamide. Calyculins were also shown to be produced by "*Ca.* Entotheonella sp.". Interestingly, the presence of *Entotheonella* in the filamentous cell fraction associated with the genus *Discodermia* raises a question whether this as-yet uncultivated symbiont is responsible for the production of *Discodermia*-derived compounds such as discodermolide. Compared with other sponge families, Theonellidae sponges have more intensively been investigated, not only for diverse structures of polyketides and peptides, but also for the biosynthetic pathways and the actual producers. For other sponges known to produce bioactive secondary metabolites, little progress has been made in terms of identifying the biosynthetic pathways or the bacteria responsible for production. Since many important bioactive substances have been reported from non-Theonellidae sponges, it is expected that similar cultivation-independent studies will be carried out for other sponges as well.

References

1. Carroll AR, Copp BR, Davis RA, Keyzers RA, Prinsep MR (2020) Nat Prod Rep 37:175
2. Hirata Y, Uemura D (1986) Pure Appl Chem 58:701
3. Hickford SJH, Blunt JW, Munro MHG (2009) Bioorg Med Chem 17:2199
4. Litaudon M, Hickford SJH, Lill RE, Lake RJ, Blunt JW, Munro MHG (1997) J Org Chem 62:1868
5. Zheng W, Seletsky BM, Palme MH, Lydon PJ, Singer LA, Chase CE, Lemelin CA, Shen Y, Davis H, Tremblay L, Towle MJ, Salvato KA, Wels BF, Aalfs KK, Kishi Y, Littlefield BA, Yu MJ (2004) Bioorg Med Chem Lett 14:5551
6. Seletsky BM, Wang Y, Hawkins LD, Palme MH, Habgood GJ, DiPietro LV, Towle MJ, Salvato KA, Wels BF, Aalfs KK, Kishi Y, Littlefield BA, Yu MJ (2004) Bioorg Med Chem Lett 14:5547
7. Huyck TK, Gradishar W, Manuguid F, Kirkpatrick P (2011) Nat Rev Drug Discov 10:173
8. Han B-N, Hong L-L, Gu B-B, Sun Y-T, Wang J, Liu J-T, Lin H-W (2019) Li Z (ed) Symbiotic microbiomes of coral reefs sponges and corals. Springer, Dordrecht, pp 329–463
9. Genta-Jouve G, Thomas OP (2013) Phytochem Rev 12:425
10. Bewley CA, Faulkner DJ (1998) Angew Chem Int Ed 37:2162
11. Piel J (2009) Nat Prod Rep 26:338
12. Webster NS, Hill RT (2001) Mar Biol 138:843

13. Friedrich AB, Fischer I, Proksch P, Hacker J, Hentschel U (2001) FEMS Microbiol Ecol 38:105
14. Stewart EJ (2012) J Bacteriol 194:4151
15. Joint I, Mühling M, Querellou J (2010) Microb Biotechnol 3:564
16. Uria A, Piel J (2009) Phytochem Rev 8:401
17. Bewley CA, Holland ND, Faulkner DJ (1996) Experientia 52:716
18. Schmidt EW, Obraztsova AY, Davidson SK, Faulkner DJ, Haygood MG (2000) Mar Biol 136:969
19. Piel J (2010) Nat Prod Rep 27:996
20. Hochmuth T, Piel J (2009) Phytochemistry 70:1841
21. Fieseler L, Hentschel U, Grozdanov L, Schirmer A, Wen G, Platzer M, Hrvatin S, Butzke D, Zimmermann K, Piel J (2007) Appl Environ Microbiol 73:2144
22. Piel J, Hui D, Wen G, Butzke D, Platzer M, Fusetani N, Matsunaga S (2004) Proc Natl Acad Sci U S A 101:16222
23. Wakimoto T, Egami Y, Nakashima Y, Wakimoto Y, Mori T, Awakawa T, Ito T, Kenmoku H, Asakawa Y, Piel J, Abe I (2014) Nat Chem Biol 10:648
24. Ueoka R, Uria AR, Reiter S, Mori T, Karbaum P, Peters EE, Helfrich EJN, Morinaka BI, Gugger M, Takeyama H, Matsunaga S, Piel J (2015) Nat Chem Biol 11:705
25. Albertsen M, Hugenholtz P, Skarshewski A, Nielsen KL, Tyson GW, Nielsen PH (2013) Nat Biotechnol 31:533
26. Uria AR, Piel J, Wakimoto T (2018) Methods Enzymol 604:287
27. Wilson MC, Mori T, Rückert C, Uria AR, Helf MJ, Takada K, Gernert C, Steffens UAE, Heycke N, Schmitt S, Rinke C, Helfrich EJN, Brachmann AO, Gurgui C, Wakimoto T, Kracht M, Crüsemann M, Hentschel U, Abe I, Matsunaga S, Kalinowski J, Takeyama H, Piel J (2014) Nature 506:58
28. Breitling R, Takano E (2016) Cold Spring Harb Perspect Biol 8:a023994
29. Morrow C, Cárdenas P (2015) Front Zool 12:7
30. Rawlings BJ (2001) Nat Prod Rep 18:190
31. Staunton J, Weissman KJ (2001) Nat Prod Rep 18:380
32. Hertweck C (2009) Angew Chem Int Ed 48:4688
33. Hertweck C (2015) Trends Biochem Sci 40:189
34. Cheng Y-Q, Tang G-L, Shen B (2003) Proc Natl Acad Sci U S A 100:3149
35. Helfrich EJN, Piel J (2016) Nat Prod Rep 33:231
36. Fischbach MA, Walsh CT (2006) Chem Rev 106:3468
37. Cane DE, Walsh CT (1999) Chem Biol 6:R319
38. Smith DRM, Uria AR, Helfrich EJN, Milbredt D, van Pée K-H, Piel J, Goss RJM (2017) ACS Chem Biol 12:1281
39. Sakemi S, Ichiba T, Kohmoto S, Saucy G, Higa T (1988) J Am Chem Soc 110:4851
40. Matsunaga S, Fusetani N, Nakao Y (1992) Tetrahedron 48:8369
41. Fusetani N, Sugawara T, Matsunaga S (1992) J Org Chem 57:3828
42. Fusetani N, Matsunaga S (1993) Chem Rev 93:1793
43. Fusetani N (1996) J Toxicol Toxin Rev 15:157
44. Kellner RLL, Dettner K (1996) Oecologia 107:293
45. Perry NB, Blunt JW, Munro MHG, Thompson AM (1990) J Org Chem 55:223
46. Cichewicz RH, Valeriote FA, Crews P (2004) Org Lett 6:1951
47. Pettit GR, Xu J-P, Chapuis J-C, Pettit RK, Tackett LP, Doubek DL, Hooper JNA, Schmidt JM (2004) J Med Chem 47:1149
48. Piel J (2002) Proc Natl Acad Sci U S A 99:14002
49. Piel J, Wen G, Platzer M, Hui D (2004) Chembiochem 5:93
50. Piel J, Höfer I, Hui D (2004) J Bacteriol 186:1280
51. Fisch KM, Gurgui C, Heycke N, van der Sar SA, Anderson SA, Webb VL, Taudien S, Platzer M, Rubio BK, Robinson SJ, Crews P, Piel J (2009) Nat Chem Biol 5:494
52. Hrvatin S, Piel J (2007) J Microbiol Methods 68:434

53. Gurgui C, Piel J (2010) Streit WR, Daniel R (eds) Metagenomics: methods and protocols. Humana Press, Totowa, pp 247–264
54. Nguyen T, Ishida K, Jenke-Kodama H, Dittmann E, Gurgui C, Hochmuth T, Taudien S, Platzer M, Hertweck C, Piel J (2008) Nat Biotechnol 26:225
55. Rust M, Helfrich EJN, Freeman MF, Nanudorn P, Field CM, Rückert C, Kündig T, Page MJ, Webb VL, Kalinowski J, Sunagawa S, Piel J (2020) Proc Natl Acad Sci U S A 117:9508
56. Pöplau P, Frank S, Morinaka BI, Piel J (2013) Angew Chem Int Ed 52:13215
57. Kato Y, Fusetani N, Matsunaga S, Hashimoto K, Sakai R, Higa T, Kashman Y (1987) Tetrahedron Lett 28:6225
58. Sakai R, Higa T, Kashman Y (1986) Chem Lett 15:1499
59. Carmely S, Kashman Y (1985) Tetrahedron Lett 26:511
60. Kitagawa I, Kobayashi M, Katori T, Yamashita M, Tanaka J, Doi M, Ishida T (1990) J Am Chem Soc 112:3710
61. Kobayashi M, Tanaka J, Katori T, Matsuura M, Kitagawa I (1989) Tetrahedron Lett 30:2963
62. Andrianasolo EH, Gross H, Goeger D, Musafija-Girt M, McPhail K, Leal RM, Mooberry SL, Gerwick WH (2005) Org Lett 7:1375
63. Bubb MR, Spector I, Bershadsky AD, Korn ED (1995) J Biol Chem 270:3463
64. Terry DR, Spector I, Higa T, Bubb MR (1997) J Biol Chem 272:7841
65. Youssef DTA, Mooberry SL (2006) J Nat Prod 69:154
66. Ishibashi M, Moore RE, Patterson GML, Xu C, Clardy J (1986) J Org Chem 51:5300
67. Carmeli S, Moore RE, Patterson GML (1990) J Nat Prod 53:1533
68. Kubanek J, Jensen PR, Keifer PA, Sullards MC, Collins DO, Fenical W (2003) Proc Natl Acad Sci U S A 100:6916
69. Kitamura M, Schupp PJ, Nakano Y, Uemura D (2009) Tetrahedron Lett 50:6606
70. Siegl A, Kamke J, Hochmuth T, Piel J, Richter M, Liang C, Dandekar T, Hentschel U (2011) ISME J 5:61
71. Binga EK, Lasken RS, Neufeld JD (2008) ISME J 2:233
72. Stepanauskas R, Sieracki ME (2007) Proc Natl Acad Sci U S A 104:9052
73. Humisto A, Jokela J, Liu L, Wahlsten M, Wang H, Permi P, Machado JP, Antunes A, Fewer DP, Sivonen K (2018) Appl Environ Microbiol 84:e02321–e02317
74. Allingham JS, Klenchin VA, Rayment I (2006) Cell Mol Life Sci 63:2119
75. Allingham JS, Zampella A, D'Auria MV, Rayment I (2005) Proc Natl Acad Sci U S A 102:14527
76. Kato Y, Fusetani N, Matsunaga S, Hashimoto K, Fujita S, Furuya T (1986) J Am Chem Soc 108:2780
77. Suganuma M, Fujiki H, Furuya-Suguri H, Yoshizawa S, Yasumoto S, Kato Y, Fusetani N, Sugimura T (1990) Cancer Res 50:3521
78. Ishihara H, Martin BL, Brautigan DL, Karaki H, Ozaki H, Kato Y, Fusetani N, Watabe S, Hashimoto K, Uemura D, Hartshorne DJ (1989) Biochem Biophys Res Comm 159:871
79. Kita A, Matsunaga S, Takai A, Kataiwa H, Wakimoto T, Fusetani N, Isobe M, Miki K (2002) Structure 10:715
80. Dumdei EJ, Blunt JW, Munro MHG, Pannell LK (1997) J Org Chem 62:2636
81. Kehraus S, König GM, Wright AD (2002) J Nat Prod 65:1056
82. Edrada RA, Ebel R, Supriyono A, Wray V, Schupp P, Steube K, van Soest R, Proksch P (2002) J Nat Prod 65:1168
83. Fu X, Schmitz FJ, Kelly-Borges M, McCready TL, Holmes CFB (1998) J Org Chem 63:7957
84. Wakimoto T, Egami Y, Abe I (2016) Nat Prod Rep 33:751
85. Wakimoto T, Matsunaga S, Takai A, Fusetani N (2002) Chem Biol 9:309
86. Zhou Y, Murphy AC, Samborskyy M, Prediger P, Dias LC, Leadlay PF (2015) Chem Biol 22:745
87. Matilla MA, Stöckmann H, Leeper FJ, Salmond GPC (2012) J Biol Chem 287:39125
88. Meoded RA, Ueoka R, Helfrich EJN, Jensen K, Magnus N, Piechulla B, Piel J (2018) Angew Chem Int Ed Engl 57:11644

89. Gallimore AR, Stark CBW, Bhatt A, Harvey BM, Demydchuk Y, Bolanos-Garcia V, Fowler DJ, Staunton J, Leadlay PF, Spencer JB (2006) Chem Biol 13:453
90. Takahashi S, Toyoda A, Sekiyama Y, Takagi H, Nogawa T, Uramoto M, Suzuki R, Koshino H, Kumano T, Panthee S, Dairi T, Ishikawa J, Ikeda H, Sakaki Y, Osada H (2011) Nat Chem Biol 7:461
91. Calderone CT, Kowtoniuk WE, Kelleher NL, Walsh CT, Dorrestein PC (2006) Proc Natl Acad Sci U S A 103:8977
92. Gunasekera SP, Gunasekera M, Longley RE, Schulte GK (1990) J Org Chem 55:4912
93. Florence GJ, Gardner NM, Paterson I (2008) Nat Prod Rep 25:342
94. Nerenberg JB, Hung DT, Somers PK, Schreiber SL (1993) J Am Chem Soc 115:12621
95. Hung DT, Nerenberg JB, Schreiber SL (1996) J Am Chem Soc 118:11054
96. Longley RE, Caddigan D, Harmody D, Gunasekera M, Gunasekera SP (1991) Transplantation 52:650
97. Longley RE, Gunasekera SP, Faherty D, Mclane J, Dumont F (1993) Ann N Y Acad Sci 696:94
98. ter Haar E, Kowalski RJ, Hamel E, Lin CM, Longley RE, Gunasekera SP, Rosenkranz HS, Day BW (1996) Biochemistry 35:243
99. Hung DT, Chen J, Schreiber SL (1996) Chem Biol 3:287
100. Bollag DM, McQueney PA, Zhu J, Hensens O, Koupal L, Liesch J, Goetz M, Lazarides E, Woods CM (1995) Cancer Res 55:2325
101. Long BH, Carboni JM, Wasserman AJ, Cornell LA, Casazza AM, Jensen PR, Lindel T, Fenical W, Fairchild CR (1998) Cancer Res 58:1111
102. Kowalski RJ, Giannakakou P, Gunasekera SP, Longley RE, Day BW, Hamel E (1997) Mol Pharmacol 52:613
103. Martello LA, McDaid HM, Regl DL, Yang CP, Meng D, Pettus TR, Kaufman MD, Arimoto H, Danishefsky SJ, Smith AB, Horwitz SB (2000) Clin Cancer Res 6:1978
104. Martello LA, LaMarche MJ, He L, Beauchamp TJ, Smith AB, Horwitz SB (2001) Chem Biol 8:843
105. Paterson I, Florence GJ (2003) Eur J Org Chem 2003:2193
106. Schirmer A, Gadkari R, Reeves CD, Ibrahim F, DeLong EF, Hutchinson CR (2005) Appl Environ Microbiol 71:4840
107. Brück WM, Sennett SH, Pomponi SA, Willenz P, McCarthy PJ (2008) ISME J 2:335
108. Erol Ö, Schäberle TF, Schmitz A, Rachid S, Gurgui C, Omari ME, Lohr F, Kehraus S, Piel J, Müller R, König GM (2010) Chembiochem 11:1253
109. Mattheus W, Gao L-J, Herdewijn P, Landuyt B, Verhaegen J, Masschelein J, Volckaert G, Lavigne R (2010) Chem Biol 17:149

Top Heterocycl Chem (2021) 58: 415–438
https://doi.org/10.1007/7081_2021_52

Published online: 19 February 2021

Synthetic Studies of Serinolipids

Hiromasa Kiyota

Contents

Abstract A series of serinol derivatives etherificated with a non-branched polyketide possessing a 6,8-dioxabicyclo[3.2.1]octane core was isolated from extracts of marine invertebrates, ascidian tunicates belongs to the family Didemnidae. Synthetic studies of these compounds, didemniserinolipids, cyclodidemniserinol trisulfate, and siladenoserinols, are described. The characteristics of each synthesis are in the construction of the bicyclic core from each precursor.

Keywords Bicyclic acetal · Cyclodidemniserinol trisulfate · Didemniserinolipid B · *Didemnum* sp. · Siladenoserinol · Total synthesis

---

H. Kiyota (✉)
Graduate School of Environmental and Life Science, Okayama University, Okayama, Japan
e-mail: kiyota@okayama-u.ac.jp

Abbreviations

2,2-DMP	2,2-Dimethoxypropane
2,6-Lut	2,6-Lutidine
Ac	Acetyl
BAIB	(Diacetoxyiodo)benzene
Bn	Benzyl
Boc	*tert*-Butoxycarbonyl
brsm	Based on recovered starting material
CSA	Camphorsulfonic acid
DBU	1,8-Diazabicyclo[5.4.0]undec-7-ene
DCC	Dicyclohexylcarbodiimide
DDQ	2,3-Dichloro-5,6-dicyano-1,4-benzoquinone
DEAD	Diethyl azodicarboxylate
DIBAL-H	Diisobutylaluminum hydride
DIPEA	Ethyl(diisopropyl)amine
DMAP	4-(Dimethylamino)pyridine
DME	1,2-Dimethoxyethane
DMF	*N*,*N*-Dimethylformamide
DMPI	Dess–Martin periodinane
DMSO	Dimethyl sulfoxide
Fmoc	9-Fluorenylmethoxycarbonyl
GPC	Glycerophosphatidylcholine
HMPA	Hexamethylphosphoramide
IBX	2-Iodoxybenzoic acid
Im	Imidazole (or imidazolyl)
KHMDS	Potassium hexamethyldisilazide
LiHMDS	Lithium hexamethyldisilazide
MCPBA	*m*-Chloroperoxybenzoic acid
MOM	Methoxymethyl
Ms	Methanesulfonyl
MS	Molecular sieves
PMB	p-Methoxybenzyl
p-NBA	*p*-Nitrobenzoic acid
Red-Al	Sodium bis(methoxyethoxy)aluminum hydride
Su	Succinimidyl
TBAF	Tetrabutylammonium fluoride
TBAI	Tetrabutylammonium iodide
TBDPS	*tert*-Buthyldiphenylsilyl
TBS	*tert*-Buthyldimethylsilyl
TEMPO	2,2,6,6-Tetramethylpiperidine *N*-oxyl
THF	Tetrahydrofuran
TMS	Trimethylsilyl
Ts	*p*-Toluenesulfonyl

1 Introduction

Marine tunicates, invertebrate animals, especially the genus *Didemnum* are rich sources of biologically active and/or structurally characteristic organic compounds [1]. Among them, a series of serinol derivatives, named didemniserinolipids, cyclodidemniserinol trisulfate, and siladenoserinols, has been isolated to date, and has attracted the attention of synthetic chemists due to characteristic structures including a non-branched polyketide chain constructing a dioxabicyclooctane and a serinol ether, a variety of functional groups such as an *N*/*O*-sulfo, a glycerophosphatidylcholine, and/or an acetyl. This chapter described the synthetic studies of these serinolipids including structure revision.

2 Synthetic Studies of Didemniserinolipid B

2.1 Isolation

Didemniserinolipids A, B, and C were isolated from a methanol extract of a marine tunicate *Didemnum* sp. collected along the coast of Sulawesi Island (Indonesia) by Jimenez's research group in 1999 [2]. The structures of these compounds were proposed as shown in Fig. 1, with the first polyketides having an ether linkage with a serinol moiety, and a 6,8-dioxabicyclo[3.2.1]octane core. Jiménez et al. elucidated the serinol fragment by a combination of ^{1}H-^{1}H COSY and ^{13}C NMR spectral data, and the relative stereochemistry of the bicyclic core was determined by HMBC and NOESY correlations (Fig. 1). Didemniserinolipid B was later found to be *O*-sulfated at C30, and its absolute configuration was also determined from the total synthesis by Ley's group (see Sect. 2.2) [3, 4].

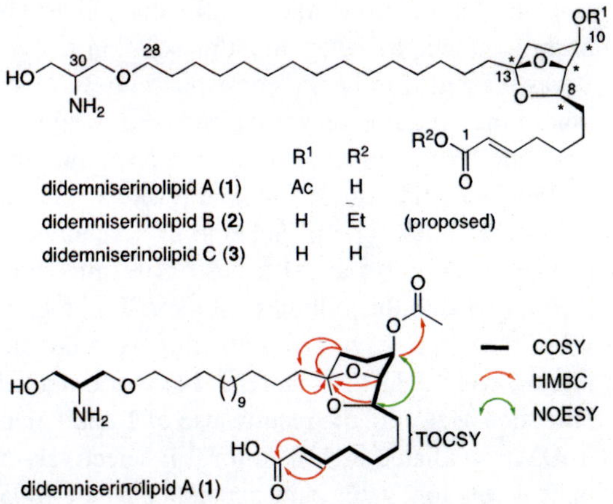

Fig. 1 Structure elucidation of didemniserinolipids A (**1**), B (**2**), and C (**3**)

Due to the attractive structural features including a non-branched long carbon chain with a dioxabicyclooctane core as well as a novel serinol sulfate, didemniserinolipid B has attracted much attention from a number of synthetic chemists.

2.2 Ley's First Total Synthesis and Structure Revision

The first synthesis and structure revision of didemniserinolipid B (**2**) was achieved by Kiyota, Ley et al. in 2002 [3]. It began with the use of both enantiomers of Garner's alcohol **4** [5] and butanediacetal building block **5** [6] and as chiral sources (Scheme 1). They presumed an (8*R*,9*R*,10*R*,13*S*)-configuration for the core part according to the related natural product (+)-2-hydroxy-*exo*-brevicomin [7], which has the same (+)-sign of specific rotation. Williamson ether synthesis using (*S*)-**4** and propargyl bromide afforded **6**, and elongation of the terminal acetylene part was done with 1,11-dibromoundecane, giving **7**. Then the remaining bromo group was substituted with dianion of triethyl phosphonoacetate to afford HWE reagent **8** in 97% yield. On the other hand, the known butanediacetal **5** derived from L-tartrate was coupled with the lithioalkyne produced from **9** to predominantly give alcohol **10** with the desired *R*-configuration (*R*/*S* = 3.4:1). The absolute configuration was confirmed by the modified Mosher method [8]. The hydroxy group was protected as a MOM ether and both of silyl protecting group were removed to give **11**. Then the less hindered terminal hydroxy group was selectively protected as a TBS ether and another was oxidized to aldehyde **12** in 73% yield. HWE reaction of **12** and phosphonate **8**, followed by reduction of the formed double bond gave **13**. Chain elongation again by HWE reaction and deprotective acetal formation furnished proposed didemniserinolipid B [(30*R*)-**2**] with the (8*R*,9*R*,10*R*,13*S*,30*R*)-configuration. Similarly, the corresponding (30*S*)-**2** was prepared from (*R*)-Garner's alcohol. ^{1}H-NMR chemical shifts for protons around the serinol moiety of the 30R and 30S isomers were different from the corresponding chemical shifts of natural **2** (Table 1).

Considering the chemical shift value of H-29 of natural **2** was still lower than that of the synthetic triacetate and the isolation of cyclodidemniserinol trisulfate, natural **2** was presumed to be the corresponding 31-*O*-sulfate. Still, the absolute configuration of the synthetic bicyclooctane core would be (8*R*,9*R*,10*R*,13*S*), the same as the synthetic compounds, because both have the same (+) sign of specific rotation.

Next, sulfation of (30*S*)-**2** and (30*R*)-**2** was examined after protection of 30-NH_2 group with Fmoc group (Scheme 2). Usual conditions using SO_3•Py or SO_3•NMe_2 in various solvents at 20 to 100°C resulted in recovery of the starting material, probably due to the bulkiness of the *N*-Fmoc group or hydrogen bonding. However, when the reaction vessel containing 10 equivalent of SO_3•Py was irradiated with microwave, 10,31-di-*O*-sulfate (**14**) was obtained in 84% after Fmoc deprotection. After considerating the results, use of 1 equivalent of SO_3•Py and 10 eq. of Na_2SO_4 in DMF irradiated at 110°C for 2 h selectively afforded the desired (30*S*)-**15**. The optical rotation value and spectral data of (30*S*)-**15** well coincided with that of

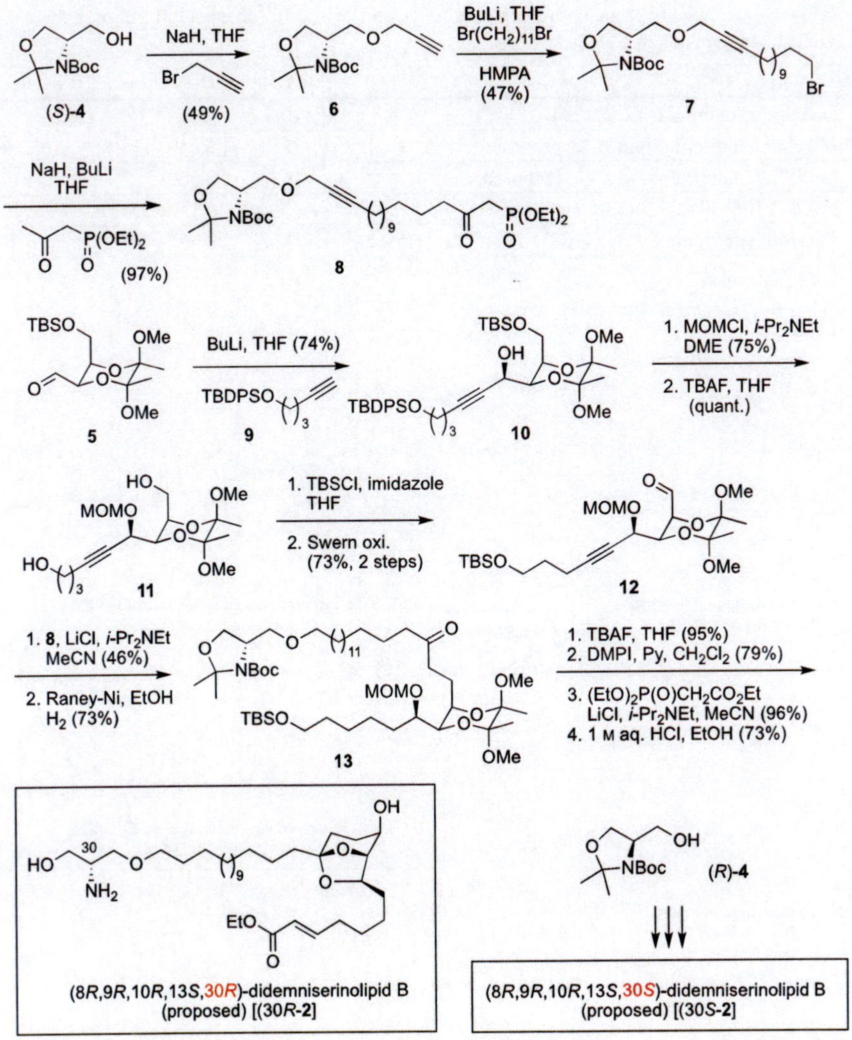

Scheme 1 Kiyota and Ley's total synthesis with structure revision (1)

natural **2**, while the ^{1}H-NMR spectral data of (30*R*)-**15** was slightly different. Accordingly, the structure of natural **2** was determined and revised to be (8*R*,9*R*,10*R*,13*S*,30*S*)-**2** 31-*O*-sulfate [(30*S*)-**15**].

Table 1 The specific rotation values in $CHCl_3$ and ^{1}H-NMR chemical shifts of natural and synthetic compounds in $CDCl_3$

	$[\alpha]_D^{25}$	H-29	H-31
Didemniserinolipid B (**2**, natural)	+10.3	4.26, 4.34	3.65
(30*R*)-didemniserinolipid B (**2**, proposed)	+21	3.64[a]	3.85[a], 3.94
(30*S*)-didemniserinolipid B (**2**, proposed)	+16	3.50	3.52, 3.62
(30*R*)-**2** 10*O*,30*N*,31*O*-triacetate	–	4.11, 4.23	4.31
(30*S*)-didemniserinolipid B (**15**, 31-*O*-sulfate, revised)	+11	3.65	4.27, 4.37

[a]Interchangeable

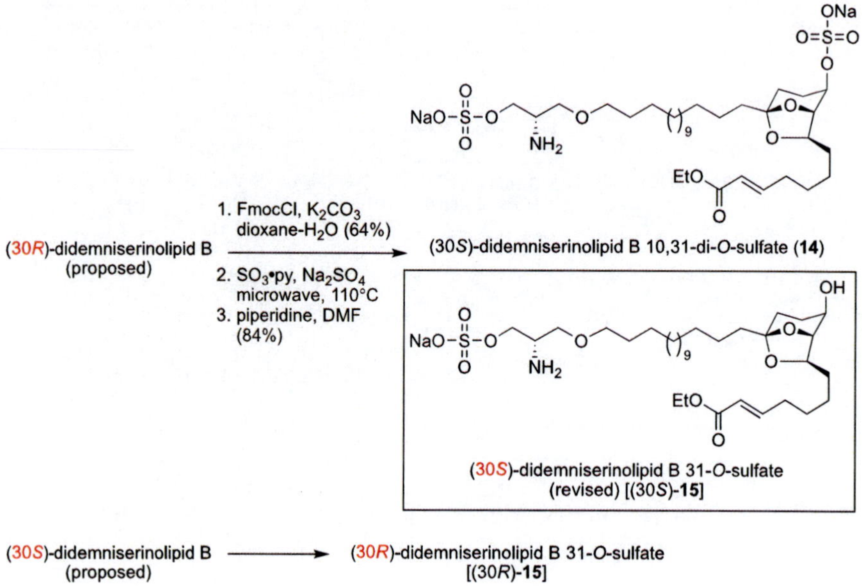

Scheme 2 Kiyota and Ley's total synthesis with structure revision (2)

2.3 Burke's Formal Total Synthesis

The second total synthesis was reported by Burke's group in 2007 [9, 10]. The retrosynthesis is shown in Scheme 3. They planned to construct the bicyclo[3.2.1] octane core **15** by a ring closing metathesis reaction of the acetal precursor **17**. Cross metathesis reaction with **18** [11] was designed for elongation of the C1-C6 side chain. The stereochemistry of the core was to be from that of L-tartrate via **19** and **20** [12].

Burke's synthesis started with ring opening of hexadecanolide **21** by dimethyl lithiomethylphosphonate to give **22** (Scheme 4). HWE reaction with phenylacetaldehyde, followed by mesylation of the hydroxy group afforded **19**, which was condensed with a symmetrical diol **20** derived from L-tartrate gave acetal

Scheme 3 Burke's retrosynthetic plan

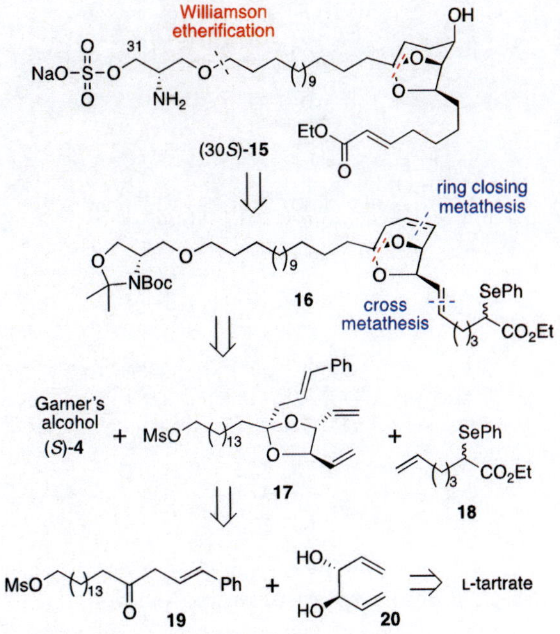

17. The key RCM reaction proceeded to afford the desired core compound **23** in 53% yield (81% brsm), however, prolonged reaction resulted in the formation of the corresponding phenyl alkene **24**. Although the phenyl ring was effective to prevent C=C migration, amounts of byproduct styrene caused the unwanted secondary reaction (due to the volatility of 2nd byproduct ethylene). This problem was solved by a relay metathesis method using 2-allyl-4-fluoro substituents on the phenyl ring (**25**). As shown in Scheme 5, metathesis first occurred at the least hindered allylic group to form **26**, which reacted with the neighboring double bond to release fluoroindane (**27**), to **28**, and finally allowing the pyran ring to give **23**. The isolated yield was up to 81%. They firstly chose mono-protected azidodiol **29** as a serinol unit [13], making **30** in 89% yield. Conversion of the *endo*-double bond to a hydroxy group was accomplished by Sharpless asymmetric dihydroxylation (**31**) and reductive desulfurylation followed by azide reduction and Fmoc protection produced the valuable intermediate **32**.

Since this method required multistep sequence and the overall yield was low, they examined an alternative synthetic route. Thus, **32** was coupled with Garner's alcohol (*S*)-**4** to give **33** which was treated with MCPBA. The resulting epoxide was reduced with $LiAlH_4$ to selectively give C10-axial alcohol **34**. Cross metathesis with ethyl 2-phenylselenenyl-6-heptenoate **18** afforded **16**. Reduction of the resulting double bond, oxidative elimination of selenoxide (**35**), acidic hydrolysis of the acetonide, and Fmoc protection of the amino group gave (30*R*)-**2**. Nature-identical (30*S*)-**15** was prepared according to Ley's procedure [3] (Scheme 6).

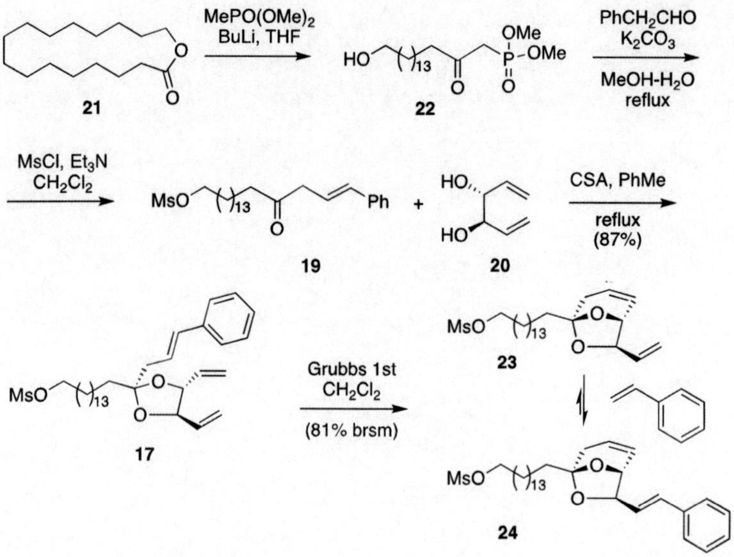

Scheme 4 Burke's formal total synthesis (1)

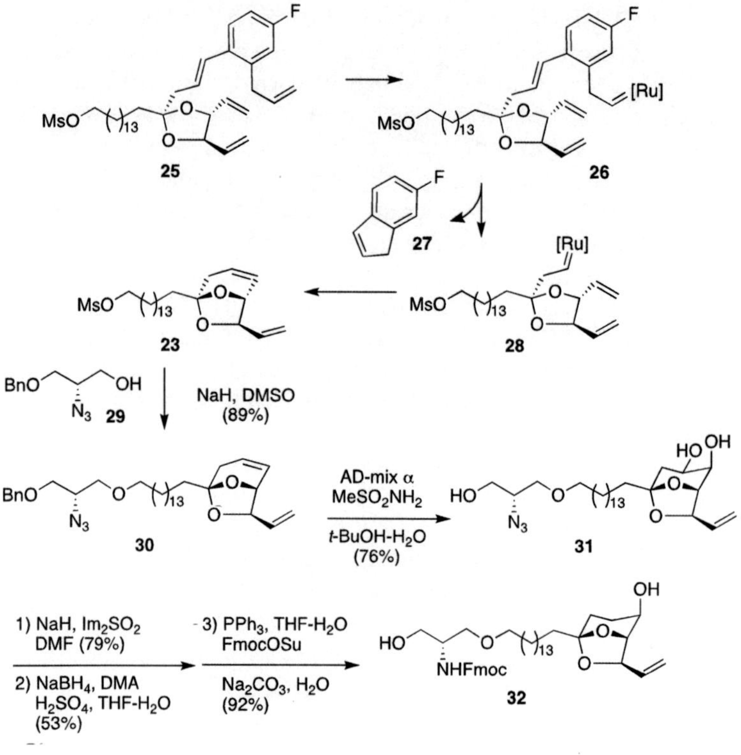

Scheme 5 Burke's formal total synthesis (2)

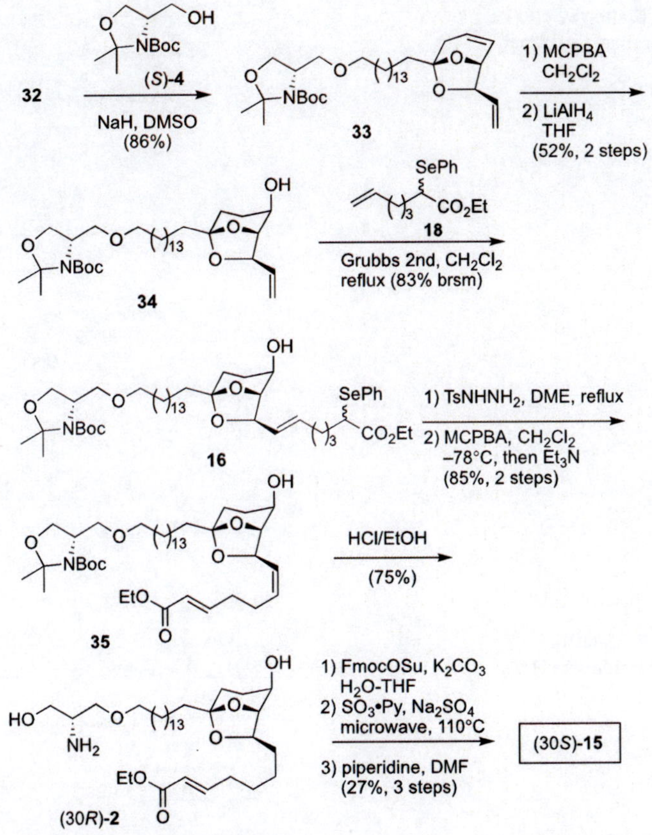

Scheme 6 Burke's formal total synthesis (3)

2.4 Ramana's Formal Total Synthesis

In 2009 Ramana and Induvadana at National Chemical Laboratory in India reported a synthesis of Burke's intermediate **35** [14, 15]. Their retrosynthetic plan is shown in Scheme 7. Similar to another groups, Williamson's etherification with Garner's alcohol (*S*)-**4** was chosen for the serinol fragment. Construction of the bicyclooctane core was designed by a double oxypalladation of alkyne-diol **36**, with a triol configuration derived from D-arabinose.

The long chain fragment **37** was prepared from propargylic ether **38** (Scheme 8). Alkylation with tetradecyl bromide and THP-deprotection afforded **39**. Then **39** was converted to terminal alkyne **40** by an acetylene zipper reaction [16]. Protection of the hydroxy group gave **37**.

Chain elongation of arabinose diacetonide **41** [17] was performed using Wittig reaction and hydrogenation followed by selective deprotection of the less hindered acetonide group to give diol **42** (Scheme 9). The less hindered hydroxy group was

Scheme 7 Retrosynthetic scheme of Ramana's formal synthesis

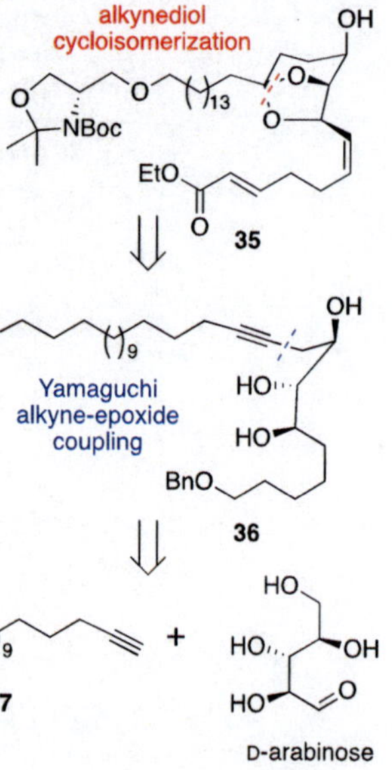

Scheme 8 Ramana's formal total synthesis (1)

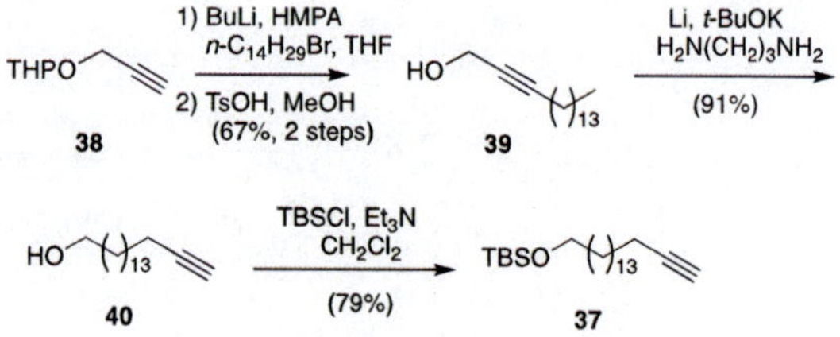

again selectively tosylated and alkaline treatment afforded epoxide **43**. Nucleophilic attack of 17-TBS-oxy-1-heptyne to **37** was accomplished using Yamaguchi's boro-alkyne method [18]. Deprotection of both acetonide and TBS groups under acidic conditions gave tetraol **36**. The key double oxypalladation succeeded using $Pd(MeCN)_2Cl_2$ as a catalyst to form the desired bicyclooctane core structure **44**. Protection of the secondary hydroxy group was achieved by successive reactions of *O*-silylation (TBS), *O*-benzylation, and de-*O*-silylation, giving **45** in 74%. After the primary hydroxy group was mesylated (**46**), Williamson ether synthesis with Garner's alcohol (*S*)-**4** furnished **47**. Chain elongation via Wittig reaction afforded the Burke's intermediate **35**.

Ramana et al. also reported a more convergent synthetic route (Scheme 10). Here, the C13-C28 fragment **48** prepared from **40** was coupled with the serinol fragment prior to Yamaguchi epoxide coupling with **43**. Conditions for the alkynediol isomerization [19] of **49** were also examined and a combination of Au and Ag catalysts produced the best result to give **50** in 85% yield. Concomitant deprotection of *N*,*O*-acetal was recovered to merge with the former route (**47**).

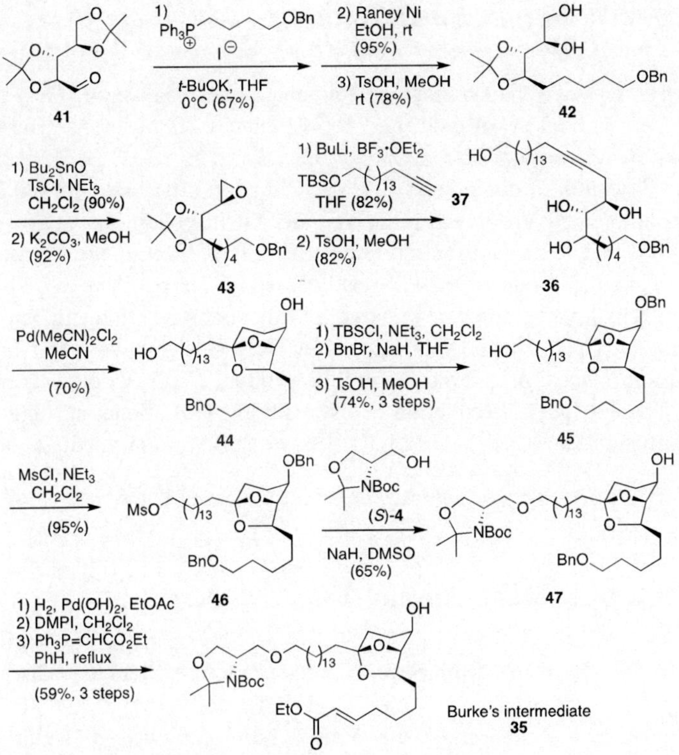

Scheme 9 Ramana's formal total synthesis (2)

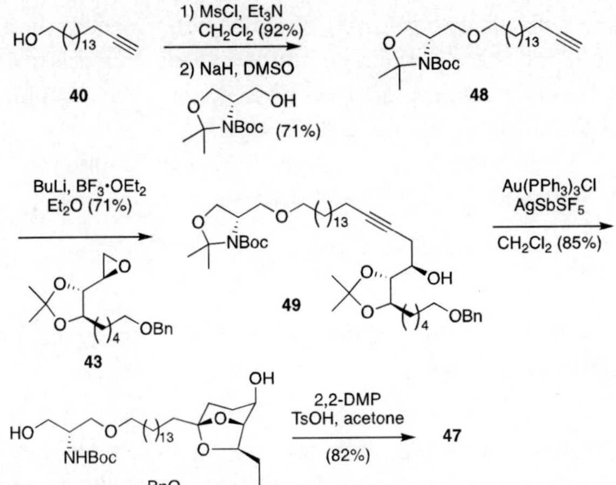

Scheme 10 Ramana's formal total synthesis (3)

2.5 Prasad's Formal Total Synthesis

Prasad et al. reported the synthesis of Ramana's intermediate **44**. They planned to use the stereochemistry of D-tartrate for C9 and C10 positions, and C8 being introduced by diastereomeric reduction.

Mono-alkylation of diamide **51** [20] derived from D-tartrate afforded ketone **52**, and the resulting keto group was reduced under Luche conditions, giving an alcohol with the desired configuration preferentially (9:1). Then rearrangement of the C9-C10 acetonide group to C8-C9 positions (**53**) was performed [21]. After the resulting C8 hydroxy group was inverted by Mitsunobu reaction and benzyl protection, methoxycarbonyl group was reduced by $NaBH_4$ to give **54**. The aliphatic long chain was introduced by IBX oxidation followed by HWE reaction with **55** to afford ketone **56** in 82% yield. Reduction of the resulting C=C bond and intramolecular acetal formation catalyzed by $FeCl_3$ [22] gave Prasad's intermediate **44** (Scheme 11).

2.6 Chandrasekhar's Formal Total Synthesis

Formal total synthesis of **1** from D-ribose was reported by Chandrasekhar's group in 2012 [23]. The synthesis began with 1-allyl ribose derivative **57** [24] (Scheme 12). Lemieux–Johnson oxidation of the double bond followed by Julia–Kocienski olefination using PT-sulfone **58** [25] afforded **59**. The acetoxy group on the other side was converted to chloro group (**60**), and reductive ring opening reaction was carried out using in situ generated $LiNH_2$ in NH_3 [26] to afford terminal alkyne **61**. Then C8-hydroxy group was inverted by Mitsunobu reaction and protected as a MOM ether **62**. They chose a Weinreb amide **64** as a coupling partner which was prepared from Garner's alcohol (*S*)-**4** via **63** in 4 steps. The coupling of **62** with **64** was successful to furnish ynone **65** in 85% yield.

The whole carbon framework was built by a similar process (**66**), and acidic treatment afforded (30*R*)-**2** as a HCl salt (Scheme 13).

2.7 Tong's Formal Total Synthesis and Synthesis of Proposed Didemniserinolipid C

Later in 2014, Ren and Tong reported the synthesis of Ley's intermediate [27, 28]. Their synthetic design included oxidative rearrangement of furan to pyrone as the key step (Scheme 14). Accordingly, introduction of a formyl substituent to ethyl 2-furanpropanoate **67** was accomplished by a Vilsmeier–Haack reaction [29], and the resulting aldehyde was coupled with PT-sulfone **68** to give **69** in 81% yield. The newly formed *E*-double bond was oxidized under Sharpless asymmetric

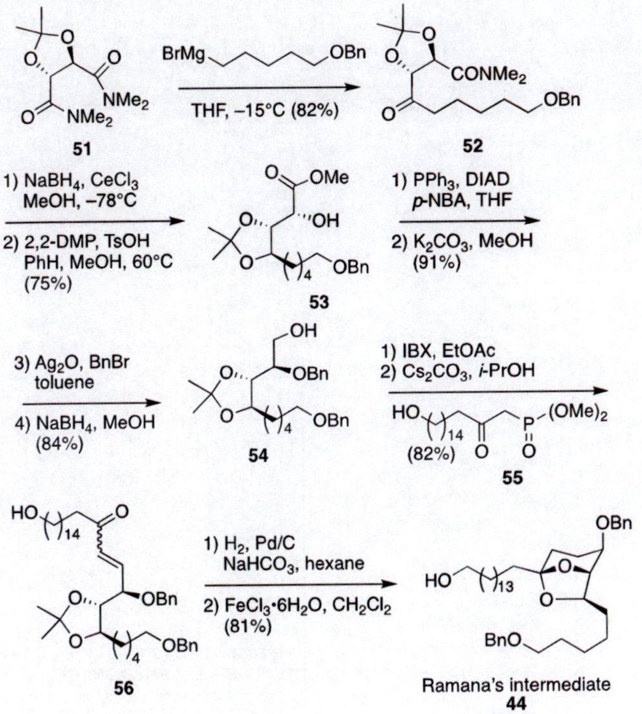

Scheme 11 Prasad's formal total synthesis

dihydroxylation conditions to afford diol **70**. Then the key Achmatowicz rearrangement [30] using MCPBA followed by acidic treatment furnished the desired bicyclic core compound **71**. After reduction of the double bond, selective reduction of the keto-carbonyl group was examined. Although the selectivity was up to 1:1 even when using bulky K-selectride, the undesired equatorial isomer eq-**72** could be recycled to ax-**72** via Dess–Martin oxidation and reduction.

After TBS-protection, the ester group was reduced by DIBAL-H to furnish **73**. Aldehyde 73 was coupled again PT-sulfone **74** to afford the desired olefin (*E*/*Z* = 2:1), which was deprotected of its PMB group and then mesylated to produce **75** (Scheme 15). The core fragment was etherificated with Garner's alcohol (*S*)-**4** to furnish **76**. Chain elongation afforded **77**, which was converted to the Ley's intermediate **78** in a similar manner, and then to the final product (*S*)-**15**.

Preparation of didemniserinolipid C (**3**) was also attempted. It was suggested that natural **3** also has a 31-*O*-sulfate group as (*S*)-**15**, but hydrolysis of the ethoxycarbonyl group of (*S*)-**15** failed. Conversion of the ethoxycarbonyl group to carboxy succeeded from **77** by reduction and oxidation, affording proposed didemniserinolipid C (**3**). The subsequent ^{1}H and ^{13}C NMR data in CD_3OD were not in agreement with those in $CDCl_3$ of natural **3**. In addition, the specific rotation

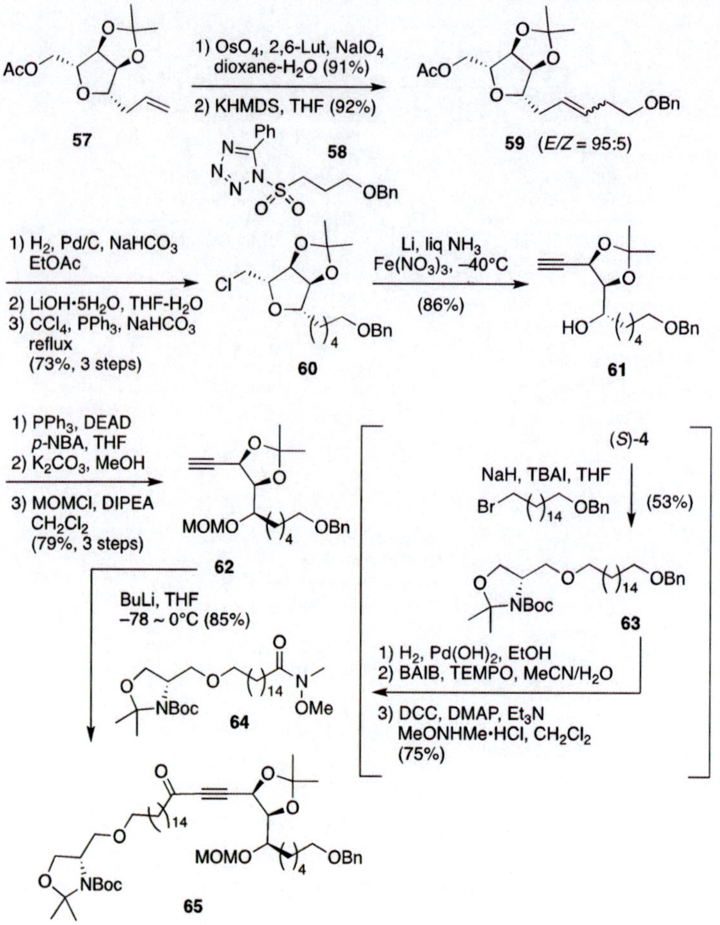

Scheme 12 Chandrasekhar's formal total synthesis (1)

value of +10.3° in $CHCl_3$ was quite different from that of natural **3** (+32.3° in MeOH). A further attempt to convert synthetic **3** did not give the presumed natural 30-*O*-sulfate.

3 Synthetic Study of Cyclodidemniserinol Trisulfate

3.1 *Isolation*

Faulkner et al. reported the isolation of cyclodidemniserinol trisulfate (**79**) from the Palauan ascidian *Didemnum guttatum* [31], with **79** showing inhibitory activity against HIV-1 integrase. Only the relative stereochemistry around the bicyclic ring

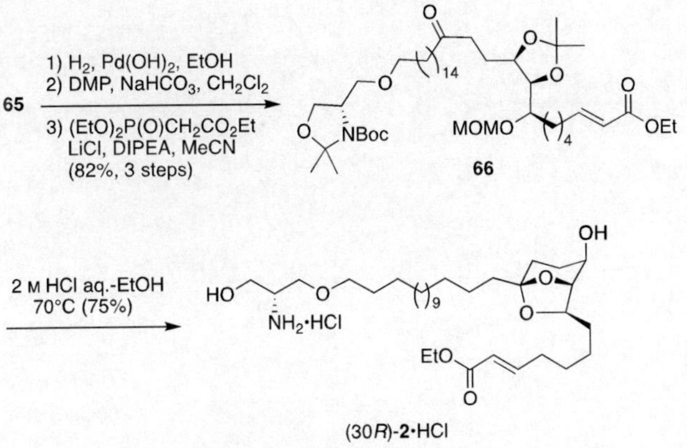

Scheme 13 Chandrasekhar's formal total synthesis (2)

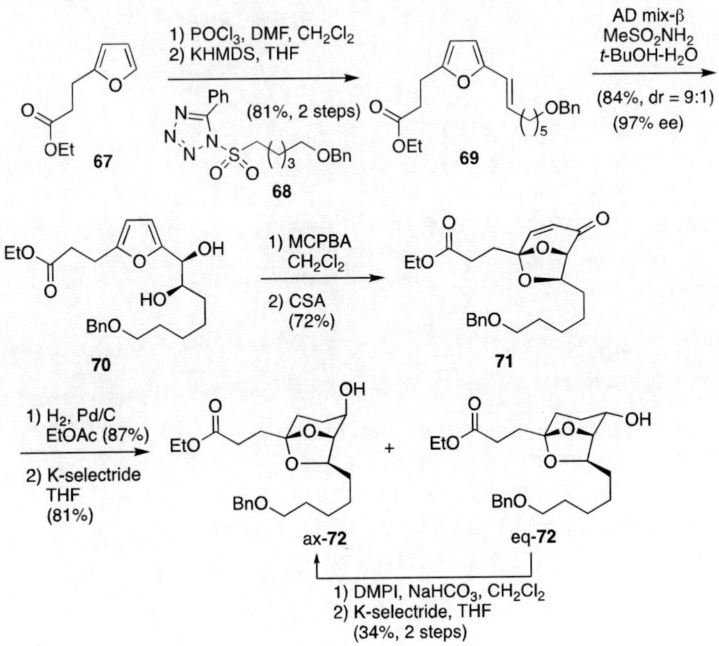

Scheme 14 Tong's formal total synthesis (1)

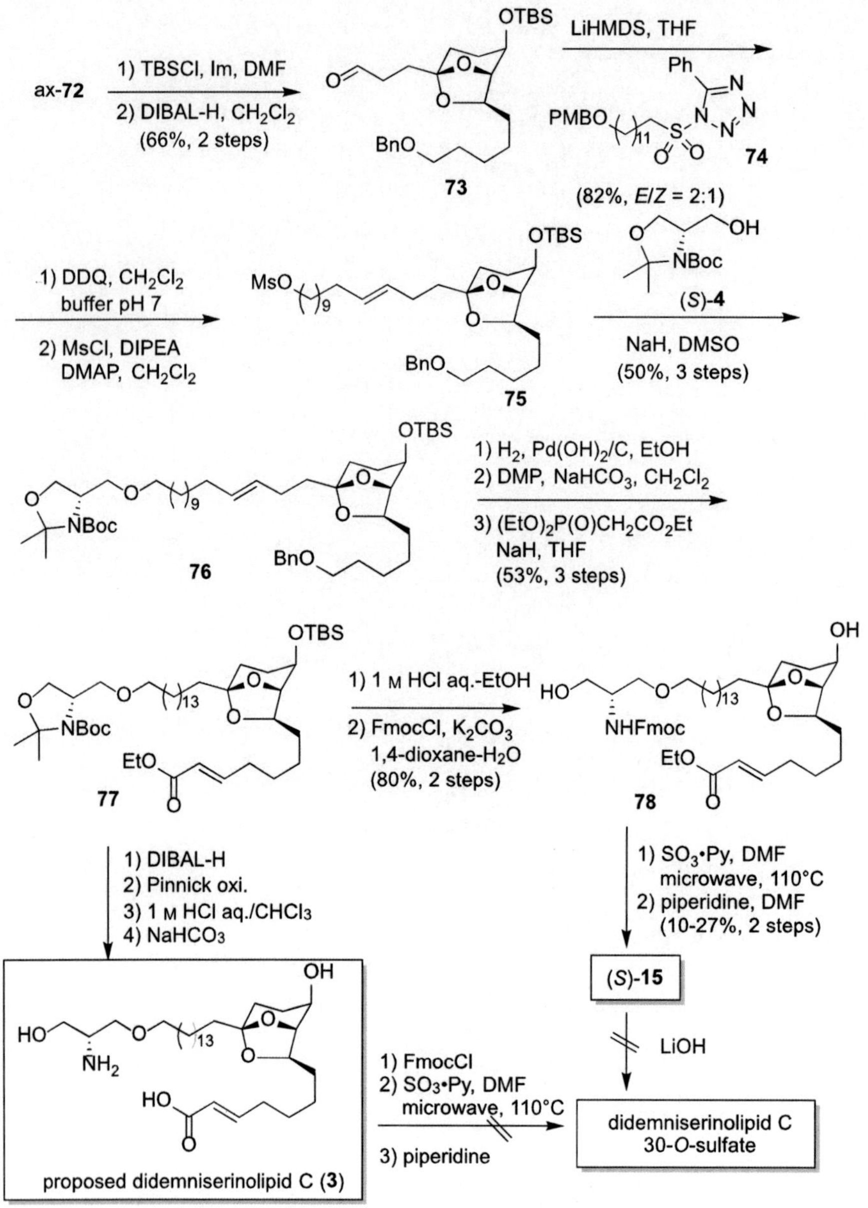

Scheme 15 Tong's formal total synthesis (2) and trial for didemniserinolipid C

Fig. 2 Cyclodidemniserinol trisulfate

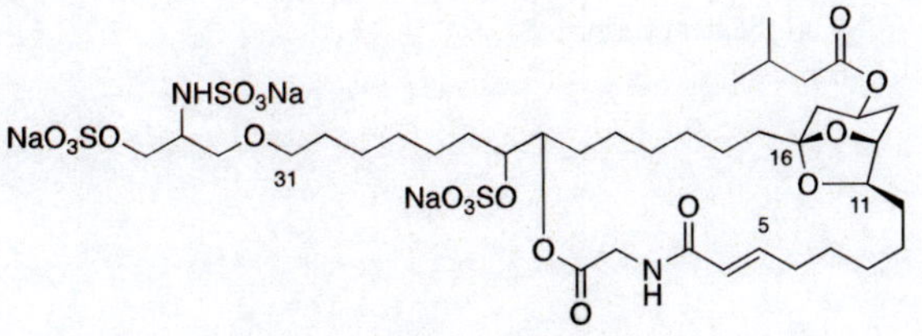

Scheme 16 Long's partial synthesis of cyclodidemniserinol trisulfate

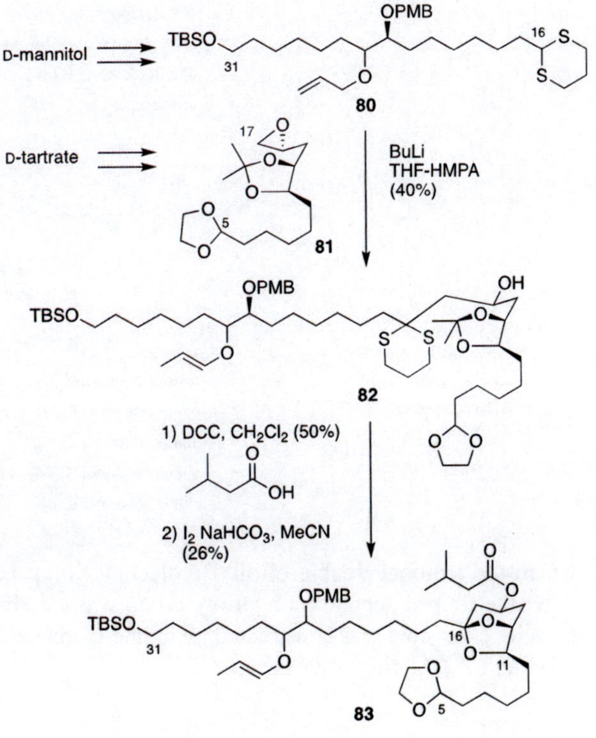

was determined (Fig. 2). The presence of three sulfate groups was observed by HRMS and IR (2,320 cm^{-1}) analyses.

3.2 *Long's Partial Synthesis*

Total synthesis of **79** has not been achieved yet, but Long's partial synthesis is to be introduced due to page limitations [32–34]. The C16-C31 fragment **80** was prepared from D-mannitol and the C5-C17 fragment **81** from D-tartrate (Scheme 16). Reaction of the lithiodithiane produced from **80** with epoxide **81** afforded **82** in 40% yield,

Fig. 3 Siladenoserinols A-L

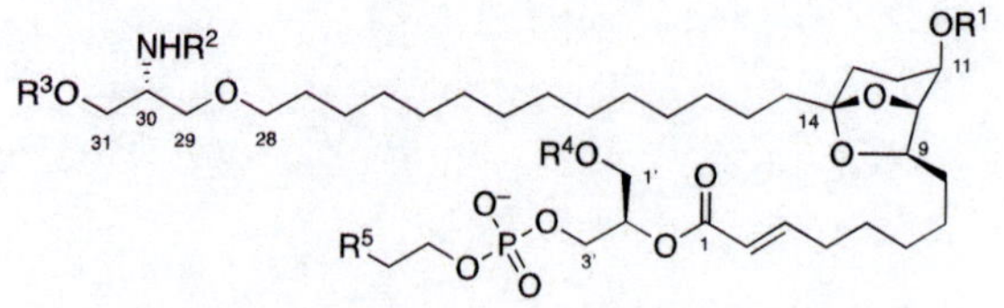

	R^1	R^2	R^3	R^4	R^5
siladenoserinol A (**84**)	Ac	SO_3H	H	Ac	$^+NMe_3$
siladenoserinol B (**85**)	Ac	SO_3H	H	Ac	$^+NH_3$
siladenoserinol C (**86**)	H	H	SO_3H	Ac	$^+NH_3$
siladenoserinol D (**87**)	Ac	SO_3H	Ac	H	$^+NMe_3$
siladenoserinol E (**88**)	H	H	SO_3H	Ac	$^+NMe_3$
siladenoserinol F (**89**)	H	SO_3H	H	Ac	$^+NMe_3$
siladenoserinol G (**90**)	Ac	H	SO_3H	Ac	$^+NMe_3$

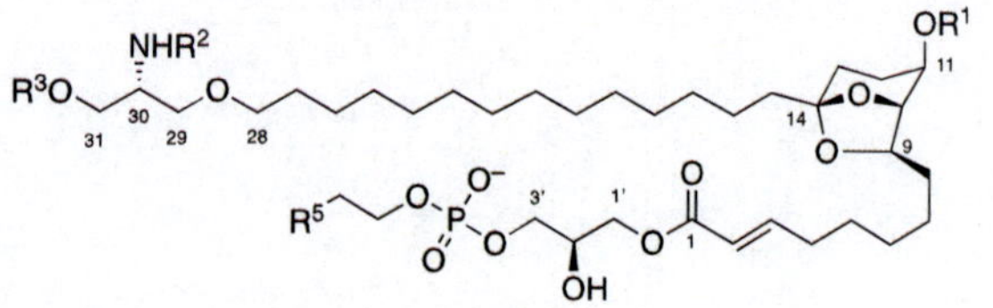

	R^1	R^2	R^3	R^4
siladenoserinol H (**91**)	Ac	SO_3H	H	$^+NMe_3$
siladenoserinol I (**92**)	Ac	H	SO_3H	$^+NMe_3$
siladenoserinol J (**93**)	H	H	SO_3H	$^+NH_3$
siladenoserinol K (**94**)	H	SO_3H	Ac	$^+NMe_3$
siladenoserinol L (**95**)	H	H	SO_3H	$^+NMe_3$

where the terminal double of allyl protecting group isomerized to a vinyl group. After the formed secondary hydroxy group was esterified with isovaleric acid, the bicyclooctane core was constructed by iodine-mediated deprotective acetalization to give the C5-C31 fragment **83**.

4 Synthetic Study of Siladenoserinols

4.1 Isolation

In 2013 Tsukamoto's group reported the isolation of a series of congeners possessing an ester linkage with glycerophosphatidylcholine, siladenoserinols A-L (Fig. 3, **84**–**95**), from a tunicate of the family Didemnidae collected in Indonesia [35]. These compounds showed inhibitory activity against p53-Hdm2 interaction. They determined the absolute configuration of the bicyclooctane core and serinol fragment by a modified Mosher method.

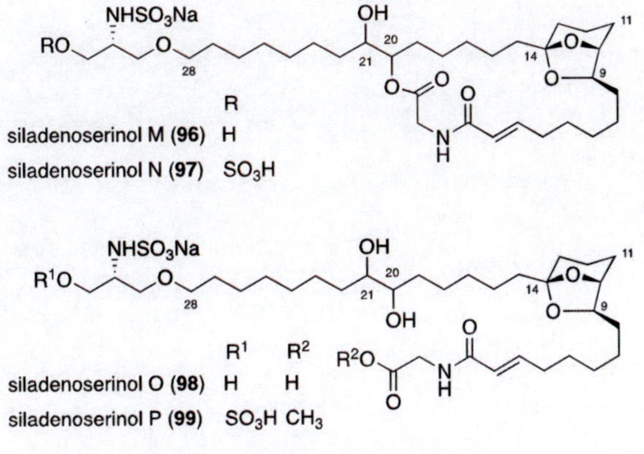

Fig. 4 Siladenoserinols M-P

Later, in 2018, the same group isolated siladenoserinols M-P (Fig. 4, **96–99**), with structures that are closely related to cyclodidemniserinol trisulfate (**79**) [36].

4.2 *Doi's First Total Synthesis of Siladenoserinol A*

In 2018, Doi's group reported the first total synthesis of siladenoserinol A (**84**) using gold-catalyzed double hydroalkoxylation as the key step [37]. They did not adopt usual esterification, but instead utilized a HWE reaction for the introduction of the GPC ester moiety, because the esterification of α,β-unsaturated acids is sluggish. This synthesis used the *R*-chirality of D-malic acid for the C11 position (Scheme 17). The benzylidene acetal moiety of **100**, prepared from D-malic acid in 5 steps, was hydrolyzed to aid triple bond reduction by Red-Al, and re-constructed again (**101**). The *E*-double bond was dihydroxylated using Sharpless AD conditions, and the resulting diol moiety was protected as an acetonide to give **102**. Elongation of the terminal hydroxy group was accomplished by tosylation, substitution by acetylide, and methoxycarboxylation. Deprotection of the acetonide group afforded alkyne diol **103**. The key double hydroalkoxylation [38] was successfully completed in 5 min in the presence of catalytic amount of $AuCl_3$ in acetonitrile, giving the bicyclic core **104**. Aldehyde prepared from DIBAL-H reduction of **104** was coupled with PT-sulfone **105** bearing a serinol moiety by a Julia–Kocienski reaction, and reduction of the resulting double bond, concomitant with removal of the benzyl group, afforded **106**. Selective oxidation of the primary hydroxy group by TEMPO and acetylation of the secondary hydroxy afforded aldehyde **107**.

The counterpart, phosphonate **108** was prepared from glycerol derivative **110** (Scheme 18) [39]. Selective acetylation of the primary hydroxy group afforded **113**

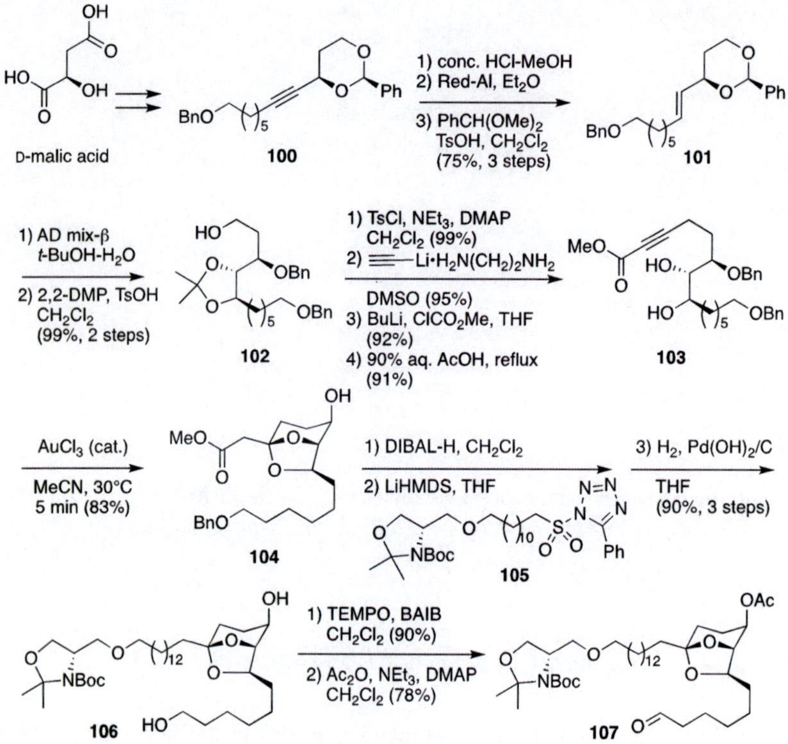

Scheme 17 Doi's first synthesis of siladenoserinol A (1)

and the secondary hydroxy was esterified with phosphonoacetic acid to give **114**. After deprotection of the benzyl group, the formed hydroxy was exchanged with the diisopropylamino group of **112**, followed by oxidation, leading to bromide **115**. Preparation of the choline residue was done by S_N2 reaction, giving **108**. HWE reaction of **107** with **108** under Masamune-Roush conditions furnished **109**. Finally, deprotection of the *N,O*-acetonide group and sulfonamidation of the resulting amino group led to the first synthesis of siladenoserinol A (**84**).

4.3 Tong's Total Synthesis of Siladenoserinols A and H

Tong et al. reported the synthesis and antibacterial activity of siladenoserinols A and H [40]. They adopted a similar strategy that they used for the synthesis of didemniserinolipid B as in Sect. 2.7. As shown in Scheme 19, compound **116** is similar to **73** where the functional group substitution position is shifted by one. **116** was subjected to Julia–Kocienski reaction with PT-sulfone **117** which is already etherified with a serinol moiety, followed by hydrogenation/hydrogenolysis and

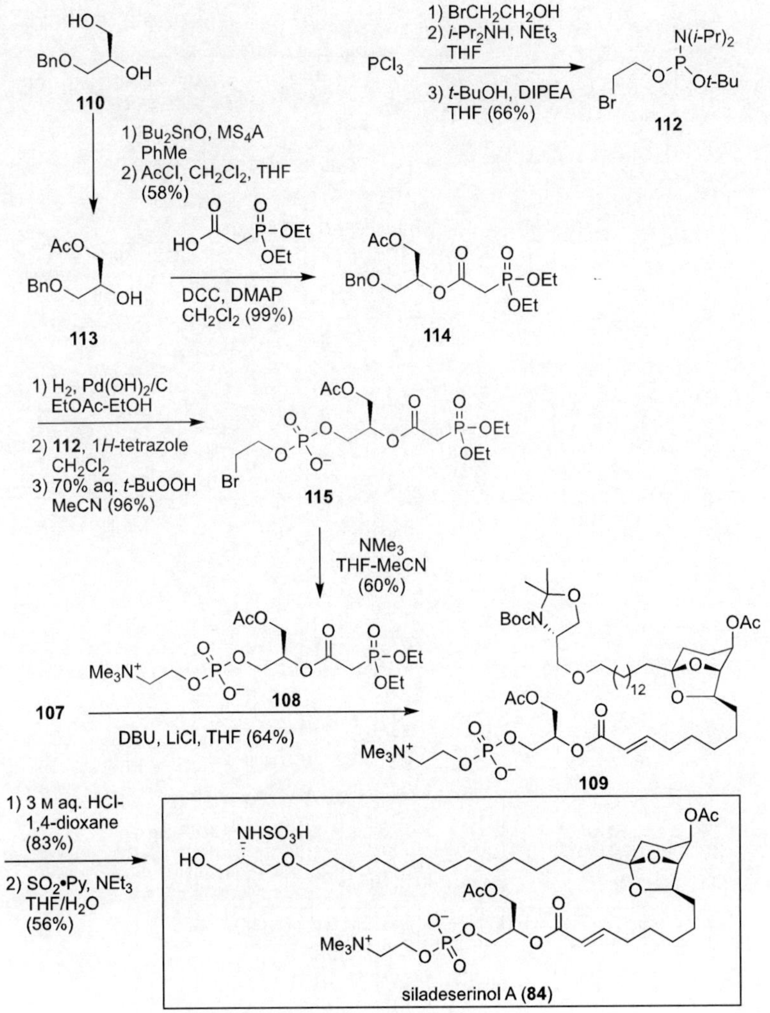

Scheme 18 Doi's first synthesis of siladenoserinol A (2)

Dess–Martin oxidation, giving aldehyde **118**. Tong et al. tried to esterify the corresponding acid **119**, however, as expected by Doi et al., esterification with GPC derivative **120** did not give any satisfied result. Thus, they followed Doi's protocol. HWE reaction of aldehyde **118** with phosphonate **121** afforded **122**. After replacement of both TIPS and TBS groups with acetyl (**123**), the acetonide group was removed under acidic conditions using 1,4-dioxane as a co-solvent, followed by sulfonamidation to furnish siladenoserinol A (**84**). Interestingly, when **123** was hydrolyzed in Et_2O, partial migration of the acyl chain occurred to afford **84** and

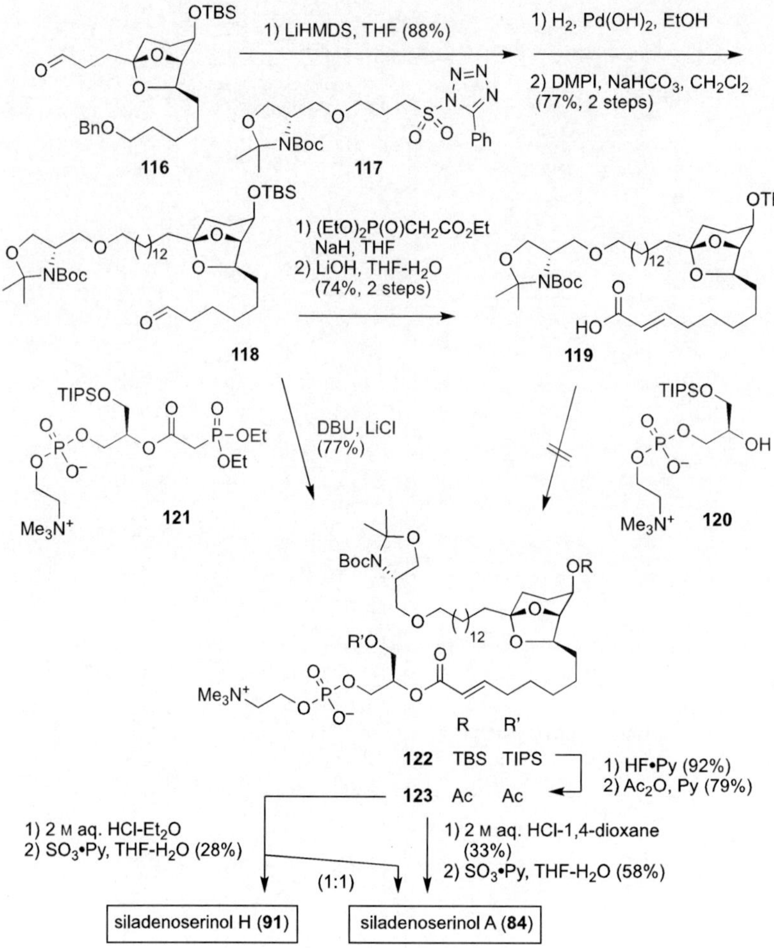

Scheme 19 Tong's synthesis of siladenoserinols A and H

siladenoserinol H (**91**) in a ratio of 1:1. These isomers were separated by reverse-phase HPLC.

5 Conclusion

Since the discovery of didemniserinolipids A-C, a series of congeners with a sulfate and/or GPC moiety has been found from the family Didemnidae. Although initial didemniserinolipids showed no significant antitumor activity, cyclodidemniserinol trisulfate was found as an HIV-1 integrase inhibitor, and siladenoserinols A-P inhibited p53-Hdm2 interaction. Further pursuit of natural serinols and synthetic studies including designed analog compounds will shed light on the development of drug leads. Novel methods to construct the bicyclooctane core is also expected.

Acknowledgements This chapter is dedicated to the late Prof. Qing-Wen Shi who passed away on 16th October 2020. He was scheduled to co-author this chapter with a different title.

References

1. Youssef DTA, Almagthali H, Shaala LA, Schmidt EW (2020) Mar Drugs 18:307
2. González N, Rodríguez J, Jiménez C (1999) J Org Chem 64:5705
3. Kiyota H, Dixon DJ, Luscombe CK, Hettstedt S, Ley SV (2002) Org Lett 4:3223
4. Kiyota H (2006) Synthesis of marine natural products with bicyclic and/or spirocyclic acetals. In: Kiyota H (ed) Marine natural products, topics in heterocyclic chemistry. Springer, Berlin, p 65
5. Garner P, Park JM (1987) J Org Chem 52:2361
6. Dixon DJ, Foster AC, Ley SV (2000) Org Lett 2:123
7. Franke W, Schroeder F, Philipp P, Meyer H, Sinwell V, Gries G (1996) Bioorg Med Chem 4:363
8. Ohtani I, Kusumi T, Kashman Y, Kakisawa H (1987) J Am Chem Soc 113:4092
9. Marvin CC, Voight EA, Burke SD (2007) Org Lett 9:5357
10. Marvin CC, Voight EA, Suh JM, Paradise CL, Burke SD (2007) J Org Chem 73:8452
11. Kraus IJ, Mandal M, Danishefsky SJ (2007) Angew Chem Int Ed 46:5576
12. Burke SD, Sametz GM (1999) Org Lett 1:71
13. He L, Wanunu M, Byun HS, Bittman R (1999) J Org Chem 64:6049
14. Ramana CV, Induvadana B (2009) Tetrahedron Lett 50:271
15. Das S, Induvadana B, Ramana CV (2013) Tetrahedron 69:1881
16. Abrams SR, Shaw AC (1993) Organic Synth 8:146
17. English JJ, Griswold JPH (1948) J Am Chem Soc 70:1390
18. Yamaguchi M, Hirao I (1983) Tetrahedron Lett 24:391
19. Alcaide B, Almendros P, Carrascosa R, Torres MR (2010) Adv Synth Catal 352:1277
20. Prasad KR, Chandrakumar A (2007) Tetrahedron 63:1798
21. Prasad KR, Chandrakumar A (2007) J Org Chem 72:6312
22. Prasad KR, Chandrakumar A (2005) Tetrahedron:Asymmetry 16:1897
23. Mahipal B, Mllikarjun K, Chandrasekhar S (2012) Tetrahedron Lett 53:45
24. Fürstner A, Radkowski K, Wirtz C, Goddard R, Lehmann CW, Mynott R (2002) J Am Chem Soc 124:7061
25. Golden JE, Aubé J (2002) Angew Chem Int Ed 41:4316
26. Yadav JS, Chander MC, Joshi BV (1988) Tetrahedron Lett 29:2737
27. Ren JG, Tong RB (2013) J Org Chem 79:6987
28. Zhang W, Tong RB (2016) Synthetic studies of 6,8-dioxabicyclo[3.2.1]octane structure including didemniserinolipid B are reviewd. J Org Chem 81:2203
29. Vilsmeier A, Haack A (1927) Chem Ber 60:119
30. Achmatowicz OJ, Bukowski P, Szechner B, Zwierzchowska Z, Zamojski A (1971) Tetrahedron 27:1973
31. Mitchell SS, Phodes D, Bushman FD, Faulkner DJ (2000) Org Lett 2:1605
32. Liu JH, Song LD, Long YQ (2009) Tetrahedron Lett 50:4587
33. Liu JH, Long YQ (2009) Tetrahedron Lett 50:4592
34. Liu JH, Jin Y, Long YQ (2010) Tetrahedron 66:1267
35. Nakamura Y, Kato H, Nishikawa T, Iwasaki N, Suwa Y, Rotinsulu H, Losung F, Maarisit W, Mangindaan REP, Morioka H, Yokosawa H, Tsukamoto S (2013) Org Lett 15:322

36. Torii M, Hirota Y, Kato H, Koyanagi Y, Kawahara T, Losung F, Mangindaan REP, Tsukamoto S (2018) Tetrahedron 74:7516
37. Yoshida M, Saito K, Kato H, Tsukamoto S, Doi T (2018) Angew Chem Int Ed 57:5147
38. Huang H, Zhou Y, Liu H (2011) Beilstein J Org Chem 7:897
39. Yamauchi K, Une F, Tabata S, Kinoshita M (1985) J Chem Soc Perkin Trans 1:765
40. Márquez-Cadena MA, Ren J, Ye W, Qian P, Tong R (2019) Org Lett 21:9704

Top Heterocycl Chem (2021) 58: 439–466
https://doi.org/10.1007/7081_2021_57

Published online: 25 June 2021

Synthesis of Chlorosulfolipid Natural Products

Philipp Sondermann and Erick M. Carreira

Contents

Abstract Chlorosulfolipids are an intriguing class of natural products that have been found across the globe as constituents of aqueous organisms. Their intriguing structure and unclear biological role have raised interest throughout various fields of chemical research, which in turn led to a number of elegant total syntheses of natural products from this family. This chapter provides an overview of methods and strategies for the preparation of chlorosulfolipids as well as challenges involved in their structure determination.

P. Sondermann and E. M. Carreira (✉)
ETH Zürich, Zürich, Switzerland
e-mail: erickm.carreira@org.chem.ethz.ch

Keywords Acyclic stereocontrol · Chlorination methods · *J*-based configurational analysis · Natural products · Nuclear magnetic resonance · Total synthesis

1 Introduction

The chlorosulfolipids comprise a relatively small class of natural products isolated from aquatic organisms. As structural motifs, they share extensive chlorination of an aliphatic backbone along with further functionalization leading to densely functionalized aliphatic structures with intricate arrays of stereogenic carbon atoms. Chlorosulfolipids have been isolated as far back as the 1960s from an array of freshwater algae, culinary mussels as well as from octacorals. Advances in structure determination of these complex lipids are intricately interwoven with new developments in spectroscopic methods. The synthesis of chlorosulfolipids has received significant attention from various research groups which in turn has led to numerous reports of elegant syntheses and novel methods.

2 Isolation of Chlorosulfolipids

Chlorosulfolipids were first reported in the 1960s when extracts of the sweet water algae *Ochromonas danica* were investigated for the presence of sulfolipids. Mass spectrometric measurements after derivatization with TMSCl along with feeding experiments of compounds containing radioactive chlorine and sulfur isotopes confirmed a new class of lipids containing both chloride and sulfate groups [1, 2]. Amongst others, extensive degradation studies and fragmentation pattern analysis allowed the determination of the structure of these lipids; however, no absolute or relative configurations were determined. After their initial discovery, chlorosulfolipids were isolated from numerous organisms and habitats. The chlorosulfolipids that have received most attention from the research community are undoubtedly Danicalipin A from the golden algae *Ochromonas danica* [3–6], Malhamensilipin A from the golden algae *Poterioochromonas malhamensis* [7], as well as Mytilipin A, B and C [8–10] from the culinary mussel *Mytilus galloprovincialis*. It is to be noted that Mytilipin A was also isolated from Formosan octoral *Dendronephthya griffini* of the coast of Taiwan [11]. This is intriguing in light of the fact that known chlorosulfolipid producing organisms such as *O. danica* and *P. malhamensis* are sweet water algae. It can be hypothesized that both the mussel *M. galloprovincialis* and the octacoral *D. griffini* may have accumulated the chlorosulfolipids as part of their diet. An excellent review on the history of chloroulfolipids has been published [12].

3 Structure Determination

The configurational assignment of chlorosulfolipids proved to be a major hurdle towards more in-depth studies of these peculiar natural products. Roughly 30 years after the initial discovery of chlorosulfolipids, Murata's *J*-based configurational assignment (JBCA) in 1999 proved to be the technique of choice to overcome this hurdle [13]. JBCA is a powerful means to determine conformations between adjacent stereocenters and with this information in hand allows deduction of relative configurations through the analysis of ${}^{2}J_{{}^{13}C-{}^{1}H}$, ${}^{3}J_{{}^{13}C-{}^{1}H}$ and ${}^{3}J_{{}^{1}H-{}^{1}H}$ coupling constants. In the case of small ${}^{3}J_{{}^{1}H-{}^{1}H}$ coupling constants between vicinal, hetereoatom-substituted methines, a set of five coupling constants allows complete assignment of relative conformation and configuration. Murata introduced the classification shown in Fig. 1 for this matter. A limitation of this approach is that for conformations A3 and B3 with large ${}^{3}J_{{}^{1}H-{}^{1}H}$ values, all five coupling constants are likely of similar size. In those cases one cannot differentiate between *syn* and *anti* configurations based solely on the coupling constants. As a solution, Murata argued that this can be resolved by different behavior of the adjacent methines in a NOESY experiment. To be precise, gauche conformation of C and C′"leads to an NOE interaction between protons H and H′". The absence (B3) or presence (A3) of an NOE interaction can thus be used to discriminate between these conformations and consequently *syn* and *anti* configurations. With this additional tool at hand JBCA allows determination of all six conformations A1-3 and B1-3 and consequently between *syn* and *anti* relative configurations solely based on spectroscopic methods.

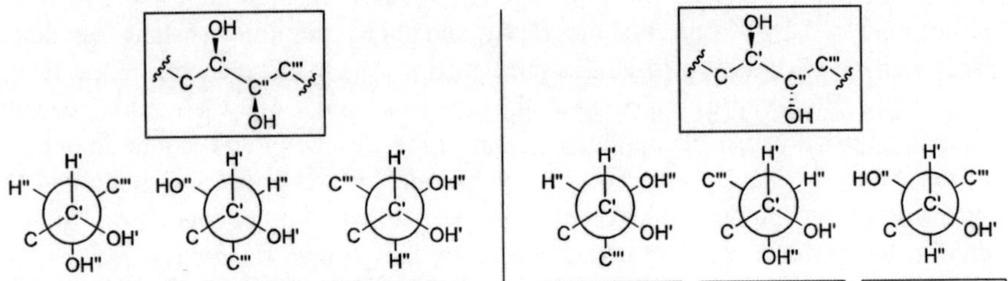

	A1	A2	A3	B1	B2	B3
$^{3}J_{H'-H''}$	small	small	large	small	small	large
$^{3}J_{H'-C'''}$	small	large	small	large	small	small
$^{3}J_{C-H''}$	small	large	small	small	large	small
$^{2}J_{C''-H'}$	small	large	large	large	small	large
$^{2}J_{C'-H''}$	small	large	large	small	large	large

Fig. 1 Classification of extracted coupling constants into categories "small" or "large" allows determination of relative conformation between two stereogenic centers when adopting conformations A1, A2, B1, or B2. For A3 and B3 conformations NOE data is necessary for unambiguous assignment

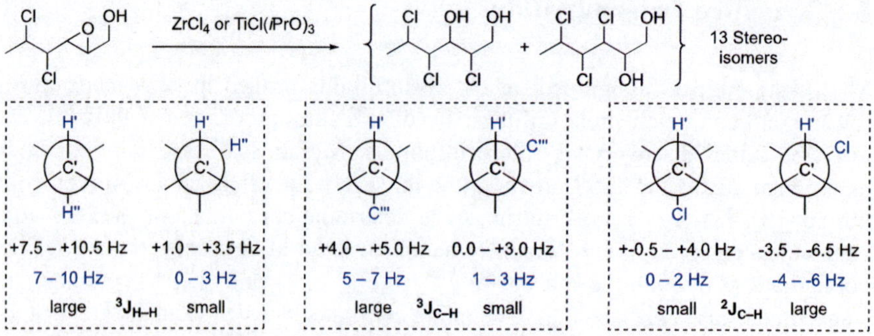

Scheme 1 Divergent synthesis of 13 different configurational and constitutional isomers for the creation of a universal coupling constant library of alkyl chlorides. Indicated coupling constants in black refer to Murata's report on oxygenated systems, blue values are typically observed in chlorinated systems

Two years after Murata's hallmark publication, Fattorusso and Ciminiello applied the new technique to a new cytotoxic chlorosulfolipid later named Mytilipin A [8]. After a report of a batch of toxic culinary mussels in which the hepatopancreas caused diarrhetic shellfish poisoning in a mouse bioassay, the group was able to not only isolate Mytilipin A, but also apply Murata's JBCA to determine the complete structure of this compound including configurational assigment. The technique had not been previously applied to chlorinated systems, but a later investigation showed it to be equally valuable for these types of systems [14]. For this, regioisomeric, crystalline trichlorohexanediols of different relative configurations were synthesized, analyzed by X-ray crystallography and their coupling constants tabulated. This allowed comparison to similar arrays in a systematic manner. While it was shown that JBCA could be applied to chlorinated systems, one should take care that slightly adjusted values of coupling constants must be taken into account: In Scheme 1, the values in black were reported by Murata for oxygenated systems and the values in blue were determined for chlorinated arrays. An application of JBCA in chlorinated systems was also demonstrated by the groups of Gerwick, Haines, and Vanderwal for the structure determination of Danicalipin A [15].

4 Synthesis of Chlorosulfolipids

Once the first complete structure of a chlorosulfolipid had been determined the field of chlorosulfolipids experienced significant interest in the synthetic community. Various groups were invested in the identification of suitable tactics and strategies to allow synthetic access to the intricate and unprecedented structures of these molecules [16]. Initial studies showed iterative Appel reactions to be unworkable in the proximity of electron-withdrawing substituents. The consecutive array of chlorines could thus likely not be assembled in this manner. Additionally, aldol additions and the likes proved to be technically difficult due to the unstable nature of the requisite α-chloroaldehydes [14, 17]. Early synthetic efforts relied consequently on diastereoselective dichlorination of olefins and epoxide openings. In part due to the interest in chlorosulfolipids the synthetic toolbox to introduce chlorine and other halogens in a stereoselective manner has vastly expanded in recent years.

4.1 Carreira's Synthesis of Mytilipin A

After the first complete structure of a chlorosulfolipid had been disclosed, early synthetic efforts by the Carreira group led to the first synthesis of a member from this family of natural products. The synthesis of Mytilipin A reported in 2009 [17] started out from dienoate **1** (Scheme 2). Reaction of this diene with Mioskowski's reagent took place at the γ,δ-olefin. The so obtained racemic dichloride **2** was further subjected to reduction with DIBAL and protection as the silyl ether. Dihydroxylation with OsO_4 followed by closure with the adduct of DABCO and Tf_2O allowed the synthesis of the *syn*-epoxide from the *trans*-olefin **3**. Subsequent deprotection under acidic conditions provided alcohol **4** that was oxidized to the aldehyde and underwent Wittig olefination with the ylide derived from phosphonium bromide **5** to yield *cis*-olefin **6**. Interestingly, epoxide opening with TMSCl predominantly led to formation of the undesired product formally derived from retentive epoxide opening **7** and only trace amounts of the desired inverted product **8**. This surprising outcome was rationalized with intermediary chloretanium (**9**) and chlorolanium (**10**) ions leading to double inversion. Further investigation later revealed retentive epoxide opening to be a more general concept when chlorides are present in proximity [18]. To address this unforeseen challenge in the context of this synthesis, the group restarted their efforts from previously synthesized **11**. Falling back to an epoxidation instead of the sequence dihydroxylation-epoxide closure led to the formation of the *anti*-epoxide. Again, oxidation and Wittig olefination now provided access to **12**. When this epoxide was subjected to Lewis acidic conditions with TMSCl retention of configuration at the allylic carbon was observed again, now as the desired outcome **14**. Dichlorination was then effected with Et_4NCl_3 with the stereoselectivity likely being dictated by the allylic strain of the allylic stereocenter. Silyl ether cleavage, oxidation to the aldehyde and Takai olefination provided the *trans*-vinylchloride **16**. Finally, treatment with $SO_3{\cdot}py$ delivered the first synthetic chlorosuflolipid, (±)-Mytilipin A (**17**).

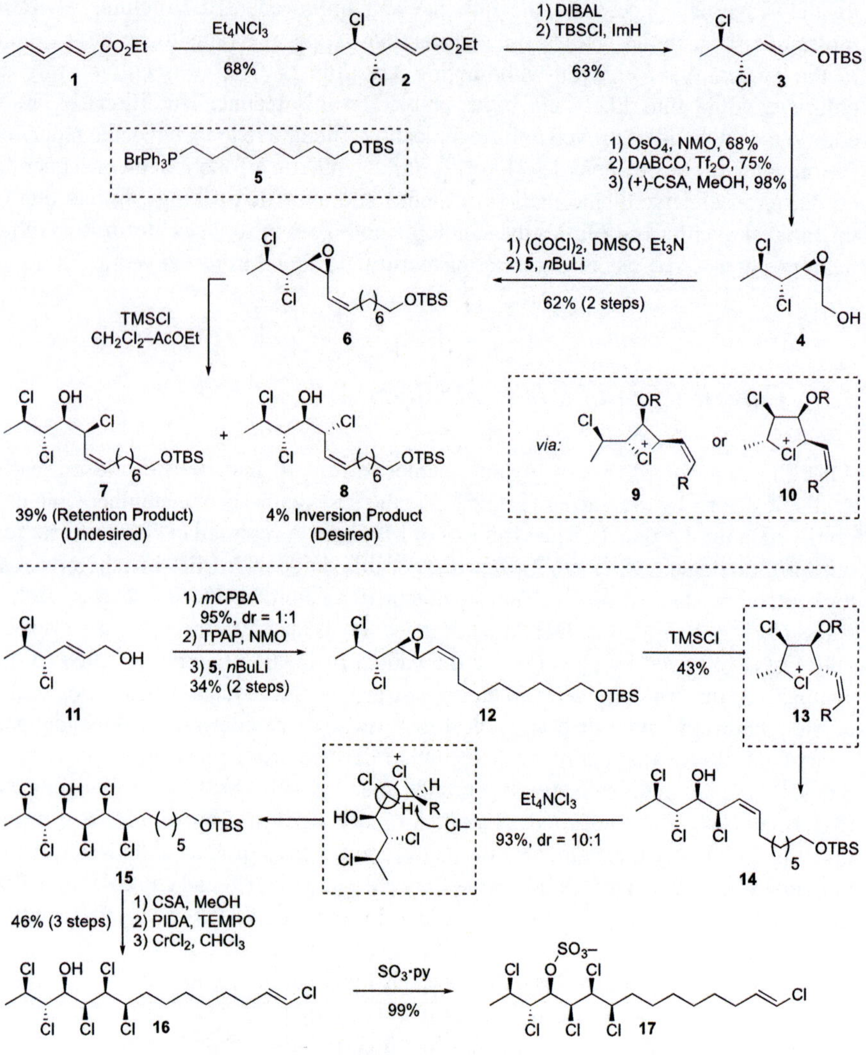

Scheme 2 Carreira's synthesis of (±)-Mytilipin A

4.2 *Vanderwal's Synthesis of Mytilipin A*

In 2013, the Vanderwal group reported a synthesis of Mytilipin A [19] that made use of the then newly developed Z-selective olefin metathesis (Scheme 3) [20]. For this effort, crotyl alcohol was dichlorinated with Mioskowski's reagent (Et_4NCl_3). Oxidation to the aldehyde was followed by an allylation reaction effected with allyl alane **20**. The formed bromohydrin **21** was not isolated, but rather converted to epoxide **22** in situ by treatment with NaOH. Following this, this olefin was reacted

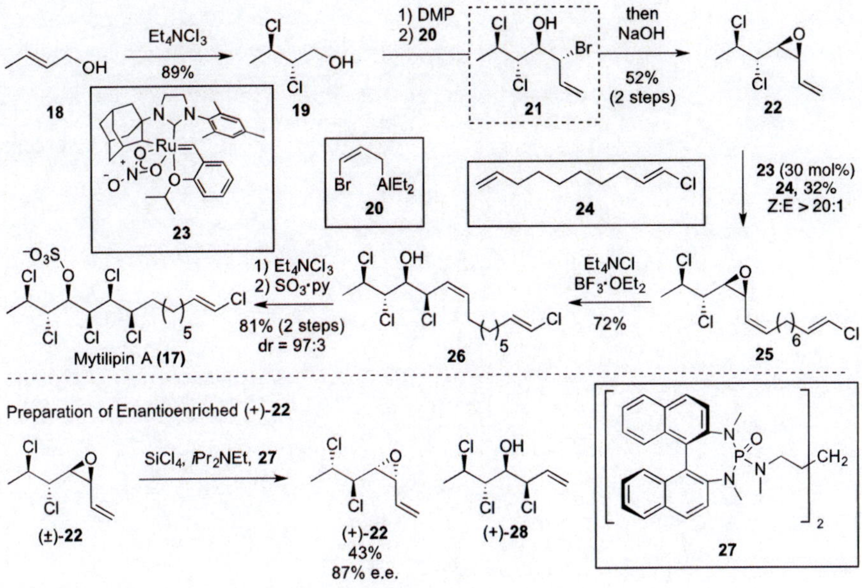

Scheme 3 Vanderwal's synthesis of (−)-Mytilipin A

with **24** in the presence of Grubbs' cyclometalated Ruthenium catalyst **23** delivering *cis*-olefin **25**, albeit in low yield. The epoxide moiety was then opened to tetrachloride **26**. In the following dichlorination the *cis*-olefin proved to be more reactive than the vinylchloride, subsequently allowing sulfation to provide (±)-Mytilipin A (**17**). Additionally, the Vanderwal group showcased that this racemic synthesis could also be carried out enantioselectively. To this effect, epoxide (+)-**22** as an intermediate synthesized en route to Mytilipin A was prepared by kinetic resolution of ± − (**22**) using Denmark procedure for epoxide openings with $SiCl_4$ and a chiral phosphoramide [21–23] (Scheme 3).

4.3 *Yoshimitsu's Synthesis of Mytilipin A*

The approaches by the Carreira and Vanderwal groups had relied on two strategies to introduce chlorides in an enantiospecific manner in their molecules: epoxide opening reactions and dichlorinations of olefins (Scheme 4). An interesting, distinct approach for the synthesis of chlorosulfolipids was presented by the group of Yoshimitsu (Scheme 4) [24]. After pivaloyl protection of enantiomerically pure epoxide **29** reaction with PPh_3 and NCS provided dichlorinated **30**. This strategy leads to different overall stereochemical outcome compared to dichlorination of an olefin: deoxygenative chlorination of *anti*-epoxides lead to 1,2-*syn* dichlorides as exemplified in the synthesis of **30** but dichlorination of *trans*-olefins leads to 1,2-*anti*

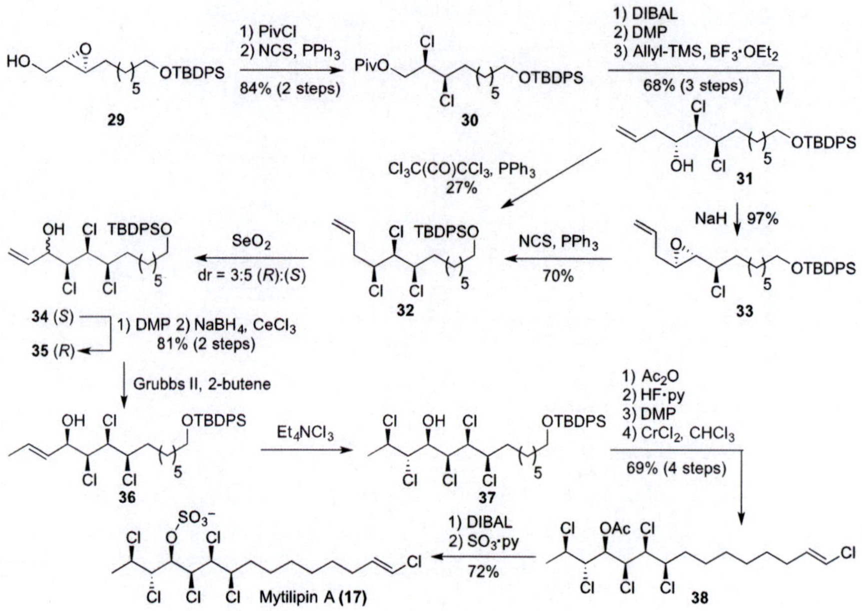

Scheme 4 Yoshimitsu's synthesis of (+)-Mytilipin A

dichlorides. The starting epoxides are readily available, e.g. by Sharpless epoxidation [25, 26]. The pivaloate was then reductively cleaved, oxidation with Dess-Martin periodinane set the stage for a Sakurai allylation, leading to **31**. A one-step conversion of the resulting alcohol in **31** to the respective chloride in **32** could only be effected in low yield. Consequently, a two-step procedure was employed, first closing the halohydrin to epoxide **33** followed by deoxygenative epoxide opening, again with NCS and PPh_3. Following this, Riley oxidation provided a diastereomeric mixture of **34** and **35**. This could be resolved, after separation, by a oxidation/reduction sequence funneling the undesired diastereomer to **35**. Cross-metathesis with 2-butene led to overall methylation of the terminal olefin, which was in turn dichlorinated with Et_4NCl_3 to give **37**. After protection of the free alcohol as the acetate, the silyl ether was converted to a C1-homologated vinylchloride by a deprotection-oxidation-Takai olefination sequence. Reductive deprotection of the acetate and sulfation then finished the synthesis of Mytilipin A (**17**).

4.4 *Burns Synthesis of Deschloromytilipin A*

The Burns group reported a synthesis of Deschloromytilipin A in 2016, a compound lacking the vinylic chloride group in comparison with Mytilipin A (Scheme 5) [27]. Key to this synthesis was an extension of a method previously developed in

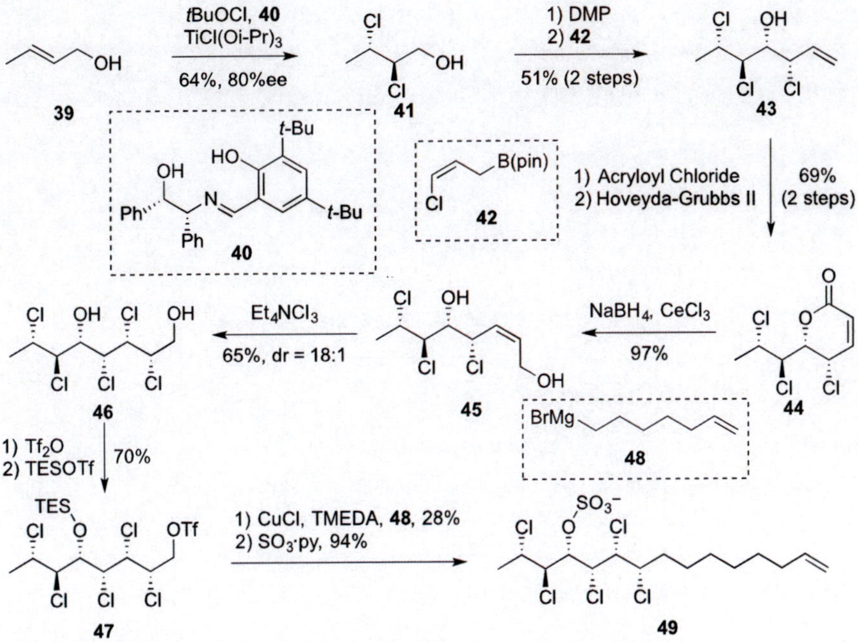

Scheme 5 Burns' synthesis of (−)-Deschloromytilipin A

the same laboratory, the asymmetric dihalogenation of allylic alcohols [28]. Using this approach, crotyl alcohol **39** was asymmetrically dichlorinated using Schiff base **40**, $TiCl(O\textit{i}Pr)_3$ and *t*BuOCl. Oxidation to the aldehyde using Dess-Martin periodinane allowed a subsequent allylation with vinylchloride **42**. Acylation of **43** with acryloyl chloride allowed subsequent ring-closing metathesis to **44**. The reduction of this compound allowed selective access to the *cis*-olefin **45**. Dichlorination of this olefin provided **46** which was triflated and silyl-protected. Copper-catalyzed sp^3–sp^3 cross coupling [29] with Grignard reagent **48** and sulfation finished the synthesis of (−)-Deschloromytilipin A.

4.5 Vanderwal's Synthesis of Danicalipin A

Shortly after the first disclosure of a synthesis of a chlorosulfolipid by the Carreira group, the Vanderwal group reported the synthesis of Danicalipin A, a chlorosulfolipid derived from the algae *Ochromonas danica* [14]. The authors had chosen a similar strategy for the stereoselective introduction of chlorine, namely an epoxide opening and two dichlorinations. Starting from dienoate **50** (Scheme 6), an initial dichlorination was succeeded by a diastereoselective dihydroxylation to **51**. The diol was activated with nosyl chloride and closed to the epoxide using K_2CO_3.

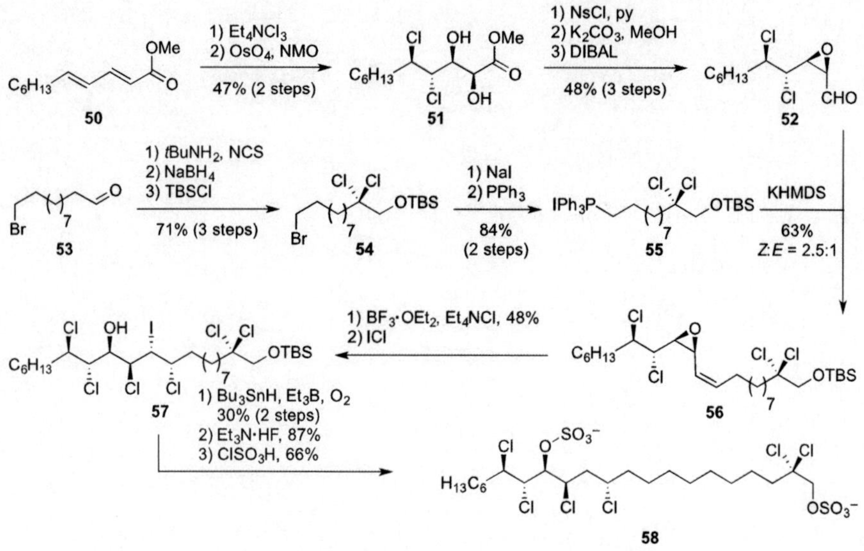

Scheme 6 Vanderwal's synthesis of (±)-Danicalipin A

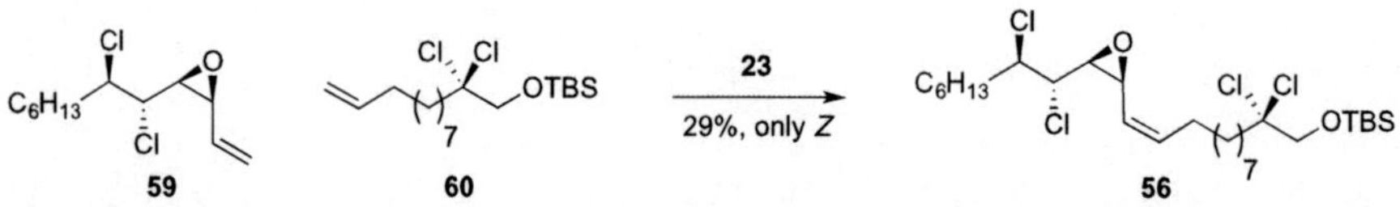

Scheme 7 Vanderwal's second generation synthesis of intermediate 118 previously used en route to (±)-Danicalipin A

Reduction to aldehyde **52** was effected with DIBAL, finishing this fragment. A second fragment was started from bromoaldehyde **53**. Enamine formation and dichlorination gave an aldehyde that was first reduced to the primary alcohol and then protected as silyl ether **54**. Transformation of the bromide to an iodide in a Finkelstein reaction and formation of a phosphonium salt gave **55** that underwent Wittig reaction with previously synthesized aldehyde **52**. The epoxide was subsequently opened at the allylic position under Lewis-acidic catalysis and the remaining olefin could be iodochlorinated with ICl. The iodide in **57** was then reductively cleaved to yield an intermediate that was deprotected and sulfated with $ClSO_3H$ to give access to synthetic (±)-Danicalipin A (**58**).

In a later publication, the Vanderwal group also made use of Grubbs' Z-selective metathesis for a formal enantioselective synthesis of Danicalipin A (Scheme 7) [21]. For this effort, **59** and **60** were subjected to the action of cyclometalated Ruthenium catalyst **23** to give **56**. The starting allylic epoxide **59** is again accessible using Denmark's kinetic resolution methodology (cf. Scheme 3).

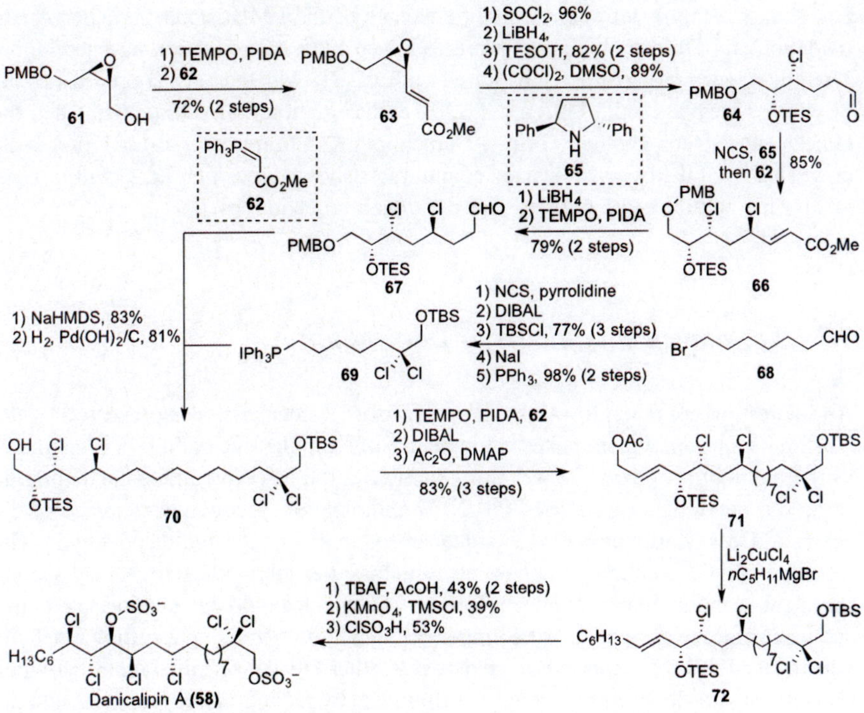

Scheme 8 Umezuwa's 2011 synthesis of (−)-Danicalipin A

4.6 *Umezuwa's Synthesis of Danicalipin A*

An additional synthesis of Danicalipin A was reported in 2011 by the group of Umezawa and Matsuda (Scheme 8) [30]. One noteworthy peculiarity about this report is that both enantiomers were synthesized, allowing comparison of their bioactivities. The route started from enantioenriched epoxide **61**. Anelli oxidation and Horner olefination with phosphonate **62** yielded *trans*-enoate **63**. $SOCl_2$ effected epoxide opening at the allylic position to give the respective halohydrin. The enoate was reduced to the saturated alcohol and the alcohol was protected as the silyl ether. Swern oxidation then yielded aldehyde **64** which could be diastereoselectively monochlorinated using Jørgsensen's diphenylpyrrolidine **65** [31] and NCS to give **66** after another Horner olefination using **62**. Again, from reduction of the enoate with $LiBH_4$ the saturated alcohol was obtained that was reoxidized in a TEMPO-catalyzed oxidation to **67**. A second fragment for the synthesis of Danicalipin was synthesized from bromoaldehyde **68**. Dichlorination was followed by reduction and protection of the aldehyde. Substitution of the primary bromide with an iodide allowed formation of the phosphonium salt **69** by addition of PPh_3. Wittig reaction of the two fragments **67** and **69** then gave an olefin that was reduced using

Pearlman's catalyst with concomitant cleavage of the PMB group. Another Anelli oxidation and Horner olefination were followed by a 1,2-reduction and aetylation. The allylic acetate was then substituted with $nC_5H_{11}MgBr$ under copper catalysis. Silyl ether cleavage with buffered TBAF, dichlorination and sulfation finished the Umezawa/Matsuda synthesis of (−)-Danicalipin A. Comparison of toxicity against brine shrimp of the two synthesized enantiomers showed a similar LC_{50} with 2.1 μg/mL for the natural and 2.4 μg/mL for the unnatural enantiomer.

4.7 *Yoshimitsu's Synthesis of Danicalipin A*

An interesting approach towards the synthesis of Danicalipin was presented by the Yoshimitsu group, which makes strategic use of a dipolar cycloaddition (Scheme 9) [32]. Enantiopure epoxide **73** was first esterified and then converted to the dichloride using the previously discussed PPh_3/NCS combination. Reductive cleavage of the ester and Dess-Martin oxidation enabled subsequent vinylation of the aldehyde. The resulting allylic alcohol was then protected as the silyl ether in **75**. A second fragment was elucidated from enol ether **76**, which could be dichlorinated in a sequence of three steps to **77**. The imine was then hydrolyzed, the resulting aldehyde was reduced and the alcohol protected as silyl ether **78**. The terminal olefin was then converted to an aldehyde by ozonolysis, followed by reduction. Appel reaction to the alkyl iodide and nucleophilic substitution with sodium nitrite provided nitroalkane

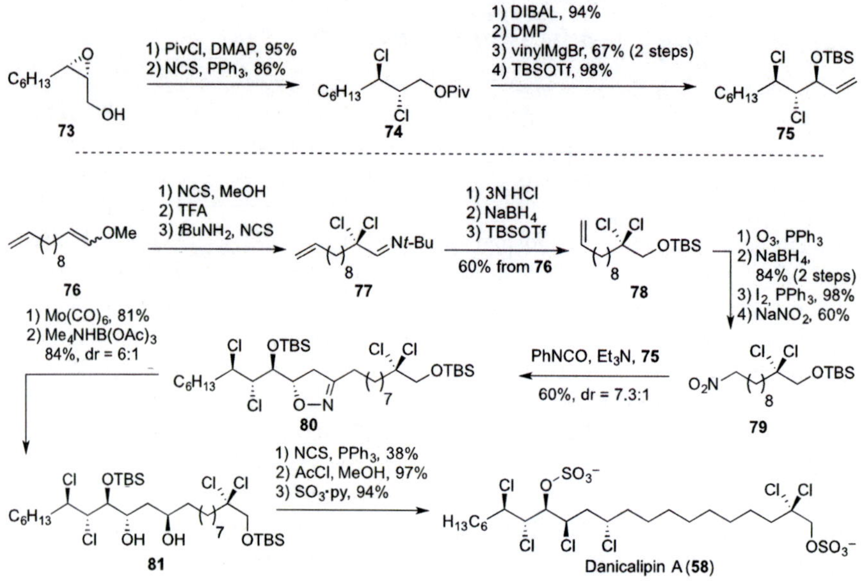

Scheme 9 Yoshimitsu's synthesis of (+)-Danicalipin A

79. Dehydration of this compound with PhNCO produced the nitrile oxide which reacted with olefin **75** in a dipolar cycloaddition to **80**. Reductive cleavage of this isoxazoline and Evans-Saksena reduction led to **81**. Two Appel reactions of the remaining free alcohols with NCS and PPh_3 were followed by silyl ether cleavage under acidic conditions and sulfation to yield synthetic (+)-Danicalipin A (58).

4.8 Carreira's Synthesis of Danicalipin A

A synthesis of (+)-Danicalipin was also reported by the Carreira group in 2015 (Scheme 10) [33]. The synthesis started from allylic alcohol **82** that was oxidized using MnO_2 to provide the desired *cis*-unsaturated aldehyde with almost no detection of the *trans* product. A subsequent allylation with the in situ generated Oehlschlager-Brown reagent [34, 35] and subsequent diastereoselective epoxidation using *m*CPBA provided **84**. The so generated homoallylic alcohol was protected as the silyl ether and the epoxide were transformed to the vicinal dichloride using deoxygenative conditions with NCS and $ClPPh_2$ to **85**. The use of sterically hindered Cy_2BH ensured the formation of the desired product in the hydroboration reaction whose product was worked up oxidatively with $NaBO_3$ to the corresponding primary alcohol. The authors report that $BH_3 \cdot THF$ instead of Cy_2BH leads to the observation of significant amounts of a dechlorinated product. After oxidation to the aldehyde, an allylation reaction provided diastereoselective access to **86**. The resulting homoallylic alcohol in **86** was transformed to the respective inverted alkyl chloride

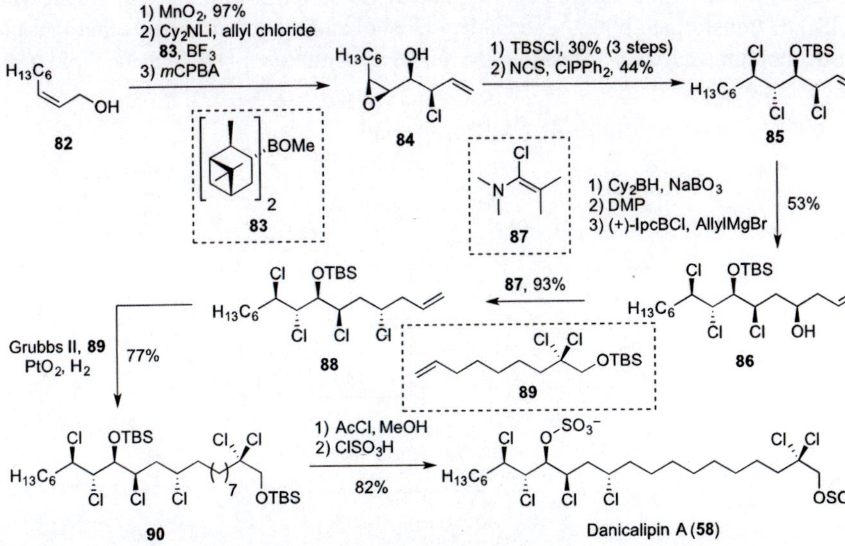

Scheme 10 Carreira's synthesis of (+)-Danicalipin A

through the use of Ghosez' reagent (**87**) [36, 37]. Grubbs' catalyst was then used to perform cross-metathesis with terminal oelfin **89**. The resulting olefin could be reduced in situ through addition of PtO_2 under a hydrogen atmosphere to **90**. The two silyl ethers could then be cleaved conveniently using acetyl chloride in methanol and the resulting compound was sulfated to deliver (+)-Danicalipin (**58**). This synthesis proved amenable to larger scales providing access to sufficient material for further biological studies including studies of self-assembly into membranes of this lipid (Scheme 10).

4.9 Burns' Synthesis of Danicalipin A

The Burns group presented a more recent synthesis of (−)-Danicalipin in 2016 (Scheme 11) [27]. The intriguing starting material for this synthesis was diborylated olefin **91**. Electrophilic substitution using ICl followed by transesterification with diol **92** selectively produced the borylated *cis*-vinylchloride **93**. This compound underwent Matteson homologation with $LiCHCl_2$ and $ZnCl_2$ to produce the insertion product **94** diastereoselectively. Metallation of **95** using *t*BuLi and addition to boronic ester **94** gave the desired allyl boronate which underwent an allylation reaction upon addition of **96** and *n*BuLi. The resulting *trans*-olefin in **97** was difunctionalized with Me_4NCl_2Br, a reaction in which it proved to have a positive effect to first treat the starting material **97** with deuterated methanol to improve the overall yield of this addition, likely by changing the H-bond strength to the adjacent alcohol in the transition state. Using conditions for radical dehalogenation with Bu_3SnH produced the desired debrominated product preferentially. Treatment with $ClSO_3H$ transformed both the secondary alcohol and the primary silyl ether into the corresponding sulfates, finishing the concise synthesis of (−)-Danicalipin (**58**) (Scheme 11).

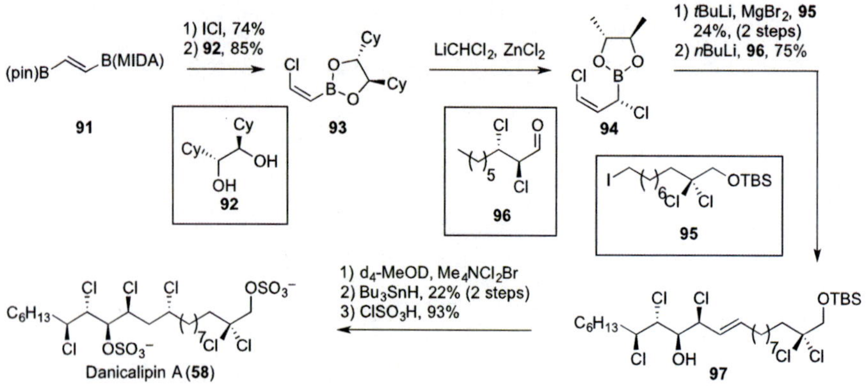

Scheme 11 Burns' synthesis of (−)-Danicalipin A

4.10 *Carreira's Synthesis of Bromodanicalipin A and Fluorodanicalipin A*

To further investigate the stereoelectronic effects of the chloride substituents on the conformation of chlorosulfolipids the Carreira group investigated the synthesis of the fluorinated and brominated analogs of Danicalipin (Scheme 12) [38]. The latter had previously been hypothesized and was later shown to be produced by *Ochromonas danica* when cultured on a bromide rich medium [6, 39]. The enantioselective synthesis of (+)-Bromodanicalipin A started from hydroxy olefin **98** which was protected using pivaloyl chloride. Hydroboration and Swern oxidation transformed the terminal olefin to the corresponding aldehyde (**99**). Reaction of this aldehyde with Br_2 in the presence of HBr delivered the desired α,α-dibromide, which was reduced and protected as the silyl ether. The pivalate protecting group was then reductively cleaved using DIBAL and transformed to the aldoxime by Swern oxidation and condensation with hydroxylamine hydrochloride. Subsequently, oxidation of the aldoxime with sodium hypochlorite gave access to the nitrile oxide, which in the presence of olefin **102** underwent dipolar cycloaddition to **103**. The

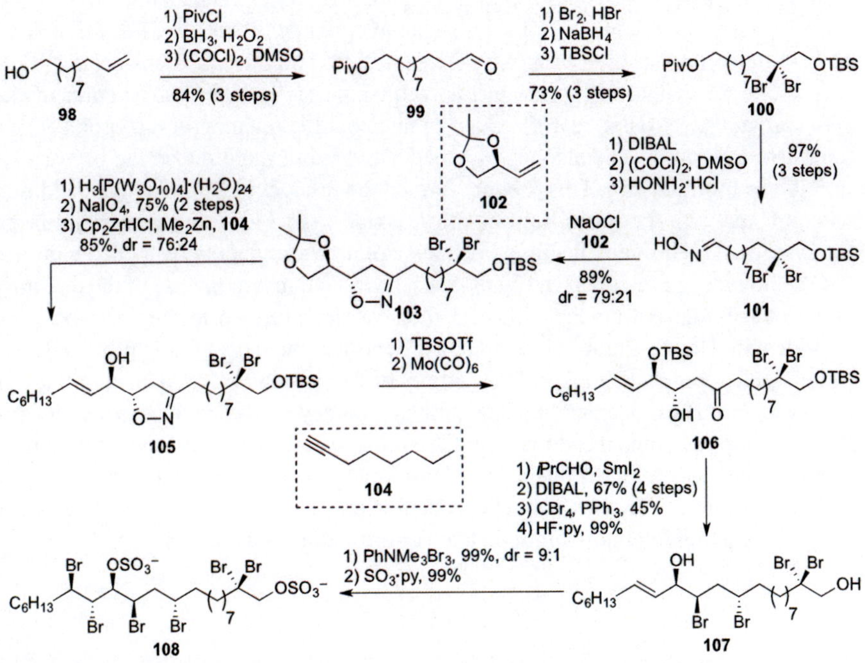

Scheme 12 Carreira's synthesis of (+)-Bromodanicalipin A

acetonide in **103** was then hydrolyzed and the resulting glycol underwent cleavage with $NaIO_4$. The so obtained aldehyde was reacted with the hydrozirconation product of **104** and Schwartz' reagent (Cp_2ZrHCl) in the presence of $ZnMe_2$ to **105**. Protection of the secondary allylic alcohol with TBSOTf and reductive cleavage of the isoxazoline delivered hydroxy ketone **106**. Diastereoselective reduction of the ketone group was achieved by an Evans-Tishchenko employing SmI_2 and isobutyraldehyde as hydride donor. The resulting ester was reductively cleaved and the 1,3-diol was brominated using two iterative Appel reactions with CBr_4 followed by desilylation to provide access to **107**. Diastereoselective dibromination could then be achieved with $PhNMe_3Br_3$ and sulfation of the two remaining alcohols completed the synthesis of (+)-Bromodanicalipin A (Scheme 12).

Along with Carreira's studies of Bromodanicalipin, a synthesis of Fluorodanicalipin A was also presented. For this effort, aldehyde **109** was difluorinated employing a strategy of first forming the enamine with racemic proline and fluorination by NFSI (**110**) followed by reduction with $NaBH_4$ to **111**. The resulting primary alcohol was protected as the benzyl ether and the terminal olefin was in turn converted to a primary alcohol with BH_3 and H_2O_2, oxidation and condensation provided aldoxime **112**. The second fragment for the synthesis of Fluorodanicalipin was synthesized from *cis*-allylic alcohol **82**. MnO_2 effected the oxidation to the aldehyde without affecting diastereomeric purity of the olefin. Addition of vinyl magnesium bromide formed a racemic allylic alcohol which also served as directing element for the subsequent epoxidation with *m*CPBA to deliver **113**. The alcohol was then benzyl protected and the epoxide was transformed to the corresponding vicinal difluoride in two sequential steps. The first reaction of the epoxide with $Et_3N{\cdot}(HF)_3$ at 150 °C led to nucleophilic opening of the epoxide by a fluoride nucleophile. The alcohol was then activated through the action of nonaflyl fluoride and formed **114**. This product formed the product of a dipolar cycloaddition when treated with the in situ formed nitrile oxide from **112**. In a strategy similar to the synthesis of Bromodanicalipin, the isoxazoline was reductively opened with Mo $(CO)_6$ and reduced to **166** in an Evans-Saksena reduction making use of the directing effect of the adjacent alcohol. The 1,3-diol was transformed to the corresponding fluoride with Deoxo-Fluor® (**117**). Finally, deprotection and sulfation provided (±)-Fluorodanicalipin A. Comparing the activity of Bromo- and Fluorodanicalipin it was observed that Bromodanicalipin has similar potency (LC_{50} = 4.7 μg/mL against Brine shrimp) as compared to natural Chlorodanicalipin whereas Fluorodanicalipin was 15 times less toxic in this assay. Additionally, conformational comparison by JBCA of Fluoro-, Bromo- and Chlorodanicalipin showed that these three lipids adopt a comparable major solution state conformation (Scheme 13).

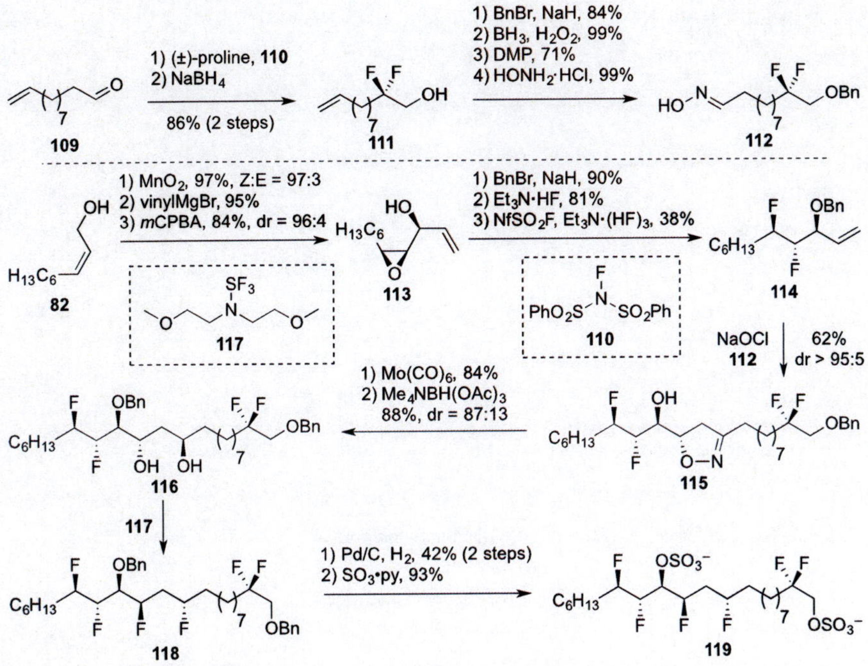

Scheme 13 Carreira's synthesis of (±)-Fluorodanicalipin A

4.11 Vanderwal's Synthesis of Malhamensilipin A

Gerwick and Vanderwal presented the first synthesis of Malhamensilipin A along with a revision of the originally proposed structure in 2009 (Scheme 14) [40]. For this effort, electron-deficient enyne **120** was selectively dihydroxylated in an asymmetric fashion using Sharpless's AD-mix. The alkyne was then partially reduced to the *cis*-olefin and the more reactive hydroxyl group was sulfonylated to **121**. The *cis*-olefin served as substrate for a dichlorination reaction with the diastereoselectivity dictated by the adjacent free hydroxyl group and the geometry of the olefin. Having served its purpose, the hydroxyl group was closed to the corresponding epoxide and the adjacent ester was partially reduced to aldehyde **122**. Wittig reaction with phosphonium ylide **123** preferentially formed *cis*-olefin **124**. The epoxide was then opened with Lewis-acidic $BF_3{\cdot}OEt_2$ at the more reactive, allylic carbon. The remaining *cis*-olefin was dichlorinated using Mioskowski's reagent (Et_4NCl_3). Transformation of the secondary alcohol and the primary silyl ether could be accomplished using $ClSO_3H$ and subsequent elimination provided exlusively

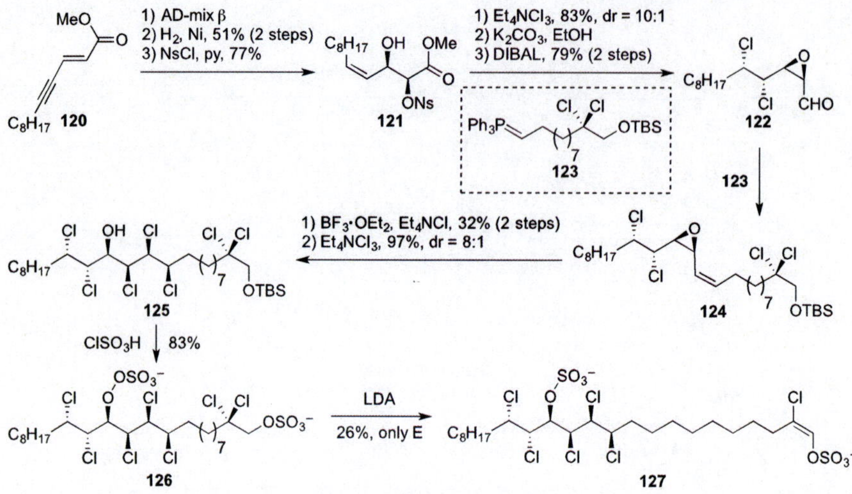

Scheme 14 Vanderwal's synthesis of (+)-Malhamensilipin A

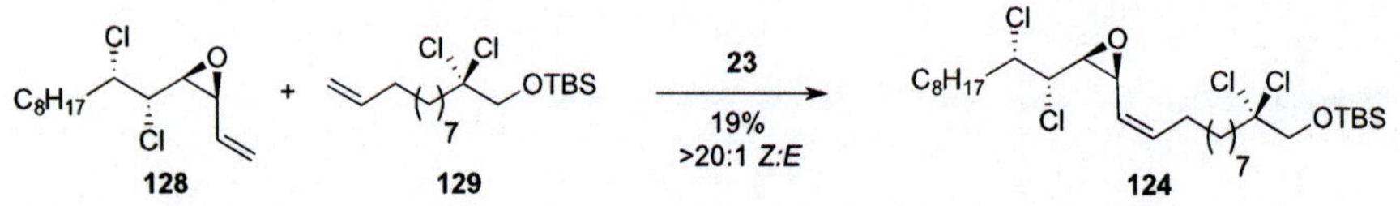

Scheme 15 Selected step from Vanderwal's second generation synthesis of (+)-Malhamensilipin A

the *E* isomer of olefin **127**, completing the synthesis of (+)-Malhamensilipin A (Scheme 14).

In addition to their initial report on the synthesis of Malhamensilipin A, the Vanderwal group also showed the applicability of *Z*-selective olefin metathesis in this setting (Scheme 15) [21]. To this effect, olefin **128** and **129** were reacted in the presence of Grubbs' cycometalated Ruthenium catalyst **23** to yield **124** in *Z*/*E* > 20:1, albeit in low yield. **124** had been previously demonstrated as an intermediate in the successful synthesis of (+)-Malhamensilipin A (**127**). The epoxide starting material **128** can in turn be accessed using Denmark's catalyst **27** (cf. Scheme 3) (Scheme 15).

4.12 Denton's Synthesis of (+)-Malhamensilipin A

A report by the Denton group from 2010 made use of multiple, previously demonstrated successful strategies for the synthesis of chlorosulfolipids in an effort to synthesize Malhamensilipin A (**127**) (Scheme 16) [41]. Starting from unsaturated aldehyde **130**, the enantioselective Scheffer-Weitz epoxidation, using the prolinol catalyst **131** originally developed by Jørgensen, led to the formation of **132** [42]. After Horner olefination with phosphonate **133** to **134**, the epoxide was transformed to vicinal dichloride **135** using catalytic triphenylphosphine oxide along with stoichiometric oxalyl chloride to form the chlorophosphonium salt. A series of ester reduction, Sharpless epoxidation and reoxidation set the stage for a Julia olefination with hetaryl sulfone **137**. The so obtained allylic epoxide **138** underwent retentive epoxide opening and epoxidation to **139**. In analogy to previous reports by Carreira, the retentive epoxide opening was rationalized by intervention of proximal chloride substitutents leading to double inversion [17, 18]. Epoxide **139** was again subjected to deoxygenative chlorination to the corresponding hexachloride, which in turn was transformed to intermediate **126** from the Vanderwal route by desilylation and sulfation (Scheme 16).

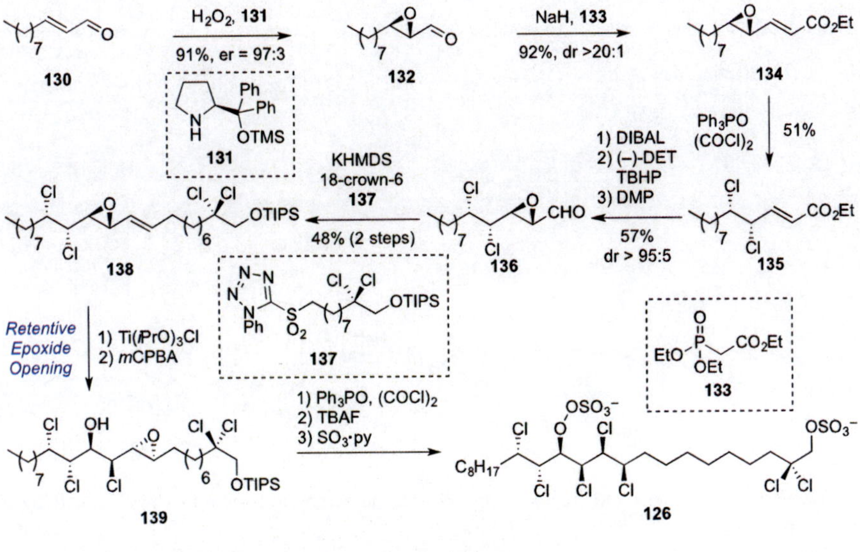

Scheme 16 Denton's synthesis of (+)-Malhamensilipin A

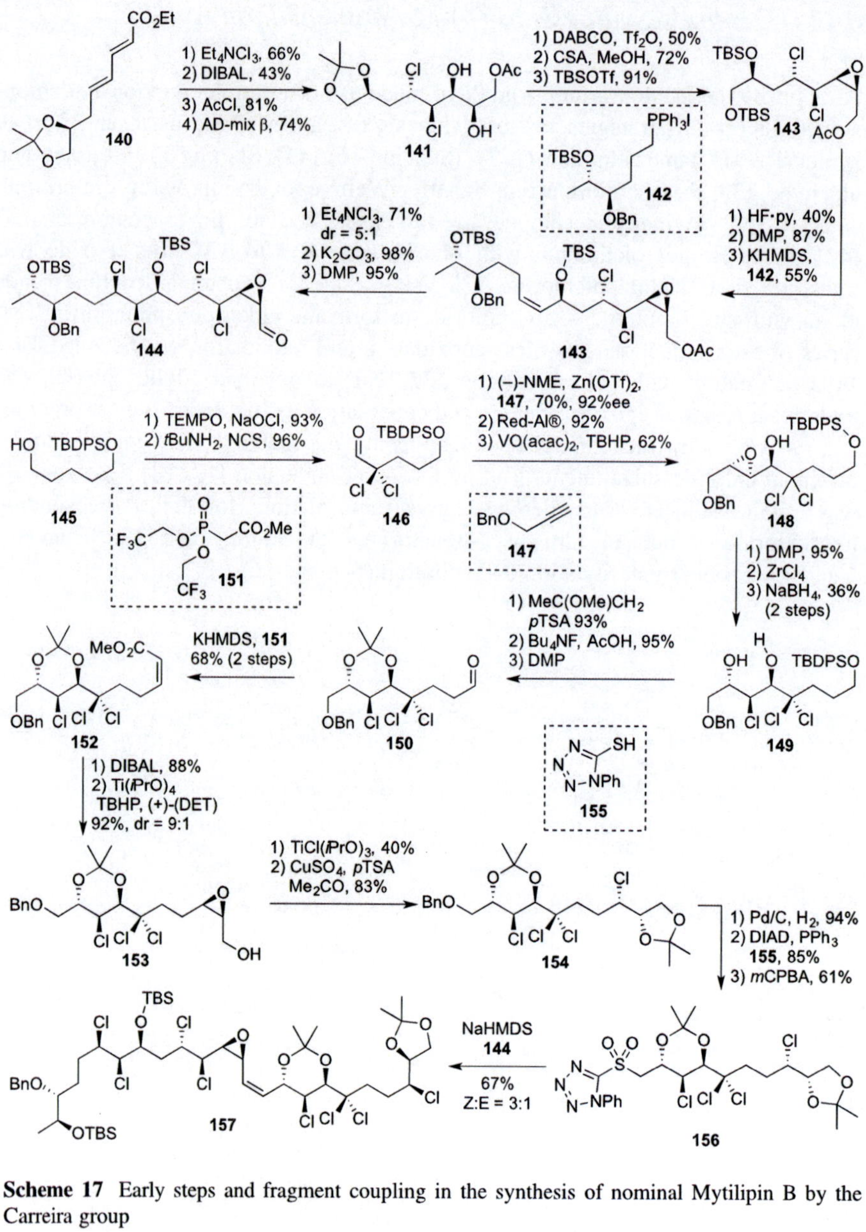

Scheme 17 Early steps and fragment coupling in the synthesis of nominal Mytilipin B by the Carreira group

4.13 Carreira's Synthesis of Nominal Mytilipin B

Carreira et al. presented a synthesis of the nominal structure of Mytilipin B in 2011 [9, 43], arguably the most complex chlorosulfolipid reported to date (Scheme 17).

Starting with **140** (Scheme 17), the distal, more reactive double bond was dichlorinated with slight diastereoselectivity induced by the proximal acetonide stereocenter followed by ester reduction, protection, and asymmetric dihydroxylation (**141**). The diol was converted to the corresponding epoxide by treatment with DABCO and Tf_2O. This sequence allowed the preparation of the *syn*-epoxide starting out from an *E*-olefin. Acetonide hydrolysis was followed by *bis*-silylation of the resulting diol. The more accessible primary alcohol was liberated with HF·py and oxidized to the aldehyde with Dess-Martin periodinane. Reaction with the phosphonium iodide **142** provided *cis*-olefin **143** in a Wittig reaction. Dichlorination with Mioskowski's reagent, acetate hydrolysis with K_2CO_3 and Dess-Martin oxidation gave fragment aldehyde **144**, finishing this fragment of the molecule. For a second fragment, the alcohol in **145** was oxidized in an Anelli oxidation and converted to the α,α-dichloroaldehyde by first transforming the aldehyde to the enamine using tBuNH$_2$ followed by treatment with NCS. Using conditions for asymmetric alkyne additions with $Zn(OTf)_2$, porpargylic ether **147** could be added to aldehyde **146**. Directed, partial reduction of the resulting propargylic alcohol with Red-Al led to formation of the *trans*-allylic alcohol, which was epoxidized with $VO(acac)_2$ to **148**. Epoxide opening suffered from low selectivity and therefore the alcohol in **148** was oxidized to the ketone, activating the position α to the ketone for regioselective opening with $ZrCl_4$. Re-reduction of the ketone then delivered desired **149**. The 1,3-*anti* diol was protected with 2,2-dimethoxypropene, the primary alcohol was deprotected with buffered TBAF and oxidized to the aldehyde with Dess-Martin periodinane. Still-Gennari olefination allowed synthesis of *cis*-enoate **150**. The ester was reduced to the allylic alcohol, which in turn was epoxidized under Sharpless's conditions to provide **153**. The epoxide moiety was then opened to the corresponding chlorohydrin, albeit with low regioselectivity. Of the separable regioisomeric products from the epoxide opening, the desired 1,2-diol was then protected as the corresponding acetonide **154** with acetone and $CuSO_4$ as desiccant. On the other side of the molecule the benzyl ether was reductively cleaved and converted to hetaryl sulfone **156** through Mitsunobu substitution with thiol **155** followed by oxidation with *m*CPBA. Fusion of this fragment **156** with previously synthesized fragment **144** was achieved in a Julia olefination using KHMDS to give *cis*-olefin **157**. The allylic epoxide was opened regioselectively with PPh_3Cl_2 and the remaining olefin was dichlorinated with Et_4NCl_3 to give **158** (Scheme 18). Deprotection of the benzyl ether and elimination with Martin's sulfurane left the base-sensitive chlorohydrin moieties of the molecule intact. Deprotection of two silyl ethers gave **159**. Selective palmitoylation of the least hindered alcohol was followed by regioselective sulfation of the resulting diol and global deprotection with TFA to give nominal Mytilipin B (**160**). However, the spectral properties of the synthesized product showed strong deviations when compared to those reported for the natural product, suggesting a potential misassignment of the structure of the Mytilipin B.

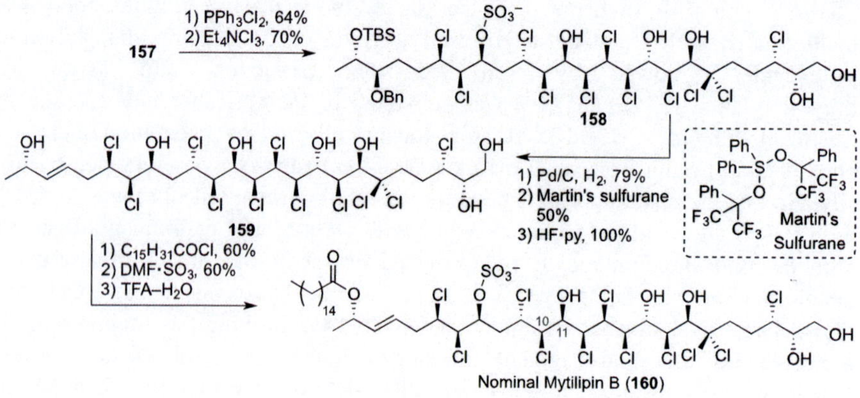

Scheme 18 Late stages in the synthesis of nominal Mytilipin B (210) by the Carreira group

4.14 Carreira's Synthesis of Revised Mytilipin B

The Carreira group subsequently investigated isolation data of Mytilipin B to revise the originally proposed structure (Scheme 19) [44]. For this effort two changes were judged most promising. The absence of an ROE signal in the spectra provided by the isolation team suggested the relative configuration between C10 and C11 (cf. **160**) should be *anti* instead of *syn* as suggested by the isolation team, leading to reassignment of the configuration at C10. The configurations at C9, C7, C6, and C5 were reassigned as well, because their configurations had been assigned relative to C10. In addition, the configuration of the allylic stereocenter at C1 was judged uncertain due to the use of a single MTPA (-α-methoxy-α-trifluoromethylphenylacetic acid) ester by the isolation team in a protocol originally developed by Riguera for MPA (α-methoxyphenylacetic acid) esters [45, 46]. During the synthetic investigations the assignment of configurations at C21 and C22 also became ambiguous. To identify the correct structure of Mytilipin B a total of four diastereomers were synthesized. For this effort, allylic alcohol **161** was epoxidized diastereoselectively to **162** using Sharpless's epoxidation conditions with (−)-DET (Scheme 19). Acetylation and protecting group exchange followed by selective deprotection of the primary silyl ether gave primary alcohol **163**. After oxidation of this primary alcohol using Dess-Martin periodinane the resulting α-silyloxy aldehyde underwent Wittig olefination using the ylide derived from phosphonium iodide **142** to give the Z-olefin. Diastereoselective dichlorination dictated by allylic strain provided a tetrachloride, which was transformed to aldehyde **165** through acetate deprotection followed by Dess-Martin oxidation. The second, sulfone fragment was prepared starting from **166**. To account for the revised configurations at C21 and C22 in Mytilipin B, (−)-DET was used to diastereoselectively epoxidize **166** to **167**. The challenging regioselective opening of the *cis*-epoxide with chloride was achieved using $MgCl_2$ in high yield. Diol **168** was protected as the acetonide with *p*TSA and $CuSO_4$ as desiccant. Benzyl ether

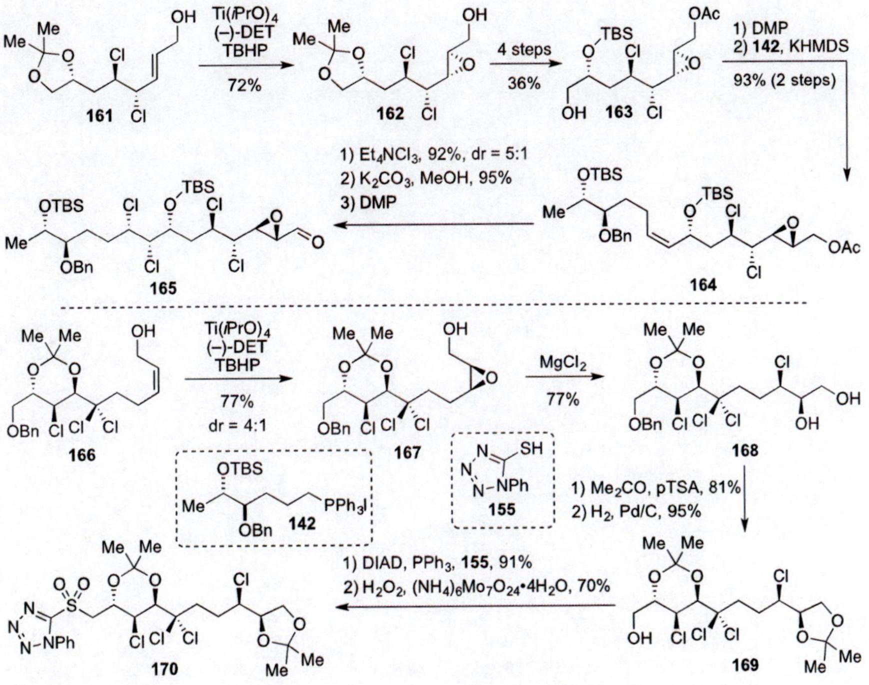

Scheme 19 Early steps in the Carreira synthesis of revised Mytilipin B

deprotection under hydrogenolytic conditions using Pd/C provided access to primary alcohol **169**, which was transformed to the hetaryl sulfone **170** by Mitsunobu reaction followed by oxidation with ammonium molybdate. The use of the latter reagent avoided deprotection of the acetonide as a side reaction that had been observed when *m*CPBA was used as the oxidant.

Sulfone **170** and aldehyde **165** were then joined using Julia olefination to deliver *Z*-alkene **171** (Scheme 20). The epoxide then underwent regioselective opening using trimethylsilyl chloride along with slow addition of hydrogen chloride in ethyl acetate. It is noteworthy that opening of the epoxide to **172** proceeded with retention of configuration, likely a consequence of anchimeric assistance by neighboring chlorides. Diastereoselective dichlorination using Mioskowski's reagent Et_4NCl_3 provided access to **173**. After debenzylation and elimination to **174**, silyl ether cleavage was followed by inversion of configuration of the allylic alcohol using Mitsunobu's reaction with 2-methyl-6-picolinic acid (**175**), yielding **176**. The use of this nucleophile allowed the use of nonbasic cleavage conditions with Cu$(OAc)_2$ and methanol and thus take into account the base-sensitive chlorohydrin moieties found in the molecule, ultimately providing access to **177**. Subsequent introduction of the palmitoyl chain, sulfation, and acetonide hydrolysis completed the synthesis of Mytilipin B (**178**). The spectral data of the obtained product matched those of the natural product. A total of four diastereomers were synthesized to enable

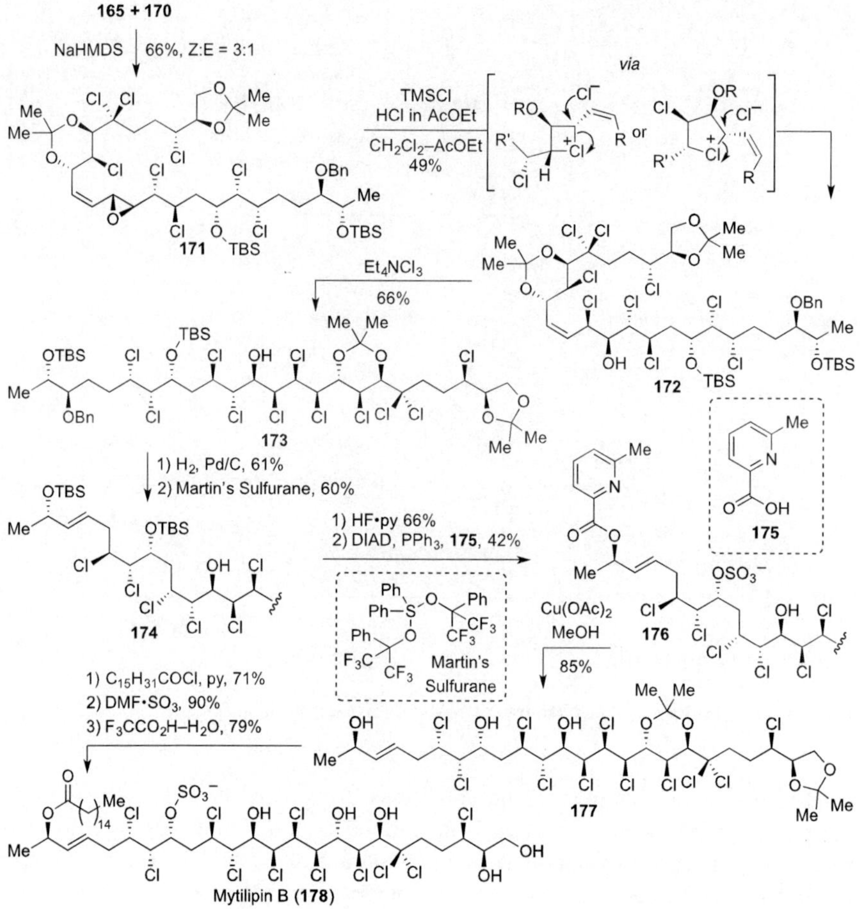

Scheme 20 Fragment coupling and final steps en route to revised Mytilipin B

confidence in the assignment. The prepared diastereomers differ in the configurations of the distal stereocenters, which are likely to have minimal impact on spectral characteristics of the respective compounds. Preparation of four permutations of the relative configurations between stereoarrays allowed direct comparison of their spectral data with those reported for the natural product to conclude the correct structure of Mytilipin B.

In summary, more than 50 years after the initial discovery of chlorosulfolipids numerous syntheses have allowed insights into the structure of chlorosulfolipids. Along the way, many synthetic strategies have been devised and methods were both developed and applied. Nonetheless, the impact of chlorosulfolipids in nature remains mysterious. The biosynthesis and in particular the enzymes involved in the stereoselective introduction of chlorine remain speculative, but would certainly enable valuable insights. Furthermore, the role the lipids play in the organism's membrane remains enigmatic, be it their ability to form bilayers or the influence on the membrane's properties. Synthetic efforts from synthetic organic groups have proven critical to enable structural characterization of this class of natural products. The development of the routes presented herein provided important contributions to synthetic methods for stereoselective synthesis. Additionally, given that the source of some of these natural products remains unknown and that isolation is only accomplished with difficulty, total synthesis is able to deliver ample material of the various natural products and analogs to enable biological studies.

References

1. Haines TH, Pousada M, Stern B, Mayers GL (1969) Microbial sulpholipids: (R)-13-chloro-1-(R)-14-docosanediol disulphate and polychlorosulpholipids in Ochromonas Danica. Biochem J 113(3):565–566. https://doi.org/10.1042/bj1130565
2. Elovson J, Vagelos PR (1969) A new class of lipids: chlorosulfolipids. Proc Natl Acad Sci U S A 62(3):957–963. https://doi.org/10.1073/pnas.62.3.957
3. Elovson J, Vagelos PR (1970) Structure of the major species of chlorosulfolipid from Ochromonas danica. 2,2,11,13,15,16-Hexachloro-N-docosane 1,14-disulfate. Biochemistry 9 (16):3110–3126. https://doi.org/10.1021/bi00818a002
4. Elovson J (1974) Biosynthesis of chlorosulfolipids in Ochromonas danica. Origin of primary and secondary hydroxyl groups determined by oxygen-18 incorporation in vivo. Biochemistry 13(10):2105–2109. https://doi.org/10.1021/bi00707a017
5. Elovson J (1974) Biosynthesis of chlorosulfolipids in Ochromonas danica. Assembly of the docosane-1,14-diol structure in vivo. Biochemistry 13(17):3483–3487. https://doi.org/10.1021/bi00714a010
6. Haines TH (1973) Halogen- and sulfur-containing lipids of ochromonas. Annu Rev Microbiol 27:403–411. https://doi.org/10.1146/annurev.mi.27.100173.002155
7. Chen JL, Proteau PJ, Roberts MA, Gerwick WH, Slate DL, Lee RH (1994) Structure of malhamensilipin A, an inhibitor of protein tyrosine kinase, from the cultured chrysophyte poterioochromonas malhamensis. J Nat Prod 57(4):524–527. https://doi.org/10.1021/np50106a015
8. Ciminiello P, Fattorusso E, Forino M, Di Rosa M, Ianaro A, Poletti R (2001) Structural elucidation of a new cytotoxin isolated from mussels of the adriatic sea. J Org Chem 66 (2):578–582. https://doi.org/10.1021/jo001437s
9. Ciminiello P, Dell'Aversano C, Fattorusso E, Forino M, Magno S, Di Rosa M, Ianaro A, Poletti R (2002) Structure and stereochemistry of a new cytotoxic polychlorinated sulfolipid from adriatic shellfish. J Am Chem Soc 124(44):13114–13120. https://doi.org/10.1021/ja0207347
10. Ciminiello P, Dell'Aversano C, Fattorusso E, Forino M, Magno S, Di Meglio P, Ianaro A, Poletti R (2004) A new cytotoxic polychlorinated sulfolipid from contaminated adriatic mussels. Tetrahedron 60(33):7093–7098. https://doi.org/10.1016/j.tet.2003.12.072
11. Chao C-H, Huang H-C, Wang G-H, Wen Z-H, Wang W-H, Chen I-M, Sheu J-H (2010) Chlorosulfolipids and the corresponding alcohols from the octocoral Dendronephthya Griffini. Chem Pharm Bull (Tokyo) 58(7):944–946. https://doi.org/10.1248/cpb.58.944
12. Bedke DK, Vanderwal CD (2011) Chlorosulfolipids: structure, synthesis, and biological relevance. Nat Prod Rep 28(1):15–25. https://doi.org/10.1039/C0NP00044B
13. Matsumori N, Kaneno D, Murata M, Nakamura H, Tachibana K (1999) Stereochemical determination of acyclic structures based on carbon−proton spin-coupling constants. A method of configuration analysis for natural products. J Org Chem 64(3):866–876. https://doi.org/10.1021/jo981810k

14. Nilewski C, Geisser RW, Ebert M-O, Carreira EM (2009) Conformational and configurational analysis in the study and synthesis of chlorinated natural products. J Am Chem Soc 131 (43):15866–15876. https://doi.org/10.1021/ja906461h
15. Bedke DK, Shibuya GM, Pereira A, Gerwick WH, Haines TH, Vanderwal CD (2009) Relative stereochemistry determination and synthesis of the major chlorosulfolipid from Ochromonas Danica. J Am Chem Soc 131(22):7570–7572. https://doi.org/10.1021/ja902138w
16. Nilewski C, Carreira EM (2012) Recent advances in the total synthesis of chlorosulfolipids. Eur J Org Chem 2012(9):1685–1698. https://doi.org/10.1002/ejoc.201101525
17. Nilewski C, Geisser RW, Carreira EM (2009) Total synthesis of a chlorosulpholipid cytotoxin associated with seafood poisoning. Nature 457(7229):573–576. https://doi.org/10.1038/nature07734
18. Shemet A, Sarlah D, Carreira EM (2015) Stereochemical studies of the opening of chloro vinyl epoxides: cyclic chloronium ions as intermediates. Org Lett 17(8):1878–1881. https://doi.org/10.1021/acs.orglett.5b00558
19. Chung W, Carlson JS, Bedke DK, Vanderwal CD (2013) A synthesis of the chlorosulfolipid mytilipin A via a longest linear sequence of seven steps. Angew Chem Int Ed 52 (38):10052–10055. https://doi.org/10.1002/anie.201304565
20. Keitz BK, Endo K, Patel PR, Herbert MB, Grubbs RH (2012) Improved ruthenium catalysts for Z-selective olefin metathesis. J Am Chem Soc 134(1):693–699. https://doi.org/10.1021/ja210225e
21. Chung W, Carlson JS, Vanderwal CD (2014) General approach to the synthesis of the chlorosulfolipids Danicalipin A, Mytilipin A, and Malhamensilipin A in enantioenriched form. J Org Chem 79(5):2226–2241. https://doi.org/10.1021/jo5000829
22. Denmark SE, Barsanti PA, Wong K-T, Stavenger RA (1998) Enantioselective ring opening of epoxides with silicon tetrachloride in the presence of a chiral Lewis base. J Org Chem 63 (8):2428–2429. https://doi.org/10.1021/jo9801420
23. Denmark SE, Barsanti PA, Beutner GL, Wilson TW (2007) Enantioselective ring opening of epoxides with silicon tetrachloride in the presence of a chiral Lewis base: mechanism studies. Adv Synth Catal 349(4–5):567–582. https://doi.org/10.1002/adsc.200600551
24. Yoshimitsu T, Fukumoto N, Nakatani R, Kojima N, Tanaka T (2010) Asymmetric total synthesis of (+)-hexachlorosulfolipid, a cytotoxin isolated from adriatic mussels. J Org Chem 75(16):5425–5437. https://doi.org/10.1021/jo100534d
25. Katsuki T, Sharpless KB (1980) The first practical method for asymmetric epoxidation. J Am Chem Soc 102(18):5974–5976. https://doi.org/10.1021/ja00538a077
26. Gao Y, Klunder JM, Hanson RM, Masamune H, Ko SY, Sharpless KB (1987) Catalytic asymmetric epoxidation and kinetic resolution: modified procedures including in situ derivatization. J Am Chem Soc 109(19):5765–5780. https://doi.org/10.1021/ja00253a032
27. Landry ML, Hu DX, McKenna GM, Burns NZ (2016) Catalytic enantioselective dihalogenation and the selective synthesis of (−)-deschloromytilipin A and (−)-Danicalipin A. J Am Chem Soc 138(15):5150–5158. https://doi.org/10.1021/jacs.6b01643
28. Hu DX, Seidl FJ, Bucher C, Burns NZ (2015) Catalytic chemo-, regio-, and enantioselective bromochlorination of allylic alcohols. J Am Chem Soc 137(11):3795–3798. https://doi.org/10.1021/jacs.5b01384
29. Yang C-T, Zhang Z-Q, Liang J, Liu J-H, Lu X-Y, Chen H-H, Liu L (2012) Copper-catalyzed cross-coupling of nonactivated secondary alkyl halides and tosylates with secondary alkyl grignard reagents. J Am Chem Soc 134(27):11124–11127. https://doi.org/10.1021/ja304848n
30. Umezawa T, Shibata M, Kaneko K, Okino T, Matsuda F (2011) Asymmetric total synthesis of Danicalipin A and evaluation of biological activity. Org Lett 13(5):904–907. https://doi.org/10.1021/ol102882a
31. Halland N, Braunton A, Bachmann S, Marigo M, Jørgensen KA (2004) Direct organocatalytic asymmetric α-chlorination of aldehydes. J Am Chem Soc 126(15):4790–4791. https://doi.org/10.1021/ja049231m

32. Yoshimitsu T, Nakatani R, Kobayashi A, Tanaka T (2011) Asymmetric total synthesis of (+)-Danicalipin A. Org Lett 13(5):908–911. https://doi.org/10.1021/ol1029518
33. Bailey AM, Wolfrum S, Carreira EM (2016) Biological investigations of (+)-Danicalipin A enabled through synthesis. Angew Chem Int Ed 55(2):639–643. https://doi.org/10.1002/anie.201509082
34. Hu S, Jayaraman S, Oehlschlager AC (1996) Diastereo- and enantioselective synthesis of Syn-α-vinylchlorohydrins and cis-vinylepoxides. J Org Chem 61(21):7513–7520. https://doi.org/10.1021/jo960875p
35. Hu S, Jayaraman S, Oehlschlager AC (1998) Diastereoselective chloroallylboration of α-chiral aldehydes. J Org Chem 63(24):8843–8849. https://doi.org/10.1021/jo980977a
36. Ghosez L, Marchand-Brynaert J (2001) 1-Chloro-N,N,2-trimethylpropenylamine. In: Encyclopedia of reagents for organic synthesis. John Wiley & Sons, Ltd, Chichester, UK
37. Munyemana F, Frisque-Hesbain A-M, Devos A, Ghosez L (1989) Synthesis of alkyl halides under neutral conditions. Tetrahedron Lett 30(23):3077–3080. https://doi.org/10.1016/S0040-4039(00)99407-3
38. Fischer S, Huwyler N, Wolfrum S, Carreira EM (2016) Synthesis and biological evaluation of bromo- and fluorodanicalipin A. Angew Chem Int Ed 55(7):2555–2558. https://doi.org/10.1002/anie.201510608
39. White AR, Duggan BM, Tsai S-C, Vanderwal CD (2016) The alga Ochromonas Danica produces bromosulfolipids. Org Lett 18(5):1124–1127. https://doi.org/10.1021/acs.orglett.6b00230
40. Pereira AR, Byrum T, Shibuya GM, Vanderwal CD, Gerwick WH (2010) Structure revision and absolute configuration of Malhamensilipin A from the freshwater chrysophyte Poterioochromonas malhamensis. J Nat Prod 73(2):279–283. https://doi.org/10.1021/np900672h
41. Saska J, Lewis W, Paton RS, Denton RM (2016) Synthesis of Malhamensilipin A exploiting iterative epoxidation/chlorination: experimental and computational analysis of epoxide-derived chloronium ions. Chem Sci 7(12):7040–7049. https://doi.org/10.1039/C6SC03012B
42. Marigo M, Franzén J, Poulsen TB, Zhuang W, Jørgensen KA (2005) Asymmetric organocatalytic epoxidation of α,β-unsaturated aldehydes with hydrogen peroxide. J Am Chem Soc 127(19):6964–6965. https://doi.org/10.1021/ja051808s
43. Nilewski C, Deprez NR, Fessard TC, Li DB, Geisser RW, Carreira EM (2011) Synthesis of Undecachlorosulfolipid A: re-evaluation of the nominal structure. Angew Chem Int Ed 50(34):7940–7943. https://doi.org/10.1002/anie.201102521
44. Sondermann P, Carreira EM (2019) Stereochemical revision, total synthesis, and solution state conformation of the complex chlorosulfolipid mytilipin B. J Am Chem Soc 141(26):10510–10519. https://doi.org/10.1021/jacs.9b05013. In this context, an overview conceptually related to this review has been discussed. See: structure elucidation of Mytilipin B, Sondermann, P. ETH Zurich, 2020, Diss. ETH No. 26830
45. Latypov SK, Seco JM, Quiñoá E, Riguera R (1996) MTPA vs MPA in the determination of the absolute configuration of chiral alcohols by ^{1}H NMR. J Org Chem 61(24):8569–8577. https://doi.org/10.1021/jo960719i
46. Latypov SK, Seco JM, Quiñoá E, Riguera R (1998) Are both the (R)- and the (S)-MPA esters really needed for the assignment of the absolute configuration of secondary alcohols by NMR? The use of a single derivative. J Am Chem Soc 120(5):877–882. https://doi.org/10.1021/ja9700055

Top Heterocycl Chem (2021) 58: 467–518
https://doi.org/10.1007/7081_2020_40

Published online: 10 October 2020

The Callipeltoside Story

James R. Frost and Steven V. Ley

Contents

Abstract In their search for novel molecules of therapeutic benefit during the 1990s, Minale and his team collected extracts from the shallow water lithistid marine sponge *Callipelta* sp. Subsequent testing of these specimens revealed promising activity in cytotoxic assays by inhibiting in vitro proliferation of KB and P388 cells. Whilst a closer inspection of the extract revealed callipeltins A–C as the major metabolites responsible for the observed activity, further analysis led to the discovery of three additional cytotoxic components: callipeltosides A, B and C (each differing in the attached sugar). At the time, these molecules represented a structurally unprecedented class of polyketides, rightly drawing the attention of the synthetic

J. R. Frost (✉)
UCB Pharma, Slough, UK
e-mail: james.frost@ucb.com

S. V. Ley (✉)
Department of Chemistry, University of Cambridge, Cambridge, UK
e-mail: svl1000@cam.ac.uk

community. Although the connectivity of these molecules could be deduced by Minale, questions surrounding the absolute stereochemistry of the sugar moieties (C1'–C8', D or L) and configuration of the *trans*-configured chlorocyclopropane with respect to the C1–C19 unit remained. Furthermore, the stereochemistry of the glycosidic linkage could not be conclusively determined. This chapter details the substantial effort of the synthetic community to elucidate the structure of these fascinating molecules from start to finish, describing nigh on 20 years of collective work by world leaders in the field.

Graphical Abstract

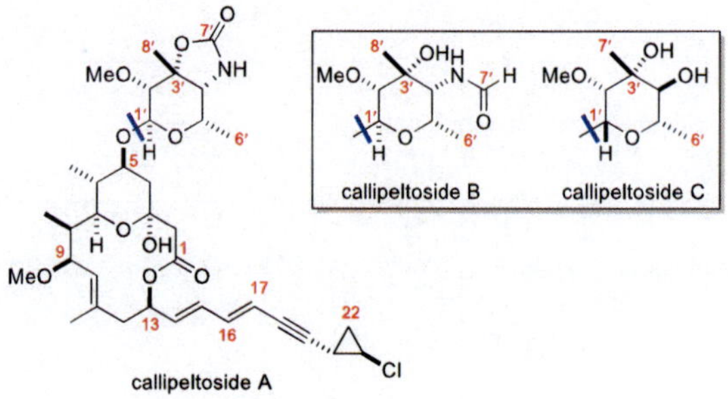

Keywords Aglycon · Callipeltoside · Carbohydrate · Chlorocyclopropane · Macrocycle

1 Isolation

Extracts from marine sponges have provided a large number of complex molecules that exhibit a wide range of biological activities. In their search for novel compounds that display antitumour and antiviral properties, Minale and his team examined an extract from the shallow water lithistid marine sponge *Callipelta* sp., located off New Caledonia [1, 2]. This specimen was found to inhibit the in vitro proliferation of P388 and KB cells whilst also providing protection against HIV. Further study of the extract indicated that callipeltins A–C were the major metabolites responsible for the biological activity [1, 2]. However, additional analysis of the dichloromethane extract of this marine sponge (2.5 kg freeze dried) revealed callipeltoside A as a minor metabolite (3.5 mg isolated) [3] and later callipeltosides B and C (both $\leq$1.0 mg) (Fig. 1) [4]. At the time of isolation, these polyketides represented an unprecedented new class of natural products, which has now been expanded upon with the disclosure of the phorbasides, aurisides, dolastatins and others [5, 6].

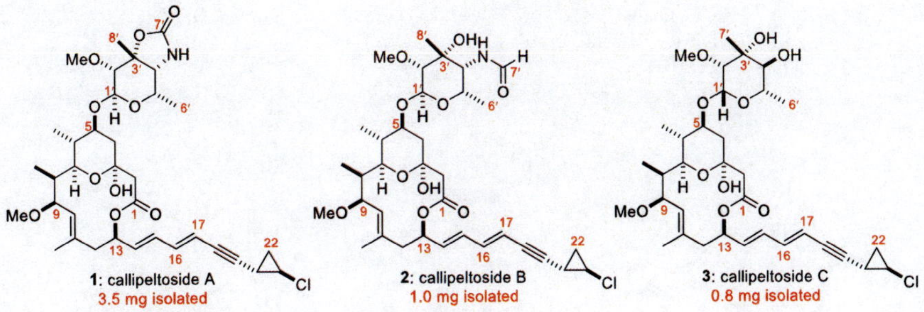

Fig. 1 The callipeltosides in their corrected forms

1.1 Structural Features and Assignment

1.1.1 Callipeltoside A

Minale and co-workers successfully deduced the connectivity of callipeltoside A through the use of COSY, HMQC and HMBC experiments. Further ROESY and nOe techniques provided an indication of the relative stereochemistry of the C1–C13 core whilst revealing the preferred chair conformation of the embedded C3–C7 pyran. Measurement of the ^{1}H coupling constant between C20 and C21 ($J_{20,21}$ = 3.1 Hz, having removed additional $J_{21,22}$ splitting) gave strong evidence for a *trans*-configured chlorocyclopropane moiety. However, the configuration of the chlorocyclopropane with respect to the rest of the molecule could not be determined [3].

The attached 4-amino-4,6-dideoxy-2-*O*,3-*C*-dimethyl-α-talopyranosyl-3,4-urethane (later named callipeltose A) sugar was also without precedent. HMBC correlations between C5 and H1′ as well as H5 and C1′ provided the point of attachment for the glycosidic linkage, with nOe experiments giving evidence for the relative stereochemistry, suggesting an α-anomeric linkage. The nOe correlations also indicated that the pyran ring adopted a six-membered boat conformation, which was imparted by the five-membered cyclic carbamate [3].

Although the connectivity of callipeltoside A had been established, the absolute configuration could not be determined, and therefore the power of total synthesis was required to confirm the structure [3]. It was later shown by Trost [7, 8] (and also Evans [9, 10], Paterson [11], Panek [12], Hoye [13] and Ley [14, 15] vide infra) that the C1–C19 fragment and callipeltose A sugar were enantiomeric to that originally reported by Minale, whilst the chlorocyclopropane (C20–C22) was incorrectly configured with respect to the rest of the molecule (Fig. 2).

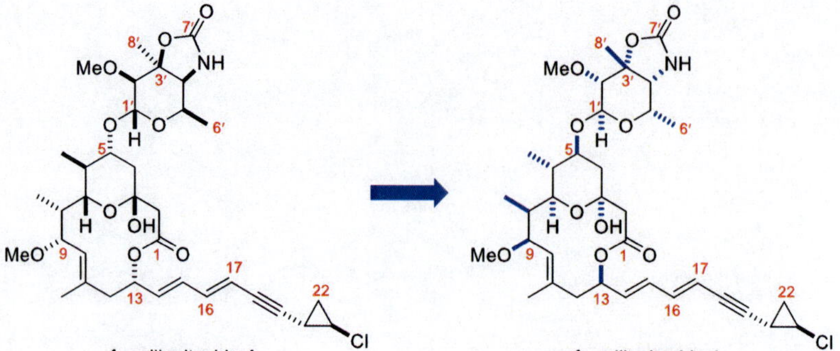

4: callipeltoside A
originally disclosed structure

1: callipeltoside A
revised structure from total synthesis

Fig. 2 Originally disclosed structure of callipeltoside A and subsequent absolute assignment following total synthesis

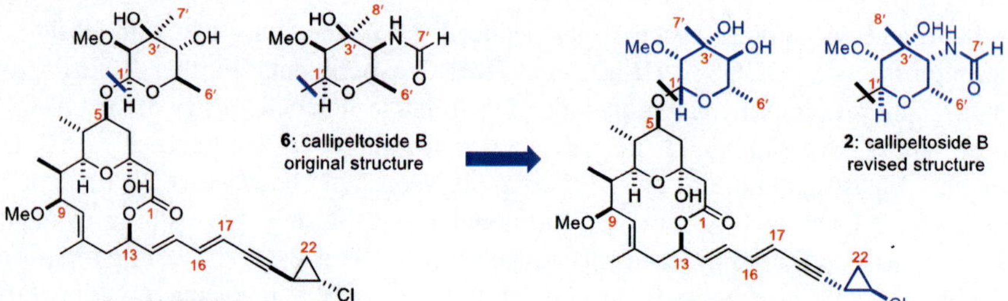

5: callipeltoside C
original structure

6: callipeltoside B
original structure

3: callipeltoside C
revised structure

2: callipeltoside B
revised structure

Fig. 3 Originally disclosed structures of callipeltosides B and C and subsequent absolute stereochemical assignment following total synthesis

1.1.2 Callipeltosides B and C

The structures of callipeltosides B and C were assigned in much the same way as callipeltoside A, with the authors suggesting them to contain 'the same macrolide portion including stereochemistry' as the parent molecule, the only difference being the attached sugar. Despite this comment, the isolation paper unusually depicts the structures of callipeltosides B and C as having an enantiomeric C1–C13 core to callipeltoside A (Fig. 3), also showing an oppositely configured chlorocyclopropane. Although Minale made no claim to know the absolute stereochemistry of callipeltosides A, B or C, the relative configuration described in the text and accompanying structures differed in their descriptions [4].

In keeping with the originally depicted structure of callipeltoside A, callipeltosides B and C were also shown to contain D-configured sugars. The configuration of these sugars was assigned on the basis that callipeltose C bore a likeness with the evalose sugar in everninomicin B [4].

The callipeltose B sugar was identified to contain an *N*-formyl group at the C4′H position, meaning that the molecule existed as 'two inseparable conformers' in a 4:1 ratio, whilst callipeltose C instead contained a secondary hydroxy group with opposite stereochemistry. Both the callipeltose B and C sugars were found to adopt chair conformations (following nOe experiments), with analysis of the ^{13}C NMR chemical shifts (extrapolated from HMQC experiments) indicating that the anomeric linkage was β-equatorial in both cases (callipeltoside B, $\delta = 99.2$; and callipeltoside C, 98.9 ppm) [4].

The first synthesis of callipeltoside C was completed by MacMillan in 2008 (Sect. 2.8), confirming that the absolute stereochemistry of the molecule in terms of the C1–C13 core, chlorocyclopropane unit and sugar configuration (L-configured rather than the original D-configuration suggested by Minale) were identical to callipeltoside A (Fig. 3). The configuration of the glycosidic linkage was tentatively suggested to be β-equatorial 'based on the isolation studies for callipeltoside B' [16].

The Ley group completed the synthesis of callipeltoside C in 2012 [14, 15], confirming the structural assignment by MacMillan [16]. Measurement of the $^{1}J_{C\text{–}H}$ coupling value of the glycosidic bond and thorough NOESY analysis of the sugar portion implied that the stereochemistry at C1′H was (*R*)-configured, in contrast to callipeltoside A. In the same work, this group also disclosed the first total synthesis of callipeltoside B, noting that the callipeltose B sugar was also L-configured, but with (*S*)-stereochemistry at the C1′H position. No further studies were performed, but it was reasonably assumed that the inversion of stereochemistry at the C1′H position for callipeltoside C is connected to the opposing configuration of the C4′H substituent (Fig. 3) [14, 15].

1.2 *Biological Activity of the Callipeltosides*

1.2.1 Studies by Minale

Preliminary studies by Minale indicated that callipeltoside A was moderately cytotoxic in assays against human bronchopulmonary non-small-cell lung carcinoma, affecting the NSCLC-N6 and P388 cell lines ($IC_{50} = 11.26$ and 15.26 μg/mL, respectively) [3]. Further research involving the NSCLC-N6 cell line showed a cell cycle-dependent effect, with cell division blocked at the G1 phase level (Table 1). However, the small amounts of isolated material prevented additional investigation into this process and hence determination of the exact mode of action [3].

Callipeltosides B and C also showed cytotoxic activity against the NSCLC-N6 cell line, albeit at a more modest level (IC_{50} values of 15.1 μg/mL and 30.0 μg/mL, respectively) [4].

Table 1 Flow cytometry assays of the NSCLC-N6 cell line treated with callipeltoside A

	Concentration (μg/mL)	Cells in the G1 phase (%)	Cells in the S phase (%)	Cells in G2/M phase (%)
Control		67.9	28.0	4.1
Callipeltoside A (1)	30	82.5	14.9	2.6
	10	77.5	20.8	1.7
	5	74.7	22.6	2.7

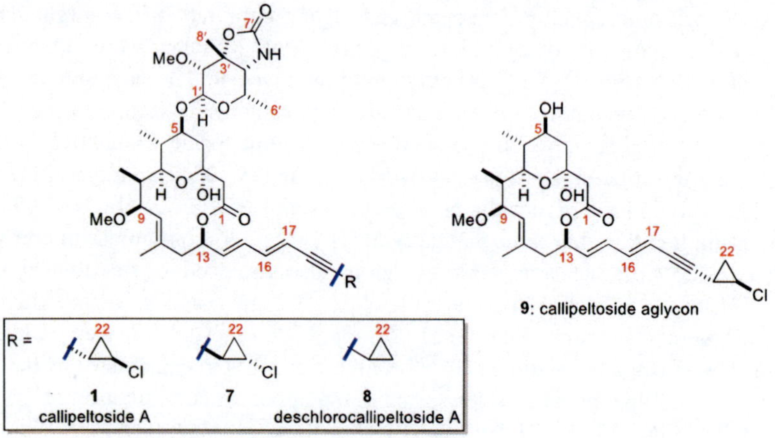

Fig. 4 Compounds synthesised and tested by Trost against the A2780 human ovarian carcinoma cell line

Table 2 Assays against the A2780 human ovarian carcinoma cell line

Entry	Compound	IC_{50} (μM)
1	**1**	20.2
2	**7**	7.0
3	**8**	17.4
4	**9**	>100

1.2.2 Studies by Trost

Further investigations into the biological activity of callipeltoside A were carried out by Trost and co-workers [8]. During their synthesis of callipeltoside A, diastereomer **7**, deschlorocallipeltoside A (**8**) and the callipeltoside aglycon (**9**) were also synthesised (Fig. 4).

Treatment of A2780 human ovarian carcinoma cells (following 48 h incubation) with callipeltoside A as well as **7**, **8** and **9** showed that the callipeltose A sugar was essential for biological activity, whilst the *trans*-chlorocyclopropane was less critical (Table 2, entries 1 and 3). Interestingly the introduction of the oppositely configured *trans*-chlorocyclopropane (**7**) resulted in improved biological activity relative to the natural product (Table 2, entry 2) [8].

2 Structural Investigations and Total Syntheses

The first total synthesis of callipeltoside A (**1**) was completed in 2002 by Trost [7, 8], following excellent initial work by Paterson who completed what proved to be the enantiomeric C1–C13 core (2001) [17].

The synthesis of callipeltoside A was also achieved thereafter by the groups of Evans (2002) [9, 10], Paterson (2003) [11], Panek (2004) [12], Hoye (2010) [13] and Ley (2012) [14, 15]. In addition, a formal synthesis of the callipeltoside aglycon has been disclosed by Marshall [18], partial syntheses by Olivo [19–21] and Yadav [22] as well as numerous contributions towards the callipeltose A sugar [23–26]. The absolute configuration of callipeltoside C (**3**) was initially determined by MacMillan (2008) [16], whilst the Ley group supported this assignment and completed the family with the first synthesis of callipeltoside B [14, 15]. Due to the page restriction imposed on this review, only the completed syntheses of these molecules and subsequent learnings will be reported.

2.1 Paterson Aglycon Synthesis (2001)

2.1.1 Retrosynthesis

Paterson's approach to the callipeltoside aglycon involved disconnection along the C1–C13O ester linkage which, in the forward direction, would be formed by Yamaguchi macrolactonisation. Since the relative configuration of the *trans*-chlorocyclopropane with respect to the rest of the molecule was in doubt, attachment of each enantiomer was necessary. Therefore, the C17–C18 bond was chosen as a further key disconnection point, revealing vinyl iodide **11** and the cyclopropyl alkyne **12**. Compound **11** was further disconnected at the pyran moiety, giving fragment **13**, which in turn could be accessed through sequential aldol reactions. This strategy therefore allows the facile synthesis of both C13 epimers and allows for the relative configuration of the macrolide ring to be resolved (Scheme 1) [17].

2.1.2 Synthesis of the C1–C17 Vinyl Iodide

The synthesis began by union of known vinyl iodide **17** and the enolate of enal **16** by Lewis acid-mediated vinylogous aldol reaction [17, 27, 28]. This resulted in an intentional 1:1 mixture of epimeric products at C13, also installing the desired trisubstituted double bond present in the natural product scaffold [17]. TBS protection of the C13 alcohol followed by boron-mediated *anti*-aldol reaction with **15** set the C8 and C9 stereocentres [17, 29–31]. Diastereoselective reduction of the ketone functionality using the Evans–Tishchenko protocol formed the remaining C7

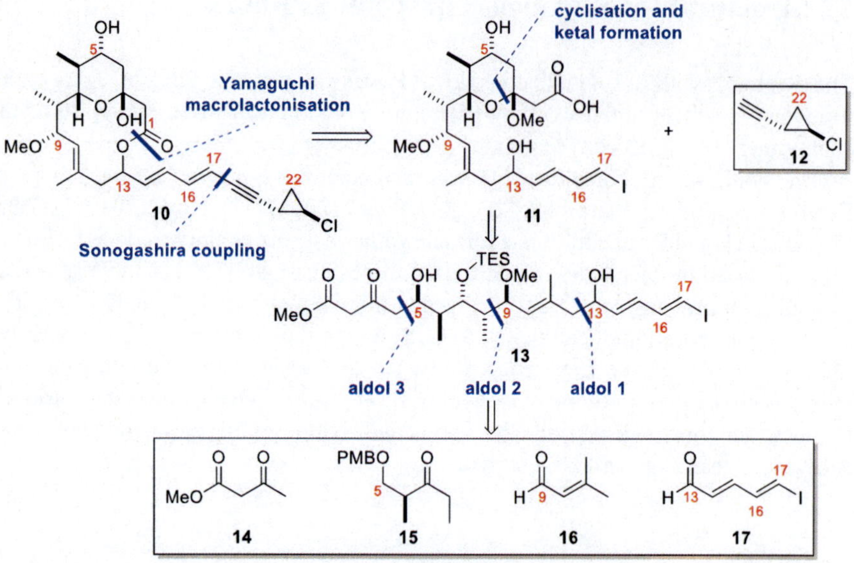

Scheme 1 Paterson's retrosynthesis of the callipeltoside aglycon

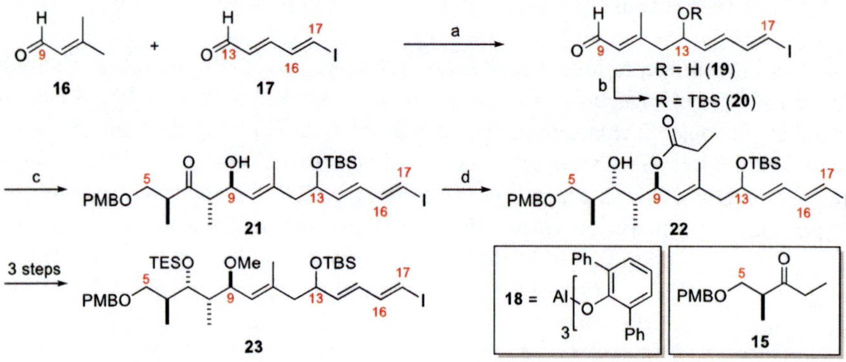

Scheme 2 Synthesis of C5–C17 fragment **23.** *Reagents and conditions*: (**a**) **16**, PhMe, −78°C; LDA, then **17**, **18**, THF, −78°C, 80%; (**b**) TBSCl, imidazole, CH_2Cl_2, RT, 87%; (**c**) (i) Cy_2BCl, Et_3N, Et_2O, −5°C, then **15**, −78°C to −27°C; (ii) CH_2Cl_2–H_2O, SiO_2, RT, 99% (over 2 steps), dr > 98.5:1.5; (**d**) SmI_2, EtCHO, THF, −20°C to −10°C, 92%, dr > 98.5:1.5

stereocentre in high selectivity. A further three synthetic manipulations provided advanced protected C5–C17 fragment **23** (Scheme 2) [17].

PMB protecting group removal and oxidation gave aldehyde **24**, which underwent Mukaiyama aldol reaction with **25** to give linear precursor **26** in good diastereoselectivity (95:5) and yield (85%). Acid-catalysed deprotection of the TES group resulted in cyclisation and methyl ketal formation to give **27** in an excellent yield of 96% (Scheme 3) [17].

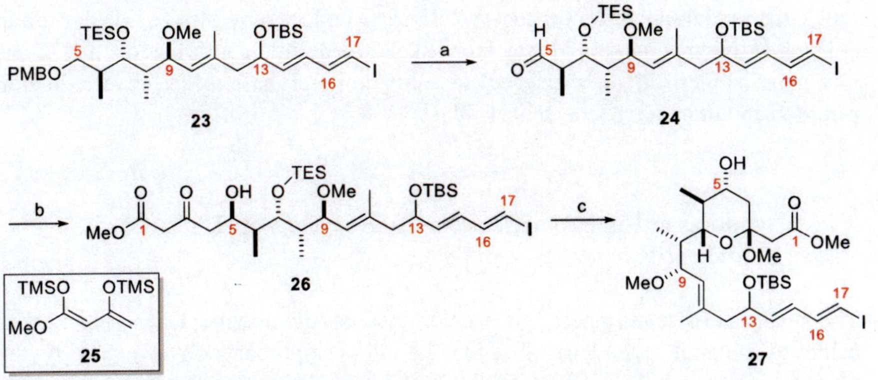

Scheme 3 Manipulation of the C1–C17 fragment and formation of the C3–C7 pyran. *Reagents and conditions*: (**a**) (i) DDQ, CH_2Cl_2–pH 7 buffer (10:1), 40°C, 88%; (ii) Dess–Martin periodinane, CH_2Cl_2, RT, 86%; (**b**) **25**, BF_3•OEt_2, CH_2Cl_2, −100°C, 85%, dr = 95:5; (**c**) PPTS, $(MeO)_3CH$, MeOH, 20°C, 96%

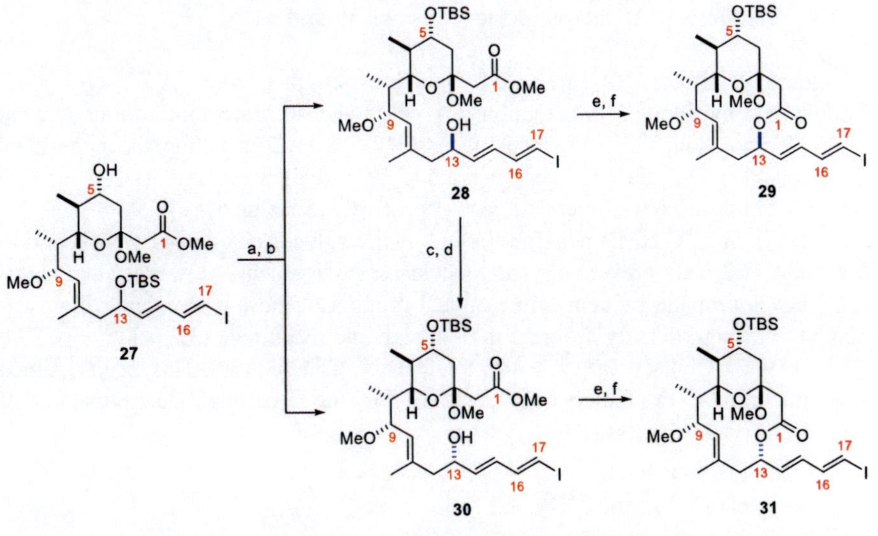

Scheme 4 Formation of the proposed callipeltoside macrocycle and the C13 epimeric material. *Reagents and conditions*: (**a**) TBSCl, imidazole, CH_2Cl_2, RT; (**b**) TBAF, THF, RT, 87% (over 2 steps); (**c**) TPAP, NMO, 4 Å MS, CH_2Cl_2, RT; (**d**) $NaBH_4$, $CeCl_3$•$7H_2O$, EtOH, −78°C to 0°C, 46% (over 2 steps); (**e**) $Ba(OH)_2$•$8H_2O$, MeOH, RT; (**f**) 2,4,6-trichlorobenzoyl chloride, Et_3N, DMAP, PhMe, 80°C, 53% (over 2 steps, **28** to **29**) and 70% (over 2 steps, **30** to **31**)

The free hydroxy was protected as the TBS-ether and the C13OH revealed thereafter following selective removal of the TBS group using TBAF. This allowed for separation of the C13 epimeric alcohols, which could then be separately saponified and the requisite 12-membered rings formed by Yamaguchi macrolactonisation (Scheme 4) [17]. With the two macrocycles in hand, the Paterson group used molecular modelling and compared the nOe data collected by Minale [3]. The group concluded that macrocycle **31** was more likely the desired conformation, and so the unwanted isomer (**28**) was recycled by oxidation and diastereoselective reduction prior to saponification (Scheme 4) [17].

2.1.3 Synthesis of Enantiomeric *trans*-Chlorocyclopropyl Alkynes 12 and *ent*-12

The synthesis of enantiomeric *trans*-chlorocyclopropyl alkynes **12** and *ent*-**12** was achieved using an asymmetric Simmons–Smith cyclopropanation developed by the Charette group [32–34]. Following this, oxidation, dibromoolefination and Corey–Fuchs reaction gave the *trans*-chlorocyclopropane **12** in short sequence (Scheme 5, one enantiomer shown) [17].

2.1.4 Synthesis of Diastereomeric Aglycons 39 and 41

Cyclopropyl alkynes **12** and *ent*-**12** were cross-coupled with the C1–C17 vinyl iodide (**31**) by Sonogashira reaction to give **38** and **40**. Each molecule underwent TBS deprotection, with treatment using PPTS in H_2O installing the hemi-ketal functionality. The diastereomeric aglycons **39** and **41** were synthesised in 54% (over 3 steps) and 67% (over 3 steps), respectively (Scheme 6) [17].

The ^{1}H and ^{13}C NMR data for diastereomeric aglycons **39** and **41** were found to be near identical, preventing any conclusive assignment. However, the *trans*-chlorocyclopropane influenced the optical rotation of these molecules, which was found to be significantly different in both sign and magnitude [**39**, $[\alpha]_D^{20} = -97.8$ ($c = 0.19$, $CHCl_3$); **41**, $[\alpha]_D^{20} = +45.8$ ($c = 0.28$, $CHCl_3$)] [17]. This result outlined the importance of the optical rotation for the structural determination of

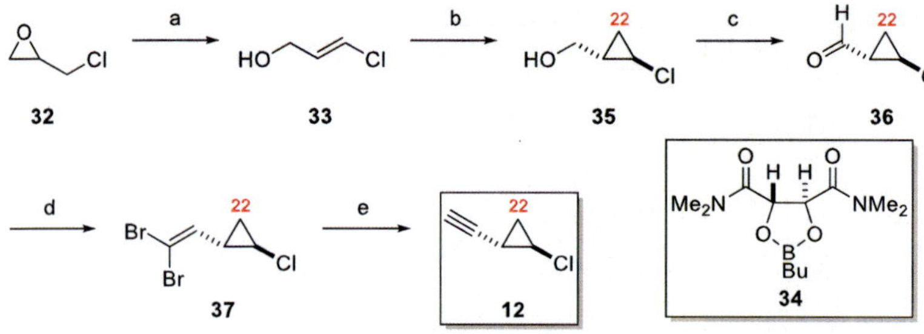

Scheme 5 Access to enantiomeric cyclopropyl alkynes (one enantiomer shown; **12**). *Reagents and conditions*: (**a**) *n*-BuLi, TMEDA, THF, −78°C to 0°C, 80%; (**b**) $ZnEt_2$, CH_2I_2, **34**, CH_2Cl_2, 0°C to RT, dr = 97.5:2.5; (**c**) $(COCl)_2$, DMSO, CH_2Cl_2, −78°C, then Et_3N, −78°C to 0°C; (**d**) Zn, pyridine, PPh_3, CBr_4, CH_2Cl_2, RT, 40% (over 3 steps); (e) *n*-BuLi, Et_2O, −78°C, no reported yield, used directly in the next step

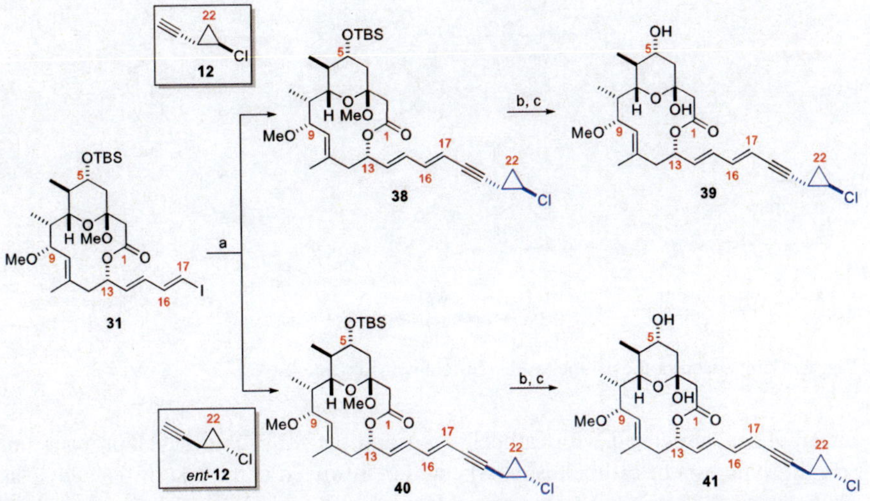

Scheme 6 Synthesis of diastereomeric aglycons **39** and **41.** *Reagents and conditions*: (**a**) **12** or *ent*-**12** [$(PPh_3)_2PdCl_2$], CuI, *i*-Pr_2NH, EtOAc, −20°C to RT; (**b**) TBAF, THF, RT; (**c**) PPTS, MeCN, H_2O, RT, 54% (over 3 steps, **31** to **39**), and 67% (over 3 steps, **31** to **41**)

callipeltoside A. However, at this point in time, attachment of the callipeltose A sugar was not reported by Paterson, and therefore the absolute configuration of callipeltoside A was not determined.

2.2 *Trost Synthesis of Deschlorocallipeltoside A (8)*

Prior to their synthesis of callipeltoside A, the Trost group completed deschlorocallipeltoside A (**8**) [35]. By targeting **8**, Trost hoped to confirm the relative stereochemistry of the C1–C22 core with respect to the callipeltose A sugar and provide a convergent strategy to this family of molecules [8, 35]. Comparison of callipeltoside A with auriside B (**42**) revealed several structural similarities (Fig. 5), suggesting that the synthetic effort should focus on the synthesis of the enantiomeric C1–C13 portion (to that disclosed by Paterson) and an L-configured sugar [8, 35, 36].

2.2.1 Retrosynthesis

Initial retrosynthetic disconnections involved the late-stage attachment of the callipeltose A sugar (**43**) by Schmidt glycosidation and Horner–Wadsworth–Emmons (HWE) coupling of the C1–C13 macrocycle with phosphonate **45** [8, 35]. Attachment of these fragments at a late stage meant that the stereochemical

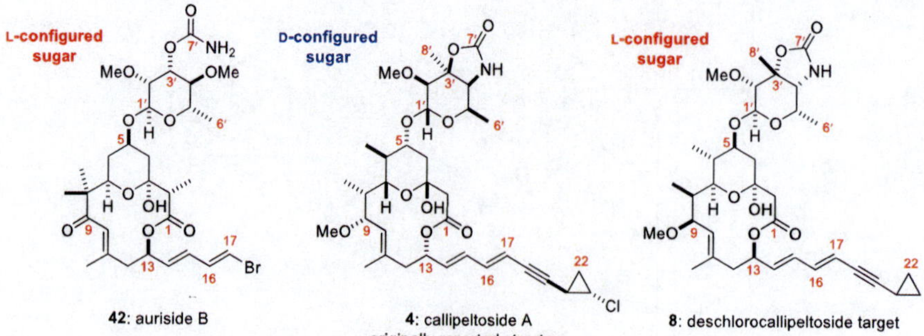

Fig. 5 Comparison of the callipeltoside scaffold with auriside B (**42**)

uncertainties associated with callipeltose A and the *trans*-chlorocyclopropane unit (for the synthesis of callipeltoside A) could be resolved at the end of the synthesis. Callipeltose A **43** was to be constructed from L-rhamnose (**46**), with phosphonate **45** by Sonogashira reaction between iodoalkyne **48** and cyclopropyl alkyne **49** followed by further synthetic manipulation [8, 35]. An acylketene cyclisation would provide the C1–C13 macrocycle, whilst a ruthenium-Alder-ene type coupling [37] and palladium-catalysed allylic alkylation [38] would be key to the synthesis of substrate **47** (Scheme 7) [8, 35].

2.2.2 Synthesis of the C1–C13 Macrocycle 44

Synthesis of the C1–C13 macrocycle began from (*S*)-configured Roche ester **50**, which was converted into alkyne **51** in 5 linear steps [8, 35]. A ruthenium-catalysed Alder-ene type reaction between **51** and **52** was implemented to construct **53**, with the trisubstituted double bond formed with complete regioselectivity (Scheme 8) [8, 35, 37]. Incorporation of the Troc protecting group was key to obtaining high yield (85%), lower catalyst loading (5 mol%) and shorter reaction time (30 min) compared to the unprotected version (62%, 10 mol%, 2 h, respectively) [8, 35]. The regioselectivity of this reaction was thought to arise from ruthenium metallacycle **58** (Scheme 9), with a dative interaction between the non-bonding methoxy electrons and ruthenium essential to reduce the equilibration of the oxidative cyclisation and enhance the rate of the β-hydride elimination step [8, 37].

The C13 stereocentre was installed via an asymmetric palladium-catalysed allylic alkylation reaction with (*R*,*R*)-diphenyl compound **61** identified as the chiral ligand of choice, providing **55** in excellent diastereoselectivity (95:5), good yield (79%) and reasonable branched-to-linear ratio (75:25) [8, 35, 38]. The proposed mechanism for this transformation is shown in Scheme 10 [8, 38]. It is postulated that following alkene coordination to the active palladium species, the corresponding palladium π-allyl species forms with loss of the OTroc group and formation of the putative cationic complex **62** [8]. It is suggested that the *anti*-π-allyl species is favoured over

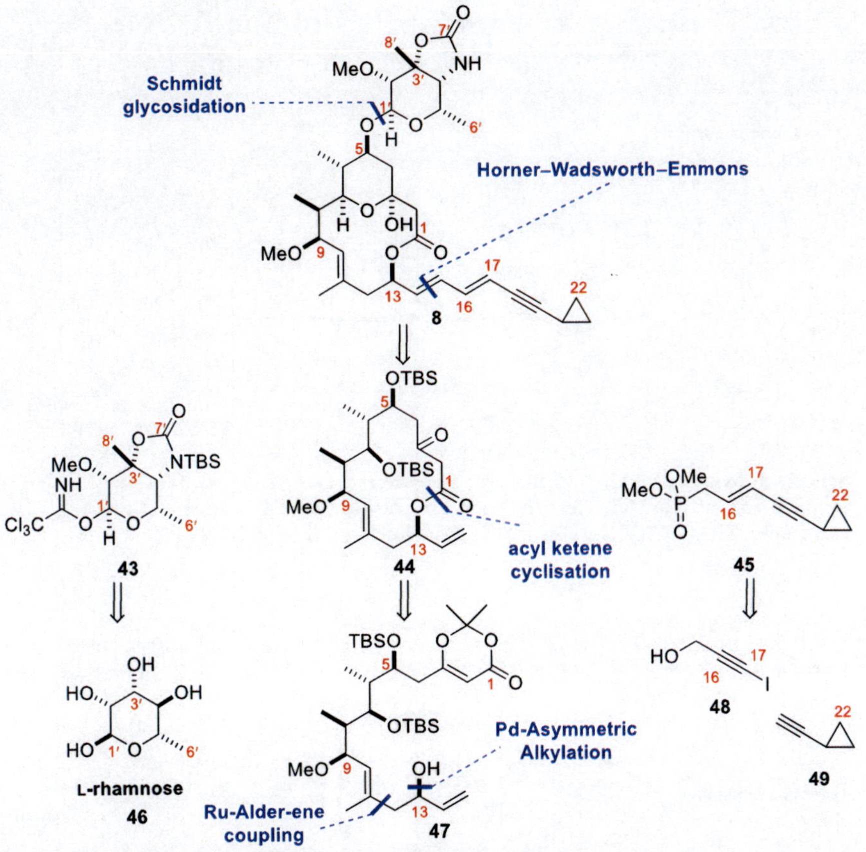

Scheme 7 Trost retrosynthesis of deschlorocallipeltoside A (**8**)

the *syn*-π-allyl form by virtue of the increased steric bulk of the substituted *syn*-π-allyl configuration [8]. The (*R*,*R*)-diphenyl ligand **61** is key for setting the C13 stereocentre, influencing the diastereofacial nucleophilic attack of *p*-methoxyphenol [8].

However, further analysis of C13 stereocentre using the *O*-methylmandelate method [39] revealed that the *undesired* C13 configuration had been set. Fortunately, this unanticipated result could be rectified by use of the enantiomeric (*S*,*S*)-diphenyl ligand **54** (see Scheme 8) in excellent diastereoselectivity (95:5) and slightly lower branched to linear regioselectivity (66:34) [8].

With the C13 stereocentre in place, routine deprotection and oxidation of the primary alcohol gave **69**. Addition of the *E*-lithium enolate of *t*-butylthiopropionate generated the C6 stereocentre in good diastereoselectivity (dr = 84:16) by means of a Cram chelation-controlled aldol reaction. Subsequent protection and reduction delivered aldehyde **70**, which was further reacted with **71** under Felkin–Anh control and protected as TBS-ether **72**. Treatment of **72** with CAN then liberated the C13

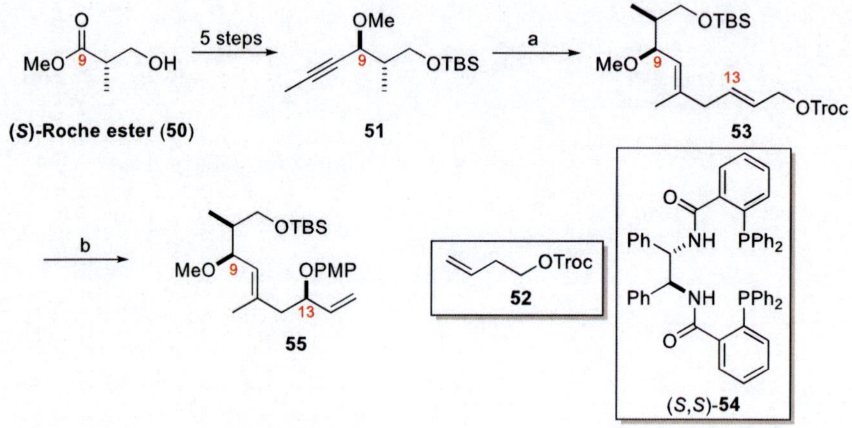

Scheme 8 Synthesis of C7–C15 fragment **55.** *Reagents and conditions*: (**a**) $CpRu(MeCN)_3PF_6$ (5 mol%), **52**, acetone, RT, 85%; (**b**) *p*-methoxyphenol, (*S*,*S*)-**54** (7.5 mol%), $[Pd_2dba_3]\cdot CHCl_3$ (2.5 mol%), TBAC, CH_2Cl_2, 79%, dr = 95:5, regioselectivity = 66:34 (2°:1°)

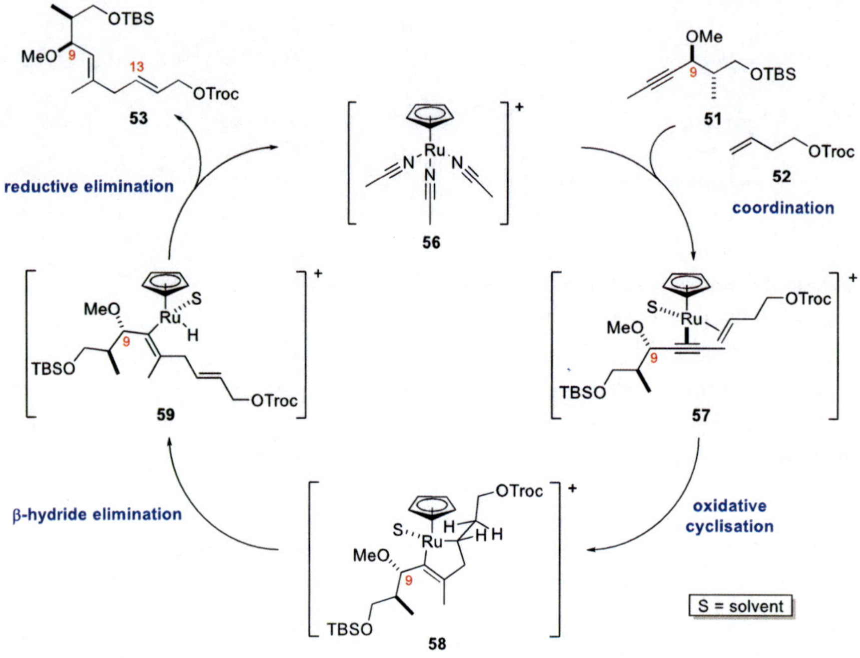

Scheme 9 Proposed mechanism for the ruthenium-catalysed Alder-ene reaction [8]

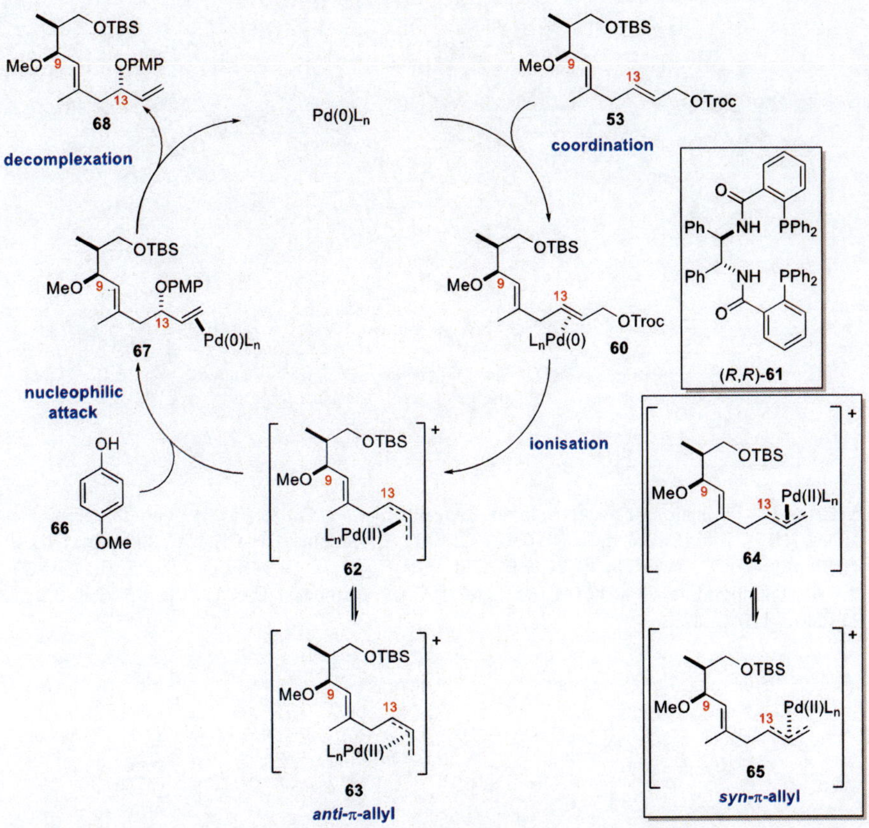

Scheme 10 Proposed catalytic cycle for the palladium-catalysed allylic alkylation using (*R*,*R*)-**61** [8]

alcohol which, upon heating, cyclised to produce the C1–C13 macrocycle (**44**) in 82% yield via the acyl ketene (Scheme 11) [8, 35].

2.2.3 Completion of the Deschlorocallipeltoside Aglycon 75

Dihydroxylation and sodium periodate-mediated cleavage converted macrocycle **44** to aldehyde **73**, with which the deschlorocallipeltoside sidechain **45**[1] could be appended via a Horner–Wadsworth–Emmons reaction [8, 35]. This was achievable in low yield (40%), with the *E*-isomer predominating in a 4:1 ratio. TBS deprotection

[1]This was synthesised by Sonogashira reaction between alkynyl iodide **48** and cyclopropyl alkyne **49** followed by stereoselective reduction, Appel reaction and displacement of the primary bromide with $P(OMe)_3$ [35].

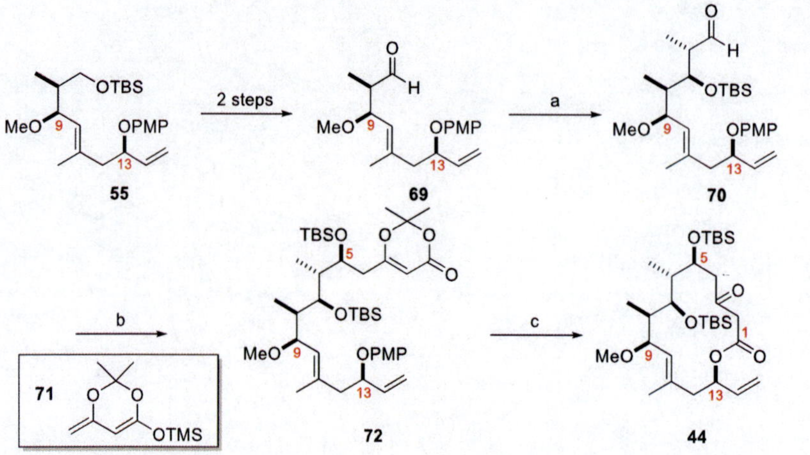

Scheme 11 Formation of macrocycle **44.** *Reagents and conditions:* (**a**) (i) *t*-butylthiopropionate, LDA, THF, −108°C, 82%, dr = 84:16; (ii) TBSOTf, 2,6-lutidine, CH_2Cl_2, 0°C, 86%; (iii) DIBAL-H, PhMe, −78°C, 79%; (**b**) (i) **71**, BF_3•OEt_2, CH_2Cl_2, −78°C, 94%, dr > 95:5; (ii) TBSOTf, 2,6-di-*t*-butylpyridine, CH_2Cl_2, 0°C, 95%; (**c**) (i) CAN, acetone–H_2O (4:1), 0°C, 82%; (ii) 0.5 mM in PhMe, 110°C, 82%

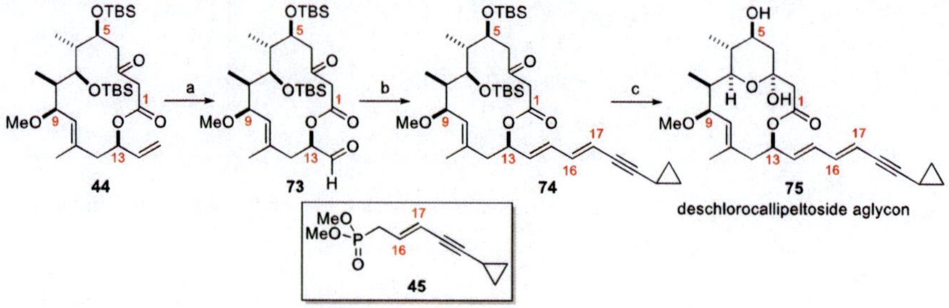

Scheme 12 Completion of the deschlorocallipeltoside aglycon **75.** *Reagents and conditions:* (**a**) OsO_4, NMO, THF–H_2O (4:1), 0°C, then $NaIO_4$, THF–H_2O, RT, 80%; (**b**) **45**, LiHMDS, THF, −78°C, 40%, *E*:*Z* = 4:1; (**c**) HF•pyridine, MeOH, RT, then PPTS, H_2O, MeCN, RT, 60%

and transketalisation with PPTS in wet MeCN then afforded the deschlorocallipeltoside aglycon **75** (Scheme 12) [35].

2.2.4 Synthesis of L-Callipeltose A (43) and Deschlorocallipeltoside A

The bicyclic ring system of callipeltose A (**76**) was synthesised according to previous methodology reported by Guiliano [25] starting from L-rhamnose. A three-step sequence then gave trichloroacetimidate **43**, in preparation for coupling

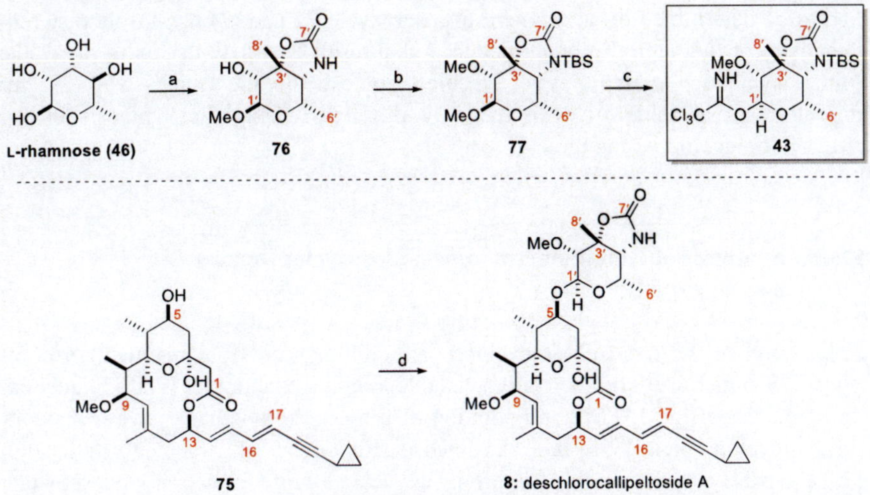

Scheme 13 Synthesis of callipeltose A (**43**) and deschlorocallipeltoside A (**8**). *Reagents and conditions*: (**a**) ref [25]; (**b**) (i) TBSOTf, 2,6-lutidine, CH_2Cl_2, RT, 55%; (ii) MeI, Ag_2O, DMF, RT, 69%; (**c**) (i) H_2SO_4, PPTS, Ac_2O, RT, 81%; (ii) K_2CO_3, MeOH, RT, 89%; (iii) Cl_3CCN, NaH, CH_2Cl_2, RT, 86%; (**d**) (i) **43**, TMSOTf, 4 Å MS, dichloroethane, RT, 80%; (ii) TBAF, AcOH, THF, RT, 95%

to the deschlorocallipeltoside aglycon (**75**) by Schmidt glycosidation (Scheme 13) [8, 35].

Deschlorocallipeltoside A was completed in two steps by attachment of trichloroacetimidate **43** to glycosyl acceptor **75** and subsequent deprotection using TBAF. The resulting ^{1}H and ^{13}C NMR data showed a striking similarity with callipeltoside A (**1**), except in the regions where the chlorine substituent would be present. This provided evidence that the relative configuration of the C1–C13 fragment with respect to the callipeltose A sugar was correct and would likely translate to the natural product following attachment of the correct *trans*-chlorocyclopropane moiety. Whilst this was a major milestone, the absolute stereochemistry of the natural product could not be inferred as the sign and magnitude of the optical rotation differed significantly [$[\alpha]_D = +45.0$ ($c = 0.50$, MeOH) verses $[\alpha]_D = -17.6$ ($c = 0.04$, MeOH) for callipeltoside A (**1**)] following omission of the chlorine substituent (see Sect. 2.1.4 for comment on the influence of the *trans*-chlorocyclopropane on the measured optical rotation) [35].

2.3 *Trost Synthesis of Callipeltoside A (2002)*

Following the synthesis of deschlorocallipeltoside A [35], Trost completed the first total synthesis of callipeltoside A [7, 8]. Prior to this, two stereochemical issues needed to be resolved: (1) the correct enantiomer of the *trans*-chlorocyclopropane had to be determined by attachment to macrocycle **73** and (2) the absolute stereochemistry of the natural product deduced as a result [7, 8]. With this in mind, the initial synthetic strategy was executed in exactly the same way as for deschlorocallipeltoside A, with the only difference being the synthesis of the *trans*-chlorocyclopropane moiety.

2.3.1 Synthesis of Enantiomeric *trans*-Chlorocyclopropanes 84 and *ent*-84

Preparation of the *trans*-chlorocyclopropane commenced from bis-menthol compound **78**, which was commercially available as either enantiomer [7, 8]. Sequential deprotonation using LiTMP and addition of bromochloromethane gave *trans*-configured cyclopropane **79** in high diastereoselectivity (>99:1) after recrystallisation [7, 8]. Hydrolysis of a single ester group and acid chloride formation gave substrate **80** which, upon submission to a modified Barton–Crich–Motherwell decarboxylation procedure [40], provided the requisite *trans*-chlorocyclopropane geometry in good yield (60%) and diastereoselectivity (dr = 97:3) [7, 8]. A further three synthetic steps provided compound **37**, which underwent a one-pot Stille reaction/elimination sequence to incorporate the C16–C17 *trans*-double bond and internal alkyne [41]. Additional manipulation by means of an Appel reaction and displacement of the resulting bromide with $P(OEt)_3$ gave phosphonate coupling partner **84** in 10 linear steps and 20% overall yield (Scheme 14, one enantiomer shown for clarity) [7, 8].

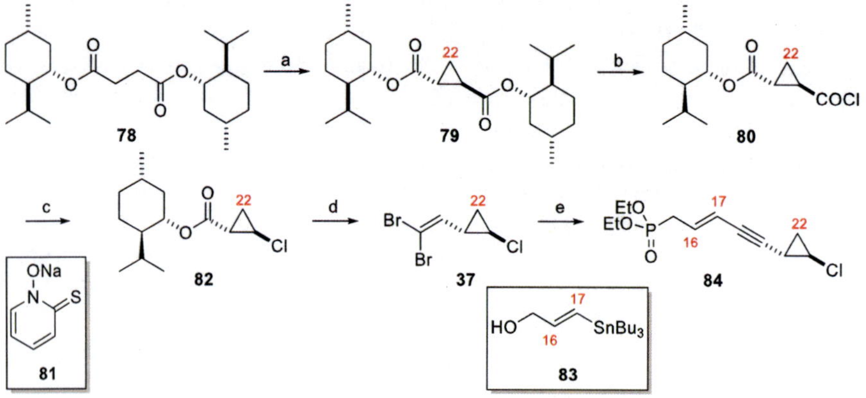

Scheme 14 Synthesis of the *trans*-chlorocyclopropane (one enantiomer shown). *Reagents and conditions*: (**a**) LiTMP, $BrCH_2Cl$, THF, −78°C, 87%, dr > 99:1 (following one crystallisation); (**b**) (i) NaOH, *i*-PrOH, 70°C; (ii) $SOCl_2$, RT, 90% (over 2 steps); (**c**) **81**, DMAP, TBAI, CCl_4 (0.02 M), RT, then AIBN, 80°C, 60%, dr = 97:3; (**d**) (i) HCl•HNMe(OMe), *i*-PrMgCl, THF, −10°C; (ii) DIBAL-H, CH_2Cl_2, −78°C; (iii) CBr_4, PPh_3, CH_2Cl_2, RT, 80% (over 3 steps); (**e**) (i) **83**, $[Pd_2dba_3]$•$CHCl_3$, $(4\text{-OMeC}_6H_4)_3P$, DIPEA, DMF, 80°C, 66%; (ii) CBr_4, PPh_3, CH_2Cl_2, −40°C, 90%; (iii) $P(OEt)_3$, 100°C, 93%

2.3.2 Completion of Callipeltoside A (1)

Phosphonate **84** and its enantiomer were appended to **73** via a Horner–Wadsworth–Emmons reaction. Subsequent deprotection and concomitant cyclisation then afforded diastereomeric aglycons **9** and **86** [7, 8], proceeding in near identical yield and selectivity to that previously reported for deschlorocallipeltoside A [35]. Once again the callipeltose A sugar **43** was attached by Schmidt glycosidation (Sect. 2.2.4), with deprotection using TBAF revealing callipeltoside diastereomers **85** and **7** (Scheme 15). Comparison of the ^{1}H and ^{13}C NMR spectra generated for these compounds with the natural isolate disappointingly revealed no significant differences [7, 8]. However, as previously noted by Paterson (Sect. 2.1.4) [17], the presence of the *trans*-chlorocyclopropane unit meant significantly different optical rotations for each diastereomer. An optical rotation of $[\alpha]_D^{22} = -19.2$ ($c = 1.0$, MeOH) was recorded for diastereomer **85**, whereas **7** was $[\alpha]_D^{22} = +156.3$ ($c = 0.55$, MeOH) [7, 8]. Comparison with the natural isolate [$[\alpha]_D^{22} = -17.6$ ($c = 0.04$, MeOH)] therefore suggested that diastereomer **85** was consistent with being callipeltoside A (**1**). Hence, the structural assignment of callipeltoside A was made possible purely by comparison of the sign and magnitude of the measured optical rotation [7, 8].

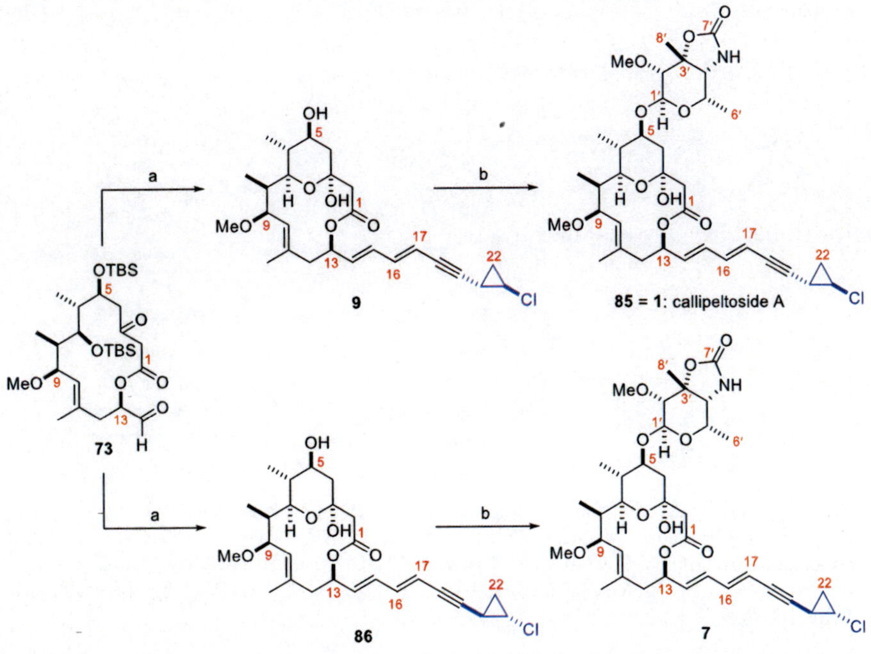

Scheme 15 Completion of callipeltoside A. *Reagents and conditions*: (**a**) (i) LiHMDS, **84** or *ent*-**84**, THF, −78°C to −40°C to RT, *E*:*Z* = 4:1; (ii) HF•pyridine, MeOH, 0°C, 50% (over 2 steps in both cases); (**b**) (i) **43**, TMSOTf, 4 Å MS, 1,2-dichloroethane, −30°C; (ii) TBAF, AcOH, THF, RT, 70% (over 2 steps in both cases)

Trost's synthesis of callipeltoside A was completed in a total of 50 manipulations, with 22 steps making up the longest linear sequence (0.47% overall yield) [7, 8]. Although the stereochemical uncertainties for callipeltoside A had been resolved, the group wanted to improve the diene-yne synthesis, as only 4:1 *E*:*Z* ratio (at C14–C15) was obtained from the earlier Horner–Wadsworth–Emmons reaction.

2.3.3 Second-Generation Synthesis of the Callipeltoside Aglycon 86

The second generation began with the interception of alkene **44** (Sect. 2.2.3), which was converted to the corresponding hemi-ketal **87** following TBS deprotection. Cross metathesis with crotonaldehyde in the presence of Grubbs' second-generation catalyst and subsequent Takai reaction provided vinyl iodide **88** in a significantly improved 8:1 *E*:*Z* ratio at C16–C17 (Scheme 16) [8]. At this point it was noted that volatile alkyne **12** (described by Paterson, Sect. 2.1.3) [17] was avoided for practical reasons, and so a one-pot reaction involving formation of the alkynyl stannane in situ and subsequent Stille coupling delivered **86**. This sequence represented an improvement on the step count and diene (*E*) selectivity of the first-generation synthesis, but was only reported with the oppositely configured *trans*-chlorocyclopropane (*ent*-**37**) (Scheme 16) [8].

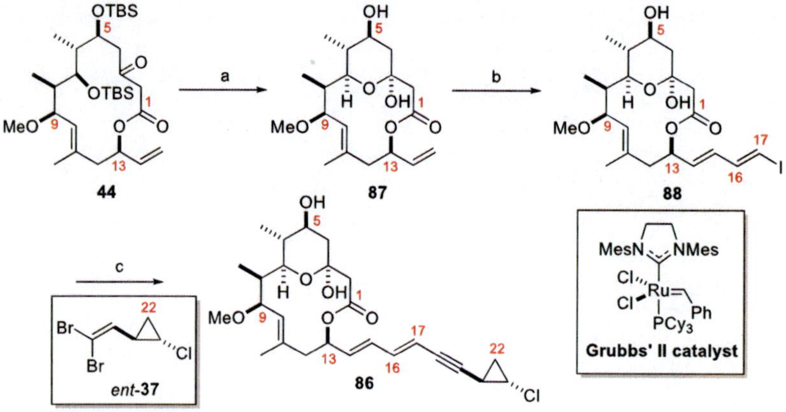

Scheme 16 Second-generation coupling of the *trans*-chlorocyclopropane. *Reagents and conditions*: (**a**) (i) HF•pyridine, MeOH, 0°C; (ii) PPTS, MeCN–H_2O (3:1), RT, 91% (over 2 steps); (**b**) crotonaldehyde, Grubbs' II, CH_2Cl_2, 40°C, then $CrCl_2$, CHI_3, dioxane–THF (4:1), 0°C, *E*:*Z* = 8:1, 84%; (**c**) *n*-BuLi, *ent*-**37**, Me_3SnCl, Et_2O, −78°C to RT, then **88**, $(MeCN)_2PdCl_2$ (3 × 10 mol%), DMF, RT, 70%

2.4 *Evans Synthesis of Callipeltoside A (2002)*

Shortly after Trost published the completion of callipeltoside A, Evans also disclosed the synthesis of the molecule [9, 10].

2.4.1 Retrosynthesis

The absolute configuration of callipeltoside A was not known at the start of the project, and so Evans applied a similar retrosynthetic approach whereby the *trans*-chlorocyclopropane was attached at a late stage. This was to be achieved by Horner–Wadsworth–Emmons reaction with sidechain **92**, enabling the stereochemistry of the sidechain to be deduced at a late stage [9, 10]. Sequential aldol reactions would be key for the synthesis of the C1–C13 fragment **91**, which would be completed by macrocyclisation to form the C13–O bond (Scheme 17).

2.4.2 Synthesis of C1–C13 Fragment 91

A copper-catalysed vinylogous aldol reaction between **95** and **96** set the (*R*)-configured C13 stereocentre and forged the C10–C11 trisubstituted double bond to give **98**

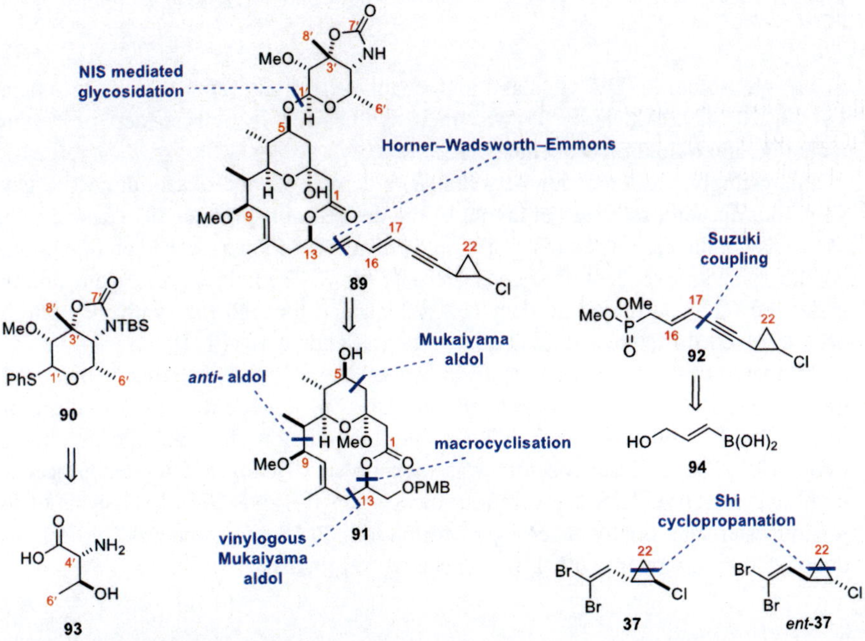

Scheme 17 Evans approach to callipeltoside A

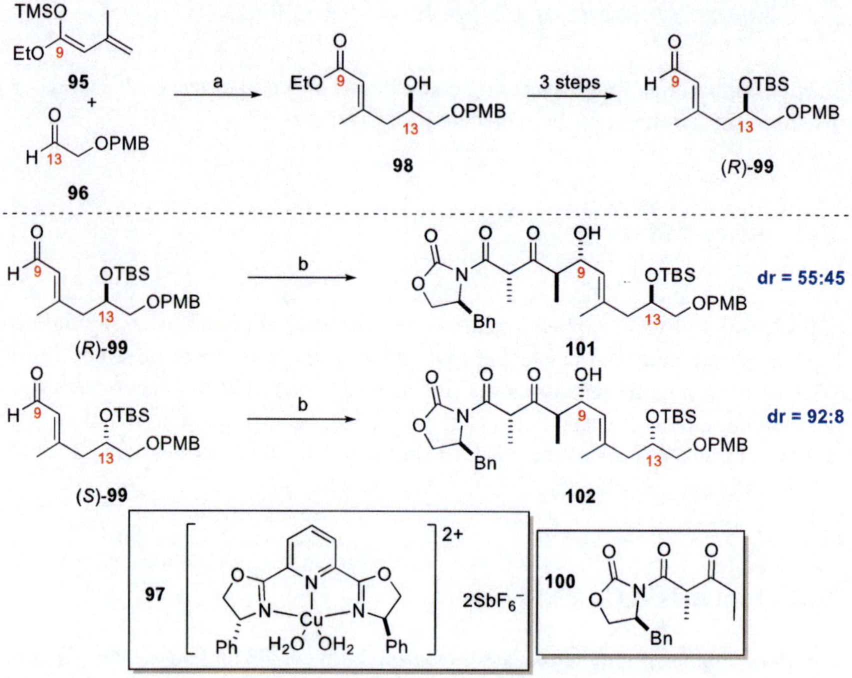

Scheme 18 Synthesis of the C5–C14 fragment. *Reagents and conditions:* **(a) 97**, CH_2Cl_2, −78°C; 1 N HCl, EtOAc, RT, 93%, er = 97.5:2.5; **(b) 100**, Cy_2BCl, $EtNMe_2$, Et_2O, 0°C to −78°C then RCHO, −78°C to −20°C

as a single isomer in 93% yield and high enantioselectivity (er = 97.5:2.5) (Scheme 18) [9, 10, 42]. Aldehyde **99** was then synthesised in a three-step protection, reduction and oxidation sequence.

Interestingly, aldol reaction between (*R*)-aldehyde **99** and oxazolidone **100** gave very poor diastereoselection in favour of the desired *anti*-product **101** (dr = 55:45). This was not the case when (*S*)-configured enantiomer **99** was subjected to the same conditions, with excellent diastereoselectivity (dr = 92:8) observed. As a result, the undesired C13 configured product **102** was carried through the synthesis with the need to invert this stereocentre at a later stage (Scheme 18) [9, 10].

Further routine manipulations followed by the addition of Chan's diene **104** under Felkin–Anh control resulted in **105** in excellent diastereoselectivity (dr > 95:5). Silylation, methanolysis and methylation then gave callipeltoside pyran **106**. The C13 stereocentre was then inverted using a four-step procedure involving selective TBS deprotection, mesylation (to give **107**), hydrolysis of the methyl ester and finally macrocycle formation. PMB-protected scaffold **91** was completed by treatment with TBAF (Scheme 19) [9, 10].

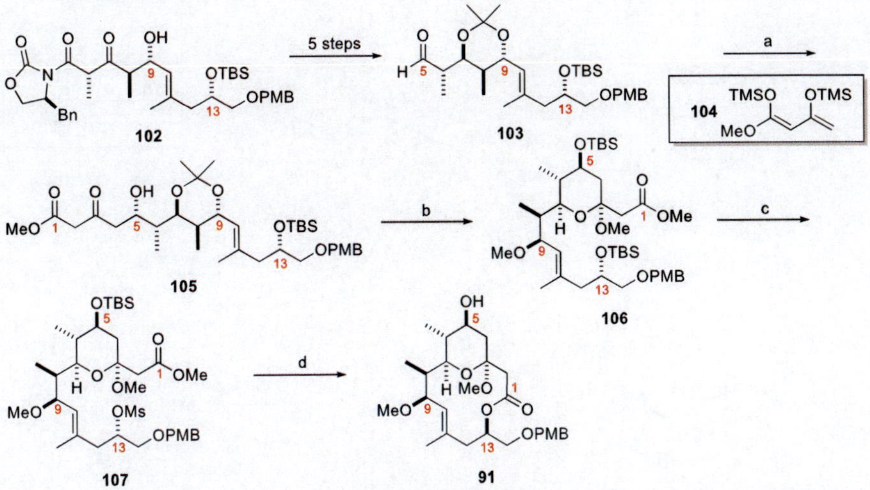

Scheme 19 Synthesis of C1–C13 macrocycle **91.** *Reagents and conditions:* (**a**) **104**, $BF_3{\bullet}OEt_2$, PhMe, −90°C, 88%, dr > 95:5; (**b**) (i) TBSOTf, 2,6-lutidine, CH_2Cl_2, −78°C; (ii) PPTS, MeOH, RT; (iii) MeOTf, 2,6-di-*t*-butylpyridine, CH_2Cl_2, RT, 50% (over 3 steps); (**c**) (i) TBAF, THF, RT; (ii) MsCl, Et_3N, DMAP, CH_2Cl_2, 0°C; (iii) LiOH, H_2O, MeOH–THF (1:1), RT, 67% (over 3 steps); (**d**) (i) Cs_2CO_3, 18-crown-6, PhMe, 110°C; (ii) TBAF, THF, RT, 66% (over 2 steps)

2.4.3 Synthesis of *trans*-Chlorocyclopropane Sidechain 116

Similar to the Trost synthesis [7, 8], Evans designed the route to the *trans*-chlorocyclopropane unit to be flexible enough such that both enantiomers could be accessed (Scheme 20, desired enantiomer shown) [9, 10, 43]. Oxidative cleavage of **108** using KIO_4 provided aldehyde **109**, from which compound **112** could be produced using either a one- or two-step procedure. Although the Takai protocol resulted in vinyl chloride **112** in good stereoselectivity (*E*:*Z* = 6.7:1), this process was not ideal since it was difficult to separate the two isomers and only resulted in poor isolated yield (< 40%). This was remedied by use of a two-step sequential homologation (to alkyne **111**) and reduction sequence using the Ohira–Bestmann reagent (**110**), followed by Masuda's one-pot hydroboration/chlorination methodology [44]. This not only resulted in better double bond selectivity (*E*:*Z* > 20:1) but also improved overall yield (70%, **111** to **112**) [9, 10, 43].

The authors comment that cyclopropanation of vinyl chloride **112** was troublesome at first, with the traditional Simmons–Smith reaction as well as the modified variants developed by Furukawa (Et_2Zn, CH_2I_2) [45] and Denmark (Et_2Zn, CH_2I_2, ZnI_2) [46] all resulting in little or no product. However, the conditions developed by Shi (Et_2Zn, CH_2I_2, TFA) [47] gave **113** in good yield (82%) and excellent diastereoselectivity (dr > 98:2) [9, 10, 43]. Deprotection, oxidative cleavage of the resulting diol and dibromoolefination gave **37** (also used by Paterson [17] and Trost [7, 8]). This underwent Suzuki reaction with boronic acid **94** under Roush-modified

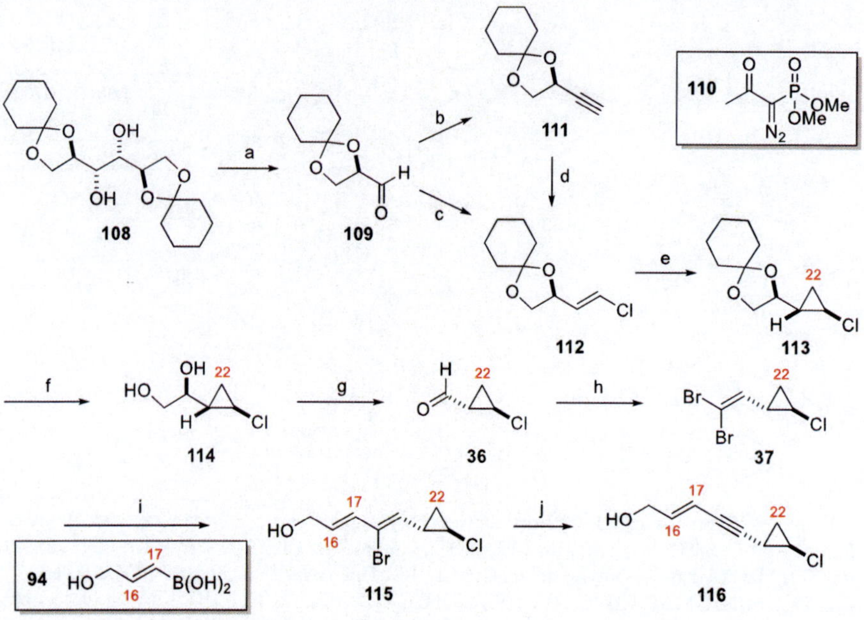

Scheme 20 Formation of *trans*-chlorocyclopropane fragment **116** (desired enantiomer). *Reagents and conditions*: (**a**) KIO_4, $KHCO_3$, H_2O–THF (3:1), RT; (**b**) **110**, K_2CO_3, MeOH, RT, 94% (over 2 steps); (**c**) $CrCl_2$, $CHCl_3$, THF, 70°C, < 40% (over 2 steps), *E*:*Z* = 6.7:1; (**d**) (i) $Thexyl_2BH$, −15°C to 0°C, THF, (ii) $CuCl_2$, H_2O, HMPA, THF, 0°C to 70°C, 70% (over 2 steps), *E*:*Z* > 20:1; (**e**) $ZnEt_2$, CF_3COOH, CH_2I_2, CH_2Cl_2, 0°C to RT, 82%, dr > 98:2; (**f**) Dowex resin, MeOH, RT, 84%; (**g**) $Pb(OAc)_4$, K_2CO_3, CH_2Cl_2, 0°C; (**h**) PPh_3, CBr_4, CH_2Cl_2, 0°C to RT, 94% (over 2 steps); (**i**) **94**, $Pd(PPh_3)_4$, TlOEt, THF–H_2O (3:1), RT, 85%; (**j**) DBU, PhMe, 110°C, 91%

conditions [48], with final elimination of HBr providing advanced fragment **116** in 33% overall yield (Scheme 20) [9, 10, 43].

2.4.4 L-Callipeltose A Sugar 90

Synthesis of the L-callipeltose A sugar began from D-threonine (**93**), which was converted into chiral substrate **118** in a three-step procedure [49]. The aldol reaction between **118** and the lithium enolate of **119** was performed in order to set the C2′ and C3′ stereocentres. However, whilst the desired C3′-stereocentre could be accessed using this method (consistent with Felkin–Anh selectivity), the undesired *syn* aldol adduct predominated (dr = 66:34), meaning poor selectivity for C2′ stereocentre formation (Scheme 21) [9, 10, 49]. This unexpected stereochemical outcome was hypothesised to be the result of chelation between the lithium and α-alkoxy substituent, thereby reinforcing the preference for the formation of an (*E*)-configured enolate. The relative configuration of the C2′ and C3′ stereocentres was determined by nOe studies following conversion of aldol adducts **120** and **122** to the

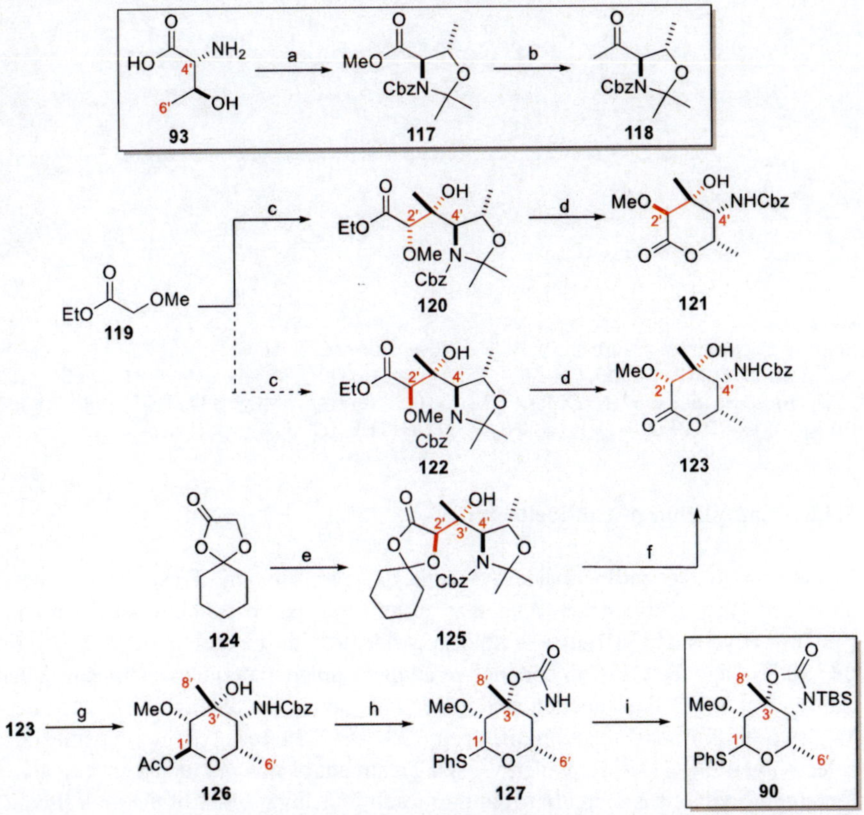

Scheme 21 Synthesis of L-callipeltose A (**90**). *Reagents and conditions*: (**a**) (i) 1 M NaOH, CbzCl, MeCN, 0°C to RT; (ii) MeI, K_2CO_3, DMF, 0°C to RT; (iii) TsOH•H_2O, 2,2-dimethoxypropane, PhH, RT, 93% (over 3 steps); (**b**) (i) *i*-PrMgCl, HN(OMe)Me•HCl, THF, −78°C to −40°C; (ii) MeMgBr, THF, −40°C to RT, 61% (over 2 steps); (**c**) LDA, then **118**, −78°C to −30°C, 56%, dr = 66:34; (**d**) HF, THF (no yield given); (**e**) LDA, then **118**, −78°C to −30°C, 80%, dr = 94:6; (**f**) (i) AcOH–H_2O (1.5:1), 70°C, 71%; (ii) Me_3OBF_4, DTBMP, CH_2Cl_2, RT, 82%; (**g**) DIBAL-H, CH_2Cl_2, −78°C, then Ac_2O, pyridine, DMAP, −78°C, 93%; (**h**) (i) BF_3•OEt_2, PhSH, CH_2Cl_2, 0°C to RT, 81%, α:β = 50:50; (ii) NaH, THF, 0°C to RT, 97%; (i) TBSOTf, 2,6-lutidine, CH_2Cl_2, 0°C to RT, 92%

corresponding lactones. In order to ensure formation of an enolate that could only exist in the (*Z*)-configuration (and therefore set the desired C2′ stereochemistry), the lithium enolate derived from **124** was studied. This resulted in formation of the correct Felkin–Anh *anti*-aldol adduct **125** in excellent yield (80%) and much-improved diastereoselectivity (dr = 94:6). With the desired C2′ stereochemistry set, conversion to lactone **123** followed by routine synthetic manipulations afforded thioglycoside **90** as a 50:50 mixture of α/β anomers (Scheme 21) [9, 10, 49].

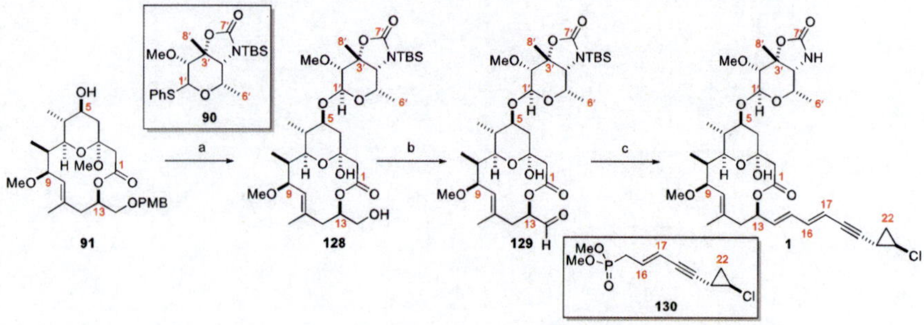

Scheme 22 Completion of callipeltoside A. *Reagents and conditions*: (**a**) (i) **90** (α/β = 50:50), NIS, TfOH, DTBMP, 4 Å MS, CH_2Cl_2, −15°C to RT; (ii) DDQ, MeOH, CH_2Cl_2–H_2O (1.3:1), RT, 83%; (**b**) SO_3•pyridine, Et_3N, DMSO, CH_2Cl_2, 0°C; (**c**) (i) LiHMDS, **130**, THF, −78°C to RT; (ii) I_2, CH_2Cl_2, RT, *E*:*Z* = 11:1; (iii) TBAF, AcOH, THF, RT, 56% (over 4 steps)

2.4.5 Completion of Callipeltoside A

In contrast to the callipeltoside A synthesis completed by Trost [7, 8], Evans appended the callipeltose A sugar prior to the *trans*-chlorocyclopropane [9, 10]. Glycosidation between thioglycoside **90** [as an anomeric mixture (α/β = 50:50)] and the C1–C13 core delivered the coupled material as a single product. Subsequent PMB deprotection and oxidation then gave aldehyde **129**. Horner–Wadsworth–Emmons reaction between **129** and **130** resulted in moderate *E*:*Z* selectivity at the C14/C15 *E*-olefin (3:1). Treatment of this mixture with a catalytic amount of iodine was thereafter found to result in isomerisation of the undesired *Z*-isomer, improving the ratio to 11:1. Final TBS deprotection then provided callipeltoside A (**1**) in 25 steps and 4.0% overall yield. The optical rotation of this synthetic material [$[\alpha]_D = -17$ ($c = 0.19$, MeOH)] was in full agreement with both the callipeltoside A natural isolate (Minale [3]) and that prepared by Trost (Scheme 22) [7, 8]. For completion, oppositely configured *trans*-chlorocyclopropane diastereomer **7** was also produced in an analogous fashion (not shown, also synthesised by Trost, Sect. 2.3.2), with the measured optical rotation having both different sign and magnitude [$[\alpha]_D = +140$ ($c = 0.05$, MeOH)] [9, 10].

2.5 Paterson Synthesis of Callipeltoside A (2003)

Paterson's synthesis of the enantiomeric callipeltoside aglycon **41** [17] (Sect. 2.1) was adapted for the synthesis of callipeltoside A. Key to this was the development of an asymmetric vinylogous Mukaiyama aldol reaction to set the C13 stereocentre. A chiral Lewis acid system requiring (*R*)-BINOL-Ti(O*i*-Pr)$_2$ [derived from (*R*)-BINOL and Ti(O*i*-Pr)$_4$] in the presence of CaH_2 provided trisubstituted alkene **133** in 94% yield and high enantioselectivity (er = 97:3). Further manipulation resulted in

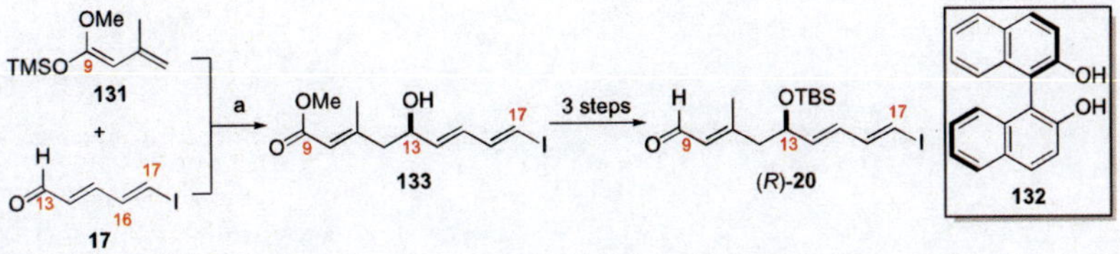

Scheme 23 Paterson's asymmetric synthesis of **(*R*)-20.** *Reagents and conditions*: (**a**) CaH_2, (*R*)-BINOL, **132**, Ti(O*i*-Pr)$_4$, THF, RT, then −78°C, aldehyde **17**, followed by silyl ketene acetal **131**, 96%, er = 97:3

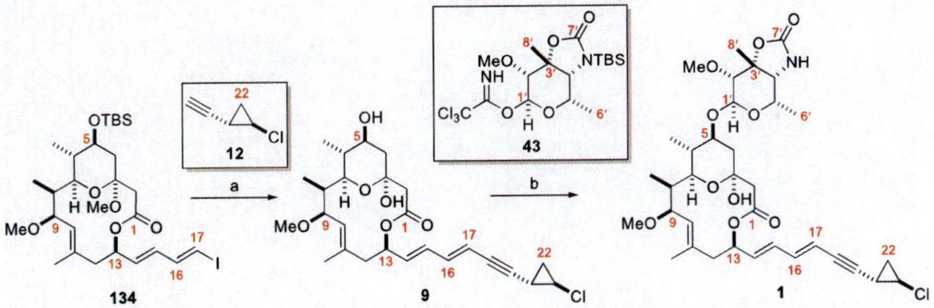

Scheme 24 Completion of callipeltoside A. *Reagents and conditions*: (**a**) (i) **12**, $Pd(PPh_3)_2Cl_2$, CuI, *i*-Pr_2NH, EtOAc, −20°C to 0°C, 83%; (ii) TFA, aq. THF, RT, 98%; (**b**) (i) **43**, TMSOTf, CH_2Cl_2, 4 Å MS, −30°C; (ii) TBAF, AcOH, THF, RT, 76% (over 2 steps)

aldehyde **(*R*)-20**, allowing for the interception of the analogous enantiomeric route (Sect. 2.1) (Scheme 23) [11].

The callipeltoside aglycon was completed in the same manner as the enantiomeric version previously shown in Sect. 2.1.4. The callipeltose A sugar was synthesised following the procedure reported by Guiliano [25] and the corresponding trichloroacetimidate coupling partner **43** attached using Schmidt glycosidation conditions (analogous to Trost; see Sect. 2.2.4) [7, 8]. TBS deprotection with TBAF then gave callipeltoside A in 23 steps (longest linear) and 4.8% overall yield (Scheme 24) [11].

2.6 *Panek Synthesis of Callipeltoside A (2004)*

2.6.1 Retrosynthesis

The fourth synthesis of callipeltoside A was completed by Panek in 2004 [12]. Since the absolute and relative stereochemistry of callipeltoside A had been confirmed at this point, late-stage installation of the *trans*-chlorocyclopropane was not required. Instead, the Panek group sought to unite fully assembled C11–C22 fragment **136** with pyran core **135** by means of a Julia–Kocienski olefination, forming the

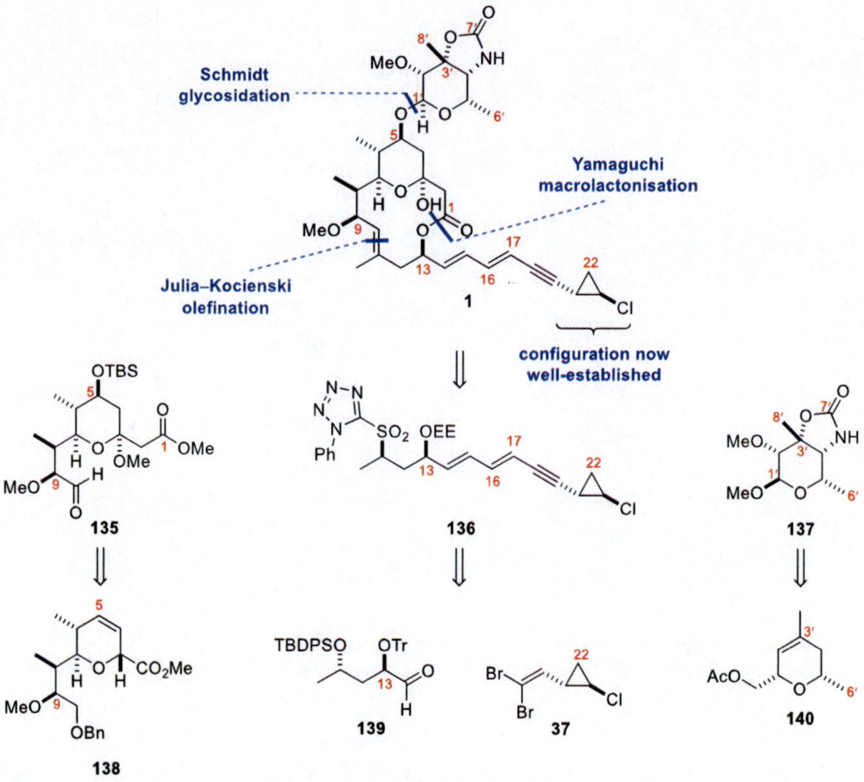

Scheme 25 Panek retrosynthesis of callipeltoside A

trisubstituted C10/C11 olefin in the process. In contrast to previous syntheses, pyran **135** and callipeltose A sugar **137** were to be derived from the relevant dihydropyrans arising from [4 + 2] annulations using chiral organosilanes (Scheme 25) [12, 50].

2.6.2 Formation of C1–C10 Pyran 135

The synthesis began with the installation of the C8 and C9 stereocentres via the *anti*-selective condensation of α-benzyloxyacetaldehyde **141** with chiral silane (*S*)-**142** in the presence of $SnCl_4$. This provided tetrahydrofuran **143** in high yield (87%) and diastereoselectivity (dr > 97:3). Treatment with $SbCl_5$ resulted in elimination and the formation of the *anti*-homoallylic alcohol, which could be converted to aldehyde **144** by methylation and ozonolysis. At this point aldehyde **144** could be reacted with chiral organosilane **145** in the presence of TfOH to afford tetrahydropyran **138** in 85% yield and as a single diastereomer (dr > 97:3). This was then converted to advanced pyranone **146** in a further three steps (Scheme 26) [12].

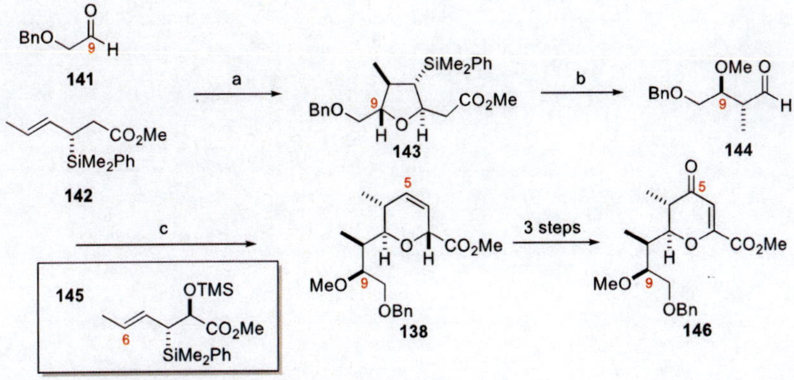

Scheme 26 Synthesis of the C1–C10 scaffold. *Reagents and conditions:* (**a**) **141**, $SnCl_4$, CH_2Cl_2, then **142**, CH_2Cl_2, −78°C, 87%, dr > 97:3; (**b**) (i) $SbCl_5$, CH_2Cl_2, −78°C to −50°C; (ii) Me_3OBF_4, proton sponge, 4 Å MS, CH_2Cl_2, RT; (iii) O_3, MeOH–CH_2Cl_2 (3:1), Me_2S, −78°C, 81% (over 3 steps); (**c**) **145**, TfOH, CH_2Cl_2, −70°C, 85%, dr > 97:3

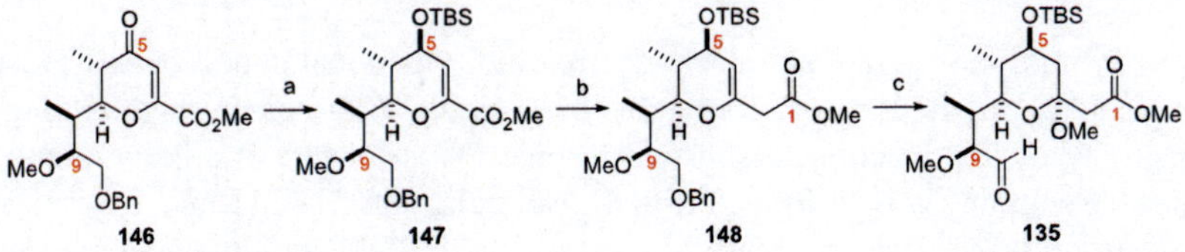

Scheme 27 Completion of pyran **135.** *Reagents and conditions:* (**a**) (i) $CeCl_3{\cdot}7H_2O$, MeOH, −78°C, then $NaBH_4$, MeOH, −78°C, 93%; (ii) TBSOTf, 2,6-lutidine, CH_2Cl_2, −78°C, 97%; (**b**) (i) LiOH, THF–MeOH–H_2O (3:1:1), RT, then 5% HCl; (ii) $(COCl)_2$, DMF, CH_2Cl_2; (iii) CH_2N_2, Et_2O, 0°C, 70% (over 3 steps); (iv) $PhCO_2Ag$, pyridine, MeOH, RT, 80%; (**c**) (i) CSA, MeOH, RT, 97%; (ii) H_2, Pd/C, EtOAc; (iii) PDC, 4 Å MS, CH_2Cl_2, RT, 88% (2 steps)

Luche reduction provided the C5 stereocentre, and the resulting hydroxy was protected as its TBS-ether (**147**). The C1–C10 scaffold was then further manipulated by ester saponification and Arndt–Eistert homologation to give **148** in four steps and 56% overall yield. Installation of the anomerically favoured ketal (MeOH/CSA), Bn deprotection and oxidation then afforded aldehyde **135** (Scheme 27) [12].

2.6.3 Synthesis of the C11–C22 Fragment

Panek chose to assemble the entire C11–C22 fragment prior to coupling with aldehyde **135**.

The synthesis again required dibromoolefin **37** as a key building block, which was synthesised as described by Paterson [11, 17] (Sect. 2.1.3). This underwent Stille reaction using the conditions developed by Shen [41] (and also used by Olivo [20] and Trost [8]) to give enyne **116**, from which phosphonate **84** was prepared in a further two-step procedure (Scheme 28). Horner–Wadsworth–Emmons reaction

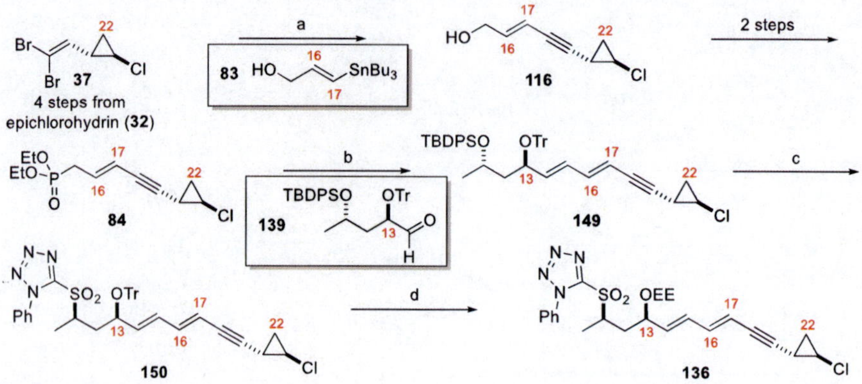

Scheme 28 Completion of the C11–C22 fragment. *Reagents and conditions*: (**a**) **83**, Pd_2dba_3, $(4\text{-}MeOC_6H_4)_3P$, DIPEA, DMF, 80°C, 68%; (**b**) LiHMDS, then **139**, THF, −78°C, 89%; (**c**) (i) TBAF, THF, RT, 91%; (ii) 1-phenyl-1*H*-tetrazole-5-thiol, DIAD, PPh_3, THF, 0°C to RT; (iii) $(NH_4)_6Mo_7O_{24}$, EtOH–acetone (2:1), RT, 66% (2 steps); (**d**) (i) TsOH, MeOH, RT; (ii) EVE, PPTS, RT, 90% (2 steps)

between phosphonate **84** and aldehyde **139** under identical conditions to Trost [8] interestingly resulted in the formation of a single *E,E*-isomer, highlighting the sensitivity of this reaction to substrate choice (Trost observed a 4:1 *E*/*Z* mixture when the coupling was performed with the callipeltoside C1–C13 macrocycle). With the entire C11–C22 carbon scaffold in place, the TBDPS group was removed and stereocentre inverted by Mitsunobu reaction with 1-phenyl-1*H*-tetrazole-5-thiol. Oxidation to the corresponding sulfone followed by protecting group manipulation gave Julia–Kocienski coupling partner **136** (Scheme 28) [12].

2.6.4 Synthesis of Callipeltose A

Organosilane methodology was also used to access the callipeltose A sugar [12, 51]. Stereoselective [4 + 2] annulation between acetaldehyde and chiral organosilane **151** (made in 5 steps, 55%) gave dihydropyran **140** in high yield and diastereoselectivity (dr = 10:1). This was converted to **152** in a further two steps, with selective reduction and methanolysis providing **153** as a single anomer. Advanced intermediate **155** was then constructed in short order by Sharpless dihydroxylation, selective methylation and carbamate formation. 1,5-C–H insertion using $Rh_2(OAc)_4$ then forged the requisite 5-membered ring present in callipeltose A in 93% yield (Scheme 29) [12, 51].

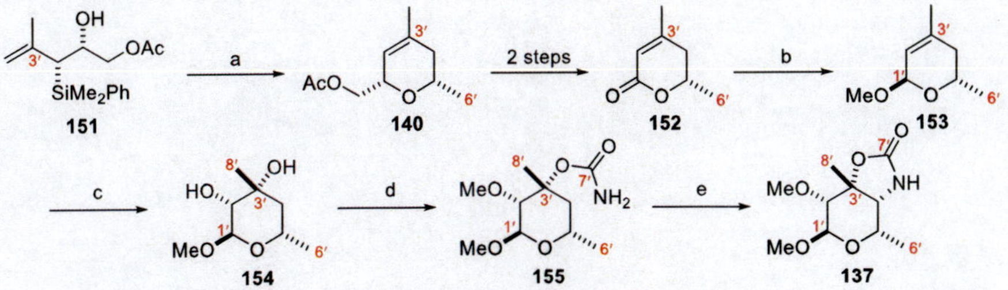

Scheme 29 Synthesis of the callipeltose A scaffold (**137**). *Reagents and conditions*: (**a**) MeCHO, TMSOTf, CH_2Cl_2, −78°C, dr = 91:9, 80%; (**b**) DIBAL-H, CH_2Cl_2, −78°C, then CSA, MeOH; (**c**) AD-mix-α, $NaHCO_3$, $MeSO_2NH_2$, OsO_4, *t*-BuOH–H_2O (1:1), RT, 65% (over 2 steps); (**d**) (i) *t*-BuOK, MeI, THF, 0°C; (ii) $Cl_3CCONCO$, CH_2Cl_2, RT, then K_2CO_3, MeOH–H_2O (10:1), 79% (over 2 steps); (**e**) $Rh_2(OAc)_4$, $PhI(OAc)_2$, MgO, PhH, 80°C, 93%.

2.6.5 Completion of Callipeltoside A

Aldehyde **135** and sulfone **136** were coupled using a Julia–Kocienski olefination to afford **156** in low yield (20%) after acidic cleavage of the ethoxyethyl ether (EE) protecting group. The yield could not be improved despite further investigation involving different bases (KHMDS, NaHMDS) and more polar solvents (DME, DMF) [12].

Hydrolysis of the ester group in **156** was followed by Yamaguchi macrolactonisation, which gave **157** as well as eliminated product **158** in a 50:50 ratio These products were separated and **157** processed by deprotection and hemi-ketal installation to give the callipeltoside aglycon (**9**).[2] This was then coupled to trichloroacetimidate **43** using the literature known Schmidt glycosidation conditions to give callipeltoside A (**1**) in 1.4% overall yield and 25 steps (longest linear) (Scheme 30) [12].

2.7 *Hoye Synthesis of Callipeltoside A (2010)*

2.7.1 Retrosynthesis

The synthesis of callipeltoside A was disclosed by Hoye in 2010 [13]. Similar disconnections involving late-stage attachment of the callipeltose A sugar via a Schmidt glycosidation and Horner–Wadsworth–Emmons reaction with phosphonate **84** were key to the assembly of callipeltoside A. However, a dual macrocyclisation/hemi-ketal formation process from acylketene precursor **161** was anticipated to form the C1–C13 macrocycle in one step. This precursor was to be accessed by alkenyllithium addition to aldehyde **163** (Scheme 31) [13, 52, 53].

---

[2]The undesired lactone (**158**) was recycled in two steps to provide additional quantities of the callipeltoside aglycon (**9**) (conditions shown in Scheme 30).

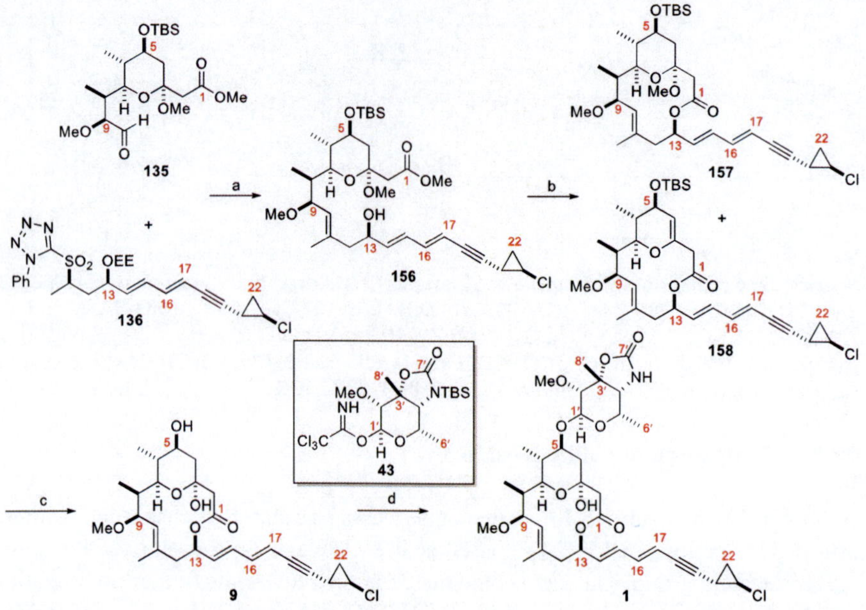

Scheme 30 Completion of callipeltoside A. *Reagents and conditions*: (**a**) (i) **136**, LiHMDS, then **135**, THF, −78°C to RT; (ii) PPTS, MeOH, 20% (over 2 steps); (**b**) (i) LiOH, MeOH–H_2O–THF, 90%; (ii) 2,4,6-$Cl_3C_6H_2COCl$, Et_3N, DMAP, PhMe, 80°C, 90% (**157** and **158**, 1:1); [**158** to **9** conversion: (i) PPh_3•HBr, H_2O–CH_2Cl_2 (11:1), RT, 97%; (ii) TBAF, THF, RT, 85%]; (**c**) (i) TBAF, THF, RT, 85%; (ii) PPTS, MeCN–H_2O (3:1), RT, 89%; (**d**) (i) **43**, TMSOTf, 4 Å MS, CH_2Cl_2, −30°C; (ii) TBAF, AcOH–THF (1:1), RT, 73% (over 2 steps)

2.7.2 Synthesis of the C1–C13 Fragment

Lithium-halogen exchange with vinyl iodide **162** followed by addition to aldehyde **163** gave a 50:50 mixture of separable C9 epimeric alcohols.[3] Methylation of the desired C9 alcohol afforded **164**, which was converted to lactonisation precursor **161** in an additional five linear steps (Scheme 32) [13].

Substrate **161** underwent the desired dual macrolactonisation/hemi-ketal formation process to afford the C1–C13 macrocyclic core **160** in 76% yield. This reaction was proposed to proceed via intermediates **165**, **166** and **167** (Scheme 33) [13, 53].

2.7.3 Completion of Callipeltoside A

Compound **168** was synthesised following a three-step protecting group manipulation. Subsequent oxidation to the aldehyde, coupling to phosphonate **84** and

[3]The undesired C13 epimer was oxidised and diastereoselectively reduced to afford additional quantities of **164**.

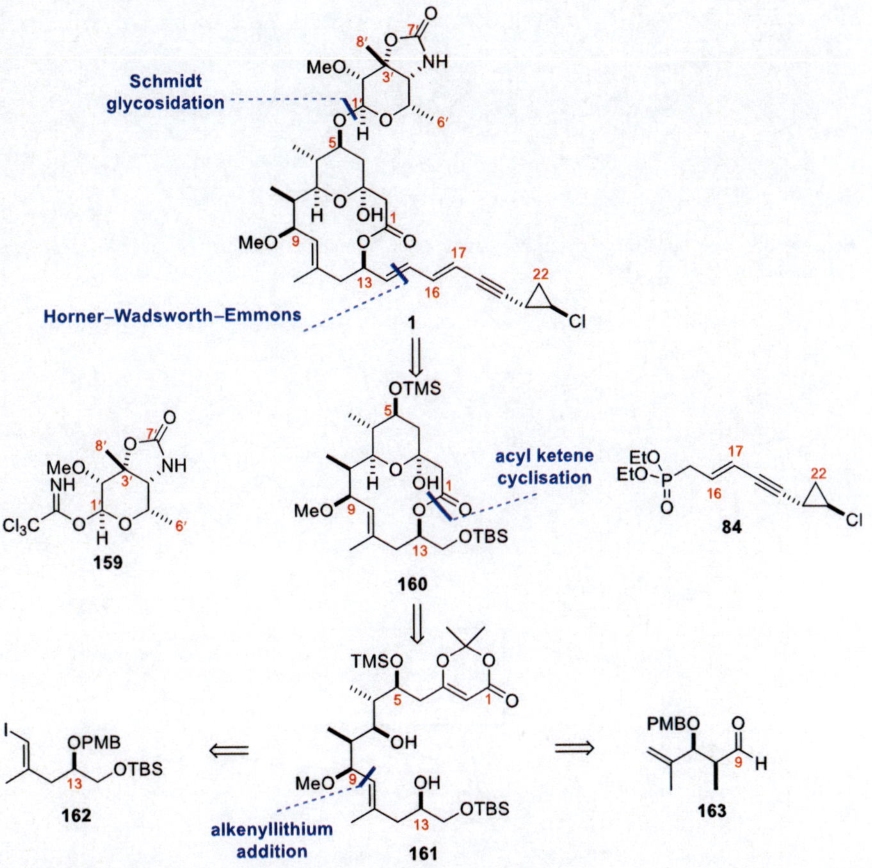

Scheme 31 Hoye's approach to callipeltoside A

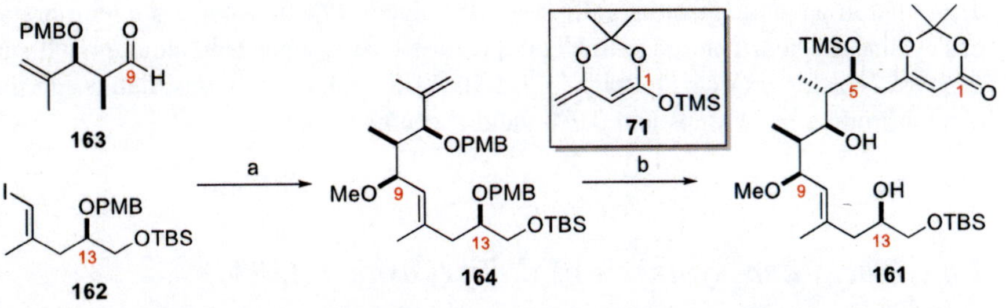

Scheme 32 Synthesis of cyclisation precursor **161.** *Reagents and conditions*: **(a)** (i) *n*-BuLi, **162**, Et_2O, −78°C, then **163**, Et_2O, dr = 50:50 (separation by SiO_2 chromatography), 34% of each; (ii) MeI, NaH, DMF, RT, 80%; **(b)** (i) 9-BBN, THF, −78°C to RT, then NaOH (10%)–H_2O_2 (30%) (6:1), 70%, dr = 87.5:12.5 (at C6, separation by SiO_2 chromatography); (ii) Dess–Martin periodinane, CH_2Cl_2, RT, 76%; (iii) **71**, BF_3•OEt_2, CH_2Cl_2, −78°C, 90%; (iv) TMSCl, Et_3N, CH_2Cl_2, RT; (v) DDQ, CH_2Cl_2–H_2O (12:1), 78% (over 2 steps)

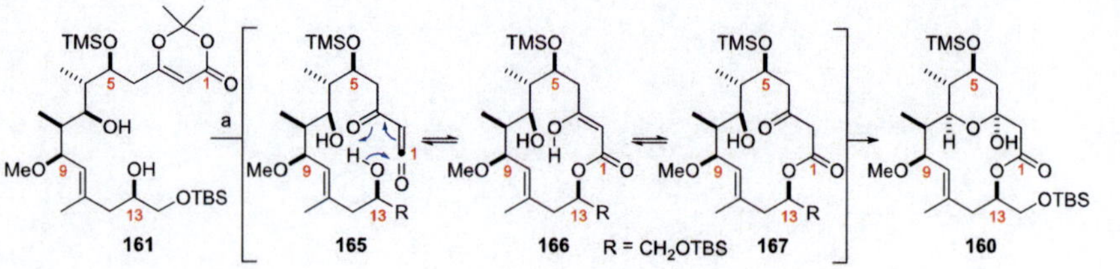

Scheme 33 Key dual macrocyclisation/hemi-ketal formation reaction [13, 53]. *Reagents and conditions*: (**a**) PhH, Δ, 76%

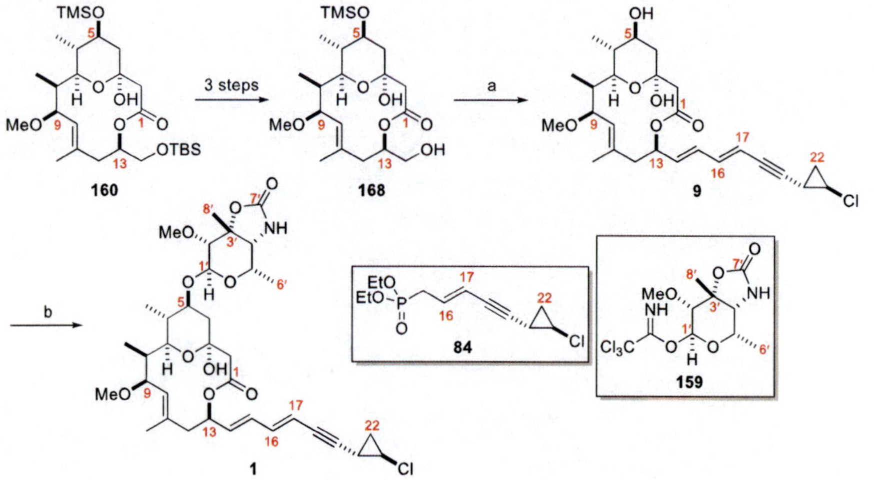

Scheme 34 Completion of callipeltoside A. *Reagents and conditions*: (**a**) (i) SO_3•pyridine, DMSO, Et_3N, CH_2Cl_2, 0°C; (ii) **84**, LiHMDS, THF, −78°C; (iii) TBAF, THF, 54% (over 3 steps); (**b**) **159**, TMSOTf, 4 Å MS, CH_2Cl_2, 0°C, 70%

deprotection gave the familiar callipeltoside aglycon (**9**). In accordance with previous syntheses, the trichloroacetimidate of callipeltose A sugar **159** (although with an exposed NH this time) was attached to **9** using Schmidt conditions, delivering the natural product in 21 steps and 0.7% yield (Scheme 34) [13].

2.8 *MacMillan Synthesis of Callipeltoside C (2008)*

2.8.1 Retrosynthesis

The first synthesis of callipeltoside C was reported in 2008 by MacMillan [16]. At this stage the relative configuration of callipeltose C with respect to the callipeltoside aglycon had not been validated, and therefore the absolute configuration of the molecule was not known [4]. In contrast to previous approaches towards these

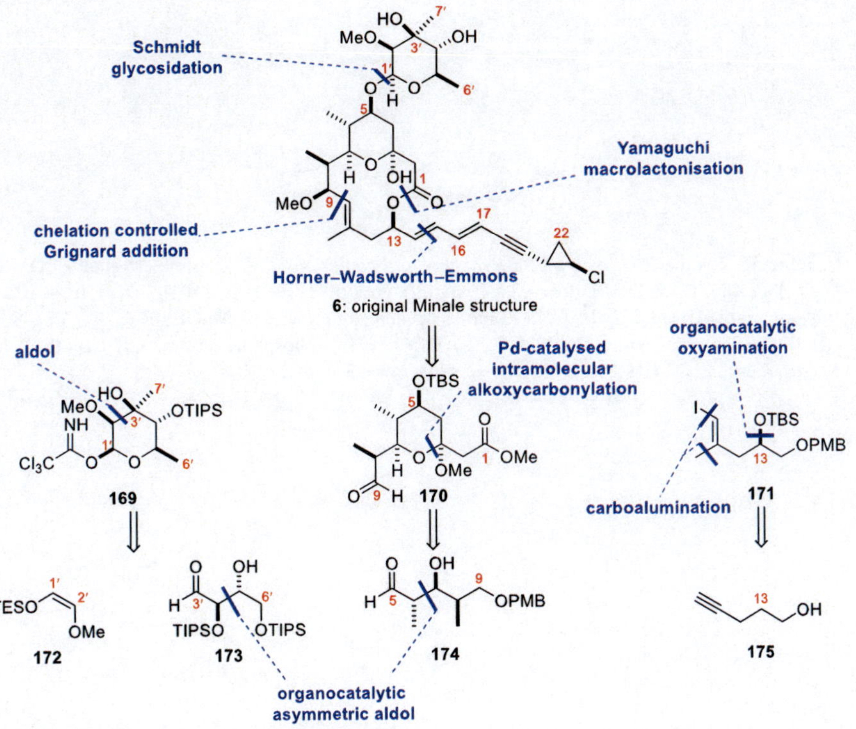

Scheme 35 Retrosynthesis of callipeltoside C

molecules, organocatalytic processes featured heavily in MacMillan's synthesis. Pyran **170** and callipeltose C sugar **169** were to be accessed using organocatalytic asymmetric aldol reactions, whilst the C13 stereocentre of **171** was to be set by an organocatalytic oxyamination reaction (Scheme 35) [16, 54–58].

2.8.2 Synthesis of Pyran 170

The synthesis began with an L-proline-catalysed organocatalytic aldol reaction between propionaldehyde (**176**) and Roche ester-derived aldehyde **177**. This gave aldehyde **174** in moderate yield (48%), but set the desired C6 and C7 stereocentres in both excellent diastereo- (dr = 92:8) and enantioselectivity (er > 99:1) [16]. The C5 stereocentre was then formed by propargylzinc addition to give the desired 1,3-*anti*-diol relationship (**180**, dr = 86:14). With the 4 contiguous C5–C8 stereocentres in place, a palladium-catalysed alkoxycarbonylation process developed by Marshall [59, 60] impressively gave advanced C1–C9 pyran **181** as a single diastereomer in one step (Scheme 36). The proposed mechanism for this transformation is described in Scheme 37 [59, 60]. Subsequent TBS protection of the C5 hydroxy, PMB removal and Parikh–Doering oxidation then gave **170** in six steps and 44% overall yield from aldehyde **177** (Scheme 36).

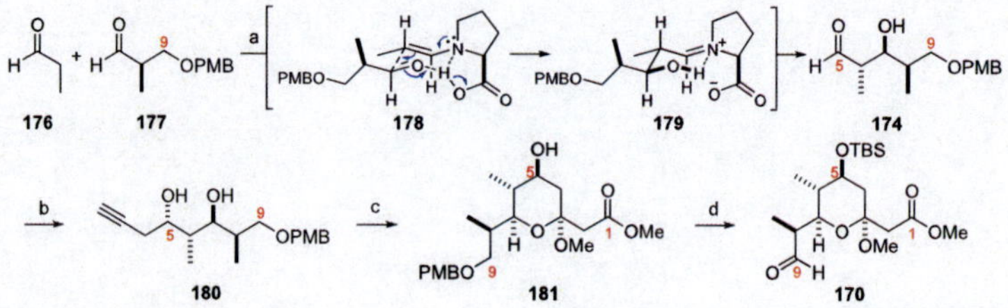

Scheme 36 Synthesis of pyran **170.** *Reagents and conditions:* (**a**) L-proline (10 mol%), DMSO, H_2O, 4°C, 48% (75% brsm), dr = 92:8 (*syn:anti*), er > 99:1; (**b**) $HCCCH_2Br$, Zn, THF, −100°C, 98%, dr = 86:14; (**c**) $PdCl_2(MeCN)_2$ (5 mol%), *p*-benzoquinone, CO, MeOH, 0°C, 75%, dr > 95:5; (**d**) (i) TBSCl, imidazole, DMF, RT; (ii) DDQ, pH 7 phosphate buffer, CH_2Cl_2, 0°C; (iii) SO_3•pyridine, Et_3N, CH_2Cl_2, DMSO, 0°C, 80% (over 3 steps)

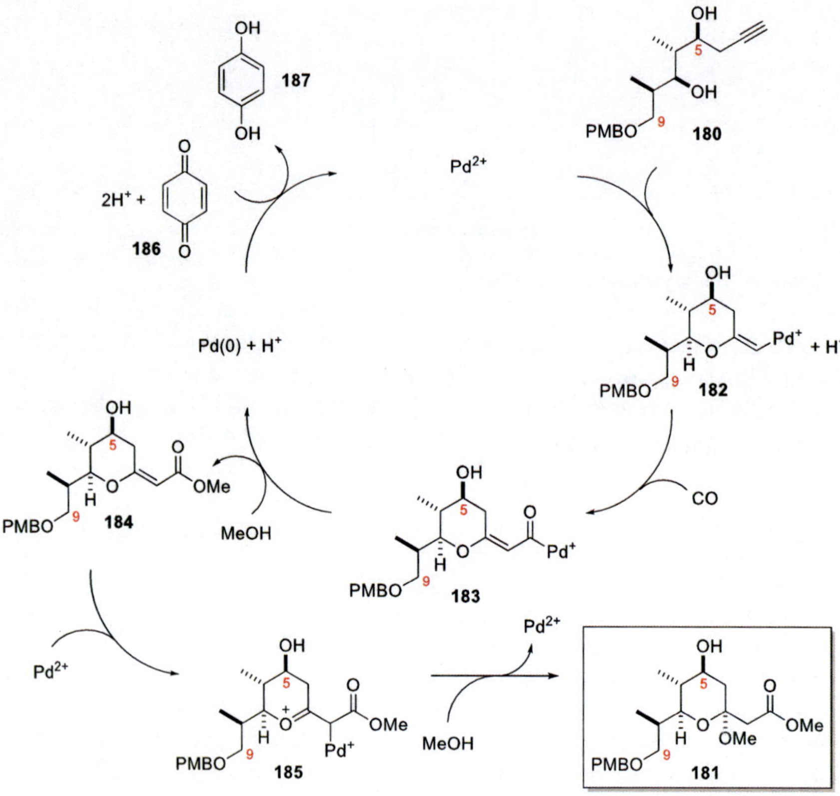

Scheme 37 Proposed mechanism for the palladium-catalysed alkoxycarbonylation reaction [59, 60]

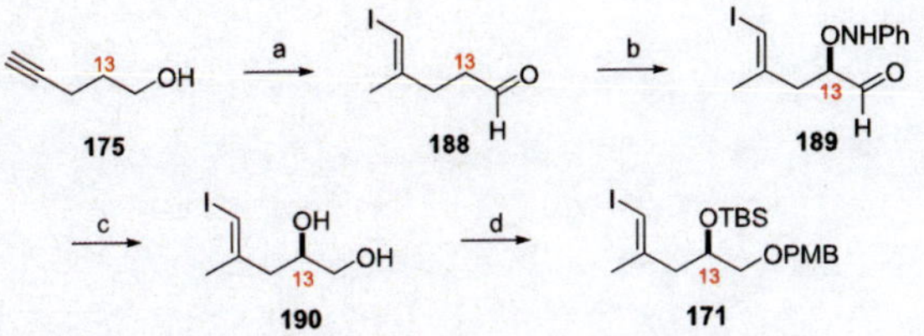

Scheme 38 Organocatalytic oxyamination approach to vinyl iodide **171.** *Reagents and conditions*: (**a**) (i) Me_3Al, Cp_2ZrCl_2, I_2, THF, −30°C; (ii) $(COCl)_2$, Et_3N, DMSO, CH_2Cl_2, −50°C, 92% (over 2 steps); (**b**) L-proline, (20 mol%), PhNO, DMSO, RT, er > 99:1; (**c**) (i) $NaBH_4$, EtOH, RT; (ii) Zn, EtOH–AcOH (3:1), 77% (over 3 steps); (**d**) (i) $Bu_2Sn(OMe)_2$, PMBCl, TBAI, PhMe, Δ; (ii) TBSCl, imidazole, DMF, RT, 80% (over 2 steps), er > 99:1

2.8.3 Formation of Vinyl Iodide 171

Carboalumination of 4-pentyn-1-ol delivered the desired trisubstituted double bond as a single regioisomer which, following treatment with iodine, gave **188**. Subsequent Swern oxidation and organocatalytic α-oxyamination then provided **189** in excellent enantioselectivity (er > 99:1) [57, 58]. Reduction of the aldehyde, N–O bond cleavage and bis-protection gave **171** in seven steps and 57% yield (Scheme 38) [16].

2.8.4 Fragment Union and Completion of the Aglycon (9)

Vinylic iodide **171** (2.2 equiv.) was converted to the corresponding Grignard and added to aldehyde **170** to provide the C9 stereocentre in both high diastereoselectivity (dr = 94:6) and yield (98%). The addition of $MgBr_2$ to substrate **170** prior to Grignard addition was found to be important for both yield and diastereoselectivity, with formation of a complex between $MgBr_2$ and the pyran and aldehyde oxygens speculated. It is also noted that an intermolecular Nozaki–Hiyama–Kishi reaction was also attempted and in contrast gave the undesired C9 alcohol configuration [16].

With **191** in hand, methylation, PMB deprotection, oxidation and Horner–Wadsworth–Emmons olefination with *trans*-chlorocyclopropane-containing phosphonate **130** gave **193** [16]. Interestingly, attachment of **130** at this stage of the synthesis resulted in high *E*:*Z* selectivity (10:1–19:1) compared to Trost [7, 8], Evans [9, 10] and Paterson [11] and once again highlights the subtleties associated with substrate choice when performing reactions under similar conditions (see Sects. 2.2.3, 2.3.2 and 2.4.5; *E*:*Z* = 4:1).

Deprotection of C13-OTBS and saponification of the methyl ester gave *seco*-acid **193** which, when subjected to Yamaguchi macrocyclisation conditions, gave eliminated lactone **158** as well as the desired product (**157**, not shown; see Sect. 2.6.5). Treatment of the mixture with aqueous PPh_3•HBr then installed the hemi-ketal functionality, whilst final TBS deprotection using TFA delivered the callipeltoside aglycon (**9**) in 18 steps and 19% overall yield (Scheme 39) [16].

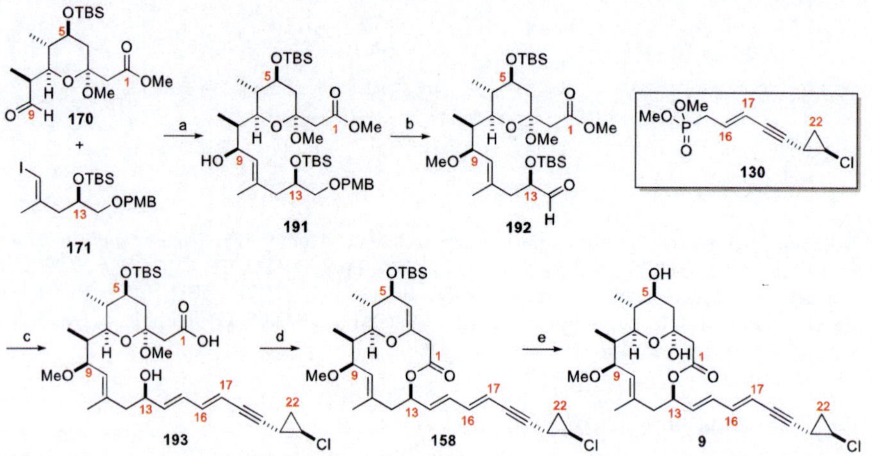

Scheme 39 Synthesis of callipeltoside aglycon (**9**). *Reagents and conditions*: (**a**) **171**, *t*-BuLi, $MgBr_2$•Et_2O, Et_2O, then **170**, CH_2Cl_2, −78°C, 98%, dr = 94:6; (**b**) (i) MeOTf, 2,6-di-*t*-butylpyridine, CH_2Cl_2, RT; (ii) DDQ, pH 7 buffer, CH_2Cl_2, 0°C; (iii) SO_3•pyridine, Et_3N, DMSO, CH_2Cl_2, −10°C, 84% (over 3 steps); (**c**) (i) **130**, LiHMDS, THF, −78°C, then **192**, THF, −78°C, 84%, *E*:*Z* = 19:1; (ii) TBAF, THF, 0°C, 100%; (iii) $Ba(OH)_2$•$8H_2O$, MeOH, RT, 84% (over 3 steps); (**d**) 2,4,6-$Cl_3C_6H_2COCl$, DIPEA, THF, RT, then addition to DMAP, PhMe, 60°C, 83%; (**e**) (i) PPh_3•HBr, H_2O, CH_2Cl_2, RT; (ii) TFA, THF–H_2O (5:1), RT, 81% (over 2 steps)

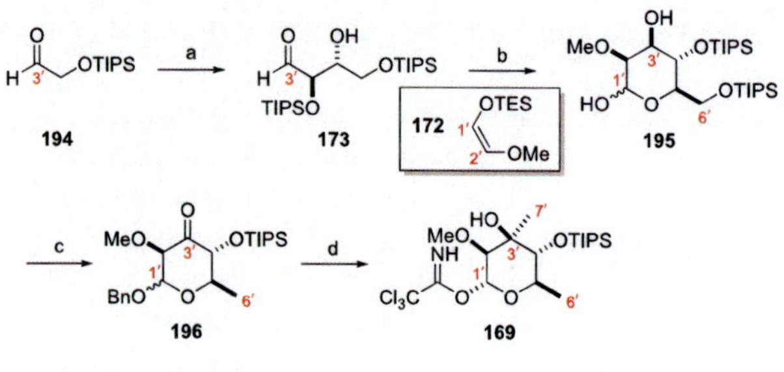

Scheme 40 Organocatalytic approach to D-configured callipeltose C. *Reagents and conditions*: (**a**) D-proline, DMF, 75%, er > 99:1; (**b**) **172**, $MgBr_2$•Et_2O, CH_2Cl_2, 47%, dr > 95:5; (**c**) (i) AcCl, BnOH, 110°C; (ii) PhOCSCl, pyridine, CH_2Cl_2; (iii) *n*-Bu_3SnH, AIBN, PhH, 120°C; (iv) Dess–Martin periodinane, CH_2Cl_2, 0°C, 47% (over 4 steps); (**d**) (i) $MgBr_2$•Et_2O, MeMgBr, CH_2Cl_2, dr > 95:5; (ii) H_2, Pd/C, EtOAc; (iii) Cl_3CCN, Cs_2CO_3, CH_2Cl_2, 86% (over 3 steps), dr = 95:5

2.8.5 Synthesis of D-Callipeltose C

The synthesis commenced with dimerisation of aldehyde **194** under organocatalytic conditions to provide **173** in good yield and excellent enantioselectivity (er > 99:1) (Scheme 40) [16, 54]. This was then subjected to the organocatalytic Mukaiyama aldol conditions developed in the MacMillan laboratory, with the Lewis acid ($MgBr_2$) and solvent (CH_2Cl_2) choice critical for the formation of the desired sugar scaffold (**195**) [16, 55, 56]. The anomeric hydroxy was then protected as its benzyl ether with concomitant removal of the primary TIPS protecting group.

Subsequent conversion of the primary alcohol to the corresponding phenyl thiocarbonate and deoxygenation under Barton–McCombie conditions provided the C6′ methyl group. Oxidation of the remaining C3′ secondary hydroxy to the ketone then gave **196**. Further manipulation by means of diastereoselective methyl Grignard addition (dr = 95:5), removal of the benzyl group and formation of the trichloroacetimidate provided the D-callipeltose C sugar coupling partner **169**. Overall, this fragment was synthesised in eight steps and 14.5% yield (Scheme 40) [16].

2.8.6 Completion of Callipeltoside C

The callipeltoside aglycon was coupled with D-callipeltose C (**169**) using the conditions developed by Tietze [16, 61] which, following TIPS deprotection, gave spectroscopic data inconsistent with the natural isolate [4]. Since previous syntheses of callipeltoside A had determined the absolute stereochemistry of the molecule, it was reasonably assumed that the problem lay with the callipeltose C sugar rather than the aglycon. As a result, the enantiomeric L-configured sugar (*ent*-**169**) was synthesised under the same conditions outlined in Scheme 40. Schmidt glycosidation and final TIPS deprotection gave compound **3**, which this time had identical spectroscopic data to naturally occurring callipeltoside C [4, 16]. This result not only confirmed the absolute stereochemistry of callipeltoside C but also suggested that both callipeltosides A and C contain L-configured sugars (Scheme 41). This was in contrast to the original structures drawn in the isolation papers [3, 4].[4] However, it is noteworthy that the glycosidic linkage attaching callipeltose C to the aglycon was only tentatively assigned, with the MacMillan group assuming it to be the β-equatorial anomer by comparison with the originally reported structure of callipeltoside B (Figs. 3, 5) [16].

2.9 Ley Syntheses of Callipeltosides A, B and C (2012)

2.9.1 Retrosynthesis

The Ley group's approach to the entire callipeltoside family involved a highly convergent strategy whereby the molecule(s) was split into three equally sized fragments: the C1–C9 pyran (**170**), C10–C22 vinylic iodide (**197**) and the relevant sugar moiety [14, 15]. The common callipeltoside aglycon would then be assembled by union of **170** and **197** via a diastereoselective Oppolzer–Radinov alkenylzinc addition [62], followed by Yamaguchi macrolactonisation. The C1–C9 pyran was to be formed using an $AuCl_3$-catalysed cyclisation, which the group had reported

[4] An optical rotation of $[\alpha]^{25}_{D} = -23.1$ ($c = 0.18$, $CDCl_3$) was measured for synthetic callipeltoside C [16]. However, no optical rotation for natural callipeltoside C was recorded due to the prohibitively small amounts isolated [4].

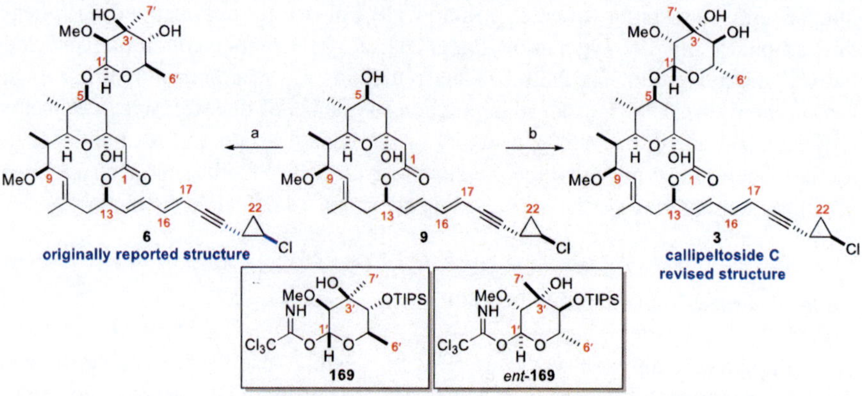

Scheme 41 Completion of callipeltoside C. *Reagents and conditions*: **(a)** (i) **169**, TMSOTf, 4 Å MS, CH_2Cl_2, −30°C; (ii) TASF, DMF, 40°C, 58% (over 2 steps); **(b)** (i) *ent*-**169**, TMSOTf, 4 Å MS, CH_2Cl_2, −30°C; (ii) TASF, DMF, 40°C, 63% (over 2 steps)

previously [63]. Since other approaches to the callipeltosides had often struggled to produce the dien-yne as a single isomer, the group also chose to address this synthetic issue. With this in mind, the C15–C16 and C17–C18 bonds were disconnected, revealing bis-stannane linchpin **201**, from which bidirectional Stille reactions could be performed. Pyranone **202** was considered to be a key building block from which each callipeltose sugar could be accessed (Scheme 42) [14, 15].

2.9.2 Synthesis of Pyran 170

Synthesis of pyran fragment **170** began with crotylation of known Roche ester-derived aldehyde **203** to give homoallylic alcohol **205** in good yield (70%) and diastereoselectivity (dr = 88:12). Conversion to the triol using catalytic osmium tetroxide followed by sodium periodate-mediated cleavage gave the requisite aldehyde, with which a chelation-controlled addition of propargyl zinc furnished diol **206** in 72% over three steps (dr = 85:15 at C5). Two further synthetic steps gave ynoate **208**, enabling all minor diastereomers to be separated (**208**, dr > 95:5) and the 1,3-*anti*-relationship to be proven by X-ray crystallography (not shown). Subsequent deprotection afforded the diol, which smoothly underwent $AuCl_3$-catalysed cyclisation in MeOH to give pyran **209** as a single, anomerically favoured diastereomer in 96% yield. Further routine synthetic manipulations gave pyran fragment **170** in 26% yield from **203** (11 linear steps) with minimal use of chromatography (Scheme 43) [14, 15].

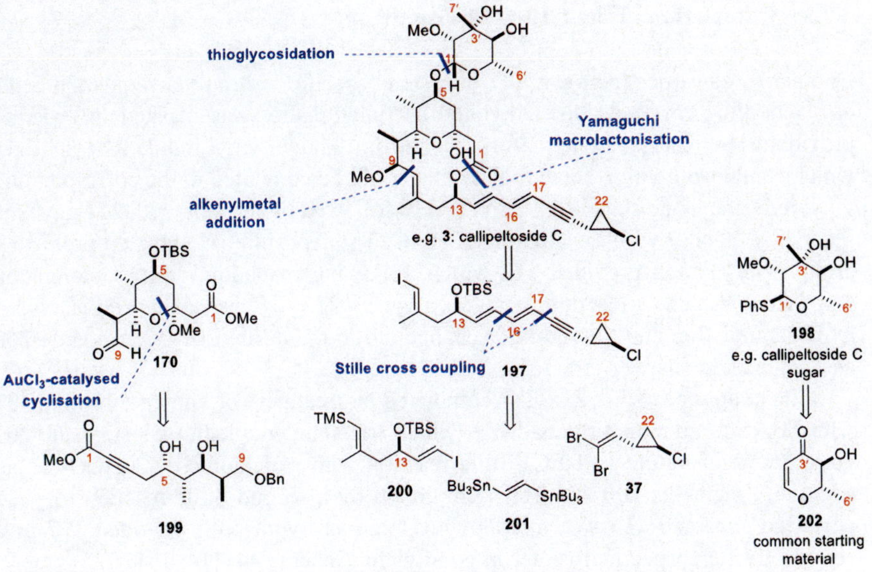

Scheme 42 Ley group approach to the callipeltosides (callipeltoside C shown for clarity)

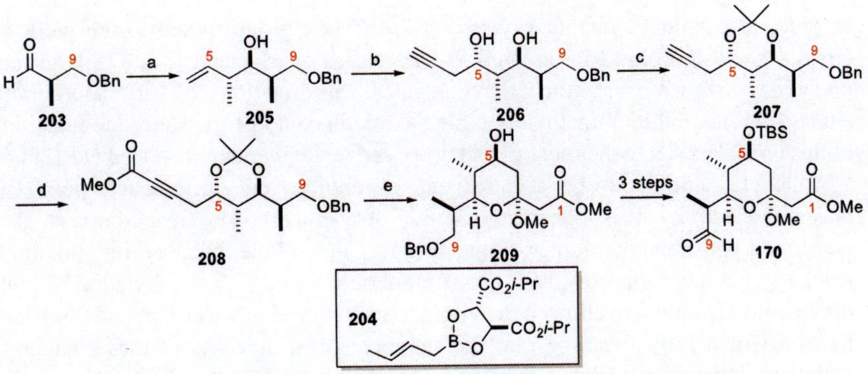

Scheme 43 Synthesis of pyran **170.** *Reagents and conditions*: (**a**) crotylborane **204**, 4 Å MS, PhMe, −78°C, 70%, dr = 88:12; (**b**) (i) OsO_4, NMO, acetone–H_2O (2:1), RT; (ii) $NaIO_4$, THF–H_2O (10:1), 0°C to RT; (iii) Zn, propargyl bromide, THF, 0°C to −100°C, 72% (over 3 steps), dr = 85:15 (at C5); (**c**) 2,2-dimethoxypropane, (±)-CSA, acetone, RT; (**d**) *n*-BuLi, THF, −40°C to −78°C, then $ClCO_2Me$, 73% (over 2 steps), dr > 95:5; (**e**) (i) QP-SA, MeOH, RT, 95%; (ii) $AuCl_3$ (2 mol%), MeOH, RT, 96%

2.9.3 Completion of the C10–C22 Fragment

In order to construct fragment **197** in a stereospecific fashion via sequential Stille cross-coupling reactions, the vinyl iodide functionality was masked in order to provide orthogonal reactivity. Work began from known vinyl iodide **210** [64, 65], which contained a vinyl silane moiety that could be converted to the corresponding vinyl iodide at a late stage. Lithiation of **210** followed by reaction with **211** gave the secondary alcohol with installation of the C13 stereocentre. A two-step protecting group manipulation provided **212** which, following oxidation, dibromoolefination and Corey–Fuchs reaction, gave alkyne **213**. Subsequent Pd-catalysed hydrostannylation and tin–iodine exchange led to the desired (*E*)-vinyl iodide **200** as a single stereoisomer [14, 15].

Stille coupling partner **215** was completed by treatment of known dibromide **37** with TBAF to give the bromoalkyne, which was then coupled to bis-stannane **201** via a low temperature (−10°C) Stille reaction with stoichiometric Ag_2CO_3 as an additive. Fragments **200** and **215** were united by a second Stille reaction to yield advanced substrate **216** which, following treatment with NIS, provided **197** in a completely stereospecific manner in good yield (Scheme 44) [14, 15].

2.9.4 Union of Fragments 170 and 197 and Macrocycle Formation

In order to couple fragments **170** and **197**, the Ley group turned to the work of Oppolzer and Radinov [62] as well as the studies of Marshall [66, 67], who had previously shown that the stereochemical information of an appropriate enantioenriched lithio *N*-methylephedrine alkoxide could be transferred to reactions of this type. The stereochemical model proposed in the literature revealed (1*R*,2*S*)-(−)-*N*-methylephedrine to be the reagent of choice; but in practice poor C9 diastereoselectivity was observed (34:66, not shown). The reaction was also performed with the enantiomeric ligand, (1*S*,2*R*)-(+)-*N*-methylephedrine, this time resulting in a much-improved C9 diastereomeric ratio of 91:9. Methylation of both diastereomeric mixtures allowed the group to intercept a known compound disclosed by MacMillan [16], meaning that the stereochemical outcome of these reactions could be straightforwardly determined. To their surprise, the diastereomerically enriched material using (1*S*,2*R*)-(+)-*N*-methylephedrine was found to be the desired compound, in contrast to the models disclosed by Oppolzer [62] and Noyori [68, 69]. It was proposed that the unexpected reversal in selectivity was due to bis-chelation between the active metal complex and the pyran oxygen, resulting in the exposure of the opposite diastereotopic face to reaction [70]. The authors note that a similar result was documented by Myers in his synthesis of the tetracycline antibiotics [70].

With the C9 stereocentre in place, the methylated derivative was deprotected and saponified to give *seco*-acid **193**. Yamaguchi macrolactonisation, and a one-pot

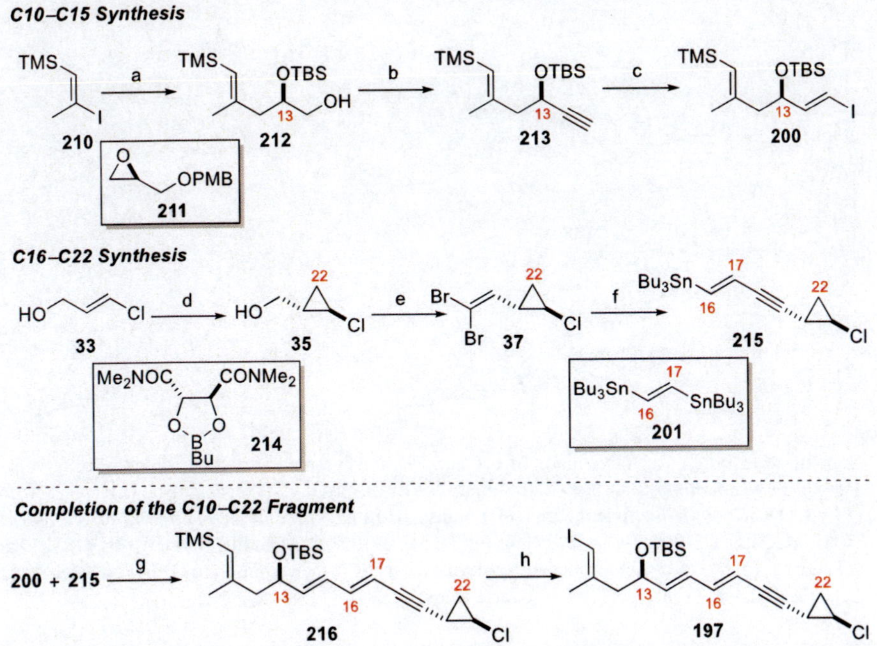

Scheme 44 Completion of the C10–C22 Fragment **197.** *Reagents and conditions*: **(a)** (i) *t*-BuLi, PhMe, −78°C, then **211**, BF_3•OEt_2, PhMe; (ii) TBSOTf, 2,6-lutidine, CH_2Cl_2, −78°C, 53% (over 2 steps); (iii) DDQ, pH 7 phosphate buffer, CH_2Cl_2, 0°C, 89%; **(b)** (i) SO_3•Py, Et_3N, DMSO, CH_2Cl_2, 0°C to RT; (ii) PPh_3, CBr_4, CH_2Cl_2, then aldehyde, 2,6-lutidine, CH_2Cl_2, 0°C, 89% (over 2 steps); (iii) *n*-BuLi, THF, −78°C, then H_2O, 85%; **(c)** (i) $Pd(PPh_3)_2Cl_2$ (3 mol%), Bu_3SnH, THF, 0°C; (ii) I_2, CH_2Cl_2, −78°C, 49% (over 2 steps); **(d)** $ZnEt_2$, CH_2I_2, CH_2Cl_2, 0°C, then **214**, **33**, CH_2Cl_2, 0°C to RT, 74%; **(e)** (i) PCC, Celite®, CH_2Cl_2, RT; (ii) PPh_3, CBr_4, CH_2Cl_2, 0°C, then aldehyde, CH_2Cl_2, 0°C to RT, 70% (over 2 steps); **(f)** (i) TBAF, DMF, 65°C; (ii) Pd_2dba_3 (10 mol%), $AsPh_3$ (40 mol%), Ag_2CO_3 (1.0 equiv.), THF, dark, −10°C, then **201**, 45% (over 2 steps); **(g)** $Pd(PFur_3)_2Cl_2$ (15 mol%), DMF, dark, RT, 63%; **(h)** NIS, MeCN, RT, 84%

deprotection/hemi-ketal formation step afforded the callipeltoside aglycon **9** (Scheme 45).

2.9.5 Preparation of Callipeltose A, B and C Thioglycosides

The synthesis of the callipeltose sugars commenced from pyranone **202**, which could be straightforwardly accessed in two steps. Nosylation of the secondary alcohol, displacement and diastereoselective methyl addition provided **218**, containing three of the stereocentres present in callipeltose A and B. Hydroxy-directed epoxidation, ring opening and selective 2° alcohol methylation gave **219**, which could be diversified to each callipeltose sugar [24]. Callipeltose A thioglycoside was completed in an additional four steps as a single anomer. Despite being able to synthesise the callipeltose B sugar, it is noted that the *N*-formyl group prevented thioglycoside

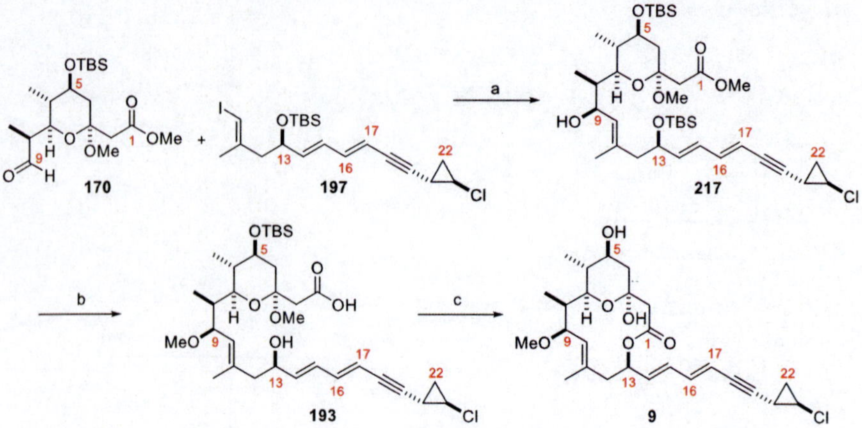

Scheme 45 Diastereoselective union of **170** and **197** and completion of the callipeltoside aglycon. *Reagents and conditions*: (**a**) 1. **197**, (1.3 equiv.), *t*-BuLi, Et_2O, −78°C; 2. $ZnBr_2$, Et_2O, −78°C to 0°C; 3. (1*S*,2*R*)-(+)-*N*-methylephedrine (1.1 equiv.), *n*-BuLi, PhMe, 0°C; 4. **170**, PhMe, 0°C, 48%; (**b**) (i) MeOTf, DTBP, CH_2Cl_2, RT, 73%; (ii) TBAF, THF, RT, 74%; (iii) $Ba(OH)_2{\bullet}8H_2O$, MeOH, RT, quant.; (**c**) (i) 2,4,6-trichlorobenzoyl chloride, Et_3N, RT, then addition to DMAP, PhMe, 80°C; (ii) TFA, THF–H_2O (5:1), RT, 58% (over 2 steps)

formation and hence coupling to the callipeltoside aglycon. As a result, compound **219** was TMS-protected and the thioglycoside formed to give callipeltose B precursor **221**, which would have to be further manipulated once attached to the aglycon (**9**) [14, 15].

The callipeltose C sugar was also synthesised from **202**, with the C3′ stereocentre once again formed by diastereoselective methyl addition, albeit this time by means of a complex-induced proximity effect from the neighbouring C4′ hydroxy [71]. TES protection, followed by the epoxidation, ring opening, methylation and thioglycosidation sequence mentioned previously, gave the callipeltose C thioglycoside (**224**) as a single anomer (Scheme 46) [14, 15].

2.9.6 Completion of Callipeltosides A, B and C

The callipeltoside aglycon was straightforwardly coupled with thioglycoside donors **220** and **224** using the conditions described by Evans [10]. Final deprotection of these two compounds then led to callipeltosides A and C, with spectroscopic data in full agreement with that previously reported [3, 4, 14, 15].

With the syntheses of both callipeltoside A and callipeltoside C completed, attention switched to callipeltoside B. *Since there had been no previous synthesis of callipeltoside B, it was assumed that like callipeltosides A and C, it also contained an* L*-configured sugar*. Attachment of azido sugar **221** (L-configured) proceeded without incident, and the azide moiety was selectively reduced using 1,3-propanedithiol [72]. Formylation of the amine using pentafluorophenyl formate

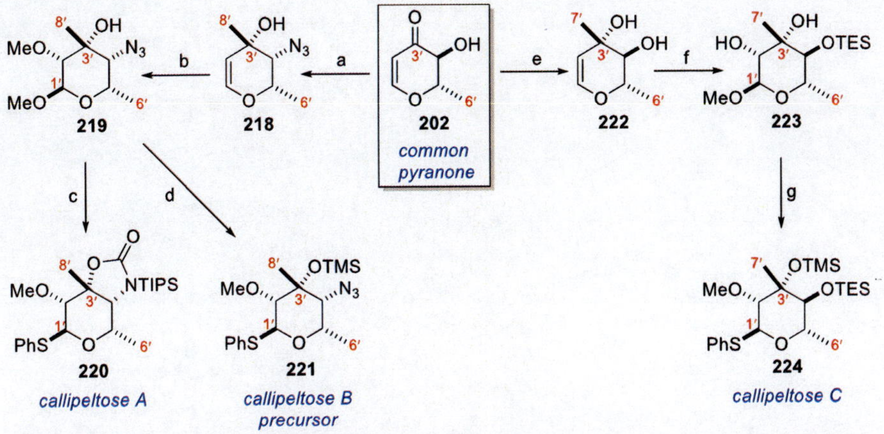

Scheme 46 Synthesis of callipeltoses A, B and C from common pyranone **202.** *Reagents and conditions*: (**a**) (i) NsCl, pyridine, CH_2Cl_2, RT, 95%; (ii) n-Bu_4NN_3, CH_2Cl_2, 0°C, 72%; (iii) MeLi, THF, −100°C, 79%; (**b**) (i) *m*-CPBA, MeOH, 0°C to RT, 52%; (ii) KO*t*-Bu, MeI, THF, 0°C, 79%; (**c**) (i) H_2, $Pd(OH)_2/C$, EtOAc, RT, 93%; (ii) triphosgene, pyridine, THF, −78°C to RT, 72%; (iii) TIPSOTf, 2,6-lutidine, CH_2Cl_2, RT, 97%; (iv) PhSH, BF_3•OEt_2, CH_2Cl_2, 0°C to RT, 80%, single anomer; (**d**) ZnI_2, TBAI, TMSSPh, DCE, 65°C, 60%, single anomer; (**e**) MeLi•LiBr, Et_2O, −78°C, 78%; (**f**) (i) TESCl, pyridine, DMAP, CH_2Cl_2, RT; (ii) *m*-CPBA, MeOH, 0°C to RT, 45% (over 2 steps); (**g**) (i) KO*t*-Bu, MeI, THF, 0°C, 81%; (ii) ZnI_2, TBAI, TMSSPh, DCE, 65°C, 86%, single anomer

(**225**) [73] and deprotection then delivered putative callipeltoside B (**2**) as a 4:1 rotameric mixture (by ^{1}H NMR) (Scheme 47). The synthetic sample was in complete agreement with the spectroscopic data for the natural isolate disclosed by Minale [4, 14, 15].

In order to ensure that the correct sugar had indeed been attached, the enantiomeric thioglycoside (*ent*-**221**, D-configured) was synthesised in analogous fashion to that shown in Scheme 46 and attached to the callipeltoside aglycon (**9**). This was then also reduced, formylated and deprotected to afford diastereomer **226**, which had a significantly different ^{1}H NMR to the natural isolate [4]. This provided further evidence that all three callipeltose sugars were L-configured, with callipeltoside B completed for the first time [14, 15].

With the entire family of molecules completed, the configuration of the glycosidic linkages in callipeltosides B and C still needed to be determined.

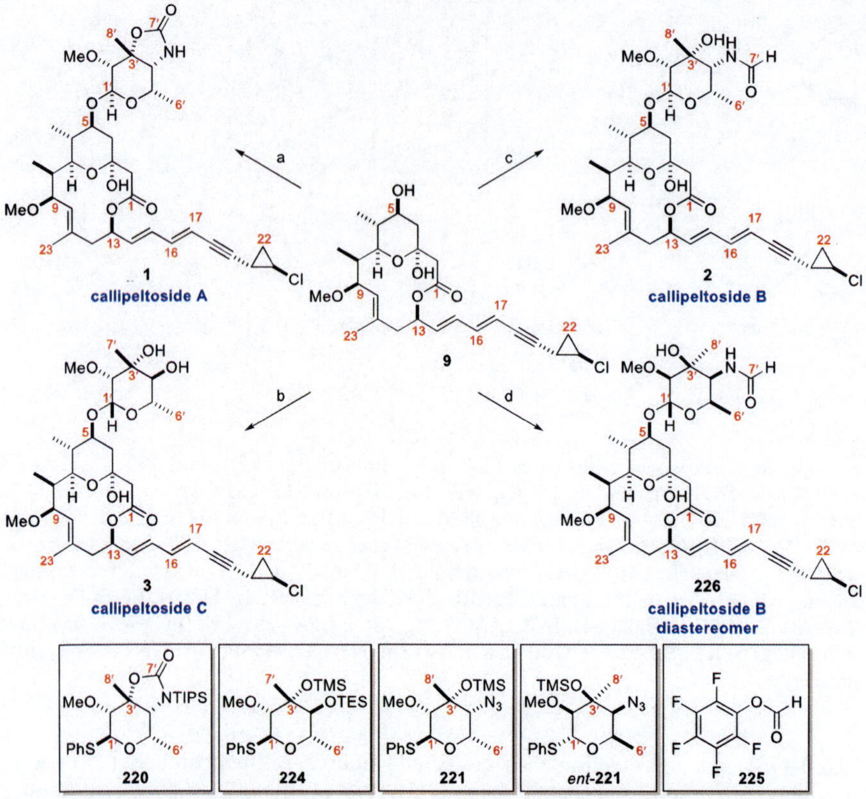

Scheme 47 Completion of callipeltosides A, B and C. *Reagents and conditions*: **(a)** (i) **220**, 4 Å MS, DTBMP, CH_2Cl_2, RT, 50 min, then −15°C, NIS, TfOH, −15°C to RT; (ii) TBAF, THF, RT, 83% (over 2 steps); **(b)** (i) **224**, 4 Å MS, DTBMP, CH_2Cl_2, RT, 50 min, then −15°C, NIS, TfOH, −15°C to RT; (ii) TASF, DMF, 40°C, 57% (over 2 steps); **(c)** (i) **221**, 4 Å MS, DTBMP, CH_2Cl_2, RT, 50 min, then −15°C, NIS, TfOH, −15°C to RT, 56%; (ii) 1,3-propandithiol, Et_3N, pyridine–H_2O (10:1), RT; (iii) **225**, $CHCl_3$, RT; (iv) TASF, DMF, 40°C, 52% (over 3 steps), 4:1 rotameric mixture (by ^{1}H NMR); **(d)** (i) *ent*-**221**, 4 Å MS, DTBMP, CH_2Cl_2, RT, 50 min, then −15°C, NIS, TfOH, −15°C to RT, 41%; (ii) 1,3-propandithiol, Et_3N, pyridine–H_2O (10:1), RT; (iii) **225**, $CHCl_3$, RT; (iv) TASF, DMF, 40°C, 57% (over 3 steps), 4:1 rotameric mixture (by ^{1}H NMR)

2.9.7 Stereochemistry of the Glycosidic Linkages of Callipeltosides B and C

Callipeltoside C

Although the Ley group had completed the second total synthesis of callipeltoside C, the configuration of the glycosidic linkage was still under question, with the MacMillan group tentatively assigning the stereochemistry by comparison with the assumed glycosidic linkage found in callipeltoside B [16]. As a result, the Ley

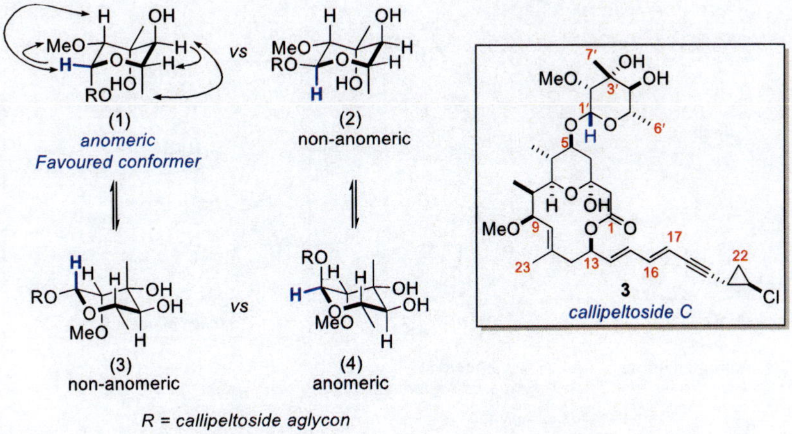

Fig. 6 Key nOe's for callipeltose C

group attempted to determine the C1′ stereochemistry by analysis of the $^1J_{\text{C–H}}$ coupling constant [74–78] and NOESY data.

$^1J_{C\text{–}H}$ Coupling Constant

Since measurement of the $^1J_{\text{C–H}}$ coupling constant [from a HSQC (heteronuclear single quantum coherence) experiment without ^{13}C decoupling] had proven to be a useful technique for carbohydrate chemists to determine whether the proton at an anomeric centre is axial or equatorial, this was assessed initially. It is noted that a value of ~170 Hz suggests an equatorial proton at C1′H, whilst ~160 Hz is indicative of an axial proton [74–78]. Unfortunately, measurement of the $^1J_{\text{C–H}}$ coupling constant gave an inconclusive result, returning a value of $^1J_{\text{C-H}} = 166.5$ Hz.

NOESY Data

As an alternative method to determine the stereochemistry of the glycosidic linkage, the Ley group conducted NOESY experiments and assessed all potential chair conformers **(1)**–**(4)** (Fig. 6; X-ray analysis of sugar intermediates suggested a chair configuration, not shown). Analysis of this averaged data indicated that structure **(1)** was the only chair conformer that accounted for all observed nOe interactions [14, 15]. With this in mind, the C1′ stereocentre was assigned to be (*R*)-configured, in agreement with the assignment previously suggested by MacMillan [16].

Callipeltoside B

The C1′H stereochemistry of callipeltoside B was analysed in exactly the same way as for callipeltoside C. Once again, the $^1J_{\text{C–H}}$ coupling value gave no indication of

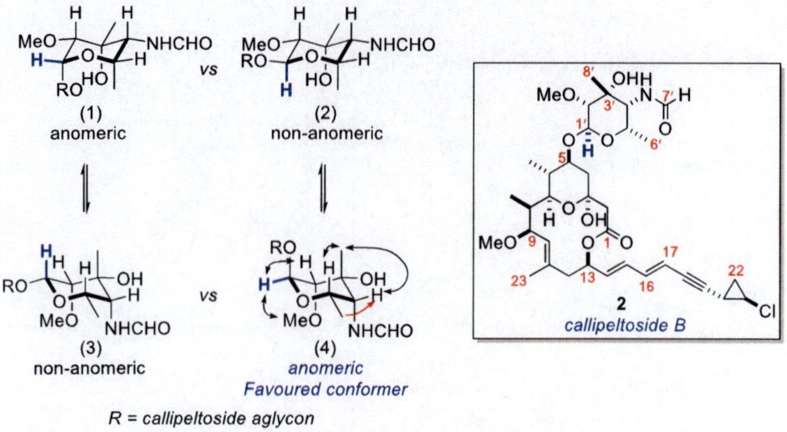

Fig. 7 Key nOe's for callipeltose B

stereochemistry, with a coupling constant of $^1J_{C\text{–}H}$ = 165.6 Hz measured. The stereochemistry was therefore inferred from NOSEY experiments (Fig. 7). This method revealed conformer (**4**) to be the only structure accounting for all observed nOe interactions. Therefore, callipeltoside B was noted to have a (*S*)-configured glycoside linkage (Fig. 7) [14, 15].

The authors note that whilst the same glycosidic linkage is present in both callipeltosides A (**1**) and B (**2**) [(*S*)-configured], callipeltoside C (**3**) is oppositely configured [(*R*)-configured]. Since each callipeltose sugar was attached under identical glycosidation conditions, it was assumed that the stereochemical course of the reaction must be influenced by the C4′ substituent. However, at present no further work has been reported to further study this [14, 15].

3 Conclusion and Final Remarks

The completion of the callipeltosides represents nearly 20 years of synthetic work, spanning across multiple research groups. Early syntheses focused on determining the configuration of the *trans*-chlorocyclopropane ring with respect to the C1–C19 core, in the process determining the absolute stereochemistry of the callipeltoside aglycon. The synthesis of callipeltoside A was accomplished shortly thereafter. These routes evolved over time, with more elegant and convergent approaches implemented once the stereochemical ambiguities associated with the aglycon had been overcome. The final pieces of the puzzle concerned the absolute configuration of the callipeltose B and C sugars, whilst the stereochemistry of the anomeric linkage of these molecules remained uncertain. The MacMillan and Ley groups later provided evidence to suggest that all members of the family contained L-configured sugars (having both also synthesised the corresponding D-configured callipeltose B and C sugars), with the latter research group further analysing the stereochemistry at C1′H to complete the study.

The different approaches towards the callipeltosides highlighted in this review stand as a testament to the power of modern synthetic technologies in their ability to supplement classical natural product structural analysis. It provides access to novel analogues and, in the programme above on the callipeltosides, makes available significantly greater quantities of material than that which is obtained by extraction from natural sources.

References

1. Zampella A, D'Auria MV, Paloma LG, Casapullo A, Minale L, Debitus C, Henin Y (1996) J Am Chem Soc 118:6202–6209
2. D'Auria MV, Zampella A, Paloma LG, Minale L (1996) Tetrahedron 52:9589–9596
3. Zampella A, D'Auria MV, Minale L, Debitus C, Roussakis C (1996) J Am Chem Soc 118:11085–11088
4. Zampella A, D'Auria MV, Minale L (1997) Tetrahedron 53:3243–3248
5. Yeung K, Paterson I (2005) Chem Rev 105:4237–4313
6. Paterson I, Findlay AD (2008) Pure Appl Chem 80:1773–1782
7. Trost BM, Dirat O, Gunzner JL (2002) Angew Chem Int Ed 41:841–843
8. Trost BM, Gunzner JL, Dirat O, Rhee YH (2002) J Am Chem Soc 124:10396–10415
9. Evans DA, Hu E, Burch JD, Jaeschke G (2002) J Am Chem Soc 124:5654–5655
10. Evans DA, Burch JD, Hu E, Jaeschke G (2008) Tetrahedron 64:4671–4699
11. Paterson I, Davies RDM, Heimann AC, Marquez R, Meyer A (2003) Org Lett 5:4477–4480
12. Huang H, Panek JS (2004) Org Lett 6:4383–4385
13. Hoye TR, Danielson M, May AE, Zhao H (2010) J Org Chem 75:7052–7060
14. Frost JR, Peason CM, Snaddon TN, Booth RA, Ley SV (2012) Angew Chem Int Ed 51:9366–9371
15. Frost JR, Pearson CM, Snaddon TN, Booth RA, Turner RM, Gold J, Shaw DM, Gaunt MJ, Ley SV (2015) Chem Eur J 21:13261–13277
16. Carpenter J, Northrup AB, Chung D, Wiener JJM, Kim S, MacMillan DWC (2008) Angew Chem Int Ed 47:3568–3572
17. Paterson I, Davies RDM, Marquez R (2001) Angew Chem Int Ed 40:603–607
18. Marshall JA, Eidam PM (2008) Org Lett 10:93–96
19. Velazquez F, Olivo HF (2000) Org Lett 2:1931–1933
20. Olivo HF, Velazquez F, Trevisan HC (2000) Org Lett 2:4055–4058
21. Romero-Ortega M, Colby DA, Olivo HF (2002) Tetrahedron Lett 43:6439–6441
22. Yadav JS, Haldar A, Maity T (2012) Eur J Org Chem:2062–2071
23. Gurjar MK, Reddy R (1998) Carbohydr Lett 3:169–172
24. Pihko AJ, Nicolaou KC, Koskinen AMP (2001) Tetrahedron Asymmetry 12:937–942
25. Smith GR, Finley JJ, Giuliano RM (1998) Carbohydr Res 308:223–227
26. Toth A, Remenyik J, Bajza I, Liptak A (2003) ARKIVOC:28–45
27. Saito S, Shiozawa M, Ito M, Yamamoto H (1998) J Am Chem Soc 120:813–814
28. Saito S, Shiozawa M, Yamamoto H (1999) Angew Chem Int Ed 38:1769–1771
29. Paterson I, Goodman JM, Isaka M (1989) Tetrahedron Lett 30:7121–7124
30. Paterson I, Perkins MV (1992) Tetrahedron Lett 33:801–804

31. Paterson I, Norcross RD, Ward RA, Romea P, Lister MA (1994) J Am Chem Soc 116:11287–11314
32. Charette AB, Juteau H (1994) J Am Chem Soc 116:2651–2652
33. Charette AB, Prescott S, Brochu C (1995) J Org Chem 60:1081–1083
34. Wang T, Liang Y, Yu Z (2011) J Am Chem Soc 133:9343–9353
35. Trost BM, Gunzner JL (2001) J Am Chem Soc 123:9449–9450
36. Sone H, Kigoshi H, Yamada K (1996) J Org Chem 61:8956–8960
37. Trost BM, Toste FD, Pinkerton AB (2001) Chem Rev 101:2067–2096
38. Trost BM, Toste FD (1999) J Am Chem Soc 121:4545–4554
39. Trost BM, Belletire JL, Godleski S, McDougal PG, Balkovec JM (1986) J Org Chem 51:2370–2374
40. Barton DHR, Crich D, Motherwell WB (1983) Tetrahedron Lett 24:4979–4982
41. Shen W, Wang L (1999) J Org Chem 64:8873–8879
42. Evans DA, Kozlowski MC, Murray JA, Burgey CS, Connell B (1999) J Am Chem Soc 121:669–685
43. Evans DA, Burch JD (2001) Org Lett 3:503–505
44. Masuda Y, Hoshi M, Arase A (1992) J Chem Soc Perkin Trans 1:2725–2726
45. Furukawa J, Kawabata N, Nishimura J (1968) Tetrahedron 24:53–58
46. Denmark S, O'Conner SP (1997) J Org Chem 62:3390–3401
47. Yang Z, Lorenz JC, Shi Y (1998) Tetrahedron Lett 39:8621–8624
48. Frank SA, Chen H, Kunz RK, Schnaderbeck MJ, Roush WR (2000) Org Lett 2:2691–2694
49. Evans DA, Hu E, Tedrow JS (2001) Org Lett 3:3133–3136
50. Masse CE, Panek JS (1995) Chem Rev 95:1293–1316
51. Huang HB, Panek JS (2003) Org Lett 5:1991–1993
52. Hoye TR, Zhao H (1999) Org Lett 1:169–172
53. Hoye TR, Danielson ME, May AE, Zhao H (2008) Angew Chem Int Ed 47:9743–9746
54. Northrup AB, MacMillan DWC (2002) J Am Chem Soc 124:6798–6799
55. Northrup AB, Mangion IK, Hettche F, MacMillan DWC (2004) Angew Chem Int Ed 43:2152–2154
56. Northrup AB, MacMillan DWC (2004) Science 303:1752–1755
57. Brown SP, Brochu MP, Sinz CJ, MacMillan DWC (2003) J Am Chem Soc 125:10808–10809
58. Zhong G (2003) Angew Chem Int Ed 42:4247–4250
59. Marshall JA, Yanik MM (2000) Tetrahedron Lett 41:4717–4721
60. Semmelhack MF, Kim C, Zhang N, Bodurow C, Sanner M, Dobler W, Meier M (1990) Pure Appl Chem 62:2035–2040
61. Tietze LF, Fischer R, Guder HJ, Goerlach A, Meumann M, Krach T (1987) Carbohydr Res 164:177–194
62. Oppolzer W, Radinov RN (1991) Tetrahedron Lett 32:5777–5780
63. Diéguez-Vázquez A, Tzschucke CC, Crecente-Campo J, McGrath S, Ley SV (2009) Eur J Org Chem:1698–1706
64. Nicolaou KC, Chakraborty TK, Piscopio AD, Minowa N, Bertinato P (1993) J Am Chem Soc 115:4419–4420
65. Nicolaou KC, Piscopio AD, Bertinato P, Chakraborty TK, Minowa N, Koide K (1995) Chem Eur J 1:318–333
66. Marshall JA, Eidam P (2004) Org Lett 6:445–448
67. Layton ME, Morales CA, Shair MD (2002) J Am Chem Soc 124:773–775
68. Pu L, Yu H (2001) Chem Rev 101:757–824
69. Noyori R, Suga S, Kawai K, Okada S, Kitamura M, Oguni N, Hayashi M, Kaneko T, Matsuda Y (1990) J Organomet Chem 382:19–37
70. Brubaker JD, Myers AG (2007) Org Lett 9:3523–3525
71. Beak P, Meyers AI (1986) Acc Chem Res 19:356–363
72. Bayley H, Standring DN, Knowles JR (1978) Tetrahedron Lett 19:3633–3634
73. Kisfaludy L, Ötvös L (1987) Synthesis:510

74. Muller N, Pritchard DE (1959) J Chem Phys 31:768–771
75. Maciel GE, McIver Jr JW, Ostlund NS, Pople JA (1970) J Am Chem Soc 92:1–11
76. Bock K, Lundt I, Pedersen C (1973) Tetrahedron Lett 13:1037–1040
77. Bock K, Wiebe L (1973) Acta Chem Scand 27:2676–2678
78. Bock K, Pedersen C (1974) J Chem Soc Perkin Trans 1:293–299